ELECTROMAGNETICS

Edward J. Rothwell
Michigan State University
East Lansing, Michigan

Michael J. Cloud
Lawrence Technological University
Southfield, Michigan

CRC Press
Boca Raton   London   New York   Washington, D.C.

Library of Congress Cataloging-in-Publication Data

Rothwell, Edward J.
    Electromagnetics / Edward J. Rothwell, Michael J. Cloud.
        p. cm.—(Electrical engineering textbook series ; 2)
    Includes bibliographical references and index.
    ISBN 0-8493-1397-X (alk. paper)
     1. Electromagnetic theory. I. Cloud, Michael J. II. Title. III. Series.

QC670 .R693 2001
530.14′1—dc21
                                                    00-065158
                                                        CIP

Visit our website at www.crcpress.com.

No claim to original U.S. Government works
International Standard Book Number 0-8493-1397-X
Library of Congress Card Number 00-065158
Printed in the United States of America 1 2 3 4 5 6 7 8 9 0
Printed on acid-free paper

In memory of Catherine Rothwell

Preface

This book is intended as a text for a first-year graduate sequence in engineering electromagnetics. Ideally such a sequence provides a transition period during which a student can solidify his or her understanding of fundamental concepts before proceeding to specialized areas of research.

The assumed background of the reader is limited to standard undergraduate topics in physics and mathematics. Worthy of explicit mention are complex arithmetic, vector analysis, ordinary differential equations, and certain topics normally covered in a "signals and systems" course (e.g., convolution and the Fourier transform). Further analytical tools, such as contour integration, dyadic analysis, and separation of variables, are covered in a self-contained mathematical appendix.

The organization of the book is in six chapters. In Chapter 1 we present essential background on the field concept, as well as information related specifically to the electromagnetic field and its sources. Chapter 2 is concerned with a presentation of Maxwell's theory of electromagnetism. Here attention is given to several useful forms of Maxwell's equations, the nature of the four field quantities and of the postulate in general, some fundamental theorems, and the wave nature of the time-varying field. The electrostatic and magnetostatic cases are treated in Chapter 3. In Chapter 4 we cover the representation of the field in the frequency domains: both temporal and spatial. Here the behavior of common engineering materials is also given some attention. The use of potential functions is discussed in Chapter 5, along with other field decompositions including the solenoidal–lamellar, transverse–longitudinal, and TE–TM types. Finally, in Chapter 6 we present the powerful integral solution to Maxwell's equations by the method of Stratton and Chu. A main mathematical appendix near the end of the book contains brief but sufficient treatments of Fourier analysis, vector transport theorems, complex-plane integration, dyadic analysis, and boundary value problems. Several subsidiary appendices provide useful tables of identities, transforms, and so on.

We would like to express our deep gratitude to those persons who contributed to the development of the book. The reciprocity-based derivation of the Stratton–Chu formula was provided by Prof. Dennis Nyquist, as was the material on wave reflection from multiple layers. The groundwork for our discussion of the Kronig–Kramers relations was provided by Michael Havrilla, and material on the time-domain reflection coefficient was developed by Jungwook Suk. We owe thanks to Prof. Leo Kempel, Dr. David Infante, and Dr. Ahmet Kizilay for carefully reading large portions of the manuscript during its preparation, and to Christopher Coleman for helping to prepare the figures. We are indebted to Dr. John E. Ross for kindly permitting us to employ one of his computer programs for scattering from a sphere and another for numerical Fourier transformation. Helpful comments and suggestions on the figures were provided by Beth Lannon–Cloud.

Thanks to Dr. C. L. Tondo of T & T Techworks, Inc., for assistance with the LaTeX macros that were responsible for the layout of the book. Finally, we would like to thank the staff members of CRC Press — Evelyn Meany, Sara Seltzer, Elena Meyers, Helena Redshaw, Jonathan Pennell, Joette Lynch, and Nora Konopka — for their guidance and support.

Contents

Chapter 1

Introductory concepts

1.1 Notation, conventions, and symbology

Any book that covers a broad range of topics will likely harbor some problems with notation and symbology. This results from having the same symbol used in different areas to represent different quantities, and also from having too many quantities to represent. Rather than invent new symbols, we choose to stay close to the standards and warn the reader about any symbol used to represent more than one distinct quantity.

The basic nature of a physical quantity is indicated by typeface or by the use of a diacritical mark. Scalars are shown in ordinary typeface: q, Φ, for example. Vectors are shown in boldface: $\mathbf{E}, \mathbf{\Pi}$. Dyadics are shown in boldface with an overbar: $\bar{\epsilon}, \bar{\mathbf{A}}$. Frequency dependent quantities are indicated by a tilde, whereas time dependent quantities are written without additional indication; thus we write $\tilde{\mathbf{E}}(\mathbf{r}, \omega)$ and $\mathbf{E}(\mathbf{r}, t)$. (Some quantities, such as impedance, are used in the frequency domain to interrelate Fourier spectra; although these quantities are frequency dependent they are seldom written in the time domain, and hence we do not attach tildes to their symbols.) We often combine diacritical marks: for example, $\tilde{\bar{\epsilon}}$ denotes a frequency domain dyadic. We distinguish carefully between phasor and frequency domain quantities. The variable ω is used for the frequency variable of the Fourier spectrum, while $\check{\omega}$ is used to indicate the constant frequency of a time harmonic signal. We thus further separate the notion of a phasor field from a frequency domain field by using a *check* to indicate a phasor field: $\check{\mathbf{E}}(\mathbf{r})$. However, there is often a simple relationship between the two, such as $\check{\mathbf{E}} = \tilde{\mathbf{E}}(\check{\omega})$.

We designate the field and source point position vectors by $\mathbf{r}$ and $\mathbf{r}'$, respectively, and the corresponding relative displacement or distance vector by $\mathbf{R}$:

$$\mathbf{R} = \mathbf{r} - \mathbf{r}'.$$

A hat designates a vector as a unit vector (e.g., $\hat{\mathbf{x}}$). The sets of coordinate variables in rectangular, cylindrical, and spherical coordinates are denoted by

$$(x, y, z), \qquad (\rho, \phi, z), \qquad (r, \theta, \phi),$$

respectively. (In the spherical system ϕ is the azimuthal angle and θ is the polar angle.) We freely use the "del" operator notation ∇ for gradient, curl, divergence, Laplacian, and so on.

The SI (MKS) system of units is employed throughout the book.

1.2   The field concept of electromagnetics

Introductory treatments of electromagnetics often stress the role of the field in force transmission: the individual fields **E** and **B** are defined via the mechanical force on a small test charge. This is certainly acceptable, but does not tell the whole story. We might, for example, be left with the impression that the EM field always arises from an interaction between charged objects. Often coupled with this is the notion that the field concept is meant merely as an aid to the calculation of force, a kind of notational convenience not placed on the same physical footing as force itself. In fact, fields are more than useful — they are fundamental. Before discussing electromagnetic fields in more detail, let us attempt to gain a better perspective on the field concept and its role in modern physical theory. Fields play a central role in any attempt to describe physical reality. They are as real as the physical substances we ascribe to everyday experience. In the words of Einstein [63],

> "It seems impossible to give an obvious qualitative criterion for distinguishing between matter and field or charge and field."

We must therefore put fields and particles of matter on the same footing: both carry energy and momentum, and both interact with the observable world.

1.2.1   Historical perspective

Early nineteenth century physical thought was dominated by the *action at a distance* concept, formulated by Newton more than 100 years earlier in his immensely successful theory of gravitation. In this view the influence of individual bodies extends across space, instantaneously affects other bodies, and remains completely unaffected by the presence of an intervening medium. Such an idea was revolutionary; until then *action by contact*, in which objects are thought to affect each other through physical contact or by contact with the intervening medium, seemed the obvious and only means for mechanical interaction. Priestly's experiments in 1766 and Coulomb's torsion-bar experiments in 1785 seemed to indicate that the force between two electrically charged objects behaves in strict analogy with gravitation: both forces obey inverse square laws and act along a line joining the objects. Oersted, Ampere, Biot, and Savart soon showed that the magnetic force on segments of current-carrying wires also obeys an inverse square law.

The experiments of Faraday in the 1830s placed doubt on whether action at a distance really describes electric and magnetic phenomena. When a material (such as a dielectric) is placed between two charged objects, the force of interaction decreases; thus, the intervening medium does play a role in conveying the force from one object to the other. To explain this, Faraday visualized "lines of force" extending from one charged object to another. The manner in which these lines were thought to interact with materials they intercepted along their path was crucial in understanding the forces on the objects. This also held for magnetic effects. Of particular importance was the number of lines passing through a certain area (the *flux*), which was thought to determine the amplitude of the effect observed in Faraday's experiments on electromagnetic induction.

Faraday's ideas presented a new world view: electromagnetic phenomena occur in the region surrounding charged bodies, and can be described in terms of the laws governing the "field" of his lines of force. Analogies were made to the stresses and strains in material objects, and it appeared that Faraday's force lines created equivalent electromagnetic stresses and strains in media surrounding charged objects. His law of induction was formulated not in terms of positions of bodies, but in terms of lines of magnetic force. Inspired by Faraday's ideas, Gauss restated Coulomb's law in terms of flux lines, and Maxwell extended the idea to time changing fields through his concept of displacement current.

In the 1860s Maxwell created what Einstein called "the most important invention since Newton's time" — a set of equations describing an entirely field-based theory of electromagnetism. These equations do not model the forces acting between bodies, as do Newton's law of gravitation and Coulomb's law, but rather describe only the dynamic, time-evolving structure of the electromagnetic field. Thus bodies are not seen to interact with each other, but rather with the (very real) electromagnetic field they create, an interaction described by a supplementary equation (the Lorentz force law). To better understand the interactions in terms of mechanical concepts, Maxwell also assigned properties of stress and energy to the field.

Using constructs that we now call the electric and magnetic fields and potentials, Maxwell synthesized all known electromagnetic laws and presented them as a system of differential and algebraic equations. By the end of the nineteenth century, Hertz had devised equations involving only the electric and magnetic fields, and had derived the laws of circuit theory (Ohm's law and Kirchoff's laws) from the field expressions. His experiments with high-frequency fields verified Maxwell's predictions of the existence of electromagnetic waves propagating at finite velocity, and helped solidify the link between electromagnetism and optics. But one problem remained: if the electromagnetic fields propagated by stresses and strains on a medium, how could they propagate through a vacuum? A substance called the *luminiferous aether*, long thought to support the transverse waves of light, was put to the task of carrying the vibrations of the electromagnetic field as well. However, the pivotal experiments of Michelson and Morely showed that the aether was fictitious, and the physical existence of the field was firmly established.

The essence of the field concept can be conveyed through a simple thought experiment. Consider two stationary charged particles in free space. Since the charges are stationary, we know that (1) another force is present to balance the Coulomb force between the charges, and (2) the momentum and kinetic energy of the system are zero. Now suppose one charge is quickly moved and returned to rest at its original position. Action at a distance would require the second charge to react immediately (Newton's third law), but by Hertz's experiments it does not. There appears to be no change in energy of the system: both particles are again at rest in their original positions. However, after a time (given by the distance between the charges divided by the speed of light) we find that the second charge does experience a change in electrical force and begins to move away from its state of equilibrium. But by doing so it has gained net kinetic energy and momentum, and the energy and momentum of the system seem larger than at the start. This can only be reconciled through field theory. If we regard the field as a physical entity, then the nonzero work required to initiate the motion of the first charge and return it to its initial state can be seen as increasing the energy of the field. A disturbance propagates at finite speed and, upon reaching the second charge, transfers energy into kinetic energy of the charge. Upon its acceleration this charge also sends out a wave of field disturbance, carrying energy with it, eventually reaching the first charge and creating a second reaction. At any given time, the net energy and momentum of the system, composed of both the bodies and the field, remain constant. We thus come to regard the electromagnetic field as a true physical entity: an entity capable of carrying energy and momentum.

1.2.2 Formalization of field theory

Before we can invoke physical laws, we must find a way to describe the *state* of the system we intend to study. We generally begin by identifying a set of *state variables* that can depict the physical nature of the system. In a mechanical theory such as Newton's law of gravitation, the state of a system of point masses is expressed in terms of the instantaneous positions and momenta of the individual particles. Hence $6N$ state variables are needed to describe the state of a system of N particles, each particle having three position coordinates and three momentum components. The time evolution of the system state is determined by a supplementary force function (e.g., gravitational attraction), the initial state (initial conditions), and Newton's second law $\mathbf{F} = d\mathbf{P}/dt$.

Descriptions using finite sets of state variables are appropriate for action-at-a-distance interpretations of physical laws such as Newton's law of gravitation or the interaction of charged particles. If Coulomb's law were taken as the force law in a mechanical description of electromagnetics, the state of a system of particles could be described completely in terms of their positions, momenta, and charges. Of course, charged particle interaction is not this simple. An attempt to augment Coulomb's force law with Ampere's force law would not account for kinetic energy loss via radiation. Hence we abandon[1] the mechanical viewpoint in favor of the field viewpoint, selecting a different set of state variables. The essence of field theory is to regard electromagnetic phenomena as affecting all of space. We shall find that we can describe the field in terms of the four vector quantities $\mathbf{E}$, $\mathbf{D}$, $\mathbf{B}$, and $\mathbf{H}$. Because these fields exist by definition at each point in space and each time t, a finite set of state variables cannot describe the system.

Here then is an important distinction between field theories and mechanical theories: the state of a field at any instant can only be described by an infinite number of state variables. Mathematically we describe fields in terms of functions of continuous variables; however, we must be careful not to confuse all quantities described as "fields" with those fields innate to a scientific field theory. For instance, we may refer to a temperature "field" in the sense that we can describe temperature as a function of space and time. However, we do *not* mean by this that temperature obeys a set of physical laws analogous to those obeyed by the electromagnetic field.

What special character, then, can we ascribe to the electromagnetic field that has meaning beyond that given by its mathematical implications? In this book, $\mathbf{E}$, $\mathbf{D}$, $\mathbf{B}$, and $\mathbf{H}$ are integral parts of a *field-theory description* of electromagnetics. In any field theory we need two types of fields: a *mediating field* generated by a source, and a field describing the source itself. In free-space electromagnetics the mediating field consists of $\mathbf{E}$ and $\mathbf{B}$, while the source field is the distribution of charge or current. An important consideration is that the source field must be independent of the mediating field that it "sources." Additionally, fields are generally regarded as unobservable: they can only be measured indirectly through interactions with observable quantities. We need a link to mechanics to observe $\mathbf{E}$ and $\mathbf{B}$: we might measure the change in kinetic energy of a particle as it interacts with the field through the Lorentz force. The Lorentz force becomes the force function in the mechanical interaction that uniquely determines the (observable) mechanical state of the particle.

A field is associated with a set of *field equations* and a set of *constitutive relations*. The field equations describe, through partial derivative operations, both the spatial distribution and temporal evolution of the field. The constitutive relations describe the effect

---

[1]Attempts have been made to formulate electromagnetic theory purely in action-at-a-distance terms, but this viewpoint has not been generally adopted [69].

of the supporting medium on the fields and are dependent upon the physical state of the medium. The state may include macroscopic effects, such as mechanical stress and thermodynamic temperature, as well as the microscopic, quantum-mechanical properties of matter.

The value of the field at any position and time in a bounded region V is then determined uniquely by specifying the sources within V, the initial state of the fields within V, and the value of the field or finitely many of its derivatives on the surface bounding V. If the boundary surface also defines a surface of discontinuity between adjacent regions of differing physical characteristics, or across discontinuous sources, then *jump conditions* may be used to relate the fields on either side of the surface.

The variety of forms of field equations is restricted by many physical principles including reference-frame invariance, conservation, causality, symmetry, and simplicity. Causality prevents the field at time $t = 0$ from being influenced by events occurring at subsequent times $t > 0$. Of course, we prefer that a field equation be mathematically robust and well-posed to permit solutions that are unique and stable.

Many of these ideas are well illustrated by a consideration of electrostatics. We can describe the electrostatic field through a mediating scalar field $\Phi(x, y, z)$ known as the electrostatic potential. The spatial distribution of the field is governed by Poisson's equation

$$\frac{\partial^2 \Phi}{\partial x^2} + \frac{\partial^2 \Phi}{\partial y^2} + \frac{\partial^2 \Phi}{\partial z^2} = -\frac{\rho}{\epsilon_0},$$

where $\rho = \rho(x, y, z)$ is the source charge density. No temporal derivatives appear, and the spatial derivatives determine the spatial behavior of the field. The function ρ represents the spatially-averaged distribution of charge that acts as the source term for the field Φ. Note that ρ incorporates no information about Φ. To uniquely specify the field at any point, we must still specify its behavior over a boundary surface. We could, for instance, specify Φ on five of the six faces of a cube and the normal derivative $\partial\Phi/\partial n$ on the remaining face. Finally, we cannot directly observe the static potential field, but we can observe its interaction with a particle. We relate the static potential field theory to the realm of mechanics via the electrostatic force $\mathbf{F} = q\mathbf{E}$ acting on a particle of charge q.

In future chapters we shall present a classical field theory for macroscopic electromagnetics. In that case the mediating field quantities are $\mathbf{E}$, $\mathbf{D}$, $\mathbf{B}$, and $\mathbf{H}$, and the source field is the current density $\mathbf{J}$.

1.3   The sources of the electromagnetic field

Electric charge is an intriguing natural entity. Human awareness of charge and its effects dates back to at least 600 BC, when the Greek philosopher Thales of Miletus observed that rubbing a piece of amber could enable the amber to attract bits of straw. Although *charging by friction* is probably still the most common and familiar manifestation of electric charge, systematic experimentation has revealed much more about the behavior of charge and its role in the physical universe. There are two kinds of charge, to which Benjamin Franklin assigned the respective names *positive* and *negative*. Franklin observed that charges of opposite kind attract and charges of the same kind repel. He also found that an increase in one kind of charge is accompanied by an increase in the other, and so first described the principle of *charge conservation*. Twentieth century physics has added dramatically to the understanding of charge:

1. Electric charge is a fundamental property of matter, as is mass or dimension.

2. Charge is *quantized*: there exists a smallest quantity (*quantum*) of charge that can be associated with matter. No smaller amount has been observed, and larger amounts always occur in integral multiples of this quantity.

3. The charge quantum is associated with the smallest subatomic particles, and these particles interact through electrical forces. In fact, matter is organized and arranged through electrical interactions; for example, our perception of physical contact is merely the macroscopic manifestation of countless charges in our fingertips pushing against charges in the things we touch.

4. Electric charge is an *invariant*: the value of charge on a particle does not depend on the speed of the particle. In contrast, the mass of a particle increases with speed.

5. Charge acts as the source of an electromagnetic field; the field is an entity that can carry energy and momentum away from the charge via propagating waves.

We begin our investigation of the properties of the electromagnetic field with a detailed examination of its source.

1.3.1   Macroscopic electromagnetics

We are interested primarily in those electromagnetic effects that can be predicted by classical techniques using continuous sources (charge and current densities). Although macroscopic electromagnetics is limited in scope, it is useful in many situations encountered by engineers. These include, for example, the determination of currents and voltages in lumped circuits, torques exerted by electrical machines, and fields radiated by antennas. Macroscopic predictions can fall short in cases where quantum effects are important: e.g., with devices such as tunnel diodes. Even so, quantum mechanics can often be coupled with classical electromagnetics to determine the macroscopic electromagnetic properties of important materials.

Electric charge is not of a continuous nature. The quantization of atomic charge — $\pm e$ for electrons and protons, $\pm e/3$ and $\pm 2e/3$ for quarks — is one of the most precisely established principles in physics (verified to 1 part in 10^{21}). The value of e itself is known to great accuracy:

$$e = 1.60217733 \times 10^{-19} \text{ Coulombs (C)}.$$

However, the discrete nature of charge is not easily incorporated into everyday engineering concerns. The strange world of the individual charge — characterized by particle spin, molecular moments, and thermal vibrations — is well described only by quantum theory. There is little hope that we can learn to describe electrical machines using such concepts. Must we therefore retreat to the macroscopic idea and ignore the discretization of charge completely? A viable alternative is to use atomic theories of matter to estimate the useful scope of macroscopic electromagnetics.

Remember, we are completely free to postulate a theory of nature whose scope may be limited. Like continuum mechanics, which treats distributions of matter as if they were continuous, macroscopic electromagnetics is regarded as valid because it is verified by experiment over a certain range of conditions. This applicability range generally corresponds to dimensions on a laboratory scale, implying a very wide range of validity for engineers.

Macroscopic effects as averaged microscopic effects. Macroscopic electromagnetics can hold in a world of discrete charges because applications usually occur over physical scales that include vast numbers of charges. Common devices, generally much larger than individual particles, "average" the rapidly varying fields that exist in the spaces between charges, and this allows us to view a source as a continuous "smear" of charge. To determine the range of scales over which the macroscopic viewpoint is valid, we must compare averaged values of microscopic fields to the macroscopic fields we measure in the lab. But if the effects of the individual charges are describable only in terms of quantum notions, this task will be daunting at best. A simple compromise, which produces useful results, is to extend the macroscopic theory right down to the microscopic level and regard discrete charges as "point" entities that produce electromagnetic fields according to Maxwell's equations. Then, in terms of scales much larger than the classical radius of an electron ($\approx 10^{-14}$ m), the expected rapid fluctuations of the fields in the spaces between charges is predicted. Finally, we ask: over what spatial scale must we average the effects of the fields and the sources in order to obtain agreement with the macroscopic equations?

In the spatial averaging approach a convenient weighting function $f(\mathbf{r})$ is chosen, and is normalized so that $\int f(\mathbf{r})\,dV = 1$. An example is the Gaussian distribution

$$f(\mathbf{r}) = (\pi a^2)^{-3/2} e^{-r^2/a^2},$$

where a is the approximate radial extent of averaging. The spatial average of a microscopic quantity $F(\mathbf{r}, t)$ is given by

$$\langle F(\mathbf{r}, t) \rangle = \int F(\mathbf{r} - \mathbf{r}', t) f(\mathbf{r}')\,dV'. \qquad (1.1)$$

The scale of validity of the macroscopic model can be found by determining the averaging radius a that produces good agreement between the averaged microscopic fields and the macroscopic fields.

The macroscopic volume charge density. At this point we do not distinguish between the "free" charge that is unattached to a molecular structure and the charge found near the surface of a conductor. Nor do we consider the dipole nature of polarizable materials or the microscopic motion associated with molecular magnetic moment or the magnetic moment of free charge. For the consideration of free-space electromagnetics, we assume charge exhibits either three degrees of freedom (*volume charge*), two degrees of freedom (*surface charge*), or one degree of freedom (*line charge*).

In typical matter, the microscopic fields vary spatially over dimensions of 10^{-10} m or less, and temporally over periods (determined by atomic motion) of 10^{-13} s or less. At the surface of a material such as a good conductor where charge often concentrates, averaging with a radius on the order of 10^{-10} m may be required to resolve the rapid variation in the distribution of individual charged particles. However, within a solid or liquid material, or within a free-charge distribution characteristic of a dense gas or an electron beam, a radius of 10^{-8} m proves useful, containing typically 10^6 particles. A diffuse gas, on the other hand, may have a particle density so low that the averaging radius takes on laboratory dimensions, and in such a case the microscopic theory must be employed even at macroscopic dimensions.

Once the averaging radius has been determined, the value of the charge density may be found via (1.1). The volume density of charge for an assortment of point sources can be written in terms of the three-dimensional Dirac delta as

$$\rho^o(\mathbf{r}, t) = \sum_i q_i \delta(\mathbf{r} - \mathbf{r}_i(t)),$$

where $\mathbf{r}_i(t)$ is the position of the charge q_i at time t. Substitution into (1.1) gives

$$\rho(\mathbf{r}, t) = \langle \rho^o(\mathbf{r}, t) \rangle = \sum_i q_i f(\mathbf{r} - \mathbf{r}_i(t)) \qquad (1.2)$$

as the averaged charge density appropriate for use in a macroscopic field theory. Because the oscillations of the atomic particles are statistically uncorrelated over the distances used in spatial averaging, the time variations of microscopic fields are not present in the macroscopic fields and temporal averaging is unnecessary. In (1.2) the time dependence of the spatially-averaged charge density is due entirely to bulk motion of the charge aggregate (macroscopic charge motion).

With the definition of macroscopic charge density given by (1.2), we can determine the total charge $Q(t)$ in any macroscopic volume region V using

$$Q(t) = \int_V \rho(\mathbf{r}, t) \, dV. \qquad (1.3)$$

We have

$$Q(t) = \sum_i q_i \int_V f(\mathbf{r} - \mathbf{r}_i(t)) \, dV = \sum_{\mathbf{r}_i(t) \in V} q_i.$$

Here we ignore the small discrepancy produced by charges lying within distance a of the boundary of V. It is common to employ a box B having volume ΔV:

$$f(\mathbf{r}) = \begin{cases} 1/\Delta V, & \mathbf{r} \in B, \\ 0, & \mathbf{r} \notin B. \end{cases}$$

In this case

$$\rho(\mathbf{r}, t) = \frac{1}{\Delta V} \sum_{\mathbf{r} - \mathbf{r}_i(t) \in B} q_i.$$

The size of B is chosen with the same considerations as to atomic scale as was the averaging radius a. Discontinuities at the edges of the box introduce some difficulties concerning charges that move in and out of the box because of molecular motion.

The macroscopic volume current density. Electric charge in motion is referred to as *electric current*. Charge motion can be associated with external forces and with microscopic fluctuations in position. Assuming charge q_i has velocity $\mathbf{v}_i(t) = d\mathbf{r}_i(t)/dt$, the charge aggregate has volume current density

$$\mathbf{J}^o(\mathbf{r}, t) = \sum_i q_i \mathbf{v}_i(t) \, \delta(\mathbf{r} - \mathbf{r}_i(t)).$$

Spatial averaging gives the macroscopic volume current density

$$\mathbf{J}(\mathbf{r}, t) = \langle \mathbf{J}^o(\mathbf{r}, t) \rangle = \sum_i q_i \mathbf{v}_i(t) f(\mathbf{r} - \mathbf{r}_i(t)). \qquad (1.4)$$

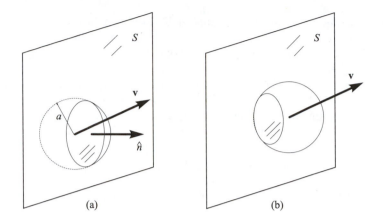

Figure 1.1: Intersection of the averaging function of a point charge with a surface S, as the charge crosses S with velocity $\mathbf{v}$: (a) at some time $t = t_1$, and (b) at $t = t_2 > t_1$. The averaging function is represented by a sphere of radius a.

Spatial averaging at time t eliminates currents associated with microscopic motions that are uncorrelated at the scale of the averaging radius (again, we do not consider the magnetic moments of particles). The assumption of a sufficiently large averaging radius leads to

$$\mathbf{J}(\mathbf{r}, t) = \rho(\mathbf{r}, t)\, \mathbf{v}(\mathbf{r}, t). \tag{1.5}$$

The total flux $I(t)$ of current through a surface S is given by

$$I(t) = \int_S \mathbf{J}(\mathbf{r}, t) \cdot \hat{\mathbf{n}}\, dS$$

where $\hat{\mathbf{n}}$ is the unit normal to S. Hence, using (1.4), we have

$$I(t) = \sum_i q_i \frac{d}{dt}(\mathbf{r}_i(t) \cdot \hat{\mathbf{n}}) \int_S f(\mathbf{r} - \mathbf{r}_i(t))\, dS$$

if $\hat{\mathbf{n}}$ stays approximately constant over the extent of the averaging function and S is not in motion. We see that the integral effectively intersects S with the averaging function surrounding each moving point charge (Figure 1.1). The time derivative of $\mathbf{r}_i \cdot \hat{\mathbf{n}}$ represents the velocity at which the averaging function is "carried across" the surface.

Electric current takes a variety of forms, each described by the relation $\mathbf{J} = \rho \mathbf{v}$. Isolated charged particles (positive and negative) and charged insulated bodies moving through space comprise *convection currents*. Negatively-charged electrons moving through the positive background lattice within a conductor comprise a *conduction current*. Empirical evidence suggests that conduction currents are also described by the relation $\mathbf{J} = \sigma \mathbf{E}$ known as *Ohm's law*. A third type of current, called *electrolytic current*, results from the flow of positive or negative ions through a fluid.

1.3.2   Impressed vs. secondary sources

In addition to the simple classification given above we may classify currents as *primary* or *secondary*, depending on the action that sets the charge in motion.

It is helpful to separate primary or "impressed" sources, which are independent of the fields they source, from secondary sources which result from interactions between the sourced fields and the medium in which the fields exist. Most familiar is the conduction current set up in a conducting medium by an externally applied electric field. The impressed source concept is particularly important in circuit theory, where independent voltage sources are modeled as providing primary voltage excitations that are independent of applied load. In this way they differ from the secondary or "dependent" sources that react to the effect produced by the application of primary sources.

In applied electromagnetics the primary source may be so distant that return effects resulting from local interaction of its impressed fields can be ignored. Other examples of primary sources include the applied voltage at the input of an antenna, the current on a probe inserted into a waveguide, and the currents producing a power-line field in which a biological body is immersed.

1.3.3   Surface and line source densities

Because they are spatially averaged effects, macroscopic sources and the fields they source cannot have true spatial discontinuities. However, it is often convenient to work with sources in one or two dimensions. Surface and line source densities are idealizations of actual, continuous macroscopic densities.

The entity we describe as a *surface charge* is a continuous volume charge distributed in a thin layer across some surface S. If the thickness of the layer is small compared to laboratory dimensions, it is useful to assign to each point $\mathbf{r}$ on the surface a quantity describing the amount of charge contained within a cylinder oriented normal to the surface and having infinitesimal cross section dS. We call this quantity the *surface charge density* $\rho_s(\mathbf{r},t)$, and write the volume charge density as

$$\rho(\mathbf{r},w,t) = \rho_s(\mathbf{r},t)f(w,\Delta),$$

where w is distance from S in the normal direction and Δ in some way parameterizes the "thickness" of the charge layer at $\mathbf{r}$. The continuous density function $f(x,\Delta)$ satisfies

$$\int_{-\infty}^{\infty} f(x,\Delta)\,dx = 1$$

and

$$\lim_{\Delta \to 0} f(x,\Delta) = \delta(x).$$

For instance, we might have

$$f(x,\Delta) = \frac{e^{-x^2/\Delta^2}}{\Delta\sqrt{\pi}}. \tag{1.6}$$

With this definition the total charge contained in a cylinder normal to the surface at $\mathbf{r}$ and having cross-sectional area dS is

$$dQ(t) = \int_{-\infty}^{\infty} [\rho_s(\mathbf{r},t)\,dS]\,f(w,\Delta)\,dw = \rho_s(\mathbf{r},t)\,dS,$$

and the total charge contained within any cylinder oriented normal to S is

$$Q(t) = \int_S \rho_s(\mathbf{r},t)\,dS. \tag{1.7}$$

We may describe a line charge as a thin "tube" of volume charge distributed along some contour Γ. The amount of charge contained between two planes normal to the contour and separated by a distance dl is described by the *line charge density* $\rho_l(\mathbf{r}, t)$. The volume charge density associated with the contour is then

$$\rho(\mathbf{r}, \rho, t) = \rho_l(\mathbf{r}, t) f_s(\rho, \Delta),$$

where ρ is the radial distance from the contour in the plane normal to Γ and $f_s(\rho, \Delta)$ is a density function with the properties

$$\int_0^\infty f_s(\rho, \Delta) 2\pi \rho \, d\rho = 1$$

and

$$\lim_{\Delta \to 0} f_s(\rho, \Delta) = \frac{\delta(\rho)}{2\pi\rho}.$$

For example, we might have

$$f_s(\rho, \Delta) = \frac{e^{-\rho^2/\Delta^2}}{\pi\Delta^2}. \tag{1.8}$$

Then the total charge contained between planes separated by a distance dl is

$$dQ(t) = \int_0^\infty [\rho_l(\mathbf{r}, t) \, dl] \, f_s(\rho, \Delta) 2\pi \rho \, d\rho = \rho_l(\mathbf{r}, t) \, dl$$

and the total charge contained between planes placed at the ends of a contour Γ is

$$Q(t) = \int_\Gamma \rho_l(\mathbf{r}, t) \, dl. \tag{1.9}$$

We may define surface and line currents similarly. A surface current is merely a volume current confined to the vicinity of a surface S. The volume current density may be represented using a surface current density function $\mathbf{J}_s(\mathbf{r}, t)$, defined at each point $\mathbf{r}$ on the surface so that

$$\mathbf{J}(\mathbf{r}, w, t) = \mathbf{J}_s(\mathbf{r}, t) f(w, \Delta).$$

Here $f(w, \Delta)$ is some appropriate density function such as (1.6), and the surface current vector obeys $\hat{\mathbf{n}} \cdot \mathbf{J}_s = 0$ where $\hat{\mathbf{n}}$ is normal to S. The total current flowing through a strip of width dl arranged perpendicular to S at $\mathbf{r}$ is

$$dI(t) = \int_{-\infty}^\infty [\mathbf{J}_s(\mathbf{r}, t) \cdot \hat{\mathbf{n}}_l(\mathbf{r}) \, dl] \, f(w, \Delta) \, dw = \mathbf{J}_s(\mathbf{r}, t) \cdot \hat{\mathbf{n}}_l(\mathbf{r}) \, dl$$

where $\hat{\mathbf{n}}_l$ is normal to the strip at $\mathbf{r}$ (and thus also tangential to S at $\mathbf{r}$). The total current passing through a strip intersecting with S along a contour Γ is thus

$$I(t) = \int_\Gamma \mathbf{J}_s(\mathbf{r}, t) \cdot \hat{\mathbf{n}}_l(\mathbf{r}) \, dl.$$

We may describe a line current as a thin "tube" of volume current distributed about some contour Γ and flowing parallel to it. The amount of current passing through a plane normal to the contour is described by the *line current density* $J_l(\mathbf{r}, t)$. The volume current density associated with the contour may be written as

$$\mathbf{J}(\mathbf{r}, \rho, t) = \hat{\mathbf{u}}(\mathbf{r}) J_l(\mathbf{r}, t) f_s(\rho, \Delta),$$

where $\hat{\mathbf{u}}$ is a unit vector along Γ, ρ is the radial distance from the contour in the plane normal to Γ, and $f_s(\rho, \Delta)$ is a density function such as (1.8). The total current passing through any plane normal to Γ at $\mathbf{r}$ is

$$I(t) = \int_0^\infty [J_l(\mathbf{r},t)\hat{\mathbf{u}}(\mathbf{r}) \cdot \hat{\mathbf{u}}(\mathbf{r})]\, f_s(\rho,\Delta) 2\pi\rho\, d\rho = J_l(\mathbf{r},t).$$

It is often convenient to employ *singular* models for continuous source densities. For instance, it is mathematically simpler to regard a surface charge as residing only in the surface S than to regard it as being distributed about the surface. Of course, the source is then discontinuous since it is zero everywhere outside the surface. We may obtain a representation of such a charge distribution by letting the thickness parameter Δ in the density functions recede to zero, thus concentrating the source into a plane or a line. We describe the limit of the density function in terms of the δ-function. For instance, the volume charge distribution for a surface charge located about the xy-plane is

$$\rho(x,y,z,t) = \rho_s(x,y,t)f(z,\Delta).$$

As $\Delta \to 0$ we have

$$\rho(x,y,z,t) = \rho_s(x,y,t) \lim_{\Delta \to 0} f(z,\Delta) = \rho_s(x,y,t)\delta(z).$$

It is a simple matter to represent singular source densities in this way as long as the surface or line is easily parameterized in terms of constant values of coordinate variables. However, care must be taken to represent the δ-function properly. For instance, the density of charge on the surface of a cone at $\theta = \theta_0$ may be described using the distance normal to this surface, which is given by $r\theta - r\theta_0$:

$$\rho(r,\theta,\phi,t) = \rho_s(r,\phi,t)\delta\left(r[\theta - \theta_0]\right).$$

Using the property $\delta(ax) = \delta(x)/a$, we can also write this as

$$\rho(r,\theta,\phi,t) = \rho_s(r,\phi,t)\frac{\delta(\theta - \theta_0)}{r}.$$

1.3.4 Charge conservation

There are four fundamental conservation laws in physics: conservation of energy, momentum, angular momentum, and charge. These laws are said to be *absolute*; they have never been observed to fail. In that sense they are true empirical laws of physics.

However, in modern physics the fundamental conservation laws have come to represent more than just observed facts. Each law is now associated with a fundamental symmetry of the universe; conversely, each known symmetry is associated with a conservation principle. For example, energy conservation can be shown to arise from the observation that the universe is symmetric with respect to time; the laws of physics do not depend on choice of time origin $t = 0$. Similarly, momentum conservation arises from the observation that the laws of physics are invariant under translation, while angular momentum conservation arises from invariance under rotation.

The law of conservation of charge also arises from a symmetry principle. But instead of being spatial or temporal in character, it is related to the invariance of electrostatic potential. Experiments show that there is no absolute potential, only potential difference. The laws of nature are invariant with respect to what we choose as the "reference" potential. This in turn is related to the invariance of Maxwell's equations under gauge transforms; the values of the electric and magnetic fields do not depend on which gauge transformation we use to relate the scalar potential Φ to the vector potential $\mathbf{A}$.

We may state the conservation of charge as follows:

The net charge in any closed system remains constant with time.

This does not mean that individual charges cannot be created or destroyed, only that the total charge in any isolated system must remain constant. Thus it is possible for a positron with charge e to annihilate an electron with charge $-e$ without changing the net charge of the system. Only if a system is not closed can its net charge be altered; since moving charge constitutes current, we can say that the total charge within a system depends on the current passing through the surface enclosing the system. This is the essence of the continuity equation. To derive this important result we consider a closed system within which the charge remains constant, and apply the Reynolds transport theorem (see § A.2).

The continuity equation. Consider a region of space occupied by a distribution of charge whose velocity is given by the vector field $\mathbf{v}$. We surround a portion of charge by a surface S and let S deform as necessary to "follow" the charge as it moves. Since S always contains precisely the same charged particles, we have an isolated system for which the time rate of change of total charge must vanish. An expression for the time rate of change is given by the Reynolds transport theorem (A.66); we have[2]

$$\frac{DQ}{Dt} = \frac{D}{Dt} \int_{V(t)} \rho \, dV = \int_{V(t)} \frac{\partial \rho}{\partial t} \, dV + \oint_{S(t)} \rho \mathbf{v} \cdot d\mathbf{S} = 0.$$

The "D/Dt" notation indicates that the volume region $V(t)$ moves with its enclosed particles. Since $\rho \mathbf{v}$ represents current density, we can write

$$\int_{V(t)} \frac{\partial \rho(\mathbf{r}, t)}{\partial t} \, dV + \oint_{S(t)} \mathbf{J}(\mathbf{r}, t) \cdot d\mathbf{S} = 0. \tag{1.10}$$

In this *large-scale form* of the continuity equation, the partial derivative term describes the time rate of change of the charge density for a fixed spatial position $\mathbf{r}$. At any time t, the time rate of change of charge density integrated over a volume is exactly compensated by the total current exiting through the surrounding surface.

We can obtain the continuity equation in *point form* by applying the divergence theorem to the second term of (1.10) to get

$$\int_{V(t)} \left[\frac{\partial \rho(\mathbf{r}, t)}{\partial t} + \nabla \cdot \mathbf{J}(\mathbf{r}, t) \right] dV = 0.$$

Since $V(t)$ is arbitrary we can set the integrand to zero to obtain

$$\frac{\partial \rho(\mathbf{r}, t)}{\partial t} + \nabla \cdot \mathbf{J}(\mathbf{r}, t) = 0. \tag{1.11}$$

---

[2]Note that in Appendix A we use the symbol $\mathbf{u}$ to represent the velocity of a material and $\mathbf{v}$ to represent the velocity of an artificial surface.

This expression involves the time derivative of ρ with $\mathbf{r}$ fixed. We can also find an expression in terms of the material derivative by using the transport equation (A.67). Enforcing conservation of charge by setting that expression to zero, we have

$$\frac{D\rho(\mathbf{r},t)}{Dt} + \rho(\mathbf{r},t)\,\nabla\cdot\mathbf{v}(\mathbf{r},t) = 0. \tag{1.12}$$

Here $D\rho/Dt$ is the time rate of change of the charge density experienced by an observer moving with the current.

We can state the large-scale form of the continuity equation in terms of a stationary volume. Integrating (1.11) over a stationary volume region V and using the divergence theorem, we find that

$$\int_V \frac{\partial\rho(\mathbf{r},t)}{\partial t}\,dV = -\oint_S \mathbf{J}(\mathbf{r},t)\cdot d\mathbf{S}.$$

Since V is not changing with time we have

$$\frac{dQ(t)}{dt} = \frac{d}{dt}\int_V \rho(\mathbf{r},t)\,dV = -\oint_S \mathbf{J}(\mathbf{r},t)\cdot d\mathbf{S}. \tag{1.13}$$

Hence any increase of total charge within V must be produced by current entering V through S.

Use of the continuity equation.   As an example, suppose that in a bounded region of space we have

$$\rho(\mathbf{r},t) = \rho_0 r e^{-\beta t}.$$

We wish to find $\mathbf{J}$ and $\mathbf{v}$, and to verify both versions of the continuity equation in point form. The spherical symmetry of ρ requires that $\mathbf{J} = \hat{\mathbf{r}}J_r$. Application of (1.13) over a sphere of radius a gives

$$4\pi\frac{d}{dt}\int_0^a \rho_0 r e^{-\beta t} r^2\,dr = -4\pi J_r(a)a^2.$$

Hence

$$\mathbf{J} = \hat{\mathbf{r}}\beta\rho_0\frac{r^2}{4}e^{-\beta t}$$

and therefore

$$\nabla\cdot\mathbf{J} = \frac{1}{r^2}\frac{\partial}{\partial r}(r^2 J_r) = \beta\rho_0 r e^{-\beta t}.$$

The velocity is

$$\mathbf{v} = \frac{\mathbf{J}}{\rho} = \hat{\mathbf{r}}\beta\frac{r}{4},$$

and we have $\nabla\cdot\mathbf{v} = 3\beta/4$. To verify the continuity equations, we compute the time derivatives

$$\frac{\partial\rho}{\partial t} = -\beta\rho_0 r e^{-\beta t},$$

$$\begin{aligned}
\frac{D\rho}{Dt} &= \frac{\partial\rho}{\partial t} + \mathbf{v}\cdot\nabla\rho \\
&= -\beta\rho_0 r e^{-\beta t} + \left(\hat{\mathbf{r}}\beta\frac{r}{4}\right)\cdot\left(\hat{\mathbf{r}}\rho_0 e^{-\beta t}\right) \\
&= -\frac{3}{4}\beta\rho_0 r e^{-\beta t}.
\end{aligned}$$

Note that the charge density decreases with time less rapidly for a moving observer than for a stationary one (3/4 as fast): the moving observer is following the charge outward, and $\rho \propto r$. Now we can check the continuity equations. First we see

$$\frac{D\rho}{Dt} + \rho \nabla \cdot \mathbf{v} = -\frac{3}{4}\beta\rho_0 r e^{-\beta t} + (\rho_0 r e^{-\beta t})\left(\frac{3}{4}\beta\right) = 0,$$

as required for a moving observer; second we see

$$\frac{\partial \rho}{\partial t} + \nabla \cdot \mathbf{J} = -\beta\rho_0 r e^{-\beta t} + \beta\rho_0 e^{-\beta t} = 0,$$

as required for a stationary observer.

The continuity equation in fewer dimensions. The continuity equation can also be used to relate current and charge on a surface or along a line. By conservation of charge we can write

$$\frac{d}{dt}\int_S \rho_s(\mathbf{r}, t)\, dS = -\oint_\Gamma \mathbf{J}_s(\mathbf{r}, t) \cdot \hat{\mathbf{m}}\, dl \tag{1.14}$$

where $\hat{\mathbf{m}}$ is the vector normal to the curve Γ and tangential to the surface S. By the surface divergence theorem (B.20), the corresponding point form is

$$\frac{\partial \rho_s(\mathbf{r}, t)}{\partial t} + \nabla_s \cdot \mathbf{J}_s(\mathbf{r}, t) = 0. \tag{1.15}$$

Here $\nabla_s \cdot \mathbf{J}_s$ is the surface divergence of the vector field $\mathbf{J}_s$. For instance, in rectangular coordinates in the $z = 0$ plane we have

$$\nabla_s \cdot \mathbf{J}_s = \frac{\partial J_{sx}}{\partial x} + \frac{\partial J_{sy}}{\partial y}.$$

In cylindrical coordinates on the cylinder $\rho = a$, we would have

$$\nabla_s \cdot \mathbf{J}_s = \frac{1}{a}\frac{\partial J_{s\phi}}{\partial \phi} + \frac{\partial J_{sz}}{\partial z}.$$

A detailed description of vector operations on a surface may be found in Tai [190], while many identities may be found in Van Bladel [202].

The equation of continuity for a line is easily established by reference to Figure 1.2. Here the net charge exiting the surface during time Δt is given by

$$\Delta t[I(u_2, t) - I(u_1, t)].$$

Thus, the rate of net increase of charge within the system is

$$\frac{dQ(t)}{dt} = \frac{d}{dt}\int \rho_l(\mathbf{r}, t)\, dl = -[I(u_2, t) - I(u_1, t)]. \tag{1.16}$$

The corresponding point form is found by letting the length of the curve approach zero:

$$\frac{\partial I(l, t)}{\partial l} + \frac{\partial \rho_l(l, t)}{\partial t} = 0, \tag{1.17}$$

where l is arc length along the curve. As an example, suppose the line current on a circular loop antenna is approximately

$$I(\phi, t) = I_0 \cos\left(\frac{\omega a}{c}\phi\right)\cos\omega t,$$

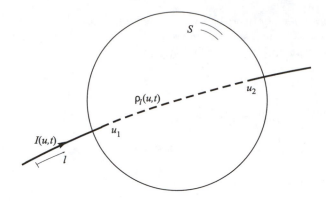

Figure 1.2: Linear form of the continuity equation.

where a is the radius of the loop, ω is the frequency of operation, and c is the speed of light. We wish to find the line charge density on the loop. Since $l = a\phi$, we can write

$$I(l, t) = I_0 \cos\left(\frac{\omega l}{c}\right) \cos\omega t.$$

Thus

$$\frac{\partial I(l, t)}{\partial l} = -I_0 \frac{\omega}{c} \sin\left(\frac{\omega l}{c}\right) \cos\omega t = -\frac{\partial \rho_l(l, t)}{\partial t}.$$

Integrating with respect to time and ignoring any constant (static) charge, we have

$$\rho(l, t) = \frac{I_0}{c} \sin\left(\frac{\omega l}{c}\right) \sin\omega t$$

or

$$\rho(\phi, t) = \frac{I_0}{c} \sin\left(\frac{\omega a}{c}\phi\right) \sin\omega t.$$

Note that we could have used the chain rule

$$\frac{\partial I(\phi, t)}{\partial l} = \frac{\partial I(\phi, t)}{\partial \phi} \frac{\partial \phi}{\partial l} \quad \text{and} \quad \frac{\partial \phi}{\partial l} = \left[\frac{\partial l}{\partial \phi}\right]^{-1} = \frac{1}{a}$$

to calculate the spatial derivative.

We can apply the volume density continuity equation (1.11) directly to surface and line distributions written in singular notation. For the loop of the previous example, we write the volume current density corresponding to the line current as

$$\mathbf{J}(\mathbf{r}, t) = \hat{\boldsymbol{\phi}}\, \delta(\rho - a)\delta(z)I(\phi, t).$$

Substitution into (1.11) then gives

$$\nabla \cdot [\hat{\boldsymbol{\phi}}\delta(\rho - a)\delta(z)I(\phi, t)] = -\frac{\partial \rho(\mathbf{r}, t)}{\partial t}.$$

The divergence formula for cylindrical coordinates gives

$$\delta(\rho - a)\delta(z)\frac{\partial I(\phi, t)}{\rho\,\partial\phi} = -\frac{\partial \rho(\mathbf{r}, t)}{\partial t}.$$

Next we substitute for $I(\phi, t)$ to get

$$-\frac{I_0}{\rho}\frac{\omega a}{c}\sin\left(\frac{\omega a}{c}\phi\right)\delta(\rho - a)\delta(z)\cos\omega t = -\frac{\partial\rho(\mathbf{r}, t)}{\partial t}.$$

Finally, integrating with respect to time and ignoring any constant term, we have

$$\rho(\mathbf{r}, t) = \frac{I_0}{c}\delta(\rho - a)\delta(z)\sin\left(\frac{\omega a}{c}\phi\right)\sin\omega t,$$

where we have set $\rho = a$ because of the presence of the factor $\delta(\rho - a)$.

1.3.5 Magnetic charge

We take for granted that electric fields are produced by electric charges, whether stationary or in motion. The smallest element of electric charge is the electric *monopole*: a single discretely charged particle from which the electric field diverges. In contrast, experiments show that magnetic fields are created only by currents or by time changing electric fields; hence, magnetic fields have moving electric charge as their source. The elemental source of magnetic field is the magnetic *dipole*, representing a tiny loop of electric current (or a spinning electric particle). The observation made in 1269 by Pierre De Maricourt, that even the smallest magnet has two poles, still holds today.

In a world filled with symmetry at the fundamental level, we find it hard to understand why there should not be a source from which the magnetic field diverges. We would call such a source *magnetic charge*, and the most fundamental quantity of magnetic charge would be exhibited by a *magnetic monopole*. In 1931 Paul Dirac invigorated the search for magnetic monopoles by making the first strong theoretical argument for their existence. Dirac showed that the existence of magnetic monopoles would imply the quantization of electric charge, and would thus provide an explanation for one of the great puzzles of science. Since that time magnetic monopoles have become important players in the "Grand Unified Theories" of modern physics, and in cosmological theories of the origin of the universe.

If magnetic monopoles are ever found to exist, there will be both positive and negatively charged particles whose motions will constitute currents. We can define a macroscopic magnetic charge density ρ_m and current density $\mathbf{J}_m$ exactly as we did with electric charge, and use conservation of magnetic charge to provide a continuity equation:

$$\nabla \cdot \mathbf{J}_m(\mathbf{r}, t) + \frac{\partial\rho_m(\mathbf{r}, t)}{\partial t} = 0. \tag{1.18}$$

With these new sources Maxwell's equations become appealingly symmetric. Despite uncertainties about the existence and physical nature of magnetic monopoles, magnetic charge and current have become an integral part of electromagnetic theory. We often use the concept of *fictitious* magnetic sources to make Maxwell's equations symmetric, and then derive various equivalence theorems for use in the solution of important problems. Thus we can put the idea of magnetic sources to use regardless of whether these sources actually exist.

1.4   Problems

1.1 Write the volume charge density for a singular surface charge located on the sphere $r = r_0$, entirely in terms of spherical coordinates. Find the total charge on the sphere.

1.2 Repeat Problem 1.1 for a charged half plane $\phi = \phi_0$.

1.3 Write the volume charge density for a singular surface charge located on the cylinder $\rho = \rho_0$, entirely in terms of cylindrical coordinates. Find the total charge on the cylinder.

1.4 Repeat Problem 1.3 for a charged half plane $\phi = \phi_0$.

Chapter 2

Maxwell's theory of electromagnetism

2.1 The postulate

In 1864, James Clerk Maxwell proposed one of the most successful theories in the history of science. In a famous memoir to the Royal Society [125] he presented nine equations summarizing all known laws on electricity and magnetism. This was more than a mere cataloging of the laws of nature. By postulating the need for an additional term to make the set of equations self-consistent, Maxwell was able to put forth what is still considered a complete theory of macroscopic electromagnetism. The beauty of Maxwell's equations led Boltzmann to ask, "Was it a god who wrote these lines ...?" [185].

Since that time authors have struggled to find the best way to present Maxwell's theory. Although it is possible to study electromagnetics from an "empirical–inductive" viewpoint (roughly following the historical order of development beginning with static fields), it is only by postulating the complete theory that we can do justice to Maxwell's vision. His concept of the existence of an electromagnetic "field" (as introduced by Faraday) is fundamental to this theory, and has become one of the most significant principles of modern science.

We find controversy even over the best way to present Maxwell's equations. Maxwell worked at a time before vector notation was completely in place, and thus chose to use scalar variables and equations to represent the fields. Certainly the true beauty of Maxwell's equations emerges when they are written in vector form, and the use of tensors reduces the equations to their underlying physical simplicity. We shall use vector notation in this book because of its wide acceptance by engineers, but we still must decide whether it is more appropriate to present the vector equations in integral or point form.

On one side of this debate, the brilliant mathematician David Hilbert felt that the fundamental natural laws should be posited as axioms, each best described in terms of integral equations [154]. This idea has been championed by Truesdell and Toupin [199]. On the other side, we may quote from the great physicist Arnold Sommerfeld: "The general development of Maxwell's theory must proceed from its differential form; for special problems the integral form may, however, be more advantageous" ([185], p. 23). Special relativity flows naturally from the point forms, with fields easily converted between moving reference frames. For stationary media, it seems to us that the only difference between the two approaches arises in how we handle discontinuities in sources and materials. If we choose to use the point forms of Maxwell's equations, then we must also postulate the boundary conditions at surfaces of discontinuity. This is pointed out clearly by Tai [192], who also notes that if the integral forms are used, then their validity across regions of discontinuity should be stated as part of the postulate.

We have decided to use the point form in this text. In doing so we follow a long history begun by Hertz in 1890 [85] when he wrote down Maxwell's differential equations as a set of axioms, recognizing the equations as the launching point for the theory of electromagnetism. Also, by postulating Maxwell's equations in point form we can take full advantage of modern developments in the theory of partial differential equations; in particular, the idea of a "well-posed" theory determines what sort of information must be specified to make the postulate useful.

We must also decide which form of Maxwell's differential equations to use as the basis of our postulate. There are several competing forms, each differing on the manner in which materials are considered. The oldest and most widely used form was suggested by Minkowski in 1908 [130]. In the Minkowski form the differential equations contain no mention of the materials supporting the fields; all information about material media is relegated to the constitutive relationships. This places simplicity of the differential equations above intuitive understanding of the behavior of fields in materials. We choose the Maxwell–Minkowski form as the basis of our postulate, primarily for ease of manipulation. But we also recognize the value of other versions of Maxwell's equations. We shall present the basic ideas behind the Boffi form, which places some information about materials into the differential equations (although constitutive relationships are still required). Missing, however, is any information regarding the velocity of a moving medium. By using the polarization and magnetization vectors $\mathbf{P}$ and $\mathbf{M}$ rather than the fields $\mathbf{D}$ and $\mathbf{H}$, it is sometimes easier to visualize the meaning of the field vectors and to understand (or predict) the nature of the constitutive relations.

The Chu and Amperian forms of Maxwell's equations have been promoted as useful alternatives to the Minkowski and Boffi forms. These include explicit information about the velocity of a moving material, and differ somewhat from the Boffi form in the physical interpretation of the electric and magnetic properties of matter. Although each of these models matter in terms of charged particles immersed in free space, magnetization in the Boffi and Amperian forms arises from electric current loops, while the Chu form employs magnetic dipoles. In all three forms polarization is modeled using electric dipoles. For a detailed discussion of the Chu and Amperian forms, the reader should consult the work of Kong [101], Tai [193], Penfield and Haus [145], or Fano, Chu and Adler [70].

Importantly, all of these various forms of Maxwell's equations produce the same values of the physical fields (at least external to the material where the fields are measurable).

We must include several other constituents, besides the field equations, to make the postulate complete. To form a complete field theory we need a source field, a mediating field, and a set of field differential equations. This allows us to mathematically describe the relationship between effect (the mediating field) and cause (the source field). In a well-posed postulate we must also include a set of constitutive relationships and a specification of some field relationship over a bounding surface and at an initial time. If the electromagnetic field is to have physical meaning, we must link it to some observable quantity such as force. Finally, to allow the solution of problems involving mathematical discontinuities we must specify certain boundary, or "jump," conditions.

2.1.1   The Maxwell–Minkowski equations

In Maxwell's macroscopic theory of electromagnetics, the source field consists of the vector field $\mathbf{J}(\mathbf{r}, t)$ (the current density) and the scalar field $\rho(\mathbf{r}, t)$ (the charge density). In

Minkowski's form of Maxwell's equations, the mediating field is the *electromagnetic field* consisting of the set of four vector fields $\mathbf{E}(\mathbf{r}, t)$, $\mathbf{D}(\mathbf{r}, t)$, $\mathbf{B}(\mathbf{r}, t)$, and $\mathbf{H}(\mathbf{r}, t)$. The field equations are the four partial differential equations referred to as the *Maxwell–Minkowski equations*

$$\nabla \times \mathbf{E}(\mathbf{r}, t) = -\frac{\partial}{\partial t}\mathbf{B}(\mathbf{r}, t), \tag{2.1}$$

$$\nabla \times \mathbf{H}(\mathbf{r}, t) = \mathbf{J}(\mathbf{r}, t) + \frac{\partial}{\partial t}\mathbf{D}(\mathbf{r}, t), \tag{2.2}$$

$$\nabla \cdot \mathbf{D}(\mathbf{r}, t) = \rho(\mathbf{r}, t), \tag{2.3}$$

$$\nabla \cdot \mathbf{B}(\mathbf{r}, t) = 0, \tag{2.4}$$

along with the continuity equation

$$\nabla \cdot \mathbf{J}(\mathbf{r}, t) = -\frac{\partial}{\partial t}\rho(\mathbf{r}, t). \tag{2.5}$$

Here (2.1) is called *Faraday's law*, (2.2) is called *Ampere's law*, (2.3) is called *Gauss's law*, and (2.4) is called the *magnetic Gauss's law*. For brevity we shall often leave the dependence on $\mathbf{r}$ and t implicit, and refer to the Maxwell–Minkowski equations as simply the "Maxwell equations," or "Maxwell's equations."

Equations (2.1)–(2.5), the point forms of the field equations, describe the relationships between the fields and their sources at each point in space where the fields are continuously differentiable (i.e., the derivatives exist and are continuous). Such points are called *ordinary points*. We shall not attempt to define the fields at other points, but instead seek conditions relating the fields across surfaces containing these points. Normally this is necessary on surfaces across which either sources or material parameters are discontinuous.

The electromagnetic fields carry SI units as follows: $\mathbf{E}$ is measured in Volts per meter (V/m), $\mathbf{B}$ is measured in Teslas (T), $\mathbf{H}$ is measured in Amperes per meter (A/m), and $\mathbf{D}$ is measured in Coulombs per square meter (C/m^2). In older texts we find the units of $\mathbf{B}$ given as Webers per square meter (Wb/m^2) to reflect the role of $\mathbf{B}$ as a flux vector; in that case the Weber (Wb = T·m^2) is regarded as a unit of magnetic flux.

The interdependence of Maxwell's equations. It is often claimed that the divergence equations (2.3) and (2.4) may be derived from the curl equations (2.1) and (2.2). While this is true, it is *not* proper to say that only the two curl equations are required to describe Maxwell's theory. This is because an additional physical assumption, not present in the two curl equations, is required to complete the derivation. Either the divergence equations must be specified, or the values of certain constants that fix the initial conditions on the fields must be specified. It is customary to specify the divergence equations and include them with the curl equations to form the complete set we now call "Maxwell's equations."

To identify the interdependence we take the divergence of (2.1) to get

$$\nabla \cdot (\nabla \times \mathbf{E}) = \nabla \cdot \left(-\frac{\partial \mathbf{B}}{\partial t}\right),$$

hence

$$\frac{\partial}{\partial t}(\nabla \cdot \mathbf{B}) = 0$$

by (B.49). This requires that $\nabla \cdot \mathbf{B}$ be constant with time, say $\nabla \cdot \mathbf{B}(\mathbf{r}, t) = C_B(\mathbf{r})$. The constant C_B must be specified as part of the postulate of Maxwell's theory, and the choice we make is subject to experimental validation. We postulate that $C_B(\mathbf{r}) = 0$, which leads us to (2.4). Note that if we can identify a time prior to which $\mathbf{B}(\mathbf{r}, t) \equiv 0$, then $C_B(\mathbf{r})$ must vanish. For this reason, $C_B(\mathbf{r}) = 0$ and (2.4) are often called the "initial conditions" for Faraday's law [159].

Next we take the divergence of (2.2) to find that

$$\nabla \cdot (\nabla \times \mathbf{H}) = \nabla \cdot \mathbf{J} + \frac{\partial}{\partial t}(\nabla \cdot \mathbf{D}).$$

Using (2.5) and (B.49), we obtain

$$\frac{\partial}{\partial t}(\rho - \nabla \cdot \mathbf{D}) = 0$$

and thus $\rho - \nabla \cdot \mathbf{D}$ must be some temporal constant $C_D(\mathbf{r})$. Again, we must postulate the value of C_D as part of the Maxwell theory. We choose $C_D(\mathbf{r}) = 0$ and thus obtain Gauss's law (2.3). If we can identify a time prior to which both $\mathbf{D}$ and ρ are everywhere equal to zero, then $C_D(\mathbf{r})$ must vanish. Hence $C_D(\mathbf{r}) = 0$ and (2.3) may be regarded as "initial conditions" for Ampere's law. Combining the two sets of initial conditions, we find that the curl equations imply the divergence equations as long as we can find a time prior to which all of the fields $\mathbf{E}, \mathbf{D}, \mathbf{B}, \mathbf{H}$ and the sources $\mathbf{J}$ and ρ are equal to zero (since all the fields are related through the curl equations, and the charge and current are related through the continuity equation). Conversely, the empirical evidence supporting the two divergence equations implies that such a time should exist.

Throughout this book we shall refer to the two curl equations as the "fundamental" Maxwell equations, and to the two divergence equations as the "auxiliary" equations. The fundamental equations describe the relationships between the fields while, as we have seen, the auxiliary equations provide a sort of initial condition. This does not imply that the auxiliary equations are of lesser importance; indeed, they are required to establish uniqueness of the fields, to derive the wave equations for the fields, and to properly describe static fields.

Field vector terminology. Various terms are used for the field vectors, sometimes harkening back to the descriptions used by Maxwell himself, and often based on the physical nature of the fields. We are attracted to Sommerfeld's separation of the fields into *entities of intensity* $(\mathbf{E}, \mathbf{B})$ and *entities of quantity* $(\mathbf{D}, \mathbf{H})$. In this system $\mathbf{E}$ is called the *electric field strength*, $\mathbf{B}$ the *magnetic field strength*, $\mathbf{D}$ the *electric excitation*, and $\mathbf{H}$ the *magnetic excitation* [185]. Maxwell separated the fields into a set $(\mathbf{E}, \mathbf{H})$ of vectors that appear within line integrals to give work-related quantities, and a set $(\mathbf{B}, \mathbf{D})$ of vectors that appear within surface integrals to give flux-related quantities; we shall see this clearly when considering the integral forms of Maxwell's equations. By this system, authors such as Jones [97] and Ramo, Whinnery, and Van Duzer [153] call $\mathbf{E}$ the *electric intensity*, $\mathbf{H}$ the *magnetic intensity*, $\mathbf{B}$ the *magnetic flux density*, and $\mathbf{D}$ the *electric flux density*.

Maxwell himself designated names for each of the vector quantities. In his classic paper "A Dynamical Theory of the Electromagnetic Field," [178] Maxwell referred to the quantity we now designate $\mathbf{E}$ as the *electromotive force*, the quantity $\mathbf{D}$ as the *electric displacement* (with a time rate of change given by his now famous "displacement current"), the quantity $\mathbf{H}$ as the *magnetic force*, and the quantity $\mathbf{B}$ as the *magnetic induction* (although he described **B** as a density of lines of magnetic force). Maxwell also included a quantity designated *electromagnetic momentum* as an integral part of his theory. We now know this as the vector potential **A**, which is not generally included as a part of the electromagnetics postulate.

Many authors follow the original terminology of Maxwell, with some slight modifications. For instance, Stratton [187] calls **E** the *electric field intensity*, **H** the *magnetic field intensity*, **D** the *electric displacement*, and **B** the *magnetic induction*. Jackson [91] calls **E** the *electric field*, **H** the *magnetic field*, **D** the *displacement*, and **B** the *magnetic induction*.

Other authors choose freely among combinations of these terms. For instance, Kong [101] calls **E** the *electric field strength*, **H** the *magnetic field strength*, **B** the *magnetic flux density*, and **D** the *electric displacement*. We do not wish to inject further confusion into the issue of nomenclature; still, we find it helpful to use as simple a naming system as possible. We shall refer to **E** as the *electric field*, **H** as the *magnetic field*, **D** as the *electric flux density* and **B** as the *magnetic flux density*. When we use the term *electromagnetic field* we imply the entire set of field vectors $(\mathbf{E}, \mathbf{D}, \mathbf{B}, \mathbf{H})$ used in Maxwell's theory.

Invariance of Maxwell's equations. Maxwell's differential equations are valid for any system in uniform relative motion with respect to the laboratory frame of reference in which we normally do our measurements. The field equations describe the relationships between the source and mediating fields *within that frame of reference*. This property was first proposed for moving material media by Minkowski in 1908 (using the term *covariance*) [130]. For this reason, Maxwell's equations expressed in the form (2.1)–(2.2) are referred to as the *Minkowski form*.

2.1.2 Connection to mechanics

Our postulate must include a connection between the abstract quantities of charge and field and a measurable physical quantity. A convenient means of linking electromagnetics to other classical theories is through mechanics. We postulate that charges experience mechanical forces given by the *Lorentz force equation*. If a small volume element dV contains a total charge $\rho\,dV$, then the force experienced by that charge when moving at velocity **v** in an electromagnetic field is

$$d\mathbf{F} = \rho\,dV\,\mathbf{E} + \rho\mathbf{v}\,dV \times \mathbf{B}. \qquad (2.6)$$

As with any postulate, we verify this equation through experiment. Note that we write the Lorentz force in terms of charge $\rho\,dV$, rather than charge density ρ, since charge is an invariant quantity under a Lorentz transformation.

The important links between the electromagnetic fields and energy and momentum must also be postulated. We postulate that the quantity

$$\mathbf{S}_{em} = \mathbf{E} \times \mathbf{H} \qquad (2.7)$$

represents the transport density of electromagnetic power, and that the quantity

$$\mathbf{g}_{em} = \mathbf{D} \times \mathbf{B} \qquad (2.8)$$

represents the transport density of electromagnetic momentum.

2.2   The well-posed nature of the postulate

It is important to investigate whether Maxwell's equations, along with the point form of the continuity equation, suffice as a useful theory of electromagnetics. Certainly we must agree that a theory is "useful" as long as it is defined as such by the scientists and engineers who employ it. In practice a theory is considered useful if it predicts accurately the behavior of nature under given circumstances, and even a theory that often fails may be useful if it is the best available. We choose here to take a more narrow view and investigate whether the theory is "well-posed."

A mathematical model for a physical problem is said to be *well-posed*, or *correctly set*, if three conditions hold:

1. the model has at least one solution (*existence*);

2. the model has at most one solution (*uniqueness*);

3. the solution is continuously dependent on the data supplied.

The importance of the first condition is obvious: if the electromagnetic model has no solution, it will be of little use to scientists and engineers. The importance of the second condition is equally obvious: if we apply two different solution methods to the same model and get two different answers, the model will not be very helpful in analysis or design work. The third point is more subtle; it is often extended in a practical sense to the following statement:

3′. Small changes in the data supplied produce equally small changes in the solution.

That is, the solution is not sensitive to errors in the data. To make sense of this we must decide which quantity is specified (the independent quantity) and which remains to be calculated (the dependent quantity). Commonly the source field (charge) is taken as the independent quantity, and the mediating (electromagnetic) field is computed from it; in such cases it can be shown that Maxwell's equations are well-posed. Taking the electromagnetic field to be the independent quantity, we can produce situations in which the computed quantity (charge or current) changes wildly with small changes in the specified fields. These situations (called *inverse problems*) are of great importance in remote sensing, where the field is measured and the properties of the object probed are thereby deduced.

At this point we shall concentrate on the "forward" problem of specifying the source field (charge) and computing the mediating field (the electromagnetic field). In this case we may question whether the first of the three conditions (existence) holds. We have twelve unknown quantities (the scalar components of the four vector fields), but only eight equations to describe them (from the scalar components of the two fundamental Maxwell equations and the two scalar auxiliary equations). With fewer equations than unknowns we cannot be sure that a solution exists, and we refer to Maxwell's equations as being *indefinite*. To overcome this problem we must specify more information in the form of constitutive relations among the field quantities $\mathbf{E}$, $\mathbf{B}$, $\mathbf{D}$, $\mathbf{H}$, and $\mathbf{J}$. When these are properly formulated, the number of unknowns and the number of equations are equal and Maxwell's equations are in *definite form*. If we provide more equations than unknowns, the solution may be non-unique. When we model the electromagnetic properties of materials we must supply precisely the right amount of information in the constitutive relations, or our postulate will not be well-posed.

Once Maxwell's equations are in definite form, standard methods for partial differential equations can be used to determine whether the electromagnetic model is well-posed. In a nutshell, the system (2.1)–(2.2) of hyperbolic differential equations is well-posed if and only if we specify $\mathbf{E}$ and $\mathbf{H}$ throughout a volume region V at some time instant and also specify, at all subsequent times,

1. the tangential component of $\mathbf{E}$ over all of the boundary surface S, or

2. the tangential component of $\mathbf{H}$ over all of S, or

3. the tangential component of $\mathbf{E}$ over part of S, and the tangential component of $\mathbf{H}$ over the remainder of S.

Proof of all three of the conditions of well-posedness is quite tedious, but a simplified uniqueness proof is often given in textbooks on electromagnetics. The procedure used by Stratton [187] is reproduced below. The interested reader should refer to Hansen [81] for a discussion of the existence of solutions to Maxwell's equations.

2.2.1   Uniqueness of solutions to Maxwell's equations

Consider a simply connected region of space V bounded by a surface S, where both V and S contain only ordinary points. The fields within V are associated with a current distribution $\mathbf{J}$, which may be internal to V (entirely or in part). By the initial conditions that imply the auxiliary Maxwell's equations, we know there is a time, say $t = 0$, prior to which the current is zero for all time, and thus by causality the fields throughout V are identically zero for all times $t < 0$. We next assume that the fields are specified throughout V at some time $t_0 > 0$, and seek conditions under which they are determined uniquely for all $t > t_0$.

Let the field set $(\mathbf{E}_1, \mathbf{D}_1, \mathbf{B}_1, \mathbf{H}_1)$ be a solution to Maxwell's equations (2.1)–(2.2) associated with the current $\mathbf{J}$ (along with an appropriate set of constitutive relations), and let $(\mathbf{E}_2, \mathbf{D}_2, \mathbf{B}_2, \mathbf{H}_2)$ be a second solution associated with $\mathbf{J}$. To determine the conditions for uniqueness of the fields, we look for a situation that results in $\mathbf{E}_1 = \mathbf{E}_2$, $\mathbf{B}_1 = \mathbf{B}_2$, and so on. The electromagnetic fields must obey

$$\nabla \times \mathbf{E}_1 = -\frac{\partial \mathbf{B}_1}{\partial t},$$

$$\nabla \times \mathbf{H}_1 = \mathbf{J} + \frac{\partial \mathbf{D}_1}{\partial t},$$

$$\nabla \times \mathbf{E}_2 = -\frac{\partial \mathbf{B}_2}{\partial t},$$

$$\nabla \times \mathbf{H}_2 = \mathbf{J} + \frac{\partial \mathbf{D}_2}{\partial t}.$$

Subtracting, we have

$$\nabla \times (\mathbf{E}_1 - \mathbf{E}_2) = -\frac{\partial (\mathbf{B}_1 - \mathbf{B}_2)}{\partial t}, \tag{2.9}$$

$$\nabla \times (\mathbf{H}_1 - \mathbf{H}_2) = \frac{\partial (\mathbf{D}_1 - \mathbf{D}_2)}{\partial t}, \tag{2.10}$$

hence defining $\mathbf{E}_0 = \mathbf{E}_1 - \mathbf{E}_2$, $\mathbf{B}_0 = \mathbf{B}_1 - \mathbf{B}_2$, and so on, we have

$$\mathbf{E}_0 \cdot (\nabla \times \mathbf{H}_0) = \mathbf{E}_0 \cdot \frac{\partial \mathbf{D}_0}{\partial t}, \tag{2.11}$$

$$\mathbf{H}_0 \cdot (\nabla \times \mathbf{E}_0) = -\mathbf{H}_0 \cdot \frac{\partial \mathbf{B}_0}{\partial t}. \tag{2.12}$$

Subtracting again, we have

$$\mathbf{E}_0 \cdot (\nabla \times \mathbf{H}_0) - \mathbf{H}_0 \cdot (\nabla \times \mathbf{E}_0) = \mathbf{H}_0 \cdot \frac{\partial \mathbf{B}_0}{\partial t} + \mathbf{E}_0 \cdot \frac{\partial \mathbf{D}_0}{\partial t},$$

hence

$$-\nabla \cdot (\mathbf{E}_0 \times \mathbf{H}_0) = \mathbf{E}_0 \cdot \frac{\partial \mathbf{D}_0}{\partial t} + \mathbf{H}_0 \cdot \frac{\partial \mathbf{B}_0}{\partial t}$$

by (B.44). Integrating both sides throughout V and using the divergence theorem on the left-hand side, we get

$$-\oint_S (\mathbf{E}_0 \times \mathbf{H}_0) \cdot d\mathbf{S} = \int_V \left(\mathbf{E}_0 \cdot \frac{\partial \mathbf{D}_0}{\partial t} + \mathbf{H}_0 \cdot \frac{\partial \mathbf{B}_0}{\partial t} \right) dV.$$

Breaking S into two arbitrary portions and using (B.6), we obtain

$$\int_{S_1} \mathbf{E}_0 \cdot (\hat{\mathbf{n}} \times \mathbf{H}_0)\, dS - \int_{S_2} \mathbf{H}_0 \cdot (\hat{\mathbf{n}} \times \mathbf{E}_0)\, dS = \int_V \left(\mathbf{E}_0 \cdot \frac{\partial \mathbf{D}_0}{\partial t} + \mathbf{H}_0 \cdot \frac{\partial \mathbf{B}_0}{\partial t} \right) dV.$$

Now if $\hat{\mathbf{n}} \times \mathbf{E}_0 = 0$ or $\hat{\mathbf{n}} \times \mathbf{H}_0 = 0$ over all of S, or some combination of these conditions holds over all of S, then

$$\int_V \left(\mathbf{E}_0 \cdot \frac{\partial \mathbf{D}_0}{\partial t} + \mathbf{H}_0 \cdot \frac{\partial \mathbf{B}_0}{\partial t} \right) dV = 0. \tag{2.13}$$

This expression implies a relationship between $\mathbf{E}_0, \mathbf{D}_0, \mathbf{B}_0,$ and $\mathbf{H}_0$. Since V is arbitrary, we see that one possibility is simply to have $\mathbf{D}_0$ and $\mathbf{B}_0$ constant with time. However, since the fields are identically zero for $t < 0$, if they are constant for all time then those constant values must be zero. Another possibility is to have one of each pair $(\mathbf{E}_0, \mathbf{D}_0)$ and $(\mathbf{H}_0, \mathbf{B}_0)$ equal to zero. Then, by (2.9) and (2.10), $\mathbf{E}_0 = 0$ implies $\mathbf{B}_0 = 0$, and $\mathbf{D}_0 = 0$ implies $\mathbf{H}_0 = 0$. Thus $\mathbf{E}_1 = \mathbf{E}_2$, $\mathbf{B}_1 = \mathbf{B}_2$, and so on, and the solution is unique throughout V. However, we cannot in general rule out more complicated relationships. The number of possibilities depends on the additional constraints on the relationship between $\mathbf{E}_0, \mathbf{D}_0, \mathbf{B}_0,$ and $\mathbf{H}_0$ that we must supply to describe the material supporting the field — i.e., the constitutive relationships. For a simple medium described by the time-constant permittivity ϵ and permeability μ, (2.13) becomes

$$\int_V \left(\mathbf{E}_0 \cdot \epsilon \frac{\partial \mathbf{E}_0}{\partial t} + \mathbf{H}_0 \cdot \mu \frac{\partial \mathbf{H}_0}{\partial t} \right) dV = 0,$$

or

$$\frac{1}{2} \frac{\partial}{\partial t} \int_V (\epsilon \mathbf{E}_0 \cdot \mathbf{E}_0 + \mu \mathbf{H}_0 \cdot \mathbf{H}_0)\, dV = 0.$$

Since the integrand is always positive or zero (and not constant with time, as mentioned above), the only possible conclusion is that $\mathbf{E}_0$ and $\mathbf{H}_0$ must both be zero, and thus the fields are unique.

When establishing more complicated constitutive relations, we must be careful to ensure that they lead to a unique solution, and that the condition for uniqueness is understood. In the case above, the assumption $\hat{\mathbf{n}} \times \mathbf{E}_0 \big|_S = 0$ implies that the tangential components of $\mathbf{E}_1$ and $\mathbf{E}_2$ are identical over S — that is, we must give specific values of these quantities on S to ensure uniqueness. A similar statement holds for the condition $\hat{\mathbf{n}} \times \mathbf{H}_0 \big|_S = 0$. Requiring that constitutive relations lead to a unique solution is known as *just setting*, and is one of several factors that must be considered, as discussed in the next section.

Uniqueness implies that the electromagnetic state of an isolated region of space may be determined without the knowledge of conditions outside the region. If we wish to solve Maxwell's equations for that region, we need know only the source density within the region and the values of the tangential fields over the bounding surface. The effects of a complicated external world are thus reduced to the specification of surface fields. This concept has numerous applications to problems in antennas, diffraction, and guided waves.

2.2.2 Constitutive relations

We now supply a set of constitutive relations to complete the conditions for well-posedness. We generally split these relations into two sets. The first describes the relationships between the electromagnetic field quantities, and the second describes mechanical interaction between the fields and resulting secondary sources. All of these relations depend on the properties of the medium supporting the electromagnetic field. Material phenomena are quite diverse, and it is remarkable that the Maxwell–Minkowski equations hold for all phenomena yet discovered. All material effects, from nonlinearity to chirality to temporal dispersion, are described by the constitutive relations.

The specification of constitutive relationships is required in many areas of physical science to describe the behavior of "ideal materials": mathematical models of actual materials encountered in nature. For instance, in continuum mechanics the constitutive equations describe the relationship between material motions and stress tensors [209]. Truesdell and Toupin [199] give an interesting set of "guiding principles" for the concerned scientist to use when constructing constitutive relations. These include consideration of *consistency* (with the basic conservation laws of nature), *coordinate invariance* (independence of coordinate system), *isotropy and aeolotropy* (dependence on, or independence of, orientation), *just setting* (constitutive parameters should lead to a unique solution), *dimensional invariance* (similarity), *material indifference* (non-dependence on the observer), and *equipresence* (inclusion of *all* relevant physical phenomena in *all* of the constitutive relations across disciplines).

The constitutive relations generally involve a set of constitutive parameters and a set of constitutive operators. The constitutive parameters may be as simple as constants of proportionality between the fields or they may be components in a dyadic relationship. The constitutive operators may be linear and integro-differential in nature, or may imply some nonlinear operation on the fields. If the constitutive parameters are spatially constant within a certain region, we term the medium *homogeneous* within that region. If the constitutive parameters vary spatially, the medium is *inhomogeneous*. If the constitutive parameters are constants with time, we term the medium *stationary*; if they are time-changing, the medium is *nonstationary*. If the constitutive operators involve time derivatives or integrals, the medium is said to be *temporally dispersive*; if space derivatives or integrals are involved, the medium is *spatially dispersive*. Examples of all these effects can be found in common materials. It is important to note that the constitutive parameters may depend on other physical properties of the material, such as temperature, mechanical stress, and isomeric state, just as the mechanical constitutive parameters of a material may depend on the electromagnetic properties (principle of equipresence).

Many effects produced by linear constitutive operators, such as those associated with temporal dispersion, have been studied primarily in the frequency domain. In this case temporal derivative and integral operations produce complex constitutive parameters. It is becoming equally important to characterize these effects directly in the time domain for use with direct time-domain field solving techniques such as the finite-difference time-domain (FDTD) method. We shall cover the very basic properties of dispersive media in this section. A detailed description of frequency-domain fields (and a discussion of complex constitutive parameters) is deferred until later in this book.

It is difficult to find a simple and consistent means for classifying materials by their electromagnetic effects. One way is to separate linear and nonlinear materials, then categorize linear materials by the way in which the fields are coupled through the constitutive relations:

1. *Isotropic* materials are those in which **D** is related to **E**, **B** is related to **H**, and the secondary source current **J** is related to **E**, with the field direction in each pair aligned.

2. In *anisotropic* materials the pairings are the same, but the fields in each pair are generally not aligned.

3. In *biisotropic* materials (such as chiral media) the fields **D** and **B** depend on both **E** and **H**, but with no realignment of **E** or **H**; for instance, **D** is given by the addition of a scalar times **E** plus a second scalar times **H**. Thus the contributions to **D** involve no changes to the directions of **E** and **H**.

4. *Bianisotropic* materials exhibit the most general behavior: **D** and **H** depend on both **E** and **B**, with an arbitrary realignment of either or both of these fields.

In 1888, Roentgen showed experimentally that a material isotropic in its own stationary reference frame exhibits bianisotropic properties when observed from a moving frame. Only recently have materials bianisotropic in their own rest frame been discovered. In 1894 Curie predicted that in a stationary material, based on symmetry, an electric field might produce magnetic effects and a magnetic field might produce electric effects. These effects, coined *magnetoelectric* by Landau and Lifshitz in 1957, were sought unsuccessfully by many experimentalists during the first half of the twentieth century. In 1959 the Soviet scientist I.E. Dzyaloshinskii predicted that, theoretically, the antiferromagnetic material chromium oxide (Cr_2O_3) should display magnetoelectric effects. The magnetoelectric effect was finally observed soon after by D.N. Astrov in a single crystal of Cr_2O_3 using a 10 kHz electric field. Since then the effect has been observed in many different materials. Recently, highly exotic materials with useful electromagnetic properties have been proposed and studied in depth, including chiroplasmas and chiroferrites [211]. As the technology of materials synthesis advances, a host of new and intriguing media will certainly be created.

The most general forms of the constitutive relations between the fields may be written in symbolic form as

$$\mathbf{D} = \mathbf{D}[\mathbf{E}, \mathbf{B}], \tag{2.14}$$
$$\mathbf{H} = \mathbf{H}[\mathbf{E}, \mathbf{B}]. \tag{2.15}$$

That is, **D** and **H** have some mathematically descriptive relationship to **E** and **B**. The specific forms of the relationships may be written in terms of dyadics as [102]

$$c\mathbf{D} = \bar{\mathbf{P}} \cdot \mathbf{E} + \bar{\mathbf{L}} \cdot (c\mathbf{B}), \tag{2.16}$$
$$\mathbf{H} = \bar{\mathbf{M}} \cdot \mathbf{E} + \bar{\mathbf{Q}} \cdot (c\mathbf{B}), \tag{2.17}$$

where each of the quantities $\bar{\mathbf{P}}, \bar{\mathbf{L}}, \bar{\mathbf{M}}, \bar{\mathbf{Q}}$ may be dyadics in the usual sense, or dyadic operators containing space or time derivatives or integrals, or some nonlinear operations on the fields. We may write these expressions as a single matrix equation

$$\begin{bmatrix} c\mathbf{D} \\ \mathbf{H} \end{bmatrix} = [\bar{\mathbf{C}}] \begin{bmatrix} \mathbf{E} \\ c\mathbf{B} \end{bmatrix} \tag{2.18}$$

where the 6×6 matrix

$$[\bar{\mathbf{C}}] = \begin{bmatrix} \bar{\mathbf{P}} & \bar{\mathbf{L}} \\ \bar{\mathbf{M}} & \bar{\mathbf{Q}} \end{bmatrix}.$$

This most general relationship between fields is the property of a bianisotropic material.

We may wonder why $\mathbf{D}$ is not related to $(\mathbf{E}, \mathbf{B}, \mathbf{H})$, $\mathbf{E}$ to $(\mathbf{D}, \mathbf{B})$, etc. The reason is that since the field pairs $(\mathbf{E}, \mathbf{B})$ and $(\mathbf{D}, \mathbf{H})$ convert identically under a Lorentz transformation, a constitutive relation that maps fields as in (2.18) is form invariant, as are the Maxwell–Minkowski equations. That is, although the constitutive parameters may vary numerically between observers moving at different velocities, the form of the relationship given by (2.18) is maintained.

Many authors choose to relate $(\mathbf{D}, \mathbf{B})$ to $(\mathbf{E}, \mathbf{H})$, often because the expressions are simpler and can be more easily applied to specific problems. For instance, in a linear, isotropic material (as shown below) $\mathbf{D}$ is directly proportional to $\mathbf{E}$ and $\mathbf{B}$ is directly proportional to $\mathbf{H}$. To provide the appropriate expression for the constitutive relations, we need only remap (2.18). This gives

$$\mathbf{D} = \bar{\epsilon} \cdot \mathbf{E} + \bar{\xi} \cdot \mathbf{H}, \tag{2.19}$$

$$\mathbf{B} = \bar{\zeta} \cdot \mathbf{E} + \bar{\mu} \cdot \mathbf{H}, \tag{2.20}$$

or

$$\begin{bmatrix} \mathbf{D} \\ \mathbf{B} \end{bmatrix} = [\bar{\mathbf{C}}_{EH}] \begin{bmatrix} \mathbf{E} \\ \mathbf{H} \end{bmatrix}, \tag{2.21}$$

where the new constitutive parameters $\bar{\epsilon}, \bar{\xi}, \bar{\zeta}, \bar{\mu}$ can be easily found from the original constitutive parameters $\bar{\mathbf{P}}, \bar{\mathbf{L}}, \bar{\mathbf{M}}, \bar{\mathbf{Q}}$. We do note, however, that in the form (2.19)–(2.20) the Lorentz invariance of the constitutive equations is not obvious.

In the following paragraphs we shall characterize some of the most common materials according to these classifications. With this approach effects such as temporal or spatial dispersion are not part of the classification process, but arise from the nature of the constitutive parameters. Hence we shall not dwell on the particulars of the constitutive parameters, but shall concentrate on the form of the constitutive relations.

Constitutive relations for fields in free space. In a vacuum the fields are related by the simple constitutive equations

$$\mathbf{D} = \epsilon_0 \mathbf{E}, \tag{2.22}$$

$$\mathbf{H} = \frac{1}{\mu_0} \mathbf{B}. \tag{2.23}$$

The quantities μ_0 and ϵ_0 are, respectively, the *free-space permeability* and *permittivity constants*. It is convenient to use three numerical quantities to describe the electromagnetic properties of free space — μ_0, ϵ_0, and the speed of light c — and interrelate them through the equation

$$c = 1/(\mu_0 \epsilon_0)^{1/2}.$$

Historically it has been the practice to define μ_0, measure c, and compute ϵ_0. In SI units

$$\mu_0 = 4\pi \times 10^{-7} \text{ H/m},$$
$$c = 2.998 \times 10^8 \text{ m/s},$$
$$\epsilon_0 = 8.854 \times 10^{-12} \text{ F/m}.$$

With the two constitutive equations we have enough information to put Maxwell's equations into definite form. Traditionally (2.22) and (2.23) are substituted into (2.1)–(2.2) to give

$$\nabla \times \mathbf{E} = -\frac{\partial \mathbf{B}}{\partial t}, \tag{2.24}$$

$$\nabla \times \mathbf{B} = \mu_0 \mathbf{J} + \mu_0 \epsilon_0 \frac{\partial \mathbf{E}}{\partial t}. \tag{2.25}$$

These are two vector equations in two vector unknowns (equivalently, six scalar equations in six scalar unknowns).

In terms of the general constitutive relation (2.18), we find that free space is isotropic with

$$\bar{\mathbf{P}} = \bar{\mathbf{Q}} = \frac{1}{\eta_0}\bar{\mathbf{I}}, \qquad \bar{\mathbf{L}} = \bar{\mathbf{M}} = 0,$$

where $\eta_0 = (\mu_0/\epsilon_0)^{1/2}$ is called the *intrinsic impedance of free space*. This emphasizes the fact that free space has, along with c, only a single empirical constant associated with it (i.e., ϵ_0 or η_0). Since no derivative or integral operators appear in the constitutive relations, free space is nondispersive.

Constitutive relations in a linear isotropic material. In a linear isotropic material there is proportionality between $\mathbf{D}$ and $\mathbf{E}$ and between $\mathbf{B}$ and $\mathbf{H}$. The constants of proportionality are the *permittivity* ϵ and the *permeability* μ. If the material is nondispersive, the constitutive relations take the form

$$\mathbf{D} = \epsilon\mathbf{E}, \qquad \mathbf{B} = \mu\mathbf{H},$$

where ϵ and μ may depend on position for inhomogeneous materials. Often the permittivity and permeability are referenced to the permittivity and permeability of free space according to

$$\epsilon = \epsilon_r\epsilon_0, \qquad \mu = \mu_r\mu_0.$$

Here the dimensionless quantities ϵ_r and μ_r are called, respectively, the *relative permittivity* and *relative permeability*.

When dealing with the Maxwell–Boffi equations (§ 2.4) the difference between the material and free space values of $\mathbf{D}$ and $\mathbf{H}$ becomes important. Thus for linear isotropic materials we often write the constitutive relations as

$$\mathbf{D} = \epsilon_0\mathbf{E} + \epsilon_0\chi_e\mathbf{E}, \tag{2.26}$$
$$\mathbf{B} = \mu_0\mathbf{H} + \mu_0\chi_m\mathbf{H}, \tag{2.27}$$

where the dimensionless quantities $\chi_e = \epsilon_r - 1$ and $\chi_m = \mu_r - 1$ are called, respectively, the *electric* and *magnetic susceptibilities* of the material. In terms of (2.18) we have

$$\bar{\mathbf{P}} = \frac{\epsilon_r}{\eta_0}\bar{\mathbf{I}}, \qquad \bar{\mathbf{Q}} = \frac{1}{\eta_0\mu_r}\bar{\mathbf{I}}, \qquad \bar{\mathbf{L}} = \bar{\mathbf{M}} = 0.$$

Generally a material will have either its electric or magnetic properties dominant. If $\mu_r = 1$ and $\epsilon_r \neq 1$ then the material is generally called a *perfect dielectric* or a *perfect insulator*, and is said to be an electric material. If $\epsilon_r = 1$ and $\mu_r \neq 1$, the material is said to be a magnetic material.

A linear isotropic material may also have conduction properties. In a *conducting material*, a constitutive relation is generally used to describe the mechanical interaction of field and charge by relating the electric field to a secondary electric current. For a nondispersive isotropic material, the current is aligned with, and proportional to, the electric field; there are no temporal operators in the constitutive relation, which is simply

$$\mathbf{J} = \sigma \mathbf{E}. \tag{2.28}$$

This is known as *Ohm's law*. Here σ is the *conductivity* of the material.

If $\mu_r \approx 1$ and σ is very small, the material is generally called a *good dielectric*. If σ is very large, the material is generally called a *good conductor*. The conditions by which we say the conductivity is "small" or "large" are usually established using the frequency response of the material. Materials that are good dielectrics over broad ranges of frequency include various glasses and plastics such as fused quartz, polyethylene, and teflon. Materials that are good conductors over broad ranges of frequency include common metals such as gold, silver, and copper.

For dispersive linear isotropic materials, the constitutive parameters become nonstationary (time dependent), and the constitutive relations involve time operators. (Note that the name *dispersive* describes the tendency for pulsed electromagnetic waves to spread out, or disperse, in materials of this type.) If we assume that the relationships given by (2.26), (2.27), and (2.28) retain their product form in the frequency domain, then by the convolution theorem we have in the time domain the constitutive relations

$$\mathbf{D}(\mathbf{r}, t) = \epsilon_0 \left(\mathbf{E}(\mathbf{r}, t) + \int_{-\infty}^{t} \chi_e(\mathbf{r}, t - t') \mathbf{E}(\mathbf{r}, t')\, dt' \right), \tag{2.29}$$

$$\mathbf{B}(\mathbf{r}, t) = \mu_0 \left(\mathbf{H}(\mathbf{r}, t) + \int_{-\infty}^{t} \chi_m(\mathbf{r}, t - t') \mathbf{H}(\mathbf{r}, t')\, dt' \right), \tag{2.30}$$

$$\mathbf{J}(\mathbf{r}, t) = \int_{-\infty}^{t} \sigma(\mathbf{r}, t - t') \mathbf{E}(\mathbf{r}, t')\, dt'. \tag{2.31}$$

These expressions were first introduced by Volterra in 1912 [199]. We see that for a linear dispersive material of this type the constitutive operators are time integrals, and that the behavior of $\mathbf{D}(t)$ depends not only on the value of $\mathbf{E}$ at time t, but on its values at all past times. Thus, in dispersive materials there is a "time lag" between the effect of the applied field and the polarization or magnetization that results. In the frequency domain, temporal dispersion is associated with complex values of the constitutive parameters, which, to describe a causal relationship, cannot be constant with frequency. The nonzero imaginary component is identified with the dissipation of electromagnetic energy as heat. Causality is implied by the upper limit being t in the convolution integrals, which indicates that $\mathbf{D}(t)$ cannot depend on future values of $\mathbf{E}(t)$. This assumption leads to a relationship between the real and imaginary parts of the frequency domain constitutive parameters as described through the Kronig–Kramers equations.

Constitutive relations for fields in perfect conductors. In a perfect electric conductor (PEC) or a perfect magnetic conductor (PMC) the fields are exactly specified as

the null field:
$$\mathbf{E} = \mathbf{D} = \mathbf{B} = \mathbf{H} = 0.$$

By Ampere's and Faraday's laws we must also have $\mathbf{J} = \mathbf{J}_m = 0$; hence, by the continuity equation, $\rho = \rho_m = 0$.

In addition to the null field, we have the condition that the tangential electric field on the surface of a PEC must be zero. Similarly, the tangential magnetic field on the surface of a PMC must be zero. This implies (§ 2.8.3) that an electric surface current may exist on the surface of a PEC but not on the surface of a PMC, while a magnetic surface current may exist on the surface of a PMC but not on the surface of a PEC.

A PEC may be regarded as the limit of a conducting material as $\sigma \to \infty$. In many practical cases, good conductors such as gold and copper can be assumed to be perfect electric conductors, which greatly simplifies the application of boundary conditions. No physical material is known to behave as a PMC, but the concept is mathematically useful for applying symmetry conditions (in which a PMC is sometimes referred to as a "magnetic wall") and for use in developing equivalence theorems.

Constitutive relations in a linear anisotropic material. In a linear anisotropic material there are relationships between $\mathbf{B}$ and $\mathbf{H}$ and between $\mathbf{D}$ and $\mathbf{E}$, but the field vectors are not aligned as in the isotropic case. We can thus write

$$\mathbf{D} = \bar{\epsilon} \cdot \mathbf{E}, \qquad \mathbf{B} = \bar{\mu} \cdot \mathbf{H}, \qquad \mathbf{J} = \bar{\sigma} \cdot \mathbf{E},$$

where $\bar{\epsilon}$ is called the *permittivity dyadic*, $\bar{\mu}$ is the *permeability dyadic*, and $\bar{\sigma}$ is the *conductivity dyadic*. In terms of the general constitutive relation (2.18) we have

$$\bar{\mathbf{P}} = c\bar{\epsilon}, \qquad \bar{\mathbf{Q}} = \frac{\bar{\mu}^{-1}}{c}, \qquad \bar{\mathbf{L}} = \bar{\mathbf{M}} = 0.$$

Many different types of materials demonstrate anisotropic behavior, including optical crystals, magnetized plasmas, and ferrites. Plasmas and ferrites are examples of *gyrotropic* media. With the proper choice of coordinate system, the frequency-domain permittivity or permeability can be written in matrix form as

$$[\tilde{\bar{\epsilon}}] = \begin{bmatrix} \epsilon_{11} & \epsilon_{12} & 0 \\ -\epsilon_{12} & \epsilon_{11} & 0 \\ 0 & 0 & \epsilon_{33} \end{bmatrix}, \qquad [\tilde{\bar{\mu}}] = \begin{bmatrix} \mu_{11} & \mu_{12} & 0 \\ -\mu_{12} & \mu_{11} & 0 \\ 0 & 0 & \mu_{33} \end{bmatrix}. \qquad (2.32)$$

Each of the matrix entries may be complex. For the special case of a *lossless* gyrotropic material, the matrices become *hermitian*:

$$[\tilde{\bar{\epsilon}}] = \begin{bmatrix} \epsilon & -j\delta & 0 \\ j\delta & \epsilon & 0 \\ 0 & 0 & \epsilon_3 \end{bmatrix}, \qquad [\tilde{\bar{\mu}}] = \begin{bmatrix} \mu & -j\kappa & 0 \\ j\kappa & \mu & 0 \\ 0 & 0 & \mu_3 \end{bmatrix}, \qquad (2.33)$$

where ϵ, ϵ_3, δ, μ, μ_3, and κ are real numbers.

Crystals have received particular attention because of their birefringent properties. A birefringent crystal can be characterized by a symmetric permittivity dyadic that has real permittivity parameters in the frequency domain; equivalently, the constitutive relations do not involve constitutive operators. A coordinate system called the *principal system*, with axes called the *principal axes*, can always be found so that the permittivity dyadic in that system is diagonal:

$$[\tilde{\bar{\epsilon}}] = \begin{bmatrix} \epsilon_x & 0 & 0 \\ 0 & \epsilon_y & 0 \\ 0 & 0 & \epsilon_z \end{bmatrix}.$$

The geometrical structure of a crystal determines the relationship between ϵ_x, ϵ_y, and ϵ_z. If $\epsilon_x = \epsilon_y < \epsilon_z$, then the crystal is *positive uniaxial* (e.g., quartz). If $\epsilon_x = \epsilon_y > \epsilon_z$, the crystal is *negative uniaxial* (e.g., calcite). If $\epsilon_x \neq \epsilon_y \neq \epsilon_z$, the crystal is *biaxial* (e.g., mica). In uniaxial crystals the z-axis is called the *optical axis*.

If the anisotropic material is dispersive, we can generalize the convolutional form of the isotropic dispersive media to obtain the constitutive relations

$$\mathbf{D}(\mathbf{r}, t) = \epsilon_0 \left(\mathbf{E}(\mathbf{r}, t) + \int_{-\infty}^{t} \bar{\chi}_e(\mathbf{r}, t - t') \cdot \mathbf{E}(\mathbf{r}, t') \, dt' \right), \tag{2.34}$$

$$\mathbf{B}(\mathbf{r}, t) = \mu_0 \left(\mathbf{H}(\mathbf{r}, t) + \int_{-\infty}^{t} \bar{\chi}_m(\mathbf{r}, t - t') \cdot \mathbf{H}(\mathbf{r}, t') \, dt' \right), \tag{2.35}$$

$$\mathbf{J}(\mathbf{r}, t) = \int_{-\infty}^{t} \bar{\sigma}(\mathbf{r}, t - t') \cdot \mathbf{E}(\mathbf{r}, t') \, dt'. \tag{2.36}$$

Constitutive relations for biisotropic materials. A biisotropic material is an isotropic magnetoelectric material. Here we have $\mathbf{D}$ related to $\mathbf{E}$ and $\mathbf{B}$, and $\mathbf{H}$ related to $\mathbf{E}$ and $\mathbf{B}$, but with no realignment of the fields as in anisotropic (or bianisotropic) materials. Perhaps the simplest example is the *Tellegen medium* devised by B.D.H. Tellegen in 1948 [196], having

$$\mathbf{D} = \epsilon \mathbf{E} + \xi \mathbf{H}, \tag{2.37}$$

$$\mathbf{B} = \xi \mathbf{E} + \mu \mathbf{H}. \tag{2.38}$$

Tellegen proposed that his hypothetical material be composed of small (but macroscopic) ferromagnetic particles suspended in a liquid. This is an example of a *synthetic* material, constructed from ordinary materials to have an exotic electromagnetic behavior. Other examples include artificial dielectrics made from metallic particles imbedded in lightweight foams [66], and *chiral materials* made from small metallic helices suspended in resins [112].

Chiral materials are also biisotropic, and have the constitutive relations

$$\mathbf{D} = \epsilon \mathbf{E} - \chi \frac{\partial \mathbf{H}}{\partial t}, \tag{2.39}$$

$$\mathbf{B} = \mu \mathbf{H} + \chi \frac{\partial \mathbf{E}}{\partial t}, \tag{2.40}$$

where the constitutive parameter χ is called the *chirality parameter*. Note the presence of temporal derivative operators. Alternatively,

$$\mathbf{D} = \epsilon (\mathbf{E} + \beta \nabla \times \mathbf{E}), \tag{2.41}$$

$$\mathbf{B} = \mu (\mathbf{H} + \beta \nabla \times \mathbf{H}), \tag{2.42}$$

by Faraday's and Ampere's laws. Chirality is a natural state of symmetry; many natural substances are chiral materials, including DNA and many sugars. The time derivatives in (2.39)–(2.40) produce rotation of the polarization of time harmonic electromagnetic waves propagating in chiral media.

Constitutive relations in nonlinear media. Nonlinear electromagnetic effects have been studied by scientists and engineers since the beginning of the era of electrical technology. Familiar examples include saturation and hysteresis in ferromagnetic materials and the behavior of p-n junctions in solid-state rectifiers. The invention of the laser extended interest in nonlinear effects to the realm of optics, where phenomena such as parametric amplification and oscillation, harmonic generation, and magneto-optic interactions have found applications in modern devices [174].

Provided that the external field applied to a nonlinear electric material is small compared to the internal molecular fields, the relationship between $\mathbf{E}$ and $\mathbf{D}$ can be expanded in a Taylor series of the electric field. For an anisotropic material exhibiting no hysteresis effects, the constitutive relation is [131]

$$D_i(\mathbf{r}, t) = \epsilon_0 E_i(\mathbf{r}, t) + \sum_{j=1}^{3} \chi_{ij}^{(1)} E_j(\mathbf{r}, t) + \sum_{j,k=1}^{3} \chi_{ijk}^{(2)} E_j(\mathbf{r}, t) E_k(\mathbf{r}, t) +$$

$$+ \sum_{j,k,l=1}^{3} \chi_{ijkl}^{(3)} E_j(\mathbf{r}, t) E_k(\mathbf{r}, t) E_l(\mathbf{r}, t) + \cdots \qquad (2.43)$$

where the index $i = 1, 2, 3$ refers to the three components of the fields $\mathbf{D}$ and $\mathbf{E}$. The first sum in (2.43) is identical to the constitutive relation for linear anisotropic materials. Thus, $\chi_{ij}^{(1)}$ is identical to the susceptibility dyadic of a linear anisotropic medium considered earlier. The quantity $\chi_{ijk}^{(2)}$ is called the *second-order susceptibility*, and is a three-dimensional matrix (or third rank tensor) describing the nonlinear electric effects quadratic in $\mathbf{E}$. Similarly $\chi_{ijkl}^{(3)}$ is called the *third-order susceptibility*, and is a four-dimensional matrix (or fourth rank tensor) describing the nonlinear electric effects cubic in $\mathbf{E}$. Numerical values of $\chi_{ijk}^{(2)}$ and $\chi_{ijkl}^{(3)}$ are given in Shen [174] for a variety of crystals.

When the material shows hysteresis effects, $\mathbf{D}$ at any point $\mathbf{r}$ and time t is due not only to the value of $\mathbf{E}$ at that point and at that time, but to the values of $\mathbf{E}$ at all points and times. That is, the material displays both temporal and spatial dispersion.

2.3   Maxwell's equations in moving frames

The essence of special relativity is that the mathematical forms of Maxwell's equations are identical in all *inertial reference frames*: frames moving with uniform velocities relative to the *laboratory frame of reference* in which we perform our measurements. This *form invariance* of Maxwell's equations is a specific example of the general physical *principle of covariance*. In the laboratory frame we write the differential equations of Maxwell's theory as

$$\nabla \times \mathbf{E}(\mathbf{r}, t) = -\frac{\partial \mathbf{B}(\mathbf{r}, t)}{\partial t},$$

$$\nabla \times \mathbf{H}(\mathbf{r}, t) = \mathbf{J}(\mathbf{r}, t) + \frac{\partial \mathbf{D}(\mathbf{r}, t)}{\partial t},$$

$$\nabla \cdot \mathbf{D}(\mathbf{r}, t) = \rho(\mathbf{r}, t),$$

$$\nabla \cdot \mathbf{B}(\mathbf{r}, t) = 0,$$

$$\nabla \cdot \mathbf{J}(\mathbf{r}, t) = -\frac{\partial \rho(\mathbf{r}, t)}{\partial t}.$$

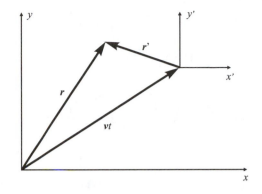

Figure 2.1: Primed coordinate system moving with velocity **v** relative to laboratory (unprimed) coordinate system.

Similarly, in an inertial frame having four-dimensional coordinates $(\mathbf{r}', t')$ we have

$$\nabla' \times \mathbf{E}'(\mathbf{r}', t') = -\frac{\partial \mathbf{B}'(\mathbf{r}', t')}{\partial t'},$$

$$\nabla' \times \mathbf{H}'(\mathbf{r}', t') = \mathbf{J}'(\mathbf{r}', t') + \frac{\partial \mathbf{D}'(\mathbf{r}', t')}{\partial t'},$$

$$\nabla' \cdot \mathbf{D}'(\mathbf{r}', t') = \rho'(\mathbf{r}', t'),$$

$$\nabla' \cdot \mathbf{B}'(\mathbf{r}', t') = 0,$$

$$\nabla' \cdot \mathbf{J}'(\mathbf{r}', t') = -\frac{\partial \rho'(\mathbf{r}', t')}{\partial t'}.$$

The primed fields measured in the moving system do *not* have the same numerical values as the unprimed fields measured in the laboratory. To convert between **E** and **E**′, **B** and **B**′, and so on, we must find a way to convert between the coordinates $(\mathbf{r}, t)$ and $(\mathbf{r}', t')$.

2.3.1 Field conversions under Galilean transformation

We shall assume that the primed coordinate system moves with constant velocity **v** relative to the laboratory frame (Figure 2.1). Prior to the early part of the twentieth century, converting between the primed and unprimed coordinate variables was intuitive and obvious: it was thought that time must be measured identically in each coordinate system, and that the relationship between the space variables can be determined simply by the displacement of the moving system at time $t = t'$. Under these assumptions, and under the further assumption that the two systems coincide at time $t = 0$, we can write

$$t' = t, \qquad x' = x - v_x t, \qquad y' = y - v_y t, \qquad z' = z - v_z t,$$

or simply

$$t' = t, \qquad \mathbf{r}' = \mathbf{r} - \mathbf{v}t.$$

This is called a *Galilean transformation*. We can use the chain rule to describe the manner in which differential operations transform, i.e., to relate derivatives with respect to the laboratory coordinates to derivatives with respect to the inertial coordinates. We have, for instance,

$$\frac{\partial}{\partial t} = \frac{\partial t'}{\partial t}\frac{\partial}{\partial t'} + \frac{\partial x'}{\partial t}\frac{\partial}{\partial x'} + \frac{\partial y'}{\partial t}\frac{\partial}{\partial y'} + \frac{\partial z'}{\partial t}\frac{\partial}{\partial z'}$$

$$= \frac{\partial}{\partial t'} - v_x \frac{\partial}{\partial x'} - v_y \frac{\partial}{\partial y'} - v_z \frac{\partial}{\partial z'}$$

$$= \frac{\partial}{\partial t'} - (\mathbf{v} \cdot \nabla'). \tag{2.44}$$

Similarly

$$\frac{\partial}{\partial x} = \frac{\partial}{\partial x'}, \qquad \frac{\partial}{\partial y} = \frac{\partial}{\partial y'}, \qquad \frac{\partial}{\partial z} = \frac{\partial}{\partial z'},$$

from which

$$\nabla \times \mathbf{A}(\mathbf{r}, t) = \nabla' \times \mathbf{A}(\mathbf{r}, t), \qquad \nabla \cdot \mathbf{A}(\mathbf{r}, t) = \nabla' \cdot \mathbf{A}(\mathbf{r}, t), \tag{2.45}$$

for each vector field $\mathbf{A}$.

Newton was aware that the laws of mechanics are invariant with respect to Galilean transformations. Do Maxwell's equations also behave in this way? Let us use the Galilean transformation to determine which relationship between the primed and unprimed fields results in form invariance of Maxwell's equations. We first examine $\nabla' \times \mathbf{E}$, the spatial rate of change of the laboratory field with respect to the inertial frame spatial coordinates:

$$\nabla' \times \mathbf{E} = \nabla \times \mathbf{E} = -\frac{\partial \mathbf{B}}{\partial t} = -\frac{\partial \mathbf{B}}{\partial t'} + (\mathbf{v} \cdot \nabla')\mathbf{B}$$

by (2.45) and (2.44). Rewriting the last term by (B.45) we have

$$(\mathbf{v} \cdot \nabla')\mathbf{B} = -\nabla' \times (\mathbf{v} \times \mathbf{B})$$

since $\mathbf{v}$ is constant and $\nabla' \cdot \mathbf{B} = \nabla \cdot \mathbf{B} = 0$, hence

$$\nabla' \times (\mathbf{E} + \mathbf{v} \times \mathbf{B}) = -\frac{\partial \mathbf{B}}{\partial t'}. \tag{2.46}$$

Similarly

$$\nabla' \times \mathbf{H} = \nabla \times \mathbf{H} = \mathbf{J} + \frac{\partial \mathbf{D}}{\partial t} = \mathbf{J} + \frac{\partial \mathbf{D}}{\partial t'} + \nabla' \times (\mathbf{v} \times \mathbf{D}) - \mathbf{v}(\nabla' \cdot \mathbf{D})$$

where $\nabla' \cdot \mathbf{D} = \nabla \cdot \mathbf{D} = \rho$ so that

$$\nabla' \times (\mathbf{H} - \mathbf{v} \times \mathbf{D}) = \frac{\partial \mathbf{D}}{\partial t'} - \rho \mathbf{v} + \mathbf{J}. \tag{2.47}$$

Also

$$\nabla' \cdot \mathbf{J} = \nabla \cdot \mathbf{J} = -\frac{\partial \rho}{\partial t} = -\frac{\partial \rho}{\partial t'} + (\mathbf{v} \cdot \nabla')\rho$$

and we may use (B.42) to write

$$(\mathbf{v} \cdot \nabla')\rho = \mathbf{v} \cdot (\nabla'\rho) = \nabla' \cdot (\rho\mathbf{v}),$$

obtaining

$$\nabla' \cdot (\mathbf{J} - \rho\mathbf{v}) = -\frac{\partial \rho}{\partial t'}. \tag{2.48}$$

Equations (2.46), (2.47), and (2.48) show that the forms of Maxwell's equations in the inertial and laboratory frames are identical provided that

$$\mathbf{E}' = \mathbf{E} + \mathbf{v} \times \mathbf{B}, \tag{2.49}$$

$$\mathbf{D}' = \mathbf{D}, \tag{2.50}$$

$$\mathbf{H}' = \mathbf{H} - \mathbf{v} \times \mathbf{D}, \tag{2.51}$$

$$\mathbf{B}' = \mathbf{B}, \tag{2.52}$$

$$\mathbf{J}' = \mathbf{J} - \rho\mathbf{v}, \tag{2.53}$$

$$\rho' = \rho. \tag{2.54}$$

That is, (2.49)–(2.54) result in form invariance of Faraday's law, Ampere's law, and the continuity equation under a Galilean transformation. These equations express the fields measured by a moving observer in terms of those measured in the laboratory frame. To convert the opposite way, we need only use the principle of relativity. Neither observer can tell whether he or she is stationary — only that the other observer is moving relative to him or her. To obtain the fields in the laboratory frame we simply change the sign on $\mathbf{v}$ and swap primed with unprimed fields in (2.49)–(2.54):

$$\mathbf{E} = \mathbf{E}' - \mathbf{v} \times \mathbf{B}', \tag{2.55}$$

$$\mathbf{D} = \mathbf{D}', \tag{2.56}$$

$$\mathbf{H} = \mathbf{H}' + \mathbf{v} \times \mathbf{D}', \tag{2.57}$$

$$\mathbf{B} = \mathbf{B}', \tag{2.58}$$

$$\mathbf{J} = \mathbf{J}' + \rho'\mathbf{v}, \tag{2.59}$$

$$\rho = \rho'. \tag{2.60}$$

According to (2.53), a moving observer interprets charge stationary in the laboratory frame as an additional current moving opposite the direction of his or her motion. This seems reasonable. However, while $\mathbf{E}$ depends on both $\mathbf{E}'$ and $\mathbf{B}'$, the field $\mathbf{B}$ is unchanged under the transformation. Why should $\mathbf{B}$ have this special status? In fact, we may uncover an inconsistency among the transformations by considering free space where (2.22) and (2.23) hold: in this case (2.49) gives

$$\mathbf{D}'/\epsilon_0 = \mathbf{D}/\epsilon_0 + \mathbf{v} \times \mu_0\mathbf{H}$$

or

$$\mathbf{D}' = \mathbf{D} + \mathbf{v} \times \mathbf{H}/c^2$$

rather than (2.50). Similarly, from (2.51) we get

$$\mathbf{B}' = \mathbf{B} - \mathbf{v} \times \mathbf{E}/c^2$$

instead of (2.52). Using these, the set of transformations becomes

$$\mathbf{E}' = \mathbf{E} + \mathbf{v} \times \mathbf{B}, \tag{2.61}$$

$$\mathbf{D}' = \mathbf{D} + \mathbf{v} \times \mathbf{H}/c^2, \tag{2.62}$$

$$\mathbf{H}' = \mathbf{H} - \mathbf{v} \times \mathbf{D}, \tag{2.63}$$

$$\mathbf{B}' = \mathbf{B} - \mathbf{v} \times \mathbf{E}/c^2, \tag{2.64}$$

$$\mathbf{J}' = \mathbf{J} - \rho\mathbf{v}, \tag{2.65}$$

$$\rho' = \rho. \tag{2.66}$$

These can also be written using dyadic notation as

$$\mathbf{E}' = \bar{\mathbf{I}} \cdot \mathbf{E} + \bar{\boldsymbol{\beta}} \cdot (c\mathbf{B}), \tag{2.67}$$

$$c\mathbf{B}' = -\bar{\boldsymbol{\beta}} \cdot \mathbf{E} + \bar{\mathbf{I}} \cdot (c\mathbf{B}), \tag{2.68}$$

and

$$c\mathbf{D}' = \bar{\mathbf{I}} \cdot (c\mathbf{D}) + \bar{\boldsymbol{\beta}} \cdot \mathbf{H}, \tag{2.69}$$

$$\mathbf{H}' = -\bar{\boldsymbol{\beta}} \cdot (c\mathbf{D}) + \bar{\mathbf{I}} \cdot \mathbf{H}, \tag{2.70}$$

where

$$[\bar{\boldsymbol{\beta}}] = \begin{bmatrix} 0 & -\beta_z & \beta_y \\ \beta_z & 0 & -\beta_x \\ -\beta_y & \beta_x & 0 \end{bmatrix}$$

with $\boldsymbol{\beta} = \mathbf{v}/c$. This set of equations is self-consistent among Maxwell's equations. However, the equations are not consistent with the assumption of a Galilean transformation of the coordinates, and thus Maxwell's equations are not covariant under a Galilean transformation. Maxwell's equations are only covariant under a Lorentz transformation as described in the next section. Expressions (2.61)–(2.64) turn out to be accurate to order v/c, hence are the results of a *first-order Lorentz transformation*. Only when v is an appreciable fraction of c do the field conversions resulting from the first-order Lorentz transformation differ markedly from those resulting from a Galilean transformation; those resulting from the true Lorentz transformation require even higher velocities to differ markedly from the first-order expressions. Engineering accuracy is often accomplished using the Galilean transformation. This pragmatic observation leads to quite a bit of confusion when considering the large-scale forms of Maxwell's equations, as we shall soon see.

2.3.2   Field conversions under Lorentz transformation

To find the proper transformation under which Maxwell's equations are covariant, we must discard our notion that time progresses the same in the primed and the unprimed frames. The proper transformation of coordinates that guarantees covariance of Maxwell's equations is the *Lorentz transformation*

$$ct' = \gamma ct - \gamma \boldsymbol{\beta} \cdot \mathbf{r}, \tag{2.71}$$

$$\mathbf{r}' = \bar{\boldsymbol{\alpha}} \cdot \mathbf{r} - \gamma \boldsymbol{\beta} ct, \tag{2.72}$$

where

$$\gamma = \frac{1}{\sqrt{1 - \beta^2}}, \qquad \bar{\boldsymbol{\alpha}} = \bar{\mathbf{I}} + (\gamma - 1)\frac{\boldsymbol{\beta}\boldsymbol{\beta}}{\beta^2}, \qquad \beta = |\boldsymbol{\beta}|.$$

This is obviously more complicated than the Galilean transformation; only as $\boldsymbol{\beta} \to 0$ are the Lorentz and Galilean transformations equivalent.

Not surprisingly, field conversions between inertial reference frames are more complicated with the Lorentz transformation than with the Galilean transformation. For simplicity we assume that the velocity of the moving frame has only an x-component: $\mathbf{v} = \hat{\mathbf{x}}v$. Later we can generalize this to any direction. Equations (2.71) and (2.72) become

$$x' = x + (\gamma - 1)x - \gamma vt, \tag{2.73}$$

$$y' = y, \tag{2.74}$$

$$z' = z, \tag{2.75}$$

$$ct' = \gamma ct - \gamma \frac{v}{c} x, \tag{2.76}$$

and the chain rule gives

$$\frac{\partial}{\partial x} = \gamma \frac{\partial}{\partial x'} - \gamma \frac{v}{c^2} \frac{\partial}{\partial t'}, \tag{2.77}$$

$$\frac{\partial}{\partial y} = \frac{\partial}{\partial y'}, \tag{2.78}$$

$$\frac{\partial}{\partial z} = \frac{\partial}{\partial z'}, \tag{2.79}$$

$$\frac{\partial}{\partial t} = -\gamma v \frac{\partial}{\partial x'} + \gamma \frac{\partial}{\partial t'}. \tag{2.80}$$

We begin by examining Faraday's law in the laboratory frame. In component form we have

$$\frac{\partial E_z}{\partial y} - \frac{\partial E_y}{\partial z} = -\frac{\partial B_x}{\partial t}, \tag{2.81}$$

$$\frac{\partial E_x}{\partial z} - \frac{\partial E_z}{\partial x} = -\frac{\partial B_y}{\partial t}, \tag{2.82}$$

$$\frac{\partial E_y}{\partial x} - \frac{\partial E_x}{\partial y} = -\frac{\partial B_z}{\partial t}. \tag{2.83}$$

These become

$$\frac{\partial E_z}{\partial y'} - \frac{\partial E_y}{\partial z'} = \gamma v \frac{\partial B_x}{\partial x'} - \gamma \frac{\partial B_x}{\partial t'}, \tag{2.84}$$

$$\frac{\partial E_x}{\partial z'} - \gamma \frac{\partial E_z}{\partial x'} + \gamma \frac{v}{c^2} \frac{\partial E_z}{\partial t'} = \gamma v \frac{\partial B_y}{\partial x'} - \gamma \frac{\partial B_y}{\partial t'}, \tag{2.85}$$

$$\gamma \frac{\partial E_y}{\partial x'} - \gamma \frac{v}{c^2} \frac{\partial E_y}{\partial t'} - \frac{\partial E_x}{\partial y'} = \gamma v \frac{\partial B_z}{\partial x'} - \gamma \frac{\partial B_z}{\partial t'}, \tag{2.86}$$

after we use (2.77)–(2.80) to convert the derivatives in the laboratory frame to derivatives with respect to the moving frame coordinates. To simplify (2.84) we consider

$$\nabla \cdot \mathbf{B} = \frac{\partial B_x}{\partial x} + \frac{\partial B_y}{\partial y} + \frac{\partial B_z}{\partial z} = 0.$$

Converting the laboratory frame coordinates to the moving frame coordinates, we have

$$\gamma \frac{\partial B_x}{\partial x'} - \gamma \frac{v}{c^2} \frac{\partial B_x}{\partial t'} + \frac{\partial B_y}{\partial y'} + \frac{\partial B_z}{\partial z'} = 0$$

or

$$-\gamma v \frac{\partial B_x}{\partial x'} = -\gamma \frac{v^2}{c^2} \frac{\partial B_x}{\partial t'} + v \frac{\partial B_y}{\partial y'} + v \frac{\partial B_z}{\partial z'}.$$

Substituting this into (2.84) and rearranging (2.85) and (2.86), we obtain

$$\frac{\partial}{\partial y'} \gamma (E_z + v B_y) - \frac{\partial}{\partial z'} \gamma (E_y - v B_z) = -\frac{\partial B_x}{\partial t'},$$

$$\frac{\partial E_x}{\partial z'} - \frac{\partial}{\partial x'} \gamma (E_z + v B_y) = -\frac{\partial}{\partial t'} \gamma \left(B_y + \frac{v}{c^2} E_z \right),$$

$$\frac{\partial}{\partial x'} \gamma (E_y - v B_z) - \frac{\partial E_x}{\partial y'} = -\frac{\partial}{\partial t'} \gamma \left(B_z - \frac{v}{c^2} E_y \right).$$

Comparison with (2.81)–(2.83) shows that form invariance of Faraday's law under the Lorentz transformation requires

$$E'_x = E_x, \qquad E'_y = \gamma(E_y - vB_z), \qquad E'_z = \gamma(E_z + vB_y),$$

and

$$B'_x = B_x, \qquad B'_y = \gamma\left(B_y + \frac{v}{c^2}E_z\right), \qquad B'_z = \gamma\left(B_z - \frac{v}{c^2}E_y\right).$$

To generalize **v** to any direction, we simply note that the components of the fields parallel to the velocity direction are identical in the moving and laboratory frames, while the components perpendicular to the velocity direction convert according to a simple cross product rule. After similar analyses with Ampere's and Gauss's laws (see Problem 2.2), we find that

$$\mathbf{E}'_\parallel = \mathbf{E}_\parallel, \qquad \mathbf{B}'_\parallel = \mathbf{B}_\parallel, \qquad \mathbf{D}'_\parallel = \mathbf{D}_\parallel, \qquad \mathbf{H}'_\parallel = \mathbf{H}_\parallel,$$

$$\mathbf{E}'_\perp = \gamma(\mathbf{E}_\perp + \boldsymbol{\beta} \times c\mathbf{B}_\perp), \tag{2.87}$$

$$c\mathbf{B}'_\perp = \gamma(c\mathbf{B}_\perp - \boldsymbol{\beta} \times \mathbf{E}_\perp), \tag{2.88}$$

$$c\mathbf{D}'_\perp = \gamma(c\mathbf{D}_\perp + \boldsymbol{\beta} \times \mathbf{H}_\perp), \tag{2.89}$$

$$\mathbf{H}'_\perp = \gamma(\mathbf{H}_\perp - \boldsymbol{\beta} \times c\mathbf{D}_\perp), \tag{2.90}$$

and

$$\mathbf{J}'_\parallel = \gamma(\mathbf{J}_\parallel - \rho\mathbf{v}), \tag{2.91}$$

$$\mathbf{J}'_\perp = \mathbf{J}_\perp, \tag{2.92}$$

$$c\rho' = \gamma(c\rho - \boldsymbol{\beta} \cdot \mathbf{J}), \tag{2.93}$$

where the symbols $\parallel$ and $\perp$ designate the components of the field parallel and perpendicular to **v**, respectively.

These conversions are self-consistent, and the Lorentz transformation is the transformation under which Maxwell's equations are covariant. If $v^2 \ll c^2$, then $\gamma \approx 1$ and to first order (2.87)–(2.93) reduce to (2.61)–(2.66). If $v/c \ll 1$, then the first-order fields reduce to the Galilean fields (2.49)–(2.54).

To convert in the opposite direction, we can swap primed and unprimed fields and change the sign on **v**:

$$\mathbf{E}_\perp = \gamma(\mathbf{E}'_\perp - \boldsymbol{\beta} \times c\mathbf{B}'_\perp), \tag{2.94}$$

$$c\mathbf{B}_\perp = \gamma(c\mathbf{B}'_\perp + \boldsymbol{\beta} \times \mathbf{E}'_\perp), \tag{2.95}$$

$$c\mathbf{D}_\perp = \gamma(c\mathbf{D}'_\perp - \boldsymbol{\beta} \times \mathbf{H}'_\perp), \tag{2.96}$$

$$\mathbf{H}_\perp = \gamma(\mathbf{H}'_\perp + \boldsymbol{\beta} \times c\mathbf{D}'_\perp), \tag{2.97}$$

and

$$\mathbf{J}_\parallel = \gamma(\mathbf{J}'_\parallel + \rho'\mathbf{v}), \tag{2.98}$$

$$\mathbf{J}_\perp = \mathbf{J}'_\perp, \tag{2.99}$$

$$c\rho = \gamma(c\rho' + \boldsymbol{\beta} \cdot \mathbf{J}'). \tag{2.100}$$

The conversion formulas can be written much more succinctly in dyadic notation:

$$\mathbf{E}' = \gamma\bar{\boldsymbol{\alpha}}^{-1} \cdot \mathbf{E} + \gamma\bar{\boldsymbol{\beta}} \cdot (c\mathbf{B}), \tag{2.101}$$

$$c\mathbf{B}' = -\gamma\bar{\boldsymbol{\beta}} \cdot \mathbf{E} + \gamma\bar{\boldsymbol{\alpha}}^{-1} \cdot (c\mathbf{B}), \tag{2.102}$$

$$c\mathbf{D}' = \gamma\bar{\boldsymbol{\alpha}}^{-1} \cdot (c\mathbf{D}) + \gamma\bar{\boldsymbol{\beta}} \cdot \mathbf{H}, \qquad (2.103)$$

$$\mathbf{H}' = -\gamma\bar{\boldsymbol{\beta}} \cdot (c\mathbf{D}) + \gamma\bar{\boldsymbol{\alpha}}^{-1} \cdot \mathbf{H}, \qquad (2.104)$$

and

$$c\rho' = \gamma(c\rho - \boldsymbol{\beta} \cdot \mathbf{J}), \qquad (2.105)$$

$$\mathbf{J}' = \bar{\boldsymbol{\alpha}} \cdot \mathbf{J} - \gamma\boldsymbol{\beta}c\rho, \qquad (2.106)$$

where $\bar{\boldsymbol{\alpha}}^{-1} \cdot \bar{\boldsymbol{\alpha}} = \bar{\mathbf{I}}$, and thus $\bar{\boldsymbol{\alpha}}^{-1} = \bar{\boldsymbol{\alpha}} - \gamma\boldsymbol{\beta}\boldsymbol{\beta}$.

Maxwell's equations are covariant under a Lorentz transformation but not under a Galilean transformation; the laws of mechanics are invariant under a Galilean transformation but not under a Lorentz transformation. How then should we analyze interactions between electromagnetic fields and particles or materials? Einstein realized that the laws of mechanics needed revision to make them Lorentz covariant: in fact, under his theory of special relativity all physical laws should demonstrate Lorentz covariance. Interestingly, charge is then Lorentz invariant, whereas mass is not (recall that invariance refers to a quantity, whereas covariance refers to the form of a natural law). We shall not attempt to describe all the ramifications of special relativity, but instead refer the reader to any of the excellent and readable texts on the subject, including those by Bohm [14], Einstein [62], and Born [18], and to the nice historical account by Miller [130]. However, we shall examine the importance of Lorentz invariants in electromagnetic theory.

Lorentz invariants. Although the electromagnetic fields are not Lorentz invariant (e.g., the numerical value of $\mathbf{E}$ measured by one observer differs from that measured by another observer in uniform relative motion), several quantities do give identical values regardless of the velocity of motion. Most fundamental are the speed of light and the quantity of electric charge which, unlike mass, is the same in all frames of reference. Other important *Lorentz invariants* include $\mathbf{E} \cdot \mathbf{B}$, $\mathbf{H} \cdot \mathbf{D}$, and the quantities

$$\mathbf{B} \cdot \mathbf{B} - \mathbf{E} \cdot \mathbf{E}/c^2,$$

$$\mathbf{H} \cdot \mathbf{H} - c^2\mathbf{D} \cdot \mathbf{D},$$

$$\mathbf{B} \cdot \mathbf{H} - \mathbf{E} \cdot \mathbf{D},$$

$$c\mathbf{B} \cdot \mathbf{D} + \mathbf{E} \cdot \mathbf{H}/c.$$

(See Problem 2.3.) To see the importance of these quantities, consider the special case of fields in empty space. If $\mathbf{E} \cdot \mathbf{B} = 0$ in one reference frame, then it is zero in all reference frames. Then if $\mathbf{B} \cdot \mathbf{B} - \mathbf{E} \cdot \mathbf{E}/c^2 = 0$ in any reference frame, the ratio of E to B is always c^2 regardless of the reference frame in which the fields are measured. This is the characteristic of a plane wave in free space.

If $\mathbf{E} \cdot \mathbf{B} = 0$ and $c^2B^2 > E^2$, then we can find a reference frame using the conversion formulas (2.101)–(2.106) (see Problem 2.5) in which the electric field is zero but the magnetic field is nonzero. In this case we call the fields *purely magnetic* in any reference frame, even if both $\mathbf{E}$ and $\mathbf{B}$ are nonzero. Similarly, if $\mathbf{E} \cdot \mathbf{B} = 0$ and $c^2B^2 < E^2$ then we can find a reference frame in which the magnetic field is zero but the electric field is nonzero. We call fields of this type *purely electric*.

The Lorentz force is not Lorentz invariant. Consider a point charge at rest in the laboratory frame. While we measure only an electric field in the laboratory frame, an inertial observer measures both electric and magnetic fields. A test charge Q in the laboratory frame experiences the Lorentz force $\mathbf{F} = Q\mathbf{E}$; in an inertial frame the same charge experiences $\mathbf{F}' = Q\mathbf{E}' + Q\mathbf{v} \times \mathbf{B}'$ (see Problem 2.6). The conversion formulas show that $\mathbf{F}$ and $\mathbf{F}'$ are not identical.

We see that both $\mathbf{E}$ and $\mathbf{B}$ are integral components of the electromagnetic field: the separation of the field into electric and magnetic components depends on the motion of the reference frame in which measurements are made. This has obvious implications when considering static electric and magnetic fields.

Derivation of Maxwell's equations from Coulomb's law. Consider a point charge at rest in the laboratory frame. If the magnetic component of force on this charge arises naturally through motion of an inertial reference frame, and if this force can be expressed in terms of Coulomb's law in the laboratory frame, then perhaps the magnetic field can be derived directly from Coulomb's and the Lorentz transformation. Perhaps it is possible to derive all of Maxwell's theory with Coulomb's law and Lorentz invariance as the only postulates.

Several authors, notably Purcell [152] and Elliott [65], have used this approach. However, Jackson [91] has pointed out that many additional assumptions are required to deduce Maxwell's equations beginning with Coulomb's law. Feynman [73] is critical of the approach, pointing out that we must introduce a vector potential which adds to the scalar potential from electrostatics in order to produce an entity that transforms according to the laws of special relativity. In addition, the assumption of Lorentz invariance seems to involve circular reasoning since the Lorentz transformation was originally introduced to make Maxwell's equations covariant. But Lucas and Hodgson [117] point out that the Lorentz transformation can be deduced from other fundamental principles (such as causality and the isotropy of space), and that the postulate of a vector potential is reasonable. Schwartz [170] gives a detailed derivation of Maxwell's equations from Coulomb's law, outlining the necessary assumptions.

Transformation of constitutive relations. Minkowski's interest in the covariance of Maxwell's equations was aimed not merely at the relationship between fields in different moving frames of reference, but at an understanding of the electrodynamics of moving media. He wished to ascertain the effect of a moving material body on the electromagnetic fields in some region of space. By proposing the covariance of Maxwell's equations in materials as well as in free space, he extended Maxwell's theory to moving material bodies.

We have seen in (2.101)–(2.104) that $(\mathbf{E}, c\mathbf{B})$ and $(c\mathbf{D}, \mathbf{H})$ convert identically under a Lorentz transformation. Since the most general form of the constitutive relations relate $c\mathbf{D}$ and $\mathbf{H}$ to the field pair $(\mathbf{E}, c\mathbf{B})$ (see § 2.2.2) as

$$\begin{bmatrix} c\mathbf{D} \\ \mathbf{H} \end{bmatrix} = [\bar{\mathbf{C}}] \begin{bmatrix} \mathbf{E} \\ c\mathbf{B} \end{bmatrix},$$

this form of the constitutive relations must be Lorentz covariant. That is, in the reference frame of a moving material we have

$$\begin{bmatrix} c\mathbf{D}' \\ \mathbf{H}' \end{bmatrix} = [\bar{\mathbf{C}}'] \begin{bmatrix} \mathbf{E}' \\ c\mathbf{B}' \end{bmatrix},$$

and should be able to convert $[\bar{\mathbf{C}}']$ to $[\bar{\mathbf{C}}]$. We should be able to find the constitutive matrix describing the relationships among the fields observed in the laboratory frame.

It is somewhat laborious to obtain the constitutive matrix $[\bar{\mathbf{C}}]$ for an arbitrary moving medium. Detailed expressions for isotropic, bianisotropic, gyrotropic, and uniaxial media are given by Kong [101]. The rather complicated expressions can be written in a more compact form if we consider the expressions for $\mathbf{B}$ and $\mathbf{D}$ in terms of the pair $(\mathbf{E}, \mathbf{H})$. For a linear isotropic material such that $\mathbf{D}' = \epsilon' \mathbf{E}'$ and $\mathbf{B}' = \mu' \mathbf{H}'$ in the moving frame, the relationships in the laboratory frame are [101]

$$\mathbf{B} = \mu' \bar{\mathbf{A}} \cdot \mathbf{H} - \mathbf{\Omega} \times \mathbf{E}, \tag{2.107}$$
$$\mathbf{D} = \epsilon' \bar{\mathbf{A}} \cdot \mathbf{E} + \mathbf{\Omega} \times \mathbf{H}, \tag{2.108}$$

where

$$\bar{\mathbf{A}} = \frac{1 - \beta^2}{1 - n^2 \beta^2} \left[\bar{\mathbf{I}} - \frac{n^2 - 1}{1 - \beta^2} \boldsymbol{\beta}\boldsymbol{\beta} \right], \tag{2.109}$$

$$\mathbf{\Omega} = \frac{n^2 - 1}{1 - n^2 \beta^2} \frac{\boldsymbol{\beta}}{c}, \tag{2.110}$$

and where $n = c(\mu'\epsilon')^{1/2}$ is the optical index of the medium. A moving material that is isotropic in its own moving reference frame is bianisotropic in the laboratory frame. If, for instance, we tried to measure the relationship between the fields of a moving isotropic fluid, but used instruments that were stationary in our laboratory (e.g., attached to our measurement bench) we would find that $\mathbf{D}$ depends not only on $\mathbf{E}$ but also on $\mathbf{H}$, and that $\mathbf{D}$ aligns with neither $\mathbf{E}$ nor $\mathbf{H}$. That a moving material isotropic in its own frame of reference is bianisotropic in the laboratory frame was known long ago. Roentgen showed experimentally in 1888 that a dielectric moving through an electric field becomes magnetically polarized, while H.A. Wilson showed in 1905 that a dielectric moving through a magnetic field becomes electrically polarized [139].

If $v^2/c^2 \ll 1$, we can consider the form of the constitutive equations for a first-order Lorentz transformation. Ignoring terms to order v^2/c^2 in (2.109) and (2.110), we obtain $\bar{\mathbf{A}} = \bar{\mathbf{I}}$ and $\mathbf{\Omega} = \mathbf{v}(n^2 - 1)/c^2$. Then, by (2.107) and (2.108),

$$\mathbf{B} = \mu' \mathbf{H} - (n^2 - 1)\frac{\mathbf{v} \times \mathbf{E}}{c^2}, \tag{2.111}$$

$$\mathbf{D} = \epsilon' \mathbf{E} + (n^2 - 1)\frac{\mathbf{v} \times \mathbf{H}}{c^2}. \tag{2.112}$$

We can also derive these from the first-order field conversion equations (2.61)–(2.64). From (2.61) and (2.62) we have

$$\mathbf{D}' = \mathbf{D} + \mathbf{v} \times \mathbf{H}/c^2 = \epsilon' \mathbf{E}' = \epsilon'(\mathbf{E} + \mathbf{v} \times \mathbf{B}).$$

Eliminating $\mathbf{B}$ via (2.64), we have

$$\mathbf{D} + \mathbf{v} \times \mathbf{H}/c^2 = \epsilon' \mathbf{E} + \epsilon' \mathbf{v} \times (\mathbf{v} \times \mathbf{E}/c^2) + \epsilon' \mathbf{v} \times \mathbf{B}' = \epsilon' \mathbf{E} + \epsilon' \mathbf{v} \times \mathbf{B}'$$

where we have neglected terms of order v^2/c^2. Since $\mathbf{B}' = \mu' \mathbf{H}' = \mu'(\mathbf{H} - \mathbf{v} \times \mathbf{D})$, we have

$$\mathbf{D} + \mathbf{v} \times \mathbf{H}/c^2 = \epsilon' \mathbf{E} + \epsilon' \mu' \mathbf{v} \times \mathbf{H} - \epsilon' \mu' \mathbf{v} \times \mathbf{v} \times \mathbf{D}.$$

Using $n^2 = c^2 \mu' \epsilon'$ and neglecting the last term since it is of order v^2/c^2, we obtain

$$\mathbf{D} = \epsilon' \mathbf{E} + (n^2 - 1)\frac{\mathbf{v} \times \mathbf{H}}{c^2},$$

which is identical to the expression (2.112) obtained by approximating the exact result to first order. Similar steps produce (2.111). In a Galilean frame where $v/c \ll 1$, the expressions reduce to $\mathbf{D} = \epsilon'\mathbf{E}$ and $\mathbf{B} = \mu'\mathbf{H}$, and the isotropy of the fields is preserved.

For a conducting medium having

$$\mathbf{J}' = \sigma'\mathbf{E}'$$

in a moving reference frame, Cullwick [48] shows that in the laboratory frame

$$\mathbf{J} = \sigma'\gamma[\bar{\mathbf{I}} - \boldsymbol{\beta}\boldsymbol{\beta}] \cdot \mathbf{E} + \sigma'\gamma c\boldsymbol{\beta} \times \mathbf{B}.$$

For $v \ll c$ we can set $\gamma \approx 1$ and see that

$$\mathbf{J} = \sigma'(\mathbf{E} + \mathbf{v} \times \mathbf{B})$$

to first order.

Constitutive relations in deforming or rotating media. The transformations discussed in the previous paragraphs hold for media in uniform relative motion. When a material body undergoes deformation or rotation, the concepts of special relativity are not directly applicable. However, authors such as Pauli [144] and Sommerfeld [185] have maintained that Minkowski's theory is *approximately* valid for deforming or rotating media if $\mathbf{v}$ is taken to be the *instantaneous* velocity at each point within the body. The reasoning is that at any instant in time each point within the body has a velocity $\mathbf{v}$ that may be associated with some inertial reference frame (generally different for each point). Thus the constitutive relations for the material at that point, within some small time interval taken about the observation time, may be assumed to be those of a stationary material, and the relations measured by an observer within the laboratory frame may be computed using the inertial frame for that point. This *instantaneous rest-frame* theory is most accurate at small accelerations $d\mathbf{v}/dt$. Van Bladel [201] outlines its shortcomings. See also Anderson [3] and Mo [132] for detailed discussions of the electromagnetic properties of material media in accelerating frames of reference.

2.4   The Maxwell–Boffi equations

In any version of Maxwell's theory, the mediating field is the electromagnetic field described by four field vectors. In Minkowski's form of Maxwell's equations we use $\mathbf{E}$, $\mathbf{D}$, $\mathbf{B}$, and $\mathbf{H}$. As an alternative consider the electromagnetic field as represented by the vector fields $\mathbf{E}$, $\mathbf{B}$, $\mathbf{P}$, and $\mathbf{M}$, and described by

$$\nabla \times \mathbf{E} = -\frac{\partial \mathbf{B}}{\partial t}, \tag{2.113}$$

$$\nabla \times (\mathbf{B}/\mu_0 - \mathbf{M}) = \mathbf{J} + \frac{\partial}{\partial t}(\epsilon_0\mathbf{E} + \mathbf{P}), \tag{2.114}$$

$$\nabla \cdot (\epsilon_0\mathbf{E} + \mathbf{P}) = \rho, \tag{2.115}$$

$$\nabla \cdot \mathbf{B} = 0. \tag{2.116}$$

These *Maxwell–Boffi equations* are named after L. Boffi, who formalized them for moving media [13]. The quantity $\mathbf{P}$ is the *polarization vector*, and $\mathbf{M}$ is the *magnetization vector*.

The use of $\mathbf{P}$ and $\mathbf{M}$ in place of $\mathbf{D}$ and $\mathbf{H}$ is sometimes called an application of the *principle of Ampere and Lorentz* [199].

Let us examine the ramification of using (2.113)–(2.116) as the basis for a postulate of electromagnetics. These equations are similar to the Maxwell–Minkowski equations used earlier; must we rebuild all the underpinning of a new postulate, or can we use our original arguments based on the Minkowski form? For instance, how do we invoke uniqueness if we no longer have the field $\mathbf{H}$? What represents the flux of energy, formerly found using $\mathbf{E} \times \mathbf{H}$? And, importantly, are (2.113)–(2.114) form invariant under a Lorentz transformation?

It turns out that the set of vector fields $(\mathbf{E}, \mathbf{B}, \mathbf{P}, \mathbf{M})$ is merely a linear mapping of the set $(\mathbf{E}, \mathbf{D}, \mathbf{B}, \mathbf{H})$. As pointed out by Tai [193], any linear mapping of the four field vectors from Minkowski's form onto any other set of four field vectors will preserve the covariance of Maxwell's equations. Boffi chose to keep $\mathbf{E}$ and $\mathbf{B}$ intact and to introduce only two new fields; he could have kept $\mathbf{H}$ and $\mathbf{D}$ instead, or used a mapping that introduced four completely new fields (as did Chu). Many authors retain $\mathbf{E}$ and $\mathbf{H}$. This is somewhat more cumbersome since these vectors do not convert as a pair under a Lorentz transformation. A discussion of the idea of field vector "pairing" appears in § 2.6.

The usefulness of the Boffi form lies in the specific mapping chosen. Comparison of (2.113)–(2.116) to (2.1)–(2.4) quickly reveals that

$$\mathbf{P} = \mathbf{D} - \epsilon_0 \mathbf{E}, \tag{2.117}$$

$$\mathbf{M} = \mathbf{B}/\mu_0 - \mathbf{H}. \tag{2.118}$$

We see that $\mathbf{P}$ is the difference between $\mathbf{D}$ in a material and $\mathbf{D}$ in free space, while $\mathbf{M}$ is the difference between $\mathbf{H}$ in free space and $\mathbf{H}$ in a material. In free space, $\mathbf{P} = \mathbf{M} = 0$.

Equivalent polarization and magnetization sources. The Boffi formulation provides a new way to regard $\mathbf{E}$ and $\mathbf{B}$. Maxwell grouped $(\mathbf{E}, \mathbf{H})$ as a pair of "force vectors" to be associated with line integrals (or curl operations in the point forms of his equations), and $(\mathbf{D}, \mathbf{B})$ as a pair of "flux vectors" associated with surface integrals (or divergence operations). That is, $\mathbf{E}$ is interpreted as belonging to the computation of "emf" as a line integral, while $\mathbf{B}$ is interpreted as a density of magnetic "flux" passing through a surface. Similarly, $\mathbf{H}$ yields the "mmf" about some closed path and $\mathbf{D}$ the electric flux through a surface. The introduction of $\mathbf{P}$ and $\mathbf{M}$ allows us to also regard $\mathbf{E}$ as a flux vector and $\mathbf{B}$ as a force vector — in essence, allowing the two fields $\mathbf{E}$ and $\mathbf{B}$ to take on the duties that required four fields in Minkowski's form. To see this, we rewrite the Maxwell–Boffi equations as

$$\nabla \times \mathbf{E} = -\frac{\partial \mathbf{B}}{\partial t},$$

$$\nabla \times \frac{\mathbf{B}}{\mu_0} = \left(\mathbf{J} + \nabla \times \mathbf{M} + \frac{\partial \mathbf{P}}{\partial t} \right) + \frac{\partial \epsilon_0 \mathbf{E}}{\partial t},$$

$$\nabla \cdot (\epsilon_0 \mathbf{E}) = (\rho - \nabla \cdot \mathbf{P}),$$

$$\nabla \cdot \mathbf{B} = 0,$$

and compare them to the Maxwell–Minkowski equations for sources in free space:

$$\nabla \times \mathbf{E} = -\frac{\partial \mathbf{B}}{\partial t},$$

$$\nabla \times \frac{\mathbf{B}}{\mu_0} = \mathbf{J} + \frac{\partial \epsilon_0 \mathbf{E}}{\partial t},$$
$$\nabla \cdot (\epsilon_0 \mathbf{E}) = \rho,$$
$$\nabla \cdot \mathbf{B} = 0.$$

The forms are preserved if we identify $\partial \mathbf{P}/\partial t$ and $\nabla \times \mathbf{M}$ as new types of current density, and $\nabla \cdot \mathbf{P}$ as a new type of charge density. We define

$$\mathbf{J}_P = \frac{\partial \mathbf{P}}{\partial t} \tag{2.119}$$

as an *equivalent polarization current* density, and

$$\mathbf{J}_M = \nabla \times \mathbf{M}$$

as an *equivalent magnetization current* density (sometimes called the *equivalent Amperian currents of magnetized matter* [199]). We define

$$\rho_P = -\nabla \cdot \mathbf{P}$$

as an *equivalent polarization charge* density (sometimes called the *Poisson–Kelvin* equivalent charge distribution [199]). Then the Maxwell–Boffi equations become simply

$$\nabla \times \mathbf{E} = -\frac{\partial \mathbf{B}}{\partial t}, \tag{2.120}$$
$$\nabla \times \frac{\mathbf{B}}{\mu_0} = (\mathbf{J} + \mathbf{J}_M + \mathbf{J}_P) + \frac{\partial \epsilon_0 \mathbf{E}}{\partial t}, \tag{2.121}$$
$$\nabla \cdot (\epsilon_0 \mathbf{E}) = (\rho + \rho_P), \tag{2.122}$$
$$\nabla \cdot \mathbf{B} = 0. \tag{2.123}$$

Here is the new view. A material can be viewed as composed of charged particles of matter immersed in free space. When these charges are properly considered as "equivalent" polarization and magnetization charges, all field effects (describable through flux and force vectors) can be handled by the two fields $\mathbf{E}$ and $\mathbf{B}$. Whereas in Minkowski's form $\mathbf{D}$ diverges from ρ, in Boffi's form $\mathbf{E}$ diverges from a *total* charge density consisting of ρ and ρ_P. Whereas in the Minkowski form $\mathbf{H}$ curls around $\mathbf{J}$, in the Boffi form $\mathbf{B}$ curls around the total current density consisting of $\mathbf{J}$, $\mathbf{J}_M$, and $\mathbf{J}_P$.

This view was pioneered by Lorentz, who by 1892 considered matter as consisting of bulk molecules in a vacuum that would respond to an applied electromagnetic field [130]. The resulting motion of the charged particles of matter then became another source term for the "fundamental" fields $\mathbf{E}$ and $\mathbf{B}$. Using this reasoning he was able to reduce the fundamental Maxwell equations to two equations in two unknowns, demonstrating a simplicity appealing to many (including Einstein). Of course, to apply this concept we must be able to describe how the charged particles respond to an applied field. Simple microscopic models of the constituents of matter are generally used: some combination of electric and magnetic dipoles, or of loops of electric and magnetic current.

The Boffi equations are mathematically appealing since they now specify both the curl and divergence of the two field quantities $\mathbf{E}$ and $\mathbf{B}$. By the Helmholtz theorem we know that a field vector is uniquely specified when both its curl and divergence are given. But this assumes that the equivalent sources produced by $\mathbf{P}$ and $\mathbf{M}$ are true source fields in the same sense as $\mathbf{J}$. We have precluded this by insisting in Chapter 1 that the source field must be independent of the mediating field it sources. If we view $\mathbf{P}$ and $\mathbf{M}$ as merely a mapping from the original vector fields of Minkowski's form, we still have four vector fields with which to contend. And with these must also be a mapping of the constitutive relationships, which now link the fields $\mathbf{E}$, $\mathbf{B}$, $\mathbf{P}$, and $\mathbf{M}$. Rather than argue the actual physical existence of the equivalent sources, we note that a real benefit of the new view is that under certain circumstances the equivalent source quantities can be determined through physical reasoning, hence we can create physical models of $\mathbf{P}$ and $\mathbf{M}$ and deduce their links to $\mathbf{E}$ and $\mathbf{B}$. We may then find it easier to understand and deduce the constitutive relationships. However we do not in general consider $\mathbf{E}$ and $\mathbf{B}$ to be in any way more "fundamental" than $\mathbf{D}$ and $\mathbf{H}$.

Covariance of the Boffi form. Because of the linear relationships (2.117) and (2.118), covariance of the Maxwell–Minkowski equations carries over to the Maxwell–Boffi equations. However, the conversion between fields in different moving reference frames will now involve $\mathbf{P}$ and $\mathbf{M}$. Since Faraday's law is unchanged in the Boffi form, we still have

$$\mathbf{E}'_{\|} = \mathbf{E}_{\|}, \tag{2.124}$$

$$\mathbf{B}'_{\|} = \mathbf{B}_{\|}, \tag{2.125}$$

$$\mathbf{E}'_{\perp} = \gamma(\mathbf{E}_{\perp} + \boldsymbol{\beta} \times c\mathbf{B}_{\perp}), \tag{2.126}$$

$$c\mathbf{B}'_{\perp} = \gamma(c\mathbf{B}_{\perp} - \boldsymbol{\beta} \times \mathbf{E}_{\perp}). \tag{2.127}$$

To see how $\mathbf{P}$ and $\mathbf{M}$ convert, we note that in the laboratory frame $\mathbf{D} = \epsilon_0 \mathbf{E} + \mathbf{P}$ and $\mathbf{H} = \mathbf{B}/\mu_0 - \mathbf{M}$, while in the moving frame $\mathbf{D}' = \epsilon_0 \mathbf{E}' + \mathbf{P}'$ and $\mathbf{H}' = \mathbf{B}'/\mu_0 - \mathbf{M}'$. Thus

$$\mathbf{P}'_{\|} = \mathbf{D}'_{\|} - \epsilon_0 \mathbf{E}'_{\|} = \mathbf{D}_{\|} - \epsilon_0 \mathbf{E}_{\|} = \mathbf{P}_{\|}$$

and

$$\mathbf{M}'_{\|} = \mathbf{B}'_{\|}/\mu_0 - \mathbf{H}'_{\|} = \mathbf{B}_{\|}/\mu_0 - \mathbf{H}_{\|} = \mathbf{M}_{\|}.$$

For the perpendicular components

$$\mathbf{D}'_{\perp} = \gamma(\mathbf{D}_{\perp} + \boldsymbol{\beta} \times \mathbf{H}_{\perp}/c) = \epsilon_0 \mathbf{E}'_{\perp} + \mathbf{P}'_{\perp} = \epsilon_0 \left[\gamma(\mathbf{E}_{\perp} + \boldsymbol{\beta} \times c\mathbf{B}_{\perp})\right] + \mathbf{P}'_{\perp};$$

substitution of $\mathbf{H}_{\perp} = \mathbf{B}_{\perp}/\mu_0 - \mathbf{M}_{\perp}$ then gives

$$\mathbf{P}'_{\perp} = \gamma(\mathbf{D}_{\perp} - \epsilon_0 \mathbf{E}_{\perp}) - \gamma\epsilon_0\boldsymbol{\beta} \times c\mathbf{B}_{\perp} + \gamma\boldsymbol{\beta} \times \mathbf{B}_{\perp}/(c\mu_0) - \gamma\boldsymbol{\beta} \times \mathbf{M}_{\perp}/c$$

or

$$c\mathbf{P}'_{\perp} = \gamma(c\mathbf{P}_{\perp} - \boldsymbol{\beta} \times \mathbf{M}_{\perp}).$$

Similarly,

$$\mathbf{M}'_{\perp} = \gamma(\mathbf{M}_{\perp} + \boldsymbol{\beta} \times c\mathbf{P}_{\perp}).$$

Hence

$$\mathbf{E}'_{\|} = \mathbf{E}_{\|}, \quad \mathbf{B}'_{\|} = \mathbf{B}_{\|}, \quad \mathbf{P}'_{\|} = \mathbf{P}_{\|}, \quad \mathbf{M}'_{\|} = \mathbf{M}_{\|}, \quad \mathbf{J}'_{\perp} = \mathbf{J}_{\perp}, \tag{2.128}$$

and

$$\mathbf{E}'_{\perp} = \gamma(\mathbf{E}_{\perp} + \boldsymbol{\beta} \times c\mathbf{B}_{\perp}), \tag{2.129}$$

$$c\mathbf{B}'_{\perp} = \gamma(c\mathbf{B}_{\perp} - \boldsymbol{\beta} \times \mathbf{E}_{\perp}), \tag{2.130}$$

$$c\mathbf{P}'_{\perp} = \gamma(c\mathbf{P}_{\perp} - \boldsymbol{\beta} \times \mathbf{M}_{\perp}), \tag{2.131}$$

$$\mathbf{M}'_{\perp} = \gamma(\mathbf{M}_{\perp} + \boldsymbol{\beta} \times c\mathbf{P}_{\perp}), \tag{2.132}$$

$$\mathbf{J}'_{\|} = \gamma(\mathbf{J}_{\|} - \rho\mathbf{v}). \tag{2.133}$$

In the case of the first-order Lorentz transformation we can set $\gamma \approx 1$ to obtain

$$\mathbf{E'} = \mathbf{E} + \mathbf{v} \times \mathbf{B}, \tag{2.134}$$

$$\mathbf{B'} = \mathbf{B} - \frac{\mathbf{v} \times \mathbf{E}}{c^2}, \tag{2.135}$$

$$\mathbf{P'} = \mathbf{P} - \frac{\mathbf{v} \times \mathbf{M}}{c^2}, \tag{2.136}$$

$$\mathbf{M'} = \mathbf{M} + \mathbf{v} \times \mathbf{P}, \tag{2.137}$$

$$\mathbf{J'} = \mathbf{J} - \rho\mathbf{v}. \tag{2.138}$$

To convert from the moving frame to the laboratory frame we simply swap primed with unprimed fields and let $\mathbf{v} \to -\mathbf{v}$.

As a simple example, consider a linear isotropic medium having

$$\mathbf{D'} = \epsilon_0 \epsilon_r' \mathbf{E'}, \qquad \mathbf{B'} = \mu_0 \mu_r' \mathbf{H'},$$

in a moving reference frame. From (2.117) we have

$$\mathbf{P'} = \epsilon_0 \epsilon_r' \mathbf{E'} - \epsilon_0 \mathbf{E'} = \epsilon_0 \chi_e' \mathbf{E'}$$

where $\chi_e' = \epsilon_r' - 1$ is the electric susceptibility of the moving material. Similarly (2.118) yields

$$\mathbf{M'} = \frac{\mathbf{B'}}{\mu_0} - \frac{\mathbf{B'}}{\mu_0 \mu_r'} = \frac{\mathbf{B'} \chi_m'}{\mu_0 \mu_r'}$$

where $\chi_m' = \mu_r' - 1$ is the magnetic susceptibility of the moving material. How are $\mathbf{P}$ and $\mathbf{M}$ related to $\mathbf{E}$ and $\mathbf{B}$ in the laboratory frame? For simplicity, we consider the first-order expressions. From (2.136) we have

$$\mathbf{P} = \mathbf{P'} + \frac{\mathbf{v} \times \mathbf{M'}}{c^2} = \epsilon_0 \chi_e' \mathbf{E'} + \frac{\mathbf{v} \times \mathbf{B'} \chi_m'}{\mu_0 \mu_r' c^2}.$$

Substituting for $\mathbf{E'}$ and $\mathbf{B'}$ from (2.134) and (2.135), and using $\mu_0 c^2 = 1/\epsilon_0$, we have

$$\mathbf{P} = \epsilon_0 \chi_e' (\mathbf{E} + \mathbf{v} \times \mathbf{B}) + \epsilon_0 \frac{\chi_m'}{\mu_r'} \mathbf{v} \times \left(\mathbf{B} - \frac{\mathbf{v} \times \mathbf{E}}{c^2} \right).$$

Neglecting the last term since it varies as v^2/c^2, we get

$$\mathbf{P} = \epsilon_0 \chi_e' \mathbf{E} + \epsilon_0 \left(\chi_e' + \frac{\chi_m'}{\mu_r'} \right) \mathbf{v} \times \mathbf{B}. \tag{2.139}$$

Similarly,

$$\mathbf{M} = \frac{\chi_m'}{\mu_0 \mu_r'} \mathbf{B} - \epsilon_0 \left(\chi_e' + \frac{\chi_m'}{\mu_r'} \right) \mathbf{v} \times \mathbf{E}. \tag{2.140}$$

2.5   Large-scale form of Maxwell's equations

We can write Maxwell's equations in a form that incorporates the spatial variation of the field in a certain region of space. To do this, we integrate the point form of Maxwell's

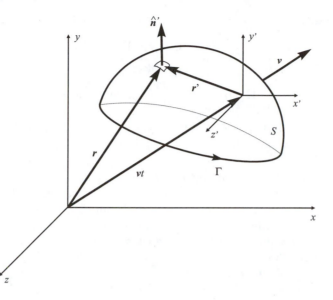

Figure 2.2: Open surface having velocity **v** relative to laboratory (unprimed) coordinate system. Surface is non-deforming.

equations over a region of space, then perform some succession of manipulations until we arrive at a form that provides us some benefit in our work with electromagnetic fields. The results are particularly useful for understanding the properties of electric and magnetic circuits, and for predicting the behavior of electrical machinery.

We shall consider two important situations: a mathematical surface that moves with constant velocity **v** and with constant shape, and a surface that moves and deforms arbitrarily.

2.5.1 Surface moving with constant velocity

Consider an open surface S moving with constant velocity **v** relative to the laboratory frame (Figure 2.2). Assume every point on the surface is an ordinary point. At any instant t we can express the relationship between the fields at points on S in either frame. In the laboratory frame we have

$$\nabla \times \mathbf{E} = -\frac{\partial \mathbf{B}}{\partial t}, \qquad \nabla \times \mathbf{H} = \frac{\partial \mathbf{D}}{\partial t} + \mathbf{J},$$

while in the moving frame

$$\nabla' \times \mathbf{E}' = -\frac{\partial \mathbf{B}'}{\partial t'}, \qquad \nabla' \times \mathbf{H}' = \frac{\partial \mathbf{D}'}{\partial t'} + \mathbf{J}'.$$

If we integrate over S and use Stokes's theorem, we get for the laboratory frame

$$\oint_\Gamma \mathbf{E} \cdot \mathbf{dl} = -\int_S \frac{\partial \mathbf{B}}{\partial t} \cdot \mathbf{dS}, \tag{2.141}$$

$$\oint_\Gamma \mathbf{H} \cdot \mathbf{dl} = \int_S \frac{\partial \mathbf{D}}{\partial t} \cdot \mathbf{dS} + \int_S \mathbf{J} \cdot \mathbf{dS}, \tag{2.142}$$

and for the moving frame

$$\oint_{\Gamma'} \mathbf{E}' \cdot \mathbf{dl}' = -\int_{S'} \frac{\partial \mathbf{B}'}{\partial t'} \cdot \mathbf{dS}', \tag{2.143}$$

$$\oint_{\Gamma'} \mathbf{H}' \cdot \mathbf{dl}' = \int_{S'} \frac{\partial \mathbf{D}'}{\partial t'} \cdot \mathbf{dS}' + \int_{S'} \mathbf{J}' \cdot \mathbf{dS}'. \tag{2.144}$$

Here boundary contour Γ has sense determined by the right-hand rule. We use the notation Γ', S', etc., to indicate that all integrations for the moving frame are computed using space and time variables in that frame. Equation (2.141) is the *integral form of Faraday's law*, while (2.142) is the *integral form of Ampere's law*.

Faraday's law states that the net circulation of $\mathbf{E}$ about a contour Γ (sometimes called the *electromotive force* or *emf*) is determined by the flux of the time-rate of change of the flux vector $\mathbf{B}$ passing through the surface bounded by Γ. Ampere's law states that the circulation of $\mathbf{H}$ (sometimes called the *magnetomotive force* or *mmf*) is determined by the flux of the current $\mathbf{J}$ plus the flux of the time-rate of change of the flux vector $\mathbf{D}$. It is the term containing $\partial \mathbf{D}/\partial t$ that Maxwell recognized as necessary to make his equations consistent; since it has units of current, it is often referred to as the *displacement current* term.

Equations (2.141)–(2.142) are the large-scale or integral forms of Maxwell's equations. They are the integral-form equivalents of the point forms, and are form invariant under Lorentz transformation. If we express the fields in terms of the moving reference frame, we can write

$$\oint_{\Gamma'} \mathbf{E}' \cdot \mathbf{dl}' = -\frac{d}{dt} \int_{S'} \mathbf{B}' \cdot \mathbf{dS}', \tag{2.145}$$

$$\oint_{\Gamma'} \mathbf{H}' \cdot \mathbf{dl}' = \frac{d}{dt} \int_{S'} \mathbf{D}' \cdot \mathbf{dS}' + \int_{S'} \mathbf{J}' \cdot \mathbf{dS}'. \tag{2.146}$$

These hold for a stationary surface, since the surface would be stationary to an observer who moves with it. We are therefore justified in removing the partial derivative from the integral. Although the surfaces and contours considered here are purely mathematical, they often coincide with actual physical boundaries. The surface may surround a moving material medium, for instance, or the contour may conform to a wire moving in an electrical machine.

We can also convert the auxiliary equations to large-scale form. Consider a volume region V surrounded by a surface S that moves with velocity $\mathbf{v}$ relative to the laboratory frame (Figure 2.3). Integrating the point form of Gauss's law over V we have

$$\int_{V} \nabla \cdot \mathbf{D} \, dV = \int_{V} \rho \, dV.$$

Using the divergence theorem and recognizing that the integral of charge density is total charge, we obtain

$$\oint_{S} \mathbf{D} \cdot \mathbf{dS} = \int_{V} \rho \, dV = Q(t) \tag{2.147}$$

where $Q(t)$ is the total charge contained within V at time t. This large-scale form of Gauss's law states that the total flux of $\mathbf{D}$ passing through a closed surface is identical to the electric charge Q contained within. Similarly,

$$\oint_{S} \mathbf{B} \cdot \mathbf{dS} = 0 \tag{2.148}$$

(The asterisk should not be confused with the notation for complex conjugate.)

Much confusion arises from the similarity between (2.153) and (2.145). In fact, these expressions are different and give different results. This is because $\mathbf{B}'$ in (2.145) is measured *in the frame of the moving circuit*, while $\mathbf{B}$ in (2.153) is measured in the frame of the laboratory. Further confusion arises from various definitions of emf. Many authors (e.g., Hermann Weyl [213]) define emf to be the circulation of $\mathbf{E}^*$. In that case the emf is equal to the negative time rate of change of the flux of the *laboratory frame* magnetic field $\mathbf{B}$ through S. Since the Lorentz force experienced by a charge q moving *with the contour* is given by $q\mathbf{E}^* = q(\mathbf{E} + \mathbf{v} \times \mathbf{B})$, this emf is the circulation of Lorentz force per unit charge along the contour. If the contour is aligned with a conducting circuit, then in some cases this emf can be given physical interpretation as the work required to move a charge around the entire circuit through the conductor against the Lorentz force. Unfortunately the usefulness of this definition of emf is lost if the time or space rate of change of the fields is so large that no true loop current can be established (hence Kirchoff's law cannot be employed). Such a problem must be treated as an electromagnetic "scattering" problem with consideration given to retardation effects. Detailed discussions of the physical interpretation of $\mathbf{E}^*$ in the definition of emf are given by Scanlon [165] and Cullwick [48].

Other authors choose to define emf as the circulation of the electric field *in the frame of the moving contour*. In this case the circulation of $\mathbf{E}'$ in (2.145) is the emf, and is related to the flux of the magnetic field *in the frame of the moving circuit*. As pointed out above, the result differs from that based on the Lorentz force. If we wish, we can also write this emf in terms of the fields expressed in the laboratory frame. To do this we must convert $\partial\mathbf{B}'/\partial t'$ to the laboratory fields using the rules for a Lorentz transformation. The result, given by Tai [194], is quite complicated and involves both the magnetic *and* electric laboratory-frame fields.

The moving-frame emf as computed from the Lorentz transformation is rarely used as a working definition of emf, mostly because circuits moving at relativistic velocities are seldom used by engineers. Unfortunately, more confusion arises for the case $v \ll c$, since for a Galilean frame the Lorentz-force and moving-frame emfs become identical. This is apparent if we use (2.52) to replace $\mathbf{B}'$ with the laboratory frame field $\mathbf{B}$, and (2.49) to replace $\mathbf{E}'$ with the combination of laboratory frame fields $\mathbf{E} + \mathbf{v} \times \mathbf{B}$. Then (2.145) becomes

$$\oint_\Gamma \mathbf{E}' \cdot \mathbf{dl} = \oint_\Gamma (\mathbf{E} + \mathbf{v} \times \mathbf{B}) \cdot \mathbf{dl} = -\frac{d}{dt} \int_S \mathbf{B} \cdot \mathbf{dS},$$

which is identical to (2.153). For circuits moving with low velocity then, the circulation of $\mathbf{E}'$ can be interpreted as work per unit charge. As an added bit of confusion, the term

$$\oint_\Gamma (\mathbf{v} \times \mathbf{B}) \cdot \mathbf{dl} = \int_S \nabla \times (\mathbf{v} \times \mathbf{B}) \cdot \mathbf{dS}$$

is sometimes called *motional emf*, since it is the component of the circulation of $\mathbf{E}^*$ that is directly attributable to the motion of the circuit.

Although less commonly done, we can also rewrite Ampere's law (2.142) using (2.151). This gives

$$\oint_\Gamma \mathbf{H} \cdot \mathbf{dl} = \int_S \mathbf{J} \cdot \mathbf{dS} + \frac{d}{dt} \int_S \mathbf{D} \cdot \mathbf{dS} - \int_S (\mathbf{v}\nabla \cdot \mathbf{D}) \cdot \mathbf{dS} + \int_S \nabla \times (\mathbf{v} \times \mathbf{D}) \cdot \mathbf{dS}.$$

Using $\nabla \cdot \mathbf{D} = \rho$ and using Stokes's theorem on the last term, we obtain

$$\oint_\Gamma (\mathbf{H} - \mathbf{v} \times \mathbf{D}) \cdot \mathbf{dl} = \frac{d}{dt} \int_S \mathbf{D} \cdot \mathbf{dS} + \int_S (\mathbf{J} - \rho\mathbf{v}) \cdot \mathbf{dS}.$$

Finally, letting $\mathbf{H}^* = \mathbf{H} - \mathbf{v} \times \mathbf{D}$ and $\mathbf{J}^* = \mathbf{J} - \rho\mathbf{v}$ we can write the *kinematic form of Ampere's law*:

$$\oint_\Gamma \mathbf{H}^* \cdot \mathbf{dl} = \frac{d}{dt} \int_S \mathbf{D} \cdot \mathbf{dS} + \int_S \mathbf{J}^* \cdot \mathbf{dS}. \qquad (2.154)$$

In a Galilean frame where we use (2.49)–(2.54), we see that (2.154) is identical to

$$\oint_\Gamma \mathbf{H}' \cdot \mathbf{dl} = \frac{d}{dt} \int_S \mathbf{D}' \cdot \mathbf{dS} + \int_S \mathbf{J}' \cdot \mathbf{dS} \qquad (2.155)$$

where the primed fields are measured in the frame of the moving contour. This equivalence does *not* hold when the Lorentz transformation is used to represent the primed fields.

Alternative form of the large-scale Maxwell equations. We can write Maxwell's equations in an alternative large-scale form involving only surface and volume integrals. This will be useful later for establishing the field jump conditions across a material or source discontinuity. Again we begin with Maxwell's equations in point form, but instead of integrating them over an open surface we integrate over a volume region V moving with velocity $\mathbf{v}$ (Figure 2.3). In the laboratory frame this gives

$$\int_V (\nabla \times \mathbf{E}) \, dV = -\int_V \frac{\partial \mathbf{B}}{\partial t} \, dV,$$

$$\int_V (\nabla \times \mathbf{H}) \, dV = \int_V \left(\frac{\partial \mathbf{D}}{\partial t} + \mathbf{J} \right) dV.$$

An application of curl theorem (B.24) then gives

$$\oint_S (\hat{\mathbf{n}} \times \mathbf{E}) \, dS = -\int_V \frac{\partial \mathbf{B}}{\partial t} \, dV, \qquad (2.156)$$

$$\oint_S (\hat{\mathbf{n}} \times \mathbf{H}) \, dS = \int_V \left(\frac{\partial \mathbf{D}}{\partial t} + \mathbf{J} \right) dV. \qquad (2.157)$$

Similar results are obtained for the fields in the moving frame:

$$\oint_{S'} (\hat{\mathbf{n}}' \times \mathbf{E}') \, dS' = -\int_{V'} \frac{\partial \mathbf{B}'}{\partial t'} \, dV',$$

$$\oint_{S'} (\hat{\mathbf{n}}' \times \mathbf{H}') \, dS' = \int_{V'} \left(\frac{\partial \mathbf{D}'}{\partial t'} + \mathbf{J}' \right) dV'.$$

These large-scale forms are an alternative to (2.141)–(2.144). They are also form-invariant under a Lorentz transformation.

An alternative to the kinematic formulation of (2.153) and (2.154) can be achieved by applying a kinematic identity for a moving volume region. If V is surrounded by a surface S that moves with velocity $\mathbf{v}$ relative to the laboratory frame, and if a vector field $\mathbf{A}$ is measured in the laboratory frame, then the vector form of the general transport theorem (A.68) states that

$$\frac{d}{dt} \int_V \mathbf{A} \, dV = \int_V \frac{\partial \mathbf{A}}{\partial t} \, dV + \oint_S \mathbf{A}(\mathbf{v} \cdot \hat{\mathbf{n}}) \, dS. \qquad (2.158)$$

Applying this to (2.156) and (2.157) we have

$$\oint_S [\hat{\mathbf{n}} \times \mathbf{E} - (\mathbf{v} \cdot \hat{\mathbf{n}})\mathbf{B}]\, dS = -\frac{d}{dt}\int_V \mathbf{B}\, dV, \tag{2.159}$$

$$\oint_S [\hat{\mathbf{n}} \times \mathbf{H} + (\mathbf{v} \cdot \hat{\mathbf{n}})\mathbf{D}]\, dS = \int_V \mathbf{J}\, dV + \frac{d}{dt}\int_V \mathbf{D}\, dV. \tag{2.160}$$

We can also apply (2.158) to the large-scale form of the continuity equation (1.10) and obtain the expression for a volume region moving with velocity $\mathbf{v}$:

$$\oint_S (\mathbf{J} - \rho\mathbf{v}) \cdot d\mathbf{S} = -\frac{d}{dt}\int_V \rho\, dV.$$

2.5.2 Moving, deforming surfaces

Because (2.151) holds for arbitrarily moving surfaces, the kinematic versions (2.153) and (2.154) hold when $\mathbf{v}$ is interpreted as an instantaneous velocity. However, if the surface and contour lie within a material body that moves relative to the laboratory frame, the constitutive equations relating $\mathbf{E}$, $\mathbf{D}$, $\mathbf{B}$, $\mathbf{H}$, and $\mathbf{J}$ in the laboratory frame differ from those relating the fields in the stationary frame of the body (if the body is not accelerating), and thus the concepts of § 2.3.2 must be employed. This is important when boundary conditions at a moving surface are needed. Particular care must be taken when the body accelerates, since the constitutive relations are then only approximate.

The representation (2.145)–(2.146) is also generally valid, provided we *define* the primed fields as those converted from laboratory fields using the Lorentz transformation with instantaneous velocity $\mathbf{v}$. Here we should use a different inertial frame for each point in the integration, and align the frame with the velocity vector $\mathbf{v}$ at the instant t. We certainly may do this since we can choose to integrate any function we wish. However, this representation may not find wide application.

We thus choose the following expressions, valid for arbitrarily moving surfaces containing only regular points, as our general forms of the large-scale Maxwell equations:

$$\oint_{\Gamma(t)} \mathbf{E}^* \cdot d\mathbf{l} = -\frac{d}{dt}\int_{S(t)} \mathbf{B} \cdot d\mathbf{S} = -\frac{d\Psi(t)}{dt},$$

$$\oint_{\Gamma(t)} \mathbf{H}^* \cdot d\mathbf{l} = \frac{d}{dt}\int_{S(t)} \mathbf{D} \cdot d\mathbf{S} + \int_{S(t)} \mathbf{J}^* \cdot d\mathbf{S},$$

where

$$\mathbf{E}^* = \mathbf{E} + \mathbf{v} \times \mathbf{B},$$
$$\mathbf{H}^* = \mathbf{H} - \mathbf{v} \times \mathbf{D},$$
$$\mathbf{J}^* = \mathbf{J} - \rho\mathbf{v},$$

and where all fields are taken to be measured in the laboratory frame with $\mathbf{v}$ the instantaneous velocity of points on the surface and contour relative to that frame. The constitutive parameters must be considered carefully if the contours and surfaces lie in a moving material medium.

Kinematic identity (2.158) is also valid for arbitrarily moving surfaces. Thus we have the following, valid for arbitrarily moving surfaces and volumes containing only regular points:

$$\oint_{S(t)} [\hat{\mathbf{n}} \times \mathbf{E} - (\mathbf{v} \cdot \hat{\mathbf{n}})\mathbf{B}] \, dS = -\frac{d}{dt} \int_{V(t)} \mathbf{B} \, dV,$$

$$\oint_{S(t)} [\hat{\mathbf{n}} \times \mathbf{H} + (\mathbf{v} \cdot \hat{\mathbf{n}})\mathbf{D}] \, dS = \int_{V(t)} \mathbf{J} \, dV + \frac{d}{dt} \int_{V(t)} \mathbf{D} \, dV.$$

We also find that the two Gauss's law expressions,

$$\oint_{S(t)} \mathbf{D} \cdot d\mathbf{S} = \int_{V(t)} \rho \, dV,$$

$$\oint_{S(t)} \mathbf{B} \cdot d\mathbf{S} = 0,$$

remain valid.

2.5.3   Large-scale form of the Boffi equations

The Maxwell–Boffi equations can be written in large-scale form using the same approach as with the Maxwell–Minkowski equations. Integrating (2.120) and (2.121) over an open surface S and applying Stokes's theorem, we have

$$\oint_{\Gamma} \mathbf{E} \cdot d\mathbf{l} = - \int_{S} \frac{\partial \mathbf{B}}{\partial t} \cdot d\mathbf{S}, \qquad (2.161)$$

$$\oint_{\Gamma} \mathbf{B} \cdot d\mathbf{l} = \mu_0 \int_{S} \left(\mathbf{J} + \mathbf{J}_M + \mathbf{J}_P + \frac{\partial \epsilon_0 \mathbf{E}}{\partial t} \right) \cdot d\mathbf{S}, \qquad (2.162)$$

for fields in the laboratory frame, and

$$\oint_{\Gamma'} \mathbf{E}' \cdot d\mathbf{l}' = - \int_{S'} \frac{\partial \mathbf{B}'}{\partial t'} \cdot d\mathbf{S}',$$

$$\oint_{\Gamma'} \mathbf{B}' \cdot d\mathbf{l}' = \mu_0 \int_{S'} \left(\mathbf{J}' + \mathbf{J}'_M + \mathbf{J}'_P + \frac{\partial \epsilon_0 \mathbf{E}'}{\partial t'} \right) \cdot d\mathbf{S}',$$

for fields in a moving frame. We see that Faraday's law is unmodified by the introduction of polarization and magnetization, hence our prior discussion of emf for moving contours remains valid. However, Ampere's law must be interpreted somewhat differently. The flux vector $\mathbf{B}$ also acts as a force vector, and its circulation is proportional to the out-flux of total current, consisting of $\mathbf{J}$ plus the equivalent magnetization and polarization currents plus the displacement current *in free space*, through the surface bounded by the circulation contour.

The large-scale forms of the auxiliary equations can be found by integrating (2.122) and (2.123) over a volume region and applying the divergence theorem. This gives

$$\oint_{S} \mathbf{E} \cdot d\mathbf{S} = \frac{1}{\epsilon_0} \int_{V} (\rho + \rho_P) \, dV,$$

$$\oint_{S} \mathbf{B} \cdot d\mathbf{S} = 0,$$

for the laboratory frame fields, and

$$\oint_{S'} \mathbf{E}' \cdot d\mathbf{S}' = \frac{1}{\epsilon_0} \int_{V'} (\rho' + \rho'_P) \, dV',$$

$$\oint_{S'} \mathbf{B}' \cdot d\mathbf{S}' = 0,$$

for the moving frame fields. Here we find the force vector $\mathbf{E}$ also acting as a flux vector, with the outflux of $\mathbf{E}$ over a closed surface proportional to the sum of the electric and polarization charges enclosed by the surface.

To provide the alternative representation, we integrate the point forms over V and use the curl theorem to obtain

$$\oint_S (\hat{\mathbf{n}} \times \mathbf{E})\, dS = -\int_V \frac{\partial \mathbf{B}}{\partial t}\, dV, \tag{2.163}$$

$$\oint_S (\hat{\mathbf{n}} \times \mathbf{B})\, dS = \mu_0 \int_V \left(\mathbf{J} + \mathbf{J}_M + \mathbf{J}_P + \frac{\partial \epsilon_0 \mathbf{E}}{\partial t} \right) dV, \tag{2.164}$$

for the laboratory frame fields, and

$$\oint_{S'} (\hat{\mathbf{n}}' \times \mathbf{E}')\, dS' = -\int_{V'} \frac{\partial \mathbf{B}'}{\partial t'}\, dV',$$

$$\oint_{S'} (\hat{\mathbf{n}}' \times \mathbf{B}')\, dS' = \mu_0 \int_{V'} \left(\mathbf{J}' + \mathbf{J}'_M + \mathbf{J}'_P + \frac{\partial \epsilon_0 \mathbf{E}'}{\partial t'} \right) dV',$$

for the moving frame fields.

The large-scale forms of the Boffi equations can also be put into kinematic form using either (2.151) or (2.158). Using (2.151) on (2.161) and (2.162) we have

$$\oint_{\Gamma(t)} \mathbf{E}^* \cdot \mathbf{dl} = -\frac{d}{dt} \int_{S(t)} \mathbf{B} \cdot \mathbf{dS}, \tag{2.165}$$

$$\oint_{\Gamma(t)} \mathbf{B}^\dagger \cdot \mathbf{dl} = \int_{S(t)} \mu_0 \mathbf{J}^\dagger \cdot \mathbf{dS} + \frac{1}{c^2} \frac{d}{dt} \int_{S(t)} \mathbf{E} \cdot \mathbf{dS}, \tag{2.166}$$

where

$$\mathbf{E}^* = \mathbf{E} + \mathbf{v} \times \mathbf{B},$$

$$\mathbf{B}^\dagger = \mathbf{B} - \frac{1}{c^2} \mathbf{v} \times \mathbf{E},$$

$$\mathbf{J}^\dagger = \mathbf{J} + \mathbf{J}_M + \mathbf{J}_P - (\rho + \rho_P)\mathbf{v}.$$

Here $\mathbf{B}^\dagger$ is equivalent to the first-order Lorentz transformation representation of the field in the moving frame (2.64). (The dagger $\dagger$ should not be confused with the symbol for the hermitian operation.) Using (2.158) on (2.163) and (2.164) we have

$$\oint_{S(t)} [\hat{\mathbf{n}} \times \mathbf{E} - (\mathbf{v} \cdot \hat{\mathbf{n}})\mathbf{B}]\, dS = -\frac{d}{dt} \int_{V(t)} \mathbf{B}\, dV, \tag{2.167}$$

and

$$\oint_{S(t)} \left[\hat{\mathbf{n}} \times \mathbf{B} + \frac{1}{c^2}(\mathbf{v} \cdot \hat{\mathbf{n}})\mathbf{E} \right] dS = \mu_0 \int_{V(t)} (\mathbf{J} + \mathbf{J}_M + \mathbf{J}_P)\, dV + \frac{1}{c^2} \frac{d}{dt} \int_{V(t)} \mathbf{E}\, dV. \tag{2.168}$$

In each case the fields are measured in the laboratory frame, and $\mathbf{v}$ is measured with respect to the laboratory frame and may vary arbitrarily over the surface or contour.

2.6   The nature of the four field quantities

Since the very inception of Maxwell's theory, its students have been distressed by the fact that while there are four electromagnetic fields $(\mathbf{E}, \mathbf{D}, \mathbf{B}, \mathbf{H})$, there are only two fundamental equations (the curl equations) to describe their interrelationship. The relegation of additional required information to constitutive equations that vary widely between classes of materials seems to lessen the elegance of the theory. While some may find elegant the separation of equations into a set expressing the basic wave nature of electromagnetism and a set describing how the fields interact with materials, the history of the discipline is one of categorizing and pairing fields as "fundamental" and "supplemental" in hopes of reducing the model to two equations in two unknowns.

Lorentz led the way in this area. With his electrical theory of matter, all material effects could be interpreted in terms of atomic charge and current immersed in free space. We have seen how the Maxwell–Boffi equations seem to eliminate the need for $\mathbf{D}$ and $\mathbf{H}$, and indeed for simple media where there is a linear relation between the remaining "fundamental" fields and the induced polarization and magnetization, it appears that only $\mathbf{E}$ and $\mathbf{B}$ are required. However, for more complicated materials that display nonlinear and bianisotropic effects we are only able to supplant $\mathbf{D}$ and $\mathbf{H}$ with two other fields $\mathbf{P}$ and $\mathbf{M}$, along with (possibly complicated) constitutive relations relating them to $\mathbf{E}$ and $\mathbf{B}$.

Even those authors who do not wish to eliminate two of the fields tend to categorize the fields into pairs based on physical arguments, implying that one or the other pair is in some way "more fundamental." Maxwell himself separated the fields into the pair $(\mathbf{E}, \mathbf{H})$ that appears within line integrals to give work and the pair $(\mathbf{B}, \mathbf{D})$ that appears within surface integrals to give flux. In what other ways might we pair the four vectors?

Most prevalent is the splitting of the fields into electric and magnetic pairs: $(\mathbf{E}, \mathbf{D})$ and $(\mathbf{B}, \mathbf{H})$. In Poynting's theorem $\mathbf{E} \cdot \mathbf{D}$ describes one component of stored energy (called "electric energy") and $\mathbf{B} \cdot \mathbf{H}$ describes another component (called "magnetic energy"). These pairs also occur in Maxwell's stress tensor. In statics, the fields decouple into electric and magnetic sets. But biisotropic and bianisotropic materials demonstrate how separation into electric and magnetic effects can become problematic.

In the study of electromagnetic waves, the ratio of E to H appears to be an important quantity, called the "intrinsic impedance." The pair $(\mathbf{E}, \mathbf{H})$ also determines the Poynting flux of power, and is required to establish the uniqueness of the electromagnetic field. In addition, constitutive relations for simple materials usually express $(\mathbf{D}, \mathbf{B})$ in terms of $(\mathbf{E}, \mathbf{H})$. Models for these materials are often conceived by viewing the fields $(\mathbf{E}, \mathbf{H})$ as interacting with the atomic structure in such a way as to produce secondary effects describable by $(\mathbf{D}, \mathbf{B})$. These considerations, along with Maxwell's categorization into a pair of work vectors and a pair of flux vectors, lead many authors to formulate electromagnetics with $\mathbf{E}$ and $\mathbf{H}$ as the "fundamental" quantities. But the pair $(\mathbf{B}, \mathbf{D})$ gives rise to electromagnetic momentum and is also perpendicular to the direction of wave propagation in an anisotropic material; in these senses, we might argue that these fields must be equally "fundamental."

Perhaps the best motivation for grouping fields comes from relativistic considerations. We have found that $(\mathbf{E}, \mathbf{B})$ transform together under a Lorentz transformation, as do $(\mathbf{D}, \mathbf{H})$. In each of these pairs we have one polar vector ($\mathbf{E}$ or $\mathbf{D}$) and one axial vector ($\mathbf{B}$ or $\mathbf{H}$). A polar vector retains its meaning under a change in handedness of the coordinate system, while an axial vector does not. The Lorentz force involves one polar vector ($\mathbf{E}$)

and one axial vector ($\mathbf{B}$) that we also call "electric" and "magnetic." If we follow the lead of some authors and *choose* to define $\mathbf{E}$ and $\mathbf{B}$ through measurements of the Lorentz force, then we recognize that $\mathbf{B}$ must be axial since it is not measured directly, but as part of the cross product $\mathbf{v} \times \mathbf{B}$ that changes its meaning if we switch from a right-hand to a left-hand coordinate system. The other polar vector ($\mathbf{D}$) and axial vector ($\mathbf{H}$) arise through the "secondary" constitutive relations. Following this reasoning we might claim that $\mathbf{E}$ and $\mathbf{B}$ are "fundamental."

Sommerfeld also associates $\mathbf{E}$ with $\mathbf{B}$ and $\mathbf{D}$ with $\mathbf{H}$. The vectors $\mathbf{E}$ and $\mathbf{B}$ are called *entities of intensity*, describing "how strong," while $\mathbf{D}$ and $\mathbf{H}$ are called *entities of quantity*, describing "how much." This is in direct analogy with stress (intensity) and strain (quantity) in materials. We might also say that the entities of intensity describe a "cause" while the entities of quantity describe an "effect." In this view $\mathbf{E}$ "induces" (causes) a polarization $\mathbf{P}$, and the field $\mathbf{D} = \epsilon_0 \mathbf{E} + \mathbf{P}$ is the result. Similarly $\mathbf{B}$ creates $\mathbf{M}$, and $\mathbf{H} = \mathbf{B}/\mu_0 - \mathbf{M}$ is the result. Interestingly, each of the terms describing energy and momentum in the electromagnetic field ($\mathbf{D} \cdot \mathbf{E}$, $\mathbf{B} \cdot \mathbf{H}$, $\mathbf{E} \times \mathbf{H}$, $\mathbf{D} \times \mathbf{B}$) involves the interaction of an entity of intensity with an entity of quantity.

Although there is a natural tendency to group things together based on conceptual similarity, there appears to be little reason to believe that any of the four field vectors are more "fundamental" than the rest. Perhaps we are fortunate that we can apply Maxwell's theory without worrying too much about such questions of underlying philosophy.

2.7 Maxwell's equations with magnetic sources

Researchers have yet to discover the "magnetic monopole": a magnetic source from which magnetic field would diverge. This has not stopped speculation on the form that Maxwell's equations might take if such a discovery were made. Arguments based on fundamental principles of physics (such as symmetry and conservation laws) indicate that in the presence of magnetic sources Maxwell's equations would assume the forms

$$\nabla \times \mathbf{E} = -\mathbf{J}_m - \frac{\partial \mathbf{B}}{\partial t}, \tag{2.169}$$

$$\nabla \times \mathbf{H} = \mathbf{J} + \frac{\partial \mathbf{D}}{\partial t}, \tag{2.170}$$

$$\nabla \cdot \mathbf{B} = \rho_m, \tag{2.171}$$

$$\nabla \cdot \mathbf{D} = \rho, \tag{2.172}$$

where $\mathbf{J}_m$ is a volume magnetic current density describing the flow of magnetic charge in exactly the same manner as $\mathbf{J}$ describes the flow of electric charge. The density of this magnetic charge is given by ρ_m and should, by analogy with electric charge density, obey a conservation law

$$\nabla \cdot \mathbf{J}_m + \frac{\partial \rho_m}{\partial t} = 0.$$

This is the magnetic source continuity equation.

It is interesting to inquire as to the units of $\mathbf{J}_m$ and ρ_m. From (2.169) we see that if $\mathbf{B}$ has units of $\mathrm{Wb/m^2}$, then $\mathbf{J}_m$ has units of $(\mathrm{Wb/s})/\mathrm{m^2}$. Similarly, (2.171) shows that ρ_m must have units of $\mathrm{Wb/m^3}$. Hence magnetic charge is measured in Wb, magnetic current in Wb/s. This gives a nice symmetry with electric sources where charge is measured in

C and current in C/s.[1] The physical symmetry is equally appealing: magnetic flux lines diverge from magnetic charge, and the total flux passing through a surface is given by the total magnetic charge contained within the surface. This is best seen by considering the large-scale forms of Maxwell's equations for stationary surfaces. We need only modify (2.145) to include the magnetic current term; this gives

$$\oint_\Gamma \mathbf{E} \cdot d\mathbf{l} = -\int_S \mathbf{J}_m \cdot d\mathbf{S} - \frac{d}{dt}\int_S \mathbf{B} \cdot d\mathbf{S}, \qquad (2.173)$$

$$\oint_\Gamma \mathbf{H} \cdot d\mathbf{l} = \int_S \mathbf{J} \cdot d\mathbf{S} + \frac{d}{dt}\int_S \mathbf{D} \cdot d\mathbf{S}. \qquad (2.174)$$

If we modify (2.148) to include magnetic charge, we get the auxiliary equations

$$\oint_S \mathbf{D} \cdot d\mathbf{S} = \int_V \rho \, dV,$$

$$\oint_S \mathbf{B} \cdot d\mathbf{S} = \int_V \rho_m \, dV.$$

Any of the large-scale forms of Maxwell's equations can be similarly modified to include magnetic current and charge. For arbitrarily moving surfaces we have

$$\oint_{\Gamma(t)} \mathbf{E}^* \cdot d\mathbf{l} = -\frac{d}{dt}\int_{S(t)} \mathbf{B} \cdot d\mathbf{S} - \int_{S(t)} \mathbf{J}_m^* \cdot d\mathbf{S},$$

$$\oint_{\Gamma(t)} \mathbf{H}^* \cdot d\mathbf{l} = \frac{d}{dt}\int_{S(t)} \mathbf{D} \cdot d\mathbf{S} + \int_{S(t)} \mathbf{J}^* \cdot d\mathbf{S},$$

where

$$\mathbf{E}^* = \mathbf{E} + \mathbf{v} \times \mathbf{B},$$
$$\mathbf{H}^* = \mathbf{H} - \mathbf{v} \times \mathbf{D},$$
$$\mathbf{J}^* = \mathbf{J} - \rho\mathbf{v},$$
$$\mathbf{J}_m^* = \mathbf{J}_m - \rho_m\mathbf{v},$$

and all fields are taken to be measured in the laboratory frame with $\mathbf{v}$ the instantaneous velocity of points on the surface and contour relative to the laboratory frame. We also have the alternative forms

$$\oint_S (\hat{\mathbf{n}} \times \mathbf{E}) \, dS = \int_V \left(-\frac{\partial \mathbf{B}}{\partial t} - \mathbf{J}_m\right) dV, \qquad (2.175)$$

$$\oint_S (\hat{\mathbf{n}} \times \mathbf{H}) \, dS = \int_V \left(\frac{\partial \mathbf{D}}{\partial t} + \mathbf{J}\right) dV, \qquad (2.176)$$

and

$$\oint_{S(t)} [\hat{\mathbf{n}} \times \mathbf{E} - (\mathbf{v} \cdot \hat{\mathbf{n}})\mathbf{B}] \, dS = -\int_{V(t)} \mathbf{J}_m \, dV - \frac{d}{dt}\int_{V(t)} \mathbf{B} \, dV, \qquad (2.177)$$

$$\oint_{S(t)} [\hat{\mathbf{n}} \times \mathbf{H} + (\mathbf{v} \cdot \hat{\mathbf{n}})\mathbf{D}] \, dS = \int_{V(t)} \mathbf{J} \, dV + \frac{d}{dt}\int_{V(t)} \mathbf{D} \, dV. \qquad (2.178)$$

---

[1]We note that if the modern unit of T is used to describe $\mathbf{B}$, then ρ_m is described using the more cumbersome units of T/m, while $\mathbf{J}_m$ is given in terms of T/s. Thus, magnetic charge is measured in Tm^2 and magnetic current in $(Tm^2)/s$.

and the two Gauss's law expressions

$$\oint_{S(t)} \mathbf{D} \cdot \hat{\mathbf{n}} \, dS = \int_{V(t)} \rho \, dV,$$

$$\oint_{S(t)} \mathbf{B} \cdot \hat{\mathbf{n}} \, dS = \int_{V(t)} \rho_m \, dV.$$

Magnetic sources also allow us to develop equivalence theorems in which difficult problems involving boundaries are replaced by simpler problems involving magnetic sources. Although these sources may not physically exist, the mathematical solutions are completely valid.

2.8 Boundary (jump) conditions

If we restrict ourselves to regions of space without spatial (jump) discontinuities in either the sources or the constitutive relations, we can find meaningful solutions to the Maxwell differential equations. We also know that for given sources, if the fields are specified on a closed boundary and at an initial time the solutions are unique. The standard approach to treating regions that do contain spatial discontinuities is to isolate the discontinuities on surfaces. That is, we introduce surfaces that serve to separate space into regions in which the differential equations are solvable and the fields are well defined. To make the solutions in adjoining regions unique, we must specify the tangential fields on each side of the adjoining surface. If we can relate the fields across the boundary, we can propagate the solution from one region to the next; in this way, information about the source in one region is effectively passed on to the solution in an adjacent region. For uniqueness, only relations between the tangential components need be specified.

We shall determine the appropriate boundary conditions (BC's) via two distinct approaches. We first model a thin source layer and consider a discontinuous surface source layer as a limiting case of the continuous thin layer. With no true discontinuity, Maxwell's differential equations hold everywhere. We then consider a true spatial discontinuity between material surfaces (with possible surface sources lying along the discontinuity). We must then isolate the region containing the discontinuity and *postulate* a field relationship that is both physically meaningful and experimentally verifiable.

We shall also consider both stationary and moving boundary surfaces, and surfaces containing magnetic as well as electric sources.

2.8.1 Boundary conditions across a stationary, thin source layer

In § 1.3.3 we discussed how in the macroscopic sense a surface source is actually a volume distribution concentrated near a surface S. We write the charge and current in terms of the point $\mathbf{r}$ on the surface and the normal distance x from the surface at $\mathbf{r}$ as

$$\rho(\mathbf{r}, x, t) = \rho_s(\mathbf{r}, t) f(x, \Delta), \qquad (2.179)$$

$$\mathbf{J}(\mathbf{r}, x, t) = \mathbf{J}_s(\mathbf{r}, t) f(x, \Delta), \qquad (2.180)$$

where $f(x, \Delta)$ is the source density function obeying

$$\int_{-\infty}^{\infty} f(x, \Delta) \, dx = 1. \qquad (2.181)$$

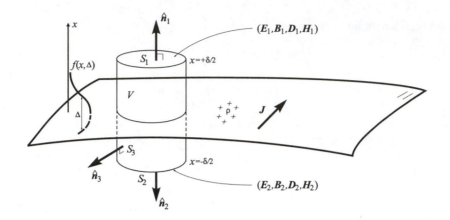

Figure 2.5: Derivation of the electromagnetic boundary conditions across a thin continuous source layer.

The parameter Δ describes the "width" of the source layer normal to the reference surface.

We use (2.156)–(2.157) to study field behavior across the source layer. Consider a volume region V that intersects the source layer as shown in Figure 2.5. Let the top and bottom surfaces be parallel to the reference surface, and label the fields on the top and bottom surfaces with subscripts 1 and 2, respectively. Since points on and within V are all regular, (2.157) yields

$$\int_{S_1} \hat{\mathbf{n}}_1 \times \mathbf{H}_1 \, dS + \int_{S_2} \hat{\mathbf{n}}_2 \times \mathbf{H}_2 \, dS + \int_{S_3} \hat{\mathbf{n}}_3 \times \mathbf{H} \, dS = \int_V \left(\mathbf{J} + \frac{\partial \mathbf{D}}{\partial t} \right) dV.$$

We now choose $\delta = k\Delta$ ($k > 1$) so that most of the source lies within V. As $\Delta \to 0$ the thin source layer recedes to a surface layer, and the volume integral of displacement current and the integral of tangential $\mathbf{H}$ over S_3 both approach zero by continuity of the fields. By symmetry $S_1 = S_2$ and $\hat{\mathbf{n}}_1 = -\hat{\mathbf{n}}_2 = \hat{\mathbf{n}}_{12}$, where $\hat{\mathbf{n}}_{12}$ is the surface normal directed into region 1 from region 2. Thus

$$\int_{S_1} \hat{\mathbf{n}}_{12} \times (\mathbf{H}_1 - \mathbf{H}_2) \, dS = \int_V \mathbf{J} \, dV. \qquad (2.182)$$

Note that

$$\int_V \mathbf{J} \, dV = \int_{S_1} \int_{-\delta/2}^{\delta/2} \mathbf{J} \, dS \, dx = \int_{-\delta/2}^{\delta/2} f(x, \Delta) \, dx \int_{S_1} \mathbf{J}_s(\mathbf{r}, t) \, dS.$$

Since we assume that the majority of the source current lies within V, the integral can be evaluated using (2.181) to give

$$\int_{S_1} [\hat{\mathbf{n}}_{12} \times (\mathbf{H}_1 - \mathbf{H}_2) - \mathbf{J}_s] \, dS = 0,$$

hence

$$\hat{\mathbf{n}}_{12} \times (\mathbf{H}_1 - \mathbf{H}_2) = \mathbf{J}_s.$$

The tangential magnetic field across a thin source distribution is discontinuous by an amount equal to the surface current density.

Similar steps with Faraday's law give

$$\hat{\mathbf{n}}_{12} \times (\mathbf{E}_1 - \mathbf{E}_2) = 0.$$

The tangential electric field is continuous across a thin source.

We can also derive conditions on the normal components of the fields, although these are not required for uniqueness. Gauss's law (2.147) applied to the volume V in Figure 2.5 gives

$$\int_{S_1} \mathbf{D}_1 \cdot \hat{\mathbf{n}}_1 \, dS + \int_{S_2} \mathbf{D}_2 \cdot \hat{\mathbf{n}}_2 \, dS + \int_{S_3} \mathbf{D} \cdot \hat{\mathbf{n}}_3 \, dS = \int_V \rho \, dV.$$

As $\Delta \to 0$, the thin source layer recedes to a surface layer. The integral of normal $\mathbf{D}$ over S_3 tends to zero by continuity of the fields. By symmetry $S_1 = S_2$ and $\hat{\mathbf{n}}_1 = -\hat{\mathbf{n}}_2 = \hat{\mathbf{n}}_{12}$. Thus

$$\int_{S_1} (\mathbf{D}_1 - \mathbf{D}_2) \cdot \hat{\mathbf{n}}_{12} \, dS = \int_V \rho \, dV. \tag{2.183}$$

The volume integral is

$$\int_V \rho \, dV = \int_{S_1} \int_{-\delta/2}^{\delta/2} \rho \, dS \, dx = \int_{-\delta/2}^{\delta/2} f(x, \Delta) \, dx \int_{S_1} \rho_s(\mathbf{r}, t) \, dS.$$

Since $\delta = k\Delta$ has been chosen so that most of the source charge lies within V, (2.181) gives

$$\int_{S_1} [(\mathbf{D}_1 - \mathbf{D}_2) \cdot \hat{\mathbf{n}}_{12} - \rho_s] \, dS = 0,$$

hence

$$(\mathbf{D}_1 - \mathbf{D}_2) \cdot \hat{\mathbf{n}}_{12} = \rho_s.$$

The normal component of $\mathbf{D}$ is discontinuous across a thin source distribution by an amount equal to the surface charge density. Similar steps with the magnetic Gauss's law yield

$$(\mathbf{B}_1 - \mathbf{B}_2) \cdot \hat{\mathbf{n}}_{12} = 0.$$

The normal component of $\mathbf{B}$ is continuous across a thin source layer.

We can follow similar steps when a thin magnetic source layer is present. When evaluating Faraday's law we must include magnetic surface current and when evaluating the magnetic Gauss's law we must include magnetic charge. However, since such sources are not physical we postpone their consideration until the next section, where appropriate boundary conditions are postulated rather than derived.

2.8.2   Boundary conditions across a stationary layer of field discontinuity

Provided that we model a surface source as a limiting case of a very thin but continuous volume source, we can derive boundary conditions across a surface layer. We might ask whether we can extend this idea to surfaces of materials where the constitutive parameters change from one region to another. Indeed, if we take Lorentz' viewpoint and visualize a material as a conglomerate of atomic charge, we should be able to apply this same idea. After all, a material should demonstrate a continuous transition (in the macroscopic

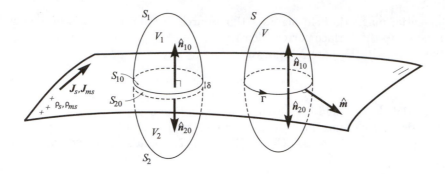

Figure 2.6: Derivation of the electromagnetic boundary conditions across a discontinuous source layer.

sense) across its boundary, and we can employ the Maxwell–Boffi equations to describe the relationship between the "equivalent" sources and the electromagnetic fields.

We should note, however, that the limiting concept is not without its critics. Stokes suggested as early as 1848 that jump conditions should never be derived from smooth solutions [199]. Let us therefore pursue the boundary conditions for a surface of true field discontinuity. This will also allow us to treat a material modeled as having a true discontinuity in its material parameters (which we can always take as a mathematical model of a more gradual transition) before we have studied in a deeper sense the physical properties of materials. This approach, taken by many textbooks, must be done carefully.

There is a logical difficulty with this approach, lying in the application of the large-scale forms of Maxwell's equations. Many authors postulate Maxwell's equations in point form, integrate to obtain the large-scale forms, then apply the large-scale forms to regions of discontinuity. Unfortunately, the large-scale forms thus obtained are only valid in the same regions where their point form antecedents were valid — discontinuities must be excluded. Schelkunoff [167] has criticized this approach, calling it a "swindle" rather than a proof, and has suggested that the proper way to handle true discontinuities is to postulate the large-scale forms of Maxwell's equations, *and* to include as part of the postulate the assumption that the large-scale forms are valid at points of field discontinuity. Does this mean we must reject our postulate of the point form Maxwell equations and reformulate everything in terms of the large-scale forms? Fortunately, no. Tai [192] has pointed out that it is still possible to postulate the point forms, as long as we also postulate appropriate boundary conditions that make the large-scale forms, as derived from the point forms, valid at surfaces of discontinuity. In essence, both approaches require an additional postulate for surfaces of discontinuity: the large scale forms require a postulate of applicability to discontinuous surfaces, and from there the boundary conditions can be derived; the point forms require a postulate of the boundary conditions that result in the large-scale forms being valid on surfaces of discontinuity. Let us examine how the latter approach works.

Consider a surface across which the constitutive relations are discontinuous, containing electric and magnetic surface currents and charges $\mathbf{J}_s$, ρ_s, $\mathbf{J}_{ms}$, and ρ_{ms} (Figure 2.6). We locate a volume region V_1 above the surface of discontinuity; this volume is bounded by a surface S_1 and another surface S_{10} which is parallel to, and a small distance $\delta/2$ above, the surface of discontinuity. A second volume region V_2 is similarly situated below the surface of discontinuity. Because these regions exclude the surface of discontinuity we can use (2.176) to get

$$\int_{S_1} \hat{\mathbf{n}} \times \mathbf{H} \, dS + \int_{S_{10}} \hat{\mathbf{n}} \times \mathbf{H} \, dS = \int_{V_1} \left(\mathbf{J} + \frac{\partial \mathbf{D}}{\partial t} \right) dV,$$

$$\int_{S_2} \hat{\mathbf{n}} \times \mathbf{H} \, dS + \int_{S_{20}} \hat{\mathbf{n}} \times \mathbf{H} \, dS = \int_{V_2} \left(\mathbf{J} + \frac{\partial \mathbf{D}}{\partial t} \right) dV.$$

Adding these we obtain

$$\int_{S_1+S_2} \hat{\mathbf{n}} \times \mathbf{H} \, dS - \int_{V_1+V_2} \left(\mathbf{J} + \frac{\partial \mathbf{D}}{\partial t} \right) dV -$$
$$- \int_{S_{10}} \hat{\mathbf{n}}_{10} \times \mathbf{H}_1 \, dS - \int_{S_{20}} \hat{\mathbf{n}}_{20} \times \mathbf{H}_2 \, dS = 0, \qquad (2.184)$$

where we have used subscripts to delineate the fields on each side of the discontinuity surface.

If δ is very small (but nonzero), then $\hat{\mathbf{n}}_{10} = -\hat{\mathbf{n}}_{20} = \hat{\mathbf{n}}_{12}$ and $S_{10} = S_{20}$. Letting $S_1 + S_2 = S$ and $V_1 + V_2 = V$, we can write (2.184) as

$$\int_S (\hat{\mathbf{n}} \times \mathbf{H}) \, dS - \int_V \left(\mathbf{J} + \frac{\partial \mathbf{D}}{\partial t} \right) dV = \int_{S_{10}} \hat{\mathbf{n}}_{12} \times (\mathbf{H}_1 - \mathbf{H}_2) \, dS. \qquad (2.185)$$

Now suppose we use the same volume region V, but let it intersect the surface of discontinuity (Figure 2.6), and suppose that the large-scale form of Ampere's law holds even if V contains points of field discontinuity. We must include the surface current in the computation. Since $\int_V \mathbf{J} \, dV$ becomes $\int_S \mathbf{J}_s \, dS$ on the surface, we have

$$\int_S (\hat{\mathbf{n}} \times \mathbf{H}) \, dS - \int_V \left(\mathbf{J} + \frac{\partial \mathbf{D}}{\partial t} \right) dV = \int_{S_{10}} \mathbf{J}_s \, dS. \qquad (2.186)$$

We wish to have this give the same value for the integrals over V and S as (2.185), which included in its derivation no points of discontinuity. This is true provided that

$$\hat{\mathbf{n}}_{12} \times (\mathbf{H}_1 - \mathbf{H}_2) = \mathbf{J}_s. \qquad (2.187)$$

Thus, under the condition (2.187) we may interpret the large-scale form of Ampere's law (as derived from the point form) as being valid for regions containing discontinuities. Note that this condition is not "derived," but must be regarded as a postulate that results in the large-scale form holding for surfaces of discontinuous field.

Similar reasoning can be used to determine the appropriate boundary condition on tangential $\mathbf{E}$ from Faraday's law. Corresponding to (2.185) we obtain

$$\int_S (\hat{\mathbf{n}} \times \mathbf{E}) \, dS - \int_V \left(-\mathbf{J}_m - \frac{\partial \mathbf{B}}{\partial t} \right) dV = \int_{S_{10}} \hat{\mathbf{n}}_{12} \times (\mathbf{E}_1 - \mathbf{E}_2) \, dS. \qquad (2.188)$$

Employing (2.175) over the region containing the field discontinuity surface we get

$$\int_S (\hat{\mathbf{n}} \times \mathbf{E}) \, dS - \int_V \left(-\mathbf{J}_m - \frac{\partial \mathbf{B}}{\partial t} \right) dV = - \int_{S_{10}} \mathbf{J}_{ms} \, dS. \qquad (2.189)$$

To have (2.188) and (2.189) produce identical results, we postulate

$$\hat{\mathbf{n}}_{12} \times (\mathbf{E}_1 - \mathbf{E}_2) = -\mathbf{J}_{ms} \qquad (2.190)$$

as the boundary condition appropriate to a surface of field discontinuity containing a magnetic surface current.

We can also postulate boundary conditions on the normal fields to make Gauss's laws valid for surfaces of discontinuous fields. Integrating (2.147) over the regions V_1 and V_2 and adding, we obtain

$$\int_{S_1+S_2} \mathbf{D} \cdot \hat{\mathbf{n}} \, dS - \int_{S_{10}} \mathbf{D}_1 \cdot \hat{\mathbf{n}}_{10} \, dS - \int_{S_{20}} \mathbf{D}_2 \cdot \hat{\mathbf{n}}_{20} \, dS = \int_{V_1+V_2} \rho \, dV.$$

As $\delta \to 0$ this becomes

$$\int_S \mathbf{D} \cdot \hat{\mathbf{n}} \, dS - \int_V \rho \, dV = \int_{S_{10}} (\mathbf{D}_1 - \mathbf{D}_2) \cdot \hat{\mathbf{n}}_{12} \, dS. \qquad (2.191)$$

If we integrate Gauss's law over the entire region V, including the surface of discontinuity, we get

$$\oint_S \mathbf{D} \cdot \hat{\mathbf{n}} \, dS = \int_V \rho \, dV + \int_{S_{10}} \rho_s \, dS. \qquad (2.192)$$

In order to get identical answers from (2.191) and (2.192), we must have

$$(\mathbf{D}_1 - \mathbf{D}_2) \cdot \hat{\mathbf{n}}_{12} = \rho_s$$

as the boundary condition appropriate to a surface of field discontinuity containing an electric surface charge. Similarly, we must postulate

$$(\mathbf{B}_1 - \mathbf{B}_2) \cdot \hat{\mathbf{n}}_{12} = \rho_{ms}$$

as the condition appropriate to a surface of field discontinuity containing a magnetic surface charge.

We can determine an appropriate boundary condition on current by using the large-scale form of the continuity equation. Applying (1.10) over each of the volume regions of Figure 2.6 and adding the results, we have

$$\int_{S_1+S_2} \mathbf{J} \cdot \hat{\mathbf{n}} \, dS - \int_{S_{10}} \mathbf{J}_1 \cdot \hat{\mathbf{n}}_{10} \, dS - \int_{S_{20}} \mathbf{J}_2 \cdot \hat{\mathbf{n}}_{20} \, dS = -\int_{V_1+V_2} \frac{\partial \rho}{\partial t} \, dV.$$

As $\delta \to 0$ we have

$$\int_S \mathbf{J} \cdot \hat{\mathbf{n}} \, dS - \int_{S_{10}} (\mathbf{J}_1 - \mathbf{J}_2) \cdot \hat{\mathbf{n}}_{12} \, dS = -\int_V \frac{\partial \rho}{\partial t} \, dV. \qquad (2.193)$$

Applying the continuity equation over the entire region V and allowing it to intersect the discontinuity surface, we get

$$\int_S \mathbf{J} \cdot \hat{\mathbf{n}} \, dS + \int_\Gamma \mathbf{J}_s \cdot \hat{\mathbf{m}} \, dl = -\int_V \frac{\partial \rho}{\partial t} \, dV - \int_{S_{10}} \frac{\partial \rho_s}{\partial t} \, dS.$$

By the two-dimensional divergence theorem (B.20) we can write this as

$$\int_S \mathbf{J} \cdot \hat{\mathbf{n}} \, dS + \int_{S_{10}} \nabla_s \cdot \mathbf{J}_s \, dS = -\int_V \frac{\partial \rho}{\partial t} \, dV - \int_{S_{10}} \frac{\partial \rho_s}{\partial t} \, dS.$$

In order for this expression to produce the same values of the integrals over S and V as in (2.193) we require

$$\nabla_s \cdot \mathbf{J}_s = -\hat{\mathbf{n}}_{12} \cdot (\mathbf{J}_1 - \mathbf{J}_2) - \frac{\partial \rho_s}{\partial t},$$

which we take as our postulate of the boundary condition on current across a surface containing discontinuities. A similar set of steps carried out using the continuity equation for magnetic sources yields

$$\nabla_s \cdot \mathbf{J}_{ms} = -\hat{\mathbf{n}}_{12} \cdot (\mathbf{J}_{m1} - \mathbf{J}_{m2}) - \frac{\partial \rho_{ms}}{\partial t}.$$

In summary, we have the following boundary conditions for fields across a surface containing discontinuities:

$$\hat{\mathbf{n}}_{12} \times (\mathbf{H}_1 - \mathbf{H}_2) = \mathbf{J}_s, \tag{2.194}$$

$$\hat{\mathbf{n}}_{12} \times (\mathbf{E}_1 - \mathbf{E}_2) = -\mathbf{J}_{ms}, \tag{2.195}$$

$$\hat{\mathbf{n}}_{12} \cdot (\mathbf{D}_1 - \mathbf{D}_2) = \rho_s, \tag{2.196}$$

$$\hat{\mathbf{n}}_{12} \cdot (\mathbf{B}_1 - \mathbf{B}_2) = \rho_{ms}, \tag{2.197}$$

and

$$\hat{\mathbf{n}}_{12} \cdot (\mathbf{J}_1 - \mathbf{J}_2) = -\nabla_s \cdot \mathbf{J}_s - \frac{\partial \rho_s}{\partial t}, \tag{2.198}$$

$$\hat{\mathbf{n}}_{12} \cdot (\mathbf{J}_{m1} - \mathbf{J}_{m2}) = -\nabla_s \cdot \mathbf{J}_{ms} - \frac{\partial \rho_{ms}}{\partial t}, \tag{2.199}$$

where $\hat{\mathbf{n}}_{12}$ points into region 1 from region 2.

2.8.3 Boundary conditions at the surface of a perfect conductor

We can easily specialize the results of the previous section to the case of perfect electric or magnetic conductors. In § 2.2.2 we saw that the constitutive relations for perfect conductors requires the null field within the material. In addition, a PEC requires zero tangential electric field, while a PMC requires zero tangential magnetic field. Using (2.194)–(2.199), we find that the boundary conditions for a perfect electric conductor are

$$\hat{\mathbf{n}} \times \mathbf{H} = \mathbf{J}_s, \tag{2.200}$$

$$\hat{\mathbf{n}} \times \mathbf{E} = 0, \tag{2.201}$$

$$\hat{\mathbf{n}} \cdot \mathbf{D} = \rho_s, \tag{2.202}$$

$$\hat{\mathbf{n}} \cdot \mathbf{B} = 0, \tag{2.203}$$

and

$$\hat{\mathbf{n}} \cdot \mathbf{J} = -\nabla_s \cdot \mathbf{J}_s - \frac{\partial \rho_s}{\partial t}, \qquad \hat{\mathbf{n}} \cdot \mathbf{J}_m = 0. \tag{2.204}$$

For a PMC the conditions are

$$\hat{\mathbf{n}} \times \mathbf{H} = 0, \tag{2.205}$$

$$\hat{\mathbf{n}} \times \mathbf{E} = -\mathbf{J}_{ms}, \tag{2.206}$$

$$\hat{\mathbf{n}} \cdot \mathbf{D} = 0, \tag{2.207}$$

$$\hat{\mathbf{n}} \cdot \mathbf{B} = \rho_{ms}, \tag{2.208}$$

and

$$\hat{\mathbf{n}} \cdot \mathbf{J}_m = -\nabla_s \cdot \mathbf{J}_{ms} - \frac{\partial \rho_{ms}}{\partial t}, \qquad \hat{\mathbf{n}} \cdot \mathbf{J} = 0. \tag{2.209}$$

We note that the normal vector $\hat{\mathbf{n}}$ points out of the conductor and into the adjacent region of nonzero fields.

2.8.4 Boundary conditions across a stationary layer of field discontinuity using equivalent sources

So far we have avoided using the physical interpretation of the equivalent sources in the Maxwell–Boffi equations so that we might investigate the behavior of fields across true discontinuities. Now that we have the appropriate boundary conditions, it is interesting to interpret them in terms of the equivalent sources.

If we put $\mathbf{H} = \mathbf{B}/\mu_0 - \mathbf{M}$ into (2.194) and rearrange, we get

$$\hat{\mathbf{n}}_{12} \times (\mathbf{B}_1 - \mathbf{B}_2) = \mu_0(\mathbf{J}_s + \hat{\mathbf{n}}_{12} \times \mathbf{M}_1 - \hat{\mathbf{n}}_{12} \times \mathbf{M}_2). \tag{2.210}$$

The terms on the right involving $\hat{\mathbf{n}}_{12} \times \mathbf{M}$ have the units of surface current and are called *equivalent magnetization surface currents*. Defining

$$\mathbf{J}_{Ms} = -\hat{\mathbf{n}} \times \mathbf{M} \tag{2.211}$$

where $\hat{\mathbf{n}}$ is directed normally outward from the material region of interest, we can rewrite (2.210) as

$$\hat{\mathbf{n}}_{12} \times (\mathbf{B}_1 - \mathbf{B}_2) = \mu_0(\mathbf{J}_s + \mathbf{J}_{Ms1} + \mathbf{J}_{Ms2}). \tag{2.212}$$

We note that J_{Ms} replaces atomic charge moving along the surface of a material with an equivalent surface current in free space.

If we substitute $\mathbf{D} = \epsilon_0\mathbf{E} + \mathbf{P}$ into (2.196) and rearrange, we get

$$\hat{\mathbf{n}}_{12} \cdot (\mathbf{E}_1 - \mathbf{E}_2) = \frac{1}{\epsilon_0}(\rho_s - \hat{\mathbf{n}}_{12} \cdot \mathbf{P}_1 + \hat{\mathbf{n}}_{12} \cdot \mathbf{P}_2). \tag{2.213}$$

The terms on the right involving $\hat{\mathbf{n}}_{12} \cdot \mathbf{P}$ have the units of surface charge and are called *equivalent polarization surface charges*. Defining

$$\rho_{Ps} = \hat{\mathbf{n}} \cdot \mathbf{P}, \tag{2.214}$$

we can rewrite (2.213) as

$$\hat{\mathbf{n}}_{12} \cdot (\mathbf{E}_1 - \mathbf{E}_2) = \frac{1}{\epsilon_0}(\rho_s + \rho_{Ps1} + \rho_{Ps2}). \tag{2.215}$$

We note that ρ_{Ps} replaces atomic charge adjacent to a surface of a material with an equivalent surface charge in free space.

In summary, the boundary conditions at a stationary surface of discontinuity written in terms of equivalent sources are

$$\hat{\mathbf{n}}_{12} \times (\mathbf{B}_1 - \mathbf{B}_2) = \mu_0(\mathbf{J}_s + \mathbf{J}_{Ms1} + \mathbf{J}_{Ms2}), \tag{2.216}$$

$$\hat{\mathbf{n}}_{12} \times (\mathbf{E}_1 - \mathbf{E}_2) = -\mathbf{J}_{ms}, \tag{2.217}$$

$$\hat{\mathbf{n}}_{12} \cdot (\mathbf{E}_1 - \mathbf{E}_2) = \frac{1}{\epsilon_0}(\rho_s + \rho_{Ps1} + \rho_{Ps2}), \tag{2.218}$$

$$\hat{\mathbf{n}}_{12} \cdot (\mathbf{B}_1 - \mathbf{B}_2) = \rho_{ms}. \tag{2.219}$$

2.8.5 Boundary conditions across a moving layer of field discontinuity

With a moving material body it is often necessary to apply boundary conditions describing the behavior of the fields across the surface of the body. If a surface of discontinuity moves with constant velocity v, the boundary conditions (2.194)–(2.199) hold as long as all fields are expressed *in the frame of the moving surface*. We can also derive boundary conditions for a deforming surface moving with arbitrary velocity by using equations (2.177)–(2.178). In this case all fields are expressed in the laboratory frame. Proceeding through the same set of steps that gave us (2.194)–(2.197), we find

$$\hat{\mathbf{n}}_{12} \times (\mathbf{H}_1 - \mathbf{H}_2) + (\hat{\mathbf{n}}_{12} \cdot \mathbf{v})(\mathbf{D}_1 - \mathbf{D}_2) = \mathbf{J}_s, \tag{2.220}$$

$$\hat{\mathbf{n}}_{12} \times (\mathbf{E}_1 - \mathbf{E}_2) - (\hat{\mathbf{n}}_{12} \cdot \mathbf{v})(\mathbf{B}_1 - \mathbf{B}_2) = -\mathbf{J}_{ms}, \tag{2.221}$$

$$\hat{\mathbf{n}}_{12} \cdot (\mathbf{D}_1 - \mathbf{D}_2) = \rho_s, \tag{2.222}$$

$$\hat{\mathbf{n}}_{12} \cdot (\mathbf{B}_1 - \mathbf{B}_2) = \rho_{ms}. \tag{2.223}$$

Note that when $\hat{\mathbf{n}}_{12} \cdot \mathbf{v} = 0$ these boundary conditions reduce to those for a stationary surface. This occurs not only when $\mathbf{v} = 0$ but also when the velocity is parallel to the surface.

The reader must be wary when employing (2.220)–(2.223). Since the fields are measured in the laboratory frame, if the constitutive relations are substituted into the boundary conditions they must also be represented in the laboratory frame. It is probable that the material parameters would be known in the rest frame of the material, in which case a conversion to the laboratory frame would be necessary.

2.9 Fundamental theorems

In this section we shall consider some of the important theorems of electromagnetics that pertain directly to Maxwell's equations. They may be derived without reference to the solutions of Maxwell's equations, and are not connected with any specialization of the equations or any specific application or geometrical configuration. In this sense these theorems are fundamental to the study of electromagnetics.

2.9.1 Linearity

Recall that a mathematical operator L is *linear* if

$$L(\alpha_1 f_1 + \alpha_2 f_2) = \alpha_1 L(f_1) + \alpha_2 L(f_2)$$

holds for any two functions $f_{1,2}$ in the domain of L and any two scalar constants $\alpha_{1,2}$. A standard observation regarding the equation

$$L(f) = s, \tag{2.224}$$

where L is a linear operator and s is a given forcing function, is that if f_1 and f_2 are solutions to

$$L(f_1) = s_1, \qquad L(f_2) = s_2, \tag{2.225}$$

respectively, and

$$s = s_1 + s_2, \tag{2.226}$$

then

$$f = f_1 + f_2 \tag{2.227}$$

is a solution to (2.224). This is the *principle of superposition*; if convenient, we can decompose s in equation (2.224) as a sum (2.226) and solve the two resulting equations (2.225) independently. The solution to (2.224) is then (2.227), "by superposition." Of course, we are free to split the right side of (2.224) into more than two terms — the method extends directly to any finite number of terms.

Because the operators $\nabla\cdot$, $\nabla\times$, and $\partial/\partial t$ are all linear, Maxwell's equations can be treated by this method. If, for instance,

$$\nabla \times \mathbf{E}_1 = -\frac{\partial \mathbf{B}_1}{\partial t}, \qquad \nabla \times \mathbf{E}_2 = -\frac{\partial \mathbf{B}_2}{\partial t},$$

then

$$\nabla \times \mathbf{E} = -\frac{\partial \mathbf{B}}{\partial t}$$

where $\mathbf{E} = \mathbf{E}_1 + \mathbf{E}_2$ and $\mathbf{B} = \mathbf{B}_1 + \mathbf{B}_2$. The motivation for decomposing terms in a particular way is often based on physical considerations; we give one example here and defer others to later sections of the book. We saw earlier that Maxwell's equations can be written in terms of both electric and (fictitious) magnetic sources as in equations (2.169)–(2.172). Let $\mathbf{E} = \mathbf{E}_e + \mathbf{E}_m$ where $\mathbf{E}_e$ is produced by electric-type sources and $\mathbf{E}_m$ is produced by magnetic-type sources, and decompose the other fields similarly. Then

$$\nabla \times \mathbf{E}_e = -\frac{\partial \mathbf{B}_e}{\partial t}, \qquad \nabla \times \mathbf{H}_e = \mathbf{J} + \frac{\partial \mathbf{D}_e}{\partial t}, \qquad \nabla \cdot \mathbf{D}_e = \rho, \qquad \nabla \cdot \mathbf{B}_e = 0,$$

with a similar equation set for the magnetic sources. We may, if desired, solve these two equation sets independently for $\mathbf{E}_e, \mathbf{D}_e, \mathbf{B}_e, \mathbf{H}_e$ and $\mathbf{E}_m, \mathbf{D}_m, \mathbf{E}_m, \mathbf{H}_m$, and then use superposition to obtain the total fields $\mathbf{E}, \mathbf{D}, \mathbf{B}, \mathbf{H}$.

2.9.2 Duality

The intriguing symmetry of Maxwell's equations leads us to an observation that can reduce the effort required to compute solutions. Consider a closed surface S enclosing a region of space that includes an electric source current $\mathbf{J}$ and a magnetic source current $\mathbf{J}_m$. The fields $(\mathbf{E}_1, \mathbf{D}_1, \mathbf{B}_1, \mathbf{H}_1)$ within the region (which may also contain arbitrary media) are described by

$$\nabla \times \mathbf{E}_1 = -\mathbf{J}_m - \frac{\partial \mathbf{B}_1}{\partial t}, \tag{2.228}$$

$$\nabla \times \mathbf{H}_1 = \mathbf{J} + \frac{\partial \mathbf{D}_1}{\partial t}, \tag{2.229}$$

$$\nabla \cdot \mathbf{D}_1 = \rho, \tag{2.230}$$

$$\nabla \cdot \mathbf{B}_1 = \rho_m. \tag{2.231}$$

Suppose we have been given a mathematical description of the sources $(\mathbf{J}, \mathbf{J}_m)$ and have solved for the field vectors $(\mathbf{E}_1, \mathbf{D}_1, \mathbf{B}_1, \mathbf{H}_1)$. Of course, we must also have been supplied with a set of boundary values and constitutive relations in order to make the solution unique. We note that if we replace the formula for $\mathbf{J}$ with the formula for $\mathbf{J}_m$ in (2.229) (and ρ with ρ_m in (2.230)) and also replace $\mathbf{J}_m$ with $-\mathbf{J}$ in (2.228) (and ρ_m with $-\rho$ in (2.231)) we get a new problem to solve, with a different solution. However, the symmetry of the equations allows us to specify the solution immediately. The new set of curl equations requires

$$\nabla \times \mathbf{E}_2 = \mathbf{J} - \frac{\partial \mathbf{B}_2}{\partial t}, \tag{2.232}$$

$$\nabla \times \mathbf{H}_2 = \mathbf{J}_m + \frac{\partial \mathbf{D}_2}{\partial t}. \tag{2.233}$$

As long as we can resolve the question of how the constitutive parameters must be altered to reflect these replacements, we can conclude by comparing (2.232) with (2.229) and (2.233) with (2.228) that the solution to these equations is merely

$$\mathbf{E}_2 = \mathbf{H}_1,$$
$$\mathbf{B}_2 = -\mathbf{D}_1,$$
$$\mathbf{D}_2 = \mathbf{B}_1,$$
$$\mathbf{H}_2 = -\mathbf{E}_1.$$

That is, if we have solved the original problem, we can use those solutions to find the new ones. This is an application of the general *principle of duality*.

Unfortunately, this approach is a little awkward since the units of the sources and fields in the two problems are different. We can make the procedure more convenient by multiplying Ampere's law by $\eta_0 - (\mu_0/\epsilon_0)^{1/2}$. Then we have

$$\nabla \times \mathbf{E} = -\mathbf{J}_m - \frac{\partial \mathbf{B}}{\partial t}, \tag{2.234}$$

$$\nabla \times (\eta_0 \mathbf{H}) = (\eta_0 \mathbf{J}) + \frac{\partial (\eta_0 \mathbf{D})}{\partial t}. \tag{2.235}$$

Thus if the original problem has solution $(\mathbf{E}_1, \eta_0 \mathbf{D}_1, \mathbf{B}_1, \eta_0 \mathbf{H}_1)$, then the dual problem with $\mathbf{J}$ replaced by $\mathbf{J}_m/\eta_0$ and $\mathbf{J}_m$ replaced by $-\eta_0 \mathbf{J}$ has solution

$$\mathbf{E}_2 = \eta_0 \mathbf{H}_1, \tag{2.236}$$
$$\mathbf{B}_2 = -\eta_0 \mathbf{D}_1, \tag{2.237}$$
$$\eta_0 \mathbf{D}_2 = \mathbf{B}_1, \tag{2.238}$$
$$\eta_0 \mathbf{H}_2 = -\mathbf{E}_1. \tag{2.239}$$

The units on the quantities in the two problems are now identical.

Of course, the constitutive parameters for the dual problem must be altered from those of the original problem to reflect the change in field quantities. From (2.19) and (2.20) we know that the most general forms of the constitutive relations (those for linear, bianisotropic media) are

$$\mathbf{D}_1 = \bar{\xi}_1 \cdot \mathbf{H}_1 + \bar{\epsilon}_1 \cdot \mathbf{E}_1, \tag{2.240}$$
$$\mathbf{B}_1 = \bar{\mu}_1 \cdot \mathbf{H}_1 + \bar{\zeta}_1 \cdot \mathbf{E}_1, \tag{2.241}$$

for the original problem, and

$$\mathbf{D}_2 = \bar{\xi}_2 \cdot \mathbf{H}_2 + \bar{\epsilon}_2 \cdot \mathbf{E}_2, \tag{2.242}$$
$$\mathbf{B}_2 = \bar{\mu}_2 \cdot \mathbf{H}_2 + \bar{\zeta}_2 \cdot \mathbf{E}_2, \tag{2.243}$$

for the dual problem. Substitution of (2.236)–(2.239) into (2.240) and (2.241) gives

$$\mathbf{D}_2 = (-\bar{\zeta}_1) \cdot \mathbf{H}_2 + \left(\frac{\bar{\mu}_1}{\eta_0^2} \right) \cdot \mathbf{E}_2, \tag{2.244}$$

$$\mathbf{B}_2 = \left(\eta_0^2 \bar{\epsilon}_1 \right) \cdot \mathbf{H}_2 + (-\bar{\xi}_1) \cdot \mathbf{E}_2. \tag{2.245}$$

Comparing (2.244) with (2.242) and (2.245) with (2.243), we conclude that

$$\bar{\zeta}_2 = -\bar{\xi}_1, \qquad \bar{\xi}_2 = -\bar{\zeta}_1, \qquad \bar{\mu}_2 = \eta_0^2 \bar{\epsilon}_1, \qquad \bar{\epsilon}_2 = \frac{\bar{\mu}_1}{\eta_0^2}.$$

As an important special case, we see that for a linear, isotropic medium specified by a permittivity ϵ and permeability μ, the dual problem is obtained by replacing ϵ_r with μ_r and μ_r with ϵ_r. The solution to the dual problem is then given by

$$\mathbf{E}_2 = \eta_0 \mathbf{H}_1, \qquad \eta_0 \mathbf{H}_2 = -\mathbf{E}_1,$$

as before. We thus see that the medium in the dual problem must have electric properties numerically equal to the magnetic properties of the medium in the original problem, and magnetic properties numerically equal to the electric properties of the medium in the original problem. This is rather inconvenient for most applications. Alternatively, we may divide Ampere's law by $\eta = (\mu/\epsilon)^{1/2}$ instead of η_0. Then the dual problem has $\mathbf{J}$ replaced by $\mathbf{J}_m/\eta$, and $\mathbf{J}_m$ replaced by $-\eta\mathbf{J}$, and the solution to the dual problem is given by

$$\mathbf{E}_2 = \eta \mathbf{H}_1, \qquad \eta \mathbf{H}_2 = -\mathbf{E}_1.$$

In this case there is no need to swap ϵ_r and μ_r, since information about these parameters is incorporated into the replacement sources.

We must also remember that to obtain a unique solution we need to specify the boundary values of the fields. In a true dual problem, the boundary values of the fields used in the original problem are used on the swapped fields in the dual problem. A typical example of this is when the condition of zero tangential electric field on a perfect electric conductor is replaced by the condition of zero tangential magnetic field on the surface of a perfect magnetic conductor. However, duality can also be used to obtain the mathematical form of the field expressions, often in a homogeneous (source-free) situation, and boundary values can be applied later to specify the solution appropriate to the problem geometry. This approach is often used to compute waveguide modal fields and the electromagnetic fields scattered from objects. In these cases a TE/TM field decomposition is employed, and duality is used to find one part of the decomposition once the other is known.

Duality of electric and magnetic point source fields.   By duality, we can sometimes use the known solution to one problem to solve a related problem by merely substituting different variables into the known mathematical expression. An example of this is the case in which we have solved for the fields produced by a certain distribution of electric sources and wish to determine the fields when the same distribution is used to describe magnetic sources.

Let us consider the case when the source distribution is that of a point current, or *Hertzian dipole*, immersed in free space. As we shall see in Chapter 5, the fields for a general source may be found by using the fields produced by these point sources. We begin by finding the fields produced by an electric dipole source at the origin aligned along the z-axis,

$$\mathbf{J} = \hat{z} I_0 \delta(\mathbf{r}),$$

then use duality to find the fields produced by a magnetic current source $\mathbf{J}_m = \hat{z} I_{m0} \delta(\mathbf{r})$.

The fields produced by the electric source must obey

$$\nabla \times \mathbf{E}_e = -\frac{\partial}{\partial t} \mu_0 \mathbf{H}_e, \qquad (2.246)$$

$$\nabla \times \mathbf{H}_e = \hat{\mathbf{z}} I_0 \delta(\mathbf{r}) + \frac{\partial}{\partial t} \epsilon_0 \mathbf{E}_e, \tag{2.247}$$

$$\nabla \cdot \epsilon_0 \mathbf{E}_e = \rho, \tag{2.248}$$

$$\nabla \cdot \mathbf{H}_e = 0, \tag{2.249}$$

while those produced by the magnetic source must obey

$$\nabla \times \mathbf{E}_m = -\hat{\mathbf{z}} I_{m0} \delta(\mathbf{r}) - \frac{\partial}{\partial t} \mu_0 \mathbf{H}_m, \tag{2.250}$$

$$\nabla \times \mathbf{H}_m = \frac{\partial}{\partial t} \epsilon_0 \mathbf{E}_m, \tag{2.251}$$

$$\nabla \cdot \mathbf{E}_m = 0, \tag{2.252}$$

$$\nabla \cdot \mu_0 \mathbf{H}_m = \rho_m. \tag{2.253}$$

We see immediately that the second set of equations is the dual of the first, as long as we scale the sources appropriately. Multiplying (2.250) by $-I_0/I_{m0}$ and (2.251) by $I_0 \eta_0^2 / I_{m0}$, we have the curl equations

$$\nabla \times \left(-\frac{I_0}{I_{m0}} \mathbf{E}_m \right) = \hat{\mathbf{z}} I_0 \delta(\mathbf{r}) + \frac{\partial}{\partial t} \left(\mu_0 \frac{I_0}{I_{m0}} \mathbf{H}_m \right), \tag{2.254}$$

$$\nabla \times \left(\frac{I_0 \eta_0^2}{I_{m0}} \mathbf{H}_m \right) = -\frac{\partial}{\partial t} \left(-\epsilon_0 \frac{I_0 \eta_0^2}{I_{m0}} \mathbf{E}_m \right). \tag{2.255}$$

Comparing (2.255) with (2.246) and (2.254) with (2.247) we see that

$$\mathbf{E}_m = -\frac{I_{m0}}{I_0} \mathbf{H}_e, \qquad \mathbf{H}_m = \frac{I_{m0}}{I_0} \frac{\mathbf{E}_e}{\eta_0^2}.$$

We note that it is impossible to have a point current source without accompanying point charge sources terminating each end of the dipole current. The point charges are required to satisfy the continuity equation, and vary in time as the moving charge that establishes the current accumulates at the ends of the dipole. From (2.247) we see that the magnetic field curls around the combination of the electric field and electric current source, while from (2.246) the electric field curls around the magnetic field, and from (2.248) diverges from the charges located at the ends of the dipole. From (2.250) we see that the electric field must curl around the combination of the magnetic field and magnetic current source, while (2.251) and (2.253) show that the magnetic field curls around the electric field and diverges from the magnetic charge.

Duality in a source-free region. Consider a closed surface S enclosing a source-free region of space. For simplicity, assume that the medium within S is linear, isotropic, and homogeneous. The fields within S are described by Maxwell's equations

$$\nabla \times \mathbf{E}_1 = -\frac{\partial}{\partial t} \mu \mathbf{H}_1, \tag{2.256}$$

$$\nabla \times \eta \mathbf{H}_1 = \frac{\partial}{\partial t} \epsilon \eta \mathbf{E}_1, \tag{2.257}$$

$$\nabla \cdot \epsilon \mathbf{E}_1 = 0, \tag{2.258}$$

$$\nabla \cdot \mu \mathbf{H}_1 = 0. \tag{2.259}$$

Under these conditions the concept of duality takes on a different face. The symmetry of the equations is such that the mathematical form of the solution for $\mathbf{E}$ is the same as that for $\eta\mathbf{H}$. That is, the fields

$$\mathbf{E}_2 = \eta\mathbf{H}_1, \tag{2.260}$$

$$\mathbf{H}_2 = -\mathbf{E}_1/\eta, \tag{2.261}$$

are also a solution to Maxwell's equations, and thus the dual problem merely involves replacing $\mathbf{E}$ by $\eta\mathbf{H}$ and $\mathbf{H}$ by $-\mathbf{E}/\eta$. However, the final forms of $\mathbf{E}$ and $\mathbf{H}$ will *not* be identical after appropriate boundary values are imposed.

This form of duality is very important for the solution of fields within waveguides or the fields scattered by objects where the sources are located outside the region where the fields are evaluated.

2.9.3   Reciprocity

The reciprocity theorem, also called the *Lorentz reciprocity theorem*, describes a specific and often useful relationship between sources and the electromagnetic fields they produce. Under certain special circumstances we find that an interaction between independent source and mediating fields called "reaction" is a spatially symmetric quantity. The reciprocity theorem is used in the study of guided waves to establish the orthogonality of guided wave modes, in microwave network theory to obtain relationships between terminal characteristics, and in antenna theory to demonstrate the equivalence of transmission and reception patterns.

Consider a closed surface S enclosing a volume V. Assume that the fields within and on S are produced by two independent source fields. The source $(\mathbf{J}_a, \mathbf{J}_{ma})$ produces the field $(\mathbf{E}_a, \mathbf{D}_a, \mathbf{B}_a, \mathbf{H}_a)$ as described by Maxwell's equations

$$\nabla \times \mathbf{E}_a = -\mathbf{J}_{ma} - \frac{\partial \mathbf{B}_a}{\partial t}, \tag{2.262}$$

$$\nabla \times \mathbf{H}_a = \mathbf{J}_a + \frac{\partial \mathbf{D}_a}{\partial t}, \tag{2.263}$$

while the source field $(\mathbf{J}_b, \mathbf{J}_{mb})$ produces the field $(\mathbf{E}_b, \mathbf{D}_b, \mathbf{B}_b, \mathbf{H}_b)$ as described by

$$\nabla \times \mathbf{E}_b = -\mathbf{J}_{mb} - \frac{\partial \mathbf{B}_b}{\partial t}, \tag{2.264}$$

$$\nabla \times \mathbf{H}_b = \mathbf{J}_b + \frac{\partial \mathbf{D}_b}{\partial t}. \tag{2.265}$$

The sources may be distributed in any way relative to S: they may lie completely inside, completely outside, or partially inside and partially outside. Material media may lie within S, and their properties may depend on position.

Let us examine the quantity

$$R \equiv \nabla \cdot (\mathbf{E}_a \times \mathbf{H}_b - \mathbf{E}_b \times \mathbf{H}_a).$$

By (B.44) we have

$$R = \mathbf{H}_b \cdot \nabla \times \mathbf{E}_a - \mathbf{E}_a \cdot \nabla \times \mathbf{H}_b - \mathbf{H}_a \cdot \nabla \times \mathbf{E}_b + \mathbf{E}_b \cdot \nabla \times \mathbf{H}_a$$

so that by Maxwell's curl equations

$$R = \left[\mathbf{H}_a \cdot \frac{\partial \mathbf{B}_b}{\partial t} - \mathbf{H}_b \cdot \frac{\partial \mathbf{B}_a}{\partial t} \right] - \left[\mathbf{E}_a \cdot \frac{\partial \mathbf{D}_b}{\partial t} - \mathbf{E}_b \cdot \frac{\partial \mathbf{D}_a}{\partial t} \right] +$$

$$+ \left[\mathbf{J}_a \cdot \mathbf{E}_b - \mathbf{J}_b \cdot \mathbf{E}_a - \mathbf{J}_{ma} \cdot \mathbf{H}_b + \mathbf{J}_{mb} \cdot \mathbf{H}_a \right].$$

The useful relationships we seek occur when the first two bracketed quantities on the right-hand side of the above expression are zero. Whether this is true depends not only on the behavior of the fields, but on the properties of the medium at the point in question. Though we have assumed that the sources of the field sets are independent, it is apparent that they must share a similar time dependence in order for the terms within each of the bracketed quantities to cancel. Of special interest is the case where the two sources are both sinusoidal in time with identical frequencies, but with differing spatial distributions. We shall consider this case in detail in § 4.10.2 after we have discussed the properties of the time harmonic field. Importantly, we will find that only certain characteristics of the constitutive parameters allow cancellation of the bracketed terms; materials with these characteristics are called *reciprocal*, and the fields they support are said to display the property of *reciprocity*. To see what this property entails, we set the bracketed terms to zero and integrate over a volume V to obtain

$$\oint_S (\mathbf{E}_a \times \mathbf{H}_b - \mathbf{E}_b \times \mathbf{H}_a) \cdot \mathbf{dS} = \int_V (\mathbf{J}_a \cdot \mathbf{E}_b - \mathbf{J}_b \cdot \mathbf{E}_a - \mathbf{J}_{ma} \cdot \mathbf{H}_b + \mathbf{J}_{mb} \cdot \mathbf{H}_a) \, dV,$$

which is the time-domain version of the *Lorentz reciprocity theorem*.

Two special cases of this theorem are important to us. If all sources lie outside S, we have *Lorentz's lemma*

$$\oint_S (\mathbf{E}_a \times \mathbf{H}_b - \mathbf{E}_b \times \mathbf{H}_a) \cdot \mathbf{dS} = 0.$$

This remarkable expression shows that a relationship exists between the fields produced by completely independent sources, and is useful for establishing waveguide mode orthogonality for time harmonic fields. If sources reside within S but the surface integral is equal to zero, we have

$$\int_V (\mathbf{J}_a \cdot \mathbf{E}_b - \mathbf{J}_b \cdot \mathbf{E}_a - \mathbf{J}_{ma} \cdot \mathbf{H}_b + \mathbf{J}_{mb} \cdot \mathbf{H}_a) \, dV = 0.$$

This occurs when the surface is bounded by a special material (such as an impedance sheet or a perfect conductor), or when the surface recedes to infinity; the expression is useful for establishing the reciprocity conditions for networks and antennas. We shall interpret it for time harmonic fields in § 4.10.2.

2.9.4   Similitude

A common approach in physical science involves the introduction of normalized variables to provide for scaling of problems along with a chance to identify certain physically significant parameters. Similarity as a general principle can be traced back to the earliest attempts to describe physical effects with mathematical equations, with serious study undertaken by Galileo. Helmholtz introduced the first systematic investigation in 1873, and the concept was rigorized by Reynolds ten years later [216]. Similitude is now considered a fundamental guiding principle in the modeling of materials [199].

The process often begins with a consideration of the fundamental differential equations. In electromagnetics we may introduce a set of dimensionless field and source variables

$$\underline{\mathbf{E}}, \quad \underline{\mathbf{D}}, \quad \underline{\mathbf{B}}, \quad \underline{\mathbf{H}}, \quad \underline{\mathbf{J}}, \quad \underline{\rho}, \qquad (2.266)$$

by setting

$$\mathbf{E} = \underline{\mathbf{E}} k_E, \quad \mathbf{B} = \underline{\mathbf{B}} k_B, \quad \mathbf{D} = \underline{\mathbf{D}} k_D,$$
$$\mathbf{H} = \underline{\mathbf{H}} k_H, \quad \mathbf{J} = \underline{\mathbf{J}} k_J, \quad \rho = \underline{\rho} k_\rho. \qquad (2.267)$$

Here we regard the quantities $k_E, k_B, \ldots$ as base units for the discussion, while the dimensionless quantities (2.266) serve to express the actual fields $\mathbf{E}, \mathbf{B}, \ldots$ in terms of these base units. Of course, the time and space variables can also be scaled: we can write

$$t = \underline{t}k_t, \qquad l = \underline{l}k_l, \tag{2.268}$$

if l is any length of interest. Again, the quantities $\underline{t}$ and $\underline{l}$ are dimensionless measure numbers used to express the actual quantities t and l relative to the chosen base amounts k_t and k_l. With (2.267) and (2.268), Maxwell's curl equations become

$$\underline{\nabla} \times \underline{\mathbf{E}} = -\frac{k_B}{k_E}\frac{k_l}{k_t}\frac{\partial \underline{\mathbf{B}}}{\partial \underline{t}}, \qquad \underline{\nabla} \times \underline{\mathbf{H}} = \frac{k_J k_l}{k_H}\underline{\mathbf{J}} + \frac{k_D}{k_H}\frac{k_l}{k_t}\frac{\partial \underline{\mathbf{D}}}{\partial \underline{t}} \tag{2.269}$$

while the continuity equation becomes

$$\underline{\nabla} \cdot \underline{\mathbf{J}} = -\frac{k_\rho}{k_J}\frac{k_l}{k_t}\frac{\partial \underline{\rho}}{\partial \underline{t}}, \tag{2.270}$$

where $\underline{\nabla}$ has been normalized by k_l. These are examples of field equations cast into dimensionless form — it is easily verified that the *similarity parameters*

$$\frac{k_B}{k_E}\frac{k_l}{k_t}, \qquad \frac{k_J k_l}{k_H}, \qquad \frac{k_D}{k_H}\frac{k_l}{k_t}, \qquad \frac{k_\rho}{k_J}\frac{k_l}{k_t}, \tag{2.271}$$

are dimensionless. The idea behind electromagnetic similitude is that a given set of normalized values $\underline{\mathbf{E}}, \underline{\mathbf{B}}, \ldots$ can satisfy equations (2.269) and (2.270) for many different physical situations, provided that the numerical values of the coefficients (2.271) are all fixed across those situations. Indeed, the differential equations would be identical.

To make this discussion a bit more concrete, let us assume a conducting linear medium where

$$\mathbf{D} = \epsilon\mathbf{E}, \qquad \mathbf{B} = \mu\mathbf{H}, \qquad \mathbf{J} = \sigma\mathbf{E},$$

and use

$$\epsilon = \underline{\epsilon}k_\epsilon, \qquad \mu = \underline{\mu}k_\mu, \qquad \sigma = \underline{\sigma}k_\sigma,$$

to express the material parameters in terms of dimensionless values $\underline{\epsilon}$, $\underline{\mu}$, and $\underline{\sigma}$. Then

$$\underline{\mathbf{D}} = \frac{k_\epsilon k_E}{k_D}\underline{\epsilon}\underline{\mathbf{E}}, \qquad \underline{\mathbf{B}} = \frac{k_\mu k_H}{k_B}\underline{\mu}\underline{\mathbf{H}}, \qquad \underline{\mathbf{J}} = \frac{k_\sigma k_E}{k_J}\underline{\sigma}\underline{\mathbf{E}},$$

and equations (2.269) become

$$\underline{\nabla} \times \underline{\mathbf{E}} = -\left(\frac{k_\mu k_l}{k_t}\frac{k_H}{k_E}\right)\underline{\mu}\frac{\partial \underline{\mathbf{H}}}{\partial \underline{t}},$$

$$\underline{\nabla} \times \underline{\mathbf{H}} = \left(k_\sigma k_l \frac{k_E}{k_H}\right)\underline{\sigma}\underline{\mathbf{E}} + \left(\frac{k_\epsilon k_l}{k_t}\frac{k_E}{k_H}\right)\underline{\epsilon}\frac{\partial \underline{\mathbf{E}}}{\partial \underline{t}}.$$

Defining

$$\alpha = \frac{k_\mu k_l}{k_t}\frac{k_H}{k_E}, \qquad \gamma = k_\sigma k_l \frac{k_E}{k_H}, \qquad \beta = \frac{k_\epsilon k_l}{k_t}\frac{k_E}{k_H},$$

we see that under the current assumptions similarity holds between two electromagnetics problems only if $\alpha\underline{\mu}$, $\gamma\underline{\sigma}$, and $\beta\underline{\epsilon}$ are numerically the same in both problems. A necessary condition for similitude, then, is that the products

$$(\alpha\underline{\mu})(\beta\underline{\epsilon}) = k_\mu k_\epsilon \left(\frac{k_l}{k_t}\right)^2 \underline{\mu}\underline{\epsilon}, \qquad (\alpha\underline{\mu})(\gamma\underline{\sigma}) = k_\mu k_\sigma \frac{k_l^2}{k_t}\underline{\mu}\underline{\sigma},$$

(which do not involve k_E or k_H) stay constant between problems. We see, for example, that we may compensate for a halving of the length scale k_l by (a) a quadrupling of the permeability μ, or (b) a simultaneous halving of the time scale k_t and doubling of the conductivity σ. A much less subtle special case is that for which $\sigma = 0$, $k_\epsilon = \epsilon_0$, $k_\mu = \mu_0$, and $\epsilon = \mu = 1$; we then have free space and must simply maintain

$$k_l/k_t = \text{constant}$$

so that the time and length scales stay proportional. In the sinusoidal steady state, for instance, the frequency would be made to vary inversely with the length scale.

2.9.5 Conservation theorems

The misconception that Poynting's theorem can be "derived" from Maxwell's equations is widespread and ingrained. We must, in fact, *postulate* the idea that the electromagnetic field can be associated with an energy flux propagating at the speed of light. Since the form of the postulate is patterned after the well-understood laws of mechanics, we begin by developing the basic equations of momentum and energy balance in mechanical systems. Then we shall see whether it is sensible to ascribe these principles to the electromagnetic field.

Maxwell's theory allows us to describe, using Maxwell's equations, the behavior of the electromagnetic fields within a (possibly) finite region V of space. The presence of any sources or material objects outside V are made known through the specification of tangential fields over the boundary of V, as required for uniqueness. Thus, the influence of external effects can always be viewed as being transported across the boundary. This is true of mechanical as well as electromagnetic effects. A charged material body can be acted on by physical contact with another body, by gravitational forces, and by the Lorentz force, each effect resulting in momentum exchange across the boundary of the object. These effects must all be taken into consideration if we are to invoke momentum conservation, resulting in a very complicated situation. This suggests that we try to decompose the problem into simpler "systems" based on physical effects.

The system concept in the physical sciences. The idea of decomposing a complicated system into simpler, self-contained systems is quite common in the physical sciences. Penfield and Haus [145] invoke this concept by introducing an *electromagnetic system* where the effects of the Lorentz force equation are considered to accompany a *mechanical system* where effects of pressure, stress, and strain are considered, and a *thermodynamic system* where the effects of heat exchange are considered. These systems can all be interrelated in a variety of ways. For instance, as a material heats up it can expand, and the resulting mechanical forces can alter the electrical properties of the material. We will follow Penfield and Haus by considering separate electromagnetic and mechanical subsystems; other systems may be added analogously.

If we separate the various systems by physical effect, we will need to know how to "reassemble the information." Two conservation theorems are very helpful in this regard: conservation of energy, and conservation of momentum. Engineers often employ these theorems to make tacit use of the system idea. For instance, when studying electromagnetic waves propagating in a waveguide, it is common practice to compute wave attenuation by calculating the Poynting flux of power into the walls of the guide. The power lost from the wave is said to "heat up the waveguide walls," which indeed it does. This is an admission that the electromagnetic system is not "closed": it requires the inclusion of a thermodynamic system in order that energy be conserved. Of course, the detailed workings of the thermodynamic system are often ignored, indicating that any thermodynamic "feedback" mechanism is weak. In the waveguide example, for instance, the heating of the metallic walls does not alter their electromagnetic properties enough to couple back into an effect on the fields in the walls or in the guide. If such effects were important, they would have to be included in the conservation theorem via the boundary fields; it is therefore reasonable to associate with these fields a "flow" of energy or momentum into V. Thus, we wish to develop conservation laws that include not only the Lorentz force effects within V, but a flow of external effects into V through its boundary surface.

To understand how external influences may effect the electromagnetic subsystem, we look to the behavior of the mechanical subsystem as an analogue. In the electromagnetic system, effects are felt both internally to a region (because of the Lorentz force effect) and through the system boundary (by the dependence of the internal fields on the boundary fields). In the mechanical and thermodynamic systems, a region of mass is affected both internally (through transfer of heat and gravitational forces) and through interactions occurring across its surface (through transfers of energy and momentum, by pressure and stress). One beauty of electromagnetic theory is that we can find a mathematical symmetry between electromagnetic and mechanical effects which parallels the above conceptual symmetry. This makes applying conservation of energy and momentum to the total system (electromagnetic, thermodynamic, and mechanical) very convenient.

Conservation of momentum and energy in mechanical systems. We begin by reviewing the interactions of material bodies in a mechanical system. For simplicity we concentrate on fluids (analogous to charge in space); the extension of these concepts to solid bodies is straightforward.

Consider a fluid with mass density ρ_m. The momentum of a small subvolume of the fluid is given by $\rho_m \mathbf{v}\, dV$, where $\mathbf{v}$ is the velocity of the subvolume. So the momentum density is $\rho_m \mathbf{v}$. Newton's second law states that a force acting throughout the subvolume results in a change in its momentum given by

$$\frac{D}{Dt}(\rho_m \mathbf{v}\, dV) = \mathbf{f}\, dV, \tag{2.272}$$

where $\mathbf{f}$ is the volume force density and the D/Dt notation shows that we are interested in the rate of change of the momentum as observed by the moving fluid element (see § A.2). Here $\mathbf{f}$ could be the weight force, for instance. Addition of the results for all elements of the fluid body gives

$$\frac{D}{Dt}\int_V \rho_m \mathbf{v}\, dV = \int_V \mathbf{f}\, dV \tag{2.273}$$

as the change in momentum for the entire body. If on the other hand the force exerted on the body is through contact with its surface, the change in momentum is

$$\frac{D}{Dt}\int_V \rho_m \mathbf{v}\, dV = \oint_S \mathbf{t}\, dS \tag{2.274}$$

where $\mathbf{t}$ is the "surface traction."

We can write the time-rate of change of momentum in a more useful form by applying the Reynolds transport theorem (A.66):

$$\frac{D}{Dt}\int_V \rho_m \mathbf{v}\, dV = \int_V \frac{\partial}{\partial t}(\rho_m \mathbf{v})\, dV + \oint_S (\rho_m \mathbf{v})\mathbf{v} \cdot d\mathbf{S}. \tag{2.275}$$

Superposing (2.273) and (2.274) and substituting into (2.275) we have

$$\int_V \frac{\partial}{\partial t}(\rho_m \mathbf{v})\, dV + \oint_S (\rho_m \mathbf{v})\mathbf{v} \cdot \mathbf{dS} = \int_V \mathbf{f}\, dV + \oint_S \mathbf{t}\, dS. \qquad (2.276)$$

If we define the dyadic quantity

$$\bar{\mathbf{T}}_k = \rho_m \mathbf{vv}$$

then (2.276) can be written as

$$\int_V \frac{\partial}{\partial t}(\rho_m \mathbf{v})\, dV + \oint_S \hat{\mathbf{n}} \cdot \bar{\mathbf{T}}_k\, dS = \int_V \mathbf{f}\, dV + \oint_S \mathbf{t}\, dS. \qquad (2.277)$$

This *principle of linear momentum* [214] can be interpreted as a large-scale form of conservation of kinetic linear momentum. Here $\hat{\mathbf{n}} \cdot \bar{\mathbf{T}}_k$ represents the flow of kinetic momentum across S, and the sum of this momentum transfer and the change of momentum within V stands equal to the forces acting internal to V and upon S.

The surface traction may be related to the surface normal $\hat{\mathbf{n}}$ through a dyadic quantity $\bar{\mathbf{T}}_m$ called the mechanical *stress tensor*:

$$\mathbf{t} = \hat{\mathbf{n}} \cdot \bar{\mathbf{T}}_m.$$

With this we may write (2.277) as

$$\int_V \frac{\partial}{\partial t}(\rho_m \mathbf{v})\, dV + \oint_S \hat{\mathbf{n}} \cdot \bar{\mathbf{T}}_k\, dS = \int_V \mathbf{f}\, dV + \oint_S \hat{\mathbf{n}} \cdot \bar{\mathbf{T}}_m\, dS$$

and apply the dyadic form of the divergence theorem (B.19) to get

$$\int_V \frac{\partial}{\partial t}(\rho_m \mathbf{v})\, dV + \int_V \nabla \cdot (\rho_m \mathbf{vv})\, dV = \int_V \mathbf{f}\, dV + \int_V \nabla \cdot \bar{\mathbf{T}}_m\, dV. \qquad (2.278)$$

Combining the volume integrals and setting the integrand to zero we have

$$\frac{\partial}{\partial t}(\rho_m \mathbf{v}) + \nabla \cdot (\rho_m \mathbf{vv}) = \mathbf{f} + \nabla \cdot \bar{\mathbf{T}}_m,$$

which is the point-form equivalent of (2.277). Note that the second term on the right-hand side is nonzero only for points residing on the surface of the body. Finally, letting $\mathbf{g}$ denote momentum density we obtain the simple expression

$$\nabla \cdot \bar{\mathbf{T}}_k + \frac{\partial \mathbf{g}_k}{\partial t} = \mathbf{f}_k, \qquad (2.279)$$

where

$$\mathbf{g}_k = \rho_m \mathbf{v}$$

is the density of kinetic momentum and

$$\mathbf{f}_k = \mathbf{f} + \nabla \cdot \bar{\mathbf{T}}_m \qquad (2.280)$$

is the total force density.

Equation (2.279) is somewhat analogous to the electric charge continuity equation (1.11). For each point of the body, the total outflux of kinetic momentum plus the time rate of change of kinetic momentum equals the total force. The resemblance to (1.11) is strong, except for the nonzero term on the right-hand side. The charge continuity equation represents a closed system: charge cannot spontaneously appear and add an extra term to the right-hand side of (1.11). On the other hand, the change in total momentum at a point can exceed that given by the momentum flowing out of the point if there is another "source" (e.g., gravity for an internal point, or pressure on a boundary point).

To obtain a momentum conservation expression that resembles the continuity equation, we must consider a "subsystem" with terms that exactly counterbalance the extra expressions on the right-hand side of (2.279). For a fluid acted on only by external pressure the sole effect enters through the traction term, and [145]

$$\nabla \cdot \bar{\mathbf{T}}_m = -\nabla p \tag{2.281}$$

where p is the pressure exerted on the fluid body. Now, using (B.63), we can write

$$-\nabla p = -\nabla \cdot \bar{\mathbf{T}}_p \tag{2.282}$$

where

$$\bar{\mathbf{T}}_p = p\bar{\mathbf{I}}$$

and $\bar{\mathbf{I}}$ is the unit dyad. Finally, using (2.282), (2.281), and (2.280) in (2.279), we obtain

$$\nabla \cdot (\bar{\mathbf{T}}_k + \bar{\mathbf{T}}_p) + \frac{\partial}{\partial t}\mathbf{g}_k = 0$$

and we have an expression for a closed system including all possible effects. Now, note that we can form the above expression as

$$\left(\nabla \cdot \bar{\mathbf{T}}_k + \frac{\partial}{\partial t}\mathbf{g}_k\right) + \left(\nabla \cdot \bar{\mathbf{T}}_p + \frac{\partial}{\partial t}\mathbf{g}_p\right) = 0 \tag{2.283}$$

where $\mathbf{g}_p = 0$ since there are no volume effects associated with pressure. This can be viewed as the sum of two closed subsystems

$$\nabla \cdot \bar{\mathbf{T}}_k + \frac{\partial}{\partial t}\mathbf{g}_k = 0, \tag{2.284}$$

$$\nabla \cdot \bar{\mathbf{T}}_p + \frac{\partial}{\partial t}\mathbf{g}_p = 0.$$

We now have the desired viewpoint. The conservation formula for the complete closed system can be viewed as a sum of formulas for open subsystems, each having the form of a conservation law for a closed system. In case we must include the effects of gravity, for instance, we need only determine $\bar{\mathbf{T}}_g$ and $\mathbf{g}_g$ such that

$$\nabla \cdot \bar{\mathbf{T}}_g + \frac{\partial}{\partial t}\mathbf{g}_g = 0$$

and add this new conservation equation to (2.283). If we can find a conservation expression of form similar to (2.284) for an "electromagnetic subsystem," we can include its effects along with the mechanical effects by merely adding together the conservation laws. We shall find just such an expression later in this section.

We stated in § 1.3 that there are four fundamental conservation principles. We have now discussed linear momentum; the principle of angular momentum follows similarly. Our next goal is to find an expression similar to (2.283) for conservation of energy. We may expect the conservation of energy expression to obey a similar law of superposition.

We begin with the fundamental definition of work: for a particle moving with velocity $\mathbf{v}$ under the influence of a force $\mathbf{f}_k$ the work is given by $\mathbf{f}_k \cdot \mathbf{v}$. Dot multiplying (2.272) by $\mathbf{v}$ and replacing $\mathbf{f}$ by $\mathbf{f}_k$ (to represent both volume and surface forces), we get

$$\mathbf{v} \cdot \frac{D}{Dt}(\rho_m \mathbf{v})\, dV = \mathbf{v} \cdot \mathbf{f}_k\, dV$$

or equivalently

$$\frac{D}{Dt}\left(\frac{1}{2}\rho_m \mathbf{v} \cdot \mathbf{v}\right) dV = \mathbf{v} \cdot \mathbf{f}_k\, dV.$$

Integration over a volume and application of the Reynolds transport theorem (A.66) then gives

$$\int_V \frac{\partial}{\partial t}\left(\frac{1}{2}\rho_m v^2\right) dV + \oint_S \hat{\mathbf{n}} \cdot \left(\mathbf{v}\frac{1}{2}\rho_m v^2\right) dS = \int_V \mathbf{f}_k \cdot \mathbf{v}\, dV.$$

Hence the sum of the time rate of change in energy internal to the body and the flow of kinetic energy across the boundary must equal the work done by internal and surface forces acting on the body. In point form,

$$\nabla \cdot \mathbf{S}_k + \frac{\partial}{\partial t} W_k = \mathbf{f}_k \cdot \mathbf{v} \qquad (2.285)$$

where

$$\mathbf{S}_k = \mathbf{v}\frac{1}{2}\rho_m v^2$$

is the density of the flow of kinetic energy and

$$W_k = \frac{1}{2}\rho_m v^2$$

is the kinetic energy density. Again, the system is not closed (the right-hand side of (2.285) is not zero) because the balancing forces are not included. As was done with the momentum equation, the effect of the work done by the pressure forces can be described in a closed-system-type equation

$$\nabla \cdot \mathbf{S}_p + \frac{\partial}{\partial t} W_p = 0. \qquad (2.286)$$

Combining (2.285) and (2.286) we have

$$\nabla \cdot (\mathbf{S}_k + \mathbf{S}_p) + \frac{\partial}{\partial t}(W_k + W_p) = 0,$$

the energy conservation equation for the closed system.

Conservation in the electromagnetic subsystem. We would now like to achieve closed-system conservation theorems for the electromagnetic subsystem so that we can add in the effects of electromagnetism. For the momentum equation, we can proceed exactly as we did with the mechanical system. We begin with

$$\mathbf{f}_{em} = \rho \mathbf{E} + \mathbf{J} \times \mathbf{B}.$$

This force term should appear on one side of the point form of the momentum conservation equation. The term on the other side must involve the electromagnetic fields, since they are the mechanism for exerting force on the charge distribution. Substituting for $\mathbf{J}$ from (2.2) and for ρ from (2.3) we have

$$\mathbf{f}_{em} = \mathbf{E}(\nabla \cdot \mathbf{D}) - \mathbf{B} \times (\nabla \times \mathbf{H}) + \mathbf{B} \times \frac{\partial \mathbf{D}}{\partial t}.$$

Using

$$\mathbf{B} \times \frac{\partial \mathbf{D}}{\partial t} = -\frac{\partial}{\partial t}(\mathbf{D} \times \mathbf{B}) + \mathbf{D} \times \frac{\partial \mathbf{B}}{\partial t}$$

and substituting from Faraday's law for $\partial \mathbf{B}/\partial t$ we have

$$-[\mathbf{E}(\nabla \cdot \mathbf{D}) - \mathbf{D} \times (\nabla \times \mathbf{E}) + \mathbf{H}(\nabla \cdot \mathbf{B}) - \mathbf{B} \times (\nabla \times \mathbf{H})] + \frac{\partial}{\partial t}(\mathbf{D} \times \mathbf{B}) = -\mathbf{f}_{em}. \quad (2.287)$$

Here we have also added the null term $\mathbf{H}(\nabla \cdot \mathbf{B})$.

The forms of (2.287) and (2.279) would be identical if the bracketed term could be written as the divergence of a dyadic function $\bar{\mathbf{T}}_{em}$. This is indeed possible for linear, homogeneous, bianisotropic media, provided that the constitutive matrix $[\bar{\mathbf{C}}_{EH}]$ in (2.21) is symmetric [101]. In that case

$$\bar{\mathbf{T}}_{em} = \frac{1}{2}(\mathbf{D} \cdot \mathbf{E} + \mathbf{B} \cdot \mathbf{H})\bar{\mathbf{I}} - \mathbf{DE} - \mathbf{BH}, \quad (2.288)$$

which is called the *Maxwell stress tensor*. Let us demonstrate this equivalence for a linear, isotropic, homogeneous material. Putting $\mathbf{D} = \epsilon \mathbf{E}$ and $\mathbf{H} = \mathbf{B}/\mu$ into (2.287) we obtain

$$\nabla \cdot \mathbf{T}_{em} = -\epsilon \mathbf{E}(\nabla \cdot \mathbf{E}) + \frac{1}{\mu}\mathbf{B} \times (\nabla \times \mathbf{B}) + \epsilon \mathbf{E} \times (\nabla \times \mathbf{E}) - \frac{1}{\mu}\mathbf{B}(\nabla \cdot \mathbf{B}). \quad (2.289)$$

Now (B.46) gives
$$\nabla(\mathbf{A} \cdot \mathbf{A}) = 2\mathbf{A} \times (\nabla \times \mathbf{A}) + 2(\mathbf{A} \cdot \nabla)\mathbf{A}$$

so that
$$\mathbf{E}(\nabla \cdot \mathbf{E}) - \mathbf{E} \times (\nabla \times \mathbf{E}) = \mathbf{E}(\nabla \cdot \mathbf{E}) + (\mathbf{E} \cdot \nabla)\mathbf{E} - \frac{1}{2}\nabla(E^2).$$

Finally, (B.55) and (B.63) give

$$\mathbf{E}(\nabla \cdot \mathbf{E}) - \mathbf{E} \times (\nabla \times \mathbf{E}) = \nabla \cdot \left(\mathbf{EE} - \frac{1}{2}\bar{\mathbf{I}}\mathbf{E} \cdot \mathbf{E}\right).$$

Substituting this expression and a similar one for $\mathbf{B}$ into (2.289) we have

$$\nabla \cdot \bar{\mathbf{T}}_{em} = \nabla \cdot \left[\frac{1}{2}(\mathbf{D} \cdot \mathbf{E} + \mathbf{B} \cdot \mathbf{H})\bar{\mathbf{I}} - \mathbf{DE} - \mathbf{BH}\right],$$

which matches (2.288).

Replacing the term in brackets in (2.287) by $\nabla \cdot \bar{\mathbf{T}}_{em}$, we get

$$\nabla \cdot \bar{\mathbf{T}}_{em} + \frac{\partial \mathbf{g}_{em}}{\partial t} = -\mathbf{f}_{em} \quad (2.290)$$

where

$$\mathbf{g}_{em} = \mathbf{D} \times \mathbf{B}.$$

Equation (2.290) is the point form of the electromagnetic conservation of momentum theorem. It is mathematically identical in form to the mechanical theorem (2.279). Integration over a volume gives the large-scale form

$$\oint_S \bar{\mathbf{T}}_{em} \cdot d\mathbf{S} + \int_V \frac{\partial \mathbf{g}_{em}}{\partial t} \, dV = -\int_V \mathbf{f}_{em} \, dV. \tag{2.291}$$

If we interpret this as we interpreted the conservation theorems from mechanics, the first term on the left-hand side represents the flow of electromagnetic momentum across the boundary of V, while the second term represents the change in momentum within V. The sum of these two quantities is exactly compensated by the total Lorentz force acting on the charges within V. Thus we identify $\mathbf{g}_{em}$ as the transport density of electromagnetic momentum.

Because (2.290) is not zero on the right-hand side, it does not represent a closed system. If the Lorentz force is the only force acting on the charges within V, then the mechanical reaction to the Lorentz force should be described by Newton's third law. Thus we have the kinematic momentum conservation formula

$$\nabla \cdot \bar{\mathbf{T}}_k + \frac{\partial \mathbf{g}_k}{\partial t} = \mathbf{f}_k = -\mathbf{f}_{em}.$$

Subtracting this expression from (2.290) we obtain

$$\nabla \cdot (\bar{\mathbf{T}}_{em} - \bar{\mathbf{T}}_k) + \frac{\partial}{\partial t}(\mathbf{g}_{em} - \mathbf{g}_k) = 0, \tag{2.292}$$

which describes momentum conservation for the closed system.

It is also possible to derive a conservation theorem for electromagnetic energy that resembles the corresponding theorem for mechanical energy. Earlier we noted that $\mathbf{v} \cdot \mathbf{f}$ represents the volume density of work produced by moving an object at velocity $\mathbf{v}$ under the action of a force $\mathbf{f}$. For the electromagnetic subsystem the work is produced by charges moving against the Lorentz force. So the volume density of work delivered to the currents is

$$w_{em} = \mathbf{v} \cdot \mathbf{f}_{em} = \mathbf{v} \cdot (\rho \mathbf{E} + \mathbf{J} \times \mathbf{B}) = (\rho \mathbf{v}) \cdot \mathbf{E} + \rho \mathbf{v} \cdot (\mathbf{v} \times \mathbf{B}). \tag{2.293}$$

Using (B.6) on the second term in (2.293) we get

$$w_{em} = (\rho \mathbf{v}) \cdot \mathbf{E} + \rho \mathbf{B} \cdot (\mathbf{v} \times \mathbf{v}).$$

The second term vanishes by definition of the cross product. This is the familiar property that the magnetic field does no work on moving charge. Hence

$$w_{em} = \mathbf{J} \cdot \mathbf{E}. \tag{2.294}$$

This important relation says that charge moving in an electric field experiences a force which results in energy transfer to (or from) the charge. We wish to write this energy transfer in terms of an energy flux vector, as we did with the mechanical subsystem.

As with our derivation of the conservation of electromagnetic momentum, we wish to relate the energy transfer to the electromagnetic fields. Substitution of $\mathbf{J}$ from (2.2) into (2.294) gives

$$w_{em} = (\nabla \times \mathbf{H}) \cdot \mathbf{E} - \frac{\partial \mathbf{D}}{\partial t} \cdot \mathbf{E},$$

hence

$$w_{em} = -\nabla \cdot (\mathbf{E} \times \mathbf{H}) + \mathbf{H} \cdot (\nabla \times \mathbf{E}) - \frac{\partial \mathbf{D}}{\partial t} \cdot \mathbf{E}$$

by (B.44). Substituting for $\nabla \times \mathbf{E}$ from (2.1) we have

$$w_{em} = -\nabla \cdot (\mathbf{E} \times \mathbf{H}) - \left[\mathbf{E} \cdot \frac{\partial \mathbf{D}}{\partial t} + \mathbf{H} \cdot \frac{\partial \mathbf{B}}{\partial t} \right].$$

This is not quite of the form (2.285) since a single term representing the time rate of change of energy density is not present. However, for a linear isotropic medium in which ϵ and μ do not depend on time (i.e., a nondispersive medium) we have

$$\mathbf{E} \cdot \frac{\partial \mathbf{D}}{\partial t} = \epsilon \mathbf{E} \cdot \frac{\partial \mathbf{E}}{\partial t} = \frac{1}{2} \epsilon \frac{\partial}{\partial t} (\mathbf{E} \cdot \mathbf{E}) = \frac{1}{2} \frac{\partial}{\partial t} (\mathbf{D} \cdot \mathbf{E}), \qquad (2.295)$$

$$\mathbf{H} \cdot \frac{\partial \mathbf{B}}{\partial t} = \mu \mathbf{H} \cdot \frac{\partial \mathbf{H}}{\partial t} = \frac{1}{2} \mu \frac{\partial}{\partial t} (\mathbf{H} \cdot \mathbf{H}) = \frac{1}{2} \frac{\partial}{\partial t} (\mathbf{H} \cdot \mathbf{B}). \qquad (2.296)$$

Using this we obtain

$$\nabla \cdot \mathbf{S}_{em} + \frac{\partial}{\partial t} W_{em} = -\mathbf{f}_{em} \cdot \mathbf{v} = -\mathbf{J} \cdot \mathbf{E} \qquad (2.297)$$

where

$$W_{em} = \frac{1}{2} (\mathbf{D} \cdot \mathbf{E} + \mathbf{B} \cdot \mathbf{H})$$

and

$$\mathbf{S}_{em} = \mathbf{E} \times \mathbf{H}. \qquad (2.298)$$

Equation (2.297) is the point form of the energy conservation theorem, also called *Poynting's theorem* after J.H. Poynting who first proposed it. The quantity $\mathbf{S}_{em}$ given in (2.298) is known as the *Poynting vector*. Integrating (2.297) over a volume and using the divergence theorem, we obtain the large-scale form

$$-\int_V \mathbf{J} \cdot \mathbf{E} \, dV = \int_V \frac{1}{2} \frac{\partial}{\partial t} (\mathbf{D} \cdot \mathbf{E} + \mathbf{B} \cdot \mathbf{H}) \, dV + \oint_S (\mathbf{E} \times \mathbf{H}) \cdot \mathbf{dS}. \qquad (2.299)$$

This also holds for a nondispersive, linear, bianisotropic medium with a symmetric constitutive matrix [101, 185].

We see that the electromagnetic energy conservation theorem (2.297) is identical in form to the mechanical energy conservation theorem (2.285). Thus, if the system is composed of just the kinetic and electromagnetic subsystems, the mechanical force exactly balances the Lorentz force, and (2.297) and (2.285) add to give

$$\nabla \cdot (\mathbf{S}_{em} + \mathbf{S}_k) + \frac{\partial}{\partial t} (W_{em} + W_k) = 0, \qquad (2.300)$$

showing that energy is conserved for the entire system.

As in the mechanical system, we identify W_{em} as the volume electromagnetic energy density in V, and $\mathbf{S}_{em}$ as the density of electromagnetic energy flowing across the boundary of V. This interpretation is somewhat controversial, as discussed below.

Interpretation of the energy and momentum conservation theorems. There has been some controversy regarding Poynting's theorem (and, equally, the momentum conservation theorem). While there is no question that Poynting's theorem is mathematically correct, we may wonder whether we are justified in associating W_{em} with W_k and $\mathbf{S}_{em}$ with $\mathbf{S}_k$ merely because of the similarities in their mathematical expressions. Certainly there is some justification for associating W_k, the kinetic energy of particles, with W_{em}, since we shall show that for static fields the term $\frac{1}{2}(\mathbf{D}\cdot\mathbf{E}+\mathbf{B}\cdot\mathbf{H})$ represents the energy required to assemble the charges and currents into a certain configuration. However, the term $\mathbf{S}_{em}$ is more problematic. In a mechanical system, $\mathbf{S}_k$ represents the flow of kinetic energy associated with moving particles — does that imply that $\mathbf{S}_{em}$ represents the flow of electromagnetic energy? That is the position generally taken, and it is widely supported by experimental evidence. However, the interpretation is not clear-cut.

If we associate $\mathbf{S}_{em}$ with the flow of electromagnetic energy at a point in space, then we must define what a flow of electromagnetic energy is. We naturally associate the flow of kinetic energy with moving particles; with what do we associate the flow of electromagnetic energy? Maxwell felt that electromagnetic energy must flow through space as a result of the mechanical stresses and strains associated with an unobserved substance called the "aether." A more modern interpretation is that the electromagnetic fields propagate as a wave through space at finite velocity; when those fields encounter a charged particle a force is exerted, work is done, and energy is "transferred" from the field to the particle. Hence the energy flow is associated with the "flow" of the electromagnetic wave.

Unfortunately, it is uncertain whether $\mathbf{E}\times\mathbf{H}$ is the appropriate quantity to associate with this flow, since only its divergence appears in Poynting's theorem. We could add any other term $\mathbf{S}'$ that satisfies $\nabla\cdot\mathbf{S}'=0$ to $\mathbf{S}_{em}$ in (2.297), and the conservation theorem would be unchanged. (Equivalently, we could add to (2.299) any term that integrates to zero over S.) There is no such ambiguity in the mechanical case because kinetic energy is rigorously defined. We are left, then, to postulate that $\mathbf{E}\times\mathbf{H}$ represents the density of energy flow associated with an electromagnetic wave (based on the symmetry with mechanics), and to look to experimental evidence as justification. In fact, experimental evidence does point to the correctness of this hypothesis, and the quantity $\mathbf{E}\times\mathbf{H}$ is widely and accurately used to compute the energy radiated by antennas, carried by waveguides, etc.

Confusion also arises regarding the interpretation of W_{em}. Since this term is so conveniently paired with the mechanical volume kinetic energy density in (2.300) it would seem that we should interpret it as an electromagnetic energy density. As such, we can think of this energy as "localized" in certain regions of space. This viewpoint has been criticized [187, 145, 69] since the large-scale form of energy conservation for a space region only requires that the total energy in the region be specified, and the integrand (energy density) giving this energy is not unique. It is also felt that energy should be associated with a "configuration" of objects (such as charged particles) and not with an arbitrary point in space. However, we retain the concept of localized energy because it is convenient and produces results consistent with experiment.

The validity of extending the static field interpretation of

$$\frac{1}{2}(\mathbf{D}\cdot\mathbf{E}+\mathbf{B}\cdot\mathbf{H})$$

as the energy "stored" by a charge and a current arrangement to the time-varying case has also been questioned. If we do extend this view to the time-varying case, Poynting's theorem suggests that every point in space somehow has an energy density associated with it, and the flow of energy from that point (via $\mathbf{S}_{em}$) must be accompanied by a change in the stored energy at that point. This again gives a very useful and intuitively satisfying point of view. Since we can associate the flow of energy with the propagation of the electromagnetic fields, we can view the fields in any region of space as having the potential to do work on charged particles in that region. If there are charged particles in that region then work is done, accompanied by a transfer of energy to the particles and a reduction in the amplitudes of the fields.

We must also remember that the association of stored electromagnetic energy density W_{em} with the mechanical energy density W_k is only possible if the medium is nondispersive. If we cannot make the assumptions that justify (2.295) and (2.296), then Poynting's theorem must take the form

$$- \int_V \mathbf{J} \cdot \mathbf{E} \, dV = \int_V \left[\mathbf{E} \cdot \frac{\partial \mathbf{D}}{\partial t} + \mathbf{H} \cdot \frac{\partial \mathbf{B}}{\partial t} \right] dV + \oint_S (\mathbf{E} \times \mathbf{H}) \cdot d\mathbf{S}. \qquad (2.301)$$

For dispersive media, the volume term on the right-hand side describes not only the stored electromagnetic energy, but also the energy dissipated within the material produced by a time lag between the field applied to the medium and the resulting polarization or magnetization of the atoms. This is clearly seen in (2.29), which shows that $\mathbf{D}(t)$ depends on the value of $\mathbf{E}$ at time t and at all past times. The stored energy and dissipative terms are hard to separate, but we can see that there must always be a stored energy term by substituting $\mathbf{D} = \epsilon_0 \mathbf{E} + \mathbf{P}$ and $\mathbf{H} = \mathbf{B}/\mu_0 - \mathbf{M}$ into (2.301) to obtain

$$- \int_V [(\mathbf{J} + \mathbf{J}_P) \cdot \mathbf{E} + \mathbf{J}_H \cdot \mathbf{H}] \, dV =$$
$$\frac{1}{2} \frac{\partial}{\partial t} \int_V (\epsilon_0 \mathbf{E} \cdot \mathbf{E} + \mu_0 \mathbf{H} \cdot \mathbf{H}) \, dV + \oint_S (\mathbf{E} \times \mathbf{H}) \cdot d\mathbf{S}. \qquad (2.302)$$

Here J_P is the equivalent polarization current (2.119) and J_H is an analogous *magnetic polarization current* given by

$$\mathbf{J}_H = \mu_0 \frac{\partial \mathbf{M}}{\partial t}.$$

In this form we easily identify the quantity

$$\frac{1}{2} (\epsilon_0 \mathbf{E} \cdot \mathbf{E} + \mu_0 \mathbf{H} \cdot \mathbf{H})$$

as the electromagnetic energy density for the fields $\mathbf{E}$ and $\mathbf{H}$ *in free space*. Any dissipation produced by polarization and magnetization lag is now handled by the interaction between the fields and equivalent current, just as $\mathbf{J} \cdot \mathbf{E}$ describes the interaction of the electric current (source and secondary) with the electric field. Unfortunately, the equivalent current interaction terms also include the additional stored energy that results from polarizing and magnetizing the material atoms, and again the effects are hard to separate.

Finally, let us consider the case of static fields. Setting the time derivative to zero in (2.299), we have

$$- \int_V \mathbf{J} \cdot \mathbf{E} \, dV = \oint_S (\mathbf{E} \times \mathbf{H}) \cdot d\mathbf{S}.$$

This shows that energy flux is required to maintain steady current flow. For instance, we need both an electromagnetic and a thermodynamic subsystem to account for energy conservation in the case of steady current flow through a resistor. The Poynting flux describes the electromagnetic energy entering the resistor and the thermodynamic flux describes the heat dissipation. For the sum of the two subsystems conservation of energy requires

$$\nabla \cdot (\mathbf{S}_{em} + \mathbf{S}_{th}) = -\mathbf{J} \cdot \mathbf{E} + P_{th} = 0.$$

To compute the heat dissipation we can use

$$P_{th} = \mathbf{J} \cdot \mathbf{E} = -\nabla \cdot \mathbf{S}_{em}$$

and thus either use the boundary fields or the fields and current internal to the resistor to find the dissipated heat.

Boundary conditions on the Poynting vector.   The large-scale form of Poynting's theorem may be used to determine the behavior of the Poynting vector on either side of a boundary surface. We proceed exactly as in § 2.8.2. Consider a surface S across which the electromagnetic sources and constitutive parameters are discontinuous (Figure 2.6). As before, let $\hat{\mathbf{n}}_{12}$ be the unit normal directed into region 1. We now simplify the notation and write $\mathbf{S}$ instead of $\mathbf{S}_{em}$. If we apply Poynting's theorem

$$\int_V \left(\mathbf{J} \cdot \mathbf{E} + \mathbf{E} \cdot \frac{\partial \mathbf{D}}{\partial t} + \mathbf{H} \cdot \frac{\partial \mathbf{B}}{\partial t} \right) dV + \oint_S \mathbf{S} \cdot \mathbf{n} \, dS = 0$$

to the two separate surfaces shown in Figure 2.6, we obtain

$$\int_V \left(\mathbf{J} \cdot \mathbf{E} + \mathbf{E} \cdot \frac{\partial \mathbf{D}}{\partial t} + \mathbf{H} \cdot \frac{\partial \mathbf{B}}{\partial t} \right) dV + \int_S \mathbf{S} \cdot \mathbf{n} \, dS = \int_{S_{10}} \hat{\mathbf{n}}_{12} \cdot (\mathbf{S}_1 - \mathbf{S}_2) \, dS. \quad (2.303)$$

If on the other hand we apply Poynting's theorem to the entire volume region including the surface of discontinuity and include the contribution produced by surface current, we get

$$\int_V \left(\mathbf{J} \cdot \mathbf{E} + \mathbf{E} \cdot \frac{\partial \mathbf{D}}{\partial t} + \mathbf{H} \cdot \frac{\partial \mathbf{B}}{\partial t} \right) dV + \int_S \mathbf{S} \cdot \mathbf{n} \, dS = - \int_{S_{10}} \mathbf{J}_s \cdot \mathbf{E} \, dS. \quad (2.304)$$

Since we are uncertain whether to use $\mathbf{E}_1$ or $\mathbf{E}_2$ in the surface term on the right-hand side, if we wish to have the integrals over V and S in (2.303) and (2.304) produce identical results we must postulate the two conditions

$$\hat{\mathbf{n}}_{12} \times (\mathbf{E}_1 - \mathbf{E}_2) = 0$$

and

$$\hat{\mathbf{n}}_{12} \cdot (\mathbf{S}_1 - \mathbf{S}_2) = -\mathbf{J}_s \cdot \mathbf{E}. \quad (2.305)$$

The first condition is merely the continuity of tangential electric field as originally postulated in § 2.8.2; it allows us to be nonspecific as to which value of $\mathbf{E}$ we use in the second condition, which is the desired boundary condition on $\mathbf{S}$.

It is interesting to note that (2.305) may also be derived directly from the two postulated boundary conditions on tangential $\mathbf{E}$ and $\mathbf{H}$. Here we write with the help of (B.6)

$$\hat{\mathbf{n}}_{12} \cdot (\mathbf{S}_1 - \mathbf{S}_2) = \hat{\mathbf{n}}_{12} \cdot (\mathbf{E}_1 \times \mathbf{H}_1 - \mathbf{E}_2 \times \mathbf{H}_2) = \mathbf{H}_1 \cdot (\hat{\mathbf{n}}_{12} \times \mathbf{E}_1) - \mathbf{H}_2 \cdot (\hat{\mathbf{n}}_{12} \times \mathbf{E}_2).$$

Since $\hat{\mathbf{n}}_{12} \times \mathbf{E}_1 = \hat{\mathbf{n}}_{12} \times \mathbf{E}_2 = \hat{\mathbf{n}}_{12} \times \mathbf{E}$, we have

$$\hat{\mathbf{n}}_{12} \cdot (\mathbf{S}_1 - \mathbf{S}_2) = (\mathbf{H}_1 - \mathbf{H}_2) \cdot (\hat{\mathbf{n}}_{12} \times \mathbf{E}) = [-\hat{\mathbf{n}}_{12} \times (\mathbf{H}_1 - \mathbf{H}_2)] \cdot \mathbf{E}.$$

Finally, using $\hat{\mathbf{n}}_{12} \times (\mathbf{H}_1 - \mathbf{H}_2) = \mathbf{J}_s$ we arrive at (2.305).

The arguments above suggest an interesting way to look at the boundary conditions. Once we identify $\mathbf{S}$ with the flow of electromagnetic energy, we may consider the condition on normal $\mathbf{S}$ as a fundamental statement of the conservation of energy. This statement implies continuity of tangential $\mathbf{E}$ in order to have an unambiguous interpretation for the meaning of the term $\mathbf{J}_s \cdot \mathbf{E}$. Then, with continuity of tangential $\mathbf{E}$ established, we can derive the condition on tangential $\mathbf{H}$ directly.

An alternative formulation of the conservation theorems. As we saw in the paragraphs above, our derivation of the conservation theorems lacks strong motivation. We manipulated Maxwell's equations until we found expressions that resembled those for mechanical momentum and energy, but in the process found that the validity of the expressions is somewhat limiting. For instance, we needed to assume a linear, homogeneous, bianisotropic medium in order to identify the Maxwell stress tensor (2.288) and the energy densities in Poynting's theorem (2.299). In the end, we were reduced to postulating the meaning of the individual terms in the conservation theorems in order for the whole to have meaning.

An alternative approach is popular in physics. It involves postulating a single Lagrangian density function for the electromagnetic field, and then applying the stationary property of the action integral. The results are precisely the same conservation expressions for linear momentum and energy as obtained from manipulating Maxwell's equations (plus the equation for conservation of angular momentum), obtained with fewer restrictions regarding the constitutive relations. This process also separates the stored energy, Maxwell stress tensor, momentum density, and Poynting vector as natural components of a tensor equation, allowing a better motivated interpretation of the meaning of these components. Since this approach is also a powerful tool in mechanics, its application is more strongly motivated than merely manipulating Maxwell's equations. Of course, some knowledge of the structure of the electromagnetic field is required to provide an appropriate postulate of the Lagrangian density. Interested readers should consult Kong [101], Jackson [91], Doughty [57], or Tolstoy [198].

2.10   The wave nature of the electromagnetic field

Throughout this chapter our goal has been a fundamental understanding of Maxwell's theory of electromagnetics. We have concentrated on developing and understanding the equations relating the field quantities, but have done little to understand the nature of the field itself. We would now like to investigate, in a very general way, the behavior of the field. We shall not attempt to solve a vast array of esoteric problems, but shall instead concentrate on a few illuminating examples.

The electromagnetic field can take on a wide variety of characteristics. Static fields differ qualitatively from those which undergo rapid time variations. Time-varying fields exhibit wave behavior and carry energy away from their sources. In the case of slow time variation this wave nature may often be neglected in favor of the nearby coupling of sources we know as the inductance effect, hence circuit theory may suffice to describe the field-source interaction. In the case of extremely rapid oscillations, particle concepts may be needed to describe the field.

The dynamic coupling between the various field vectors in Maxwell's equations provides a means of characterizing the field. Static fields are characterized by decoupling of the electric and magnetic fields. Quasistatic fields exhibit some coupling, but the wave characteristic of the field is ignored. Tightly coupled fields are dominated by the wave effect, but may still show a static-like spatial distribution near the source. Any such "near-zone" effects are generally ignored for fields at light-wave frequencies, and the particle nature of light must often be considered.

2.10.1 Electromagnetic waves

An early result of Maxwell's theory was the prediction and later verification by Heinrich Hertz of the existence of electromagnetic waves. We now know that nearly any time-varying source produces waves, and that these waves have certain important properties. An electromagnetic wave is a propagating electromagnetic field that travels with finite velocity as a disturbance through a medium. The field itself is the disturbance, rather than merely representing a physical displacement or other effect on the medium. This fact is fundamental for understanding how electromagnetic waves can travel through a true vacuum. Many specific characteristics of the wave, such as velocity and polarization, depend on the properties of the medium through which it propagates. The evolution of the disturbance also depends on these properties: we say that a material exhibits "dispersion" if the disturbance undergoes a change in its temporal behavior as the wave progresses. As waves travel they carry energy and momentum away from their source. This energy may be later returned to the source or delivered to some distant location. Waves are also capable of transferring energy to, or withdrawing energy from, the medium through which they propagate. When energy is carried outward from the source never to return, we refer to the process as "electromagnetic radiation." The effects of radiated fields can be far-reaching; indeed, radio astronomers observe waves that originated at the very edges of the universe.

Light is an electromagnetic phenomenon, and many of the familiar characteristics of light that we recognize from our everyday experience may be applied to all electromagnetic waves. For instance, radio waves bend (or "refract") in the ionosphere much as light waves bend while passing through a prism. Microwaves reflect from conducting surfaces in the same way that light waves reflect from a mirror; detecting these reflections forms the basis of radar. Electromagnetic waves may also be "confined" by reflecting boundaries to form waves standing in one or more directions. With this concept we can use waveguides or transmission lines to guide electromagnetic energy from spot to spot, or to concentrate it in the cavity of a microwave oven.

The manifestations of electromagnetic waves are so diverse that no one book can possibly describe the entire range of phenomena or application. In this section we shall merely introduce the reader to some of the most fundamental concepts of electromagnetic wave behavior. In the process we shall also introduce the three most often studied types of traveling electromagnetic waves: plane waves, spherical waves, and cylindrical waves. In later sections we shall study some of the complicated interactions of these waves with objects and boundaries, in the form of guided waves and scattering problems.

Mathematically, electromagnetic waves arise as a subset of solutions to Maxwell's equations. These solutions obey the electromagnetic "wave equation," which may be derived from Maxwell's equations under certain circumstances. Not all electromagnetic fields satisfy the wave equation. Obviously, time-invariant fields cannot represent evolving wave disturbances, and must obey the static field equations. Time-varying fields in certain metals may obey the diffusion equation rather than the wave equation, and must thereby exhibit different behavior. In the study of quasistatic fields we often ignore the displacement current term in Maxwell's equations, producing solutions that are most important near the sources of the fields and having little associated radiation. When the displacement term is significant we produce solutions with the properties of waves.

2.10.2   Wave equation for bianisotropic materials

In deriving electromagnetic wave equations we transform the first-order coupled partial differential equations we know as Maxwell's equations into uncoupled second-order equations. That is, we perform a set of operations (and make appropriate assumptions) to reduce the set of four differential equations in the four unknown fields $\mathbf{E}$, $\mathbf{D}$, $\mathbf{B}$, and $\mathbf{H}$, into a set of differential equations each involving a single unknown (usually $\mathbf{E}$ or $\mathbf{H}$). It is possible to derive wave equations for $\mathbf{E}$ and $\mathbf{H}$ even for the most general cases of inhomogeneous, bianisotropic media, as long as the constitutive parameters $\bar{\mu}$ and $\bar{\xi}$ are constant with time. Substituting the constitutive relations (2.19)–(2.20) into the Maxwell–Minkowski curl equations (2.169)–(2.170) we get

$$\nabla \times \mathbf{E} = -\frac{\partial}{\partial t}(\bar{\zeta} \cdot \mathbf{E} + \bar{\mu} \cdot \mathbf{H}) - \mathbf{J}_m, \tag{2.306}$$

$$\nabla \times \mathbf{H} = \frac{\partial}{\partial t}(\bar{\epsilon} \cdot \mathbf{E} + \bar{\xi} \cdot \mathbf{H}) + \mathbf{J}. \tag{2.307}$$

Separate equations for $\mathbf{E}$ and $\mathbf{H}$ are facilitated by introducing a new dyadic operator $\bar{\nabla}$, which when dotted with a vector field $\mathbf{V}$ gives the curl:

$$\bar{\nabla} \cdot \mathbf{V} = \nabla \times \mathbf{V}. \tag{2.308}$$

It is easy to verify that in rectangular coordinates $\bar{\nabla}$ is

$$[\bar{\nabla}] = \begin{bmatrix} 0 & -\partial/\partial z & \partial/\partial y \\ \partial/\partial z & 0 & -\partial/\partial x \\ -\partial/\partial y & \partial/\partial x & 0 \end{bmatrix}.$$

With this notation, Maxwell's curl equations (2.306)–(2.307) become simply

$$\left(\bar{\nabla} + \frac{\partial}{\partial t}\bar{\zeta}\right) \cdot \mathbf{E} = -\frac{\partial}{\partial t}\bar{\mu} \cdot \mathbf{H} - \mathbf{J}_m, \tag{2.309}$$

$$\left(\bar{\nabla} - \frac{\partial}{\partial t}\bar{\xi}\right) \cdot \mathbf{H} = \frac{\partial}{\partial t}\bar{\epsilon} \cdot \mathbf{E} + \mathbf{J}. \tag{2.310}$$

Obtaining separate equations for $\mathbf{E}$ and $\mathbf{H}$ is straightforward. Defining the inverse dyadic $\bar{\mu}^{-1}$ through

$$\bar{\mu} \cdot \bar{\mu}^{-1} = \bar{\mu}^{-1} \cdot \bar{\mu} = \bar{\mathbf{I}},$$

we can write (2.309) as

$$\frac{\partial}{\partial t}\mathbf{H} = -\bar{\mu}^{-1} \cdot \left(\bar{\nabla} + \frac{\partial}{\partial t}\bar{\zeta}\right) \cdot \mathbf{E} - \bar{\mu}^{-1} \cdot \mathbf{J}_m \tag{2.311}$$

where we have assumed that $\bar{\mu}$ is independent of time. Assuming that $\bar{\xi}$ is also independent of time, we can differentiate (2.310) with respect to time to obtain

$$\left(\bar{\nabla} - \frac{\partial}{\partial t}\bar{\xi}\right) \cdot \frac{\partial \mathbf{H}}{\partial t} = \frac{\partial^2}{\partial t^2}\left(\bar{\epsilon} \cdot \mathbf{E}\right) + \frac{\partial \mathbf{J}}{\partial t}.$$

Substituting $\partial\mathbf{H}/\partial t$ from (2.311) and rearranging, we get

$$\left[\left(\bar{\nabla} - \frac{\partial}{\partial t}\bar{\xi}\right) \cdot \bar{\mu}^{-1} \cdot \left(\bar{\nabla} + \frac{\partial}{\partial t}\bar{\zeta}\right) + \frac{\partial^2}{\partial t^2}\bar{\epsilon}\right] \cdot \mathbf{E} = -\left(\bar{\nabla} - \frac{\partial}{\partial t}\bar{\xi}\right) \cdot \bar{\mu}^{-1} \cdot \mathbf{J}_m - \frac{\partial\mathbf{J}}{\partial t}.$$

(2.312)

This is the general wave equation for $\mathbf{E}$. Using an analogous set of steps, and assuming $\bar{\epsilon}$ and $\bar{\zeta}$ are independent of time, we can find

$$\left[\left(\bar{\nabla} + \frac{\partial}{\partial t}\bar{\zeta}\right) \cdot \bar{\epsilon}^{-1} \cdot \left(\bar{\nabla} - \frac{\partial}{\partial t}\bar{\xi}\right) + \frac{\partial^2}{\partial t^2}\bar{\mu}\right] \cdot \mathbf{H} = \left(\bar{\nabla} + \frac{\partial}{\partial t}\bar{\zeta}\right) \cdot \bar{\epsilon}^{-1} \cdot \mathbf{J} - \frac{\partial\mathbf{J}_m}{\partial t}.$$

(2.313)

This is the wave equation for $\mathbf{H}$. The case in which the constitutive parameters are time-dependent will be handled using frequency domain techniques in later chapters.

Wave equations for anisotropic, isotropic, and homogeneous media are easily obtained from (2.312) and (2.313) as special cases. For example, the wave equations for a homogeneous, isotropic medium can be found by setting $\bar{\zeta} = \bar{\xi} = 0$, $\bar{\mu} = \mu\bar{\mathbf{I}}$, and $\bar{\epsilon} = \epsilon\bar{\mathbf{I}}$:

$$\frac{1}{\mu}\bar{\nabla} \cdot (\bar{\nabla} \cdot \mathbf{E}) + \epsilon\frac{\partial^2\mathbf{E}}{\partial t^2} = -\frac{1}{\mu}\bar{\nabla} \cdot \mathbf{J}_m - \frac{\partial\mathbf{J}}{\partial t},$$

$$\frac{1}{\epsilon}\bar{\nabla} \cdot (\bar{\nabla} \cdot \mathbf{H}) + \mu\frac{\partial^2\mathbf{H}}{\partial t^2} = \frac{1}{\epsilon}\bar{\nabla} \cdot \mathbf{J} - \frac{\partial\mathbf{J}_m}{\partial t}.$$

Returning to standard curl notation we find that these become

$$\nabla \times (\nabla \times \mathbf{E}) + \mu\epsilon\frac{\partial^2\mathbf{E}}{\partial t^2} = -\nabla \times \mathbf{J}_m - \mu\frac{\partial\mathbf{J}}{\partial t},$$

(2.314)

$$\nabla \times (\nabla \times \mathbf{H}) + \mu\epsilon\frac{\partial^2\mathbf{H}}{\partial t^2} = \nabla \times \mathbf{J} - \epsilon\frac{\partial\mathbf{J}_m}{\partial t}.$$

(2.315)

In each of the wave equations it appears that operations on the electromagnetic fields have been separated from operations on the source terms. However, we have not yet invoked any coupling between the fields and sources associated with secondary interactions. That is, we need to separate the impressed sources, which are independent of the fields they source, with secondary sources resulting from interactions between the sourced fields and the medium in which the fields exist. The simple case of an isotropic conducting medium will be discussed below.

Wave equation using equivalent sources. An alternative approach for studying wave behavior in general media is to use the Maxwell–Boffi form of the field equations

$$\nabla \times \mathbf{E} = -\frac{\partial\mathbf{B}}{\partial t},$$

(2.316)

$$\nabla \times \frac{\mathbf{B}}{\mu_0} = (\mathbf{J} + \mathbf{J}_M + \mathbf{J}_P) + \frac{\partial\epsilon_0\mathbf{E}}{\partial t},$$

(2.317)

$$\nabla \cdot (\epsilon_0\mathbf{E}) = (\rho + \rho_P),$$

(2.318)

$$\nabla \cdot \mathbf{B} = 0.$$

(2.319)

Taking the curl of (2.316) we have

$$\nabla \times (\nabla \times \mathbf{E}) = -\frac{\partial}{\partial t}\nabla \times \mathbf{B}.$$

Substituting for $\nabla \times \mathbf{B}$ from (2.317) we then obtain

$$\nabla \times (\nabla \times \mathbf{E}) + \mu_0 \epsilon_0 \frac{\partial^2 \mathbf{E}}{\partial t^2} = -\mu_0 \frac{\partial}{\partial t}(\mathbf{J} + \mathbf{J}_M + \mathbf{J}_P), \qquad (2.320)$$

which is the wave equation for $\mathbf{E}$. Taking the curl of (2.317) and substituting from (2.316) we obtain the wave equation

$$\nabla \times (\nabla \times \mathbf{B}) + \mu_0 \epsilon_0 \frac{\partial^2 \mathbf{B}}{\partial t^2} = \mu_0 \nabla \times (\mathbf{J} + \mathbf{J}_M + \mathbf{J}_P) \qquad (2.321)$$

for $\mathbf{B}$. Solution of the wave equations is often facilitated by writing the curl-curl operation in terms of the vector Laplacian. Using (B.47), and substituting for the divergence from (2.318) and (2.319), we can write the wave equations as

$$\nabla^2 \mathbf{E} - \mu_0 \epsilon_0 \frac{\partial^2 \mathbf{E}}{\partial t^2} = \frac{1}{\epsilon_0} \nabla(\rho + \rho_P) + \mu_0 \frac{\partial}{\partial t}(\mathbf{J} + \mathbf{J}_M + \mathbf{J}_P), \qquad (2.322)$$

$$\nabla^2 \mathbf{B} - \mu_0 \epsilon_0 \frac{\partial^2 \mathbf{B}}{\partial t^2} = -\mu_0 \nabla \times (\mathbf{J} + \mathbf{J}_M + \mathbf{J}_P). \qquad (2.323)$$

The simplicity of these equations relative to (2.312) and (2.313) is misleading. We have not considered the constitutive equations relating the polarization $\mathbf{P}$ and magnetization $\mathbf{M}$ to the fields, nor have we considered interactions leading to secondary sources.

2.10.3   Wave equation in a conducting medium

As an example of the type of wave equation that arises when secondary sources are included, consider a homogeneous isotropic conducting medium described by permittivity ϵ, permeability μ, and conductivity σ. In a conducting medium we must separate the source field into a causative impressed term $\mathbf{J}^i$ that is independent of the fields it sources, and a secondary term $\mathbf{J}^s$ that is an effect of the sourced fields. In an isotropic conducting medium the effect is described by Ohm's law $\mathbf{J}^s = \sigma \mathbf{E}$. Writing the total current as $\mathbf{J} = \mathbf{J}^i + \mathbf{J}^s$, and assuming that $\mathbf{J}_m = 0$, we write the wave equation (2.314) as

$$\nabla \times (\nabla \times \mathbf{E}) + \mu \epsilon \frac{\partial^2 \mathbf{E}}{\partial t^2} = -\mu \frac{\partial (\mathbf{J}^i + \sigma \mathbf{E})}{\partial t}. \qquad (2.324)$$

Using (B.47) and substituting $\nabla \cdot \mathbf{E} = \rho/\epsilon$, we can write the wave equation for $\mathbf{E}$ as

$$\nabla^2 \mathbf{E} - \mu \sigma \frac{\partial \mathbf{E}}{\partial t} - \mu \epsilon \frac{\partial^2 \mathbf{E}}{\partial t^2} = \mu \frac{\partial \mathbf{J}^i}{\partial t} + \frac{1}{\epsilon} \nabla \rho. \qquad (2.325)$$

Substituting $\mathbf{J} = \mathbf{J}^i + \sigma \mathbf{E}$ into (2.315) and using (B.47), we obtain

$$\nabla(\nabla \cdot \mathbf{H}) - \nabla^2 \mathbf{H} + \mu \epsilon \frac{\partial^2 \mathbf{H}}{\partial t^2} = \nabla \times \mathbf{J}^i + \sigma \nabla \times \mathbf{E}.$$

Since $\nabla \times \mathbf{E} = -\partial \mathbf{B}/\partial t$ and $\nabla \cdot \mathbf{H} = \nabla \cdot \mathbf{B}/\mu = 0$, we have

$$\nabla^2 \mathbf{H} - \mu \sigma \frac{\partial \mathbf{H}}{\partial t} - \mu \epsilon \frac{\partial^2 \mathbf{H}}{\partial t^2} = -\nabla \times \mathbf{J}^i. \qquad (2.326)$$

This is the wave equation for $\mathbf{H}$.

2.10.4 Scalar wave equation for a conducting medium

In many applications, particularly those involving planar boundary surfaces, it is convenient to decompose the vector wave equation into cartesian components. Using $\nabla^2 \mathbf{V} = \hat{\mathbf{x}} \nabla^2 V_x + \hat{\mathbf{y}} \nabla^2 V_y + \hat{\mathbf{z}} \nabla^2 V_z$ in (2.325) and in (2.326), we find that the rectangular components of $\mathbf{E}$ and $\mathbf{H}$ must obey the scalar wave equation

$$\nabla^2 \psi(\mathbf{r}, t) - \mu\sigma \frac{\partial \psi(\mathbf{r}, t)}{\partial t} - \mu\epsilon \frac{\partial^2 \psi(\mathbf{r}, t)}{\partial t^2} = s(\mathbf{r}, t). \qquad (2.327)$$

For the electric field wave equation we have

$$\psi = E_\alpha, \qquad s = \mu \frac{\partial J^i_\alpha}{\partial t} + \frac{1}{\epsilon} \hat{\boldsymbol{\alpha}} \cdot \nabla\rho,$$

where $\alpha = x, y, z$. For the magnetic field wave equations we have

$$\psi = H_\alpha, \qquad s = \hat{\boldsymbol{\alpha}} \cdot (-\nabla \times \mathbf{J}^i).$$

2.10.5 Fields determined by Maxwell's equations vs. fields determined by the wave equation

Although we derive the wave equations directly from Maxwell's equations, we may wonder whether the solutions to second-order differential equations such as (2.314)–(2.315) are necessarily the same as the solutions to the first-order Maxwell equations. Hansen and Yaghjian [81] show that if all information about the fields is supplied by the sources $\mathbf{J}(\mathbf{r}, t)$ and $\rho(\mathbf{r}, t)$, rather than by specification of field values on boundaries, the solutions to Maxwell's equations and the wave equations are equivalent as long as the second derivatives of the quantities

$$\nabla \cdot \mathbf{E}(\mathbf{r}, t) - \rho(\mathbf{r}, t)/\epsilon, \qquad \nabla \cdot \mathbf{H}(\mathbf{r}, t),$$

are continuous functions of $\mathbf{r}$ and t. If boundary values are supplied in an attempt to guarantee uniqueness, then solutions to the wave equation and to Maxwell's equations may differ. This is particularly important when comparing numerical solutions obtained directly from Maxwell's equations (using the FDTD method, say) to solutions obtained from the wave equation. "Spurious" solutions having no physical significance are a continual plague for engineers who employ numerical techniques. The interested reader should see Jiang [94].

We note that these conclusions do not hold for static fields. The conditions for equivalence of the first-order and second-order static field equations are considered in § 3.2.4.

2.10.6 Transient uniform plane waves in a conducting medium

We can learn a great deal about the wave nature of the electromagnetic field by solving the wave equation (2.325) under simple circumstances. In Chapter 5 we shall solve for the field produced by an arbitrary distribution of impressed sources, but here we seek a simple solution to the homogeneous form of the equation. This allows us to study the phenomenology of wave propagation without worrying about the consequences of specific source functions. We shall also assume a high degree of symmetry so that we are not bogged down in details about the vector directions of the field components.

We seek a solution of the wave equation in which the fields are invariant over a chosen planar surface. The resulting fields are said to comprise a *uniform plane wave*. Although we can envision a uniform plane wave as being created by a uniform surface source of doubly-infinite extent, plane waves are also useful as models for spherical waves over localized regions of the wavefront.

We choose the plane of field invariance to be the xy-plane and later generalize the resulting solution to any planar surface by a simple rotation of the coordinate axes. Since the fields vary with z only we choose to write the wave equation (2.325) in rectangular coordinates, giving for a source-free region of space[2]

$$\hat{\mathbf{x}}\frac{\partial^2 E_x(z,t)}{\partial z^2} + \hat{\mathbf{y}}\frac{\partial^2 E_y(z,t)}{\partial z^2} + \hat{\mathbf{z}}\frac{\partial^2 E_z(z,t)}{\partial z^2} - \mu\sigma\frac{\partial \mathbf{E}(z,t)}{\partial t} - \mu\epsilon\frac{\partial^2 \mathbf{E}(z,t)}{\partial t^2} = 0. \quad (2.328)$$

If we return to Maxwell's equations, we soon find that not all components of $\mathbf{E}$ are present in the plane-wave solution. Faraday's law states that

$$\nabla \times \mathbf{E}(z,t) = -\hat{\mathbf{x}}\frac{\partial E_y(z,t)}{\partial z} + \hat{\mathbf{y}}\frac{\partial E_x(z,t)}{\partial z} = \hat{\mathbf{z}} \times \frac{\partial \mathbf{E}(z,t)}{\partial z} = -\mu\frac{\partial \mathbf{H}(z,t)}{\partial t}. \quad (2.329)$$

We see that $\partial H_z/\partial t = 0$, hence H_z must be constant with respect to time. Because a nonzero constant field component would not exhibit wave-like behavior, we can only have $H_z = 0$ in our wave solution. Similarly, Ampere's law in a homogeneous conducting region free from impressed sources states that

$$\nabla \times \mathbf{H}(z,t) = \mathbf{J} + \frac{\partial \mathbf{D}(z,t)}{\partial t} = \sigma\mathbf{E}(z,t) + \epsilon\frac{\partial \mathbf{E}(z,t)}{\partial t}$$

or

$$-\hat{\mathbf{x}}\frac{\partial H_y(z,t)}{\partial z} + \hat{\mathbf{y}}\frac{\partial H_x(z,t)}{\partial z} = \hat{\mathbf{z}} \times \frac{\partial \mathbf{H}(z,t)}{\partial z} = \sigma\mathbf{E}(z,t) + \epsilon\frac{\partial \mathbf{E}(z,t)}{\partial t}. \quad (2.330)$$

This implies that

$$\sigma E_z(z,t) + \epsilon\frac{\partial E_z(z,t)}{\partial t} = 0,$$

which is a differential equation for E_z with solution

$$E_z(z,t) = E_0(z)\,e^{-\frac{\sigma}{\epsilon}t}.$$

Since we are interested only in wave-type solutions, we choose $E_z = 0$.

Hence $E_z = H_z = 0$, and thus both $\mathbf{E}$ and $\mathbf{H}$ are perpendicular to the z-direction. Using (2.329) and (2.330), we also see that

$$\frac{\partial}{\partial t}(\mathbf{E} \cdot \mathbf{H}) = \mathbf{E} \cdot \frac{\partial \mathbf{H}}{\partial t} + \mathbf{H} \cdot \frac{\partial \mathbf{E}}{\partial t}$$

$$= -\frac{1}{\mu}\mathbf{E} \cdot \left(\hat{\mathbf{z}} \times \frac{\partial \mathbf{E}}{\partial z}\right) - \mathbf{H} \cdot \left(\frac{\sigma}{\epsilon}\mathbf{E}\right) + \frac{1}{\epsilon}\mathbf{H} \cdot \left(\hat{\mathbf{z}} \times \frac{\partial \mathbf{H}}{\partial z}\right)$$

or

$$\left(\frac{\partial}{\partial t} + \frac{\sigma}{\epsilon}\right)(\mathbf{E} \cdot \mathbf{H}) = \frac{1}{\mu}\hat{\mathbf{z}} \cdot \left(\mathbf{E} \times \frac{\partial \mathbf{E}}{\partial z}\right) - \frac{1}{\epsilon}\hat{\mathbf{z}} \cdot \left(\mathbf{H} \times \frac{\partial \mathbf{H}}{\partial z}\right).$$

We seek solutions of the type $\mathbf{E}(z,t) = \hat{\mathbf{p}}E(z,t)$ and $\mathbf{H}(z,t) = \hat{\mathbf{q}}H(z,t)$, where $\hat{\mathbf{p}}$ and $\hat{\mathbf{q}}$ are constant unit vectors. Under this condition we have $\mathbf{E} \times \partial\mathbf{E}/\partial z = 0$ and $\mathbf{H} \times \partial\mathbf{H}/\partial z = 0$, giving

$$\left(\frac{\partial}{\partial t} + \frac{\sigma}{\epsilon}\right)(\mathbf{E} \cdot \mathbf{H}) = 0.$$

---

[2]The term "source free" applied to a conducting region implies that the region is devoid of impressed sources *and*, because of the relaxation effect, has no free charge. See the discussion in Jones [97].

Thus we also have $\mathbf{E} \cdot \mathbf{H} = 0$, and find that $\mathbf{E}$ must be perpendicular to $\mathbf{H}$. So $\mathbf{E}$, $\mathbf{H}$, and $\hat{\mathbf{z}}$ comprise a mutually orthogonal triplet of vectors. A wave having this property is said to be *TEM to the z-direction* or simply TEM_z. Here "TEM" stands for *transverse electromagnetic*, indicating the orthogonal relationship between the field vectors and the z-direction. Note that

$$\hat{\mathbf{p}} \times \hat{\mathbf{q}} = \pm \hat{\mathbf{z}}.$$

The constant direction described by $\hat{\mathbf{p}}$ is called the *polarization* of the plane wave.

We are now ready to solve the source-free wave equation (2.328). If we dot both sides of the homogeneous expression by $\hat{\mathbf{p}}$ we obtain

$$\hat{\mathbf{p}} \cdot \hat{\mathbf{x}} \frac{\partial^2 E_x}{\partial z^2} + \hat{\mathbf{p}} \cdot \hat{\mathbf{y}} \frac{\partial^2 E_y}{\partial z^2} - \mu\sigma \frac{\partial(\hat{\mathbf{p}} \cdot \mathbf{E})}{\partial t} - \mu\epsilon \frac{\partial^2(\hat{\mathbf{p}} \cdot \mathbf{E})}{\partial t^2} = 0.$$

Noting that

$$\hat{\mathbf{p}} \cdot \hat{\mathbf{x}} \frac{\partial^2 E_x}{\partial z^2} + \hat{\mathbf{p}} \cdot \hat{\mathbf{y}} \frac{\partial^2 E_y}{\partial z^2} = \frac{\partial^2}{\partial z^2}(\hat{\mathbf{p}} \cdot \hat{\mathbf{x}} E_x + \hat{\mathbf{p}} \cdot \hat{\mathbf{y}} E_y) = \frac{\partial^2}{\partial z^2}(\hat{\mathbf{p}} \cdot \mathbf{E}),$$

we have the wave equation

$$\frac{\partial^2 E(z,t)}{\partial z^2} - \mu\sigma \frac{\partial E(z,t)}{\partial t} - \mu\epsilon \frac{\partial^2 E(z,t)}{\partial t^2} = 0. \tag{2.331}$$

Similarly, dotting both sides of (2.326) with $\hat{\mathbf{q}}$ and setting $\mathbf{J}^i = 0$ we obtain

$$\frac{\partial^2 H(z,t)}{\partial z^2} - \mu\sigma \frac{\partial H(z,t)}{\partial t} - \mu\epsilon \frac{\partial^2 H(z,t)}{\partial t^2} = 0. \tag{2.332}$$

In a source-free homogeneous conducting region $\mathbf{E}$ and $\mathbf{H}$ satisfy identical wave equations.

Solutions are considered in § A.1. There we solve for the total field for all z, t given the value of the field and its derivative over the $z = 0$ plane. This solution can be directly applied to find the total field of a plane wave reflected by a perfect conductor. Let us begin by considering the lossless case where $\sigma = 0$, and assuming the region $z < 0$ contains a perfect electric conductor. The conditions on the field in the $z = 0$ plane are determined by the required boundary condition on a perfect conductor: the tangential electric field must vanish. From (2.330) we see that since $\mathbf{E} \perp \hat{\mathbf{z}}$, requiring

$$\left. \frac{\partial H(z,t)}{\partial z} \right|_{z=0} = 0 \tag{2.333}$$

gives $\mathbf{E}(0,t) = 0$ and thus satisfies the boundary condition. Writing

$$H(0,t) = H_0 f(t), \qquad \left. \frac{\partial H(z,t)}{\partial z} \right|_{z=0} = H_0 g(t) = 0, \tag{2.334}$$

and setting $\Omega = 0$ in (A.41) we obtain the solution to (2.332):

$$H(z,t) = \frac{H_0}{2} f\left(t - \frac{z}{v}\right) + \frac{H_0}{2} f\left(t + \frac{z}{v}\right), \tag{2.335}$$

where $v = 1/(\mu\epsilon)^{1/2}$. Since we designate the vector direction of $\mathbf{H}$ as $\hat{\mathbf{q}}$, the vector field is

$$\mathbf{H}(z,t) = \hat{\mathbf{q}} \frac{H_0}{2} f\left(t - \frac{z}{v}\right) + \hat{\mathbf{q}} \frac{H_0}{2} f\left(t + \frac{z}{v}\right). \tag{2.336}$$

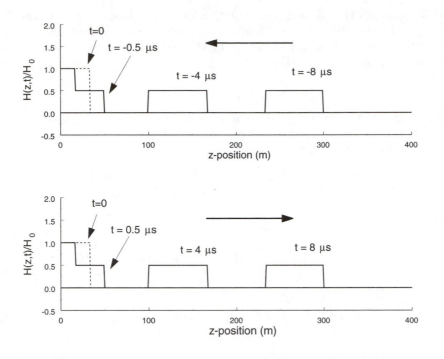

Figure 2.7: Propagation of a transient plane wave in a lossless medium.

From (2.329) we also have the solution for $\mathbf{E}(z,t)$:

$$\mathbf{E}(z,t) = \hat{\mathbf{p}}\frac{v\mu H_0}{2}f\left(t - \frac{z}{v}\right) - \hat{\mathbf{p}}\frac{v\mu H_0}{2}f\left(t + \frac{z}{v}\right), \qquad (2.337)$$

where

$$\hat{\mathbf{p}} \times \hat{\mathbf{q}} = \hat{\mathbf{z}}.$$

The boundary conditions $E(0,t) = 0$ and $H(0,t) = H_0 f(t)$ are easily verified by substitution.

This solution displays the quintessential behavior of electromagnetic waves. We may interpret the term $f(t + z/v)$ as a wave field disturbance, propagating at velocity v in the $-z$-direction, incident from $z > 0$ upon the conductor. The term $f(t - z/v)$ represents a wave field disturbance propagating in the $+z$-direction with velocity v, reflected from the conductor. By "propagating" we mean that if we increment time, the disturbance will occupy a spatial position determined by incrementing z by vt. For free space where $v = 1/(\mu_0\epsilon_0)^{1/2}$, the velocity of propagation is the speed of light c.

A specific example should serve to clarify our interpretation of the wave solution. Taking $\mu = \mu_0$ and $\epsilon = 81\epsilon_0$, representing typical constitutive values for fresh water, we can plot (2.335) as a function of position for fixed values of time. The result is shown in Figure 2.7, where we have chosen

$$f(t) = \text{rect}(t/\tau) \qquad (2.338)$$

with $\tau = 1$ μs. We see that the disturbance is spatially distributed as a rectangular pulse of extent $L = 2v\tau = 66.6$ m, where $v = 3.33 \times 10^7$ m/s is the wave velocity, and where 2τ is the temporal duration of the pulse. At $t = -8$ μs the leading edge of the pulse is at $z = 233$ m, while at -4 μs the pulse has traveled a distance $z = vt = (3.33 \times 10^7) \times (4 \times 10^{-6}) = 133$ m in the $-z$-direction, and the leading edge is thus at 100 m. At $t = -1$ μs the leading edge strikes the conductor and begins to induce a current in the conductor surface. This current sets up the reflected wave, which begins to travel in the opposite $(+z)$ direction. At $t = -0.5$ μs a portion of the wave has begun to travel in the $+z$-direction while the trailing portion of the disturbance continues to travel in the $-z$-direction. At $t = 1$ μs the wave has been completely reflected from the surface, and thus consists only of the component traveling in the $+z$-direction. Note that if we plot the total field in the $z = 0$ plane, the sum of the forward and backward traveling waves produces the pulse waveform (2.338) as expected.

Using the expressions for **E** and **H** we can determine many interesting characteristics of the wave. We see that the terms $f(t \pm z/v)$ represent the components of the waves traveling in the $\mp z$-directions, respectively. If we were to isolate these waves from each other (by, for instance, measuring them as functions of time at a position where they do not overlap) we would find from (2.336) and (2.337) that the ratio of E to H for a wave traveling in either direction is

$$\left| \frac{E(z,t)}{H(z,t)} \right| = v\mu = (\mu/\epsilon)^{1/2},$$

independent of the time and position of the measurement. This ratio, denoted by η and carrying units of ohms, is called the *intrinsic impedance* of the medium through which the wave propagates. Thus, if we let $E_0 = \eta H_0$ we can write

$$\mathbf{E}(z,t) = \hat{\mathbf{p}} \frac{E_0}{2} f\left(t - \frac{z}{v}\right) - \hat{\mathbf{p}} \frac{E_0}{2} f\left(t + \frac{z}{v}\right). \tag{2.339}$$

We can easily determine the current induced in the conductor by applying the boundary condition (2.200):

$$\mathbf{J}_s = \hat{\mathbf{n}} \times \mathbf{H}|_{z=0} = \hat{\mathbf{z}} \times [H_0 \hat{\mathbf{q}} f(t)] = -\hat{\mathbf{p}} H_0 f(t). \tag{2.340}$$

We can also determine the pressure exerted on the conductor due to the Lorentz force interaction between the fields and the induced current. The total force on the conductor can be computed by integrating the Maxwell stress tensor (2.288) over the xy-plane[3]:

$$\mathbf{F}_{em} = -\int_S \bar{\mathbf{T}}_{em} \cdot d\mathbf{S}.$$

The surface traction is

$$\mathbf{t} = \bar{\mathbf{T}}_{em} \cdot \hat{\mathbf{n}} = \left[\frac{1}{2}(\mathbf{D} \cdot \mathbf{E} + \mathbf{B} \cdot \mathbf{H})\bar{\mathbf{I}} - \mathbf{DE} - \mathbf{BH} \right] \cdot \hat{\mathbf{z}}.$$

Since **E** and **H** are both normal to $\hat{\mathbf{z}}$, the last two terms in this expression are zero. Also, the boundary condition on **E** implies that it vanishes in the xy-plane. Thus

$$\mathbf{t} = \frac{1}{2}(\mathbf{B} \cdot \mathbf{H})\hat{\mathbf{z}} = \hat{\mathbf{z}} \frac{\mu}{2} H^2(t).$$

---

[3]We may neglect the momentum term in (2.291), which is small compared to the stress tensor term. See Problem 2.20.

With $H_0 = E_0/\eta$ we have

$$\mathbf{t} = \hat{\mathbf{z}} \frac{E_0^2}{2\eta^2} \mu f^2(t). \tag{2.341}$$

As a numerical example, consider a high-altitude nuclear electromagnetic pulse (HEMP) generated by the explosion of a large nuclear weapon in the upper atmosphere. Such an explosion could generate a transient electromagnetic wave of short (sub-microsecond) duration with an electric field amplitude of 50,000 V/m in air [200]. Using (2.341), we find that the wave would exert a peak pressure of $P = |\mathbf{t}| = .011$ Pa $= 1.6 \times 10^{-6}$ lb/in^2 if reflected from a perfect conductor at normal incidence. Obviously, even for this extreme field level the pressure produced by a transient electromagnetic wave is quite small. However, from (2.340) we find that the current induced in the conductor would have a peak value of 133 A/m. Even a small portion of this current could destroy a sensitive electronic circuit if it were to leak through an opening in the conductor. This is an important concern for engineers designing circuitry to be used in high-field environments, and demonstrates why the concepts of current and voltage can often supersede the concept of force in terms of importance.

Finally, let us see how the terms in the Poynting power balance theorem relate. Consider a cubic region V bounded by the planes $z = z_1$ and $z = z_2$, $z_2 > z_1$. We choose the field waveform $f(t)$ and locate the planes so that we can isolate either the forward or backward traveling wave. Since there is no current in V, Poynting's theorem (2.299) becomes

$$\frac{1}{2} \frac{\partial}{\partial t} \int_V (\epsilon \mathbf{E} \cdot \mathbf{E} + \mu \mathbf{H} \cdot \mathbf{H}) \, dV = - \oint_S (\mathbf{E} \times \mathbf{H}) \cdot d\mathbf{S}.$$

Consider the wave traveling in the $-z$-direction. Substitution from (2.336) and (2.337) gives the time-rate of change of stored energy as

$$S_{\text{cube}}(t) = \frac{1}{2} \frac{\partial}{\partial t} \int_V \left[\epsilon E^2(z,t) + \mu H^2(z,t) \right] dV$$

$$= \frac{1}{2} \frac{\partial}{\partial t} \int_x \int_y dx\, dy \int_{z_1}^{z_2} \left[\epsilon \frac{(v\mu)^2 H_0^2}{4} f^2 \left(t + \frac{z}{v} \right) + \mu \frac{H_0^2}{4} f^2 \left(t + \frac{z}{v} \right) \right] dz$$

$$= \frac{1}{2} \frac{\partial}{\partial t} \mu \frac{H_0^2}{2} \int_x \int_y dx\, dy \int_{z_1}^{z_2} f^2 \left(t + \frac{z}{v} \right) dz.$$

Integration over x and y gives the area A of the cube face. Putting $u = t + z/v$ we see that

$$S = A \mu \frac{H_0^2}{4} \frac{\partial}{\partial t} \int_{t+z_1/v}^{t+z_2/v} f^2(u) v \, du.$$

Leibnitz' rule for differentiation (A.30) then gives

$$S_{\text{cube}}(t) = A \frac{\mu v H_0^2}{4} \left[f^2 \left(t + \frac{z_2}{v} \right) - f^2 \left(t + \frac{z_1}{v} \right) \right]. \tag{2.342}$$

Again substituting for $E(t + z/v)$ and $H(t + z/v)$ we can write

$$S_{\text{cube}}(t) = - \oint_S (\mathbf{E} \times \mathbf{H}) \cdot d\mathbf{S}$$

$$= - \int_x \int_y \frac{v\mu H_0^2}{4} f^2 \left(t + \frac{z_1}{v} \right) (-\hat{\mathbf{p}} \times \hat{\mathbf{q}}) \cdot (-\hat{\mathbf{z}}) \, dx \, dy -$$

$$- \int_x \int_y \frac{v\mu H_0^2}{4} f^2 \left(t + \frac{z_2}{v} \right) (-\hat{\mathbf{p}} \times \hat{\mathbf{q}}) \cdot (\hat{\mathbf{z}}) \, dx \, dy.$$

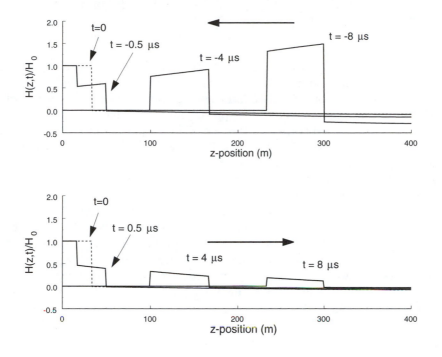

Figure 2.8: Propagation of a transient plane wave in a dissipative medium.

The second term represents the energy change in V produced by the backward traveling wave entering the cube by passing through the plane at $z = z_2$, while the first term represents the energy change in V produced by the wave exiting the cube by passing through the plane $z = z_1$. Contributions from the sides, top, and bottom are zero since $\mathbf{E} \times \mathbf{H}$ is perpendicular to $\hat{\mathbf{n}}$ over those surfaces. Since $\hat{\mathbf{p}} \times \hat{\mathbf{q}} = \hat{\mathbf{z}}$, we get

$$S_{\text{cube}}(t) = A \frac{\mu v H_0^2}{4} \left[f^2 \left(t + \frac{z_2}{v} \right) - f^2 \left(t + \frac{z_1}{v} \right) \right],$$

which matches (2.342) and thus verifies Poynting's theorem. We may interpret this result as follows. The propagating electromagnetic disturbance carries energy through space. The energy within any region is associated with the field in that region, and can change with time as the propagating wave carries a flux of energy across the boundary of the region. The energy continues to propagate even if the source is changed or is extinguished altogether. That is, the behavior of the leading edge of the disturbance is determined by causality — it is affected by obstacles it encounters, but not by changes in the source that occur after the leading edge has been established.

When propagating through a dissipative region a plane wave takes on a somewhat different character. Again applying the conditions (2.333) and (2.334), we obtain from (A.41) the solution to the wave equation (2.332):

$$H(z,t) = \frac{H_0}{2} e^{-\frac{\Omega}{v}z} f \left(t - \frac{z}{v} \right) + \frac{H_0}{2} e^{\frac{\Omega}{v}z} f \left(t + \frac{z}{v} \right) -$$

$$-\frac{z\Omega^2 H_0}{2v}e^{-\Omega t}\int_{t-\frac{z}{v}}^{t+\frac{z}{v}} f(u)e^{\Omega u}\frac{J_1\left(\frac{\Omega}{v}\sqrt{z^2-(t-u)^2v^2}\right)}{\frac{\Omega}{v}\sqrt{z^2-(t-u)^2v^2}}\,du \qquad (2.343)$$

where $\Omega = \sigma/2\epsilon$. The first two terms resemble those for the lossless case, modified by an exponential damping factor. This accounts for the loss in amplitude that must accompany the transfer of energy from the propagating wave to joule loss (heat) within the conducting medium. The remaining term appears only when the medium is lossy, and results in an extension of the disturbance through the medium because of the currents induced by the passing wavefront. This "wake" follows the leading edge of the disturbance as is shown clearly in Figure 2.8. Here we have repeated the calculation of Figure 2.7, but with $\sigma = 2 \times 10^{-4}$, approximating the conductivity of fresh water. As the wave travels to the left it attenuates and leaves a trailing remnant behind. Upon reaching the conductor it reflects much as in the lossless case, resulting in a time dependence at $z = 0$ given by the finite-duration rectangular pulse (2.338). In order for the pulse to be of finite duration, the wake left by the reflected pulse must exactly cancel the wake associated with the incident pulse that continues to arrive after the reflection. As the reflected pulse sweeps forward, the wake is obliterated everywhere behind.

If we were to verify the Poynting theorem for a dissipative medium (which we shall not attempt because of the complexity of the computation), we would need to include the $\mathbf{E} \cdot \mathbf{J}$ term. Here $\mathbf{J}$ is the induced conduction current and the integral of $\mathbf{E} \cdot \mathbf{J}$ accounts for the joule loss within a region V balanced by the difference in Poynting energy flux carried into and out of V.

Once we have the fields for a wave propagating along the z-direction, it is a simple matter to generalize these results to any propagation direction. Assume that $\hat{\mathbf{u}}$ is normal to the surface of a plane over which the fields are invariant. Then $u = \hat{\mathbf{u}} \cdot \mathbf{r}$ describes the distance from the origin along the direction $\hat{\mathbf{u}}$. We need only replace z by $\hat{\mathbf{u}} \cdot \mathbf{r}$ in any of the expressions obtained above to determine the fields of a plane wave propagating in the u-direction. We must also replace the orthogonality condition $\hat{\mathbf{p}} \times \hat{\mathbf{q}} = \hat{\mathbf{z}}$ with

$$\hat{\mathbf{p}} \times \hat{\mathbf{q}} = \hat{\mathbf{u}}.$$

For instance, the fields associated with a wave propagating through a lossless medium in the positive u-direction are, from (2.336)–(2.337),

$$\mathbf{H}(\mathbf{r},t) = \hat{\mathbf{q}}\frac{H_0}{2}f\left(t - \frac{\hat{\mathbf{u}} \cdot \mathbf{r}}{v}\right), \qquad \mathbf{E}(\mathbf{r},t) = \hat{\mathbf{p}}\frac{v\mu H_0}{2}f\left(t - \frac{\hat{\mathbf{u}} \cdot \mathbf{r}}{v}\right).$$

2.10.7 Propagation of cylindrical waves in a lossless medium

Much as we envisioned a uniform plane wave arising from a uniform planar source, we can imagine a uniform cylindrical wave arising from a uniform line source. Although this line source must be infinite in extent, uniform cylindrical waves (unlike plane waves) display the physical behavior of diverging from their source while carrying energy outwards to infinity.

A *uniform cylindrical wave* has fields that are invariant over a cylindrical surface: $\mathbf{E}(\mathbf{r},t) = \mathbf{E}(\rho,t)$, $\mathbf{H}(\mathbf{r},t) = \mathbf{H}(\rho,t)$. For simplicity, we shall assume that waves propagate in a homogeneous, isotropic, linear, and lossless medium described by permittivity ϵ and permeability μ. From Maxwell's equations we find that requiring the fields to be independent of ϕ and z puts restrictions on the remaining vector components. Faraday's law states

$$\nabla \times \mathbf{E}(\rho, t) = -\hat{\phi}\frac{\partial E_z(\rho, t)}{\partial \rho} + \hat{z}\frac{1}{\rho}\frac{\partial}{\partial \rho}[\rho E_\phi(\rho, t)] = -\mu\frac{\partial \mathbf{H}(\rho, t)}{\partial t}. \qquad (2.344)$$

Equating components we see that $\partial H_\rho/\partial t = 0$, and because our interest lies in wave solutions we take $H_\rho = 0$. Ampere's law in a homogeneous lossless region free from impressed sources states in a similar manner

$$\nabla \times \mathbf{H}(\rho, t) = -\hat{\phi}\frac{\partial H_z(\rho, t)}{\partial \rho} + \hat{z}\frac{1}{\rho}\frac{\partial}{\partial \rho}[\rho H_\phi(\rho, t)] = \epsilon\frac{\partial \mathbf{E}(\rho, t)}{\partial t}. \qquad (2.345)$$

Equating components we find that $E_\rho = 0$. Since $E_\rho = H_\rho = 0$, both $\mathbf{E}$ and $\mathbf{H}$ are perpendicular to the ρ-direction. Note that if there is only a z-component of $\mathbf{E}$ then there is only a ϕ-component of $\mathbf{H}$. This case, termed *electric polarization*, results in

$$\frac{\partial E_z(\rho, t)}{\partial \rho} = \mu\frac{\partial H_\phi(\rho, t)}{\partial t}.$$

Similarly, if there is only a z-component of $\mathbf{H}$ then there is only a ϕ-component of $\mathbf{E}$. This case, termed *magnetic polarization*, results in

$$-\frac{\partial H_z(\rho, t)}{\partial \rho} = \epsilon\frac{\partial E_\phi(\rho, t)}{\partial t}.$$

Since $\mathbf{E} = \hat{\phi}E_\phi + \hat{z}E_z$ and $\mathbf{H} = \hat{\phi}H_\phi + \hat{z}H_z$, we can always decompose a cylindrical electromagnetic wave into cases of electric and magnetic polarization. In each case the resulting field is TEM$_\rho$ since the vectors $\mathbf{E}, \mathbf{H}, \hat{\rho}$ are mutually orthogonal.

Wave equations for E_z in the electric polarization case and for H_z in the magnetic polarization case can be found in the usual manner. Taking the curl of (2.344) and substituting from (2.345) we find

$$\nabla \times (\nabla \times \mathbf{E}) = -\hat{z}\frac{1}{\rho}\frac{\partial}{\partial \rho}\left(\rho\frac{\partial E_z}{\partial \rho}\right) - \hat{\phi}\frac{\partial}{\partial \rho}\left(\frac{1}{\rho}\frac{\partial}{\partial \rho}[\rho E_\phi]\right)$$

$$= -\frac{1}{v^2}\frac{\partial^2 \mathbf{E}}{\partial t^2} = -\frac{1}{v^2}\left[\hat{z}\frac{\partial^2 E_z}{\partial t^2} + \hat{\phi}\frac{\partial^2 E_\phi}{\partial t^2}\right]$$

where $v = 1/(\mu\epsilon)^{1/2}$. Noting that $E_\phi = 0$ for the electric polarization case we obtain the wave equation for E_z. A similar set of steps beginning with the curl of (2.345) gives an identical equation for H_z. Thus

$$\frac{1}{\rho}\frac{\partial}{\partial \rho}\left(\rho\frac{\partial}{\partial \rho}\begin{bmatrix} E_z \\ H_z \end{bmatrix}\right) - \frac{1}{v^2}\frac{\partial^2}{\partial t^2}\begin{bmatrix} E_z \\ H_z \end{bmatrix} = 0. \qquad (2.346)$$

We can obtain a solution for (2.346) in much the same way as we do for the wave equations in § A.1. We begin by substituting for $E_z(\rho, t)$ in terms of its temporal Fourier representation

$$E_z(\rho, t) = \frac{1}{2\pi}\int_{-\infty}^{\infty}\tilde{E}_z(\rho, \omega)e^{j\omega t}\, d\omega$$

to obtain

$$\frac{1}{2\pi}\int_{-\infty}^{\infty}\left[\frac{1}{\rho}\frac{\partial}{\partial \rho}\left(\rho\frac{\partial}{\partial \rho}\tilde{E}_z(\rho, \omega)\right) + \frac{\omega^2}{v^2}\tilde{E}_z(\rho, \omega)\right]e^{j\omega t}\, d\omega = 0.$$

The Fourier integral theorem implies that the integrand is zero. Then, expanding out the ρ derivatives, we find that $\tilde{E}_z(\rho, \omega)$ obeys the ordinary differential equation

$$\frac{d^2 \tilde{E}_z}{d\rho^2} + \frac{1}{\rho}\frac{d\tilde{E}_z}{d\rho} + k^2 \tilde{E}_z = 0$$

where $k = \omega/v$. This is merely Bessel's differential equation (A.124). It is a second-order equation with two independent solutions chosen from the list

$$J_0(k\rho), \quad Y_0(k\rho), \quad H_0^{(1)}(k\rho), \quad H_0^{(2)}(k\rho).$$

We find that $J_0(k\rho)$ and $Y_0(k\rho)$ are useful for describing standing waves between boundaries while $H_0^{(1)}(k\rho)$ and $H_0^{(2)}(k\rho)$ are useful for describing waves propagating in the ρ-direction. Of these, $H_0^{(1)}(k\rho)$ represents waves traveling inward while $H_0^{(2)}(k\rho)$ represents waves traveling outward. Concentrating on the outward traveling wave we find that

$$\tilde{E}_z(\rho, \omega) = \tilde{A}(\omega)\left[-j\frac{\pi}{2}H_0^{(2)}(k\rho)\right] = \tilde{A}(\omega)\tilde{g}(\rho, \omega).$$

Here $A(t) \leftrightarrow \tilde{A}(\omega)$ is the disturbance waveform, assumed to be a real, causal function. To make $E_z(\rho, t)$ real we require that the inverse transform of $\tilde{g}(\rho, \omega)$ be real. This requires the inclusion of the $-j\pi/2$ factor in $\tilde{g}(\rho, \omega)$. Inverting we have

$$E_z(\rho, t) = A(t) * g(\rho, t) \tag{2.347}$$

where $g(\rho, t) \leftrightarrow (-j\pi/2)H_0^{(2)}(k\rho)$.

The inverse transform needed to obtain $g(\rho, t)$ may be found in Campbell [26]:

$$g(\rho, t) = \mathcal{F}^{-1}\left\{-j\frac{\pi}{2}H_0^{(2)}\left(\omega\frac{\rho}{v}\right)\right\} = \frac{U\left(t - \frac{\rho}{v}\right)}{\sqrt{t^2 - \frac{\rho^2}{v^2}}},$$

where $U(t)$ is the unit step function defined in (A.5). Substituting this into (2.347) and writing the convolution in integral form we have

$$E_z(\rho, t) = \int_{-\infty}^{\infty} A(t - t')\frac{U(t' - \rho/v)}{\sqrt{t'^2 - \rho^2/v^2}}\, dt'.$$

The change of variable $x = t' - \rho/v$ then gives

$$E_z(\rho, t) = \int_0^{\infty} \frac{A(t - x - \rho/v)}{\sqrt{x^2 + 2x\rho/v}}\, dx. \tag{2.348}$$

Those interested in the details of the inverse transform should see Chew [33].

As an example, consider a lossless medium with $\mu_r = 1$, $\epsilon_r = 81$, and a waveform

$$A(t) = E_0[U(t) - U(t - \tau)]$$

where $\tau = 2$ μs. This situation is the same as that in the plane wave example above, except that the pulse waveform begins at $t = 0$. Substituting for $A(t)$ into (2.348) and using the integral

$$\int \frac{dx}{\sqrt{x}\sqrt{x + a}} = 2\ln\left[\sqrt{x} + \sqrt{x + a}\right]$$

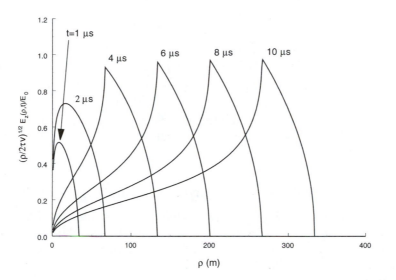

Figure 2.9: Propagation of a transient cylindrical wave in a lossless medium.

we can write the electric field in closed form as

$$E_z(\rho, t) = 2E_0 \ln \left[\frac{\sqrt{x_2} + \sqrt{x_2 + 2\rho/v}}{\sqrt{x_1} + \sqrt{x_1 + 2\rho/v}} \right], \qquad (2.349)$$

where $x_2 = \max[0, t - \rho/v]$ and $x_1 = \max[0, t - \rho/v - \tau]$. The field is plotted in Figure 2.9 for various values of time. Note that the leading edge of the disturbance propagates outward at a velocity v and a wake trails behind the disturbance. This wake is similar to that for a plane wave in a dissipative medium, but it exists in this case even though the medium is lossless. We can think of the wave as being created by a line source of infinite extent, pulsed by the disturbance waveform. Although current changes simultaneously everywhere along the line, it takes the disturbance longer to propagate to an observation point in the $z = 0$ plane from source points $z \neq 0$ than from the source point at $z = 0$. Thus, the field at an arbitrary observation point ρ arrives from different source points at different times. If we look at Figure 2.9 we note that there is always a nonzero field near $\rho = 0$ (or any value of $\rho < vt$) regardless of the time, since at any given t the disturbance is arriving from some point along the line source.

We also see in Figure 2.9 that as ρ becomes large the peak value of the propagating disturbance approaches a certain value. This value occurs at $t_m = \rho/v + \tau$ or, equivalently, $\rho_m = v(t - \tau)$. If we substitute this value into (2.349) we find that

$$E_z(\rho, t_m) = 2E_0 \ln \left[\sqrt{\frac{\tau}{2\rho/v}} + \sqrt{1 + \frac{\tau}{2\rho/v}} \right].$$

For large values of ρ/v,

$$E_z(\rho, t_m) \approx 2E_0 \ln\left[1 + \sqrt{\frac{\tau}{2\rho/v}}\right].$$

Using $\ln(1 + x) \approx x$ when $x \ll 1$, we find that

$$E_z(\rho, t_m) \approx E_0\sqrt{\frac{2\tau v}{\rho}}.$$

Thus, as $\rho \to \infty$ we have $\mathbf{E} \times \mathbf{H} \sim 1/\rho$ and the flux of energy passing through a cylindrical surface of area $\rho\, d\phi\, dz$ is independent of ρ. This result is similar to that seen for spherical waves where $\mathbf{E} \times \mathbf{H} \sim 1/r^2$.

2.10.8   Propagation of spherical waves in a lossless medium

In the previous section we found solutions that describe uniform cylindrical waves dependent only on the radial variable ρ. It turns out that similar solutions are not possible in spherical coordinates; fields that only depend on r cannot satisfy Maxwell's equations since, as shown in § 2.10.9, a source having the appropriate symmetry for the production of uniform spherical waves in fact produces no field at all external to the region it occupies. As we shall see in Chapter 5, the fields produced by localized sources are in general quite complex. However, certain solutions that are only slightly nonuniform may be found, and these allow us to investigate the most important properties of spherical waves. We shall find that spherical waves diverge from a localized point source and expand outward with finite velocity, carrying energy away from the source.

Consider a homogeneous, lossless, source-free region of space characterized by permittivity ϵ and permeability μ. We seek solutions to the wave equation that are TEM$_r$ in spherical coordinates ($H_r = E_r = 0$), and independent of the azimuthal angle ϕ. Thus we may write

$$\mathbf{E}(\mathbf{r}, t) = \hat{\boldsymbol{\theta}}E_\theta(r, \theta, t) + \hat{\boldsymbol{\phi}}E_\phi(r, \theta, t),$$
$$\mathbf{H}(\mathbf{r}, t) = \hat{\boldsymbol{\theta}}H_\theta(r, \theta, t) + \hat{\boldsymbol{\phi}}H_\phi(r, \theta, t).$$

Maxwell's equations show that not all of these vector components are required. Faraday's law states that

$$\nabla \times \mathbf{E}(r, \theta, t) = \hat{\mathbf{r}}\frac{1}{r\sin\theta}\frac{\partial}{\partial\theta}[\sin\theta E_\phi(r, \theta, t)] - \hat{\boldsymbol{\theta}}\frac{1}{r}\frac{\partial}{\partial r}[rE_\phi(r, \theta, t)] + \hat{\boldsymbol{\phi}}\frac{1}{r}\frac{\partial}{\partial r}[rE_\theta(r, \theta, t)]$$
$$= -\mu\frac{\partial\mathbf{H}(r, \theta, t)}{\partial t}. \tag{2.350}$$

Since we require $H_r = 0$ we must have

$$\frac{\partial}{\partial\theta}[\sin\theta E_\phi(r, \theta, t)] = 0.$$

This implies that either $E_\phi \sim 1/\sin\theta$ or $E_\phi = 0$. We shall choose $E_\phi = 0$ and investigate whether the resulting fields satisfy the remaining Maxwell equations.

In a source-free region of space we have $\nabla \cdot \mathbf{D} = \epsilon\nabla \cdot \mathbf{E} = 0$. Since we now have only a θ-component of the electric field, this requires

$$\frac{1}{r}\frac{\partial}{\partial\theta}E_\theta(r, \theta, t) + \frac{\cot\theta}{r}E_\theta(r, \theta, t) = 0.$$

From this we see that when $E_\phi = 0$ the component E_θ must obey

$$E_\theta(r,\theta,t) = \frac{f_E(r,t)}{\sin\theta}.$$

By (2.350) there is only a ϕ-component of magnetic field, and it must obey $H_\phi(r,\theta,t) = f_H(r,t)/\sin\theta$ where

$$-\mu\frac{\partial}{\partial t}f_H(r,t) = \frac{1}{r}\frac{\partial}{\partial r}[rf_E(r,t)]. \tag{2.351}$$

Thus the spherical wave has the property $\mathbf{E} \perp \mathbf{H} \perp \mathbf{r}$, and is TEM to the r-direction.

We can obtain a wave equation for E_θ by taking the curl of (2.350) and substituting from Ampere's law:

$$\nabla \times (\nabla \times \mathbf{E}) = -\hat{\boldsymbol{\theta}}\frac{1}{r}\frac{\partial^2}{\partial r^2}[rE_\theta] = \nabla \times \left[-\mu\frac{\partial}{\partial t}\mathbf{H}\right] = -\mu\frac{\partial}{\partial t}\left[\sigma\mathbf{E} + \epsilon\frac{\partial}{\partial t}\mathbf{E}\right].$$

This gives

$$\frac{\partial^2}{\partial r^2}[rf_E(r,t)] - \mu\sigma\frac{\partial}{\partial t}[rf_E(r,t)] - \mu\epsilon\frac{\partial^2}{\partial t^2}[rf_E(r,t)] = 0, \tag{2.352}$$

which is the desired wave equation for $\mathbf{E}$. Proceeding similarly we find that H_ϕ obeys

$$\frac{\partial^2}{\partial r^2}[rf_H(r,t)] - \mu\sigma\frac{\partial}{\partial t}[rf_H(r,t)] - \mu\epsilon\frac{\partial^2}{\partial t^2}[rf_H(r,t)] = 0. \tag{2.353}$$

We see that the wave equation for rf_E is identical to that for the plane wave field E_z (2.331). Thus, we can use the solution obtained in § A.1, as we did with the plane wave, with a few subtle differences. First, we cannot have $r < 0$. Second, we do not anticipate a solution representing a wave traveling in the $-r$-direction — i.e., a wave converging toward the origin. (In other situations we might need such a solution in order to form a standing wave between two spherical boundary surfaces, but here we are only interested in the basic propagating behavior of spherical waves.) Thus, we choose as our solution the term (A.45) and find for a lossless medium where $\Omega = 0$

$$E_\theta(r,\theta,t) = \frac{1}{r\sin\theta}A\left(t - \frac{r}{v}\right). \tag{2.354}$$

From (2.351) we see that

$$H_\phi = \frac{1}{\mu v}\frac{1}{r\sin\theta}A\left(t - \frac{r}{v}\right). \tag{2.355}$$

Since $\mu v = (\mu/\epsilon)^{1/2} = \eta$, we can also write this as

$$\mathbf{H} = \frac{\hat{\mathbf{r}} \times \mathbf{E}}{\eta}.$$

We note that our solution is not appropriate for unbounded space since the fields have a singularity at $\theta = 0$. Thus we must exclude the z-axis. This can be accomplished by using PEC cones of angles θ_1 and θ_2, $\theta_2 > \theta_1$. Because the electric field $\mathbf{E} = \hat{\boldsymbol{\theta}}E_\theta$ is normal to these cones, the boundary condition that tangential $\mathbf{E}$ vanishes is satisfied.

It is informative to see how the terms in the Poynting power balance theorem relate for a spherical wave. Consider the region between the spherical surfaces $r = r_1$ and $r = r_2$, $r_2 > r_1$. Since there is no current within the volume region, Poynting's theorem (2.299) becomes

$$\frac{1}{2}\frac{\partial}{\partial t}\int_V (\epsilon\mathbf{E} \cdot \mathbf{E} + \mu\mathbf{H} \cdot \mathbf{H})\,dV = -\oint_S (\mathbf{E} \times \mathbf{H}) \cdot d\mathbf{S}. \tag{2.356}$$

From (2.354) and (2.355), the time-rate of change of stored energy is

$$
\begin{aligned}
P_{\text{sphere}}(t) &= \frac{1}{2}\frac{\partial}{\partial t}\int_V \left[\epsilon E^2(r,\theta,t) + \mu H^2(r,\theta,t)\right] dV \\
&= \frac{1}{2}\frac{\partial}{\partial t}\int_0^{2\pi} d\phi \int_{\theta_1}^{\theta_2}\frac{d\theta}{\sin\theta}\int_{r_1}^{r_2}\left[\epsilon\frac{1}{r^2}A^2\left(t-\frac{r}{v}\right) + \mu\frac{1}{r^2}\frac{1}{(v\mu)^2}A^2\left(t-\frac{r}{v}\right)\right]r^2\,dr \\
&= 2\pi\epsilon F\frac{\partial}{\partial t}\int_{r_1}^{r_2}A^2\left(t-\frac{r}{v}\right)dr
\end{aligned}
$$

where

$$
F = \ln\left[\frac{\tan(\theta_2/2)}{\tan(\theta_1/2)}\right].
$$

Putting $u = t - r/v$ we see that

$$
P_{\text{sphere}}(t) = -2\pi\epsilon F\frac{\partial}{\partial t}\int_{t-r_1/v}^{t-r_2/v}A^2(u)v\,du.
$$

An application of Leibnitz' rule for differentiation (A.30) gives

$$
P_{\text{sphere}}(t) = -\frac{2\pi}{\eta}F\left[A^2\left(t-\frac{r_2}{v}\right) - A^2\left(t-\frac{r_1}{v}\right)\right]. \tag{2.357}
$$

Next we find the Poynting flux term:

$$
\begin{aligned}
P_{\text{sphere}}(t) &= -\oint_S (\mathbf{E}\times\mathbf{H})\cdot d\mathbf{S} \\
&= -\int_0^{2\pi} d\phi \int_{\theta_1}^{\theta_2}\left[\frac{1}{r_1}A\left(t-\frac{r_1}{v}\right)\hat{\boldsymbol{\theta}}\right]\times\left[\frac{1}{r_1}\frac{1}{\mu v}A\left(t-\frac{r_1}{v}\right)\hat{\boldsymbol{\phi}}\right]\cdot(-\hat{\mathbf{r}})r_1^2\frac{d\theta}{\sin\theta} - \\
&\quad -\int_0^{2\pi} d\phi \int_{\theta_1}^{\theta_2}\left[\frac{1}{r_2}A\left(t-\frac{r_2}{v}\right)\hat{\boldsymbol{\theta}}\right]\times\left[\frac{1}{r_2}\frac{1}{\mu v}A\left(t-\frac{r_2}{v}\right)\hat{\boldsymbol{\phi}}\right]\cdot\hat{\mathbf{r}}r_2^2\frac{d\theta}{\sin\theta}.
\end{aligned}
$$

The first term represents the power carried by the traveling wave into the volume region by passing through the spherical surface at $r = r_1$, while the second term represents the power carried by the wave out of the region by passing through the surface $r = r_2$. Integration gives

$$
P_{\text{sphere}}(t) = -\frac{2\pi}{\eta}F\left[A^2\left(t-\frac{r_2}{v}\right) - A^2\left(t-\frac{r_1}{v}\right)\right], \tag{2.358}
$$

which matches (2.357), thus verifying Poynting's theorem.

It is also interesting to compute the total energy passing through a surface of radius r_0. From (2.358) we see that the flux of energy (power density) passing outward through the surface $r = r_0$ is

$$
P_{\text{sphere}}(t) = \frac{2\pi}{\eta}FA^2\left(t-\frac{r_0}{v}\right).
$$

The total energy associated with this flux can be computed by integrating over all time: we have

$$
E = \frac{2\pi}{\eta}F\int_{-\infty}^{\infty}A^2\left(t-\frac{r_0}{v}\right)dt = \frac{2\pi}{\eta}F\int_{-\infty}^{\infty}A^2(u)\,du
$$

after making the substitution $u = t - r_0/v$. The total energy passing through a spherical surface is independent of the radius of the sphere. This is an important property of spherical waves. The $1/r$ dependence of the electric and magnetic fields produces a power density that decays with distance in precisely the right proportion to compensate for the r^2-type increase in the surface area through which the power flux passes.

2.10.9 Nonradiating sources

Not all time-dependent sources produce electromagnetic waves. In fact, certain local-ized source distributions produce no fields external to the region containing the sources. Such distributions are said to be *nonradiating*, and the fields they produce (within their source regions) lack wave characteristics.

Let us consider a specific example involving two concentric spheres. The inner sphere, carrying a uniformly distributed total charge $-Q$, is rigid and has a fixed radius a; the outer sphere, carrying uniform charge $+Q$, is a flexible balloon that can be stretched to any radius $b = b(t)$. The two surfaces are initially stationary, some external force being required to hold them in place. Now suppose we apply a time-varying force that results in $b(t)$ changing from $b(t_1) = b_1$ to $b(t_2) = b_2 > b_1$. This creates a radially directed time-varying current $\hat{\mathbf{r}} J_r(\mathbf{r}, t)$. By symmetry J_r depends only on r and produces a field $\mathbf{E}$ that depends only on r and is directed radially. An application of Gauss's law over a sphere of radius $r_0 > b_2$, which contains zero total charge, gives

$$4\pi r_0^2 E_r(r_0, t) = 0,$$

hence $\mathbf{E}(\mathbf{r}, t) = 0$ for $r > r_0$ and all time t. So $\mathbf{E} = 0$ external to the current distribution and no outward traveling wave is produced. Gauss's law also shows that $\mathbf{E} = 0$ inside the rigid sphere, while between the spheres

$$\mathbf{E}(\mathbf{r}, t) = -\hat{\mathbf{r}} \frac{Q}{4\pi\epsilon_0 r^2}.$$

Now work is certainly required to stretch the balloon and overcome the Lorentz force between the two charged surfaces. But an application of Poynting's theorem over a surface enclosing both spheres shows that no energy is carried away by an electromagnetic wave. Where does the expended energy go? The presence of only two nonzero terms in Poynting's theorem clearly indicates that the power term $\int_V \mathbf{E} \cdot \mathbf{J} \, dV$ corresponding to the external work must be balanced exactly by a change in stored energy. As the radius of the balloon increases, so does the region of nonzero field as well as the stored energy.

In free space any current source expressible in the form

$$\mathbf{J}(\mathbf{r}, t) = \nabla \left(\frac{\partial \psi(\mathbf{r}, t)}{\partial t} \right) \tag{2.359}$$

and localized to a volume region V, such as the current in the example above, is nonra-diating. Indeed, Ampere's law states that

$$\nabla \times \mathbf{H} = \epsilon_0 \frac{\partial \mathbf{E}}{\partial t} + \nabla \left(\frac{\partial \psi(\mathbf{r}, t)}{\partial t} \right) \tag{2.360}$$

for $\mathbf{r} \in V$; taking the curl we have

$$\nabla \times (\nabla \times \mathbf{H}) = \epsilon_0 \frac{\partial \nabla \times \mathbf{E}}{\partial t} + \nabla \times \nabla \left(\frac{\partial \psi(\mathbf{r}, t)}{\partial t} \right).$$

But the second term on the right is zero, so

$$\nabla \times (\nabla \times \mathbf{H}) = \epsilon_0 \frac{\partial \nabla \times \mathbf{E}}{\partial t}$$

and this equation holds for all $\mathbf{r}$. By Faraday's law we can rewrite it as

$$\left((\nabla \times \nabla \times) + \frac{1}{c^2} \frac{\partial^2}{\partial t^2} \right) \mathbf{H}(\mathbf{r}, t) = 0.$$

So **H** obeys the homogeneous wave equation everywhere, and $\mathbf{H} = 0$ follows from causality. The laws of Ampere and Faraday may also be combined with (2.359) to show that

$$\left((\nabla \times \nabla \times) + \frac{1}{c^2} \frac{\partial^2}{\partial t^2} \right) \left[\mathbf{E}(\mathbf{r}, t) + \frac{1}{\epsilon_0} \nabla \psi(\mathbf{r}, t) \right] = 0$$

for all **r**. By causality

$$\mathbf{E}(\mathbf{r}, t) = -\frac{1}{\epsilon_0} \nabla \psi(\mathbf{r}, t) \qquad (2.361)$$

everywhere. But since $\psi(\mathbf{r}, t) = 0$ external to V, we must also have $\mathbf{E} = 0$ there. Note that $\mathbf{E} = -\nabla \psi / \epsilon_0$ is consistent with Ampere's law (2.360) provided that $\mathbf{H} = 0$ everywhere.

We see that sources having spherical symmetry such that

$$\mathbf{J}(\mathbf{r}, t) = \hat{\mathbf{r}} J_r(r, t) = \nabla \left(\frac{\partial \psi(r, t)}{\partial t} \right) = \hat{\mathbf{r}} \frac{\partial^2 \psi(r, t)}{\partial r \partial t}$$

obey (2.359) and are therefore nonradiating. Hence the fields associated with any outward traveling spherical wave must possess some angular variation. This holds, for example, for the fields far removed from a time-varying source of finite extent.

As pointed out by Lindell [113], nonradiating sources are not merely hypothetical. The outflowing currents produced by a highly symmetric nuclear explosion in outer space or in a homogeneous atmosphere would produce no electromagnetic field outside the source region. The large electromagnetic-pulse effects discussed in § 2.10.6 are due to inhomogeneities in the earth's atmosphere. We also note that the fields produced by a radiating source $\mathbf{J}^r(\mathbf{r}, t)$ do not change external to the source if we superpose a nonradiating component $\mathbf{J}^{nr}(\mathbf{r}, t)$ to create a new source $\mathbf{J} = \mathbf{J}^{nr} + \mathbf{J}^r$. We say that the two sources $\mathbf{J}$ and $\mathbf{J}^r$ are *equivalent* for the region V external to the sources. This presents difficulties in remote sensing where investigators are often interested in reconstructing an unknown source by probing the fields external to (and usually far away from) the source region. Unique reconstruction is possible only if the fields within the source region are also measured.

For the time harmonic case, Devaney and Wolf [54] provide the most general possible form for a nonradiating source. See § 4.11.9 for details.

2.11   Problems

2.1 Consider the constitutive equations (2.16)–(2.17) relating **E**, **D**, **B**, and **H** in a bianisotropic medium. Using the definition for **P** and **M**, show that the constitutive equations relating **E**, **B**, **P**, and **M** are

$$\mathbf{P} = \left(\frac{1}{c} \bar{\mathbf{P}} - \epsilon_0 \bar{\mathbf{I}} \right) \cdot \mathbf{E} + \bar{\mathbf{L}} \cdot \mathbf{B},$$

$$\mathbf{M} = -\bar{\mathbf{M}} \cdot \mathbf{E} - \left(c \bar{\mathbf{Q}} - \frac{1}{\mu_0} \bar{\mathbf{I}} \right) \cdot \mathbf{B}.$$

Also find the constitutive equations relating **E**, **H**, **P**, and **M**.

2.2 Consider Ampere's law and Gauss's law written in terms of rectangular components in the laboratory frame of reference. Assume that an inertial frame moves with velocity $\mathbf{v} = \hat{x}v$ with respect to the laboratory frame. Using the Lorentz transformation given by (2.73)–(2.76), show that

$$c\mathbf{D}'_\perp = \gamma(c\mathbf{D}_\perp + \boldsymbol{\beta} \times \mathbf{H}_\perp),$$
$$\mathbf{H}'_\perp = \gamma(\mathbf{H}_\perp - \boldsymbol{\beta} \times c\mathbf{D}_\perp),$$
$$\mathbf{J}'_\parallel = \gamma(\mathbf{J}_\parallel - \rho\mathbf{v}),$$
$$\mathbf{J}'_\perp = \mathbf{J}_\perp,$$
$$c\rho' = \gamma(c\rho - \boldsymbol{\beta} \cdot \mathbf{J}),$$

where "$\perp$" means perpendicular to the direction of the velocity and "$\parallel$" means parallel to the direction of the velocity.

2.3 Show that the following quantities are invariant under Lorentz transformation:

(a) $\mathbf{E} \cdot \mathbf{B}$,

(b) $\mathbf{H} \cdot \mathbf{D}$,

(c) $\mathbf{B} \cdot \mathbf{B} - \mathbf{E} \cdot \mathbf{E}/c^2$,

(d) $\mathbf{H} \cdot \mathbf{H} - c^2\mathbf{D} \cdot \mathbf{D}$,

(e) $\mathbf{B} \cdot \mathbf{H} - \mathbf{E} \cdot \mathbf{D}$,

(f) $c\mathbf{B} \cdot \mathbf{D} + \mathbf{E} \cdot \mathbf{H}/c$.

2.4 Show that if $c^2B^2 > E^2$ holds in one reference frame, then it holds in all other reference frames. Repeat for the inequality $c^2B^2 < E^2$.

2.5 Show that if $\mathbf{E} \cdot \mathbf{B} = 0$ and $c^2B^2 > E^2$ holds in one reference frame, then a reference frame may be found such that $\mathbf{E} = 0$. Show that if $\mathbf{E} \cdot \mathbf{B} = 0$ and $c^2B^2 < E^2$ holds in one reference frame, then a reference frame may be found such that $\mathbf{B} = 0$.

2.6 A test charge Q at rest in the laboratory frame experiences a force $\mathbf{F} = Q\mathbf{E}$ as measured by an observer in the laboratory frame. An observer in an inertial frame measures a force on the charge given by $\mathbf{F}' = Q\mathbf{E}' + Q\mathbf{v} \times \mathbf{B}'$. Show that $\mathbf{F} \neq \mathbf{F}'$ and find the formula for converting between $\mathbf{F}$ and $\mathbf{F}'$.

2.7 Consider a material moving with velocity $\mathbf{v}$ with respect to the laboratory frame of reference. When the fields are measured in the moving frame, the material is found to be isotropic with $\mathbf{D}' = \epsilon'\mathbf{E}'$ and $\mathbf{B}' = \mu'\mathbf{H}'$. Show that the fields measured in the laboratory frame are given by (2.107) and (2.108), indicating that the material is bianisotropic when measured in the laboratory frame.

2.8 Show that by assuming $v^2/c^2 \ll 1$ in (2.61)–(2.64) we may obtain (2.111).

2.9 Derive the following expressions that allow us to convert the value of the magnetization measured in the laboratory frame of reference to the value measured in a moving frame:

$$\mathbf{M}'_\perp = \gamma(\mathbf{M}_\perp + \boldsymbol{\beta} \times c\mathbf{P}_\perp), \qquad \mathbf{M}'_\parallel = \mathbf{M}_\parallel.$$

2.10 Beginning with the expressions (2.61)–(2.64) for the field conversions under a first-order Lorentz transformation, show that

$$\mathbf{P}' = \mathbf{P} - \frac{\mathbf{v} \times \mathbf{M}}{c^2}, \qquad \mathbf{M}' = \mathbf{M} + \mathbf{v} \times \mathbf{P}.$$

2.11 Consider a simple isotropic material moving through space with velocity $\mathbf{v}$ relative to the laboratory frame. The relative permittivity and permeability of the material measured in the moving frame are ϵ_r' and μ_r', respectively. Show that the magnetization as measured in the laboratory frame is related to the laboratory frame electric field and magnetic flux density as

$$\mathbf{M} = \frac{\chi_m'}{\mu_0 \mu_r'}\mathbf{B} - \epsilon_0 \left(\chi_e' + \frac{\chi_m'}{\mu_r'} \right) \mathbf{v} \times \mathbf{E}$$

when a first-order Lorentz transformation is used. Here $\chi_e' = \epsilon_r' - 1$ and $\chi_m' = \mu_r' - 1$.

2.12 Consider a simple isotropic material moving through space with velocity $\mathbf{v}$ relative to the laboratory frame. The relative permittivity and permeability of the material measured in the moving frame are ϵ_r' and μ_r', respectively. Derive the formulas for the magnetization and polarization in the laboratory frame in terms of $\mathbf{E}$ and $\mathbf{B}$ measured in the laboratory frame by using the Lorentz transformations (2.128) and (2.129)–(2.132). Show that these expressions reduce to (2.139) and (2.140) under the assumption of a first-order Lorentz transformation ($v^2/c^2 \ll 1$).

2.13 Derive the kinematic form of the large-scale Maxwell–Boffi equations (2.165) and (2.166). Derive the alternative form of the large-scale Maxwell–Boffi equations (2.167) and (2.168).

2.14 Modify the kinematic form of the Maxwell–Boffi equations (2.165)–(2.166) to account for the presence of magnetic sources. Repeat for the alternative forms (2.167)–(2.168).

2.15 Consider a thin magnetic source distribution concentrated near a surface S. The magnetic charge and current densities are given by

$$\rho_m(\mathbf{r}, x, t) = \rho_{ms}(\mathbf{r}, t) f(x, \Delta), \qquad \mathbf{J}_m(\mathbf{r}, x, t) = \mathbf{J}_{ms}(\mathbf{r}, t) f(x, \Delta),$$

where $f(x, \Delta)$ satisfies

$$\int_{-\infty}^{\infty} f(x, \Delta)\, dx = 1.$$

Let $\Delta \to 0$ and derive the boundary conditions on $(\mathbf{E}, \mathbf{D}, \mathbf{B}, \mathbf{H})$ across S.

2.16 Beginning with the kinematic forms of Maxwell's equations (2.177)–(2.178), derive the boundary conditions for a moving surface

$$\hat{\mathbf{n}}_{12} \times (\mathbf{H}_1 - \mathbf{H}_2) + (\hat{\mathbf{n}}_{12} \cdot \mathbf{v})(\mathbf{D}_1 - \mathbf{D}_2) = \mathbf{J}_s,$$
$$\hat{\mathbf{n}}_{12} \times (\mathbf{E}_1 - \mathbf{E}_2) - (\hat{\mathbf{n}}_{12} \cdot \mathbf{v})(\mathbf{B}_1 - \mathbf{B}_2) = -\mathbf{J}_{ms}.$$

2.17 Beginning with Maxwell's equations and the constitutive relationships for a bianisotropic medium (2.19)–(2.20), derive the wave equation for $\mathbf{H}$ (2.313). Specialize the result for the case of an anisotropic medium.

2.18 Consider an isotropic but inhomogeneous material, so that

$$\mathbf{D}(\mathbf{r}, t) = \epsilon(\mathbf{r})\mathbf{E}(\mathbf{r}, t), \qquad \mathbf{B}(\mathbf{r}, t) = \mu(\mathbf{r})\mathbf{H}(\mathbf{r}, t).$$

Show that the wave equations for the fields within this material may be written as

$$\nabla^2 \mathbf{E} - \mu\epsilon\frac{\partial^2 \mathbf{E}}{\partial t^2} + \nabla\left[\mathbf{E} \cdot \left(\frac{\nabla\epsilon}{\epsilon}\right)\right] - (\nabla \times \mathbf{E}) \times \left(\frac{\nabla\mu}{\mu}\right) = \mu\frac{\partial \mathbf{J}}{\partial t} + \nabla\left(\frac{\rho}{\epsilon}\right),$$

$$\nabla^2 \mathbf{H} - \mu\epsilon\frac{\partial^2 \mathbf{H}}{\partial t^2} + \nabla\left[\mathbf{H} \cdot \left(\frac{\nabla\mu}{\mu}\right)\right] - (\nabla \times \mathbf{H}) \times \left(\frac{\nabla\epsilon}{\epsilon}\right) = -\nabla \times \mathbf{J} - \mathbf{J} \times \left(\frac{\nabla\epsilon}{\epsilon}\right).$$

2.19 Consider a homogeneous, isotropic material in which $\mathbf{D} = \epsilon\mathbf{E}$ and $\mathbf{B} = \mu\mathbf{H}$. Using the definitions of the equivalent sources, show that the wave equations (2.322)–(2.323) are equivalent to (2.314)–(2.315).

2.20 When we calculate the force on a conductor produced by an incident plane wave, we often neglect the momentum term

$$\frac{\partial}{\partial t}(\mathbf{D} \times \mathbf{B}).$$

Compute this term for the plane wave field (2.336) in free space at the surface of the conductor and compare to the term obtained from the Maxwell stress tensor (2.341). What is the relative difference in amplitude?

2.21 When a material is only slightly conducting, and thus Ω is very small, we often neglect the third term in the plane wave solution (2.343). Reproduce the plot of Figure 2.8 with this term omitted and compare. Discuss how the omitted term affects the shape of the propagating waveform.

2.22 A total charge Q is evenly distributed over a spherical surface. The surface expands outward at constant velocity so that the radius of the surface is $b = vt$ at time t. (a) Use Gauss's law to find $\mathbf{E}$ everywhere as a function of time. (b) Show that $\mathbf{E}$ may be found from a potential function

$$\psi(\mathbf{r}, t) = \frac{Q}{4\pi r}(r - vt)U(r - vt)$$

according to (2.361). Here $U(t)$ is the unit step function. (c) Write down the form of $\mathbf{J}$ for the expanding sphere and show that since it may be found from (2.359) it is a nonradiating source.

Chapter 3

The static electromagnetic field

3.1 Static fields and steady currents

Perhaps the most carefully studied area of electromagnetics is that in which the fields are time-invariant. This area, known generally as *statics*, offers (1) the most direct opportunities for solution of the governing equations, and (2) the clearest physical pictures of the electromagnetic field. We therefore devote the present chapter to a treatment of static fields. We begin to seek and examine specific solutions to the field equations; however, our selection of examples is shaped by a search for insight into the behavior of the field itself, rather than a desire to catalog the solutions of numerous statics problems.

We note at the outset that a static field is physically sensible only as a limiting case of a time-varying field as the latter approaches a time-invariant equilibrium, and then only in local regions. The static field equations we shall study thus represent an idealized model of the physical fields.

If we examine the Maxwell–Minkowski equations (2.1)–(2.4) and set the time derivatives to zero, we obtain the *static field Maxwell equations*

$$\nabla \times \mathbf{E}(\mathbf{r}) = 0, \tag{3.1}$$

$$\nabla \cdot \mathbf{D}(\mathbf{r}) = \rho(\mathbf{r}), \tag{3.2}$$

$$\nabla \times \mathbf{H}(\mathbf{r}) = \mathbf{J}(\mathbf{r}), \tag{3.3}$$

$$\nabla \cdot \mathbf{B}(\mathbf{r}) = 0. \tag{3.4}$$

We note that if the fields are to be everywhere time-invariant, then the sources $\mathbf{J}$ and ρ must also be everywhere time-invariant. Under this condition the dynamic coupling between the fields described by Maxwell's equations disappears; any connection between $\mathbf{E}$, $\mathbf{D}$, $\mathbf{B}$, and $\mathbf{H}$ imposed by the time-varying nature of the field is gone. For static fields we also require that any dynamic coupling between fields in the constitutive relations vanish. In this *static field limit* we cannot derive the divergence equations from the curl equations, since we can no longer use the initial condition argument that the fields were identically zero prior to some time.

The static field equations are useful for approximating many physical situations in which the fields rapidly settle to a local, macroscopically-static state. This may occur so rapidly and so completely that, in a practical sense, the static equations describe the fields within our ability to measure and to compute. Such is the case when a capacitor is rapidly charged using a battery in series with a resistor; for example, a 1 pF capacitor charging through a 1 Ω resistor reaches 99.99% of its total charge static limit within 10 ps.

3.1.1   Decoupling of the electric and magnetic fields

For the remainder of this chapter we shall assume that there is no coupling between **E** and **H** or between **D** and **B** in the constitutive relations. Then the static equations decouple into two independent sets of equations in terms of two independent sets of fields. The static electric field set (**E**,**D**) is described by the equations

$$\nabla \times \mathbf{E}(\mathbf{r}) = 0, \tag{3.5}$$

$$\nabla \cdot \mathbf{D}(\mathbf{r}) = \rho(\mathbf{r}). \tag{3.6}$$

Integrating these over a stationary contour and surface, respectively, we have the large-scale forms

$$\oint_\Gamma \mathbf{E} \cdot \mathbf{dl} = 0, \tag{3.7}$$

$$\oint_S \mathbf{D} \cdot \mathbf{dS} = \int_V \rho \, dV. \tag{3.8}$$

The static magnetic field set (**B**,**H**) is described by

$$\nabla \times \mathbf{H}(\mathbf{r}) = \mathbf{J}(\mathbf{r}), \tag{3.9}$$

$$\nabla \cdot \mathbf{B}(\mathbf{r}) = 0, \tag{3.10}$$

or, in large-scale form,

$$\oint_\Gamma \mathbf{H} \cdot \mathbf{dl} = \int_S \mathbf{J} \cdot \mathbf{dS}, \tag{3.11}$$

$$\oint_S \mathbf{B} \cdot \mathbf{dS} = 0. \tag{3.12}$$

We can also specialize the Maxwell–Boffi equations to static form. Assuming that the fields, sources, and equivalent sources are time-invariant, the electrostatic field **E**(**r**) is described by the point-form equations

$$\nabla \times \mathbf{E} = 0, \tag{3.13}$$

$$\nabla \cdot \mathbf{E} = \frac{1}{\epsilon_0} \left(\rho - \nabla \cdot \mathbf{P} \right), \tag{3.14}$$

or the equivalent large-scale equations

$$\oint_\Gamma \mathbf{E} \cdot \mathbf{dl} = 0, \tag{3.15}$$

$$\oint_S \mathbf{E} \cdot \mathbf{dS} = \frac{1}{\epsilon_0} \int_V \left(\rho - \nabla \cdot \mathbf{P} \right) dV. \tag{3.16}$$

Similarly, the magnetostatic field **B** is described by

$$\nabla \times \mathbf{B} = \mu_0 \left(\mathbf{J} + \nabla \times \mathbf{M} \right), \tag{3.17}$$

$$\nabla \cdot \mathbf{B} = 0, \tag{3.18}$$

or

$$\oint_\Gamma \mathbf{B} \cdot \mathbf{dl} = \mu_0 \int_S \left(\mathbf{J} + \nabla \times \mathbf{M} \right) \cdot \mathbf{dS}, \tag{3.19}$$

$$\oint_S \mathbf{B} \cdot \mathbf{dS} = 0. \tag{3.20}$$

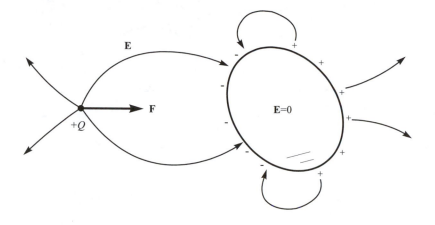

Figure 3.1: Positive point charge in the vicinity of an insulated, uncharged conductor.

It is important to note that any separation of the electromagnetic field into independent static electric and magnetic portions is illusory. As we mentioned in § 2.3.2, the electric and magnetic components of the EM field depend on the motion of the observer. An observer stationary with respect to a single charge measures only a static electric field, while an observer in uniform motion with respect to the charge measures both electric and magnetic fields.

3.1.2 Static field equilibrium and conductors

Suppose we could arrange a group of electric charges into a static configuration in free space. The charges would produce an electric field, resulting in a force on the distribution via the Lorentz force law, and hence would begin to move. Regardless of how we arrange the charges they cannot maintain their original static configuration without the help of some mechanical force to counterbalance the electrical force. This is a statement of Earnshaw's theorem, discussed in detail in § 3.4.2.

The situation is similar for charges within and on electric conductors. A conductor is a material having many charges free to move under external influences, both electric and non-electric. In a metallic conductor, electrons move against a background lattice of positive charges. An *uncharged conductor* is neutral: the amount of negative charge carried by the electrons is equal to the positive charge in the background lattice. The distribution of charges in an uncharged conductor is such that the macroscopic electric field is zero inside and outside the conductor. When the conductor is exposed to an additional electric field, the electrons move under the influence of the Lorentz force, creating a *conduction current*. Rather than accelerating indefinitely, conduction electrons experience collisions with the lattice, thereby giving up their kinetic energy. Macroscopically, the charge motion can be described in terms of a time-average velocity, hence a macroscopic current density can be assigned to the density of moving charge. The relationship between the applied, or "impressed," field and the resulting current density is given by *Ohm's law*; in a linear, isotropic, nondispersive material this is

$$\mathbf{J}(\mathbf{r}, t) = \sigma(\mathbf{r})\mathbf{E}(\mathbf{r}, t). \tag{3.21}$$

The conductivity σ describes the impediment to charge motion through the lattice: the

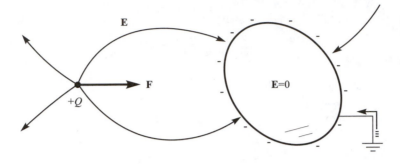

Figure 3.2: Positive point charge near a grounded conductor.

higher the conductivity, the farther an electron may move on average before undergoing
a collision.

Let us examine how a state of equilibrium is established in a conductor. We shall consider several important situations. First, suppose we bring a positively charged particle
into the vicinity of a neutral, insulated conductor (we say that a conductor is "insulated"
if no means exists for depositing excess charge onto the conductor). The Lorentz force
on the free electrons in the conductor results in their motion toward the particle (Figure
3.1). A reaction force **F** attracts the particle to the conductor. If the particle and the
conductor are both held rigidly in space by an external mechanical force, the electrons
within the conductor continue to move toward the surface. In a metal, when these electrons reach the surface and try to continue further they experience a rapid reversal in the
direction of the Lorentz force, drawing them back toward the surface. A sufficiently large
force (described by the *work function* of the metal) will be able to draw these charges
from the surface, but anything less will permit the establishment of a stable equilibrium
at the surface. If σ is large then equilibrium is established quickly, and a nonuniform
static charge distribution appears on the conductor surface. The electric field within the
conductor must settle to zero at equilibrium, since a nonzero field would be associated
with a current $\mathbf{J} = \sigma\mathbf{E}$. In addition, the component of the field tangential to the surface
must be zero or the charge would be forced to move along the surface. *At equilibrium,
the field within and tangential to a conductor must be zero.* Note also that equilibrium
cannot be established without external forces to hold the conductor and particle in place.

Next, suppose we bring a positively charged particle into the vicinity of a *grounded*
(rather than insulated) conductor as in Figure 3.2. Use of the term "grounded" means
that the conductor is attached via a filamentary conductor to a remote reservoir of charge
known as *ground*; in practical applications the earth acts as this charge reservoir. Charges
are drawn from or returned to the reservoir, without requiring any work, in response to
the Lorentz force on the charge within the conducting body. As the particle approaches,
negative charge is drawn to the body and then along the surface until a static equilibrium
is re-established. Unlike the insulated body, the grounded conductor in equilibrium has
excess negative charge, the amount of which depends on the proximity of the particle.
Again, both particle and conductor must be held in place by external mechanical forces,
and the total field produced by both the static charge on the conductor and the particle
must be zero at points interior to the conductor.

Finally, consider the process whereby excess charge placed inside a conducting body
redistributes as equilibrium is established. We assume an isotropic, homogeneous conducting body with permittivity ϵ and conductivity σ. An initially static charge with

density $\rho_0(\mathbf{r})$ is introduced at time $t = 0$. The charge density must obey the continuity equation

$$\nabla \cdot \mathbf{J}(\mathbf{r}, t) = -\frac{\partial \rho(\mathbf{r}, t)}{\partial t};$$

since $\mathbf{J} = \sigma \mathbf{E}$, we have

$$\sigma \nabla \cdot \mathbf{E}(\mathbf{r}, t) = -\frac{\partial \rho(\mathbf{r}, t)}{\partial t}.$$

By Gauss's law, $\nabla \cdot \mathbf{E}$ can be eliminated:

$$\frac{\sigma}{\epsilon} \rho(\mathbf{r}, t) = -\frac{\partial \rho(\mathbf{r}, t)}{\partial t}.$$

Solving this differential equation for the unknown $\rho(\mathbf{r}, t)$ we have

$$\rho(\mathbf{r}, t) = \rho_0(\mathbf{r}) e^{-\sigma t/\epsilon}. \tag{3.22}$$

The charge density within a homogeneous, isotropic conducting body decreases exponentially with time, regardless of the original charge distribution and shape of the body. Of course, the total charge must be constant, and thus charge within the body travels to the surface where it distributes itself in such a way that the field internal to the body approaches zero at equilibrium. The rate at which the volume charge dissipates is determined by the *relaxation time* ϵ/σ; for copper (a good conductor) this is an astonishingly small 10^{-19} s. Even distilled water, a relatively poor conductor, has $\epsilon/\sigma = 10^{-6}$ s. Thus we see how rapidly static equilibrium can be approached.

3.1.3 Steady current

Since time-invariant fields must arise from time-invariant sources, we have from the continuity equation

$$\nabla \cdot \mathbf{J}(\mathbf{r}) = 0. \tag{3.23}$$

In large-scale form this is

$$\oint_S \mathbf{J} \cdot d\mathbf{S} = 0. \tag{3.24}$$

A current with the property (3.23) is said to be a *steady current*. By (3.24), a steady current must be completely lineal (and infinite in extent) or must form closed loops. However, if a current forms loops then the individual moving charges must undergo acceleration (from the change in direction of velocity). Since a single accelerating particle radiates energy in the form of an electromagnetic wave, we might expect a large steady loop current to produce a great deal of radiation. In fact, if we superpose the fields produced by the many particles comprising a steady current, we find that a steady current produces no radiation [91]. Remarkably, to obtain this result we must consider the exact relativistic fields, and thus our finding is precise within the limits of our macroscopic assumptions.

If we try to create a steady current in free space, the flowing charges will tend to disperse because of the Lorentz force from the field set up by the charges, and the resulting current will not form closed loops. A beam of electrons or ions will produce both an electric field (because of the nonzero net charge of the beam) and a magnetic field (because of the current). At nonrelativistic particle speeds, the electric field produces an outward force on the charges that is much greater than the inward (or *pinch*) force produced by the magnetic field. Application of an additional, external force will allow the creation of a *collimated beam* of charge, as occurs in an electron tube where a series of permanent magnets can be used to create a beam of steady current.

More typically, steady currents are created using wire conductors to guide the moving charge. When an external force, such as the electric field created by a battery, is applied to an uncharged conductor, the free electrons will begin to move through the positive lattice, forming a current. Each electron moves only a short distance before colliding with the positive lattice, and if the wire is bent into a loop the resulting macroscopic current will be steady in the sense that the temporally and spatially averaged microscopic current will obey $\nabla \cdot \mathbf{J} = 0$. We note from the examples above that any charges attempting to leave the surface of the wire are drawn back by the electrostatic force produced by the resulting imbalance in electrical charge. For conductors, the "drift" velocity associated with the moving electrons is proportional to the applied field:

$$\mathbf{u}_d = -\mu_e \mathbf{E}$$

where μ_e is the *electron mobility*. The mobility of copper $(3.2 \times 10^{-3} \text{ m}^2/\text{V} \cdot \text{s})$ is such that an applied field of 1 V/m results in a drift velocity of only a third of a centimeter per second.

Integral properties of a steady current.   Steady currents obey several useful integral properties. To develop these properties we need an integral identity. Let $f(\mathbf{r})$ and $g(\mathbf{r})$ be scalar functions, continuous and with continuous derivatives in a volume region V. Let $\mathbf{J}$ represent a steady current field of finite extent, completely contained within V. We begin by using (B.42) to expand

$$\nabla \cdot (fg\mathbf{J}) = fg(\nabla \cdot \mathbf{J}) + \mathbf{J} \cdot \nabla(fg).$$

Noting that $\nabla \cdot \mathbf{J} = 0$ and using (B.41), we get

$$\nabla \cdot (fg\mathbf{J}) = (f\mathbf{J}) \cdot \nabla g + (g\mathbf{J}) \cdot \nabla f.$$

Now let us integrate over V and employ the divergence theorem:

$$\oint_S (fg)\mathbf{J} \cdot d\mathbf{S} = \int_V [(f\mathbf{J}) \cdot \nabla g + (g\mathbf{J}) \cdot \nabla f] \, dV.$$

Since $\mathbf{J}$ is contained entirely within S, we must have $\hat{\mathbf{n}} \cdot \mathbf{J} = 0$ everywhere on S. Hence

$$\int_V [(f\mathbf{J}) \cdot \nabla g + (g\mathbf{J}) \cdot \nabla f] \, dV = 0. \tag{3.25}$$

We can obtain a useful relation by letting $f = 1$ and $g = x_i$ in (3.25), where $(x, y, z) = (x_1, x_2, x_3)$. This gives

$$\int_V J_i(\mathbf{r}) \, dV = 0, \tag{3.26}$$

where $J_1 = J_x$ and so on. Hence the volume integral of any rectangular component of $\mathbf{J}$ is zero. Similarly, letting $f = g = x_i$ we find that

$$\int_V x_i J_i(\mathbf{r}) \, dV = 0. \tag{3.27}$$

With $f = x_i$ and $g = x_j$ we obtain

$$\int_V [x_i \mathbf{J}_j(\mathbf{r}) + x_j \mathbf{J}_i(\mathbf{r})] \, dV = 0. \tag{3.28}$$

3.2 Electrostatics

3.2.1 The electrostatic potential and work

The equation

$$\oint_{\Gamma} \mathbf{E} \cdot \mathbf{dl} = 0 \tag{3.29}$$

satisfied by the electrostatic field $\mathbf{E}(\mathbf{r})$ is particularly interesting. A field with zero circulation is said to be *conservative*. To see why, let us examine the work required to move a particle of charge Q around a closed path in the presence of $\mathbf{E}(\mathbf{r})$. Since work is the line integral of force and $\mathbf{B} = 0$, the work expended by the external system moving the charge against the Lorentz force is

$$W = -\oint_{\Gamma} (Q\mathbf{E} + Q\mathbf{v} \times \mathbf{B}) \cdot \mathbf{dl} = -Q\oint_{\Gamma} \mathbf{E} \cdot \mathbf{dl} = 0.$$

This property is analogous to the conservation property for a classical gravitational field: any potential energy gained by raising a point mass is lost when the mass is lowered.

Direct experimental verification of the electrostatic conservative property is difficult, aside from the fact that the motion of Q may alter $\mathbf{E}$ by interacting with the sources of $\mathbf{E}$. By moving Q with nonuniform velocity (i.e., with acceleration at the beginning of the loop, direction changes in transit, and deceleration at the end) we observe a radiative loss of energy, and this energy cannot be regained by the mechanical system providing the motion. To avoid this problem we may assume that the charge is moved so slowly, or in such small increments, that it does not radiate. We shall use this concept later to determine the "assembly energy" in a charge distribution.

The electrostatic potential. By the point form of (3.29),

$$\nabla \times \mathbf{E}(\mathbf{r}) = 0,$$

we can introduce a scalar field $\Phi = \Phi(\mathbf{r})$ such that

$$\mathbf{E}(\mathbf{r}) = -\nabla\Phi(\mathbf{r}). \tag{3.30}$$

The function Φ carries units of volts and is known as the *electrostatic potential*. Let us consider the work expended by an external agent in moving a charge between points P_1 at $\mathbf{r}_1$ and P_2 at $\mathbf{r}_2$:

$$W_{21} = -Q\int_{P_1}^{P_2} -\nabla\Phi(\mathbf{r}) \cdot \mathbf{dl} = Q\int_{P_1}^{P_2} d\Phi(\mathbf{r}) = Q\left[\Phi(\mathbf{r}_2) - \Phi(\mathbf{r}_1)\right].$$

The work W_{21} is clearly independent of the path taken between P_1 and P_2; the quantity

$$V_{21} = \frac{W_{21}}{Q} = \Phi(\mathbf{r}_2) - \Phi(\mathbf{r}_1) = -\int_{P_1}^{P_2} \mathbf{E} \cdot \mathbf{dl}, \tag{3.31}$$

called the *potential difference*, has an obvious physical meaning as work per unit charge required to move a particle against an electric field between two points.

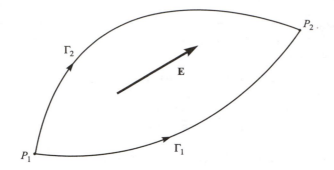

Figure 3.3: Demonstration of path independence of the electric field line integral.

Of course, the large-scale form (3.29) also implies the path-independence of work in the electrostatic field. Indeed, we may pass an arbitrary closed contour Γ through P_1 and P_2 and then split it into two pieces Γ_1 and Γ_2 as shown in Figure 3.3. Since

$$-Q \oint_{\Gamma_1 - \Gamma_2} \mathbf{E} \cdot \mathbf{dl} = -Q \int_{\Gamma_1} \mathbf{E} \cdot \mathbf{dl} + Q \int_{\Gamma_2} \mathbf{E} \cdot \mathbf{dl} = 0,$$

we have

$$-Q \int_{\Gamma_1} \mathbf{E} \cdot \mathbf{dl} = -Q \int_{\Gamma_2} \mathbf{E} \cdot \mathbf{dl}$$

as desired.

We sometimes refer to $\Phi(\mathbf{r})$ as the *absolute electrostatic potential*. Choosing a suitable reference point P_0 at location $\mathbf{r}_0$ and writing the potential difference as

$$V_{21} = [\Phi(\mathbf{r}_2) - \Phi(\mathbf{r}_0)] - [\Phi(\mathbf{r}_1) - \Phi(\mathbf{r}_0)],$$

we can justify calling $\Phi(\mathbf{r})$ the *absolute potential referred to P_0*. Note that P_0 might describe a locus of points, rather than a single point, since many points can be at the same potential. Although we can choose any reference point without changing the resulting value of $\mathbf{E}$ found from (3.30), for simplicity we often choose $\mathbf{r}_0$ such that $\Phi(\mathbf{r}_0) = 0$.

Several properties of the electrostatic potential make it convenient for describing static electric fields. We know that, at equilibrium, the electrostatic field within a conducting body must vanish. By (3.30) the potential at all points within the body must therefore have the same constant value. It follows that the surface of a conductor is an *equipotential surface*: a surface for which $\Phi(\mathbf{r})$ is constant.

As an infinite reservoir of charge that can be tapped through a filamentary conductor, the entity we call "ground" must also be an equipotential object. If we connect a conductor to ground, we have seen that charge may flow freely onto the conductor. Since no work is expended, "grounding" a conductor obviously places the conductor at the same absolute potential as ground. For this reason, ground is often assigned the role as the potential reference with an absolute potential of zero volts. Later we shall see that for sources of finite extent ground must be located at infinity.

3.2.2 Boundary conditions

Boundary conditions for the electrostatic field. The boundary conditions found for the dynamic electric field remain valid in the electrostatic case. Thus

$$\hat{\mathbf{n}}_{12} \times (\mathbf{E}_1 - \mathbf{E}_2) = 0 \tag{3.32}$$

and

$$\hat{\mathbf{n}}_{12} \cdot (\mathbf{D}_1 - \mathbf{D}_2) = \rho_s. \tag{3.33}$$

Here $\hat{\mathbf{n}}_{12}$ points into region 1 from region 2. Because the static curl and divergence equations are independent, so are the boundary conditions (3.32) and (3.33).

For a linear and isotropic dielectric where $\mathbf{D} = \epsilon\mathbf{E}$, equation (3.33) becomes

$$\hat{\mathbf{n}}_{12} \cdot (\epsilon_1 \mathbf{E}_1 - \epsilon_2 \mathbf{E}_2) = \rho_s. \tag{3.34}$$

Alternatively, using $\mathbf{D} = \epsilon_0 \mathbf{E} + \mathbf{P}$ we can write (3.33) as

$$\hat{\mathbf{n}}_{12} \cdot (\mathbf{E}_1 - \mathbf{E}_2) = \frac{1}{\epsilon_0} \left(\rho_s + \rho_{Ps1} + \rho_{Ps2} \right) \tag{3.35}$$

where

$$\rho_{Ps} = \hat{\mathbf{n}} \cdot \mathbf{P}$$

is the polarization surface charge with $\hat{\mathbf{n}}$ pointing outward from the material body.

We can also write the boundary conditions in terms of the electrostatic potential. With $\mathbf{E} = -\nabla\Phi$, equation (3.32) becomes

$$\Phi_1(\mathbf{r}) = \Phi_2(\mathbf{r}) \tag{3.36}$$

for all points $\mathbf{r}$ on the surface. Actually Φ_1 and Φ_2 may differ by a constant; because this constant is eliminated when the gradient is taken to find $\mathbf{E}$, it is generally ignored. We can write (3.35) as

$$\epsilon_0 \left(\frac{\partial \Phi_1}{\partial n} - \frac{\partial \Phi_2}{\partial n} \right) = -\rho_s - \rho_{Ps1} - \rho_{Ps2}$$

where the normal derivative is taken in the $\hat{\mathbf{n}}_{12}$ direction. For a linear, isotropic dielectric (3.33) becomes

$$\epsilon_1 \frac{\partial \Phi_1}{\partial n} - \epsilon_2 \frac{\partial \Phi_2}{\partial n} = -\rho_s. \tag{3.37}$$

Again, we note that (3.36) and (3.37) are independent.

Boundary conditions for steady electric current. The boundary condition on the normal component of current found in § 2.8.2 remains valid in the steady current case. Assume that the boundary exists between two linear, isotropic conducting regions having constitutive parameters (ϵ_1, σ_1) and (ϵ_2, σ_2), respectively. By (2.198) we have

$$\hat{\mathbf{n}}_{12} \cdot (\mathbf{J}_1 - \mathbf{J}_2) = -\nabla_s \cdot \mathbf{J}_s \tag{3.38}$$

where $\hat{\mathbf{n}}_{12}$ points into region 1 from region 2. A surface current will not appear on the boundary between two regions having finite conductivity, although a surface charge may accumulate there during the transient period when the currents are established [31]. If charge is influenced to move from the surface, it will move into the adjacent regions,

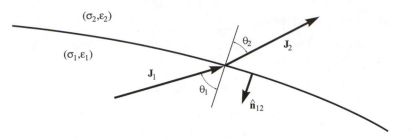

Figure 3.4: Refraction of steady current at a material interface.

rather than along the surface, and a new charge will replace it, supplied by the current. Thus, for finite conducting regions (3.38) becomes

$$\hat{\mathbf{n}}_{12} \cdot (\mathbf{J}_1 - \mathbf{J}_2) = 0. \tag{3.39}$$

A boundary condition on the tangential component of current can also be found. Substituting $\mathbf{E} = \mathbf{J}/\sigma$ into (3.32) we have

$$\hat{\mathbf{n}}_{12} \times \left(\frac{\mathbf{J}_1}{\sigma_1} - \frac{\mathbf{J}_2}{\sigma_2} \right) = 0.$$

We can also write this as

$$\frac{J_{1t}}{\sigma_1} = \frac{J_{2t}}{\sigma_2} \tag{3.40}$$

where

$$\mathbf{J}_{1t} = \hat{\mathbf{n}}_{12} \times \mathbf{J}_1, \qquad \mathbf{J}_{2t} = \hat{\mathbf{n}}_{12} \times \mathbf{J}_2.$$

We may combine the boundary conditions for the normal components of current and electric field to better understand the behavior of current at a material boundary. Substituting $\mathbf{E} = \mathbf{J}/\sigma$ into (3.34) we have

$$\frac{\epsilon_1}{\sigma_1} J_{1n} - \frac{\epsilon_2}{\sigma_2} J_{2n} = \rho_s \tag{3.41}$$

where $J_{1n} = \hat{\mathbf{n}}_{12} \cdot \mathbf{J}_1$ and $J_{2n} = \hat{\mathbf{n}}_{12} \cdot \mathbf{J}_2$. Combining (3.41) with (3.39), we have

$$\rho_s = J_{1n} \left(\frac{\epsilon_1}{\sigma_1} - \frac{\epsilon_2}{\sigma_2} \right) = E_{1n} \left(\epsilon_1 - \frac{\sigma_1}{\sigma_2} \epsilon_2 \right) = J_{2n} \left(\frac{\epsilon_1}{\sigma_1} - \frac{\epsilon_2}{\sigma_2} \right) = E_{2n} \left(\epsilon_1 \frac{\sigma_2}{\sigma_1} - \epsilon_2 \right)$$

where

$$E_{1n} = \hat{\mathbf{n}}_{12} \cdot \mathbf{E}_1, \qquad E_{2n} = \hat{\mathbf{n}}_{12} \cdot \mathbf{E}_2.$$

Unless $\epsilon_1 \sigma_2 - \sigma_1 \epsilon_2 = 0$, a surface charge will exist on the interface between dissimilar current-carrying conductors.

We may also combine the vector components of current on each side of the boundary to determine the effects of the boundary on current direction (Figure 3.4). Let $\theta_{1,2}$ denote the angle between $\mathbf{J}_{1,2}$ and $\hat{\mathbf{n}}_{12}$ so that

$$J_{1n} = J_1 \cos \theta_1, \qquad J_{1t} = J_1 \sin \theta_1$$
$$J_{2n} = J_2 \cos \theta_2, \qquad J_{2t} = J_2 \sin \theta_2.$$

Then $J_1 \cos\theta_1 = J_2 \cos\theta_2$ by (3.39), while $\sigma_2 J_1 \sin\theta_1 = \sigma_1 J_2 \sin\theta_2$ by (3.40). Hence

$$\sigma_2 \tan\theta_1 = \sigma_1 \tan\theta_2. \tag{3.42}$$

It is interesting to consider the case of current incident from a conducting material onto an insulating material. If region 2 is an insulator, then $J_{2n} = J_{2t} = 0$; by (3.39) we have $J_{1n} = 0$. But (3.40) does not require $J_{1t} = 0$; with $\sigma_2 = 0$ the right-hand side of (3.40) is indeterminate and thus J_{1t} may be nonzero. In other words, when current moving through a conductor approaches an insulating surface, it bends and flows tangential to the surface. This concept is useful in explaining how wires guide current.

Interestingly, (3.42) shows that when $\sigma_2 \ll \sigma_1$ we have $\theta_2 \to 0$; current passing from a conducting region into a slightly-conducting region does so normally.

3.2.3 Uniqueness of the electrostatic field

In § 2.2.1 we found that the electromagnetic field is unique within a region V when the tangential component of $\mathbf{E}$ is specified over the surrounding surface. Unfortunately, this condition is not appropriate in the electrostatic case. We should remember that an additional requirement for uniqueness of solution to Maxwell's equations is that the field be specified throughout V at some time t_0. For a static field this would completely determine $\mathbf{E}$ without need for the surface field!

Let us determine conditions for uniqueness beginning with the static field equations. Consider a region V surrounded by a surface S. Static charge may be located entirely or partially within V, or entirely outside V, and produces a field within V. The region may also contain any arrangement of conductors or other materials. Suppose $(\mathbf{D}_1, \mathbf{E}_1)$ and $(\mathbf{D}_2, \mathbf{E}_2)$ represent solutions to the static field equations within V with source $\rho(\mathbf{r})$. We wish to find conditions that guarantee both $\mathbf{E}_1 = \mathbf{E}_2$ and $\mathbf{D}_1 = \mathbf{D}_2$.

Since $\nabla \cdot \mathbf{D}_1 = \rho$ and $\nabla \cdot \mathbf{D}_2 = \rho$, the difference field $\mathbf{D}_0 = \mathbf{D}_2 - \mathbf{D}_1$ obeys the homogeneous equation

$$\nabla \cdot \mathbf{D}_0 = 0. \tag{3.43}$$

Consider the quantity

$$\nabla \cdot (\mathbf{D}_0 \Phi_0) = \Phi_0 (\nabla \cdot \mathbf{D}_0) + \mathbf{D}_0 \cdot (\nabla \Phi_0)$$

where $\mathbf{E}_0 = \mathbf{E}_2 - \mathbf{E}_1 = -\nabla\Phi_0 = -\nabla(\Phi_2 - \Phi_1)$. We integrate over V and use the divergence theorem and (3.43) to obtain

$$\oint_S \Phi_0 \left[\mathbf{D}_0 \cdot \hat{\mathbf{n}}\right] dS = \int_V \mathbf{D}_0 \cdot (\nabla\Phi_0)\, dV = -\int_V \mathbf{D}_0 \cdot \mathbf{E}_0\, dV. \tag{3.44}$$

Now suppose that $\Phi_0 = 0$ everywhere on S, or that $\hat{\mathbf{n}} \cdot \mathbf{D}_0 = 0$ everywhere on S, or that $\Phi_0 = 0$ over part of S and $\hat{\mathbf{n}} \cdot \mathbf{D}_0 = 0$ elsewhere on S. Then

$$\int_V \mathbf{D}_0 \cdot \mathbf{E}_0\, dV = 0. \tag{3.45}$$

Since V is arbitrary, either $\mathbf{D}_0 = 0$ or $\mathbf{E}_0 = 0$. Assuming $\mathbf{E}$ and $\mathbf{D}$ are linked by the constitutive relations, we have $\mathbf{E}_1 = \mathbf{E}_2$ and $\mathbf{D}_1 = \mathbf{D}_2$.

Hence the fields within V are unique provided that either Φ, the normal component of $\mathbf{D}$, or some combination of the two, is specified over S. We often use a multiply-connected surface to exclude conductors. By (3.33) we see that specification of the normal component of $\mathbf{D}$ on a conductor is equivalent to specification of the surface charge density. Thus we must specify the potential or surface charge density over all conducting surfaces.

One other condition results in zero on the left-hand side of (3.44). If S recedes to infinity and Φ_0 and $\mathbf{D}_0$ decrease sufficiently fast, then (3.45) still holds and uniqueness is guaranteed. If $\mathbf{D}, \mathbf{E} \sim 1/r^2$ as $r \to \infty$, then $\Phi \sim 1/r$ and the surface integral in (3.44) tends to zero since the area of an expanding sphere increases only as r^2. We shall find later in this section that for sources of finite extent the fields do indeed vary inversely with distance squared from the source, hence we may allow S to expand and encompass all space.

For the case in which conducting bodies are immersed in an infinite homogeneous medium and the static fields must be determined throughout all space, a multiply-connected surface is used with one part receding to infinity and the remaining parts surrounding the conductors. Here uniqueness is guaranteed by specifying the potentials or charges on the surfaces of the conducting bodies.

3.2.4   Poisson's and Laplace's equations

For computational purposes it is often convenient to deal with the differential versions

$$\nabla \times \mathbf{E}(\mathbf{r}) = 0, \tag{3.46}$$

$$\nabla \cdot \mathbf{D}(\mathbf{r}) = \rho(\mathbf{r}), \tag{3.47}$$

of the electrostatic field equations. We must supplement these with constitutive relations between $\mathbf{E}$ and $\mathbf{D}$; at this point we focus our attention on linear, isotropic materials for which

$$\mathbf{D}(\mathbf{r}) = \epsilon(\mathbf{r})\mathbf{E}(\mathbf{r}).$$

Using this in (3.47) along with $\mathbf{E} = -\nabla\Phi$ (justified by (3.46)), we can write

$$\nabla \cdot [\epsilon(\mathbf{r})\nabla\Phi(\mathbf{r})] = -\rho(\mathbf{r}). \tag{3.48}$$

This is *Poisson's equation*. The corresponding homogeneous equation

$$\nabla \cdot [\epsilon(\mathbf{r})\nabla\Phi(\mathbf{r})] = 0, \tag{3.49}$$

holding at points $\mathbf{r}$ where $\rho(\mathbf{r}) = 0$, is *Laplace's equation*. Equations (3.48) and (3.49) are valid for inhomogeneous media. By (B.42) we can write

$$\nabla\Phi(\mathbf{r}) \cdot \nabla\epsilon(\mathbf{r}) + \epsilon(\mathbf{r})\nabla \cdot [\nabla\Phi(\mathbf{r})] = -\rho(\mathbf{r}).$$

For a homogeneous medium, $\nabla\epsilon = 0$; since $\nabla \cdot (\nabla\Phi) \equiv \nabla^2\Phi$, we have

$$\nabla^2\Phi(\mathbf{r}) = -\rho(\mathbf{r})/\epsilon \tag{3.50}$$

in such a medium. Correspondingly,

$$\nabla^2\Phi(\mathbf{r}) = 0$$

at points where $\rho(\mathbf{r}) = 0$.

Poisson's and Laplace's equations can be solved by separation of variables, Fourier transformation, conformal mapping, and numerical techniques such as the finite difference and moment methods. In Appendix A we consider the separation of variables solution to Laplace's equation in three major coordinate systems for a variety of problems. For an introduction to numerical techniques the reader is referred to the books by Sadiku [162], Harrington [82], and Peterson et al. [146]. Solution to Poisson's equation is often undertaken using the method of Green's functions, which we shall address later in this section. We shall also consider the solution to Laplace's equation for bodies immersed in an applied, or "impressed," field.

Uniqueness of solution to Poisson's equation. Before attempting any solutions, we must ask two very important questions. How do we know that solving the second-order differential equation produces the same values for $\mathbf{E} = -\nabla\Phi$ as solving the first-order equations directly for $\mathbf{E}$? And, if these solutions are the same, what are the conditions for uniqueness of solution to Poisson's and Laplace's equations? To answer the first question, a sufficient condition is to have Φ twice differentiable. We shall not attempt to prove this, but shall instead show that the condition for uniqueness of the second-order equations is the same as that for the first-order equations.

Consider a region of space V surrounded by a surface S. Static charge may be located entirely or partially within V, or entirely outside V, and produces a field within V. This region may also contain any arrangement of conductors or other materials. Now, assume that Φ_1 and Φ_2 represent solutions to the static field equations within V with source $\rho(\mathbf{r})$. We wish to find conditions under which $\Phi_1 = \Phi_2$.

Since we have

$$\nabla \cdot [\epsilon(\mathbf{r})\nabla\Phi_1(\mathbf{r})] = -\rho(\mathbf{r}), \qquad \nabla \cdot [\epsilon(\mathbf{r})\nabla\Phi_2(\mathbf{r})] = -\rho(\mathbf{r}),$$

the difference field $\Phi_0 = \Phi_2 - \Phi_1$ obeys

$$\nabla \cdot [\epsilon(\mathbf{r})\nabla\Phi_0(\mathbf{r})] = 0. \tag{3.51}$$

That is, Φ_0 obeys Laplace's equation. Now consider the quantity

$$\nabla \cdot (\epsilon\Phi_0\nabla\Phi_0) = \epsilon|\nabla\Phi_0|^2 + \Phi_0\nabla \cdot (\epsilon\nabla\Phi_0).$$

Integration over V and use of the divergence theorem and (3.51) gives

$$\oint_S \Phi_0(\mathbf{r})\,[\epsilon(\mathbf{r})\nabla\Phi_0(\mathbf{r})] \cdot d\mathbf{S} = \int_V \epsilon(\mathbf{r})|\nabla\Phi_0(\mathbf{r})|^2\,dV.$$

As with the first order equations, we see that specifying either $\Phi(\mathbf{r})$ or $\epsilon(\mathbf{r})\nabla\Phi(\mathbf{r})\cdot\hat{\mathbf{n}}$ over S results in $\Phi_0(\mathbf{r}) = 0$ throughout V, hence $\Phi_1 = \Phi_2$. As before, specifying $\epsilon(\mathbf{r})\nabla\Phi(\mathbf{r})\cdot\hat{\mathbf{n}}$ for a conducting surface is equivalent to specifying the surface charge on S.

Integral solution to Poisson's equation: the static Green's function. The method of Green's functions is one of the most useful techniques for solving Poisson's equation. We seek a solution for a single point source, then use Green's second identity to write the solution for an arbitrary charge distribution in terms of a superposition integral.

We seek the solution to Poisson's equation for a region of space V as shown in Figure 3.5. The region is assumed homogeneous with permittivity ϵ, and its surface is multiply-connected, consisting of a bounding surface S_B and any number of closed surfaces internal to V. We denote by S the composite surface consisting of S_B and the N internal surfaces S_n, $n = 1, \ldots, N$. The internal surfaces are used to exclude material bodies, such as the

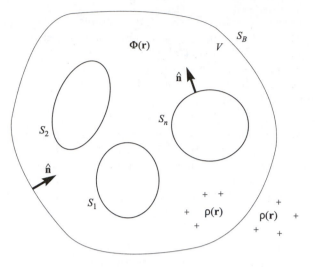

Figure 3.5: Computation of potential from known sources and values on bounding surfaces.

plates of a capacitor, which may be charged and on which the potential is assumed to be known. To solve for $\Phi(\mathbf{r})$ within V we must know the potential produced by a point source. This potential, called the *Green's function*, is denoted $G(\mathbf{r}|\mathbf{r}')$; it has two arguments because it satisfies Poisson's equation at $\mathbf{r}$ when the source is located at $\mathbf{r}'$:

$$\nabla^2 G(\mathbf{r}|\mathbf{r}') = -\delta(\mathbf{r} - \mathbf{r}'). \tag{3.52}$$

Later we shall demonstrate that in all cases of interest to us the Green's function is symmetric in its arguments:

$$G(\mathbf{r}'|\mathbf{r}) = G(\mathbf{r}|\mathbf{r}'). \tag{3.53}$$

This property of G is known as *reciprocity*.

Our development rests on the mathematical result (B.30) known as *Green's second identity*. We can derive this by subtracting the identities

$$\nabla \cdot (\phi \nabla \psi) = \phi \nabla \cdot (\nabla \psi) + (\nabla \phi) \cdot (\nabla \psi),$$
$$\nabla \cdot (\psi \nabla \phi) = \psi \nabla \cdot (\nabla \phi) + (\nabla \psi) \cdot (\nabla \phi),$$

to obtain

$$\nabla \cdot (\phi \nabla \psi - \psi \nabla \phi) = \phi \nabla^2 \psi - \psi \nabla^2 \phi.$$

Integrating this over a volume region V with respect to the dummy variable $\mathbf{r}'$ and using the divergence theorem, we obtain

$$\int_V [\phi(\mathbf{r}')\nabla'^2 \psi(\mathbf{r}') - \psi(\mathbf{r}')\nabla'^2 \phi(\mathbf{r}')]\, dV' = -\oint_S [\phi(\mathbf{r}')\nabla'\psi(\mathbf{r}') - \psi(\mathbf{r}')\nabla'\phi(\mathbf{r}')] \cdot d\mathbf{S}'.$$

The negative sign on the right-hand side occurs because $\hat{\mathbf{n}}$ is an *inward* normal to V. Finally, since $\partial \psi(\mathbf{r}')/\partial n' = \hat{\mathbf{n}}' \cdot \nabla' \psi(\mathbf{r}')$, we have

$$\int_V [\phi(\mathbf{r}')\nabla'^2 \psi(\mathbf{r}') - \psi(\mathbf{r}')\nabla'^2 \phi(\mathbf{r}')]\, dV' = -\oint_S \left[\phi(\mathbf{r}')\frac{\partial \psi(\mathbf{r}')}{\partial n'} - \psi(\mathbf{r}')\frac{\partial \phi(\mathbf{r}')}{\partial n'} \right] dS'$$

as desired.

To solve for Φ in V we shall make some seemingly unmotivated substitutions into this identity. First note that by (3.52) and (3.53) we can write

$$\nabla'^2 G(\mathbf{r}|\mathbf{r}') = -\delta(\mathbf{r}' - \mathbf{r}). \tag{3.54}$$

We now set $\phi(\mathbf{r}') = \Phi(\mathbf{r}')$ and $\psi(\mathbf{r}') = G(\mathbf{r}|\mathbf{r}')$ to obtain

$$\int_V [\Phi(\mathbf{r}')\nabla'^2 G(\mathbf{r}|\mathbf{r}') - G(\mathbf{r}|\mathbf{r}')\nabla'^2 \Phi(\mathbf{r}')]\, dV' =$$
$$-\oint_S \left[\Phi(\mathbf{r}')\frac{\partial G(\mathbf{r}|\mathbf{r}')}{\partial n'} - G(\mathbf{r}|\mathbf{r}')\frac{\partial \Phi(\mathbf{r}')}{\partial n'}\right] dS', \tag{3.55}$$

hence

$$\int_V \left[\Phi(\mathbf{r}')\delta(\mathbf{r}' - \mathbf{r}) - G(\mathbf{r}|\mathbf{r}')\frac{\rho(\mathbf{r}')}{\epsilon}\right] dV' = \oint_S \left[\Phi(\mathbf{r}')\frac{\partial G(\mathbf{r}|\mathbf{r}')}{\partial n'} - G(\mathbf{r}|\mathbf{r}')\frac{\partial \Phi(\mathbf{r}')}{\partial n'}\right] dS'.$$

By the sifting property of the Dirac delta

$$\Phi(\mathbf{r}) = \int_V G(\mathbf{r}|\mathbf{r}')\frac{\rho(\mathbf{r}')}{\epsilon}\, dV' + \oint_{S_B} \left[\Phi(\mathbf{r}')\frac{\partial G(\mathbf{r}|\mathbf{r}')}{\partial n'} - G(\mathbf{r}|\mathbf{r}')\frac{\partial \Phi(\mathbf{r}')}{\partial n'}\right] dS' +$$
$$+ \sum_{n=1}^{N} \oint_{S_n} \left[\Phi(\mathbf{r}')\frac{\partial G(\mathbf{r}|\mathbf{r}')}{\partial n'} - G(\mathbf{r}|\mathbf{r}')\frac{\partial \Phi(\mathbf{r}')}{\partial n'}\right] dS'. \tag{3.56}$$

With this we may compute the potential anywhere within V in terms of the charge density within V and the values of the potential and its normal derivative over S. We must simply determine $G(\mathbf{r}|\mathbf{r}')$ first.

Let us take a moment to specialize (3.56) to the case of unbounded space. Provided that the sources are of finite extent, as $S_B \to \infty$ we shall find that

$$\Phi(\mathbf{r}) = \int_V G(\mathbf{r}|\mathbf{r}')\frac{\rho(\mathbf{r}')}{\epsilon}\, dV' + \sum_{n=1}^{N} \oint_{S_n} \left[\Phi(\mathbf{r}')\frac{\partial G(\mathbf{r}|\mathbf{r}')}{\partial n'} - G(\mathbf{r}|\mathbf{r}')\frac{\partial \Phi(\mathbf{r}')}{\partial n'}\right] dS'.$$

A useful derivative identity. Many differential operations on the displacement vector $\mathbf{R} = \mathbf{r} - \mathbf{r}'$ occur in the study of electromagnetics. The identities

$$\nabla R = -\nabla' R = \hat{\mathbf{R}}, \qquad \nabla\left(\frac{1}{R}\right) = -\nabla'\left(\frac{1}{R}\right) = -\frac{\hat{\mathbf{R}}}{R^2}, \tag{3.57}$$

for example, follow from direct differentiation of the rectangular coordinate representation

$$\mathbf{R} = \hat{\mathbf{x}}(x - x') + \hat{\mathbf{y}}(y - y') + \hat{\mathbf{z}}(z - z').$$

The identity

$$\nabla^2\left(\frac{1}{R}\right) = -4\pi\delta(\mathbf{r} - \mathbf{r}'), \tag{3.58}$$

crucial to potential theory, is more difficult to establish. We shall prove the equivalent version

$$\nabla'^2\left(\frac{1}{R}\right) = -4\pi\delta(\mathbf{r}' - \mathbf{r})$$

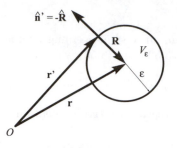

Figure 3.6: Geometry for establishing the singular property of $\nabla^2(1/R)$.

by showing that

$$\int_V f(\mathbf{r}')\nabla'^2 \left(\frac{1}{R}\right) dV' = \begin{cases} -4\pi f(\mathbf{r}), & \mathbf{r} \in V, \\ 0, & \mathbf{r} \notin V, \end{cases} \qquad (3.59)$$

holds for any continuous function $f(\mathbf{r})$. By direct differentiation we have

$$\nabla'^2 \left(\frac{1}{R}\right) = 0 \text{ for } \mathbf{r}' \neq \mathbf{r},$$

hence the second part of (3.59) is established. This also shows that if $\mathbf{r} \in V$ then the domain of integration in (3.59) can be restricted to a sphere of arbitrarily small radius ε centered at $\mathbf{r}$ (Figure 3.6). The result we seek is found in the limit as $\varepsilon \to 0$. Thus we are interested in computing

$$\int_V f(\mathbf{r}')\nabla'^2 \left(\frac{1}{R}\right) dV' = \lim_{\varepsilon \to 0} \int_{V_\varepsilon} f(\mathbf{r}')\nabla'^2 \left(\frac{1}{R}\right) dV'.$$

Since f is continuous at $\mathbf{r}' = \mathbf{r}$, we have by the mean value theorem

$$\int_V f(\mathbf{r}')\nabla'^2 \left(\frac{1}{R}\right) dV' = f(\mathbf{r}) \lim_{\varepsilon \to 0} \int_{V_\varepsilon} \nabla'^2 \left(\frac{1}{R}\right) dV'.$$

The integral over V_ε can be computed using $\nabla'^2(1/R) = \nabla' \cdot \nabla'(1/R)$ and the divergence theorem:

$$\int_{V_\varepsilon} \nabla'^2 \left(\frac{1}{R}\right) dV' = \int_{S_\varepsilon} \hat{\mathbf{n}}' \cdot \nabla' \left(\frac{1}{R}\right) dS',$$

where S_ε bounds V_ε. Noting that $\hat{\mathbf{n}}' = -\hat{\mathbf{R}}$, using (3.57), and writing the integral in spherical coordinates $(\varepsilon, \theta, \phi)$ centered at the point $\mathbf{r}$, we have

$$\int_V f(\mathbf{r}')\nabla'^2 \left(\frac{1}{R}\right) dV' = f(\mathbf{r}) \lim_{\varepsilon \to 0} \int_0^{2\pi} \int_0^\pi -\hat{\mathbf{R}} \cdot \left(\frac{\hat{\mathbf{R}}}{\varepsilon^2}\right) \varepsilon^2 \sin\theta \, d\theta \, d\phi = -4\pi f(\mathbf{r}).$$

Hence the first part of (3.59) is also established.

The Green's function for unbounded space. In view of (3.58), one solution to (3.52) is

$$G(\mathbf{r}|\mathbf{r}') = \frac{1}{4\pi|\mathbf{r} - \mathbf{r}'|}. \qquad (3.60)$$

This simple Green's function is generally used to find the potential produced by charge in unbounded space. Here $N = 0$ (no internal surfaces) and $S_B \to \infty$. Thus

$$\Phi(\mathbf{r}) = \int_V G(\mathbf{r}|\mathbf{r}') \frac{\rho(\mathbf{r}')}{\epsilon} \, dV' + \lim_{S_B \to \infty} \oint_{S_B} \left[\Phi(\mathbf{r}') \frac{\partial G(\mathbf{r}|\mathbf{r}')}{\partial n'} - G(\mathbf{r}|\mathbf{r}') \frac{\partial \Phi(\mathbf{r}')}{\partial n'} \right] dS'.$$

We have seen that the Green's function varies inversely with distance from the source, and thus expect that, as a superposition of point-source potentials, $\Phi(\mathbf{r})$ will also vary inversely with distance from a source of finite extent as that distance becomes large with respect to the size of the source. The normal derivatives then vary inversely with distance squared. Thus, each term in the surface integrand will vary inversely with distance cubed, while the surface area itself varies with distance squared. The result is that the surface integral vanishes as the surface recedes to infinity, giving

$$\Phi(\mathbf{r}) = \int_V G(\mathbf{r}|\mathbf{r}') \frac{\rho(\mathbf{r}')}{\epsilon} \, dV'.$$

By (3.60) we then have

$$\Phi(\mathbf{r}) = \frac{1}{4\pi\epsilon} \int_V \frac{\rho(\mathbf{r}')}{|\mathbf{r} - \mathbf{r}'|} \, dV' \tag{3.61}$$

where the integration is performed over all of space. Since

$$\lim_{\mathbf{r} \to \infty} \Phi(\mathbf{r}) = 0,$$

points at infinity are a convenient reference for the absolute potential.

Later we shall need to know the amount of work required to move a charge Q from infinity to a point P located at $\mathbf{r}$. If a potential field is produced by charge located in unbounded space, moving an additional charge into position requires the work

$$W_{21} = -Q \int_\infty^P \mathbf{E} \cdot d\mathbf{l} = Q[\Phi(\mathbf{r}) - \Phi(\infty)] = Q\Phi(\mathbf{r}). \tag{3.62}$$

Coulomb's law. We can obtain $\mathbf{E}$ from (3.61) by direct differentiation. We have

$$\mathbf{E}(\mathbf{r}) = -\frac{1}{4\pi\epsilon} \nabla \int_V \frac{\rho(\mathbf{r}')}{|\mathbf{r} - \mathbf{r}'|} \, dV' = -\frac{1}{4\pi\epsilon} \int_V \rho(\mathbf{r}') \nabla \left(\frac{1}{|\mathbf{r} - \mathbf{r}'|} \right) \, dV',$$

hence

$$\mathbf{E}(\mathbf{r}) = \frac{1}{4\pi\epsilon} \int_V \rho(\mathbf{r}') \frac{\mathbf{r} - \mathbf{r}'}{|\mathbf{r} - \mathbf{r}'|^3} \, dV' \tag{3.63}$$

by (3.57). So Coulomb's law follows from the two fundamental postulates of electrostatics (3.5) and (3.6).

Green's function for unbounded space: two dimensions. We define the two-dimensional Green's function as the potential at a point $\mathbf{r} = \boldsymbol{\rho} + \hat{\mathbf{z}}z$ produced by a z-directed line source of constant density located at $\mathbf{r}' = \boldsymbol{\rho}'$. Perhaps the simplest way to compute this is to first find $\mathbf{E}$ produced by a line source on the z-axis. By (3.63) we have

$$\mathbf{E}(\mathbf{r}) = \frac{1}{4\pi\epsilon} \int_\Gamma \rho_l(z') \frac{\mathbf{r} - \mathbf{r}'}{|\mathbf{r} - \mathbf{r}'|^3} \, dl'.$$

Then, since $\mathbf{r} = \hat{\mathbf{z}}z + \hat{\boldsymbol{\rho}}\rho$, $\mathbf{r}' = \hat{\mathbf{z}}z'$, and $dl' = dz'$, we have

$$\mathbf{E}(\rho) = \frac{\rho_l}{4\pi\epsilon} \int_{-\infty}^{\infty} \frac{\hat{\boldsymbol{\rho}}\rho + \hat{\mathbf{z}}(z - z')}{[\rho^2 + (z - z')^2]^{3/2}} \, dz'.$$

Carrying out the integration we find that $\mathbf{E}$ has only a ρ-component which varies only with ρ:

$$\mathbf{E}(\rho) = \hat{\boldsymbol{\rho}}\frac{\rho_l}{2\pi\epsilon\rho}. \tag{3.64}$$

The absolute potential referred to a radius ρ_0 can be found by computing the line integral of $\mathbf{E}$ from ρ to ρ_0:

$$\Phi(\rho) = -\frac{\rho_l}{2\pi\epsilon} \int_{\rho_0}^{\rho} \frac{d\rho'}{\rho'} = \frac{\rho_l}{2\pi\epsilon} \ln\left(\frac{\rho_0}{\rho}\right).$$

We may choose any reference point ρ_0 except $\rho_0 = 0$ or $\rho_0 = \infty$. This choice is equivalent to the addition of an arbitrary constant, hence we can also write

$$\Phi(\rho) = \frac{\rho_l}{2\pi\epsilon} \ln\left(\frac{1}{\rho}\right) + C. \tag{3.65}$$

The potential for a general two-dimensional charge distribution in unbounded space is by superposition

$$\Phi(\rho) = \int_{S_T} \frac{\rho_T(\rho')}{\epsilon} G(\rho|\rho') \, dS', \tag{3.66}$$

where the Green's function is the potential of a unit line source located at ρ':

$$G(\rho|\rho') = \frac{1}{2\pi} \ln\left(\frac{\rho_0}{|\rho - \rho'|}\right). \tag{3.67}$$

Here S_T denotes the transverse (xy) plane, and ρ_T denotes the two-dimensional charge distribution (C/m^2) within that plane.

   We note that the potential field (3.66) of a two-dimensional source decreases logarithmically with distance. Only the potential produced by a source of finite extent decreases inversely with distance.

Dirichlet and Neumann Green's functions.   The unbounded space Green's function may be inconvenient for expressing the potential in a region having internal surfaces. In fact, (3.56) shows that to use this function we would be forced to specify both Φ and its normal derivative over all surfaces. This, of course, would exceed the actual requirements for uniqueness.

   Many functions can satisfy (3.52). For instance,

$$G(\mathbf{r}|\mathbf{r}') = \frac{A}{|\mathbf{r} - \mathbf{r}'|} + \frac{B}{|\mathbf{r} - \mathbf{r}_i|} \tag{3.68}$$

satisfies (3.52) if $\mathbf{r}_i \notin V$. Evaluation of (3.55) with the Green's function (3.68) reproduces the general formulation (3.56) since the Laplacian of the second term in (3.68) is identically zero in V. In fact, we can add any function to the free-space Green's function, provided that the additional term obeys Laplace's equation within V:

$$G(\mathbf{r}|\mathbf{r}') = \frac{A}{|\mathbf{r} - \mathbf{r}'|} + F(\mathbf{r}|\mathbf{r}'), \qquad \nabla'^2 F(\mathbf{r}|\mathbf{r}') = 0. \tag{3.69}$$

A good choice for $G(\mathbf{r}|\mathbf{r}')$ will minimize the effort required to evaluate $\Phi(\mathbf{r})$. Examining (3.56) we notice two possibilities. If we demand that

$$G(\mathbf{r}|\mathbf{r}') = 0 \text{ for all } \mathbf{r}' \in S \tag{3.70}$$

then the surface integral terms in (3.56) involving $\partial\Phi/\partial n'$ will vanish. The Green's function satisfying (3.70) is known as the *Dirichlet Green's function*. Let us designate it by G_D and use reciprocity to write (3.70) as

$$G_D(\mathbf{r}|\mathbf{r}') = 0 \text{ for all } \mathbf{r} \in S.$$

The resulting specialization of (3.56),

$$\Phi(\mathbf{r}) = \int_V G_D(\mathbf{r}|\mathbf{r}')\frac{\rho(\mathbf{r}')}{\epsilon}\, dV' + \oint_{S_B} \Phi(\mathbf{r}')\frac{\partial G_D(\mathbf{r}|\mathbf{r}')}{\partial n'}\, dS' +$$
$$+ \sum_{n=1}^{N} \oint_{S_n} \Phi(\mathbf{r}')\frac{\partial G_D(\mathbf{r}|\mathbf{r}')}{\partial n'}\, dS', \tag{3.71}$$

requires the specification of Φ (but not its normal derivative) over the boundary surfaces. In case S_B and S_n surround and are adjacent to perfect conductors, the Dirichlet boundary condition has an important physical meaning. The corresponding Green's function is the potential at point $\mathbf{r}$ produced by a point source at $\mathbf{r}'$ in the presence of the conductors when the conductors are grounded — i.e., held at zero potential. Then we must specify the actual constant potentials on the conductors to determine Φ everywhere within V using (3.71). The additional term $F(\mathbf{r}|\mathbf{r}')$ in (3.69) accounts for the potential produced by surface charges on the grounded conductors.

By analogy with (3.70) it is tempting to try to define another electrostatic Green's function according to

$$\frac{\partial G(\mathbf{r}|\mathbf{r}')}{\partial n'} = 0 \text{ for all } \mathbf{r}' \in S. \tag{3.72}$$

But this choice is not permissible if V is a finite-sized region. Let us integrate (3.54) over V and employ the divergence theorem and the sifting property to get

$$\oint_S \frac{\partial G(\mathbf{r}|\mathbf{r}')}{\partial n'}\, dS' = -1; \tag{3.73}$$

in conjunction with this, equation (3.72) would imply the false statement $0 = -1$. Suppose instead that we introduce a Green's function according to

$$\frac{\partial G(\mathbf{r}|\mathbf{r}')}{\partial n'} = -\frac{1}{A} \text{ for all } \mathbf{r}' \in S. \tag{3.74}$$

where A is the total area of S. This choice avoids a contradiction in (3.73); it does not nullify any terms in (3.56), but does reduce the surface integral terms involving Φ to constants. Taken together, these terms all comprise a single additive constant on the right-hand side; although the corresponding potential $\Phi(\mathbf{r})$ is thereby determined only to within this additive constant, the value of $\mathbf{E}(\mathbf{r}) = -\nabla\Phi(\mathbf{r})$ will be unaffected. By reciprocity we can rewrite (3.74) as

$$\frac{\partial G_N(\mathbf{r}|\mathbf{r}')}{\partial n} = -\frac{1}{A} \text{ for all } \mathbf{r} \in S. \tag{3.75}$$

The Green's function G_N so defined is known as the *Neumann Green's function*. Observe that if V is not finite-sized then $A \to \infty$ and according to (3.74) the choice (3.72) becomes allowable.

Finding the Green's function that obeys one of the boundary conditions for a given geometry is often a difficult task. Nevertheless, certain canonical geometries make the Green's function approach straightforward and simple. Such is the case in image theory, when a charge is located near a simple conducting body such as a ground screen or a sphere. In these cases the function $F(\mathbf{r}|\mathbf{r}')$ consists of a single correction term as in (3.68). We shall consider these simple cases in examples to follow.

Reciprocity of the static Green's function.   It remains to show that

$$G(\mathbf{r}|\mathbf{r}') = G(\mathbf{r}'|\mathbf{r})$$

for any of the Green's functions introduced above. The unbounded-space Green's function is reciprocal by inspection; $|\mathbf{r} - \mathbf{r}'|$ is unaffected by interchanging $\mathbf{r}$ and $\mathbf{r}'$. However, we can give a more general treatment covering this case as well as the Dirichlet and Neumann cases. We begin with

$$\nabla^2 G(\mathbf{r}|\mathbf{r}') = -\delta(\mathbf{r} - \mathbf{r}').$$

In Green's second identity let

$$\phi(\mathbf{r}) = G(\mathbf{r}|\mathbf{r}_a), \qquad\qquad \psi(\mathbf{r}) = G(\mathbf{r}|\mathbf{r}_b),$$

where $\mathbf{r}_a$ and $\mathbf{r}_b$ are arbitrary points, and integrate over the unprimed coordinates. We have

$$\int_V [G(\mathbf{r}|\mathbf{r}_a)\nabla^2 G(\mathbf{r}|\mathbf{r}_b) - G(\mathbf{r}|\mathbf{r}_b)\nabla^2 G(\mathbf{r}|\mathbf{r}_a)]\,dV =$$
$$- \oint_S \left[G(\mathbf{r}|\mathbf{r}_a)\frac{\partial G(\mathbf{r}|\mathbf{r}_b)}{\partial n} - G(\mathbf{r}|\mathbf{r}_b)\frac{\partial G(\mathbf{r}|\mathbf{r}_a)}{\partial n} \right]\,dS.$$

If G is the unbounded-space Green's function, the surface integral must vanish since $S_B \to \infty$. It must also vanish under Dirichlet or Neumann boundary conditions. Since

$$\nabla^2 G(\mathbf{r}|\mathbf{r}_a) = -\delta(\mathbf{r} - \mathbf{r}_a), \qquad\qquad \nabla^2 G(\mathbf{r}|\mathbf{r}_b) = -\delta(\mathbf{r} - \mathbf{r}_b),$$

we have

$$\int_V [G(\mathbf{r}|\mathbf{r}_a)\delta(\mathbf{r} - \mathbf{r}_b) - G(\mathbf{r}|\mathbf{r}_b)\delta(\mathbf{r} - \mathbf{r}_a)]\,dV = 0,$$

hence

$$G(\mathbf{r}_b|\mathbf{r}_a) = G(\mathbf{r}_a|\mathbf{r}_b)$$

by the sifting property. By the arbitrariness of $\mathbf{r}_a$ and $\mathbf{r}_b$, reciprocity is established.

Electrostatic shielding.   The Dirichlet Green's function can be used to explain *electrostatic shielding*. We consider a closed, grounded, conducting shell with charge outside but not inside (Figure 3.7). By (3.71) the potential at points inside the shell is

$$\Phi(\mathbf{r}) = \oint_{S_B} \Phi(\mathbf{r}')\frac{\partial G_D(\mathbf{r}|\mathbf{r}')}{\partial n'}\,dS',$$

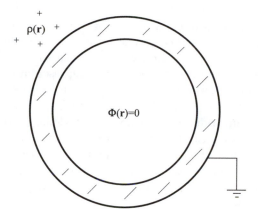

Figure 3.7: Electrostatic shielding by a conducting shell.

where S_B is tangential to the inner surface of the shell and we have used $\rho = 0$ within the shell. Because $\Phi(\mathbf{r}') = 0$ for all $\mathbf{r}'$ on S_B, we have

$$\Phi(\mathbf{r}) = 0$$

everywhere in the region enclosed by the shell. This result is independent of the charge outside the shell, and the interior region is "shielded" from the effects of that charge.

Conversely, consider a grounded conducting shell with charge contained inside. If we surround the outside of the shell by a surface S_1 and let S_B recede to infinity, then (3.71) becomes

$$\Phi(\mathbf{r}) = \lim_{S_B \to \infty} \oint_{S_B} \Phi(\mathbf{r}') \frac{\partial G_D(\mathbf{r}|\mathbf{r}')}{\partial n'} \, dS' + \oint_{S_1} \Phi(\mathbf{r}') \frac{\partial G_D(\mathbf{r}|\mathbf{r}')}{\partial n'} \, dS'.$$

Again there is no charge in V (since the charge lies completely inside the shell). The contribution from S_B vanishes. Since S_1 lies adjacent to the outer surface of the shell, $\Phi(\mathbf{r}') \equiv 0$ on S_1. Thus $\Phi(\mathbf{r}) = 0$ for all points outside the conducting shell.

Example solution to Poisson's equation: planar layered media. For simple geometries Poisson's equation may be solved as part of a *boundary value problem* (§ A.4). Occasionally such a solution has an appealing interpretation as the superposition of potentials produced by the physical charge and its "images." We shall consider here the case of planar media and subsequently use the results to predict the potential produced by charge near a conducting sphere.

Consider a layered dielectric medium where various regions of space are separated by planes at constant values of z. Material region i occupies volume region V_i and has permittivity ϵ_i; it may or may not contain source charge. The solution to Poisson's equation is given by (3.56). The contribution

$$\Phi^p(\mathbf{r}) = \int_V G(\mathbf{r}|\mathbf{r}') \frac{\rho(\mathbf{r}')}{\epsilon} \, dV'$$

produced by sources within V is known as the *primary potential*. The term

$$\Phi^s(\mathbf{r}) = \oint_S \left[\Phi(\mathbf{r}') \frac{\partial G(\mathbf{r}|\mathbf{r}')}{\partial n'} - G(\mathbf{r}|\mathbf{r}') \frac{\partial \Phi(\mathbf{r}')}{\partial n'} \right] dS',$$

on the other hand, involves an integral over the surface fields and is known as the *secondary potential*. This term is linked to effects outside V. Since the "sources" of Φ^s (i.e., the surface fields) lie on the boundary of V, Φ^s satisfies Laplace's equation within V. We may therefore use other, more convenient, representations of Φ^s provided they satisfy Laplace's equation. However, as solutions to a homogeneous equation they are of indefinite form until linked to appropriate boundary values.

Since the geometry is invariant in the x and y directions, we represent each potential function in terms of a 2-D Fourier transform over these variables. We leave the z dependence intact so that we may apply boundary conditions directly in the spatial domain. The transform representations of the Green's functions for the primary and secondary potentials are derived in Appendix A. From (A.55) we see that the primary potential within region V_i can be written as

$$\Phi_i^p(\mathbf{r}) = \int_{V_i} G^p(\mathbf{r}|\mathbf{r}')\frac{\rho(\mathbf{r}')}{\epsilon_i}\, dV' \tag{3.76}$$

where

$$G^p(\mathbf{r}|\mathbf{r}') = \frac{1}{4\pi|\mathbf{r}-\mathbf{r}'|} = \frac{1}{(2\pi)^2}\int_{-\infty}^{\infty}\frac{e^{-k_\rho|z-z'|}}{2k_\rho}e^{j\mathbf{k}_\rho\cdot(\mathbf{r}-\mathbf{r}')}\, d^2 k_\rho \tag{3.77}$$

is the primary Green's function with $\mathbf{k}_\rho = \hat{\mathbf{x}}k_x + \hat{\mathbf{y}}k_y$, $k_\rho = |\mathbf{k}_\rho|$, and $d^2 k_\rho = dk_x\, dk_y$. We also find in (A.56) that a solution of Laplace's equation can be written as

$$\Phi^s(\mathbf{r}) = \frac{1}{(2\pi)^2}\int_{-\infty}^{\infty}\left[A(\mathbf{k}_\rho)e^{k_\rho z} + B(\mathbf{k}_\rho)e^{-k_\rho z}\right]e^{j\mathbf{k}_\rho\cdot\mathbf{r}}\, d^2 k_\rho \tag{3.78}$$

where $A(\mathbf{k}_\rho)$ and $B(\mathbf{k}_\rho)$ must be found by the application of appropriate boundary conditions.

As a simple example, consider a charge distribution $\rho(\mathbf{r})$ in free space above a grounded conducting plane located at $z = 0$. We wish to find the potential in the region $z > 0$ using the Fourier transform representation of the potentials. The total potential is a sum of primary and secondary terms:

$$\Phi(x, y, z) = \int_V \left[\frac{1}{(2\pi)^2}\int_{-\infty}^{\infty}\frac{e^{-k_\rho|z-z'|}}{2k_\rho}e^{j\mathbf{k}_\rho\cdot(\mathbf{r}-\mathbf{r}')}\, d^2 k_\rho\right]\frac{\rho(\mathbf{r}')}{\epsilon_0}\, dV' +$$

$$+ \frac{1}{(2\pi)^2}\int_{-\infty}^{\infty}\left[B(\mathbf{k}_\rho)e^{-k_\rho z}\right]e^{j\mathbf{k}_\rho\cdot\mathbf{r}}\, d^2 k_\rho,$$

where the integral is over the region $z > 0$. Here we have set $A(\mathbf{k}_\rho) = 0$ because $e^{k_\rho z}$ grows with increasing z. Since the plane is grounded we must have $\Phi(x, y, 0) = 0$. Because $z < z'$ when we apply this condition, we have $|z - z'| = z' - z$ and thus

$$\Phi(x, y, 0) = \frac{1}{(2\pi)^2}\int_{-\infty}^{\infty}\left[\int_V \frac{\rho(\mathbf{r}')}{\epsilon_0}\frac{e^{-k_\rho z'}}{2k_\rho}e^{-j\mathbf{k}_\rho\cdot\mathbf{r}'}\, dV' + B(\mathbf{k}_\rho)\right]e^{j\mathbf{k}_\rho\cdot\mathbf{r}}\, d^2 k_\rho = 0.$$

Invoking the Fourier integral theorem we find

$$B(\mathbf{k}_\rho) = -\int_V \frac{\rho(\mathbf{r}')}{\epsilon_0}\frac{e^{-k_\rho z'}}{2k_\rho}e^{-j\mathbf{k}_\rho\cdot\mathbf{r}'}\, dV',$$

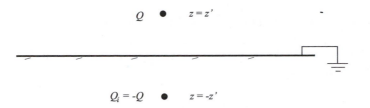

Figure 3.8: Construction of electrostatic Green's function for a ground plane.

hence the total potential is

$$\Phi(x,y,z) = \int_V \left[\frac{1}{(2\pi)^2} \int_{-\infty}^{\infty} \frac{e^{-k_\rho|z-z'|} - e^{-k_\rho(z+z')}}{2k_\rho} e^{j\mathbf{k}_\rho \cdot (\mathbf{r}-\mathbf{r}')}\, d^2k_\rho \right] \frac{\rho(\mathbf{r}')}{\epsilon_0}\, dV'$$

$$= \int_V G(\mathbf{r}|\mathbf{r}') \frac{\rho(\mathbf{r}')}{\epsilon_0}\, dV'$$

where $G(\mathbf{r}|\mathbf{r}')$ is the Green's function for the region above a grounded planar conductor. We can interpret this Green's function as a sum of the primary Green's function (3.77) and a secondary Green's function

$$G^s(\mathbf{r}|\mathbf{r}') = -\frac{1}{(2\pi)^2} \int_{-\infty}^{\infty} \frac{e^{-k_\rho(z+z')}}{2k_\rho} e^{j\mathbf{k}_\rho \cdot (\mathbf{r}-\mathbf{r}')}\, d^2k_\rho. \qquad (3.79)$$

For $z > 0$ the term $z + z'$ can be replaced by $|z + z'|$. Then, comparing (3.79) with (3.77), we see that

$$G^s(\mathbf{r}\,|\,x',y',z') = -G^p(\mathbf{r}\,|\,x',y',-z') = -\frac{1}{4\pi|\mathbf{r} - \mathbf{r}_i'|} \qquad (3.80)$$

where $\mathbf{r}_i' = \hat{\mathbf{x}}x' + \hat{\mathbf{y}}y' - \hat{\mathbf{z}}z'$. Because the Green's function is the potential of a point charge, we may interpret the secondary Green's function as produced by a negative unit charge placed in a position $-z'$ immediately beneath the positive unit charge that produces G^p (Figure 3.8). This secondary charge is the "image" of the primary charge. That two such charges would produce a null potential on the ground plane is easily verified.

As a more involved example, consider a charge distribution $\rho(\mathbf{r})$ above a planar interface separating two homogeneous dielectric media. Region 1 occupies $z > 0$ and has permittivity ϵ_1, while region 2 occupies $z < 0$ and has permittivity ϵ_2. In region 1 we can write the total potential as a sum of primary and secondary components, discarding the term that grows with z:

$$\Phi_1(x,y,z) = \int_V \left[\frac{1}{(2\pi)^2} \int_{-\infty}^{\infty} \frac{e^{-k_\rho|z-z'|}}{2k_\rho} e^{j\mathbf{k}_\rho \cdot (\mathbf{r}-\mathbf{r}')}\, d^2k_\rho \right] \frac{\rho(\mathbf{r}')}{\epsilon_1}\, dV' +$$

$$+ \frac{1}{(2\pi)^2} \int_{-\infty}^{\infty} \left[B(\mathbf{k}_\rho)e^{-k_\rho z} \right] e^{j\mathbf{k}_\rho \cdot \mathbf{r}}\, d^2k_\rho. \qquad (3.81)$$

With no source in region 2, the potential there must obey Laplace's equation and therefore consists of only a secondary component:

$$\Phi_2(\mathbf{r}) = \frac{1}{(2\pi)^2} \int_{-\infty}^{\infty} \left[A(\mathbf{k}_\rho)e^{k_\rho z} \right] e^{j\mathbf{k}_\rho \cdot \mathbf{r}}\, d^2k_\rho. \qquad (3.82)$$

To determine A and B we impose (3.36) and (3.37). By (3.36) we have

$$\frac{1}{(2\pi)^2} \int_{-\infty}^{\infty} \left[\int_V \frac{\rho(\mathbf{r}')}{\epsilon_1} \frac{e^{-k_\rho z'}}{2k_\rho} e^{-j\mathbf{k}_\rho \cdot \mathbf{r}'} \, dV' + B(\mathbf{k}_\rho) - A(\mathbf{k}_\rho) \right] e^{j\mathbf{k}_\rho \cdot \mathbf{r}} \, d^2 k_\rho = 0,$$

hence

$$\int_V \frac{\rho(\mathbf{r}')}{\epsilon_1} \frac{e^{-k_\rho z'}}{2k_\rho} e^{-j\mathbf{k}_\rho \cdot \mathbf{r}'} \, dV' + B(\mathbf{k}_\rho) - A(\mathbf{k}_\rho) = 0$$

by the Fourier integral theorem. Applying (3.37) at $z = 0$ with $\hat{\mathbf{n}}_{12} = \hat{\mathbf{z}}$, and noting that there is no excess surface charge, we find

$$\int_V \rho(\mathbf{r}') \frac{e^{-k_\rho z'}}{2k_\rho} e^{-j\mathbf{k}_\rho \cdot \mathbf{r}'} \, dV' - \epsilon_1 B(\mathbf{k}_\rho) - \epsilon_2 A(\mathbf{k}_\rho) = 0.$$

The solutions

$$A(\mathbf{k}_\rho) = \frac{2\epsilon_1}{\epsilon_1 + \epsilon_2} \int_V \frac{\rho(\mathbf{r}')}{\epsilon_1} \frac{e^{-k_\rho z'}}{2k_\rho} e^{-j\mathbf{k}_\rho \cdot \mathbf{r}'} \, dV',$$

$$B(\mathbf{k}_\rho) = \frac{\epsilon_1 - \epsilon_2}{\epsilon_1 + \epsilon_2} \int_V \frac{\rho(\mathbf{r}')}{\epsilon_1} \frac{e^{-k_\rho z'}}{2k_\rho} e^{-j\mathbf{k}_\rho \cdot \mathbf{r}'} \, dV',$$

are then substituted into (3.81) and (3.82) to give

$$\Phi_1(\mathbf{r}) = \int_V \left[\frac{1}{(2\pi)^2} \int_{-\infty}^{\infty} \frac{e^{-k_\rho |z-z'|} + \frac{\epsilon_1 - \epsilon_2}{\epsilon_1 + \epsilon_2} e^{-k_\rho(z+z')}}{2k_\rho} e^{j\mathbf{k}_\rho \cdot (\mathbf{r}-\mathbf{r}')} \, d^2 k_\rho \right] \frac{\rho(\mathbf{r}')}{\epsilon_1} \, dV'$$

$$= \int_V G_1(\mathbf{r}|\mathbf{r}') \frac{\rho(\mathbf{r}')}{\epsilon_1} \, dV',$$

$$\Phi_2(\mathbf{r}) = \int_V \left[\frac{1}{(2\pi)^2} \int_{-\infty}^{\infty} \frac{2\epsilon_2}{\epsilon_1 + \epsilon_2} \frac{e^{-k_\rho(z'-z)}}{2k_\rho} e^{j\mathbf{k}_\rho \cdot (\mathbf{r}-\mathbf{r}')} \, d^2 k_\rho \right] \frac{\rho(\mathbf{r}')}{\epsilon_2} \, dV'$$

$$= \int_V G_2(\mathbf{r}|\mathbf{r}') \frac{\rho(\mathbf{r}')}{\epsilon_2} \, dV'.$$

Since $z' > z$ for all points in region 2, we can replace $z' - z$ by $|z - z'|$ in the formula for Φ_2.

As with the previous example, let us compare the result to the form of the primary Green's function (3.77). We see that

$$G_1(\mathbf{r}|\mathbf{r}') = \frac{1}{4\pi|\mathbf{r} - \mathbf{r}'|} + \frac{\epsilon_1 - \epsilon_2}{\epsilon_1 + \epsilon_2} \frac{1}{4\pi|\mathbf{r} - \mathbf{r}_1'|},$$

$$G_2(\mathbf{r}|\mathbf{r}') = \frac{2\epsilon_2}{\epsilon_1 + \epsilon_2} \frac{1}{4\pi|\mathbf{r} - \mathbf{r}_2'|},$$

where $\mathbf{r}_1' = \hat{\mathbf{x}} x' + \hat{\mathbf{y}} y' - \hat{\mathbf{z}} z'$ and $\mathbf{r}_2' = \hat{\mathbf{x}} x' + \hat{\mathbf{y}} y' + \hat{\mathbf{z}} z'$. So we can also write

$$\Phi_1(\mathbf{r}) = \frac{1}{4\pi} \int_V \left[\frac{1}{|\mathbf{r} - \mathbf{r}'|} + \frac{\epsilon_1 - \epsilon_2}{\epsilon_1 + \epsilon_2} \frac{1}{|\mathbf{r} - \mathbf{r}_1'|} \right] \frac{\rho(\mathbf{r}')}{\epsilon_1} \, dV',$$

$$\Phi_2(\mathbf{r}) = \frac{1}{4\pi} \int_V \left[\frac{2\epsilon_2}{\epsilon_1 + \epsilon_2} \frac{1}{|\mathbf{r} - \mathbf{r}_2'|} \right] \frac{\rho(\mathbf{r}')}{\epsilon_2} \, dV'.$$

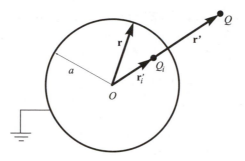

Figure 3.9: Green's function for a grounded conducting sphere.

Note that $\Phi_2 \to \Phi_1$ as $\epsilon_2 \to \epsilon_1$.

There is an image interpretation for the secondary Green's functions. The secondary Green's function for region 1 appears as a potential produced by an image of the primary charge located at $-z'$ in an infinite medium of permittivity ϵ_1, and with an amplitude of $(\epsilon_1 - \epsilon_2)/(\epsilon_1 + \epsilon_2)$ times the primary charge. The Green's function in region 2 is produced by an image charge located at z' (i.e., at the location of the primary charge) in an infinite medium of permittivity ϵ_2 with an amplitude of $2\epsilon_2/(\epsilon_1 + \epsilon_2)$ times the primary charge.

Example solution to Poisson's equation: conducting sphere. As an example involving a nonplanar geometry, consider the potential produced by a source near a grounded conducting sphere in free space (Figure 3.9). Based on our experience with planar layered media, we hypothesize that the secondary potential will be produced by an image charge; hence we try the simple Green's function

$$G^s(\mathbf{r}|\mathbf{r}') = \frac{A(\mathbf{r}')}{4\pi|\mathbf{r} - \mathbf{r}'_i|}$$

where the amplitude A and location $\mathbf{r}'_i$ of the image are to be determined. We further assume, based on our experience with planar problems, that the image charge will reside inside the sphere along a line joining the origin to the primary charge. Since $\mathbf{r} = a\hat{\mathbf{r}}$ for all points on the sphere, the total Green's function must obey the Dirichlet condition

$$G(\mathbf{r}|\mathbf{r}')|_{r=a} = \left.\frac{1}{4\pi|\mathbf{r} - \mathbf{r}'|}\right|_{r=a} + \left.\frac{A(\mathbf{r}')}{4\pi|\mathbf{r} - \mathbf{r}'_i|}\right|_{r=a} = \frac{1}{4\pi|a\hat{\mathbf{r}} - r'\hat{\mathbf{r}}'|} + \frac{A(\mathbf{r}')}{4\pi|a\hat{\mathbf{r}} - r'_i\hat{\mathbf{r}}'|} = 0$$

in order to have the potential, given by (3.56), vanish on the sphere surface. Factoring a from the first denominator and r'_i from the second we obtain

$$\frac{1}{4\pi a|\hat{\mathbf{r}} - \frac{r'}{a}\hat{\mathbf{r}}'|} + \frac{A(\mathbf{r}')}{4\pi r'_i|\frac{a}{r'_i}\hat{\mathbf{r}} - \hat{\mathbf{r}}'|} = 0.$$

Now $|k\hat{\mathbf{r}} - k'\hat{\mathbf{r}}'| = k^2 + k'^2 - 2kk'\cos\gamma$ where γ is the angle between $\hat{\mathbf{r}}$ and $\hat{\mathbf{r}}'$ and k, k' are constants; this means that $|k\hat{\mathbf{r}} - \hat{\mathbf{r}}'| = |\hat{\mathbf{r}} - k\hat{\mathbf{r}}'|$. Hence as long as we choose

$$\frac{r'}{a} = \frac{a}{r'_i}, \qquad \frac{A}{r'_i} = -\frac{1}{a},$$

the total Green's function vanishes everywhere on the surface of the sphere. The image charge is therefore located within the sphere at $\mathbf{r}'_i = a^2\mathbf{r}'/r'^2$ and has amplitude $A = $

$-a/r'$. (Note that both the location and amplitude of the image depend on the location of the primary charge.) With this Green's function and (3.71), the potential of an arbitrary source placed near a grounded conducting sphere is

$$\Phi(\mathbf{r}) = \int_V \frac{\rho(\mathbf{r}')}{\epsilon} \frac{1}{4\pi} \left[\frac{1}{|\mathbf{r} - \mathbf{r}'|} - \frac{a/r'}{|\mathbf{r} - \frac{a^2}{r'^2}\mathbf{r}'|} \right] dV'.$$

The Green's function may be used to compute the surface charge density induced on the sphere by a unit point charge: it is merely necessary to find the normal component of electric field from the gradient of $\Phi(\mathbf{r})$. We leave this as an exercise for the reader, who may then integrate the surface charge and thereby show that the total charge induced on the sphere is equal to the image charge. So the total charge induced on a grounded sphere by a point charge q at a point $r = r'$ is $Q = -qa/r'$.

It is possible to find the total charge induced on the sphere without finding the image charge first. This is an application of Green's reciprocation theorem (§ 3.4.4). According to (3.211), if we can find the potential V_P at a point $\mathbf{r}$ produced by the sphere when it is isolated and carrying a total charge Q_0, then the total charge Q induced on the grounded sphere in the vicinity of a point charge q placed at $\mathbf{r}$ is given by

$$Q = -qV_P/V_1$$

where V_1 is the potential of the isolated sphere. We can apply this formula by noting that an isolated sphere carrying charge Q_0 produces a field $\mathbf{E}(\mathbf{r}) = \hat{\mathbf{r}}Q_0/4\pi\epsilon r^2$. Integration from a radius r to infinity gives the potential referred to infinity: $\Phi(\mathbf{r}) = Q_0/4\pi\epsilon r$. So the potential of the isolated sphere is $V_1 = Q_0/4\pi\epsilon a$, while the potential at radius r' is $V_P = Q_0/4\pi\epsilon r'$. Substitution gives $Q = -qa/r'$ as before.

3.2.5   Force and energy

Maxwell's stress tensor.   The electrostatic version of Maxwell's stress tensor can be obtained from (2.288) by setting $\mathbf{B} = \mathbf{H} = 0$:

$$\bar{\mathbf{T}}_e = \frac{1}{2}(\mathbf{D} \cdot \mathbf{E})\bar{\mathbf{I}} - \mathbf{DE}. \tag{3.83}$$

The total electric force on the charges in a region V bounded by the surface S is given by the relation

$$\mathbf{F}_e = -\oint_S \bar{\mathbf{T}}_e \cdot \mathbf{dS} = \int_V \mathbf{f}_e \, dV$$

where $\mathbf{f}_e = \rho\mathbf{E}$ is the electric force volume density.

In particular, suppose that S is adjacent to a solid conducting body embedded in a dielectric having permittivity $\epsilon(\mathbf{r})$. Since all the charge is at the surface of the conductor, the force within V acts directly on the surface. Thus, $-\bar{\mathbf{T}}_e \cdot \hat{\mathbf{n}}$ is the surface force density (*traction*) $\mathbf{t}$. Using $\mathbf{D} = \epsilon\mathbf{E}$, and remembering that the fields are normal to the conductor, we find that

$$\bar{\mathbf{T}}_e \cdot \hat{\mathbf{n}} = \frac{1}{2}\epsilon E_n^2\hat{\mathbf{n}} - \epsilon\mathbf{EE} \cdot \hat{\mathbf{n}} = -\frac{1}{2}\epsilon E_n^2\hat{\mathbf{n}} = -\frac{1}{2}\rho_s\mathbf{E}.$$

The surface force density is perpendicular to the surface.

As a simple but interesting example, consider the force acting on a rigid conducting sphere of radius a carrying total charge Q in a homogeneous medium. At equilibrium the charge is distributed uniformly with surface density $\rho_s = Q/4\pi a^2$, producing a field $\mathbf{E} = \hat{\mathbf{r}} Q/4\pi\epsilon r^2$ external to the sphere. Hence a force density

$$\mathbf{t} = \frac{1}{2}\hat{\mathbf{r}}\frac{Q^2}{\epsilon(4\pi a^2)^2}$$

acts at each point on the surface. This would cause the sphere to expand outward if the structural integrity of the material were to fail. Integration over the entire sphere yields

$$\mathbf{F} = \frac{1}{2}\frac{Q^2}{\epsilon(4\pi a^2)^2}\int_S \hat{\mathbf{r}}\, dS = 0.$$

However, integration of $\mathbf{t}$ over the upper hemisphere yields

$$\mathbf{F} = \frac{1}{2}\frac{Q^2}{\epsilon(4\pi a^2)^2}\int_0^{2\pi}\int_0^{\pi/2}\hat{\mathbf{r}}a^2\sin\theta\, d\theta\, d\phi.$$

Substitution of $\hat{\mathbf{r}} = \hat{\mathbf{x}}\sin\theta\cos\phi + \hat{\mathbf{y}}\sin\theta\sin\phi + \hat{\mathbf{z}}\cos\theta$ leads immediately to $F_x = F_y = 0$, but the z-component is

$$F_z = \frac{1}{2}\frac{Q^2}{\epsilon(4\pi a^2)^2}\int_0^{2\pi}\int_0^{\pi/2}a^2\cos\theta\sin\theta\, d\theta\, d\phi = \frac{Q^2}{32\epsilon\pi a^2}.$$

This result can also be obtained by integrating $-\bar{\mathbf{T}}_e \cdot \hat{\mathbf{n}}$ over the entire xy-plane with $\hat{\mathbf{n}} = -\hat{\mathbf{z}}$. Since $-\bar{\mathbf{T}}_e \cdot (-\hat{\mathbf{z}}) = \hat{\mathbf{z}}\frac{\epsilon}{2}\mathbf{E}\cdot\mathbf{E}$ we have

$$\mathbf{F} = \hat{\mathbf{z}}\frac{1}{2}\frac{Q^2}{(4\pi\epsilon)^2}\int_0^{2\pi}\int_a^{\infty}\frac{r\, dr\, d\phi}{r^4} = \hat{\mathbf{z}}\frac{Q^2}{32\epsilon\pi a^2}.$$

As a more challenging example, consider two identical line charges parallel to the z-axis and located at $x = \pm d/2$, $y = 0$ in free space. We can find the force on one line charge due to the other by integrating Maxwell's stress tensor over the yz-plane. From (3.64) we find that the total electric field on the yz-plane is

$$\mathbf{E}(y,z) = \frac{y}{y^2 + (d/2)^2}\frac{\rho_l}{\pi\epsilon_0}\hat{\mathbf{y}}$$

where ρ_l is the line charge density. The force density for either line charge is $-\bar{\mathbf{T}}_e \cdot \hat{\mathbf{n}}$, where we use $\hat{\mathbf{n}} = \pm\hat{\mathbf{x}}$ to obtain the force on the charge at $x = \mp d/2$. The force density for the charge at $x = -d/2$ is

$$\bar{\mathbf{T}}_e \cdot \hat{\mathbf{n}} = \frac{1}{2}(\mathbf{D}\cdot\mathbf{E})\bar{\mathbf{I}}\cdot\hat{\mathbf{x}} - \mathbf{D}\mathbf{E}\cdot\hat{\mathbf{x}} = \frac{\epsilon_0}{2}\left[\frac{y}{y^2 + (d/2)^2}\frac{\rho_l}{\pi\epsilon_0}\right]^2\hat{\mathbf{x}}$$

and the total force is

$$\mathbf{F}_- = -\int_{-\infty}^{\infty}\int_{-\infty}^{\infty}\frac{\rho_l^2}{2\pi^2\epsilon_0}\frac{y^2}{[y^2 + (d/2)^2]^2}\hat{\mathbf{x}}\, dy\, dz.$$

On a per unit length basis the force is

$$\frac{\mathbf{F}_-}{l} = -\hat{\mathbf{x}}\frac{\rho_l^2}{2\pi^2\epsilon_0}\int_{-\infty}^{\infty}\frac{y^2}{[y^2 + (d/2)^2]^2}\, dy = -\hat{\mathbf{x}}\frac{\rho_l^2}{2\pi d\epsilon_0}.$$

Note that the force is repulsive as expected.

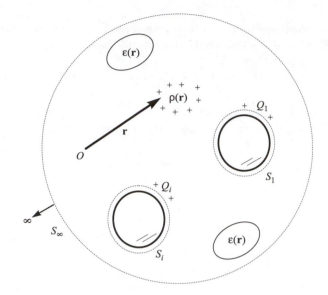

Figure 3.10: Computation of electrostatic stored energy via the assembly energy of a charge distribution.

Electrostatic stored energy. In § 2.9.5 we considered the energy relations for the electromagnetic field. Those relations remain valid in the static case. Since our interpretation of the dynamic relations was guided in part by our knowledge of the energy stored in a static field, we must, for completeness, carry out a study of that effect here.

The energy of a static configuration is taken to be the work required to assemble the configuration from a chosen starting point. For a configuration of static charges, the *stored electric energy* is the energy required to assemble the configuration, starting with all charges removed to infinite distance (the assumed zero potential reference). If the assembled charges are not held in place by an external mechanical force they will move, thereby converting stored electric energy into other forms of energy (e.g., kinetic energy and radiation).

By (3.62), the work required to move a point charge q from a reservoir at infinity to a point P at $\mathbf{r}$ in a potential field Φ is

$$W = q\Phi(\mathbf{r}).$$

If instead we have a continuous charge density ρ present, and wish to increase this to $\rho + \delta\rho$ by bringing in a small quantity of charge $\delta\rho$, a total work

$$\delta W = \int_{V_\infty} \delta\rho(\mathbf{r})\Phi(\mathbf{r})\,dV \tag{3.84}$$

is required, and the potential field is increased to $\Phi + \delta\Phi$. Here V_∞ denotes all of space. (We could restrict the integral to the region containing the charge, but we shall find it helpful to extend the domain of integration to all of space.)

Now consider the situation shown in Figure 3.10. Here we have charge in the form of both volume densities and surface densities on conducting bodies. Also present may be linear material bodies. We can think of assembling the charge in two distinctly different ways. We could, for instance, bring small portions of charge (or point charges) together to form the distribution ρ. Or, we could slowly build up ρ by adding infinitesimal, but spatially identical, distributions. That is, we can create the distribution ρ from a zero initial state by repeatedly adding a charge distribution

$$\delta\rho(\mathbf{r}) = \rho(\mathbf{r})/N,$$

where N is a large number. Whenever we add $\delta\rho$ we must perform the work given by (3.84), but we also increase the potential proportionately (remembering that all materials are assumed linear). At each step, more work is required. The total work is

$$W = \sum_{n=1}^{N} \int_{V_\infty} \delta\rho(\mathbf{r})[(n-1)\delta\Phi(\mathbf{r})]\, dV = \left[\sum_{n=1}^{N}(n-1)\right] \int_{V_\infty} \frac{\rho(\mathbf{r})}{N} \frac{\Phi(\mathbf{r})}{N}\, dV. \qquad (3.85)$$

We must use an infinite number of steps so that no energy is lost to radiation at any step (since the charge we add each time is infinitesimally small). Using

$$\sum_{n=1}^{N}(n-1) = N(N-1)/2,$$

(3.85) becomes

$$W = \frac{1}{2}\int_{V_\infty} \rho(\mathbf{r})\Phi(\mathbf{r})\, dV \qquad (3.86)$$

as $N \to \infty$. Finally, since some assembled charge will be in the form of a volume density and some in the form of the surface density on conductors, we can generalize (3.86) to

$$W = \frac{1}{2}\int_{V'} \rho(\mathbf{r})\Phi(\mathbf{r})\, dV + \frac{1}{2}\sum_{i=1}^{I} Q_i V_i. \qquad (3.87)$$

Here V' is the region outside the conductors, Q_i is the total charge on the ith conductor ($i = 1, \ldots, I$), and V_i is the absolute potential (referred to infinity) of the ith conductor.

An intriguing property of electrostatic energy is that the charges on the conductors will arrange themselves, while seeking static equilibrium, into a minimum-energy configuration (Thomson's theorem).

In keeping with our field-centered view of electromagnetics, we now wish to write the energy (3.86) entirely in terms of the field vectors $\mathbf{E}$ and $\mathbf{D}$. Since $\rho = \nabla \cdot \mathbf{D}$ we have

$$W = \frac{1}{2}\int_{V_\infty} [\nabla \cdot \mathbf{D}(\mathbf{r})]\Phi(\mathbf{r})\, dV.$$

Then, by (B.42),

$$W = \frac{1}{2}\int_{V_\infty} \nabla \cdot [\Phi(\mathbf{r})\mathbf{D}(\mathbf{r})]\, dV - \frac{1}{2}\int_{V_\infty} \mathbf{D}(\mathbf{r}) \cdot [\nabla\Phi(\mathbf{r})]\, dV.$$

Use of the divergence theorem and (3.30) leads to

$$W = \frac{1}{2}\oint_{S_\infty} \Phi(\mathbf{r})\mathbf{D}(\mathbf{r}) \cdot d\mathbf{S} + \frac{1}{2}\int_{V_\infty} \mathbf{D}(\mathbf{r}) \cdot \mathbf{E}(\mathbf{r})\, dV$$

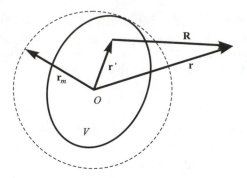

Figure 3.11: Multipole expansion.

where S_∞ is the bounding surface that recedes toward infinity to encompass all of space. Because $\Phi \sim 1/r$ and $D \sim 1/r^2$ as $r \to \infty$, the integral over S_∞ tends to zero and

$$W = \frac{1}{2} \int_{V_\infty} \mathbf{D}(\mathbf{r}) \cdot \mathbf{E}(\mathbf{r}) \, dV. \qquad (3.88)$$

Hence we may compute the assembly energy in terms of the fields supported by the charge ρ.

It is significant that the assembly energy W is identical to the term within the time derivative in Poynting's theorem (2.299). Hence our earlier interpretation, that this term represents the time-rate of change of energy "stored" in the electric field, has a firm basis. Of course, the assembly energy is a static concept, and our generalization to dynamic fields is purely intuitive. We also face similar questions regarding the meaning of energy density, and whether energy can be "localized" in space. The discussions in § 2.9.5 still apply.

3.2.6 Multipole expansion

Consider an arbitrary but spatially localized charge distribution of total charge Q in an unbounded homogeneous medium (Figure 3.11). We have already obtained the potential (3.61) of the source; as we move the observation point away, Φ should decrease in a manner roughly proportional to $1/r$. The actual variation depends on the nature of the charge distribution and can be complicated. Often this dependence is dominated by a specific inverse power of distance for observation points far from the source, and we can investigate it by expanding the potential in powers of $1/r$. Although such *multipole expansions* of the potential are rarely used to perform actual computations, they can provide insight into both the behavior of static fields and the physical meaning of the polarization vector $\mathbf{P}$.

Let us place our origin of coordinates somewhere within the charge distribution, as shown in Figure 3.11, and expand the Green's function spatial dependence in a three-dimensional Taylor series about the origin:

$$\frac{1}{R} = \sum_{n=0}^{\infty} \frac{1}{n!} (\mathbf{r}' \cdot \nabla')^n \frac{1}{R}\bigg|_{\mathbf{r}'=0} = \frac{1}{r} + (\mathbf{r}' \cdot \nabla') \frac{1}{R}\bigg|_{\mathbf{r}'=0} + \frac{1}{2}(\mathbf{r}' \cdot \nabla')^2 \frac{1}{R}\bigg|_{\mathbf{r}'=0} + \cdots, \qquad (3.89)$$

where $R = |\mathbf{r} - \mathbf{r}'|$. Convergence occurs if $|\mathbf{r}| > |\mathbf{r}'|$. In the notation $(\mathbf{r}' \cdot \nabla')^n$ we interpret a power on a derivative operator as the order of the derivative. Substituting (3.89) into

(3.61) and writing the derivatives in Cartesian coordinates we obtain

$$\Phi(\mathbf{r}) = \frac{1}{4\pi\epsilon} \int_V \rho(\mathbf{r}') \left[\frac{1}{R}\Big|_{\mathbf{r}'=0} + (\mathbf{r}' \cdot \nabla') \frac{1}{R}\Big|_{\mathbf{r}'=0} + \frac{1}{2}(\mathbf{r}' \cdot \nabla')^2 \frac{1}{R}\Big|_{\mathbf{r}'=0} + \cdots \right] dV'. \quad (3.90)$$

For the second term we can use (3.57) to write

$$(\mathbf{r}' \cdot \nabla') \frac{1}{R}\Big|_{\mathbf{r}'=0} = \mathbf{r}' \cdot \left(\nabla' \frac{1}{R}\right)\Big|_{\mathbf{r}'=0} = \mathbf{r}' \cdot \left(\frac{\hat{\mathbf{R}}}{R^2}\right)\Big|_{\mathbf{r}'=0} = \mathbf{r}' \cdot \frac{\hat{\mathbf{r}}}{r^2}. \quad (3.91)$$

The third term is complicated. Let us denote (x, y, z) by (x_1, x_2, x_3) and perform an expansion in rectangular coordinates:

$$(\mathbf{r}' \cdot \nabla')^2 \frac{1}{R}\Big|_{\mathbf{r}'=0} = \sum_{i=1}^{3} \sum_{j=1}^{3} x_i' x_j' \frac{\partial^2}{\partial x_i' \partial x_j'} \frac{1}{R}\Big|_{\mathbf{r}'=0}.$$

It turns out [172] that this can be written as

$$(\mathbf{r}' \cdot \nabla')^2 \frac{1}{R}\Big|_{\mathbf{r}'=0} = \frac{1}{r^3} \hat{\mathbf{r}} \cdot (3\mathbf{r}'\mathbf{r}' - r'^2\bar{\mathbf{I}}) \cdot \hat{\mathbf{r}}.$$

Substitution into (3.90) gives

$$\Phi(r) = \frac{Q}{4\pi\epsilon r} + \frac{\hat{\mathbf{r}} \cdot \mathbf{p}}{4\pi\epsilon r^2} + \frac{1}{2} \frac{\hat{\mathbf{r}} \cdot \bar{\mathbf{Q}} \cdot \hat{\mathbf{r}}}{4\pi\epsilon r^3} + \cdots, \quad (3.92)$$

which is the multipole expansion for $\Phi(r)$. It converges for all $r > r_m$ where r_m is the radius of the smallest sphere completely containing the charge centered at $\mathbf{r}' = 0$ (Figure 3.11). In (3.92) the terms Q, $\mathbf{p}$, $\bar{\mathbf{Q}}$, and so on are called the *multipole moments* of $\rho(\mathbf{r})$. The first moment is merely the total charge

$$Q = \int_V \rho(\mathbf{r}') \, dV'.$$

The second moment is the *electric dipole moment vector*

$$\mathbf{p} = \int_V \mathbf{r}' \rho(\mathbf{r}') \, dV'.$$

The third moment is the *electric quadrupole moment dyadic*

$$\bar{\mathbf{Q}} = \int_V (3\mathbf{r}'\mathbf{r}' - r'^2\bar{\mathbf{I}}) \rho(\mathbf{r}') \, dV'.$$

The expansion (3.92) allows us to identify the dominant power of r for $r \gg r_m$. The first nonzero term in (3.92) dominates the potential at points far from the source. Interestingly, the first nonvanishing moment is independent of the location of the origin of $\mathbf{r}'$, while all subsequent higher moments depend on the location of the origin [91]. We can see this most easily through a few simple examples.

For a single point charge q located at $\mathbf{r}_0$ we can write $\rho(\mathbf{r}) = q\delta(\mathbf{r} - \mathbf{r}_0)$. The first moment of ρ is

$$Q = \int_V q\delta(\mathbf{r}' - \mathbf{r}_0) \, dV' = q.$$

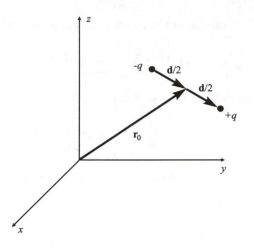

Figure 3.12: A dipole distribution.

Note that this is independent of $\mathbf{r}_0$. The second moment

$$\mathbf{p} = \int_V \mathbf{r}' q\delta(\mathbf{r}' - \mathbf{r}_0)\, dV' = q\mathbf{r}_0$$

depends on $\mathbf{r}_0$, as does the third moment

$$\bar{\mathbf{Q}} = \int_V (3\mathbf{r}'\mathbf{r}' - r'^2\bar{\mathbf{I}})q\delta(\mathbf{r}' - \mathbf{r}_0)\, dV' = q(3\mathbf{r}_0\mathbf{r}_0 - r_0^2\bar{\mathbf{I}}).$$

If $\mathbf{r}_0 = 0$ then only the first moment is nonzero; that this must be the case is obvious from (3.61).

For the dipole of Figure 3.12 we can write

$$\rho(\mathbf{r}) = -q\delta(\mathbf{r} - \mathbf{r}_0 + \mathbf{d}/2) + q\delta(\mathbf{r} - \mathbf{r}_0 - \mathbf{d}/2).$$

In this case

$$Q = -q + q = 0, \qquad \mathbf{p} = q\mathbf{d}, \qquad \bar{\mathbf{Q}} = q[3(\mathbf{r}_0\mathbf{d} + \mathbf{d}\mathbf{r}_0) - 2(\mathbf{r}_0 \cdot \mathbf{d})\bar{\mathbf{I}}].$$

Only the first nonzero moment, in this case $\mathbf{p}$, is independent of $\mathbf{r}_0$. For $\mathbf{r}_0 = 0$ the only nonzero multipole moment would be the dipole moment $\mathbf{p}$. If the dipole is aligned along the z-axis with $\mathbf{d} = d\hat{\mathbf{z}}$ and $\mathbf{r}_0 = 0$, then the exact potential is

$$\Phi(\mathbf{r}) = \frac{1}{4\pi\epsilon}\frac{p\cos\theta}{r^2}.$$

By (3.30) we have

$$\mathbf{E}(\mathbf{r}) = \frac{1}{4\pi\epsilon}\frac{p}{r^3}(\hat{\mathbf{r}}2\cos\theta + \hat{\boldsymbol{\theta}}\sin\theta), \tag{3.93}$$

which is the classic result for the electric field of a dipole.

Finally, consider the quadrupole shown in Figure 3.13. The charge density is

$$\rho(\mathbf{r}) = -q\delta(\mathbf{r} - \mathbf{r}_0) + q\delta(\mathbf{r} - \mathbf{r}_0 - \mathbf{d}_1) + q\delta(\mathbf{r} - \mathbf{r}_0 - \mathbf{d}_2) - q\delta(\mathbf{r} - \mathbf{r}_0 - \mathbf{d}_1 - \mathbf{d}_2).$$

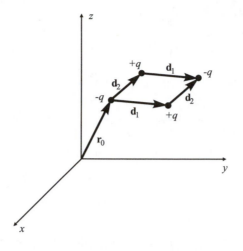

Figure 3.13: A quadrupole distribution.

Carrying through the details, we find that the first two moments of ρ vanish, while the third is given by

$$\bar{\mathbf{Q}} = q[-3(\mathbf{d}_1\mathbf{d}_2 + \mathbf{d}_2\mathbf{d}_1) + 2(\mathbf{d}_1 \cdot \mathbf{d}_2)\bar{\mathbf{I}}].$$

As expected, it is independent of $\mathbf{r}_0$.

It is tedious to carry (3.92) beyond the quadrupole term using the Taylor expansion. Another approach is to expand $1/R$ in spherical harmonics. Referring to Appendix E.3 we find that

$$\frac{1}{|\mathbf{r} - \mathbf{r}'|} = 4\pi \sum_{n=0}^{\infty} \sum_{m=-n}^{n} \frac{1}{2n+1} \frac{r'^n}{r^{n+1}} Y_{nm}^*(\theta', \phi') Y_{nm}(\theta, \phi)$$

(see Jackson [91] or Arfken [5] for a detailed derivation). This expansion converges for $|\mathbf{r}| > |\mathbf{r}_m|$. Substitution into (3.61) gives

$$\Phi(\mathbf{r}) = \frac{1}{\epsilon} \sum_{n=0}^{\infty} \frac{1}{r^{n+1}} \left[\frac{1}{2n+1} \sum_{m=-n}^{n} q_{nm} Y_{nm}(\theta, \phi) \right] \tag{3.94}$$

where

$$q_{nm} = \int_V \rho(\mathbf{r}') r'^n Y_{nm}^*(\theta', \phi') \, dV'.$$

We can now identify any inverse power of r in the multipole expansion, but at the price of dealing with a double summation. For a charge distribution with axial symmetry (no ϕ-variation), only the coefficient q_{n0} is nonzero. The relation

$$Y_{n0}(\theta, \phi) = \sqrt{\frac{2n+1}{4\pi}} P_n(\cos\theta)$$

allows us to simplify (3.94) and obtain

$$\Phi(\mathbf{r}) = \frac{1}{4\pi\epsilon} \sum_{n=0}^{\infty} \frac{1}{r^{n+1}} q_n P_n(\cos\theta) \tag{3.95}$$

where

$$q_n = 2\pi \int_{r'} \int_{\theta'} \rho(r', \theta') r'^n P_n(\cos \theta') r'^2 \sin \theta' \, d\theta' \, dr'.$$

As a simple example consider a spherical distribution of charge given by

$$\rho(\mathbf{r}) = \frac{3Q}{\pi a^3} \cos \theta, \qquad r \leq a.$$

This can be viewed as two adjacent hemispheres carrying total charges $\pm Q$. Since $\cos \theta = P_1(\cos \theta)$, we compute

$$q_n = 2\pi \int_0^a \int_0^\pi \frac{3Q}{\pi a^3} P_1(\cos \theta') r'^n P_n(\cos \theta') r'^2 \sin \theta' \, d\theta' \, dr'$$

$$= 2\pi \frac{3Q}{\pi a^3} \frac{a^{n+3}}{n+3} \int_0^\pi P_1(\cos \theta) P_n(\cos \theta') \sin \theta' \, d\theta'.$$

Using the orthogonality relation (E.123) we find

$$q_n = 2\pi \frac{3Q}{\pi a^3} \frac{a^{n+3}}{n+3} \delta_{1n} \frac{2}{2n+1}.$$

Hence the only nonzero coefficient is $q_1 = Qa$ and

$$\Phi(\mathbf{r}) = \frac{1}{4\pi\epsilon} \frac{1}{r^2} Qa P_1(\cos \theta) = \frac{Qa}{4\pi\epsilon r^2} \cos \theta.$$

This is the potential of a dipole having moment $\mathbf{p} = \hat{\mathbf{z}} Qa$. Thus we could replace the sphere with point charges $\mp Q$ at $z = \mp a/2$ without changing the field for $r > a$.

Physical interpretation of the polarization vector in a dielectric. We have used the Maxwell–Minkowski equations to determine the electrostatic potential of a charge distribution in the presence of a dielectric medium. Alternatively, we can use the Maxwell–Boffi equations

$$\nabla \times \mathbf{E} = 0, \tag{3.96}$$

$$\nabla \cdot \mathbf{E} = \frac{1}{\epsilon_0} (\rho - \nabla \cdot \mathbf{P}). \tag{3.97}$$

Equation (3.96) allows us to define a scalar potential through (3.30). Substitution into (3.97) gives

$$\nabla^2 \Phi(\mathbf{r}) = -\frac{1}{\epsilon_0} [\rho(\mathbf{r}) + \rho_P(\mathbf{r})] \tag{3.98}$$

where $\rho_P = -\nabla \cdot \mathbf{P}$. This has the form of Poisson's equation (3.50), but with charge density term $\rho(\mathbf{r}) + \rho_P(\mathbf{r})$. Hence the solution is

$$\Phi(\mathbf{r}) = \frac{1}{4\pi\epsilon_0} \int_V \frac{\rho(\mathbf{r}') - \nabla' \cdot \mathbf{P}(\mathbf{r}')}{|\mathbf{r} - \mathbf{r}'|} \, dV'.$$

To this we must add any potential produced by surface sources such as ρ_s. If there is a discontinuity in the dielectric region, there is also a surface polarization source $\rho_{Ps} = \hat{\mathbf{n}} \cdot \mathbf{P}$

according to (3.35). Separating the volume into regions with bounding surfaces S_i across which the permittivity is discontinuous, we may write

$$\Phi(\mathbf{r}) = \frac{1}{4\pi\epsilon_0} \int_V \frac{\rho(\mathbf{r}')}{|\mathbf{r}-\mathbf{r}'|}\, dV' + \frac{1}{4\pi\epsilon_0} \int_S \frac{\rho_s(\mathbf{r}')}{|\mathbf{r}-\mathbf{r}'|}\, dS' +$$

$$+ \sum_i \left[\frac{1}{4\pi\epsilon_0} \int_{V_i} \frac{-\nabla'\cdot\mathbf{P}(\mathbf{r}')}{|\mathbf{r}-\mathbf{r}'|}\, dV' + \frac{1}{4\pi\epsilon_0} \oint_{S_i} \frac{\hat{\mathbf{n}}'\cdot\mathbf{P}(\mathbf{r}')}{|\mathbf{r}-\mathbf{r}'|}\, dS' \right], \qquad (3.99)$$

where $\hat{\mathbf{n}}$ points outward from region i. Using the divergence theorem on the fourth term and employing (B.42), we obtain

$$\Phi(\mathbf{r}) = \frac{1}{4\pi\epsilon_0} \int_V \frac{\rho(\mathbf{r}')}{|\mathbf{r}-\mathbf{r}'|}\, dV' + \frac{1}{4\pi\epsilon_0} \int_S \frac{\rho_s(\mathbf{r}')}{|\mathbf{r}-\mathbf{r}'|}\, dS' +$$

$$+ \sum_i \left[\frac{1}{4\pi\epsilon_0} \int_{V_i} \mathbf{P}(\mathbf{r}')\cdot\nabla'\left(\frac{1}{|\mathbf{r}-\mathbf{r}'|}\right) dV' \right].$$

Since $\nabla'(1/R) = \hat{\mathbf{R}}/R^2$, the third term is a sum of integrals of the form

$$\frac{1}{4\pi\epsilon} \int_{V_i} \mathbf{P}(\mathbf{r}')\cdot\frac{\hat{\mathbf{R}}}{R^2}\, dV.$$

Comparing this to the second term of (3.92), we see that this integral represents a volume superposition of dipole terms where $\mathbf{P}$ is a volume density of dipole moments.

Thus, a dielectric with permittivity ϵ is equivalent to a volume distribution of dipoles in free space. No higher-order moments are required, and no zero-order moments are needed since any net charge is included in ρ. Note that we have arrived at this conclusion based only on Maxwell's equations and the assumption of a linear, isotropic relationship between $\mathbf{D}$ and $\mathbf{E}$. Assuming our macroscopic theory is correct, we are tempted to make assumptions about the behavior of matter on a microscopic level (e.g., atoms exposed to fields are polarized and their electron clouds are displaced from their positively charged nuclei), but this area of science is better studied from the viewpoints of particle physics and quantum mechanics.

Potential of an azimuthally-symmetric charged spherical surface. In several of our example problems we shall be interested in evaluating the potential of a charged spherical surface. When the charge is azimuthally-symmetric, the potential is particularly simple.

We will need the value of the integral

$$F(\mathbf{r}) = \frac{1}{4\pi} \int_S \frac{f(\theta')}{|\mathbf{r}-\mathbf{r}'|}\, dS' \qquad (3.100)$$

where $\mathbf{r} = r\hat{\mathbf{r}}$ describes an arbitrary observation point and $\mathbf{r}' = a\hat{\mathbf{r}}'$ identifies the source point on the surface of the sphere of radius a. The integral is most easily done using the expansion (E.200) for $|\mathbf{r}-\mathbf{r}'|^{-1}$ in spherical harmonics. We have

$$F(\mathbf{r}) = a^2 \sum_{n=0}^{\infty} \sum_{m=-n}^{n} \frac{Y_{nm}(\theta,\phi)}{2n+1} \frac{r_<^n}{r_>^{n+1}} \int_{-\pi}^{\pi} \int_0^{\pi} f(\theta') Y_{nm}^*(\theta',\phi') \sin\theta'\, d\theta'\, d\phi'$$

where $r_< = \min\{r, a\}$ and $r_> = \max\{r, a\}$. Using orthogonality of the exponentials we find that only the $m = 0$ terms contribute:

$$F(\mathbf{r}) = 2\pi a^2 \sum_{n=0}^{\infty} \frac{Y_{n0}(\theta, \phi)}{2n + 1} \frac{r_<^n}{r_>^{n+1}} \int_0^{\pi} f(\theta') Y_{n0}^*(\theta', \phi') \sin\theta' \, d\theta'.$$

Finally, since

$$Y_{n0} = \sqrt{\frac{2n + 1}{4\pi}} P_n(\cos\theta)$$

we have

$$F(\mathbf{r}) = \frac{1}{2} a^2 \sum_{n=0}^{\infty} P_n(\cos\theta) \frac{r_<^n}{r_>^{n+1}} \int_0^{\pi} f(\theta') P_n(\cos\theta') \sin\theta' \, d\theta'. \qquad (3.101)$$

As an example, suppose $f(\theta) = \cos\theta = P_1(\cos\theta)$. Then

$$F(\mathbf{r}) = \frac{1}{2} a^2 \sum_{n=0}^{\infty} P_n(\cos\theta) \frac{r_<^n}{r_>^{n+1}} \int_0^{\pi} P_1(\cos\theta') P_n(\cos\theta') \sin\theta' \, d\theta'.$$

The orthogonality of the Legendre polynomials can be used to show that

$$\int_0^{\pi} P_1(\cos\theta') P_n(\cos\theta') \sin\theta' \, d\theta' = \frac{2}{3} \delta_{1n},$$

hence

$$F(\mathbf{r}) = \frac{a^2}{3} \cos\theta \frac{r_<}{r_>^2}. \qquad (3.102)$$

3.2.7   Field produced by a permanently polarized body

Certain materials, called *electrets*, exhibit polarization in the absence of an external electric field. A permanently polarized material produces an electric field both internal and external to the material, hence there must be a charge distribution to support the fields. We can interpret this charge as being caused by the permanent separation of atomic charge within the material, but if we are only interested in the macroscopic field then we need not worry about the microscopic implications of such materials. Instead, we can use the Maxwell–Boffi equations and find the potential produced by the material by using (3.99). Thus, the field of an electret with known polarization $\mathbf{P}$ occupying volume region V in free space is dipolar in nature and is given by

$$\Phi(\mathbf{r}) = \frac{1}{4\pi\epsilon_0} \int_V \frac{-\nabla' \cdot \mathbf{P}(\mathbf{r}')}{|\mathbf{r} - \mathbf{r}'|} \, dV' + \frac{1}{4\pi\epsilon_0} \oint_S \frac{\hat{\mathbf{n}}' \cdot \mathbf{P}(\mathbf{r}')}{|\mathbf{r} - \mathbf{r}'|} \, dS'$$

where $\hat{\mathbf{n}}$ points out of the volume region V.

As an example, consider a material sphere of radius a, permanently polarized along its axis with uniform polarization $\mathbf{P}(\mathbf{r}) = \hat{\mathbf{z}} P_0$. We have the equivalent source densities

$$\rho_p = -\nabla \cdot \mathbf{P} = 0, \qquad \rho_{Ps} = \hat{\mathbf{n}} \cdot \mathbf{P} = \hat{\mathbf{r}} \cdot \hat{\mathbf{z}} P_0 = P_0 \cos\theta.$$

Then

$$\Phi(\mathbf{r}) = \frac{1}{4\pi\epsilon_0} \oint_S \frac{\rho_{Ps}(\mathbf{r}')}{|\mathbf{r} - \mathbf{r}'|} \, dS' = \frac{1}{4\pi\epsilon_0} \oint_S \frac{P_0 \cos\theta'}{|\mathbf{r} - \mathbf{r}'|} \, dS'.$$

The integral takes the form (3.100), hence by (3.102) the solution is

$$\Phi(\mathbf{r}) = P_0 \frac{a^2}{3\epsilon_0} \cos\theta \frac{r_<}{r_>^2}. \tag{3.103}$$

If we are interested only in the potential for $r > a$, we can use the multipole expansion (3.95) to obtain

$$\Phi(\mathbf{r}) = \frac{1}{4\pi\epsilon_0} \sum_{n=0}^{\infty} \frac{1}{r^{n+1}} q_n P_n(\cos\theta), \qquad r > a$$

where

$$q_n = 2\pi \int_0^\pi \rho_{Ps}(\theta') a^n P_n(\cos\theta') a^2 \sin\theta' \, d\theta'.$$

Substituting for ρ_{Ps} and remembering that $\cos\theta = P_1(\cos\theta)$, we have

$$q_n = 2\pi a^{n+2} P_0 \int_0^\pi P_1(\cos\theta') P_n(\cos\theta') \sin\theta' \, d\theta'.$$

Using the orthogonality relation (E.123) we find

$$q_n = 2\pi a^{n+2} P_0 \delta_{1n} \frac{2}{2n+1}.$$

Therefore the only nonzero coefficient is

$$q_1 = \frac{4\pi a^3 P_0}{3}$$

and

$$\Phi(\mathbf{r}) = \frac{1}{4\pi\epsilon_0} \frac{1}{r^2} \frac{4\pi a^3 P_0}{3} P_1(\cos\theta) = \frac{P_0 a^3}{3\epsilon_0 r^2} \cos\theta, \qquad r > a.$$

This is a dipole field, and matches (3.103) as expected.

3.2.8 Potential of a dipole layer

Surface charge layers sometimes occur in bipolar form, such as in the membrane surrounding an animal cell. These can be modeled as a dipole layer consisting of parallel surface charges of opposite sign.

Consider a surface S located in free space. Parallel to this surface, and a distance $\Delta/2$ below, is located a surface charge layer of density $\rho_s(\mathbf{r}) = P_s(\mathbf{r})$. Also parallel to S, but a distance $\Delta/2$ above, is a surface charge layer of density $\rho_s(\mathbf{r}) = -P_s(\mathbf{r})$. We define the *surface dipole moment density* D_s as

$$D_s(\mathbf{r}) = \Delta P_s(\mathbf{r}). \tag{3.104}$$

Letting the position vector $\mathbf{r}_0'$ point to the surface S we can write the potential (3.61) produced by the two charge layers as

$$\Phi(\mathbf{r}) = \frac{1}{4\pi\epsilon_0} \int_{S+} P_s(\mathbf{r}') \frac{1}{|\mathbf{r} - \mathbf{r}_0' - \hat{\mathbf{n}}'\frac{\Delta}{2}|} \, dS' - \frac{1}{4\pi\epsilon_0} \int_{S-} P_s(\mathbf{r}') \frac{1}{|\mathbf{r} - \mathbf{r}_0' + \hat{\mathbf{n}}'\frac{\Delta}{2}|} \, dS'.$$

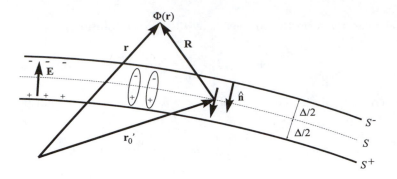

Figure 3.14: A dipole layer.

We are interested in the case in which the two charge layers collapse onto the surface S, and wish to compute the potential produced by a given dipole moment density. When $\Delta \to 0$ we have $\mathbf{r}_0' \to \mathbf{r}'$ and may write

$$\Phi(\mathbf{r}) = \lim_{\Delta \to 0} \frac{1}{4\pi\epsilon_0} \int_S \frac{D_s(\mathbf{r}')}{\Delta} \left[\frac{1}{|\mathbf{R} - \hat{\mathbf{n}}'\frac{\Delta}{2}|} - \frac{1}{|\mathbf{R} + \hat{\mathbf{n}}'\frac{\Delta}{2}|} \right] dS',$$

where $\mathbf{R} = \mathbf{r} - \mathbf{r}'$. By the binomial theorem, the limit of the term in brackets can be written as

$$\lim_{\Delta \to 0} \left(\left[R^2 + \left(\frac{\Delta}{2}\right)^2 - 2\mathbf{R} \cdot \hat{\mathbf{n}}'\frac{\Delta}{2} \right]^{-\frac{1}{2}} - \left[R^2 + \left(\frac{\Delta}{2}\right)^2 + 2\mathbf{R} \cdot \hat{\mathbf{n}}'\frac{\Delta}{2} \right]^{-\frac{1}{2}} \right)$$

$$= \lim_{\Delta \to 0} \left(R^{-1} \left[1 + \frac{\hat{\mathbf{R}} \cdot \hat{\mathbf{n}}'}{R}\frac{\Delta}{2} \right] - R^{-1} \left[1 - \frac{\hat{\mathbf{R}} \cdot \hat{\mathbf{n}}'}{R}\frac{\Delta}{2} \right] \right) = \Delta\hat{\mathbf{n}}' \cdot \frac{\mathbf{R}}{R^3}.$$

Thus

$$\Phi(\mathbf{r}) = \frac{1}{4\pi\epsilon_0} \int_S \mathbf{D}_s(\mathbf{r}') \cdot \frac{\mathbf{R}}{R^3} \, dS' \qquad (3.105)$$

where $\mathbf{D}_s = \hat{\mathbf{n}} D_s$ is the surface vector dipole moment density. The potential of a dipole layer decreases more rapidly ($\sim 1/r^2$) than that of a unipolar charge layer. We saw similar behavior in the dipole term of the multipole expansion (3.92) for a general charge distribution.

We can use (3.105) to study the behavior of the potential across a dipole layer. As we approach the layer from above, the greatest contribution to Φ comes from the charge region immediately beneath the observation point. Assuming that the surface dipole moment density is continuous beneath the point, we can compute the difference in the fields across the layer at point $\mathbf{r}$ by replacing the arbitrary surface layer by a disk of constant surface dipole moment density $\mathbf{D}_0 = \mathbf{D}_s(\mathbf{r})$. For simplicity we center the disk at $z = 0$ in the xy-plane as shown in Figure 3.15 and compute the potential difference ΔV across the layer; i.e., $\Delta V = \Phi(h) - \Phi(-h)$ on the disk axis as $h \to 0$. Using (3.105) along with $\mathbf{r}' = \pm h\hat{\mathbf{z}} - \rho'\hat{\boldsymbol{\rho}}'$, we obtain

$$\Delta V = \lim_{h \to 0} \left[\frac{1}{4\pi\epsilon_0} \int_0^{2\pi} \int_0^a [\hat{\mathbf{z}} D_0] \cdot \frac{\hat{\mathbf{z}} h - \hat{\boldsymbol{\rho}}' \rho'}{(h^2 + \rho'^2)^{3/2}} \rho' \, d\rho' \, d\phi' - \right.$$

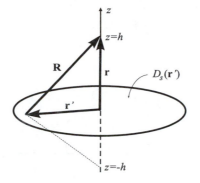

Figure 3.15: Auxiliary disk for studying the potential distribution across a dipole layer.

$$-\frac{1}{4\pi\epsilon_0}\int_0^{2\pi}\int_0^a [\hat{\mathbf{z}}D_0]\cdot\frac{-\hat{\mathbf{z}}h-\hat{\boldsymbol{\rho}}'\rho'}{(h^2+\rho'^2)^{3/2}}\rho'\,d\rho'\,d\phi'\Bigg]$$

where a is the disk radius. Integration yields

$$\Delta V = \frac{D_0}{2\epsilon_0}\lim_{h\to 0}\left[\frac{-2}{\sqrt{1+\left(\frac{a}{h}\right)^2}}+2\right]=\frac{D_0}{\epsilon_0},$$

independent of a. Generalizing this to an arbitrary surface dipole moment density, we find that the boundary condition on the potential is given by

$$\Phi_2(\mathbf{r}) - \Phi_1(\mathbf{r}) = \frac{D_s(\mathbf{r})}{\epsilon_0} \qquad (3.106)$$

where "1" denotes the positive side of the dipole moments and "2" the negative side. Physically, the potential difference in (3.106) is produced by the line integral of $\mathbf{E}$ "internal" to the dipole layer. Since there is no field internal to a unipolar surface layer, V is continuous across a surface containing charge ρ_s but having $D_s = 0$.

3.2.9   Behavior of electric charge density near a conducting edge

Sharp corners are often encountered in the application of electrostatics to practical geometries. The behavior of the charge distribution near these corners must be understood in order to develop numerical techniques for solving more complicated problems. We can use a simple model of a corner if we restrict our interest to the region near the edge. Consider the intersection of two planes as shown in Figure 3.16. The region near the intersection represents the corner we wish to study. We assume that the planes are held at zero potential and that the charge on the surface is induced by a two-dimensional charge distribution $\rho(\mathbf{r})$, or by a potential difference between the edge and another conductor far removed from the edge.

We can find the potential in the region near the edge by solving Laplace's equation in cylindrical coordinates. This problem is studied in Appendix A where the separation of variables solution is found to be either (A.127) or (A.128). Using (A.128) and enforcing $\Phi = 0$ at both $\phi = 0$ and $\phi = \beta$, we obtain the null solution. Hence the solution must take the form (A.127):

$$\Phi(\rho,\phi) = [A_\phi \sin(k_\phi\phi) + B_\phi \cos(k_\phi\phi)][a_\rho\rho^{-k_\phi}+b_\rho\rho^{k_\phi}]. \qquad (3.107)$$

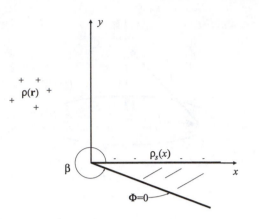

Figure 3.16: A conducting edge.

Since the origin is included we cannot have negative powers of ρ and must put $a_\rho = 0$. The boundary condition $\Phi(\rho, 0) = 0$ requires $B_\phi = 0$. The condition $\Phi(\rho, \beta) = 0$ then requires $\sin(k_\phi \beta) = 0$, which holds only if $k_\phi = n\pi/\beta$, $n = 1, 2, \ldots$. The general solution for the potential near the edge is therefore

$$\Phi(\rho, \phi) = \sum_{n=1}^{N} A_n \sin\left(\frac{n\pi}{\beta}\phi\right) \rho^{n\pi/\beta} \qquad (3.108)$$

where the constants A_n depend on the excitation source or system of conductors. (Note that if the corner is held at potential $V_0 \neq 0$, we must merely add V_0 to the solution.) The charge on the conducting surfaces can be computed from the boundary condition on normal $\mathbf{D}$. Using (3.30) we have

$$E_\phi = -\frac{1}{\rho}\frac{\partial}{\partial \phi} \sum_{n=1}^{N} A_n \sin\left(\frac{n\pi}{\beta}\phi\right) \rho^{n\pi/\beta} = -\sum_{n=1}^{N} A_n \frac{n\pi}{\beta} \cos\left(\frac{n\pi}{\beta}\phi\right) \rho^{(n\pi/\beta)-1},$$

hence

$$\rho_s(x) = -\epsilon \sum_{n=1}^{N} A_n \frac{n\pi}{\beta} x^{(n\pi/\beta)-1}$$

on the surface at $\phi = 0$. Near the edge, at small values of x, the variation of ρ_s is dominated by the lowest power of x. (Here we ignore those special excitation arrangements that produce $A_1 = 0$.) Thus

$$\rho_s(x) \sim x^{(\pi/\beta)-1}.$$

The behavior of the charge clearly depends on the wedge angle β. For a sharp edge (half plane) we put $\beta = 2\pi$ and find that the field varies as $x^{-1/2}$. This *square-root edge singularity* is very common on thin plates, fins, etc., and means that charge tends to accumulate near the edge of a flat conducting surface. For a right-angle corner where $\beta = 3\pi/2$, there is the somewhat weaker singularity $x^{-1/3}$. When $\beta = \pi$, the two surfaces fold out into an infinite plane and the charge, not surprisingly, is invariant with x to lowest order near the folding line. When $\beta < \pi$ the corner becomes interior and we find that the charge density varies with a positive power of distance from the edge. For very sharp interior angles the power is large, meaning that little charge accumulates on the inner surfaces near an interior corner.

3.2.10 Solution to Laplace's equation for bodies immersed in an impressed field

An important class of problems is based on the idea of placing a body into an existing electric field, assuming that the field arises from sources so remote that the introduction of the body does not alter the original field. The pre-existing field is often referred to as the *applied* or *impressed field*, and the solution external to the body is usually formulated as the sum of the applied field and a *secondary* or *scattered field* that satisfies Laplace's equation. This total field differs from the applied field, and must satisfy the appropriate boundary condition on the body. If the body is a conductor then the total potential must be constant everywhere on the boundary surface. If the body is a solid homogeneous dielectric then the total potential field must be continuous across the boundary.

As an example, consider a dielectric sphere of permittivity ϵ and radius a, centered at the origin and immersed in a constant electric field $\mathbf{E}_0(\mathbf{r}) = E_0\hat{\mathbf{z}}$. By (3.30) the applied potential field is $\Phi_0(\mathbf{r}) = -E_0 z = -E_0 r \cos\theta$ (to within a constant). Outside the sphere $(r > a)$ we write the total potential field as

$$\Phi_2(\mathbf{r}) = \Phi_0(\mathbf{r}) + \Phi^s(\mathbf{r})$$

where $\Phi^s(\mathbf{r})$ is the secondary or scattered potential. Since Φ^s must satisfy Laplace's equation, we can write it as a separation of variables solution (§ A.4). By azimuthal symmetry the potential has an r-dependence as in (A.146), and a θ-dependence as in (A.142) with $B_\theta = 0$ and $m = 0$. Thus Φ^s has a representation identical to (A.147), except that we cannot use terms that are unbounded as $r \to \infty$. We therefore use

$$\Phi^s(r, \theta) = \sum_{n=0}^{\infty} B_n r^{-(n+1)} P_n(\cos\theta). \tag{3.109}$$

The potential inside the sphere also obeys Laplace's equation, so we can use the same form (A.147) while discarding terms unbounded at the origin. Thus

$$\Phi_1(r, \theta) = \sum_{n=0}^{\infty} A_n r^n P_n(\cos\theta) \tag{3.110}$$

for $r < a$. To find the constants A_n and B_n we apply (3.36) and (3.37) to the total field. Application of (3.36) at $r = a$ gives

$$-E_0 a \cos\theta + \sum_{n=0}^{\infty} B_n a^{-(n+1)} P_n(\cos\theta) = \sum_{n=0}^{\infty} A_n a^n P_n(\cos\theta).$$

Multiplying through by $P_m(\cos\theta)\sin\theta$, integrating from $\theta = 0$ to $\theta = \pi$, and using the orthogonality relationship (E.123), we obtain

$$-E_0 a + a^{-2}B_1 = A_1 a, \tag{3.111}$$

$$B_n a^{-(n+1)} = A_n a^n, \quad n \neq 1, \tag{3.112}$$

where we have used $P_1(\cos\theta) = \cos\theta$. Next, since $\rho_s = 0$, equation (3.37) requires that

$$\epsilon_1 \frac{\partial \Phi_1(\mathbf{r})}{\partial r} = \epsilon_2 \frac{\partial \Phi_2(\mathbf{r})}{\partial r}$$

at $r = a$. This gives

$$-\epsilon_0 E_0 \cos\theta + \epsilon_0 \sum_{n=0}^{\infty} [-(n+1)B_n]a^{-n-2}P_n(\cos\theta) = \epsilon \sum_{n=0}^{\infty} [nA_n]a^{n-1}P_n(\cos\theta).$$

By orthogonality of the Legendre functions we have

$$-\epsilon_0 E_0 - 2\epsilon_0 B_1 a^{-3} = \epsilon A_1, \tag{3.113}$$

$$-\epsilon_0(n+1)B_n a^{-n-2} = \epsilon n A_n a^{n-1}, \quad n \neq 1. \tag{3.114}$$

Equations (3.112) and (3.114) cannot hold simultaneously unless $A_n = B_n = 0$ for $n \neq 1$. Solving (3.111) and (3.113) we have

$$A_1 = -E_0 \frac{3\epsilon_0}{\epsilon + 2\epsilon_0}, \qquad B_1 = E_0 a^3 \frac{\epsilon - \epsilon_0}{\epsilon + 2\epsilon_0}.$$

Hence

$$\Phi_1(\mathbf{r}) = -E_0 \frac{3\epsilon_0}{\epsilon + 2\epsilon_0} r\cos\theta = -E_0 z \frac{3\epsilon_0}{\epsilon + 2\epsilon_0}, \tag{3.115}$$

$$\Phi_2(\mathbf{r}) = -E_0 r\cos\theta + E_0 \frac{a^3}{r^2} \frac{\epsilon - \epsilon_0}{\epsilon + 2\epsilon_0} \cos\theta. \tag{3.116}$$

Interestingly, the electric field

$$\mathbf{E}_1(\mathbf{r}) = -\nabla\Phi_1(\mathbf{r}) = \hat{\mathbf{z}} E_0 \frac{3\epsilon_0}{\epsilon + 2\epsilon_0}$$

inside the sphere is constant with position and is aligned with the applied external field. However, it is weaker than the applied field since $\epsilon > \epsilon_0$. To explain this, we compute the polarization charge within and on the sphere. Using $\mathbf{D} = \epsilon\mathbf{E} = \epsilon_0\mathbf{E} + \mathbf{P}$ we have

$$\mathbf{P}_1 = \hat{\mathbf{z}}(\epsilon - \epsilon_0)E_0 \frac{3\epsilon_0}{\epsilon + 2\epsilon_0}. \tag{3.117}$$

The volume polarization charge density $-\nabla \cdot \mathbf{P}$ is zero, while the polarization surface charge density is

$$\rho_{Ps} = \hat{\mathbf{r}} \cdot \mathbf{P} = (\epsilon - \epsilon_0)E_0 \frac{3\epsilon_0}{\epsilon + 2\epsilon_0} \cos\theta.$$

Hence the secondary electric field can be attributed to an induced surface polarization charge, and is in a direction opposing the applied field. According to the Maxwell–Boffi viewpoint we should be able to replace the sphere by the surface polarization charge immersed in free space, and use the formula (3.61) to reproduce (3.115) and (3.116). This is left as an exercise for the reader.

3.3   Magnetostatics

The large-scale forms of the magnetostatic field equations are

$$\oint_\Gamma \mathbf{H} \cdot \mathbf{dl} = \int_S \mathbf{J} \cdot \mathbf{dS}, \tag{3.118}$$

$$\oint_S \mathbf{B} \cdot \mathbf{dS} = 0, \tag{3.119}$$

while the point forms are

$$\nabla \times \mathbf{H}(\mathbf{r}) = \mathbf{J}(\mathbf{r}), \tag{3.120}$$
$$\nabla \cdot \mathbf{B}(\mathbf{r}) = 0. \tag{3.121}$$

Note the interesting dichotomy between the electrostatic field equations and the magnetostatic field equations. Whereas the electrostatic field exhibits zero curl and a divergence proportional to the source (charge), the magnetostatic field has zero divergence and a curl proportional to the source (current). Because the vector relationship between the magnetostatic field and its source is of a more complicated nature than the scalar relationship between the electrostatic field and its source, more effort is required to develop a strong understanding of magnetic phenomena. Also, it must always be remembered that although the equations describing the electrostatic and magnetostatic field sets decouple, the phenomena themselves remain linked. Since current is moving charge, electrical phenomena are associated with the establishment of the current that supports a magnetostatic field. We know, for example, that in order to have current in a wire an electric field must be present to drive electrons through the wire.

The magnetic scalar potential. Under certain conditions the equations of magnetostatics have the same form as those of electrostatics. If $\mathbf{J} = 0$ in a region V, the magnetostatic equations are

$$\nabla \times \mathbf{H}(\mathbf{r}) = 0, \tag{3.122}$$
$$\nabla \cdot \mathbf{B}(\mathbf{r}) = 0; \tag{3.123}$$

compare with (3.5)–(3.6) when $\rho = 0$. Using (3.122) we can define a *magnetic scalar potential* Φ_m:

$$\mathbf{H} = -\nabla \Phi_m. \tag{3.124}$$

The negative sign is chosen for consistency with (3.30). We can then define a magnetic potential difference between two points as

$$V_{m21} = -\int_{P_1}^{P_2} \mathbf{H} \cdot d\mathbf{l} = -\int_{P_1}^{P_2} -\nabla \Phi_m(\mathbf{r}) \cdot d\mathbf{l} = \int_{P_1}^{P_2} d\Phi_m(\mathbf{r}) = \Phi_m(\mathbf{r}_2) - \Phi_m(\mathbf{r}_1).$$

Unlike the electrostatic potential difference, V_{m21} is not unique. Consider Figure 3.17, which shows a plane passing through the cross-section of a wire carrying total current I. Although there is no current within the region V (external to the wire), equation (3.118) still gives

$$\int_{\Gamma_2} \mathbf{H} \cdot d\mathbf{l} - \int_{\Gamma_3} \mathbf{H} \cdot d\mathbf{l} = I.$$

Thus

$$\int_{\Gamma_2} \mathbf{H} \cdot d\mathbf{l} = \int_{\Gamma_3} \mathbf{H} \cdot d\mathbf{l} + I,$$

and the integral $\int_\Gamma \mathbf{H} \cdot d\mathbf{l}$ is not path-independent. However,

$$\int_{\Gamma_1} \mathbf{H} \cdot d\mathbf{l} = \int_{\Gamma_2} \mathbf{H} \cdot d\mathbf{l}$$

since no current passes through the surface bounded by $\Gamma_1 - \Gamma_2$. So we can artificially impose uniqueness by demanding that no path cross a cut such as that indicated by the line L in the figure.

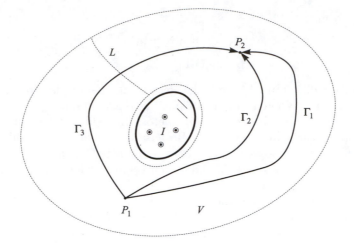

Figure 3.17: Magnetic potential.

Because V_{m21} is not unique, the field $\mathbf{H}$ is nonconservative. In point form this is shown by the fact that $\nabla \times \mathbf{H}$ is not identically zero. We are not too concerned about energy-related implications of the nonconservative nature of $\mathbf{H}$; the electric point charge has no magnetic analogue that might fail to conserve potential energy if moved around in a magnetic field.

Assuming a linear, isotropic region where $\mathbf{B}(\mathbf{r}) = \mu(\mathbf{r})\mathbf{H}(\mathbf{r})$, we can substitute (3.124) into (3.123) and expand to obtain

$$\nabla \mu(\mathbf{r}) \cdot \nabla \Phi_m(\mathbf{r}) + \mu(\mathbf{r}) \nabla^2 \Phi_m(\mathbf{r}) = 0.$$

For a homogeneous medium this reduces to Laplace's equation

$$\nabla^2 \Phi_m = 0.$$

We can also obtain an analogue to Poisson's equation of electrostatics if we use

$$\mathbf{B} = \mu_0(\mathbf{H} + \mathbf{M}) = -\mu_0 \nabla \Phi_m + \mu_0 \mathbf{M}$$

in (3.123); we have

$$\nabla^2 \Phi_m = -\rho_M \tag{3.125}$$

where

$$\rho_M = -\nabla \cdot \mathbf{M}$$

is called the *equivalent magnetization charge density*. This form can be used to describe fields of permanent magnets in the absence of $\mathbf{J}$. Comparison with (3.98) shows that ρ_M is analogous to the polarization charge ρ_P.

Since Φ_m obeys Poisson's equation, the details regarding uniqueness and the construction of solutions follow from those of the electrostatic case. If we include the possibility of a surface density of magnetization charge, then the integral solution for Φ_m in unbounded space is

$$\Phi_m(\mathbf{r}) = \frac{1}{4\pi} \int_V \frac{\rho_M(\mathbf{r}')}{|\mathbf{r} - \mathbf{r}'|}\, dV' + \frac{1}{4\pi} \int_S \frac{\rho_{Ms}(\mathbf{r}')}{|\mathbf{r} - \mathbf{r}'|}\, dS'. \tag{3.126}$$

Here ρ_{Ms}, the surface density of magnetization charge, is identified as $\hat{\mathbf{n}} \cdot \mathbf{M}$ in the boundary condition (3.152).

3.3.1 The magnetic vector potential

Although the magnetic scalar potential is useful for describing fields of permanent magnets and for solving certain boundary value problems, it does not include the effects of source current. A second type of potential function, called the *magnetic vector potential,* can be used with complete generality to describe the magnetostatic field. Because $\nabla \cdot \mathbf{B} = 0$, we can write by (B.49)

$$\mathbf{B}(\mathbf{r}) = \nabla \times \mathbf{A}(\mathbf{r}) \tag{3.127}$$

where $\mathbf{A}$ is the vector potential. Now $\mathbf{A}$ is not determined by (3.127) alone, since the gradient of any scalar field can be added to $\mathbf{A}$ without changing the value of $\nabla \times \mathbf{A}$. Such "gauge transformations" are discussed in Chapter 5, where we find that $\nabla \cdot \mathbf{A}$ must also be specified for uniqueness of $\mathbf{A}$.

The vector potential can be used to develop a simple formula for the magnetic flux passing through an open surface S:

$$\Psi_m = \int_S \mathbf{B} \cdot d\mathbf{S} = \int_S (\nabla \times \mathbf{A}) \cdot d\mathbf{S} = \oint_\Gamma \mathbf{A} \cdot d\mathbf{l}, \tag{3.128}$$

where Γ is the contour bounding S.

In the linear isotropic case where $\mathbf{B} = \mu \mathbf{H}$ we can find a partial differential equation for $\mathbf{A}$ by substituting (3.127) into (3.120). Using (B.43) we have

$$\nabla \times \left[\frac{1}{\mu(\mathbf{r})} \nabla \times \mathbf{A}(\mathbf{r}) \right] = \mathbf{J}(\mathbf{r}),$$

hence

$$\frac{1}{\mu(\mathbf{r})} \nabla \times [\nabla \times \mathbf{A}(\mathbf{r})] - [\nabla \times \mathbf{A}(\mathbf{r})] \times \nabla \left(\frac{1}{\mu(\mathbf{r})} \right) = \mathbf{J}(\mathbf{r}).$$

In a homogeneous region we have

$$\nabla \times (\nabla \times \mathbf{A}) = \mu \mathbf{J} \tag{3.129}$$

or

$$\nabla(\nabla \cdot \mathbf{A}) - \nabla^2 \mathbf{A} = \mu \mathbf{J} \tag{3.130}$$

by (B.47). As mentioned above we must eventually specify $\nabla \cdot \mathbf{A}$. Although the choice is arbitrary, certain selections make the computation of $\mathbf{A}$ both mathematically tractable and physically meaningful. The "Coulomb gauge condition" $\nabla \cdot \mathbf{A} = 0$ reduces (3.130) to

$$\nabla^2 \mathbf{A} = -\mu \mathbf{J}. \tag{3.131}$$

The vector potential concept can also be applied to the Maxwell–Boffi magnetostatic equations

$$\nabla \times \mathbf{B} = \mu_0 (\mathbf{J} + \nabla \times \mathbf{M}), \tag{3.132}$$
$$\nabla \cdot \mathbf{B} = 0. \tag{3.133}$$

By (3.133) we may still define $\mathbf{A}$ through (3.127). Substituting this into (3.132) we have, under the Coulomb gauge,

$$\nabla^2 \mathbf{A} = -\mu_0 [\mathbf{J} + \mathbf{J}_M] \tag{3.134}$$

where $\mathbf{J}_M = \nabla \times \mathbf{M}$ is the magnetization current density.

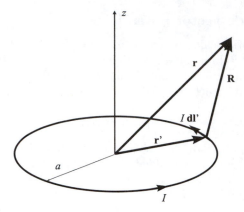

Figure 3.18: Circular loop of wire.

The differential equations (3.131) and (3.134) are vector versions of Poisson's equation, and may be solved quite easily for unbounded space by decomposing the vector source into rectangular components. For instance, dotting (3.131) with $\hat{\mathbf{x}}$ we find that

$$\nabla^2 A_x = -\mu J_x.$$

This scalar version of Poisson's equation has solution

$$A_x(\mathbf{r}) = \frac{\mu}{4\pi} \int_V \frac{J_x(\mathbf{r}')}{|\mathbf{r} - \mathbf{r}'|}\, dV'$$

in unbounded space. Repeating this for each component and assembling the results, we obtain the solution for the vector potential in an unbounded homogeneous medium:

$$\mathbf{A}(\mathbf{r}) = \frac{\mu}{4\pi} \int_V \frac{\mathbf{J}(\mathbf{r}')}{|\mathbf{r} - \mathbf{r}'|}\, dV'. \tag{3.135}$$

Any surface sources can be easily included through a surface integral:

$$\mathbf{A}(\mathbf{r}) = \frac{\mu}{4\pi} \int_V \frac{\mathbf{J}(\mathbf{r}')}{|\mathbf{r} - \mathbf{r}'|}\, dV' + \frac{\mu}{4\pi} \int_S \frac{\mathbf{J}_s(\mathbf{r}')}{|\mathbf{r} - \mathbf{r}'|}\, dS'. \tag{3.136}$$

In unbounded free space containing materials represented by $\mathbf{M}$, we have

$$\mathbf{A}(\mathbf{r}) = \frac{\mu_0}{4\pi} \int_V \frac{\mathbf{J}(\mathbf{r}') + \mathbf{J}_M(\mathbf{r}')}{|\mathbf{r} - \mathbf{r}'|}\, dV' + \frac{\mu_0}{4\pi} \int_S \frac{\mathbf{J}_s(\mathbf{r}') + \mathbf{J}_{Ms}(\mathbf{r}')}{|\mathbf{r} - \mathbf{r}'|}\, dV' \tag{3.137}$$

where $\mathbf{J}_{Ms} = -\hat{\mathbf{n}} \times \mathbf{M}$ is the surface density of magnetization current as described in (3.153). It may be verified directly from (3.137) that $\nabla \cdot \mathbf{A} = 0$.

Field of a circular loop. Consider a circular loop of line current of radius a in unbounded space (Figure 3.18). Using $\mathbf{J}(\mathbf{r}') = I\hat{\boldsymbol{\phi}}'\delta(z')\delta(\rho' - a)$ and noting that $\mathbf{r} = \rho\hat{\boldsymbol{\rho}} + z\hat{\mathbf{z}}$ and $\mathbf{r}' = a\hat{\boldsymbol{\rho}}'$, we can write (3.136) as

$$\mathbf{A}(\mathbf{r}) = \frac{\mu I}{4\pi} \int_0^{2\pi} \hat{\boldsymbol{\phi}}' \frac{a\, d\phi'}{[\rho^2 + a^2 + z^2 - 2a\rho\cos(\phi - \phi')]^{1/2}}.$$

Because $\hat{\boldsymbol{\phi}}' = -\hat{\mathbf{x}} \cos\phi' + \hat{\mathbf{y}} \sin\phi'$ we find that

$$\mathbf{A}(\mathbf{r}) = \frac{\mu I a}{4\pi} \hat{\boldsymbol{\phi}} \int_0^{2\pi} \frac{\cos\phi'}{\left[\rho^2 + a^2 + z^2 - 2a\rho\cos\phi'\right]^{1/2}} \, d\phi'.$$

We put the integral into standard form by setting $\phi' = \pi - 2x$:

$$\mathbf{A}(\mathbf{r}) = -\frac{\mu I a}{4\pi} \hat{\boldsymbol{\phi}} \int_{-\pi/2}^{\pi/2} \frac{1 - 2\sin^2 x}{\left[\rho^2 + a^2 + z^2 + 2a\rho(1 - 2\sin^2 x)\right]^{1/2}} 2\, dx.$$

Letting

$$k^2 = \frac{4a\rho}{(a+\rho)^2 + z^2}, \qquad F^2 = (a+\rho)^2 + z^2,$$

we have

$$\mathbf{A}(\mathbf{r}) = -\frac{\mu I a}{4\pi} \hat{\boldsymbol{\phi}} \frac{4}{F} \int_0^{\pi/2} \frac{1 - 2\sin^2 x}{[1 - k^2 \sin^2 x]^{1/2}} \, dx.$$

Then, since

$$\frac{1 - 2\sin^2 x}{[1 - k^2\sin^2 x]^{1/2}} = \frac{k^2 - 2}{k^2}[1 - k^2\sin^2 x]^{-1/2} + \frac{2}{k^2}[1 - k^2\sin^2 x]^{1/2},$$

we have

$$\mathbf{A}(\mathbf{r}) = \hat{\boldsymbol{\phi}}\frac{\mu I}{\pi k}\sqrt{\frac{a}{\rho}}\left[\left(1 - \frac{1}{2}k^2\right)K(k^2) - E(k^2)\right]. \qquad (3.138)$$

Here

$$K(k^2) = \int_0^{\pi/2} \frac{du}{[1 - k^2\sin^2 u]^{1/2}}, \qquad E(k^2) = \int_0^{\pi/2} [1 - k^2\sin^2 u]^{1/2}\, du,$$

are complete elliptic integrals of the first and second kinds, respectively.

We have $k^2 \ll 1$ when the observation point is far from the loop ($r^2 = \rho^2 + z^2 \gg a^2$). Using the expansions [47]

$$K(k^2) = \frac{\pi}{2}\left[1 + \frac{1}{4}k^2 + \frac{9}{64}k^4 + \cdots\right], \qquad E(k^2) = \frac{\pi}{2}\left[1 - \frac{1}{4}k^2 - \frac{3}{64}k^4 - \cdots\right],$$

in (3.138) and keeping the first nonzero term, we find that

$$\mathbf{A}(\mathbf{r}) \approx \hat{\boldsymbol{\phi}}\frac{\mu I}{4\pi r^2}(\pi a^2)\sin\theta. \qquad (3.139)$$

Defining the *magnetic dipole moment* of the loop as

$$\mathbf{m} = \hat{\mathbf{z}} I \pi a^2,$$

we can write (3.139) as

$$\mathbf{A}(\mathbf{r}) = \frac{\mu}{4\pi}\frac{\mathbf{m} \times \hat{\mathbf{r}}}{r^2}. \qquad (3.140)$$

Generalization to an arbitrarily-oriented circular loop with center located at $\mathbf{r}_0$ is accomplished by writing $\mathbf{m} = \hat{\mathbf{n}} I A$ where A is the loop area and $\hat{\mathbf{n}}$ is normal to the loop in the right-hand sense. Then

$$\mathbf{A}(\mathbf{r}) = \frac{\mu}{4\pi}\mathbf{m} \times \frac{\mathbf{r} - \mathbf{r}_0}{|\mathbf{r} - \mathbf{r}_0|^3}.$$

We shall find, upon investigating the general multipole expansion of **A** below, that this holds for any planar loop.

The magnetic field of the loop can be found by direct application of (3.127). For the case $r^2 \gg a^2$ we take the curl of (3.139) and find that

$$\mathbf{B}(\mathbf{r}) = \frac{\mu}{4\pi} \frac{m}{r^3} (\hat{\mathbf{r}}\, 2 \cos\theta + \hat{\boldsymbol{\theta}}\, \sin\theta). \qquad (3.141)$$

Comparison with (3.93) shows why we often refer to a small loop as a *magnetic dipole*. But (3.141) is approximate, and since there are no magnetic monopoles we cannot construct an exact magnetic analogue to the electric dipole. On the other hand, we shall find below that the multipole expansion of a finite-extent steady current begins with the dipole term (since the current must form closed loops). We may regard small loops as the elemental units of steady current from which all other currents may be constructed.

3.3.2 Multipole expansion

It is possible to derive a general multipole expansion for **A** analogous to (3.94). But the vector nature of **A** requires that we use vector spherical harmonics, hence the result is far more complicated than (3.94). A simpler approach yields the first few terms and requires only the Taylor expansion of $1/R$. Consider a steady current localized near the origin and contained within a sphere of radius r_m. We substitute the expansion (3.89) into (3.135) to obtain

$$\mathbf{A}(\mathbf{r}) = \frac{\mu}{4\pi} \int_V \mathbf{J}(\mathbf{r}') \left[\frac{1}{R}\Big|_{\mathbf{r}'=0} + (\mathbf{r}' \cdot \nabla') \frac{1}{R}\Big|_{\mathbf{r}'=0} + \frac{1}{2}(\mathbf{r}' \cdot \nabla')^2 \frac{1}{R}\Big|_{\mathbf{r}'=0} + \cdots \right] dV', \quad (3.142)$$

which we view as

$$\mathbf{A}(\mathbf{r}) = \mathbf{A}^{(0)}(\mathbf{r}) + \mathbf{A}^{(1)}(\mathbf{r}) + \mathbf{A}^{(2)}(\mathbf{r}) + \cdots.$$

The first term is merely

$$\mathbf{A}^{(0)}(\mathbf{r}) = \frac{\mu}{4\pi r} \int_V \mathbf{J}(\mathbf{r}')\, dV' = \frac{\mu}{4\pi r} \sum_{i=1}^{3} \hat{\mathbf{x}}_i \int_V J_i(\mathbf{r}')\, dV'$$

where $(x, y, z) = (x_1, x_2, x_3)$. However, by (3.26) each of the integrals is zero and we have

$$\mathbf{A}^{(0)}(\mathbf{r}) = 0;$$

the leading term in the multipole expansion of **A** for a general steady current distribution vanishes.

Using (3.91) we can write the second term as

$$\mathbf{A}^{(1)}(\mathbf{r}) = \frac{\mu}{4\pi r^3} \int_V \mathbf{J}(\mathbf{r}') \sum_{i=1}^{3} x_i x_i'\, dV' = \frac{\mu}{4\pi r^3} \sum_{j=1}^{3} \hat{\mathbf{x}}_j \sum_{i=1}^{3} x_i \int_V x_i' J_j(\mathbf{r}')\, dV'. \qquad (3.143)$$

By adding the null relation (3.28) we can write

$$\int_V x_i' J_j\, dV' = \int_V x_i' J_j\, dV' + \int_V [x_i' J_j + x_j' J_i]\, dV' = 2 \int_V x_i' J_j\, dV' + \int_V x_j' J_i\, dV'$$

or

$$\int_V x_i' J_j\, dV' = \frac{1}{2} \int_V [x_i' J_j - x_j' J_i]\, dV'. \qquad (3.144)$$

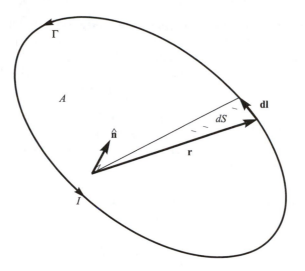

Figure 3.19: A planar wire loop.

By this and (3.143) the second term in the multipole expansion is

$$\mathbf{A}^{(1)}(\mathbf{r}) = \frac{\mu}{4\pi r^3}\frac{1}{2}\int_V \sum_{j=1}^{3}\hat{\mathbf{x}}_j \sum_{i=1}^{3} x_i[x_i' J_j - x_j' J_i]\,dV' = -\frac{\mu}{4\pi r^3}\frac{1}{2}\int_V \mathbf{r} \times [\mathbf{r}' \times \mathbf{J}(\mathbf{r}')]\,dV'.$$

Defining the *dipole moment vector*

$$\mathbf{m} = \frac{1}{2}\int_V \mathbf{r} \times \mathbf{J}(\mathbf{r})\,dV \tag{3.145}$$

we have

$$\mathbf{A}^{(1)}(\mathbf{r}) = \frac{\mu}{4\pi}\mathbf{m} \times \left(\frac{\hat{\mathbf{r}}}{r^2}\right) = -\frac{\mu}{4\pi}\mathbf{m} \times \nabla\frac{1}{r}. \tag{3.146}$$

This is the *dipole moment potential* for the steady current $\mathbf{J}$. Since steady currents of finite extent consist of loops, the dipole component is generally the first nonzero term in the expansion of $\mathbf{A}$. Higher-order components may be calculated, but extension of (3.142) beyond the dipole term is quite tedious and will not be attempted.

As an example let us compute the dipole moment of the planar but otherwise arbitrary loop shown in Figure 3.19. Specializing (3.145) for a line current we have

$$\mathbf{m} = \frac{I}{2}\oint_\Gamma \mathbf{r} \times \mathbf{dl}.$$

Examining Figure 3.19, we see that

$$\frac{1}{2}\mathbf{r} \times \mathbf{dl} = \hat{\mathbf{n}}\,dS$$

where dS is the area of the sector swept out by $\mathbf{r}$ as it moves along $\mathbf{dl}$, and $\hat{\mathbf{n}}$ is the normal to the loop in the right-hand sense. Thus

$$\mathbf{m} = \hat{\mathbf{n}}IA \tag{3.147}$$

where A is the area of the loop.

Physical interpretation of M in a magnetic material. In (3.137) we presented an expression for the vector potential produced by a magnetized material in terms of equivalent magnetization surface and volume currents. Suppose a magnetized medium is separated into volume regions with bounding surfaces across which the permeability is discontinuous. With $\mathbf{J}_M = \nabla \times \mathbf{M}$ and $\mathbf{J}_{Ms} = -\hat{\mathbf{n}} \times \mathbf{M}$ we obtain

$$\mathbf{A}(\mathbf{r}) = \frac{\mu_0}{4\pi} \int_V \frac{\mathbf{J}(\mathbf{r}')}{|\mathbf{r}-\mathbf{r}'|}\,dV' + \frac{\mu_0}{4\pi} \int_S \frac{\mathbf{J}_s(\mathbf{r}')}{|\mathbf{r}-\mathbf{r}'|}\,dS' +$$
$$+ \sum_i \frac{\mu_0}{4\pi} \left[\int_{V_i} \frac{\nabla' \times \mathbf{M}(\mathbf{r}')}{|\mathbf{r}-\mathbf{r}'|}\,dV' + \int_{S_i} \frac{-\hat{\mathbf{n}}' \times \mathbf{M}(\mathbf{r}')}{|\mathbf{r}-\mathbf{r}'|}\,dS' \right]. \tag{3.148}$$

Here $\hat{\mathbf{n}}$ points outward from region V_i. Using the curl theorem on the fourth term and employing the vector identity (B.43), we have

$$\mathbf{A}(\mathbf{r}) = \frac{\mu_0}{4\pi} \int_V \frac{\mathbf{J}(\mathbf{r}')}{|\mathbf{r}-\mathbf{r}'|}\,dV' + \frac{\mu_0}{4\pi} \int_S \frac{\mathbf{J}_s(\mathbf{r}')}{|\mathbf{r}-\mathbf{r}'|}\,dS' +$$
$$+ \sum_i \left[\frac{\mu_0}{4\pi} \int_{V_i} \mathbf{M}(\mathbf{r}') \times \nabla' \left(\frac{1}{|\mathbf{r}-\mathbf{r}'|} \right) dV' \right]. \tag{3.149}$$

But $\nabla'(1/R) = \hat{\mathbf{R}}/R^2$, hence the third term is a sum of integrals of the form

$$\frac{\mu_0}{4\pi} \int_{V_i} \mathbf{M}(\mathbf{r}') \times \frac{\hat{\mathbf{R}}}{R^2}\,dV'.$$

Comparison with (3.146) shows that this integral represents a volume superposition of dipole moments where $\mathbf{M}$ is a volume density of magnetic dipole moments. Hence a magnetic material with permeability μ is equivalent to a volume distribution of magnetic dipoles in free space. As with our interpretation of the polarization vector in a dielectric, we base this conclusion only on Maxwell's equations and the assumption of a linear, isotropic relationship between $\mathbf{B}$ and $\mathbf{H}$.

3.3.3 Boundary conditions for the magnetostatic field

The boundary conditions found for the dynamic magnetic field remain valid in the magnetostatic case. Hence

$$\hat{\mathbf{n}}_{12} \times (\mathbf{H}_1 - \mathbf{H}_2) = \mathbf{J}_s \tag{3.150}$$

and

$$\hat{\mathbf{n}}_{12} \cdot (\mathbf{B}_1 - \mathbf{B}_2) = 0, \tag{3.151}$$

where $\hat{\mathbf{n}}_{12}$ points into region 1 from region 2. Since the magnetostatic curl and divergence equations are independent, so are the boundary conditions (3.150) and (3.151). We can also write (3.151) in terms of equivalent sources by (2.118):

$$\hat{\mathbf{n}}_{12} \cdot (\mathbf{H}_1 - \mathbf{H}_2) = \rho_{Ms1} + \rho_{Ms2}, \tag{3.152}$$

where $\rho_{Ms} = \hat{\mathbf{n}} \cdot \mathbf{M}$ is called the *equivalent magnetization surface charge density*. Here $\hat{\mathbf{n}}$ points outward from the material body.

For a linear, isotropic material described by $\mathbf{B} = \mu\mathbf{H}$, equation (3.150) becomes

$$\hat{\mathbf{n}}_{12} \times \left(\frac{\mathbf{B}_1}{\mu_1} - \frac{\mathbf{B}_2}{\mu_2} \right) = \mathbf{J}_s.$$

With (2.118) we can also write (3.150) as

$$\hat{\mathbf{n}}_{12} \times (\mathbf{B}_1 - \mathbf{B}_2) = \mu_0 \left(\mathbf{J}_s + \mathbf{J}_{Ms1} + \mathbf{J}_{Ms2} \right) \tag{3.153}$$

where $\mathbf{J}_{Ms} = -\hat{\mathbf{n}} \times \mathbf{M}$ is the equivalent magnetization surface current density.

We may also write the boundary conditions in terms of the scalar or vector potential. Using $\mathbf{H} = -\nabla \Phi_m$, we can write (3.150) as

$$\Phi_{m1}(\mathbf{r}) = \Phi_{m2}(\mathbf{r}) \tag{3.154}$$

provided that the surface current $\mathbf{J}_s = 0$. As was the case with (3.36), the possibility of an additive constant here is generally ignored. To write (3.151) in terms of Φ_m we first note that $\mathbf{B}/\mu_0 - \mathbf{M} = -\nabla \Phi_m$; substitution into (3.151) gives

$$\frac{\partial \Phi_{m1}}{\partial n} - \frac{\partial \Phi_{m2}}{\partial n} = -\rho_{Ms1} - \rho_{Ms2} \tag{3.155}$$

where the normal derivative is taken in the direction of $\hat{\mathbf{n}}_{12}$. For a linear isotropic material where $\mathbf{B} = \mu \mathbf{H}$ we have

$$\mu_1 \frac{\partial \Phi_{m1}}{\partial n} = \mu_2 \frac{\partial \Phi_{m2}}{\partial n}. \tag{3.156}$$

Note that (3.154) and (3.156) are independent.

Boundary conditions on $\mathbf{A}$ may be derived using the approach of § 2.8.2. Consider Figure 2.6. Here the surface may carry either an electric surface current $\mathbf{J}_s$ or an equivalent magnetization current $\mathbf{J}_{Ms}$, and thus may be a surface of discontinuity between differing magnetic media. If we integrate $\nabla \times \mathbf{A}$ over the volume regions V_1 and V_2 and add the results we find that

$$\int_{V_1} \nabla \times \mathbf{A} \, dV + \int_{V_2} \nabla \times \mathbf{A} \, dV = \int_{V_1+V_2} \mathbf{B} \, dV.$$

By the curl theorem

$$\int_{S_1+S_2} \hat{\mathbf{n}} \times \mathbf{A} \, dS + \int_{S_{10}} -\hat{\mathbf{n}}_{10} \times \mathbf{A}_1 \, dS + \int_{S_{20}} -\hat{\mathbf{n}}_{20} \times \mathbf{A}_2 \, dS = \int_{V_1+V_2} \mathbf{B} \, dV$$

where $\mathbf{A}_1$ is the field on the surface S_{10} and $\mathbf{A}_2$ is the field on S_{20}. As $\delta \to 0$ the surfaces S_1 and S_2 combine to give S. Also S_{10} and S_{20} coincide, as do the normals $\hat{\mathbf{n}}_{10} = -\hat{\mathbf{n}}_{20} = \hat{\mathbf{n}}_{12}$. Thus

$$\int_S (\hat{\mathbf{n}} \times \mathbf{A}) \, dS - \int_V \mathbf{B} \, dV = \int_{S_{10}} \hat{\mathbf{n}}_{12} \times (\mathbf{A}_1 - \mathbf{A}_2) \, dS. \tag{3.157}$$

Now let us integrate over the entire volume region V including the surface of discontinuity. This gives

$$\int_S (\hat{\mathbf{n}} \times \mathbf{A}) \, dS - \int_V \mathbf{B} \, dV = 0,$$

and for agreement with (3.157) we must have

$$\hat{\mathbf{n}}_{12} \times (\mathbf{A}_1 - \mathbf{A}_2) = 0. \tag{3.158}$$

A similar development shows that

$$\hat{\mathbf{n}}_{12} \cdot (\mathbf{A}_1 - \mathbf{A}_2) = 0. \tag{3.159}$$

Therefore $\mathbf{A}$ is continuous across a surface carrying electric or magnetization current.

3.3.4   Uniqueness of the magnetostatic field

Because the uniqueness conditions established for the dynamic field do not apply to magnetostatics, we begin with the magnetostatic field equations. Consider a region of space V bounded by a surface S. There may be source currents and magnetic materials both inside and outside V. Assume $(\mathbf{B}_1, \mathbf{H}_1)$ and $(\mathbf{B}_2, \mathbf{H}_2)$ are solutions to the magnetostatic field equations with source $\mathbf{J}$. We seek conditions under which $\mathbf{B}_1 = \mathbf{B}_2$ and $\mathbf{H}_1 = \mathbf{H}_2$.

The difference field $\mathbf{H}_0 = \mathbf{H}_2 - \mathbf{H}_1$ obeys $\nabla \times \mathbf{H}_0 = 0$. Using (B.44) we examine the quantity

$$\nabla \cdot (\mathbf{A}_0 \times \mathbf{H}_0) = \mathbf{H}_0 \cdot (\nabla \times \mathbf{A}_0) - \mathbf{A}_0 \cdot (\nabla \times \mathbf{H}_0) = \mathbf{H}_0 \cdot (\nabla \times \mathbf{A}_0)$$

where $\mathbf{A}_0$ is defined by $\mathbf{B}_0 = \mathbf{B}_2 - \mathbf{B}_1 = \nabla \times \mathbf{A}_0 = \nabla \times (\mathbf{A}_2 - \mathbf{A}_1)$. Integrating over V we obtain

$$\oint_S (\mathbf{A}_0 \times \mathbf{H}_0) \cdot d\mathbf{S} = \int_V \mathbf{H}_0 \cdot (\nabla \times \mathbf{A}_0)\, dV = \int_V \mathbf{H}_0 \cdot \mathbf{B}_0 \, dV.$$

Then, since $(\mathbf{A}_0 \times \mathbf{H}_0) \cdot \hat{\mathbf{n}} = -\mathbf{A}_0 \cdot (\hat{\mathbf{n}} \times \mathbf{H}_0)$, we have

$$-\oint_S \mathbf{A}_0 \cdot (\hat{\mathbf{n}} \times \mathbf{H}_0)\, dS = \int_V \mathbf{H}_0 \cdot \mathbf{B}_0 \, dV. \tag{3.160}$$

If $\mathbf{A}_0 = 0$ or $\hat{\mathbf{n}} \times \mathbf{H}_0 = 0$ everywhere on S, or $\mathbf{A}_0 = 0$ on part of S and $\hat{\mathbf{n}} \times \mathbf{H}_0 = 0$ on the remainder, then

$$\int_V \mathbf{H}_0 \cdot \mathbf{B}_0 \, dS = 0. \tag{3.161}$$

So $\mathbf{H}_0 = 0$ or $\mathbf{B}_0 = 0$ by arbitrariness of V. Assuming $\mathbf{H}$ and $\mathbf{B}$ are linked by the constitutive relations, we have $\mathbf{H}_1 = \mathbf{H}_2$ and $\mathbf{B}_1 = \mathbf{B}_2$. The fields within V are unique provided that $\mathbf{A}$, the tangential component of $\mathbf{H}$, or some combination of the two, is specified over the bounding surface S.

One other condition will cause the left-hand side of (3.160) to vanish. If S recedes to infinity then, provided that the potential functions vanish sufficiently fast, the condition (3.161) still holds and uniqueness is guaranteed. Equation (3.135) shows that $\mathbf{A} \sim 1/r$ as $\mathbf{r} \to \infty$, hence $\mathbf{B}, \mathbf{H} \sim 1/r^2$. So uniqueness is ensured by the specification of $\mathbf{J}$ in unbounded space.

3.3.5   Integral solution for the vector potential

We have used the scalar Green's theorem to find a solution for the electrostatic potential within a region V in terms of the source charge in V and the values of the potential and its normal derivative on the boundary surface S. Analogously, we may find $\mathbf{A}$ within V in terms of the source current in V and the values of $\mathbf{A}$ and its derivatives on S. The vector relationship between $\mathbf{B}$ and $\mathbf{A}$ complicates the derivation somewhat, requiring Green's second identity for vector fields.

Let $\mathbf{P}$ and $\mathbf{Q}$ be continuous with continuous first and second derivatives throughout V and on S. The divergence theorem shows that

$$\int_V \nabla \cdot [\mathbf{P} \times (\nabla \times \mathbf{Q})]\, dV = \int_S [\mathbf{P} \times (\nabla \times \mathbf{Q})] \cdot d\mathbf{S}.$$

By virtue of (B.44) we have

$$\int_V [(\nabla \times \mathbf{Q}) \cdot (\nabla \times \mathbf{P}) - \mathbf{P} \cdot (\nabla \times \{\nabla \times \mathbf{Q}\})] \, dV = \int_S [\mathbf{P} \times (\nabla \times \mathbf{Q})] \cdot d\mathbf{S}.$$

We now interchange $\mathbf{P}$ and $\mathbf{Q}$ and subtract the result from the above, obtaining

$$\int_V [\mathbf{Q} \cdot (\nabla \times \{\nabla \times \mathbf{P}\}) - \mathbf{P} \cdot (\nabla \times \{\nabla \times \mathbf{Q}\})] \, dV =$$

$$\int_S [\mathbf{P} \times (\nabla \times \mathbf{Q}) - \mathbf{Q} \times (\nabla \times \mathbf{P})] \cdot d\mathbf{S}. \tag{3.162}$$

Note that $\hat{\mathbf{n}}$ points outward from V. This is *Green's second identity for vector fields.*

Now assume that V contains a magnetic material of uniform permeability μ and set

$$\mathbf{P} = \mathbf{A}(\mathbf{r}'), \qquad \mathbf{Q} = \frac{\mathbf{c}}{R},$$

in (3.162) written in terms of primed coordinates. Here $\mathbf{c}$ is a constant vector, nonzero but otherwise arbitrary. We first examine the volume integral terms. Note that

$$\nabla' \times (\nabla' \times \mathbf{Q}) = \nabla' \times \left(\nabla' \times \frac{\mathbf{c}}{R}\right) = -\nabla'^2 \left(\frac{\mathbf{c}}{R}\right) + \nabla' \left[\nabla' \cdot \left(\frac{\mathbf{c}}{R}\right)\right].$$

By (B.53) and (3.58) we have

$$\nabla'^2 \left(\frac{\mathbf{c}}{R}\right) = \frac{1}{R}\nabla'^2 \mathbf{c} + \mathbf{c}\nabla'^2 \left(\frac{1}{R}\right) + 2\left(\nabla'\frac{1}{R} \cdot \nabla'\right)\mathbf{c} = \mathbf{c}\nabla'^2 \left(\frac{1}{R}\right) = -\mathbf{c}4\pi\delta(\mathbf{r} - \mathbf{r}'),$$

hence

$$\mathbf{P} \cdot [\nabla' \times (\nabla' \times \mathbf{Q})] = 4\pi \mathbf{c} \cdot \mathbf{A}\delta(\mathbf{r} - \mathbf{r}') + \mathbf{A} \cdot \nabla' \left[\nabla' \cdot \left(\frac{\mathbf{c}}{R}\right)\right].$$

Since $\nabla \cdot \mathbf{A} = 0$ the second term on the right-hand side can be rewritten using (B.42):

$$\nabla' \cdot (\psi \mathbf{A}) = \mathbf{A} \cdot (\nabla'\psi) + \psi\nabla' \cdot \mathbf{A} = \mathbf{A} \cdot (\nabla'\psi).$$

Thus

$$\mathbf{P} \cdot [\nabla' \times (\nabla' \times \mathbf{Q})] = 4\pi \mathbf{c} \cdot \mathbf{A}\delta(\mathbf{r} - \mathbf{r}') + \nabla' \cdot \left[\mathbf{A}\left\{\mathbf{c} \cdot \nabla'\left(\frac{1}{R}\right)\right\}\right],$$

where we have again used (B.42). The other volume integral term can be found by substituting from (3.129):

$$\mathbf{Q} \cdot [\nabla' \times (\nabla' \times \mathbf{P})] = \mu\frac{1}{R}\mathbf{c} \cdot \mathbf{J}(\mathbf{r}').$$

Next we investigate the surface integral terms. Consider

$$\hat{\mathbf{n}}' \cdot [\mathbf{P} \times (\nabla' \times \mathbf{Q})] = \hat{\mathbf{n}}' \cdot \left\{\mathbf{A} \times \left[\nabla' \times \left(\frac{\mathbf{c}}{R}\right)\right]\right\}$$

$$= \hat{\mathbf{n}}' \cdot \left\{\mathbf{A} \times \left[\frac{1}{R}\nabla' \times \mathbf{c} - \mathbf{c} \times \nabla'\left(\frac{1}{R}\right)\right]\right\}$$

$$= -\hat{\mathbf{n}}' \cdot \left\{\mathbf{A} \times \left[\mathbf{c} \times \nabla'\left(\frac{1}{R}\right)\right]\right\}.$$

This can be put in slightly different form by the use of (B.8). Note that

$$(\mathbf{A} \times \mathbf{B}) \cdot (\mathbf{C} \times \mathbf{D}) = \mathbf{A} \cdot [\mathbf{B} \times (\mathbf{C} \times \mathbf{D})]$$
$$= (\mathbf{C} \times \mathbf{D}) \cdot (\mathbf{A} \times \mathbf{B})$$
$$= \mathbf{C} \cdot [\mathbf{D} \times (\mathbf{A} \times \mathbf{B})],$$

hence

$$\hat{\mathbf{n}}' \cdot [\mathbf{P} \times (\nabla' \times \mathbf{Q})] = -\mathbf{c} \cdot \left[\nabla' \left(\frac{1}{R} \right) \times (\hat{\mathbf{n}}' \times \mathbf{A}) \right].$$

The other surface term is given by

$$\hat{\mathbf{n}}' \cdot [\mathbf{Q} \times (\nabla' \times \mathbf{P})] = \hat{\mathbf{n}}' \cdot \left[\frac{\mathbf{c}}{R} \times (\nabla' \times \mathbf{A}) \right] = \hat{\mathbf{n}}' \cdot \left(\frac{\mathbf{c}}{R} \times \mathbf{B} \right) = -\frac{\mathbf{c}}{R} \cdot (\hat{\mathbf{n}}' \times \mathbf{B}).$$

We can now substitute each of the terms into (3.162) and obtain

$$\mu\mathbf{c} \cdot \int_V \frac{\mathbf{J}(\mathbf{r}')}{R} \, dV' - 4\pi\mathbf{c} \cdot \int_V \mathbf{A}(\mathbf{r}')\delta(\mathbf{r} - \mathbf{r}') \, dV' - \mathbf{c} \cdot \oint_S [\hat{\mathbf{n}}' \cdot \mathbf{A}(\mathbf{r}')]\nabla' \left(\frac{1}{R} \right) \, dS'$$
$$= -\mathbf{c} \cdot \oint_S \nabla' \left(\frac{1}{R} \right) \times [\hat{\mathbf{n}}' \times \mathbf{A}(\mathbf{r}')] \, dS' + \mathbf{c} \cdot \oint_S \frac{1}{R}\hat{\mathbf{n}}' \times \mathbf{B}(\mathbf{r}') \, dS'.$$

Since $\mathbf{c}$ is arbitrary we can remove the dot products to obtain a vector equation. Then

$$\mathbf{A}(\mathbf{r}) = \frac{\mu}{4\pi} \int_V \frac{\mathbf{J}(\mathbf{r}')}{R} \, dV' - \frac{1}{4\pi} \oint_S \left\{ [\hat{\mathbf{n}}' \times \mathbf{A}(\mathbf{r}')] \times \nabla' \left(\frac{1}{R} \right) + \right.$$
$$\left. + \frac{1}{R}\hat{\mathbf{n}}' \times \mathbf{B}(\mathbf{r}') + [\hat{\mathbf{n}}' \cdot \mathbf{A}(\mathbf{r}')]\nabla' \left(\frac{1}{R} \right) \right\} \, dS'. \tag{3.163}$$

We have expressed $\mathbf{A}$ in a closed region in terms of the sources within the region and the values of $\mathbf{A}$ and $\mathbf{B}$ on the surface. While uniqueness requires specification of *either* $\mathbf{A}$ or $\hat{\mathbf{n}} \times \mathbf{B}$ on S, the expression (3.163) includes *both* quantities. This is similar to (3.56) for electrostatic fields, which required both the scalar potential and its normal derivative.

The reader may be troubled by the fact that we require $\mathbf{P}$ and $\mathbf{Q}$ to be somewhat well behaved, then proceed to involve the singular function $\mathbf{c}/R$ and integrate over the singularity. We choose this approach to simplify the presentation; a more rigorous approach which excludes the singular point with a small sphere also gives (3.163). This approach was used in § 3.2.4 to establish (3.58). The interested reader should see Stratton [187] for details on the application of this technique to obtain (3.163).

It is interesting to note that as $S \to \infty$ the surface integral vanishes since $\mathbf{A} \sim 1/r$ and $\mathbf{B} \sim 1/r^2$, and we recover (3.135). Moreover, (3.163) returns the null result when evaluated at points outside S (see Stratton [187]). We shall see this again when studying the integral solutions for electrodynamic fields in § 6.1.3.

Finally, with

$$\mathbf{Q} = \nabla' \left(\frac{1}{R} \right) \times \mathbf{c}$$

we can find an integral expression for $\mathbf{B}$ within an enclosed region, representing a generalization of the Biot–Savart law (Problem 3.20). However, this case will be covered in the more general development of § 6.1.1.

The Biot–Savart law. We can obtain an expression for **B** in unbounded space by performing the curl operation directly on the vector potential:

$$\mathbf{B}(\mathbf{r}) = \nabla \times \frac{\mu}{4\pi} \int_V \frac{\mathbf{J}(\mathbf{r}')}{|\mathbf{r}-\mathbf{r}'|}\, dV' = \frac{\mu}{4\pi} \int_V \nabla \times \frac{\mathbf{J}(\mathbf{r}')}{|\mathbf{r}-\mathbf{r}'|}\, dV'.$$

Using (B.43) and $\nabla \times \mathbf{J}(\mathbf{r}') = 0$, we have

$$\mathbf{B}(\mathbf{r}) = -\frac{\mu}{4\pi} \int_V \mathbf{J} \times \nabla \frac{1}{|\mathbf{r}-\mathbf{r}'|}\, dV'.$$

The *Biot–Savart law*

$$\mathbf{B}(\mathbf{r}) = \frac{\mu}{4\pi} \int_V \mathbf{J}(\mathbf{r}') \times \frac{\hat{\mathbf{R}}}{R^2}\, dV' \tag{3.164}$$

follows from (3.57).

For the case of a line current we can replace $\mathbf{J}\,dV'$ by $I\mathbf{dl}'$ and obtain

$$\mathbf{B}(\mathbf{r}) = I\frac{\mu}{4\pi} \int_\Gamma \mathbf{dl}' \times \frac{\hat{\mathbf{R}}}{R^2}. \tag{3.165}$$

For an infinitely long line current on the z-axis we have

$$\mathbf{B}(\mathbf{r}) = I\frac{\mu}{4\pi} \int_{-\infty}^{\infty} \hat{\mathbf{z}} \times \frac{\hat{\mathbf{z}}(z-z') + \hat{\rho}\rho}{[(z-z')^2 + \rho^2]^{3/2}}\, dz' = \hat{\phi}\frac{\mu I}{2\pi\rho}. \tag{3.166}$$

This same result follows from taking $\nabla \times \mathbf{A}$ after direct computation of $\mathbf{A}$, or from direct application of the large-scale form of Ampere's law.

3.3.6 Force and energy

Ampere force on a system of currents. If a steady current $\mathbf{J}(\mathbf{r})$ occupying a region V is exposed to a magnetic field, the force on the moving charge is given by the Lorentz force law

$$d\mathbf{F}(\mathbf{r}) = \mathbf{J}(\mathbf{r}) \times \mathbf{B}(\mathbf{r}). \tag{3.167}$$

This can be integrated to give the total force on the current distribution:

$$\mathbf{F} = \int_V \mathbf{J}(\mathbf{r}) \times \mathbf{B}(\mathbf{r})\, dV. \tag{3.168}$$

It is apparent that the charge flow comprising a steady current must be constrained in some way, or the Lorentz force will accelerate the charge and destroy the steady nature of the current. This constraint is often provided by a conducting wire.

As an example, consider an infinitely long wire of circular cross-section centered on the z-axis in free space. If the wire carries a total current I uniformly distributed over the cross-section, then within the wire $\mathbf{J} = \hat{\mathbf{z}}I/(\pi a^2)$ where a is the wire radius. The resulting field can be found through direct integration using (3.164), or by the use of symmetry and either (3.118) or (3.120). Since $\mathbf{B}(\mathbf{r}) = \hat{\phi}B_\phi(\rho)$, equation (3.118) shows that

$$\int_0^{2\pi} B_\phi(\rho)\rho\, d\phi = \begin{cases} \frac{\mu_0 I}{a^2}\rho^2, & \rho \leq a \\ \mu_0 I, & \rho \geq a. \end{cases}$$

Thus

$$\mathbf{B}(\mathbf{r}) = \begin{cases} \hat{\phi}\,\mu_0 I \rho/2\pi a^2, & \rho \le a, \\ \hat{\phi}\,\mu_0 I/2\pi\rho, & \rho \ge a. \end{cases} \tag{3.169}$$

The force density within the wire,

$$d\mathbf{F} = \mathbf{J} \times \mathbf{B} = -\hat{\rho}\frac{\mu_0 I^2 \rho}{2\pi^2 a^4},$$

is directed inward and tends to compress the wire. Integration over the wire volume gives $\mathbf{F} = 0$ because

$$\int_0^{2\pi} \hat{\rho}\,d\phi = 0;$$

however, a section of the wire may experience a net force. For instance, we can compute the force on one half of the wire split down its axis by using $\hat{\rho} = \hat{\mathbf{x}}\cos\phi + \hat{\mathbf{y}}\sin\phi$ to obtain $F_x = 0$ and

$$F_y = -\frac{\mu_0 I^2}{2\pi^2 a^4}\int dz \int_0^a \rho^2\,d\rho \int_0^\pi \sin\phi\,d\phi = -\frac{\mu_0 I^2}{3\pi^2 a}\int dz.$$

The force per unit length

$$\frac{\mathbf{F}}{l} = -\hat{\mathbf{y}}\frac{\mu_0 I^2}{3\pi^2 a} \tag{3.170}$$

is directed toward the other half as expected.

If the wire takes the form of a loop carrying current I, then (3.167) becomes

$$d\mathbf{F}(\mathbf{r}) = I\,d\mathbf{l}(\mathbf{r}) \times \mathbf{B}(\mathbf{r}) \tag{3.171}$$

and the total force acting is

$$\mathbf{F} = I\oint_\Gamma d\mathbf{l}(\mathbf{r}) \times \mathbf{B}(\mathbf{r}).$$

We can write the force on $\mathbf{J}$ in terms of the current producing $\mathbf{B}$. Assuming this latter current $\mathbf{J}'$ occupies region V', the Biot–Savart law (3.164) yields

$$\mathbf{F} = \frac{\mu}{4\pi}\int_V \mathbf{J}(\mathbf{r}) \times \int_{V'} \mathbf{J}(\mathbf{r}') \times \frac{\mathbf{r}-\mathbf{r}'}{|\mathbf{r}-\mathbf{r}'|^3}\,dV'\,dV. \tag{3.172}$$

This can be specialized to describe the force between line currents. Assume current 1, following a path Γ_1 along the direction $d\mathbf{l}$, carries current I_1, while current 2, following path Γ_2 along the direction $d\mathbf{l}'$, carries current I_2. Then the force on current 1 is

$$\mathbf{F}_1 = I_1 I_2 \frac{\mu}{4\pi}\oint_{\Gamma_1}\oint_{\Gamma_2} d\mathbf{l} \times \left(d\mathbf{l}' \times \frac{\mathbf{r}-\mathbf{r}'}{|\mathbf{r}-\mathbf{r}'|^3}\right).$$

This equation, known as *Ampere's force law*, can be written in a better form for computational purposes. We use (B.7) and $\nabla(1/R)$ from (3.57):

$$\mathbf{F}_1 = I_1 I_2 \frac{\mu}{4\pi}\oint_{\Gamma_2} d\mathbf{l}' \oint_{\Gamma_1} d\mathbf{l} \cdot \nabla'\left(\frac{1}{|\mathbf{r}-\mathbf{r}'|}\right) - I_1 I_2 \frac{\mu}{4\pi}\oint_{\Gamma_1}\oint_{\Gamma_2}(d\mathbf{l}\cdot d\mathbf{l}')\frac{\mathbf{r}-\mathbf{r}'}{|\mathbf{r}-\mathbf{r}'|^3}. \tag{3.173}$$

The first term involves an integral of a perfect differential about a closed path, producing a null result. Thus

$$\mathbf{F}_1 = -I_1 I_2 \frac{\mu}{4\pi}\oint_{\Gamma_1}\oint_{\Gamma_2}(d\mathbf{l}\cdot d\mathbf{l}')\frac{\mathbf{r}-\mathbf{r}'}{|\mathbf{r}-\mathbf{r}'|^3}. \tag{3.174}$$

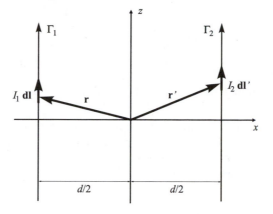

Figure 3.20: Parallel, current carrying wires.

As a simple example, consider parallel wires separated by a distance d (Figure 3.20). In this case

$$\mathbf{F}_1 = -I_1 I_2 \frac{\mu}{4\pi} \int \left[\int_{-\infty}^{\infty} \frac{-d\hat{\mathbf{x}} + (z - z')\hat{\mathbf{z}}}{[d^2 + (z - z')^2]^{3/2}} \, dz' \right] dz = I_1 I_2 \frac{\mu}{2\pi d} \hat{\mathbf{x}} \int dz$$

so the force per unit length is

$$\frac{\mathbf{F}_1}{l} = \hat{\mathbf{x}} I_1 I_2 \frac{\mu}{2\pi d}. \tag{3.175}$$

The force is attractive if $I_1 I_2 \geq 0$ (i.e., if the currents flow in the same direction).

Maxwell's stress tensor. The magnetostatic version of the stress tensor can be obtained from (2.288) by setting $\mathbf{E} = \mathbf{D} = 0$:

$$\bar{\mathbf{T}}_m = \frac{1}{2}(\mathbf{B} \cdot \mathbf{H})\bar{\mathbf{I}} - \mathbf{B}\mathbf{H}. \tag{3.176}$$

The total magnetic force on the current in a region V surrounded by surface S is given by

$$\mathbf{F}_m = -\oint_S \bar{\mathbf{T}}_m \cdot d\mathbf{S} = \int_V \mathbf{f}_m \, dV$$

where $\mathbf{f}_m = \mathbf{J} \times \mathbf{B}$ is the magnetic force volume density.

Let us compute the force between two parallel wires carrying identical currents in free space (let $I_1 = I_2 = I$ in Figure 3.20) and compare the result with (3.175). The force on the wire at $x = -d/2$ can be computed by integrating $\bar{\mathbf{T}}_m \cdot \hat{\mathbf{n}}$ over the yz-plane with $\hat{\mathbf{n}} = \hat{\mathbf{x}}$. Using (3.166) we see that in this plane the total magnetic field is

$$\mathbf{B} = -\hat{\mathbf{x}} \mu_0 \frac{I}{\pi} \frac{y}{y^2 + d^2/4}.$$

Therefore

$$\bar{\mathbf{T}}_m \cdot \hat{\mathbf{n}} = \frac{1}{2} B_x \frac{B_x}{\mu_0} \hat{\mathbf{x}} - \hat{\mathbf{x}} B_x \frac{B_x}{\mu_0} = -\mu_0 \frac{I^2}{2\pi^2} \frac{y^2}{[y^2 + d^2/4]^2} \hat{\mathbf{x}}$$

and by integration

$$\mathbf{F}_1 = \mu_0 \frac{I^2}{2\pi^2} \hat{\mathbf{x}} \int dz \int_{-\infty}^{\infty} \frac{y^2}{[y^2 + d^2/4]^2}\, dy = I^2 \frac{\mu_0}{2\pi d} \hat{\mathbf{x}} \int dz.$$

The resulting force per unit length agrees with (3.175) when $I_1 = I_2 = I$.

Torque in a magnetostatic field.   The torque exerted on a current-carrying conductor immersed in a magnetic field plays an important role in many engineering applications. If a rigid body is exposed to a force field of volume density $d\mathbf{F}(\mathbf{r})$, the torque on that body about a certain origin is given by

$$\mathbf{T} = \int_V \mathbf{r} \times d\mathbf{F}\, dV \qquad (3.177)$$

where integration is performed over the body and $\mathbf{r}$ extends from the origin of torque. If the force arises from the interaction of a current with a magnetostatic field, then $d\mathbf{F} = \mathbf{J} \times \mathbf{B}$ and

$$\mathbf{T} = \int_V \mathbf{r} \times (\mathbf{J} \times \mathbf{B})\, dV. \qquad (3.178)$$

For a line current we can replace $\mathbf{J}\, dV$ with $I\mathbf{dl}$ to obtain

$$\mathbf{T} = I \int_\Gamma \mathbf{r} \times (\mathbf{dl} \times \mathbf{B}).$$

If $\mathbf{B}$ is uniform then by (B.7) we have

$$\mathbf{T} = \int_V [\mathbf{J}(\mathbf{r} \cdot \mathbf{B}) - \mathbf{B}(\mathbf{r} \cdot \mathbf{J})]\, dV.$$

The second term can be written as

$$\int_V \mathbf{B}(\mathbf{r} \cdot \mathbf{J})\, dV = \mathbf{B} \sum_{i=1}^{3} \int_V x_i J_i\, dV = 0$$

where $(x_1, x_2, x_3) = (x, y, z)$, and where we have employed (3.27). Thus

$$\mathbf{T} = \int_V \mathbf{J}(\mathbf{r} \cdot \mathbf{B})\, dV = \sum_{j=1}^{3} \hat{\mathbf{x}}_j \int_V J_j \sum_{i=1}^{3} x_i B_i\, dV = \sum_{i=1}^{3} B_i \sum_{j=1}^{3} \hat{\mathbf{x}}_j \int_V J_j x_i\, dV.$$

We can replace the integral using (3.144) to get

$$\mathbf{T} = \frac{1}{2} \int_V \sum_{j=1}^{3} \hat{\mathbf{x}}_j \sum_{i=1}^{3} B_i [x_i J_j - x_j J_i]\, dV = -\frac{1}{2} \int_V \mathbf{B} \times (\mathbf{r} \times \mathbf{J})\, dV.$$

Since $\mathbf{B}$ is uniform we have, by (3.145),

$$\mathbf{T} = \mathbf{m} \times \mathbf{B} \qquad (3.179)$$

where $\mathbf{m}$ is the dipole moment. For a planar loop we can use (3.147) to obtain

$$\mathbf{T} = IA\hat{\mathbf{n}} \times \mathbf{B}.$$

Joule's law. In § 2.9.5 we showed that when a moving charge interacts with an electric field in a volume region V, energy is transferred between the field and the charge. If the source of that energy is outside V, the energy is carried into V as an energy flux over the boundary surface S. The energy balance described by Poynting's theorem (2.299) also holds for static fields supported by steady currents: we must simply recognize that we have no time-rate of change of stored energy. Thus

$$- \int_V \mathbf{J} \cdot \mathbf{E} \, dV = \oint_S (\mathbf{E} \times \mathbf{H}) \cdot d\mathbf{S}. \tag{3.180}$$

The term

$$P = - \int_V \mathbf{J} \cdot \mathbf{E} \, dV \tag{3.181}$$

describes the rate at which energy is supplied to the fields by the current within V; we have $P > 0$ if there are sources within V that result in energy transferred to the fields, and $P < 0$ if there is energy transferred to the currents. The latter case occurs when there are conducting materials in V. Within these conductors

$$P = - \int_V \sigma \mathbf{E} \cdot \mathbf{E} \, dV. \tag{3.182}$$

Here $P < 0$; energy is transferred from the fields to the currents, and from the currents into heat (i.e., into lattice vibrations via collisions). Equation (3.182) is called *Joule's law*, and the transfer of energy from the fields into heat is *Joule heating*. Joule's law is the power relationship for a conducting material.

An important example involves a straight section of conducting wire having circular cross-section. Assume a total current I is uniformly distributed over the cross-section of the wire, and that the wire is centered on the z-axis and extends between the planes $z = 0, L$. Let the potential difference between the ends be V. Using (3.169) we see that at the surface of the wire

$$\mathbf{H} = \hat{\boldsymbol{\phi}} \frac{I}{2\pi a}, \qquad \mathbf{E} = \hat{\mathbf{z}} \frac{V}{L}.$$

The corresponding Poynting flux $\mathbf{E} \times \mathbf{H}$ is $-\hat{\boldsymbol{\rho}}$-directed, implying that energy flows into wire volume through the curved side surface. We can verify (3.180):

$$- \int_V \mathbf{J} \cdot \mathbf{E} \, dV = \int_0^L \int_0^{2\pi} \int_0^a \hat{\mathbf{z}} \frac{I}{\pi a^2} \cdot \hat{\mathbf{z}} \frac{V}{L} \rho \, d\rho \, d\phi \, dz = -IV,$$

$$\oint_S (\mathbf{E} \times \mathbf{H}) \cdot d\mathbf{S} = \int_0^{2\pi} \int_0^L \left(-\hat{\boldsymbol{\rho}} \frac{IV}{2\pi a L} \right) \cdot \hat{\boldsymbol{\rho}} a \, d\phi \, dz = -IV.$$

Stored magnetic energy. We have shown that the energy stored in a static charge distribution may be regarded as the "assembly energy" required to bring charges from infinity against the Coulomb force. By proceeding very slowly with this assembly, we are able to avoid any complications resulting from the motion of the charges.

Similarly, we may equate the energy stored in a steady current distribution to the energy required for its assembly from current filaments[1] brought in from infinity. However, the calculation of assembly energy is more complicated in this case: moving a current

---

[1] Recall that a flux tube of a vector field is bounded by streamlines of the field. A current filament is a flux tube of current having vanishingly small, but nonzero, cross-section.

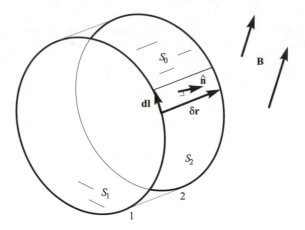

Figure 3.21: Calculation of work to move a filamentary loop in an applied magnetic field.

filament into the vicinity of existing filaments changes the total magnetic flux passing through the existing loops, regardless of how slowly we assemble the filaments. As described by Faraday's law, this change in flux must be associated with an induced emf, which will tend to change the current flowing in the filament (and any existing filaments) unless energy is expended to keep the current constant (by the application of a battery emf in the opposite direction). We therefore regard the assembly energy as consisting of two parts: (1) the energy required to bring a filament with *constant* current from infinity against the Ampere force, and (2) the energy required to keep the current in this filament, and any existing filaments, constant. We ignore the energy required to keep the steady current flowing through an isolated loop (i.e., the energy needed to overcome Joule losses).

We begin by computing the amount of energy required to bring a filament with current I from infinity to a given position within an applied magnetostatic field $\mathbf{B}(\mathbf{r})$. In this first step we assume that the field is supported by localized sources, hence vanishes at infinity, and that it will not be altered by the motion of the filament. The force on each small segment of the filament is given by Ampere's force law (3.171), and the total force is found by integration. Suppose an external agent displaces the filament incrementally from a starting position 1 to an ending position 2 along a vector $\delta \mathbf{r}$ as shown in Figure 3.21. The work required is

$$\delta W = -(I\mathbf{dl} \times \mathbf{B}) \cdot \delta \mathbf{r} = (I\mathbf{dl} \times \delta \mathbf{r}) \cdot \mathbf{B}$$

for each segment of the wire. Figure 3.21 shows that $\mathbf{dl} \times \delta \mathbf{r}$ describes a small patch of surface area between the starting and ending positions of the filament, hence $-(\mathbf{dl} \times \delta \mathbf{r}) \cdot \mathbf{B}$ is the *outward* flux of $\mathbf{B}$ through the patch. Integrating over all segments comprising the filament, we obtain

$$\Delta W = I \oint_{\Gamma} (\mathbf{dl} \times \delta \mathbf{r}) \cdot \mathbf{B} = -I \int_{S_0} \mathbf{B} \cdot \mathbf{dS}$$

for the total work required to displace the entire filament through $\delta \mathbf{r}$; here the surface S_0 is described by the superposition of all patches. If S_1 and S_2 are the surfaces bounded by the filament in its initial and final positions, respectively, then S_1, S_2, and S_0 taken together form a closed surface. The outward flux of $\mathbf{B}$ through this surface is

$$\oint_{S_0+S_1+S_2} \mathbf{B} \cdot \mathbf{dS} = 0$$

so that

$$\Delta W = -I \int_{S_0} \mathbf{B} \cdot \mathbf{dS} = I \int_{S_1+S_2} \mathbf{B} \cdot \mathbf{dS}$$

where $\hat{\mathbf{n}}$ is outward from the closed surface. Finally, let $\Psi_{1,2}$ be the flux of $\mathbf{B}$ through $S_{1,2}$ in the direction determined by $\mathbf{dl}$ and the right-hand rule. Then

$$\Delta W = -I(\Psi_2 - \Psi_1) = -I\Delta\Psi. \tag{3.183}$$

Now suppose that the initial position of the filament is at infinity. We bring the filament into a final position within the field $\mathbf{B}$ through a succession of small displacements, each requiring work (3.183). By superposition over all displacements, the total work is $W = -I(\Psi - \Psi_\infty)$ where Ψ_∞ and Ψ are the fluxes through the filament in its initial and final positions, respectively. However, since the source of the field is localized, we know that $\mathbf{B}$ is zero at infinity. Therefore $\Psi_\infty = 0$ and

$$W = -I\Psi = -I \int_S \mathbf{B} \cdot \hat{\mathbf{n}} \, dS \tag{3.184}$$

where $\hat{\mathbf{n}}$ is determined from $\mathbf{dl}$ in the right-hand sense.

Now let us find the work required to position two current filaments in a field-free region of space, starting with both filaments at infinity. Assume filament 1 carries current I_1 and filament 2 carries current I_2, and that we hold these currents constant as we move the filaments into position. We can think of assembling these filaments in two ways: by placing filament 1 first, or by placing filament 2 first. In either case, placing the first filament requires no work since (3.184) is zero. The work required to place the second filament is $W_1 = -I_1\Psi_1$ if filament 2 is placed first, where Ψ_1 is the flux passing through filament 1 in its final position, caused by the presence of filament 2. If filament 1 is placed first, the work required is $W_2 = -I_2\Psi_2$. Since the work cannot depend on which loop is placed first, we have $W_1 = W_2 = W$ where we can use either $W = -I_1\Psi_1$ or $W = -I_2\Psi_2$. It is even more convenient, as we shall see, to average these values and use

$$W = -\frac{1}{2}\left(I_1\Psi_1 + I_2\Psi_2\right). \tag{3.185}$$

We must determine the energy required to keep the currents constant as we move the filaments into position. When moving the first filament into place there is no induced emf, since no applied field is yet present. However, when moving the second filament into place we will change the flux linked by *both* the first and second loops. This change of flux will induce an emf in each of the loops, and this will change the current. To keep the current constant we must supply an opposing emf. Let dW_{emf}/dt be the rate of work required to keep the current constant. Then by (2.153) and (3.181) we have

$$\frac{dW_{\text{emf}}}{dt} = -\int_V \mathbf{J} \cdot \mathbf{E} \, dV = -I \int \mathbf{E} \cdot \mathbf{dl} = -I\frac{d\Psi}{dt}.$$

Integrating, we find the total work ΔW required to keep the current constant in either loop as the flux through the loop is changed by an amount $\Delta\Psi$:

$$\Delta W_{emf} = I\Delta\Psi.$$

So the total work required to keep I_1 constant as the loops are moved from infinity (where the flux is zero) to their final positions is $I_1\Psi_1$. Similarly, a total work $I_2\Psi_2$ is required to keep I_2 constant during the same process. Adding these to (3.185), the work required to position the loops, we obtain the complete assembly energy

$$W = \frac{1}{2}\left(I_1\Psi_1 + I_2\Psi_2\right)$$

for two filaments. The extension to N filaments is

$$W_m = \frac{1}{2}\sum_{n=1}^{N} I_n\Psi_n. \tag{3.186}$$

Consequently, the energy of a single current filament is

$$W_m = \frac{1}{2}I\Psi. \tag{3.187}$$

We may interpret this as the "assembly energy" required to bring the single loop into existence by bringing vanishingly small loops (magnetic dipoles) in from infinity. We may also interpret it as the energy required to establish the current in this single filament against the back emf. That is, if we establish I by slowly increasing the current from zero in N small steps $\Delta I = I/N$, an energy $\Psi_n\Delta I$ will be required at each step. Since Ψ_n increases proportionally to I, we have

$$W_m = \sum_{n=1}^{N} \frac{I}{N}\left[(n-1)\frac{\Psi}{N}\right]$$

where Ψ is the flux when the current is fully established. Since $\sum_{n=1}^{N}(n-1) = N(N-1)/2$ we obtain

$$W_m = \frac{1}{2}I\Psi \tag{3.188}$$

as $N \to \infty$.

A volume current $\mathbf{J}$ can be treated as though it were composed of N current filaments. Equations (3.128) and (3.186) give

$$W_m = \frac{1}{2}\sum_{n=1}^{N} I_n \oint_{\Gamma_n} \mathbf{A}\cdot\mathbf{dl}.$$

Since the total current is

$$I = \int_{CS} \mathbf{J}\cdot\mathbf{dS} = \sum_{n=1}^{N} I_n$$

where CS denotes the cross-section of the steady current, we have as $N \to \infty$

$$W_m = \frac{1}{2}\int_V \mathbf{A}\cdot\mathbf{J}\,dV. \tag{3.189}$$

Alternatively, using (3.135), we may write

$$W_m = \frac{1}{2}\int_V\int_V \frac{\mathbf{J}(\mathbf{r})\cdot\mathbf{J}(\mathbf{r}')}{|\mathbf{r}-\mathbf{r}'|}\,dV\,dV'.$$

Note the similarity between (3.189) and (3.86). We now manipulate (3.189) into a form involving only the electromagnetic fields. By Ampere's law

$$W_m = \frac{1}{2} \int_V \mathbf{A} \cdot (\nabla \times \mathbf{H}) \, dV.$$

Using (B.44) and the divergence theorem we can write

$$W_m = \frac{1}{2} \oint_S (\mathbf{H} \times \mathbf{A}) \cdot d\mathbf{S} + \frac{1}{2} \int_V \mathbf{H} \cdot (\nabla \times \mathbf{A}) \, dV.$$

We now let S expand to infinity. This does not change the value of W_m since we do not enclose any more current; however, since $\mathbf{A} \sim 1/r$ and $\mathbf{H} \sim 1/r^2$, the surface integral vanishes. Thus, remembering that $\nabla \times \mathbf{A} = \mathbf{B}$, we have

$$W_m = \frac{1}{2} \int_{V_\infty} \mathbf{H} \cdot \mathbf{B} \, dV \tag{3.190}$$

where V_∞ denotes all of space.

Although we do not provide a derivation, (3.190) is also valid within linear materials. For nonlinear materials, the total energy required to build up a magnetic field from $\mathbf{B}_1$ to $\mathbf{B}_2$ is

$$W_m = \frac{1}{2} \int_{V_\infty} \left[\int_{\mathbf{B}_1}^{\mathbf{B}_2} \mathbf{H} \cdot d\mathbf{B} \right] dV. \tag{3.191}$$

This accounts for the work required to drive a ferromagnetic material through its hysteresis loop. Readers interested in a complete derivation of (3.191) should consult Stratton [187].

As an example, consider two thin-walled, coaxial, current-carrying cylinders having radii a, b ($b > a$). The intervening region is a linear magnetic material having permeability μ. Assume that the inner and outer conductors carry total currents I in the $\pm z$ directions, respectively. From the large-scale form of Ampere's law we find that

$$\mathbf{H} = \begin{cases} 0, & \rho \leq a, \\ \hat{\phi} \, I/2\pi\rho, & a \leq \rho \leq b, \\ 0, & \rho > b, \end{cases} \tag{3.192}$$

hence by (3.190)

$$W_m = \frac{1}{2} \int dz \int_0^{2\pi} \int_a^b \frac{\mu I^2}{(2\pi\rho)^2} \rho \, d\rho \, d\phi,$$

and the stored energy is

$$\frac{W_m}{l} = \mu \frac{I^2}{4\pi} \ln\left(\frac{b}{a}\right) \tag{3.193}$$

per unit length.

Suppose instead that the inner cylinder is solid and that current is spread uniformly throughout. Then the field between the cylinders is still given by (3.192) but within the inner conductor we have

$$\mathbf{H} = \hat{\phi} \frac{I\rho}{2\pi a^2}$$

by (3.169). Thus, to (3.193) we must add the energy

$$\frac{W_{m,\text{inside}}}{l} = \frac{1}{2} \int_0^{2\pi} \int_0^a \frac{\mu_0 I^2 \rho^2}{(2\pi a^2)^2} \rho \, d\rho \, d\phi = \frac{\mu_0 I^2}{16\pi}$$

stored within the solid wire. The result is

$$\frac{W_m}{l} = \frac{\mu_0 I^2}{4\pi} \left[\mu_r \ln \left(\frac{b}{a} \right) + \frac{1}{4} \right].$$

3.3.7   Magnetic field of a permanently magnetized body

We now have the tools necessary to compute the magnetic field produced by a permanent magnet (a body with permanent magnetization $\mathbf{M}$). As an example, we shall find the field due to a uniformly magnetized sphere in three different ways: by computing the vector potential integral and taking the curl, by computing the scalar potential integral and taking the gradient, and by finding the scalar potential using separation of variables and applying the boundary condition across the surface of the sphere.

Consider a magnetized sphere of radius a, residing in free space and having permanent magnetization

$$\mathbf{M}(\mathbf{r}) = M_0 \hat{\mathbf{z}}.$$

The equivalent magnetization current and charge densities are given by

$$\mathbf{J}_M = \nabla \times \mathbf{M} = 0, \tag{3.194}$$

$$\mathbf{J}_{Ms} = -\hat{\mathbf{n}} \times \mathbf{M} = -\hat{\mathbf{r}} \times M_0 \hat{\mathbf{z}} = M_0 \hat{\boldsymbol{\phi}} \sin \theta, \tag{3.195}$$

and

$$\rho_M = -\nabla \cdot \mathbf{M} = 0, \tag{3.196}$$

$$\rho_{Ms} = \hat{\mathbf{n}} \cdot \mathbf{M} = \hat{\mathbf{r}} \cdot M_0 \hat{\mathbf{z}} = M_0 \cos \theta. \tag{3.197}$$

The vector potential is produced by the equivalent magnetization surface current. Using (3.137) we find that

$$\mathbf{A}(\mathbf{r}) = \frac{\mu_0}{4\pi} \int_S \frac{\mathbf{J}_{Ms}}{|\mathbf{r} - \mathbf{r}'|} \, dS' = \frac{\mu_0}{4\pi} \int_{-\pi}^{\pi} \int_0^{\pi} \frac{M_0 \hat{\boldsymbol{\phi}}' \sin \theta'}{|\mathbf{r} - \mathbf{r}'|} \sin \theta' \, d\theta' \, d\phi'.$$

Since $\hat{\boldsymbol{\phi}}' = -\hat{\mathbf{x}} \sin \phi' + \hat{\mathbf{y}} \cos \phi'$, the rectangular components of $\mathbf{A}$ are

$$\left\{ \begin{matrix} -A_x \\ A_y \end{matrix} \right\} = \frac{\mu_0}{4\pi} \int_{-\pi}^{\pi} \int_0^{\pi} \frac{M_0 \begin{matrix} \sin \phi' \\ \cos \phi' \end{matrix} \sin \theta'}{|\mathbf{r} - \mathbf{r}'|} a^2 \sin \theta' \, d\theta' \, d\phi'. \tag{3.198}$$

The integrals are most easily computed via the spherical harmonic expansion (E.200) for the inverse distance $|\mathbf{r} - \mathbf{r}'|^{-1}$:

$$\left\{ \begin{matrix} -A_x \\ A_y \end{matrix} \right\} = \mu_0 M_0 a^2 \sum_{n=0}^{\infty} \sum_{m=-n}^{n} \frac{Y_{nm}(\theta, \phi)}{2n+1} \frac{r_<^n}{r_>^{n+1}} \int_{-\pi}^{\pi} \int_0^{\pi} \begin{matrix} \sin \phi' \\ \cos \phi' \end{matrix} \sin^2 \theta' Y_{nm}^*(\theta', \phi') \, d\theta' \, d\phi'.$$

Because the ϕ' variation is $\sin \phi'$ or $\cos \phi'$, all terms in the sum vanish except $n = 1$, $m = \pm 1$. Since

$$Y_{1,-1}(\theta, \phi) = \sqrt{\frac{3}{8\pi}} \sin \theta e^{-j\phi}, \qquad Y_{1,1}(\theta, \phi) = -\sqrt{\frac{3}{8\pi}} \sin \theta e^{j\phi},$$

we have

$$\left\{ \begin{array}{c} -A_x \\ A_y \end{array} \right\} = \mu_0 M_0 \frac{a^2}{3} \frac{r_<}{r_>^2} \frac{3}{8\pi} \sin\theta \int_0^\pi \sin^3\theta' \, d\theta' \cdot$$

$$\cdot \left[e^{-j\phi} \int_{-\pi}^\pi \frac{\sin\phi'}{\cos\phi'} e^{j\phi'} \, d\phi' + e^{j\phi} \int_{-\pi}^\pi \frac{\sin\phi'}{\cos\phi'} e^{-j\phi'} \, d\phi' \right].$$

Carrying out the integrals we find that

$$\left\{ \begin{array}{c} -A_x \\ A_y \end{array} \right\} = \mu_0 M_0 \frac{a^2}{3} \frac{r_<}{r_>^2} \sin\theta \left\{ \begin{array}{c} \sin\phi \\ \cos\phi \end{array} \right\}$$

or

$$\mathbf{A} = \mu_0 M_0 \frac{a^2}{3} \frac{r_<}{r_>^2} \sin\theta \hat{\boldsymbol{\phi}}.$$

Finally, $\mathbf{B} = \nabla \times \mathbf{A}$ gives

$$\mathbf{B} = \left\{ \begin{array}{ll} \frac{2\mu_0 M_0}{3} \hat{\mathbf{z}}, & r < a, \\ \frac{\mu_0 M_0 a^3}{3r^3} \left(\hat{\mathbf{r}} \, 2\cos\theta + \hat{\boldsymbol{\theta}} \sin\theta \right), & r > a. \end{array} \right. \tag{3.199}$$

Hence $\mathbf{B}$ within the sphere is uniform and in the same direction as $\mathbf{M}$, while $\mathbf{B}$ outside the sphere has the form of the magnetic dipole field with moment

$$m = \left(\frac{4}{3}\pi a^3 \right) M_0.$$

We can also compute $\mathbf{B}$ by first finding the scalar potential through direct computation of the integral (3.126). Substituting for ρ_{Ms} from (3.197), we have

$$\Phi_m(\mathbf{r}) = \frac{1}{4\pi} \int_S \frac{\rho_{Ms}(\mathbf{r}')}{|\mathbf{r} - \mathbf{r}'|} \, dS' = \frac{1}{4\pi} \int_{-\pi}^\pi \int_0^\pi \frac{M_0 \cos\theta'}{|\mathbf{r} - \mathbf{r}'|} \sin\theta' \, d\theta' \, d\phi'.$$

This integral has the form of (3.100) with $f(\theta) = M_0 \cos\theta$. Thus, from (3.102),

$$\Phi_m(\mathbf{r}) = M_0 \frac{a^2}{3} \cos\theta \frac{r_<}{r_>^2}. \tag{3.200}$$

The magnetic field $\mathbf{H}$ is then

$$\mathbf{H} = -\nabla\Phi_m = \left\{ \begin{array}{ll} -\frac{M_0}{3} \hat{\mathbf{z}}, & r < a, \\ \frac{M_0 a^3}{3r^3} \left(\hat{\mathbf{r}} \, 2\cos\theta + \hat{\boldsymbol{\theta}} \sin\theta \right), & r > a. \end{array} \right.$$

Inside the sphere $\mathbf{B}$ is given by $\mathbf{B} = \mu_0(\mathbf{H} + \mathbf{M})$, while outside the sphere it is merely $\mathbf{B} = \mu_0 \mathbf{H}$. These observations lead us again to (3.199).

Since the scalar potential obeys Laplace's equation both inside and outside the sphere, as a last approach to the problem we shall write Φ_m in terms of the separation of variables solution discussed in § A.4. We can repeat our earlier arguments for the dielectric sphere in an impressed electric field (§ 3.2.10). Copying equations (3.109) and (3.110), we can write for $r \le a$

$$\Phi_{m1}(r, \theta) = \sum_{n=0}^\infty A_n r^n P_n(\cos\theta), \tag{3.201}$$

and for $r \geq a$

$$\Phi_{m2}(r,\theta) = \sum_{n=0}^{\infty} B_n r^{-(n+1)} P_n(\cos\theta). \qquad (3.202)$$

The boundary condition (3.154) at $r = a$ requires that

$$\sum_{n=0}^{\infty} A_n a^n P_n(\cos\theta) = \sum_{n=0}^{\infty} B_n a^{-(n+1)} P_n(\cos\theta);$$

upon application of the orthogonality of the Legendre functions, this becomes

$$A_n a^n = B_n a^{-(n+1)}. \qquad (3.203)$$

We can write (3.155) as

$$-\frac{\partial \Phi_{m1}}{\partial r} + \frac{\partial \Phi_{m2}}{\partial r} = -\rho_{Ms}$$

so that at $r = a$

$$-\sum_{n=0}^{\infty} A_n n a^{n-1} P_n(\cos\theta) - \sum_{n=0}^{\infty} B_n(n+1)a^{-(n+2)} P_n(\cos\theta) = -M_0 \cos\theta.$$

After application of orthogonality this becomes

$$A_1 + 2B_1 a^{-3} = M_0, \qquad (3.204)$$
$$n a^{n-1} A_n = -(n+1)B_n a^{-(n+2)}, \qquad n \neq 1. \qquad (3.205)$$

Solving (3.203) and (3.204) simultaneously for $n = 1$ we find that

$$A_1 = \frac{M_0}{3}, \qquad B_1 = \frac{M_0}{3} a^3.$$

We also see that (3.203) and (3.205) are inconsistent unless $A_n = B_n = 0$, $n \neq 1$. Substituting these results into (3.201) and (3.202), we have

$$\Phi_m = \begin{cases} \frac{M_0}{3} r \cos\theta, & r \leq a, \\ \frac{M_0}{3} \frac{a^3}{r^2} \cos\theta, & r \geq a, \end{cases}$$

which is (3.200).

3.3.8   Bodies immersed in an impressed magnetic field: magnetostatic shielding

A highly permeable enclosure can provide partial shielding from external magnetostatic fields. Consider a spherical shell of highly permeable material (Figure 3.22); assume it is immersed in a uniform impressed field $\mathbf{H}_0 = H_0 \hat{\mathbf{z}}$. We wish to determine the internal field and the factor by which it is reduced from the external applied field. Because there are no sources (the applied field is assumed to be created by sources far removed), we may use magnetic scalar potentials to represent the fields everywhere. We may represent the scalar potentials using a separation of variables solution to Laplace's equation, with a contribution only from the $n = 1$ term in the series. In region 1 we have both scattered

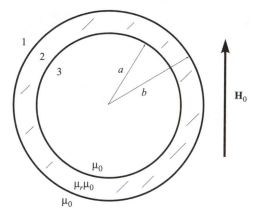

Figure 3.22: Spherical shell of magnetic material.

and applied potentials, where the applied potential is just $\Phi_0 = -H_0 z = -H_0 r \cos\theta$, since $\mathbf{H}_0 = -\nabla\Phi_0 = H_0\hat{z}$. We have

$$\Phi_1(\mathbf{r}) = A_1 r^{-2} \cos\theta - H_0 r \cos\theta, \qquad (3.206)$$

$$\Phi_2(\mathbf{r}) = (B_1 r^{-2} + C_1 r) \cos\theta, \qquad (3.207)$$

$$\Phi_3(\mathbf{r}) = D_1 r \cos\theta. \qquad (3.208)$$

We choose (3.109) for the scattered potential in region 1 so that it decays as $\mathbf{r} \to \infty$, and (3.110) for the scattered potential in region 3 so that it remains finite at $r = 0$. In region 2 we have no restrictions and therefore include both contributions. The coefficients A_1, B_1, C_1, D_1 are found by applying the appropriate boundary conditions at $r = a$ and $r = b$. By continuity of the scalar potential across each boundary we have

$$A_1 b^{-2} - H_0 b = B_1 b^{-2} + C_1 b,$$
$$B_1 a^{-2} + C_1 a = D_1 a.$$

By (3.156), the quantity $\mu \partial\Phi/\partial r$ is also continuous at $r = a$ and $r = b$; this gives two more equations:

$$\mu_0(-2A_1 b^{-3} - H_0) = \mu(-2B_1 b^{-3} + C_1),$$
$$\mu(-2B_1 a^{-3} + C_1) = \mu_0 D_1.$$

Simultaneous solution yields

$$D_1 = -\frac{9\mu_r}{K} H_0$$

where

$$K = (2 + \mu_r)(1 + 2\mu_r) - 2(a/b)^3(\mu_r - 1)^2.$$

Substituting this into (3.208) and using $\mathbf{H} = -\nabla\Phi_m$, we find that

$$\mathbf{H} = \kappa H_0 \hat{z}$$

within the enclosure, where $\kappa = 9\mu_r/K$. This field is uniform and, since $\kappa < 1$ for $\mu_r > 1$, it is weaker than the applied field. For $\mu_r \gg 1$ we have $K \approx 2\mu_r^2[1 - (a/b)^3]$. Denoting the shell thickness by $\Delta = b - a$, we find that $K \approx 6\mu_r^2 \Delta/a$ when $\Delta/a \ll 1$. Thus

$$\kappa = \frac{3}{2} \frac{1}{\mu_r \frac{\Delta}{a}}$$

describes the coefficient of shielding for a highly permeable spherical enclosure, valid when $\mu_r \gg 1$ and $\Delta/a \ll 1$. A shell for which $\mu_r = 10,000$ and $a/b = 0.99$ can reduce the enclosure field to 0.15% of the applied field.

3.4   Static field theorems

3.4.1   Mean value theorem of electrostatics

The average value of the electrostatic potential over a sphere is equal to the potential at the center of the sphere, provided that the sphere encloses no electric charge. To see this, write

$$\Phi(\mathbf{r}) = \frac{1}{4\pi\epsilon} \int_V \frac{\rho(\mathbf{r}')}{R} \, dV' + \frac{1}{4\pi} \oint_S \left[-\Phi(\mathbf{r}')\frac{\hat{\mathbf{R}}}{R^2} + \frac{\nabla'\Phi(\mathbf{r}')}{R} \right] \cdot d\mathbf{S}';$$

put $\rho \equiv 0$ in V, and use the obvious facts that if S is a sphere centered at point $\mathbf{r}$ then (1) R is constant on S and (2) $\hat{\mathbf{n}}' = -\hat{\mathbf{R}}$:

$$\Phi(\mathbf{r}) = \frac{1}{4\pi R^2} \oint_S \Phi(\mathbf{r}') \, dS' - \frac{1}{4\pi R} \oint_S \mathbf{E}(\mathbf{r}') \cdot d\mathbf{S}'.$$

The last term vanishes by Gauss's law, giving the desired result.

3.4.2   Earnshaw's theorem

It is impossible for a charge to rest in stable equilibrium under the influence of electrostatic forces alone. This is an easy consequence of the mean value theorem of electrostatics, which precludes the existence of a point where Φ can assume a maximum or a minimum.

3.4.3   Thomson's theorem

Static charge on a system of perfect conductors distributes itself so that the electric stored energy is a minimum. Figure 3.23 shows a system of n conducting bodies held at potentials $\Phi_1, \ldots, \Phi_n$. Suppose the potential field associated with the actual distribution of charge on these bodies is Φ, giving

$$W_e = \frac{\epsilon}{2} \int_V \mathbf{E} \cdot \mathbf{E} \, dV = \frac{\epsilon}{2} \int_V \nabla\Phi \cdot \nabla\Phi \, dV$$

for the actual stored energy. Now assume a slightly different charge distribution, resulting in a new potential $\Phi' = \Phi + \delta\Phi$ that satisfies the same boundary conditions (i.e., assume $\delta\Phi = 0$ on each conducting body). The stored energy associated with this hypothetical situation is

$$W_e' = W_e + \delta W_e = \frac{\epsilon}{2} \int_V \nabla(\Phi + \delta\Phi) \cdot \nabla(\Phi + \delta\Phi) \, dV$$

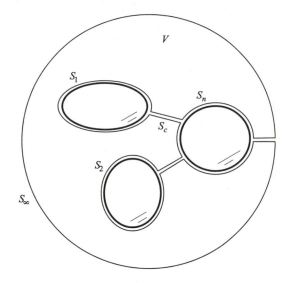

Figure 3.23: System of conductors used to derive Thomson's theorem.

so that

$$\delta W_e = \epsilon \int_V \nabla\Phi \cdot \nabla(\delta\Phi)\, dV + \frac{\epsilon}{2}\int_V |\nabla(\delta\Phi)|^2\, dV;$$

Thomson's theorem will be proved if we can show that

$$\int_V \nabla\Phi \cdot \nabla(\delta\Phi)\, dV = 0, \qquad (3.209)$$

because then we shall have

$$\delta W_e = \frac{\epsilon}{2}\int_V |\nabla(\delta\Phi)|^2\, dV \geq 0.$$

To establish (3.209), we use Green's first identity

$$\int_V (\nabla u \cdot \nabla v + u\nabla^2 v)\, dV = \oint_S u\nabla v \cdot d\mathbf{S}$$

with $u = \delta\Phi$ and $v = \Phi$:

$$\int_V \nabla\Phi \cdot \nabla(\delta\Phi)\, dV = \oint_S \delta\Phi\, \nabla\Phi \cdot d\mathbf{S}.$$

Here S is composed of (1) the exterior surfaces S_k ($k = 1, \ldots, n$) of the n bodies, (2) the surfaces S_c of the "cuts" that are introduced in order to keep V a simply-connected region (a condition for the validity of Green's identity), and (3) the sphere S_∞ of very large radius r. Thus

$$\int_V \nabla\Phi \cdot \nabla(\delta\Phi)\, dV = \sum_{k=1}^{n}\int_{S_k} \delta\Phi\, \nabla\Phi \cdot d\mathbf{S} + \int_{S_c} \delta\Phi\, \nabla\Phi \cdot d\mathbf{S} + \int_{S_\infty} \delta\Phi\, \nabla\Phi \cdot d\mathbf{S}.$$

The first term on the right vanishes because $\delta\Phi = 0$ on each S_k. The second term vanishes because the contributions from opposite sides of each cut cancel (note that $\hat{\mathbf{n}}$ occurs in pairs that are oppositely directed). The third term vanishes because $\Phi \sim 1/r$, $\nabla\Phi \sim 1/r^2$, and $dS \sim r^2$ where $r \to \infty$ for points on S_∞.

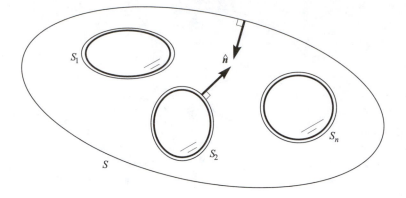

Figure 3.24: System of conductors used to derive Green's reciprocation theorem.

3.4.4   Green's reciprocation theorem

Consider a system of n conducting bodies as in Figure 3.24. An associated mathematical surface S_t consists of the exterior surfaces $S_1, \ldots, S_n$ of the n bodies, taken together with a surface S that enclosed all of the bodies. Suppose Φ and Φ' are electrostatic potentials produced by two distinct distributions of stationary charge over the set of conductors. Then $\nabla^2 \Phi = 0 = \nabla^2 \Phi'$ and Green's second identity gives

$$\oint_{S_t} \left(\Phi \frac{\partial \Phi'}{\partial n} - \Phi' \frac{\partial \Phi}{\partial n} \right) dS = 0$$

or

$$\sum_{k=1}^{n} \int_{S_k} \Phi \frac{\partial \Phi'}{\partial n} \, dS + \int_S \Phi \frac{\partial \Phi'}{\partial n} \, dS = \sum_{k=1}^{n} \int_{S_k} \Phi' \frac{\partial \Phi}{\partial n} \, dS + \int_S \Phi' \frac{\partial \Phi}{\partial n} \, dS.$$

Now let S be a sphere of very large radius R so that at points on S we have

$$\Phi, \Phi' \sim \frac{1}{R}, \qquad \frac{\partial \Phi}{\partial n}, \frac{\partial \Phi'}{\partial n} \sim \frac{1}{R^2}, \qquad dS \sim R^2;$$

as $R \to \infty$ then,

$$\sum_{k=1}^{n} \int_{S_k} \Phi \frac{\partial \Phi'}{\partial n} \, dS = \sum_{k=1}^{n} \int_{S_k} \Phi' \frac{\partial \Phi}{\partial n} \, dS.$$

Furthermore, the conductors are equipotentials so that

$$\sum_{k=1}^{n} \Phi_k \int_{S_k} \frac{\partial \Phi'}{\partial n} \, dS = \sum_{k=1}^{n} \Phi'_k \int_{S_k} \frac{\partial \Phi}{\partial n} \, dS$$

and we therefore have

$$\sum_{k=1}^{n} q'_k \Phi_k = \sum_{k=1}^{n} q_k \Phi'_k \qquad (3.210)$$

where the kth conductor ($k = 1, \ldots, n$) has potential Φ_k when it carries charge q_k, and has potential Φ'_k when it carries charge q'_k. This is *Green's reciprocation theorem*. A classic application is to determine the charge induced on a grounded conductor by

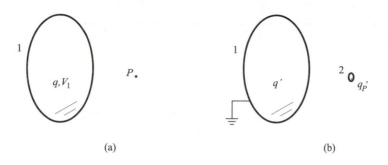

(a)   (b)

Figure 3.25: Application of Green's reciprocation theorem. (a) The "unprimed situation" permits us to determine the potential V_P at point P produced by a charge q placed on body 1. Here V_1 is the potential of body 1. (b) In the "primed situation" we ground body 1 and induce a charge q' by bringing a point charge q'_P into proximity.

a nearby point charge. This is accomplished as follows. Let the conducting body of interest be designated as body 1, and model the nearby point charge q_P as a very small conducting body designated as body 2 and located at point P in space. Take

$$q_1 = q, \qquad q_2 = 0, \qquad \Phi_1 - V_1, \qquad \Phi_2 - V_P,$$

and

$$q'_1 = q', \qquad q'_2 = q'_P, \qquad \Phi'_1 = 0, \qquad \Phi'_2 = V'_P,$$

giving the two situations shown in Figure 3.25. Substitution into Green's reciprocation theorem

$$q'_1 \Phi_1 + q'_2 \Phi_2 = q_1 \Phi'_1 + q_2 \Phi'_2$$

gives $q'V_1 + q'_P V_P = 0$ so that

$$q' = -q'_P V_P / V_1. \tag{3.211}$$

3.5 Problems

3.1 The z-axis carries a line charge of nonuniform density $\rho_l(z)$. Show that the electric field in the plane $z = 0$ is given by

$$\mathbf{E}(\rho, \phi) = \frac{1}{4\pi\epsilon} \left[\hat{\boldsymbol{\rho}} \rho \int_{-\infty}^{\infty} \frac{\rho_l(z') \, dz'}{(\rho^2 + z'^2)^{3/2}} - \hat{\mathbf{z}} \int_{-\infty}^{\infty} \frac{\rho_l(z') z' \, dz'}{(\rho^2 + z'^2)^{3/2}} \right].$$

Compute $\mathbf{E}$ when $\rho_l = \rho_0 \operatorname{sgn}(z)$, where $\operatorname{sgn}(z)$ is the signum function (A.6).

3.2 The ring $\rho = a$, $z = 0$, carries a line charge of nonuniform density $\rho_l(\phi)$. Show that the electric field at an arbitrary point on the z-axis is given by

$$\mathbf{E}(z) = \frac{-a^2}{4\pi\epsilon(a^2 + z^2)^{3/2}} \left[\hat{\mathbf{x}} \int_0^{2\pi} \rho_l(\phi') \cos \phi' \, d\phi' + \hat{\mathbf{y}} \int_0^{2\pi} \rho_l(\phi') \sin \phi' \, d\phi' \right] +$$

$$+ \hat{\mathbf{z}} \frac{az}{4\pi\epsilon(a^2 + z^2)^{3/2}} \int_0^{2\pi} \rho_l(\phi') \, d\phi'.$$

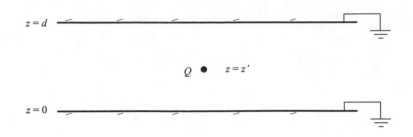

Figure 3.26: Geometry for computing Green's function for parallel plates.

Compute $\mathbf{E}$ when $\rho_l(\phi) = \rho_0 \sin\phi$. Repeat for $\rho_l(\phi) = \rho_0 \cos^2\phi$.

3.3 The plane $z = 0$ carries a surface charge of nonuniform density $\rho_s(\rho, \phi)$. Show that at an arbitrary point on the z-axis the rectangular components of $\mathbf{E}$ are given by

$$E_x(z) = -\frac{1}{4\pi\epsilon} \int_0^\infty \int_0^{2\pi} \frac{\rho_s(\rho', \phi')\,\rho'^2 \cos\phi'\,d\phi'\,d\rho'}{(\rho'^2 + z^2)^{3/2}},$$

$$E_y(z) = -\frac{1}{4\pi\epsilon} \int_0^\infty \int_0^{2\pi} \frac{\rho_s(\rho', \phi')\,\rho'^2 \sin\phi'\,d\phi'\,d\rho'}{(\rho'^2 + z^2)^{3/2}},$$

$$E_z(z) = \frac{z}{4\pi\epsilon} \int_0^\infty \int_0^{2\pi} \frac{\rho_s(\rho', \phi')\,\rho'\,d\phi'\,d\rho'}{(\rho'^2 + z^2)^{3/2}}.$$

Compute $\mathbf{E}$ when $\rho_s(\rho, \phi) = \rho_0 U(\rho - a)$ where $U(\rho)$ is the unit step function (A.5). Repeat for $\rho_s(\rho, \phi) = \rho_0[1 - U(\rho - a)]$.

3.4 The sphere $r = a$ carries a surface charge of nonuniform density $\rho_s(\theta)$. Show that the electric intensity at an arbitrary point on the z-axis is given by

$$\mathbf{E}(z) = \hat{\mathbf{z}}\frac{a^2}{2\epsilon} \int_0^\pi \frac{\rho_s(\theta')(z - a\cos\theta')\sin\theta'\,d\theta'}{(a^2 + z^2 - 2az\cos\theta')^{3/2}}.$$

Compute $\mathbf{E}(z)$ when $\rho_s(\theta) = \rho_0$, a constant. Repeat for $\rho_s(\theta) = \rho_0\cos^2\theta$.

3.5 Beginning with the postulates for the electrostatic field

$$\nabla \times \mathbf{E} = 0, \qquad \nabla \cdot \mathbf{D} = \rho,$$

use the technique of § 2.8.2 to derive the boundary conditions (3.32)–(3.33).

3.6 A material half space of permittivity ϵ_1 occupies the region $z > 0$, while a second material half space of permittivity ϵ_2 occupies $z < 0$. Find the polarization surface charge densities and compute the total induced polarization charge for a point charge Q located at $z = h$.

3.7 Consider a point charge between two grounded conducting plates as shown in Figure 3.26. Write the Green's function as the sum of primary and secondary terms and apply the boundary conditions to show that the secondary Green's function is

$$G^s(\mathbf{r}|\mathbf{r}') = \frac{1}{(2\pi)^2} \int_{-\infty}^\infty \int_{-\infty}^\infty \left[-e^{-k_\rho(d-z)}\frac{\sinh k_\rho z'}{\sinh k_\rho d} - e^{-k_\rho z}\frac{\sinh k_\rho(d - z')}{\sinh k_\rho d} \right] \frac{e^{-j\mathbf{k}_\rho\cdot\mathbf{r}'}}{2k_\rho}\,d^2k_\rho.$$

$$(3.212)$$

3.8 Use the expansion

$$\frac{1}{\sinh k_\rho d} = \operatorname{csch} k_\rho d = 2 \sum_{n=0}^{\infty} e^{-(2n+1)k_\rho d}$$

to show that the secondary Green's function for parallel conducting plates (3.212) may be written as an infinite sequence of images of the primary point charge. Identify the geometrical meaning of each image term.

3.9 Find the Green's functions for a dielectric slab of thickness d placed over a perfectly conducting ground plane located at $z = 0$.

3.10 Find the Green's functions for a dielectric slab of thickness $2d$ immersed in free space and centered on the $z = 0$ plane. Compare to the Green's function found in Problem 3.9.

3.11 Referring to the system of Figure 3.9, find the charge density on the surface of the sphere and integrate to show that the total charge is equal to the image charge.

3.12 Use the method of Green's functions to find the potential inside a conducting sphere for ρ inside the sphere.

3.13 Solve for the total potential and electric field of a grounded conducting sphere centered at the origin within a uniform impressed electric field $\mathbf{E} = E_0 \hat{\mathbf{z}}$. Find total charge induced on the sphere.

3.14 Consider a spherical cavity of radius a centered at the origin within a homogeneous dielectric material of permittivity $\epsilon = \epsilon_0 \epsilon_r$. Solve for total potential and electric field inside the cavity in the presence of an impressed field $\mathbf{E} = E_0 \hat{\mathbf{z}}$. Show that the field in the cavity is stronger than the applied field, and explain this using polarization surface charge.

3.15 Find the field of a point charge Q located at $z = d$ above a perfectly conducting ground plane at $z = 0$. Use the boundary condition to find the charge density on the plane and integrate to show that the total charge is $-Q$. Integrate Maxwell's stress tensor over the surface of the ground plane and show that the force on the ground plane is the same as the force on the image charge found from Coulomb's law.

3.16 Consider in free space a point charge $-q$ at $\mathbf{r} = \mathbf{r}_0 + \mathbf{d}$, a point charge $-q$ at $\mathbf{r} = \mathbf{r}_0 - \mathbf{d}$, and a point charge $2q$ at $\mathbf{r}_0$. Find the first three multipole moments and the resulting potential produced by this charge distribution.

3.17 A spherical charge distribution of radius a in free space has the density

$$\rho(\mathbf{r}) = \frac{Q}{\pi a^3} \cos 2\theta.$$

Compute the multipole moments for the charge distribution and find the resulting potential. Find a suitable arrangement of point charges that will produce the same potential field for $r > a$ as produced by the spherical charge.

3.18 Compute the magnetic flux density $\mathbf{B}$ for the circular wire loop of Figure 3.18 by (a) using the Biot–Savart law (3.165), and (b) computing the curl of (3.138).

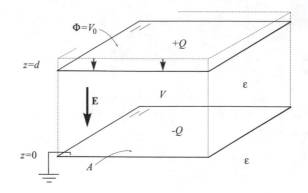

Figure 3.27: Parallel plate capacitor.

3.19 Two circular current-carrying wires are arranged coaxially along the z-axis. Loop 1 has radius a_1, carries current I_1, and is centered in the $z = 0$ plane. Loop 2 has radius a_2, carries current I_2, and is centered in the $z = d$ plane. Find the force between the loops.

3.20 Choose $\mathbf{Q} = \nabla' \left(\frac{1}{R}\right) \times \mathbf{c}$ in (3.162) and derive the following expression for $\mathbf{B}$:

$$
\mathbf{B}(\mathbf{r}) = \frac{\mu}{4\pi} \int_V \mathbf{J}(\mathbf{r}') \times \nabla' \left(\frac{1}{R}\right) dV' -
$$
$$
- \frac{1}{4\pi} \oint_S \left[[\hat{\mathbf{n}}' \times \mathbf{B}(\mathbf{r}')] \times \nabla' \left(\frac{1}{R}\right) + [\hat{\mathbf{n}}' \cdot \mathbf{B}(\mathbf{r}')] \nabla' \left(\frac{1}{R}\right) \right] dS',
$$

where $\hat{\mathbf{n}}$ is the normal vector outward from V. Compare to the Stratton–Chu formula (6.8).

3.21 Compute the curl of (3.163) to obtain the integral expression for $\mathbf{B}$ given in Problem 3.20. Compare to the Stratton–Chu formula (6.8).

3.22 Obtain (3.170) by integration of Maxwell's stress tensor over the xz-plane.

3.23 Consider two thin conducting parallel plates embedded in a region of permittivity ϵ (Figure 3.27). The bottom plate is connected to ground, and we apply an excess charge $+Q$ to the top plate (and thus $-Q$ is drawn onto the bottom plate.) Neglecting fringing, (a) solve Laplace's equation to show that

$$
\Phi(z) = \frac{Q}{A\epsilon} z.
$$

Use (3.87) to show that

$$
W = \frac{Q^2 d}{2A\epsilon}.
$$

(b) Verify W using (3.88). (c) Use $\mathbf{F} = -\hat{\mathbf{z}} dW/dz$ to show that the force on the top plate is

$$
\mathbf{F} = -\hat{\mathbf{z}} \frac{Q^2}{2A\epsilon}.
$$

(d) Verify $\mathbf{F}$ by integrating Maxwell's stress tensor over a closed surface surrounding the top plate.

3.24 Consider two thin conducting parallel plates embedded in a region of permittivity ϵ (Figure 3.27). The bottom plate is connected to ground, and we apply a potential V_0 to the top plate using a battery. Neglecting fringing, (a) solve Laplace's equation to show that

$$\Phi(z) = \frac{V_0}{d} z.$$

Use (3.87) to show that

$$W = \frac{V_0^2 A \epsilon}{2d}.$$

(b) Verify W using (3.88). (c) Use $\mathbf{F} = -\hat{\mathbf{z}} dW/dz$ to show that the force on the top plate is

$$\mathbf{F} = -\hat{\mathbf{z}} \frac{V_0^2 A \epsilon}{2d^2}.$$

(d) Verify $\mathbf{F}$ by integrating Maxwell's stress tensor over a closed surface surrounding the top plate.

3.25 A group of N perfectly conducting bodies is arranged in free space. Body n is held at potential V_n with respect to ground, and charge Q_n is induced upon its surface. By linearity we may write

$$Q_m = \sum_{n=1}^{N} c_{mn} V_n$$

where the c_{mn} are called the *capacitance coefficients*. Using Green's reciprocation theorem, demonstrate that $c_{mn} = c_{nm}$. Hint: Use (3.210). Choose one set of voltages so that $V_k = 0$, $k \neq n$, and place V_n at some potential, say $V_n = V_0$, producing the set of charges $\{Q_k\}$. For the second set choose $V_k' = 0$, $k \neq m$, and $V_m = V_0$, producing $\{Q_k'\}$.

3.26 For the set of conductors of Problem 3.25, show that we may write

$$Q_m = C_{mm} V_m + \sum_{k \neq m} C_{mk}(V_m - V_k)$$

where

$$C_{mn} = -c_{mn}, \quad m \neq n, \qquad C_{mm} = \sum_{k=1}^{N} c_{mk}.$$

Here C_{mm}, called the *self capacitance*, describes the interaction between the mth conductor and ground, while C_{mn}, called the *mutual capacitance*, describes the interaction between the mth and nth conductors.

3.27 For the set of conductors of Problem 3.25, show that the stored electric energy is given by

$$W = \frac{1}{2} \sum_{m=1}^{N} \sum_{n=1}^{N} c_{mn} V_n V_m.$$

3.28 A group of N wires is arranged in free space as shown in Figure 3.28. Wire n carries a steady current I_n, and a flux Ψ_n passes through the surface defined by its contour Γ_n. By linearity we may write

$$\Psi_m = \sum_{n=1}^{N} L_{mn} I_n$$

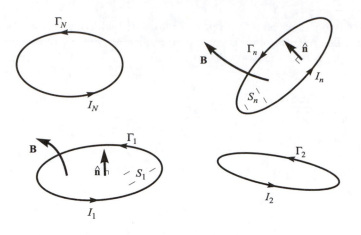

Figure 3.28: A system of current-carrying wires.

where the L_{mn} are called the *coefficients of inductance*. Derive *Neumann's formula*

$$L_{mn} = \frac{\mu_0}{4\pi} \oint_{\Gamma_n} \oint_{\Gamma_m} \frac{d\mathbf{l} \cdot d\mathbf{l}'}{|\mathbf{r} - \mathbf{r}'|},$$

and thereby demonstrate the reciprocity relation $L_{mn} = L_{nm}$.

3.29 For the group of wires shown in Figure 3.28, show that the stored magnetic energy is given by

$$W = \frac{1}{2} \sum_{m=1}^{N} \sum_{n=1}^{N} L_{mn} I_n I_m.$$

3.30 Prove the *minimum heat generation theorem*: steady electric currents distribute themselves in a conductor in such a way that the dissipated power is a minimum. Hint: Let $\mathbf{J}$ be the actual distribution of current in a conducting body, and let the power it dissipates be P. Let $\mathbf{J}' = \mathbf{J} + \delta\mathbf{J}$ be any other current distribution, and let the power it dissipates be $P' = P + \delta P$. Show that

$$\delta P = \frac{1}{2} \int_V \frac{1}{\sigma} |\delta\mathbf{J}|^2 \, dV \geq 0.$$

Chapter 4

Temporal and spatial frequency domain representation

4.1 Interpretation of the temporal transform

When a field is represented by a continuous superposition of elemental components, the resulting decomposition can simplify computation and provide physical insight. Such representation is usually accomplished through the use of an integral transform. Although several different transforms are used in electromagnetics, we shall concentrate on the powerful and efficient Fourier transform.

Let us consider the Fourier transform of the electromagnetic field. The field depends on x, y, z, t, and we can transform with respect to any or all of these variables. However, a consideration of units leads us to consider a transform over t separately. Let $\psi(\mathbf{r}, t)$ represent any rectangular component of the electric or magnetic field. Then the temporal transform will be designated by $\tilde{\psi}(\mathbf{r}, \omega)$:

$$\psi(\mathbf{r}, t) \leftrightarrow \tilde{\psi}(\mathbf{r}, \omega).$$

Here ω is the transform variable. The transform field $\tilde{\psi}$ is calculated using (A.1):

$$\tilde{\psi}(\mathbf{r}, \omega) = \int_{-\infty}^{\infty} \psi(\mathbf{r}, t)\, e^{-j\omega t}\, dt. \tag{4.1}$$

The inverse transform is, by (A.2),

$$\psi(\mathbf{r}, t) = \frac{1}{2\pi} \int_{-\infty}^{\infty} \tilde{\psi}(\mathbf{r}, \omega)\, e^{j\omega t}\, d\omega. \tag{4.2}$$

Since $\tilde{\psi}$ is complex it may be written in amplitude–phase form:

$$\tilde{\psi}(\mathbf{r}, \omega) = |\tilde{\psi}(\mathbf{r}, \omega)| e^{j\xi^{\psi}(\mathbf{r}, \omega)},$$

where we take $-\pi < \xi^{\psi}(\mathbf{r}, \omega) \le \pi$.

Since $\psi(\mathbf{r}, t)$ must be real, (4.1) shows that

$$\tilde{\psi}(\mathbf{r}, -\omega) = \tilde{\psi}^{*}(\mathbf{r}, \omega). \tag{4.3}$$

Furthermore, the transform of the derivative of ψ may be found by differentiating (4.2). We have

$$\frac{\partial}{\partial t}\psi(\mathbf{r}, t) = \frac{1}{2\pi} \int_{-\infty}^{\infty} j\omega\tilde{\psi}(\mathbf{r}, \omega)\, e^{j\omega t}\, d\omega,$$

hence

$$\frac{\partial}{\partial t}\psi(\mathbf{r},t) \;\leftrightarrow\; j\omega\tilde{\psi}(\mathbf{r},\omega). \tag{4.4}$$

By virtue of (4.2), any electromagnetic field component can be decomposed into a continuous, weighted superposition of elemental temporal terms $e^{j\omega t}$. Note that the weighting factor $\tilde{\psi}(\mathbf{r},\omega)$, often called the *frequency spectrum* of $\psi(\mathbf{r},t)$, is not arbitrary because $\psi(\mathbf{r},t)$ must obey a scalar wave equation such as (2.327). For a source-free region of space we have

$$\left(\nabla^2 - \mu\sigma\frac{\partial}{\partial t} - \mu\epsilon\frac{\partial^2}{\partial t^2}\right)\frac{1}{2\pi}\int_{-\infty}^{\infty}\tilde{\psi}(\mathbf{r},\omega)\,e^{j\omega t}\,d\omega = 0.$$

Differentiating under the integral sign we have

$$\frac{1}{2\pi}\int_{-\infty}^{\infty}\left[\left(\nabla^2 - j\omega\mu\sigma + \omega^2\mu\epsilon\right)\tilde{\psi}(\mathbf{r},\omega)\right]e^{j\omega t}\,d\omega = 0,$$

hence by the Fourier integral theorem

$$\left(\nabla^2 + k^2\right)\tilde{\psi}(\mathbf{r},\omega) = 0 \tag{4.5}$$

where

$$k = \omega\sqrt{\mu\epsilon}\sqrt{1 - j\frac{\sigma}{\omega\epsilon}}$$

is the *wavenumber*. Equation (4.5) is called the *scalar Helmholtz equation*, and represents the wave equation in the temporal frequency domain.

4.2   The frequency-domain Maxwell equations

If the region of interest contains sources, we can return to Maxwell's equations and represent all quantities using the temporal inverse Fourier transform. We have, for example,

$$\mathbf{E}(\mathbf{r},t) = \frac{1}{2\pi}\int_{-\infty}^{\infty}\tilde{\mathbf{E}}(\mathbf{r},\omega)\,e^{j\omega t}\,d\omega$$

where

$$\tilde{\mathbf{E}}(\mathbf{r},\omega) = \sum_{i=1}^{3}\hat{\mathbf{i}}_i\tilde{E}_i(\mathbf{r},\omega) = \sum_{i=1}^{3}\hat{\mathbf{i}}_i|\tilde{E}_i(\mathbf{r},\omega)|e^{j\xi_i^E(\mathbf{r},\omega)}. \tag{4.6}$$

All other field quantities will be written similarly with an appropriate superscript on the phase. Substitution into Ampere's law gives

$$\nabla \times \frac{1}{2\pi}\int_{-\infty}^{\infty}\tilde{\mathbf{H}}(\mathbf{r},\omega)\,e^{j\omega t}\,d\omega = \frac{\partial}{\partial t}\frac{1}{2\pi}\int_{-\infty}^{\infty}\tilde{\mathbf{D}}(\mathbf{r},\omega)\,e^{j\omega t}\,d\omega + \frac{1}{2\pi}\int_{-\infty}^{\infty}\tilde{\mathbf{J}}(\mathbf{r},\omega)\,e^{j\omega t}\,d\omega,$$

hence

$$\frac{1}{2\pi}\int_{-\infty}^{\infty}[\nabla \times \tilde{\mathbf{H}}(\mathbf{r},\omega) - j\omega\tilde{\mathbf{D}}(\mathbf{r},\omega) - \tilde{\mathbf{J}}(\mathbf{r},\omega)]e^{j\omega t}\,d\omega = 0$$

after we differentiate under the integral signs and combine terms. So

$$\nabla \times \tilde{\mathbf{H}} = j\omega\tilde{\mathbf{D}} + \tilde{\mathbf{J}} \tag{4.7}$$

by the Fourier integral theorem. This version of Ampere's law involves only the frequency-domain fields. By similar reasoning we have

$$\nabla \times \tilde{\mathbf{E}} = -j\omega\tilde{\mathbf{B}}, \tag{4.8}$$

$$\nabla \cdot \tilde{\mathbf{D}} = \tilde{\rho}, \tag{4.9}$$

$$\nabla \cdot \tilde{\mathbf{B}}(\mathbf{r}, \omega) = 0, \tag{4.10}$$

and

$$\nabla \cdot \tilde{\mathbf{J}} + j\omega\tilde{\rho} = 0.$$

Equations (4.7)–(4.10) govern the temporal spectra of the electromagnetic fields. We may manipulate them to obtain wave equations, and apply the boundary conditions from the following section. After finding the frequency-domain fields we may find the temporal fields by Fourier inversion. The frequency-domain equations involve one fewer derivative (the time derivative has been replaced by multiplication by $j\omega$), hence may be easier to solve. However, the inverse transform may be difficult to compute.

4.3 Boundary conditions on the frequency-domain fields

Several boundary conditions on the source and mediating fields were derived in § 2.8.2. For example, we found that the tangential electric field must obey

$$\hat{\mathbf{n}}_{12} \times \mathbf{E}_1(\mathbf{r}, t) - \hat{\mathbf{n}}_{12} \times \mathbf{E}_2(\mathbf{r}, t) = -\mathbf{J}_{ms}(\mathbf{r}, t).$$

The technique of the previous section gives us

$$\hat{\mathbf{n}}_{12} \times [\tilde{\mathbf{E}}_1(\mathbf{r}, \omega) - \tilde{\mathbf{E}}_2(\mathbf{r}, \omega)] = -\tilde{\mathbf{J}}_{ms}(\mathbf{r}, \omega)$$

as the condition satisfied by the frequency-domain electric field. The remaining boundary conditions are treated similarly. Let us summarize the results, including the effects of fictitious magnetic sources:

$$\hat{\mathbf{n}}_{12} \times (\tilde{\mathbf{H}}_1 - \tilde{\mathbf{H}}_2) = \tilde{\mathbf{J}}_s,$$

$$\hat{\mathbf{n}}_{12} \times (\tilde{\mathbf{E}}_1 - \tilde{\mathbf{E}}_2) = -\tilde{\mathbf{J}}_{ms},$$

$$\hat{\mathbf{n}}_{12} \cdot (\tilde{\mathbf{D}}_1 - \tilde{\mathbf{D}}_2) = \tilde{\rho}_s,$$

$$\hat{\mathbf{n}}_{12} \cdot (\tilde{\mathbf{B}}_1 - \tilde{\mathbf{B}}_2) = \tilde{\rho}_{ms},$$

and

$$\hat{\mathbf{n}}_{12} \cdot (\tilde{\mathbf{J}}_1 - \tilde{\mathbf{J}}_2) = -\nabla_s \cdot \tilde{\mathbf{J}}_s - j\omega\tilde{\rho}_s,$$

$$\hat{\mathbf{n}}_{12} \cdot (\tilde{\mathbf{J}}_{m1} - \tilde{\mathbf{J}}_{m2}) = -\nabla_s \cdot \tilde{\mathbf{J}}_{ms} - j\omega\tilde{\rho}_{ms}.$$

Here $\hat{\mathbf{n}}_{12}$ points into region 1 from region 2.

4.4   Constitutive relations in the frequency domain and the Kronig–Kramers relations

All materials are to some extent dispersive. If a field applied to a material undergoes a sufficiently rapid change, there is a time lag in the response of the polarization or magnetization of the atoms. It has been found that such materials have constitutive relations involving products in the frequency domain, and that the frequency-domain constitutive parameters are complex, frequency-dependent quantities. We shall restrict ourselves to the special case of anisotropic materials and refer the reader to Kong [101] and Lindell [113] for the more general case. For anisotropic materials we write

$$\tilde{\mathbf{P}} = \epsilon_0 \tilde{\bar{\chi}}_e \cdot \tilde{\mathbf{E}}, \tag{4.11}$$

$$\tilde{\mathbf{M}} = \tilde{\bar{\chi}}_m \cdot \tilde{\mathbf{H}}, \tag{4.12}$$

$$\tilde{\mathbf{D}} = \tilde{\bar{\epsilon}} \cdot \tilde{\mathbf{E}} = \epsilon_0 [\bar{\mathbf{I}} + \tilde{\bar{\chi}}_e] \cdot \tilde{\mathbf{E}}, \tag{4.13}$$

$$\tilde{\mathbf{B}} = \tilde{\bar{\mu}} \cdot \tilde{\mathbf{H}} = \mu_0 [\bar{\mathbf{I}} + \tilde{\bar{\chi}}_m] \cdot \tilde{\mathbf{H}}, \tag{4.14}$$

$$\tilde{\mathbf{J}} = \tilde{\bar{\sigma}} \cdot \tilde{\mathbf{E}}. \tag{4.15}$$

By the convolution theorem and the assumption of causality we immediately obtain the dyadic versions of (2.29)–(2.31):

$$\mathbf{D}(\mathbf{r}, t) = \epsilon_0 \left(\mathbf{E}(\mathbf{r}, t) + \int_{-\infty}^{t} \bar{\chi}_e(\mathbf{r}, t - t') \cdot \mathbf{E}(\mathbf{r}, t') \, dt' \right),$$

$$\mathbf{B}(\mathbf{r}, t) = \mu_0 \left(\mathbf{H}(\mathbf{r}, t) + \int_{-\infty}^{t} \bar{\chi}_m(\mathbf{r}, t - t') \cdot \mathbf{H}(\mathbf{r}, t') \, dt' \right),$$

$$\mathbf{J}(\mathbf{r}, t) = \int_{-\infty}^{t} \bar{\sigma}(\mathbf{r}, t - t') \cdot \mathbf{E}(\mathbf{r}, t') \, dt'.$$

These describe the essential behavior of a dispersive material. The susceptances and conductivity, describing the response of the atomic structure to an applied field, depend not only on the present value of the applied field but on all past values as well.

Now since $\mathbf{D}(\mathbf{r}, t)$, $\mathbf{B}(\mathbf{r}, t)$, and $\mathbf{J}(\mathbf{r}, t)$ are all real, so are the entries in the dyadic matrices $\bar{\epsilon}(\mathbf{r}, t)$, $\bar{\mu}(\mathbf{r}, t)$, and $\bar{\sigma}(\mathbf{r}, t)$. Thus, applying (4.3) to each entry we must have

$$\tilde{\bar{\chi}}_e(\mathbf{r}, -\omega) = \tilde{\bar{\chi}}_e^*(\mathbf{r}, \omega), \quad \tilde{\bar{\chi}}_m(\mathbf{r}, -\omega) = \tilde{\bar{\chi}}_m^*(\mathbf{r}, \omega), \quad \tilde{\bar{\sigma}}(\mathbf{r}, -\omega) = \tilde{\bar{\sigma}}^*(\mathbf{r}, \omega), \tag{4.16}$$

and hence

$$\tilde{\bar{\epsilon}}(\mathbf{r}, -\omega) = \tilde{\bar{\epsilon}}^*(\mathbf{r}, \omega), \qquad \tilde{\bar{\mu}}(\mathbf{r}, -\omega) = \tilde{\bar{\mu}}^*(\mathbf{r}, \omega). \tag{4.17}$$

If we write the constitutive parameters in terms of real and imaginary parts as

$$\tilde{\epsilon}_{ij} = \tilde{\epsilon}'_{ij} + j\tilde{\epsilon}''_{ij}, \qquad \tilde{\mu}_{ij} = \tilde{\mu}'_{ij} + j\tilde{\mu}''_{ij}, \qquad \tilde{\sigma}_{ij} = \tilde{\sigma}'_{ij} + j\tilde{\sigma}''_{ij},$$

these conditions become

$$\tilde{\epsilon}'_{ij}(\mathbf{r}, -\omega) = \tilde{\epsilon}'_{ij}(\mathbf{r}, \omega), \qquad \tilde{\epsilon}''_{ij}(\mathbf{r}, -\omega) = -\tilde{\epsilon}''_{ij}(\mathbf{r}, \omega),$$

and so on. Therefore the real parts of the constitutive parameters are even functions of frequency, and the imaginary parts are odd functions of frequency.

In most instances, the presence of an imaginary part in the constitutive parameters implies that the material is either *dissipative (lossy)*, transforming some of the electromagnetic energy in the fields into thermal energy, or *active*, transforming the chemical or mechanical energy of the material into energy in the fields. We investigate this further in § 4.5 and § 4.8.3.

We can also write the constitutive equations in amplitude–phase form. Letting

$$\tilde{\epsilon}_{ij} = |\tilde{\epsilon}_{ij}|e^{j\xi^{\epsilon}_{ij}}, \qquad \tilde{\mu}_{ij} = |\tilde{\mu}_{ij}|e^{j\xi^{\mu}_{ij}}, \qquad \tilde{\sigma}_{ij} = |\tilde{\sigma}_{ij}|e^{j\xi^{\sigma}_{ij}},$$

and using the field notation (4.6), we can write (4.13)–(4.15) as

$$\tilde{D}_i = |\tilde{D}_i|e^{j\xi^{D}_i} = \sum_{j=1}^{3} |\tilde{\epsilon}_{ij}||\tilde{E}_j|e^{j[\xi^{E}_j + \xi^{\epsilon}_{ij}]}, \tag{4.18}$$

$$\tilde{B}_i = |\tilde{B}_i|e^{j\xi^{B}_i} = \sum_{j=1}^{3} |\tilde{\mu}_{ij}||\tilde{H}_j|e^{j[\xi^{H}_j + \xi^{\mu}_{ij}]}, \tag{4.19}$$

$$\tilde{J}_i = |\tilde{J}_i|e^{j\xi^{J}_i} = \sum_{j=1}^{3} |\tilde{\sigma}_{ij}||\tilde{E}_j|e^{j[\xi^{E}_j + \xi^{\sigma}_{ij}]}. \tag{4.20}$$

Here we remember that the amplitudes and phases may be functions of both $\mathbf{r}$ and ω. For isotropic materials these reduce to

$$\tilde{D}_i = |\tilde{D}_i|e^{j\xi^{D}_i} = |\tilde{\epsilon}||\tilde{E}_i|e^{j(\xi^{E}_i + \xi^{\epsilon})}, \tag{4.21}$$

$$\tilde{B}_i = |\tilde{B}_i|e^{j\xi^{B}_i} = |\tilde{\mu}||\tilde{H}_i|e^{j(\xi^{H}_i + \xi^{\mu})}, \tag{4.22}$$

$$\tilde{J}_i = |\tilde{J}_i|e^{j\xi^{J}_i} = |\tilde{\sigma}||\tilde{E}_i|e^{j(\xi^{E}_i + \xi^{\sigma})}. \tag{4.23}$$

4.4.1 The complex permittivity

As mentioned above, dissipative effects may be associated with complex entries in the permittivity matrix. Since conduction effects can also lead to dissipation, the permittivity and conductivity matrices are often combined to form a *complex permittivity*. Writing the current as a sum of impressed and secondary conduction terms ($\tilde{\mathbf{J}} = \tilde{\mathbf{J}}^i + \tilde{\mathbf{J}}^c$) and substituting (4.13) and (4.15) into Ampere's law, we find

$$\nabla \times \tilde{\mathbf{H}} = \tilde{\mathbf{J}}^i + \bar{\tilde{\sigma}} \cdot \tilde{\mathbf{E}} + j\omega\bar{\tilde{\epsilon}} \cdot \tilde{\mathbf{E}}.$$

Defining the complex permittivity

$$\bar{\tilde{\epsilon}}^c(\mathbf{r}, \omega) = \frac{\bar{\tilde{\sigma}}(\mathbf{r}, \omega)}{j\omega} + \bar{\tilde{\epsilon}}(\mathbf{r}, \omega), \tag{4.24}$$

we have

$$\nabla \times \tilde{\mathbf{H}} = \tilde{\mathbf{J}}^i + j\omega\bar{\tilde{\epsilon}}^c \cdot \tilde{\mathbf{E}}.$$

Using the complex permittivity we can include the effects of conduction current by merely replacing the total current with the impressed current. Since Faraday's law is unaffected, any equation (such as the wave equation) derived previously using total current retains its form with the same substitution.

By (4.16) and (4.17) the complex permittivity obeys

$$\bar{\tilde{\epsilon}}^c(\mathbf{r}, -\omega) = \bar{\tilde{\epsilon}}^{c*}(\mathbf{r}, \omega) \tag{4.25}$$

or

$$\tilde{\epsilon}_{ij}^{c\prime}(\mathbf{r}, -\omega) = \tilde{\epsilon}_{ij}^{c\prime}(\mathbf{r}, \omega), \qquad \tilde{\epsilon}_{ij}^{c\prime\prime}(\mathbf{r}, -\omega) = -\tilde{\epsilon}_{ij}^{c\prime\prime}(\mathbf{r}, \omega).$$

For an isotropic material it takes the particularly simple form

$$\tilde{\epsilon}^c = \frac{\tilde{\sigma}}{j\omega} + \tilde{\epsilon} = \frac{\tilde{\sigma}}{j\omega} + \epsilon_0 + \epsilon_0 \tilde{\chi}_e, \tag{4.26}$$

and we have

$$\tilde{\epsilon}^{c\prime}(\mathbf{r}, -\omega) = \tilde{\epsilon}^{c\prime}(\mathbf{r}, \omega), \qquad \tilde{\epsilon}^{c\prime\prime}(\mathbf{r}, -\omega) = -\tilde{\epsilon}^{c\prime\prime}(\mathbf{r}, \omega). \tag{4.27}$$

4.4.2   High and low frequency behavior of constitutive parameters

At low frequencies the permittivity reduces to the electrostatic permittivity. Since $\tilde{\epsilon}^\prime$ is even in ω and $\tilde{\epsilon}^{\prime\prime}$ is odd, we have for small ω

$$\tilde{\epsilon}^\prime \sim \epsilon_0 \epsilon_r, \qquad \tilde{\epsilon}^{\prime\prime} \sim \omega.$$

If the material has some dc conductivity σ_0, then for low frequencies the complex permittivity behaves as

$$\tilde{\epsilon}^{c\prime} \sim \epsilon_0 \epsilon_r, \qquad \tilde{\epsilon}^{c\prime\prime} \sim \sigma_0/\omega. \tag{4.28}$$

If $\mathbf{E}$ or $\mathbf{H}$ changes very rapidly, there may be no polarization or magnetization effect at all. This occurs at frequencies so high that the atomic structure of the material cannot respond to the rapidly oscillating applied field. Above some frequency then, we can assume $\tilde{\chi}_e = 0$ and $\tilde{\chi}_m = 0$ so that

$$\tilde{\mathbf{P}} = 0, \qquad \tilde{\mathbf{M}} = 0,$$

and

$$\tilde{\mathbf{D}} = \epsilon_0 \tilde{\mathbf{E}}, \qquad \tilde{\mathbf{B}} = \mu_0 \tilde{\mathbf{H}}.$$

In our simple models of dielectric materials (§ 4.6) we find that as ω becomes large

$$\tilde{\epsilon}^\prime - \epsilon_0 \sim 1/\omega^2, \qquad \tilde{\epsilon}^{\prime\prime} \sim 1/\omega^3. \tag{4.29}$$

Our assumption of a macroscopic model of matter provides a fairly strict upper frequency limit to the range of validity of the constitutive parameters. We must assume that the wavelength of the electromagnetic field is large compared to the size of the atomic structure. This limit suggests that permittivity and permeability might remain meaningful even at optical frequencies, and for dielectrics this is indeed the case since the values of $\tilde{\mathbf{P}}$ remain significant. However, $\tilde{\mathbf{M}}$ becomes insignificant at much lower frequencies, and at optical frequencies we may use $\tilde{\mathbf{B}} = \mu_0 \tilde{\mathbf{H}}$ [107].

4.4.3   The Kronig–Kramers relations

The principle of causality is clearly implicit in (2.29)–(2.31). We shall demonstrate that causality leads to explicit relationships between the real and imaginary parts of the frequency-domain constitutive parameters. For simplicity we concentrate on the isotropic case and merely note that the present analysis may be applied to all the dyadic components of an anisotropic constitutive parameter. We also concentrate on the complex permittivity and extend the results to permeability by induction.

The implications of causality on the behavior of the constitutive parameters in the time domain can be easily identified. Writing (2.29) and (2.31) after setting $u = t - t'$ and then $u = t'$, we have

$$\mathbf{D}(\mathbf{r}, t) = \epsilon_0 \mathbf{E}(\mathbf{r}, t) + \epsilon_0 \int_0^\infty \chi_e(\mathbf{r}, t') \mathbf{E}(\mathbf{r}, t - t') \, dt',$$

$$\mathbf{J}(\mathbf{r}, t) = \int_0^\infty \sigma(\mathbf{r}, t') \mathbf{E}(\mathbf{r}, t - t') \, dt'.$$

We see that there is no contribution from values of $\chi_e(\mathbf{r}, t)$ or $\sigma(\mathbf{r}, t)$ for times $t < 0$. So we can write

$$\mathbf{D}(\mathbf{r}, t) = \epsilon_0 \mathbf{E}(\mathbf{r}, t) + \epsilon_0 \int_{-\infty}^\infty \chi_e(\mathbf{r}, t') \mathbf{E}(\mathbf{r}, t - t') \, dt',$$

$$\mathbf{J}(\mathbf{r}, t) = \int_{-\infty}^\infty \sigma(\mathbf{r}, t') \mathbf{E}(\mathbf{r}, t - t') \, dt',$$

with the additional assumption

$$\chi_e(\mathbf{r}, t) = 0, \quad t < 0, \qquad \sigma(\mathbf{r}, t) = 0, \quad t < 0. \tag{4.30}$$

By (4.30) we can write the frequency-domain complex permittivity (4.26) as

$$\tilde{\epsilon}^c(\mathbf{r}, \omega) - \epsilon_0 = \frac{1}{j\omega} \int_0^\infty \sigma(\mathbf{r}, t') e^{-j\omega t'} \, dt' + \epsilon_0 \int_0^\infty \chi_e(\mathbf{r}, t') e^{-j\omega t'} \, dt'. \tag{4.31}$$

In order to derive the Kronig–Kramers relations we must understand the behavior of $\tilde{\epsilon}^c(\mathbf{r}, \omega) - \epsilon_0$ in the complex ω-plane. Writing $\omega = \omega_r + j\omega_i$, we need to establish the following two properties.

Property 1: The function $\tilde{\epsilon}^c(\mathbf{r}, \omega) - \epsilon_0$ is analytic in the lower half-plane ($\omega_i < 0$) except at $\omega = 0$ where it has a simple pole.

We can establish the analyticity of $\tilde{\sigma}(\mathbf{r}, \omega)$ by integrating over any closed contour in the lower half-plane. We have

$$\oint_\Gamma \tilde{\sigma}(\mathbf{r}, \omega) \, d\omega = \oint_\Gamma \left[\int_0^\infty \sigma(\mathbf{r}, t') e^{-j\omega t'} \, dt' \right] d\omega = \int_0^\infty \sigma(\mathbf{r}, t') \left[\oint_\Gamma e^{-j\omega t'} \, d\omega \right] dt'. \tag{4.32}$$

Note that an exchange in the order of integration in the above expression is only valid for ω in the lower half-plane where $\lim_{t' \to \infty} e^{-j\omega t'} = 0$. Since the function $f(\omega) = e^{-j\omega t'}$ is analytic in the lower half-plane, its closed contour integral is zero by the Cauchy–Goursat theorem. Thus, by (4.32) we have

$$\oint_\Gamma \tilde{\sigma}(\mathbf{r}, \omega) \, d\omega = 0.$$

Then, since $\tilde{\sigma}$ may be assumed to be continuous in the lower half-plane for a physical medium, and since its closed path integral is zero for all possible paths Γ, it is by Morera's theorem [110] analytic in the lower half-plane. By similar reasoning $\chi_e(\mathbf{r}, \omega)$ is analytic in the lower half-plane. Since the function $1/\omega$ has a simple pole at $\omega = 0$, the composite function $\tilde{\epsilon}^c(\mathbf{r}, \omega) - \epsilon_0$ given by (4.31) is analytic in the lower half-plane excluding $\omega = 0$ where it has a simple pole.

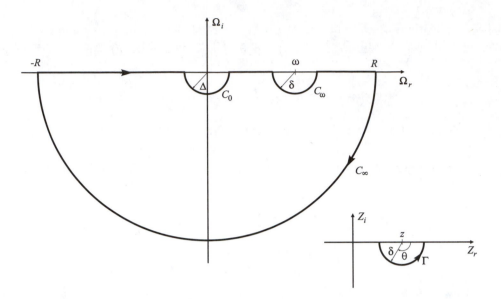

Figure 4.1: Complex integration contour used to establish the Kronig–Kramers relations.

Property 2:   We have

$$\lim_{\omega \to \pm\infty} \tilde{\epsilon}^c(\mathbf{r}, \omega) - \epsilon_0 = 0.$$

To establish this property we need the *Riemann–Lebesgue lemma* [142], which states that if $f(t)$ is absolutely integrable on the interval (a, b) where a and b are finite or infinite constants, then

$$\lim_{\omega \to \pm\infty} \int_a^b f(t)e^{-j\omega t}\, dt = 0.$$

From this we see that

$$\lim_{\omega \to \pm\infty} \frac{\tilde{\sigma}(\mathbf{r}, \omega)}{j\omega} = \lim_{\omega \to \pm\infty} \frac{1}{j\omega} \int_0^\infty \sigma(\mathbf{r}, t')e^{-j\omega t'}\, dt' = 0,$$

$$\lim_{\omega \to \pm\infty} \epsilon_0 \chi_e(\mathbf{r}, \omega) = \lim_{\omega \to \pm\infty} \epsilon_0 \int_0^\infty \chi_e(\mathbf{r}, t')e^{-j\omega t'}\, dt' = 0,$$

and thus

$$\lim_{\omega \to \pm\infty} \tilde{\epsilon}^c(\mathbf{r}, \omega) - \epsilon_0 = 0.$$

To establish the Kronig–Kramers relations we examine the integral

$$\oint_\Gamma \frac{\tilde{\epsilon}^c(\mathbf{r}, \Omega) - \epsilon_0}{\Omega - \omega}\, d\Omega$$

where Γ is the contour shown in Figure 4.1. Since the points $\Omega = 0, \omega$ are excluded, the integrand is analytic everywhere within and on Γ, hence the integral vanishes by the Cauchy–Goursat theorem. By Property 2 we have

$$\lim_{R \to \infty} \int_{C_\infty} \frac{\tilde{\epsilon}^c(\mathbf{r}, \Omega) - \epsilon_0}{\Omega - \omega}\, d\Omega = 0,$$

hence

$$\int_{C_0+C_\omega} \frac{\tilde{\epsilon}^c(\mathbf{r},\Omega)-\epsilon_0}{\Omega-\omega}\,d\Omega + \text{P.V.} \int_{-\infty}^{\infty} \frac{\tilde{\epsilon}^c(\mathbf{r},\Omega)-\epsilon_0}{\Omega-\omega}\,d\Omega = 0. \qquad (4.33)$$

Here "P.V." indicates that the integral is computed in the Cauchy principal value sense (see Appendix A). To evaluate the integrals over C_0 and C_ω, consider a function $f(Z)$ analytic in the lower half of the Z-plane ($Z = Z_r + jZ_i$). If the point z lies on the real axis as shown in Figure 4.1, we can calculate the integral

$$F(z) = \lim_{\delta\to 0} \int_{\Gamma} \frac{f(Z)}{Z-z}\,dZ$$

through the parameterization $Z - z = \delta e^{j\theta}$. Since $dZ = j\delta e^{j\theta}\,d\theta$ we have

$$F(z) = \lim_{\delta\to 0} \int_{-\pi}^{0} \frac{f\left(z+\delta e^{j\theta}\right)}{\delta e^{j\theta}} \left[j\delta e^{j\theta}\right] d\theta = jf(z) \int_{-\pi}^{0} d\theta = j\pi f(z).$$

Replacing Z by Ω and z by 0 we can compute

$$\lim_{\Delta\to 0} \int_{C_0} \frac{\tilde{\epsilon}^c(\mathbf{r},\Omega)-\epsilon_0}{\Omega-\omega}\,d\Omega$$

$$= \lim_{\Delta\to 0} \int_{C_0} \frac{\left[\frac{1}{j}\int_0^\infty \sigma(\mathbf{r},t')e^{-j\Omega t'}\,dt' + \Omega\epsilon_0\int_0^\infty \chi_e(\mathbf{r},t')e^{-j\Omega t'}\,dt'\right]\frac{1}{\Omega-\omega}}{\Omega}\,d\Omega$$

$$= -\frac{\pi\int_0^\infty \sigma(\mathbf{r},t')\,dt'}{\omega}.$$

We recognize

$$\int_0^\infty \sigma(\mathbf{r},t')\,dt' = \sigma_0(\mathbf{r})$$

as the dc conductivity and write

$$\lim_{\Delta\to 0} \int_{C_0} \frac{\tilde{\epsilon}^c(\mathbf{r},\Omega)-\epsilon_0}{\Omega-\omega}\,d\Omega = -\frac{\pi\sigma_0(\mathbf{r})}{\omega}.$$

If we replace Z by Ω and z by ω we get

$$\lim_{\delta\to 0} \int_{C_\omega} \frac{\tilde{\epsilon}^c(\mathbf{r},\Omega)-\epsilon_0}{\Omega-\omega}\,d\Omega = j\pi\tilde{\epsilon}^c(\mathbf{r},\omega) - j\pi\epsilon_0.$$

Substituting these into (4.33) we have

$$\tilde{\epsilon}^c(\mathbf{r},\omega)-\epsilon_0 = -\frac{1}{j\pi}\,\text{P.V.} \int_{-\infty}^{\infty} \frac{\tilde{\epsilon}^c(\mathbf{r},\Omega)-\epsilon_0}{\Omega-\omega}\,d\Omega + \frac{\sigma_0(\mathbf{r})}{j\omega}. \qquad (4.34)$$

If we write $\tilde{\epsilon}^c(\mathbf{r},\omega) = \tilde{\epsilon}^{c\prime}(\mathbf{r},\omega) + j\tilde{\epsilon}^{c\prime\prime}(\mathbf{r},\omega)$ and equate real and imaginary parts in (4.34) we find that

$$\tilde{\epsilon}^{c\prime}(\mathbf{r},\omega)-\epsilon_0 = -\frac{1}{\pi}\,\text{P.V.} \int_{-\infty}^{\infty} \frac{\tilde{\epsilon}^{c\prime\prime}(\mathbf{r},\Omega)}{\Omega-\omega}\,d\Omega, \qquad (4.35)$$

$$\tilde{\epsilon}^{c\prime\prime}(\mathbf{r},\omega) = \frac{1}{\pi}\,\text{P.V.} \int_{-\infty}^{\infty} \frac{\tilde{\epsilon}^{c\prime}(\mathbf{r},\Omega)-\epsilon_0}{\Omega-\omega}\,d\Omega - \frac{\sigma_0(\mathbf{r})}{\omega}. \qquad (4.36)$$

These are the *Kronig–Kramers relations,* named after R. de L. Kronig and H.A. Kramers who derived them independently. The expressions show that causality requires the real and imaginary parts of the permittivity to depend upon each other through the Hilbert transform pair [142].

It is often more convenient to write the Kronig–Kramers relations in a form that employs only positive frequencies. This can be accomplished using the even–odd behavior of the real and imaginary parts of $\tilde{\epsilon}^c$. Breaking the integrals in (4.35)–(4.36) into the ranges $(-\infty, 0)$ and $(0, \infty)$, and substituting from (4.27), we can show that

$$\tilde{\epsilon}^{c\prime}(\mathbf{r}, \omega) - \epsilon_0 = -\frac{2}{\pi} \, \text{P.V.} \int_0^\infty \frac{\Omega \tilde{\epsilon}^{c\prime\prime}(\mathbf{r}, \Omega)}{\Omega^2 - \omega^2} \, d\Omega, \qquad (4.37)$$

$$\tilde{\epsilon}^{c\prime\prime}(\mathbf{r}, \omega) = \frac{2\omega}{\pi} \, \text{P.V.} \int_0^\infty \frac{\tilde{\epsilon}^{c\prime}(\mathbf{r}, \Omega)}{\Omega^2 - \omega^2} \, d\Omega - \frac{\sigma_0(\mathbf{r})}{\omega}. \qquad (4.38)$$

The symbol P.V. in this case indicates that values of the integrand around both $\Omega = 0$ and $\Omega = \omega$ must be excluded from the integration. The details of the derivation of (4.37)–(4.38) are left as an exercise. We shall use (4.37) in § 4.6 to demonstrate the Kronig–Kramers relationship for a model of complex permittivity of an actual material.

We cannot specify $\tilde{\epsilon}^{c\prime}$ arbitrarily; for a passive medium $\tilde{\epsilon}^{c\prime\prime}$ must be zero or negative at all values of ω, and (4.36) will not necessarily return these required values. However, if we have a good measurement or physical model for $\tilde{\epsilon}^{c\prime\prime}$, as might come from studies of the absorbing properties of the material, we can approximate the real part of the permittivity using (4.35). We shall demonstrate this using simple models for permittivity in § 4.6.

The Kronig–Kramers properties hold for μ as well. We must for practical reasons consider the fact that magnetization becomes unimportant at a much lower frequency than does polarization, so that the infinite integrals in the Kronig–Kramers relations should be truncated at some upper frequency ω_{max}. If we use a model or measured values of $\tilde{\mu}^{\prime\prime}$ to determine $\tilde{\mu}^\prime$, the form of the relation (4.37) should be [107]

$$\tilde{\mu}^\prime(\mathbf{r}, \omega) - \mu_0 = -\frac{2}{\pi} \, \text{P.V.} \int_0^{\omega_{\text{max}}} \frac{\Omega \tilde{\mu}^{\prime\prime}(\mathbf{r}, \Omega)}{\Omega^2 - \omega^2} d\Omega,$$

where ω_{max} is the frequency at which magnetization ceases to be important, and above which $\tilde{\mu} = \mu_0$.

4.5   Dissipated and stored energy in a dispersive medium

Let us write down Poynting's power balance theorem for a dispersive medium. Writing $\mathbf{J} = \mathbf{J}^i + \mathbf{J}^c$ we have (§ 2.9.5)

$$-\mathbf{J}^i \cdot \mathbf{E} = \mathbf{J}^c \cdot \mathbf{E} + \nabla \cdot [\mathbf{E} \times \mathbf{H}] + \left[\mathbf{E} \cdot \frac{\partial \mathbf{D}}{\partial t} + \mathbf{H} \cdot \frac{\partial \mathbf{B}}{\partial t} \right]. \qquad (4.39)$$

We cannot express this in terms of the time rate of change of a stored energy density because of the difficulty in interpreting the term

$$\mathbf{E} \cdot \frac{\partial \mathbf{D}}{\partial t} + \mathbf{H} \cdot \frac{\partial \mathbf{B}}{\partial t} \qquad (4.40)$$

when the constitutive parameters have the form (2.29)–(2.31). Physically, this term describes both the energy stored in the electromagnetic field *and* the energy dissipated by the material because of time lags between the application of $\mathbf{E}$ and $\mathbf{H}$ and the polarization or magnetization of the atoms (and thus the response fields $\mathbf{D}$ and $\mathbf{B}$). In principle this term can also be used to describe *active* media that transfer mechanical or chemical energy of the material into field energy.

Instead of attempting to interpret (4.40), we concentrate on the physical meaning of

$$-\nabla \cdot \mathbf{S}(\mathbf{r}, t) = -\nabla \cdot [\mathbf{E}(\mathbf{r}, t) \times \mathbf{H}(\mathbf{r}, t)].$$

We shall postulate that this term describes the net flow of electromagnetic energy into the point $\mathbf{r}$ at time t. Then (4.39) shows that in the absence of impressed sources the energy flow must act to (1) increase or decrease the stored energy density at $\mathbf{r}$, (2) dissipate energy in ohmic losses through the term involving $\mathbf{J}^c$, or (3) dissipate (or provide) energy through the term (4.40). Assuming linearity we may write

$$-\nabla \cdot \mathbf{S}(\mathbf{r}, t) = \frac{\partial}{\partial t} w_e(\mathbf{r}, t) + \frac{\partial}{\partial t} w_m(\mathbf{r}, t) + \frac{\partial}{\partial t} w_Q(\mathbf{r}, t), \qquad (4.41)$$

where the terms on the right-hand side represent the time rates of change of, respectively, stored electric, stored magnetic, and dissipated energies.

4.5.1 Dissipation in a dispersive material

Although we may, in general, be unable to separate the individual terms in (4.41), we can examine these terms under certain conditions. For example, consider a field that builds from zero starting from time $t = -\infty$ and then decays back to zero at $t = \infty$. Then by direct integration[1]

$$-\int_{-\infty}^{\infty} \nabla \cdot \mathbf{S}(t)\, dt = w_{em}(t = \infty) - w_{em}(t = -\infty) + w_Q(t = \infty) - w_Q(t = -\infty)$$

where $w_{em} = w_e + w_m$ is the volume density of stored electromagnetic energy. This stored energy is zero at $t = \pm\infty$ since the fields are zero at those times. Thus,

$$\Delta w_Q = -\int_{-\infty}^{\infty} \nabla \cdot \mathbf{S}(t)\, dt = w_Q(t = \infty) - w_Q(t = -\infty)$$

represents the volume density of the net energy dissipated by a lossy medium (or supplied by an active medium). We may thus classify materials according to the scheme

$$\Delta w_Q = 0, \qquad \text{lossless,}$$
$$\Delta w_Q > 0, \qquad \text{lossy,}$$
$$\Delta w_Q \geq 0, \qquad \text{passive,}$$
$$\Delta w_Q < 0, \qquad \text{active.}$$

For an anisotropic material with the constitutive relations

$$\tilde{\mathbf{D}} = \tilde{\bar{\epsilon}} \cdot \tilde{\mathbf{E}}, \qquad \tilde{\mathbf{B}} = \tilde{\bar{\mu}} \cdot \tilde{\mathbf{H}}, \qquad \tilde{\mathbf{J}}^c = \tilde{\bar{\sigma}} \cdot \tilde{\mathbf{E}},$$

---

[1]Note that in this section we suppress the $\mathbf{r}$-dependence of most quantities for clarity of presentation.

we find that dissipation is associated with negative imaginary parts of the constitutive parameters. To see this we write

$$\mathbf{E}(\mathbf{r}, t) = \frac{1}{2\pi} \int_{-\infty}^{\infty} \tilde{\mathbf{E}}(\mathbf{r}, \omega) e^{j\omega t} \, d\omega, \qquad \mathbf{D}(\mathbf{r}, t) = \frac{1}{2\pi} \int_{-\infty}^{\infty} \tilde{\mathbf{D}}(\mathbf{r}, \omega') e^{j\omega' t} \, d\omega',$$

and thus find

$$\mathbf{J}^c \cdot \mathbf{E} + \mathbf{E} \cdot \frac{\partial \mathbf{D}}{\partial t} = \frac{1}{(2\pi)^2} \int_{-\infty}^{\infty} \int_{-\infty}^{\infty} \tilde{\mathbf{E}}(\omega) \cdot \bar{\tilde{\epsilon}}^c(\omega') \cdot \tilde{\mathbf{E}}(\omega') e^{j(\omega+\omega')t} j\omega' \, d\omega \, d\omega'$$

where $\bar{\tilde{\epsilon}}^c$ is the complex dyadic permittivity (4.24). Then

$$\Delta w_Q = \frac{1}{(2\pi)^2} \int_{-\infty}^{\infty} \int_{-\infty}^{\infty} \left[\tilde{\mathbf{E}}(\omega) \cdot \bar{\tilde{\epsilon}}^c(\omega') \cdot \tilde{\mathbf{E}}(\omega') + \tilde{\mathbf{H}}(\omega) \cdot \bar{\tilde{\mu}}(\omega') \cdot \tilde{\mathbf{H}}(\omega') \right] \cdot$$
$$\cdot \left[\int_{-\infty}^{\infty} e^{j(\omega+\omega')t} \, dt \right] j\omega' \, d\omega \, d\omega'. \tag{4.42}$$

Using (A.4) and integrating over ω we obtain

$$\Delta w_Q = \frac{1}{2\pi} \int_{-\infty}^{\infty} \left[\tilde{\mathbf{E}}(-\omega') \cdot \bar{\tilde{\epsilon}}^c(\omega') \cdot \tilde{\mathbf{E}}(\omega') + \tilde{\mathbf{H}}(-\omega') \cdot \bar{\tilde{\mu}}(\omega') \cdot \tilde{\mathbf{H}}(\omega') \right] j\omega' \, d\omega'. \tag{4.43}$$

Let us examine (4.43) more closely for the simple case of an isotropic material for which

$$\Delta w_Q = \frac{1}{2\pi} \int_{-\infty}^{\infty} \left\{ [j\tilde{\epsilon}^{c\prime}(\omega') - \tilde{\epsilon}^{c\prime\prime}(\omega')] \tilde{\mathbf{E}}(-\omega') \cdot \tilde{\mathbf{E}}(\omega') + \right.$$
$$\left. + [j\tilde{\mu}'(\omega') - \tilde{\mu}''(\omega')] \tilde{\mathbf{H}}(-\omega') \cdot \tilde{\mathbf{H}}(\omega') \right\} \omega' \, d\omega'.$$

Using the frequency symmetry property for complex permittivity (4.17) (which also holds for permeability), we find that for isotropic materials

$$\tilde{\epsilon}^{c\prime}(\mathbf{r}, \omega) = \tilde{\epsilon}^{c\prime}(\mathbf{r}, -\omega), \qquad \tilde{\epsilon}^{c\prime\prime}(\mathbf{r}, \omega) = -\tilde{\epsilon}^{c\prime\prime}(\mathbf{r}, -\omega), \tag{4.44}$$
$$\tilde{\mu}'(\mathbf{r}, \omega) = \tilde{\mu}'(\mathbf{r}, -\omega), \qquad \tilde{\mu}''(\mathbf{r}, \omega) = -\tilde{\mu}''(\mathbf{r}, -\omega). \tag{4.45}$$

Thus, the products of ω' and the real parts of the constitutive parameters are odd functions, while for the imaginary parts these products are even. Since the dot products of the vector fields are even functions, we find that the integrals of the terms containing the real parts of the constitutive parameters vanish, leaving

$$\Delta w_Q = 2\frac{1}{2\pi} \int_0^{\infty} \left[-\tilde{\epsilon}^{c\prime\prime} |\tilde{\mathbf{E}}|^2 - \tilde{\mu}'' |\tilde{\mathbf{H}}|^2 \right] \omega \, d\omega. \tag{4.46}$$

Here we have used (4.3) in the form

$$\tilde{\mathbf{E}}(\mathbf{r}, -\omega) = \tilde{\mathbf{E}}^*(\mathbf{r}, \omega), \qquad \tilde{\mathbf{H}}(\mathbf{r}, -\omega) = \tilde{\mathbf{H}}^*(\mathbf{r}, \omega). \tag{4.47}$$

Equation (4.46) leads us to associate the imaginary parts of the constitutive parameters with dissipation. Moreover, a lossy isotropic material for which $\Delta w_Q > 0$ must have at least one of $\epsilon^{c\prime\prime}$ and μ'' less than zero over some range of positive frequencies, while an active isotropic medium must have at least one of these greater than zero. In general, we speak of a lossy material as having negative imaginary constitutive parameters:

$$\tilde{\epsilon}^{c\prime\prime} < 0, \qquad \tilde{\mu}^{\prime\prime} < 0, \qquad \omega > 0. \tag{4.48}$$

A *lossless* medium must have

$$\tilde{\epsilon}^{\prime\prime} = \tilde{\mu}^{\prime\prime} = \tilde{\sigma} = 0$$

for all ω.

Things are not as simple in the more general anisotropic case. An integration of (4.42) over ω' instead of ω produces

$$\Delta w_Q = -\frac{1}{2\pi} \int_{-\infty}^{\infty} \left[\tilde{\mathbf{E}}(\omega) \cdot \bar{\bar{\epsilon}}^c(-\omega) \cdot \tilde{\mathbf{E}}(-\omega) + \tilde{\mathbf{H}}(\omega) \cdot \bar{\bar{\mu}}(-\omega) \cdot \tilde{\mathbf{H}}(-\omega) \right] j\omega \, d\omega.$$

Adding half of this expression to half of (4.43) and using (4.25), (4.17), and (4.47), we obtain

$$\Delta w_Q = \frac{1}{4\pi} \int_{-\infty}^{\infty} \left[\tilde{\mathbf{E}}^* \cdot \bar{\bar{\epsilon}}^c \cdot \tilde{\mathbf{E}} - \tilde{\mathbf{E}} \cdot \bar{\bar{\epsilon}}^{c*} \cdot \tilde{\mathbf{E}}^* + \tilde{\mathbf{H}}^* \cdot \bar{\bar{\mu}} \cdot \tilde{\mathbf{H}} - \tilde{\mathbf{H}} \cdot \bar{\bar{\mu}}^* \cdot \tilde{\mathbf{H}}^* \right] j\omega \, d\omega.$$

Finally, using the dyadic identity (A.76), we have

$$\Delta w_Q = \frac{1}{4\pi} \int_{-\infty}^{\infty} \left[\tilde{\mathbf{E}}^* \cdot \left(\bar{\bar{\epsilon}}^c - \bar{\bar{\epsilon}}^{c\dagger} \right) \cdot \tilde{\mathbf{E}} + \tilde{\mathbf{H}}^* \cdot \left(\bar{\bar{\mu}} - \bar{\bar{\mu}}^{\dagger} \right) \cdot \tilde{\mathbf{H}} \right] j\omega \, d\omega$$

where the dagger (†) denotes the hermitian (conjugate-transpose) operation. The condition for a lossless anisotropic material is

$$\bar{\bar{\epsilon}}^c = \bar{\bar{\epsilon}}^{c\dagger}, \qquad \bar{\bar{\mu}} = \bar{\bar{\mu}}^{\dagger}, \tag{4.49}$$

or

$$\tilde{\epsilon}_{ij} = \tilde{\epsilon}_{ji}^*, \qquad \tilde{\mu}_{ij} = \tilde{\mu}_{ji}^*, \qquad \tilde{\sigma}_{ij} = \tilde{\sigma}_{ji}^*. \tag{4.50}$$

These relationships imply that in the lossless case the diagonal entries of the constitutive dyadics are purely real.

Equations (4.50) show that complex entries in a permittivity or permeability matrix do not necessarily imply loss. For example, we will show in § 4.6.2 that an electron plasma exposed to a z-directed dc magnetic field has a permittivity of the form

$$[\bar{\bar{\epsilon}}] = \begin{bmatrix} \epsilon & -j\delta & 0 \\ j\delta & \epsilon & 0 \\ 0 & 0 & \epsilon_z \end{bmatrix}$$

where ϵ, ϵ_z, and δ are real functions of space and frequency. Since $\bar{\bar{\epsilon}}$ is hermitian it describes a lossless plasma. Similarly, a gyrotropic medium such as a ferrite exposed to a z-directed magnetic field has a permeability dyadic

$$[\bar{\bar{\mu}}] = \begin{bmatrix} \mu & -j\kappa & 0 \\ j\kappa & \mu & 0 \\ 0 & 0 & \mu_0 \end{bmatrix},$$

which also describes a lossless material.

4.5.2 Energy stored in a dispersive material

In the previous section we were able to isolate the dissipative effects for a dispersive material under special circumstances. It is not generally possible, however, to isolate a term describing the stored energy. The Kronig–Kramers relations imply that if the constitutive parameters of a material are frequency-dependent, they must have both real and imaginary parts; such a material, if isotropic, must be lossy. So dispersive materials are generally lossy and must have both dissipative and energy-storage characteristics. However, many materials have frequency ranges called *transparency ranges* over which $\tilde{\epsilon}^{c\prime\prime}$ and $\tilde{\mu}^{\prime\prime}$ are small compared to $\tilde{\epsilon}^{c\prime}$ and $\tilde{\mu}^{\prime}$. If we restrict our interest to these ranges, we may approximate the material as lossless and compute a stored energy. An important special case involves a monochromatic field oscillating at a frequency within this range.

To study the energy stored by a monochromatic field in a dispersive material we must consider the transient period during which energy accumulates in the fields. The assumption of a purely sinusoidal field variation would not include the effects described by the temporal constitutive relations (2.29)–(2.30), which show that as the field builds the energy must be added with a time lag. Instead we shall assume fields with the temporal variation

$$\mathbf{E}(\mathbf{r}, t) = f(t) \sum_{i=1}^{3} \hat{\mathbf{i}}_i |E_i(\mathbf{r})| \cos[\omega_0 t + \xi_i^E(\mathbf{r})] \tag{4.51}$$

where $f(t)$ is an appropriate function describing the build-up of the sinusoidal field. To compute the stored energy of a sinusoidal wave we must parameterize $f(t)$ so that we may drive it to unity as a limiting case of the parameter. A simple choice is

$$f(t) = e^{-\alpha^2 t^2} \leftrightarrow \tilde{F}(\omega) = \sqrt{\frac{\pi}{\alpha^2}} e^{-\frac{\omega^2}{4\alpha^2}}. \tag{4.52}$$

Note that since $f(t)$ approaches unity as $\alpha \to 0$, we have the generalized Fourier transform relation

$$\lim_{\alpha \to 0} \tilde{F}(\omega) = 2\pi\delta(\omega). \tag{4.53}$$

Substituting (4.51) into the Fourier transform formula (4.1) we find that

$$\tilde{\mathbf{E}}(\mathbf{r}, \omega) = \frac{1}{2} \sum_{i=1}^{3} \hat{\mathbf{i}}_i |E_i(\mathbf{r})| e^{j\xi_i^E(\mathbf{r})} \tilde{F}(\omega - \omega_0) + \frac{1}{2} \sum_{i=1}^{3} \hat{\mathbf{i}}_i |E_i(\mathbf{r})| e^{-j\xi_i^E(\mathbf{r})} \tilde{F}(\omega + \omega_0).$$

We can simplify this by defining

$$\check{\mathbf{E}}(\mathbf{r}) = \sum_{i=1}^{3} \hat{\mathbf{i}}_i |E_i(\mathbf{r})| e^{j\xi_i^E(\mathbf{r})} \tag{4.54}$$

as the *phasor* vector field to obtain

$$\tilde{\mathbf{E}}(\mathbf{r}, \omega) = \frac{1}{2} \left[\check{\mathbf{E}}(\mathbf{r}) \tilde{F}(\omega - \omega_0) + \check{\mathbf{E}}^*(\mathbf{r}) \tilde{F}(\omega + \omega_0) \right]. \tag{4.55}$$

We shall discuss the phasor concept in detail in § 4.7.

The field $\mathbf{E}(\mathbf{r}, t)$ is shown in Figure 4.2 as a function of t, while $\tilde{\mathbf{E}}(\mathbf{r}, \omega)$ is shown in Figure 4.2 as a function of ω. As α becomes small the spectrum of $\mathbf{E}(\mathbf{r}, t)$ concentrates around $\omega = \pm\omega_0$. We assume the material is transparent for all values α of interest so

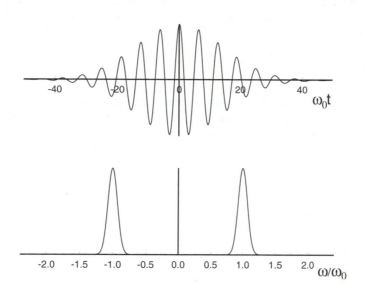

Figure 4.2: Temporal (top) and spectral magnitude (bottom) dependences of $\mathbf{E}$ used to compute energy stored in a dispersive material.

that we may treat ϵ as real. Then, since there is no dissipation, we conclude that the term (4.40) represents the time rate of change of stored energy at time t, including the effects of field build-up. Hence the interpretation[2]

$$\mathbf{E} \cdot \frac{\partial \mathbf{D}}{\partial t} = \frac{\partial w_e}{\partial t}, \qquad \mathbf{H} \cdot \frac{\partial \mathbf{B}}{\partial t} = \frac{\partial w_m}{\partial t}.$$

We shall concentrate on the electric field term and later obtain the magnetic field term by induction.

Since for periodic signals it is more convenient to deal with the time-averaged stored energy than with the instantaneous stored energy, we compute the time average of $w_e(\mathbf{r}, t)$ over the period of the sinusoid centered at the time origin. That is, we compute

$$\langle w_e \rangle = \frac{1}{T} \int_{-T/2}^{T/2} w_e(t) \, dt \tag{4.56}$$

where $T = 2\pi/\omega_0$. With $\alpha \to 0$, this time-average value is accurate for all periods of the sinusoidal wave.

Because the most expedient approach to the computation of (4.56) is to employ the Fourier spectrum of $\mathbf{E}$, we use

$$\mathbf{E}(\mathbf{r}, t) = \frac{1}{2\pi} \int_{-\infty}^{\infty} \tilde{\mathbf{E}}(\mathbf{r}, \omega) e^{j\omega t} \, d\omega = \frac{1}{2\pi} \int_{-\infty}^{\infty} \tilde{\mathbf{E}}^*(\mathbf{r}, \omega') e^{-j\omega' t} \, d\omega',$$

$$\frac{\partial \mathbf{D}(\mathbf{r}, t)}{\partial t} = \frac{1}{2\pi} \int_{-\infty}^{\infty} (j\omega) \tilde{\mathbf{D}}(\mathbf{r}, \omega) e^{j\omega t} \, d\omega = \frac{1}{2\pi} \int_{-\infty}^{\infty} (-j\omega') \tilde{\mathbf{D}}^*(\mathbf{r}, \omega') e^{-j\omega' t} \, d\omega'.$$

---

[2]Note that in this section we suppress the $\mathbf{r}$-dependence of most quantities for clarity of presentation.

We have obtained the second form of each of these expressions using the property (4.3) for the transform of a real function, and by using the change of variables $\omega' = -\omega$. Multiplying the two forms of the expressions and adding half of each, we find that

$$\frac{\partial w_e}{\partial t} = \frac{1}{2} \int_{-\infty}^{\infty} \frac{d\omega}{2\pi} \int_{-\infty}^{\infty} \frac{d\omega'}{2\pi} \left[j\omega \tilde{\mathbf{E}}^*(\omega') \cdot \tilde{\mathbf{D}}(\omega) - j\omega' \tilde{\mathbf{E}}(\omega) \cdot \tilde{\mathbf{D}}^*(\omega') \right] e^{-j(\omega'-\omega)t}. \quad (4.57)$$

Now let us consider a dispersive isotropic medium described by the constitutive relations $\tilde{\mathbf{D}} = \tilde{\epsilon}\tilde{\mathbf{E}}$, $\tilde{\mathbf{B}} = \tilde{\mu}\tilde{\mathbf{H}}$. Since the imaginary parts of $\tilde{\epsilon}$ and $\tilde{\mu}$ are associated with power dissipation in the medium, we shall approximate $\tilde{\epsilon}$ and $\tilde{\mu}$ as purely real. Then (4.57) becomes

$$\frac{\partial w_e}{\partial t} = \frac{1}{2} \int_{-\infty}^{\infty} \frac{d\omega}{2\pi} \int_{-\infty}^{\infty} \frac{d\omega'}{2\pi} \tilde{\mathbf{E}}^*(\omega') \cdot \tilde{\mathbf{E}}(\omega) \left[j\omega \tilde{\epsilon}(\omega) - j\omega' \tilde{\epsilon}(\omega') \right] e^{-j(\omega'-\omega)t}.$$

Substitution from (4.55) now gives

$$\frac{\partial w_e}{\partial t} = \frac{1}{8} \int_{-\infty}^{\infty} \frac{d\omega}{2\pi} \int_{-\infty}^{\infty} \frac{d\omega'}{2\pi} \left[j\omega \tilde{\epsilon}(\omega) - j\omega' \tilde{\epsilon}(\omega') \right] \cdot$$

$$\cdot \left[\check{\mathbf{E}} \cdot \check{\mathbf{E}}^* \tilde{F}(\omega - \omega_0)\tilde{F}(\omega' - \omega_0) + \check{\mathbf{E}} \cdot \check{\mathbf{E}}^* \tilde{F}(\omega + \omega_0)\tilde{F}(\omega' + \omega_0) + \right.$$

$$\left. + \check{\mathbf{E}} \cdot \check{\mathbf{E}} \tilde{F}(\omega - \omega_0)\tilde{F}(\omega' + \omega_0) + \check{\mathbf{E}}^* \cdot \check{\mathbf{E}}^* \tilde{F}(\omega + \omega_0)\tilde{F}(\omega' - \omega_0) \right] e^{-j(\omega'-\omega)t}.$$

Let $\omega \to -\omega$ wherever the term $\tilde{F}(\omega + \omega_0)$ appears, and $\omega' \to -\omega'$ wherever the term $\tilde{F}(\omega' + \omega_0)$ appears. Since $\tilde{F}(-\omega) = \tilde{F}(\omega)$ and $\tilde{\epsilon}(-\omega) = \tilde{\epsilon}(\omega)$, we find that

$$\frac{\partial w_e}{\partial t} = \frac{1}{8} \int_{-\infty}^{\infty} \frac{d\omega}{2\pi} \int_{-\infty}^{\infty} \frac{d\omega'}{2\pi} \tilde{F}(\omega - \omega_0)\tilde{F}(\omega' - \omega_0) \cdot$$

$$\cdot \left[\check{\mathbf{E}} \cdot \check{\mathbf{E}}^* [j\omega \tilde{\epsilon}(\omega) - j\omega' \tilde{\epsilon}(\omega')] e^{j(\omega-\omega')t} + \check{\mathbf{E}} \cdot \check{\mathbf{E}}^* [j\omega' \tilde{\epsilon}(\omega') - j\omega \tilde{\epsilon}(\omega)] e^{j(\omega'-\omega)t} + \right.$$

$$\left. + \check{\mathbf{E}} \cdot \check{\mathbf{E}} [j\omega \tilde{\epsilon}(\omega) + j\omega' \tilde{\epsilon}(\omega')] e^{j(\omega+\omega')t} + \check{\mathbf{E}}^* \cdot \check{\mathbf{E}}^* [-j\omega \tilde{\epsilon}(\omega) - j\omega' \tilde{\epsilon}(\omega')] e^{-j(\omega+\omega')t} \right].$$

$$(4.58)$$

For small α the spectra are concentrated near $\omega = \omega_0$ or $\omega' = \omega_0$. For terms involving the difference in the permittivities we can expand $g(\omega) = \omega \tilde{\epsilon}(\omega)$ in a Taylor series about ω_0 to obtain the approximation

$$\omega \tilde{\epsilon}(\omega) \approx \omega_0 \tilde{\epsilon}(\omega_0) + (\omega - \omega_0)g'(\omega_0)$$

where

$$g'(\omega_0) = \left. \frac{\partial[\omega \tilde{\epsilon}(\omega)]}{\partial \omega} \right|_{\omega=\omega_0}.$$

This is not required for terms involving a sum of permittivities since these will not tend to cancel. For such terms we merely substitute $\omega = \omega_0$ or $\omega' = \omega_0$. With these (4.58) becomes

$$\frac{\partial w_e}{\partial t} = \frac{1}{8} \int_{-\infty}^{\infty} \frac{d\omega}{2\pi} \int_{-\infty}^{\infty} \frac{d\omega'}{2\pi} \tilde{F}(\omega - \omega_0)\tilde{F}(\omega' - \omega_0) \cdot$$

$$\cdot \left[\check{\mathbf{E}} \cdot \check{\mathbf{E}}^* g'(\omega_0)[j(\omega - \omega')] e^{j(\omega-\omega')t} + \check{\mathbf{E}} \cdot \check{\mathbf{E}}^* g'(\omega_0)[j(\omega' - \omega)] e^{j(\omega'-\omega)t} + \right.$$

$$\left. + \check{\mathbf{E}} \cdot \check{\mathbf{E}} \tilde{\epsilon}(\omega_0)[j(\omega + \omega')] e^{j(\omega+\omega')t} + \check{\mathbf{E}}^* \cdot \check{\mathbf{E}}^* \tilde{\epsilon}(\omega_0)[-j(\omega + \omega')] e^{-j(\omega+\omega')t} \right].$$

By integration

$$w_e(t) = \frac{1}{8} \int_{-\infty}^{\infty} \frac{d\omega}{2\pi} \int_{-\infty}^{\infty} \frac{d\omega'}{2\pi} \tilde{F}(\omega - \omega_0)\tilde{F}(\omega' - \omega_0) \cdot$$

$$\cdot \left[\check{\mathbf{E}} \cdot \check{\mathbf{E}}^* g'(\omega_0)e^{j(\omega-\omega')t} + \check{\mathbf{E}} \cdot \check{\mathbf{E}}^* g'(\omega_0)e^{j(\omega'-\omega)t} + \right.$$

$$\left. + \check{\mathbf{E}} \cdot \check{\mathbf{E}}\tilde{\epsilon}(\omega_0)e^{j(\omega+\omega')t} + \check{\mathbf{E}}^* \cdot \check{\mathbf{E}}^*\tilde{\epsilon}(\omega_0)e^{-j(\omega+\omega')t} \right].$$

Our last step is to compute the time-average value of w_e and let $\alpha \to 0$. Applying (4.56) we find

$$\langle w_e \rangle = \frac{1}{8} \int_{-\infty}^{\infty} \frac{d\omega}{2\pi} \int_{-\infty}^{\infty} \frac{d\omega'}{2\pi} \tilde{F}(\omega - \omega_0)\tilde{F}(\omega' - \omega_0) \cdot$$

$$\cdot \left[2\check{\mathbf{E}} \cdot \check{\mathbf{E}}^* g'(\omega_0) \operatorname{sinc}\left([\omega - \omega']\frac{\pi}{\omega_0}\right) + \{\check{\mathbf{E}}^* \cdot \check{\mathbf{E}}^* + \check{\mathbf{E}} \cdot \check{\mathbf{E}}\} \tilde{\epsilon}(\omega_0) \operatorname{sinc}\left([\omega + \omega']\frac{\pi}{\omega_0}\right) \right]$$

where $\operatorname{sinc}(x)$ is defined in (A.9) and we note that $\operatorname{sinc}(-x) = \operatorname{sinc}(x)$. Finally we let $\alpha \to 0$ and use (4.53) to replace $\tilde{F}(\omega)$ by a δ-function. Upon integration these δ-functions set $\omega = \omega_0$ and $\omega' = \omega_0$. Since $\operatorname{sinc}(0) = 1$ and $\operatorname{sinc}(2\pi) = 0$, the time-average stored electric energy density becomes simply

$$\langle w_e \rangle = \frac{1}{4}|\check{\mathbf{E}}|^2 \frac{\partial[\omega\tilde{\epsilon}]}{\partial\omega}\bigg|_{\omega=\omega_0}. \tag{4.59}$$

Similarly,

$$\langle w_m \rangle = \frac{1}{4}|\check{\mathbf{H}}|^2 \frac{\partial[\omega\tilde{\mu}]}{\partial\omega}\bigg|_{\omega=\omega_0}.$$

This approach can also be applied to anisotropic materials to give

$$\langle w_e \rangle = \frac{1}{4}\check{\mathbf{E}}^* \cdot \frac{\partial[\omega\tilde{\bar{\epsilon}}]}{\partial\omega}\bigg|_{\omega=\omega_0} \cdot \check{\mathbf{E}}, \tag{4.60}$$

$$\langle w_m \rangle = \frac{1}{4}\check{\mathbf{H}}^* \cdot \frac{\partial[\omega\tilde{\bar{\mu}}]}{\partial\omega}\bigg|_{\omega=\omega_0} \cdot \check{\mathbf{H}}. \tag{4.61}$$

See Collin [39] for details. For the case of a lossless, nondispersive material where the constitutive parameters are frequency independent, we can use (4.49) and (A.76) to simplify this and obtain

$$\langle w_e \rangle = \frac{1}{4}\check{\mathbf{E}}^* \cdot \bar{\epsilon} \cdot \check{\mathbf{E}} = \frac{1}{4}\check{\mathbf{E}} \cdot \check{\mathbf{D}}^*, \tag{4.62}$$

$$\langle w_m \rangle = \frac{1}{4}\check{\mathbf{H}}^* \cdot \bar{\mu} \cdot \check{\mathbf{H}} = \frac{1}{4}\check{\mathbf{H}} \cdot \check{\mathbf{B}}^*, \tag{4.63}$$

in the anisotropic case and

$$\langle w_e \rangle = \frac{1}{4}\epsilon|\check{\mathbf{E}}|^2 = \frac{1}{4}\check{\mathbf{E}} \cdot \check{\mathbf{D}}^*, \tag{4.64}$$

$$\langle w_m \rangle = \frac{1}{4}\mu|\check{\mathbf{H}}|^2 = \frac{1}{4}\check{\mathbf{H}} \cdot \check{\mathbf{B}}^*, \tag{4.65}$$

in the isotropic case. Here $\check{\mathbf{E}}$, $\check{\mathbf{D}}$, $\check{\mathbf{B}}$, $\check{\mathbf{H}}$ are all phasor fields as defined by (4.54).

4.5.3   The energy theorem

A convenient expression for the time-average stored energies (4.60) and (4.61) is found by manipulating the frequency-domain Maxwell equations. Beginning with the complex conjugates of the two frequency-domain curl equations for anisotropic media,

$$\nabla \times \tilde{\mathbf{E}}^* = j\omega \bar{\tilde{\mu}}^* \cdot \tilde{\mathbf{H}}^*,$$
$$\nabla \times \tilde{\mathbf{H}}^* = \tilde{\mathbf{J}}^* - j\omega \bar{\tilde{\epsilon}}^* \cdot \tilde{\mathbf{E}}^*,$$

we differentiate with respect to frequency:

$$\nabla \times \frac{\partial \tilde{\mathbf{E}}^*}{\partial \omega} = j \frac{\partial [\omega \bar{\tilde{\mu}}^*]}{\partial \omega} \cdot \tilde{\mathbf{H}}^* + j\omega \bar{\tilde{\mu}}^* \cdot \frac{\partial \tilde{\mathbf{H}}^*}{\partial \omega}, \tag{4.66}$$

$$\nabla \times \frac{\partial \tilde{\mathbf{H}}^*}{\partial \omega} = \frac{\partial \tilde{\mathbf{J}}^*}{\partial \omega} - j \frac{\partial [\omega \bar{\tilde{\epsilon}}^*]}{\partial \omega} \cdot \tilde{\mathbf{E}}^* - j\omega \bar{\tilde{\epsilon}}^* \cdot \frac{\partial \tilde{\mathbf{E}}^*}{\partial \omega}. \tag{4.67}$$

These terms also appear as a part of the expansion

$$\nabla \cdot \left[\tilde{\mathbf{E}} \times \frac{\partial \tilde{\mathbf{H}}^*}{\partial \omega} + \frac{\partial \tilde{\mathbf{E}}^*}{\partial \omega} \times \tilde{\mathbf{H}} \right] =$$

$$\frac{\partial \tilde{\mathbf{H}}^*}{\partial \omega} \cdot [\nabla \times \tilde{\mathbf{E}}] - \tilde{\mathbf{E}} \cdot \nabla \times \frac{\partial \tilde{\mathbf{H}}^*}{\partial \omega} + \tilde{\mathbf{H}} \cdot \nabla \times \frac{\partial \tilde{\mathbf{E}}^*}{\partial \omega} - \frac{\partial \tilde{\mathbf{E}}^*}{\partial \omega} \cdot [\nabla \times \tilde{\mathbf{H}}]$$

where we have used (B.44). Substituting from (4.66)–(4.67) and eliminating $\nabla \times \tilde{\mathbf{E}}$ and $\nabla \times \tilde{\mathbf{H}}$ by Maxwell's equations we have

$$\frac{1}{4} \nabla \cdot \left(\tilde{\mathbf{E}} \times \frac{\partial \tilde{\mathbf{H}}^*}{\partial \omega} + \frac{\partial \tilde{\mathbf{E}}^*}{\partial \omega} \times \tilde{\mathbf{H}} \right) =$$

$$j \frac{1}{4} \omega \left(\tilde{\mathbf{E}} \cdot \bar{\tilde{\epsilon}}^* \cdot \frac{\partial \tilde{\mathbf{E}}^*}{\partial \omega} - \frac{\partial \tilde{\mathbf{E}}^*}{\partial \omega} \cdot \bar{\tilde{\epsilon}} \cdot \tilde{\mathbf{E}} \right) + j \frac{1}{4} \omega \left(\tilde{\mathbf{H}} \cdot \bar{\tilde{\mu}}^* \cdot \frac{\partial \tilde{\mathbf{H}}^*}{\partial \omega} - \frac{\partial \tilde{\mathbf{H}}^*}{\partial \omega} \cdot \bar{\tilde{\mu}} \cdot \tilde{\mathbf{H}} \right) +$$

$$+ j \frac{1}{4} \left(\tilde{\mathbf{E}} \cdot \frac{\partial [\omega \bar{\tilde{\epsilon}}^*]}{\partial \omega} \cdot \tilde{\mathbf{E}}^* + \tilde{\mathbf{H}} \cdot \frac{\partial [\omega \bar{\tilde{\mu}}^*]}{\partial \omega} \cdot \tilde{\mathbf{H}}^* \right) - \frac{1}{4} \left(\tilde{\mathbf{E}} \cdot \frac{\partial \tilde{\mathbf{J}}^*}{\partial \omega} + \tilde{\mathbf{J}} \cdot \frac{\partial \tilde{\mathbf{E}}^*}{\partial \omega} \right).$$

Let us assume that the sources and fields are narrowband, centered on ω_0, and that ω_0 lies within a transparency range so that within the band the material may be considered lossless. Invoking from (4.49) the facts that $\bar{\tilde{\epsilon}} = \bar{\tilde{\epsilon}}^\dagger$ and $\bar{\tilde{\mu}} = \bar{\tilde{\mu}}^\dagger$, we find that the first two terms on the right are zero. Integrating over a volume and taking the complex conjugate of both sides we obtain

$$\frac{1}{4} \oint_S \left(\tilde{\mathbf{E}}^* \times \frac{\partial \tilde{\mathbf{H}}}{\partial \omega} + \frac{\partial \tilde{\mathbf{E}}}{\partial \omega} \times \tilde{\mathbf{H}}^* \right) \cdot d\mathbf{S} =$$

$$-j \frac{1}{4} \int_V \left(\tilde{\mathbf{E}}^* \cdot \frac{\partial [\omega \bar{\tilde{\epsilon}}]}{\partial \omega} \cdot \tilde{\mathbf{E}} + \tilde{\mathbf{H}}^* \cdot \frac{\partial [\omega \bar{\tilde{\mu}}]}{\partial \omega} \cdot \tilde{\mathbf{H}} \right) dV - \frac{1}{4} \int_V \left(\tilde{\mathbf{E}}^* \cdot \frac{\partial \tilde{\mathbf{J}}}{\partial \omega} + \tilde{\mathbf{J}}^* \cdot \frac{\partial \tilde{\mathbf{E}}}{\partial \omega} \right) dV.$$

Evaluating each of the terms at $\omega = \omega_0$ and using (4.60)–(4.61) we have

$$\frac{1}{4} \oint_S \left(\tilde{\mathbf{E}}^* \times \frac{\partial \tilde{\mathbf{H}}}{\partial \omega} + \frac{\partial \tilde{\mathbf{E}}}{\partial \omega} \times \tilde{\mathbf{H}}^* \right) \bigg|_{\omega = \omega_0} \cdot d\mathbf{S} =$$

$$-j \left[\langle W_e \rangle + \langle W_m \rangle \right] - \frac{1}{4} \int_V \left(\tilde{\mathbf{E}}^* \cdot \frac{\partial \tilde{\mathbf{J}}}{\partial \omega} + \tilde{\mathbf{J}}^* \cdot \frac{\partial \tilde{\mathbf{E}}}{\partial \omega} \right) \bigg|_{\omega = \omega_0} dV \tag{4.68}$$

where $\langle W_e \rangle + \langle W_m \rangle$ is the total time-average electromagnetic energy stored in the volume region V. This is known as the *energy theorem*. We shall use it in § 4.11.3 to determine the velocity of energy transport for a plane wave.

4.6 Some simple models for constitutive parameters

Thus far our discussion of electromagnetic fields has been restricted to macroscopic phenomena. Although we recognize that matter is composed of microscopic constituents, we have chosen to describe materials using constitutive relationships whose parameters, such as permittivity, conductivity, and permeability, are viewed in the macroscopic sense. By performing experiments on the laboratory scale we can measure the constitutive parameters to the precision required for engineering applications.

At some point it becomes useful to establish models of the macroscopic behavior of materials based on microscopic considerations, formulating expressions for the constitutive parameters using atomic descriptors such as number density, atomic charge, and molecular dipole moment. These models allow us to predict the behavior of broad classes of materials, such as dielectrics and conductors, over wide ranges of frequency and field strength.

Accurate models for the behavior of materials under the influence of electromagnetic fields must account for many complicated effects, including those best described by quantum mechanics. However, many simple models can be obtained using classical mechanics and field theory. We shall investigate several of the most useful of these, and in the process try to gain a feeling for the relationship between the field applied to a material and the resulting polarization or magnetization of the underlying atomic structure.

For simplicity we shall consider only homogeneous materials. The fundamental atomic descriptor of "number density," N, is thus taken to be independent of position and time. The result may be more generally applicable since we may think of an inhomogeneous material in terms of the spatial variation of constitutive parameters originally determined assuming homogeneity. However, we shall not attempt to study the microscopic conditions that give rise to inhomogeneities.

4.6.1 Complex permittivity of a non-magnetized plasma

A plasma is an ionized gas in which the charged particles are free to move under the influence of an applied field and through particle-particle interactions. A plasma differs from other materials in that there is no atomic lattice restricting the motion of the particles. However, even in a gas the interactions between the particles and the fields give rise to a polarization effect, causing the permittivity of the gas to differ from that of free space. In addition, exposing the gas to an external field will cause a secondary current to flow as a result of the Lorentz force on the particles. As the moving particles collide with one another they relinquish their momentum, an effect describable in terms of a conductivity. In this section we shall perform a simple analysis to determine the complex permittivity of a non-magnetized plasma.

To make our analysis tractable, we shall make several assumptions.

1. We assume that the plasma is *neutral*: i.e., that the free electrons and positive ions are of equal number and distributed in like manner. If the particles are sufficiently

dense to be considered in the macroscopic sense, then there is no net field produced by the gas and thus no electromagnetic interaction between the particles. We also assume that the plasma is homogeneous and that the number density of the electrons N (number of electrons per m^3) is independent of time and position. In contrast to this are *electron beams*, whose properties differ significantly from neutral plasmas because of bunching of electrons by the applied field [148].

2. We ignore the motion of the positive ions in the computation of the secondary current, since the ratio of the mass of an ion to that of an electron is at least as large as the ratio of a proton to an electron ($m_p/m_e = 1837$) and thus the ions accelerate much more slowly.

3. We assume that the applied field is that of an electromagnetic wave. In § 2.10.6 we found that for a wave in free space the ratio of magnetic to electric field is $|\mathbf{H}|/|\mathbf{E}| = \sqrt{\epsilon_0/\mu_0}$, so that

$$\frac{|\mathbf{B}|}{|\mathbf{E}|} = \mu_0\sqrt{\frac{\epsilon_0}{\mu_0}} = \sqrt{\mu_0\epsilon_0} = \frac{1}{c}.$$

Thus, in the Lorentz force equation we may approximate the force on an electron as

$$\mathbf{F} = -q_e(\mathbf{E} + \mathbf{v} \times \mathbf{B}) \approx -q_e\mathbf{E}$$

as long as $v \ll c$. Here q_e is the *unsigned* charge on an electron, $q_e = 1.6021 \times 10^{-19}$ C. Note that when an external static magnetic field accompanies the field of the wave, as is the case in the earth's ionosphere for example, we cannot ignore the magnetic component of the Lorentz force. This case will be considered in § 4.6.2.

4. We assume that the mechanical interactions between particles can be described using a *collision frequency* ν, which describes the rate at which a directed plasma velocity becomes random in the absence of external forces.

With these assumptions we can write the equation of motion for the plasma medium. Let $\mathbf{v}(\mathbf{r}, t)$ represent the macroscopic velocity of the plasma medium. Then, by Newton's second law, the force acting at each point on the medium is balanced by the time-rate of change in momentum at that point. Because of collisions, the total change in momentum density is described by

$$\mathbf{F}(\mathbf{r}, t) = -Nq_e\mathbf{E}(\mathbf{r}, t) = \frac{d\wp(\mathbf{r}, t)}{dt} + \nu\wp(\mathbf{r}, t) \qquad (4.69)$$

where

$$\wp(\mathbf{r}, t) = Nm_e\mathbf{v}(\mathbf{r}, t)$$

is the volume density of momentum. Note that if there is no externally-applied electromagnetic force, then (4.69) becomes

$$\frac{d\wp(\mathbf{r}, t)}{dt} + \nu\wp(\mathbf{r}, t) = 0.$$

Hence

$$\wp(\mathbf{r}, t) = \wp_0(\mathbf{r})e^{-\nu t},$$

and we see that ν describes the rate at which the electron velocities move toward a random state, producing a macroscopic plasma velocity $\mathbf{v}$ of zero.

The time derivative in (4.69) is the total derivative as defined in (A.58):

$$\frac{d\wp(\mathbf{r},t)}{dt} = \frac{\partial\wp(\mathbf{r},t)}{\partial t} + (\mathbf{v}\cdot\nabla)\wp(\mathbf{r},t). \tag{4.70}$$

The second term on the right accounts for the time-rate of change of momentum per-ceived as the observer moves through regions of spatially-changing momentum. Since the electron velocity is induced by the electromagnetic field, we anticipate that for a sinusoidal wave the spatial variation will be on the order of the wavelength of the field: $\lambda = 2\pi c/\omega$. Thus, while the first term in (4.70) is proportional to ω, the second term is proportional to $\omega v/c$ and can be neglected for non-relativistic particle velocities. Then, writing $\mathbf{E}(\mathbf{r},t)$ and $\mathbf{v}(\mathbf{r},t)$ as inverse Fourier transforms, we see that (4.69) yields

$$-q_e\tilde{\mathbf{E}} = j\omega m_e\tilde{\mathbf{v}} + m_e\nu\tilde{\mathbf{v}} \tag{4.71}$$

and thus

$$\tilde{\mathbf{v}} = -\frac{\frac{q_e}{m_e}\tilde{\mathbf{E}}}{\nu + j\omega}. \tag{4.72}$$

The secondary current associated with the moving electrons is (since q_e is unsigned)

$$\tilde{\mathbf{J}}^s = -Nq_e\tilde{\mathbf{v}} = \frac{\epsilon_0\omega_p^2}{\omega^2 + \nu^2}(\nu - j\omega)\tilde{\mathbf{E}} \tag{4.73}$$

where

$$\omega_p^2 = \frac{Nq_e^2}{\epsilon_0 m_e} \tag{4.74}$$

is called the *plasma frequency*.

The frequency-domain Ampere's law for primary and secondary currents in free space is merely

$$\nabla \times \tilde{\mathbf{H}} = \tilde{\mathbf{J}}^i + \tilde{\mathbf{J}}^s + j\omega\epsilon_0\tilde{\mathbf{E}}.$$

Substitution from (4.73) gives

$$\nabla \times \tilde{\mathbf{H}} = \tilde{\mathbf{J}}^i + \frac{\epsilon_0\omega_p^2\nu}{\omega^2 + \nu^2}\tilde{\mathbf{E}} + j\omega\epsilon_0\left[1 - \frac{\omega_p^2}{\omega^2 + \nu^2}\right]\tilde{\mathbf{E}}.$$

We can determine the material properties of the plasma by realizing that the above expression can be written as

$$\nabla \times \tilde{\mathbf{H}} = \tilde{\mathbf{J}}^i + \tilde{\mathbf{J}}^s + j\omega\tilde{\mathbf{D}}$$

with the constitutive relations

$$\tilde{\mathbf{J}}^s = \tilde{\sigma}\tilde{\mathbf{E}}, \qquad \tilde{\mathbf{D}} = \tilde{\epsilon}\tilde{\mathbf{E}}.$$

Here we identify the conductivity of the plasma as

$$\tilde{\sigma}(\omega) = \frac{\epsilon_0\omega_p^2\nu}{\omega^2 + \nu^2} \tag{4.75}$$

and the permittivity as

$$\tilde{\epsilon}(\omega) = \epsilon_0\left[1 - \frac{\omega_p^2}{\omega^2 + \nu^2}\right].$$

We can also write Ampere's law as

$$\nabla \times \tilde{\mathbf{H}} = \tilde{\mathbf{J}}^i + j\omega\tilde{\epsilon}^c\tilde{\mathbf{E}}$$

where $\tilde{\epsilon}^c$ is the complex permittivity

$$\tilde{\epsilon}^c(\omega) = \tilde{\epsilon}(\omega) + \frac{\tilde{\sigma}(\omega)}{j\omega} = \epsilon_0\left[1 - \frac{\omega_p^2}{\omega^2 + \nu^2}\right] - j\frac{\epsilon_0\omega_p^2\nu}{\omega(\omega^2 + \nu^2)}. \tag{4.76}$$

If we wish to describe the plasma in terms of a polarization vector, we merely use $\tilde{\mathbf{D}} = \epsilon_0\tilde{\mathbf{E}} + \tilde{\mathbf{P}} = \tilde{\epsilon}\tilde{\mathbf{E}}$ to obtain the polarization vector $\tilde{\mathbf{P}} = (\tilde{\epsilon} - \epsilon_0)\tilde{\mathbf{E}} = \epsilon_0\tilde{\chi}_e\tilde{\mathbf{E}}$, where $\tilde{\chi}_e$ is the electric susceptibility

$$\tilde{\chi}_e(\omega) = -\frac{\omega_p^2}{\omega^2 + \nu^2}.$$

We note that $\tilde{\mathbf{P}}$ is directed opposite the applied field $\tilde{\mathbf{E}}$, resulting in $\tilde{\epsilon} < \epsilon_0$.

The plasma is dispersive since both its permittivity and conductivity depend on ω. As $\omega \to 0$ we have $\tilde{\epsilon}^{c\prime} \to \epsilon_0\epsilon_r$ where $\epsilon_r = 1 - \omega_p^2/\nu^2$, and also $\tilde{\epsilon}^{c\prime\prime} \sim 1/\omega$, as remarked in (4.28). As $\omega \to \infty$ we have $\tilde{\epsilon}^{c\prime} - \epsilon_0 \sim 1/\omega^2$ and $\tilde{\epsilon}^{c\prime\prime} \sim 1/\omega^3$, as mentioned in (4.29). When a transient plane wave propagates through a dispersive medium, the frequency dependence of the constitutive parameters tends to cause spreading of the waveshape.

We see that the plasma conductivity (4.75) is proportional to the collision frequency ν, and that, since $\tilde{\epsilon}^{c\prime\prime} < 0$ by the arguments of § 4.5, the plasma must be lossy. Loss arises from the transfer of electromagnetic energy into heat through electron collisions. If there are no collisions ($\nu = 0$), there is no mechanism for the transfer of energy into heat, and the conductivity of a lossless (or "collisionless") plasma reduces to zero as expected.

In a lowloss plasma ($\nu \to 0$) we may determine the time-average stored electromagnetic energy for sinusoidal excitation at frequency $\check{\omega}$. We must be careful to use (4.59), which holds for materials with dispersion. If we apply the simpler formula (4.64), we find that for $\nu \to 0$

$$\langle w_e \rangle = \frac{1}{4}\epsilon_0|\check{\mathbf{E}}|^2 - \frac{1}{4}\epsilon_0|\check{\mathbf{E}}|^2\frac{\omega_p^2}{\check{\omega}^2}.$$

For those excitation frequencies obeying $\check{\omega} < \omega_p$ we have $\langle w_e \rangle < 0$, implying that the material is active. Since there is no mechanism for the plasma to produce energy, this is obviously not valid. But an application of (4.59) gives

$$\langle w_e \rangle = \frac{1}{4}|\check{\mathbf{E}}|^2\frac{\partial}{\partial\omega}\left[\epsilon_0\omega\left(1 - \frac{\omega_p^2}{\omega^2}\right)\right]\Bigg|_{\omega=\check{\omega}} = \frac{1}{4}\epsilon_0|\check{\mathbf{E}}|^2 + \frac{1}{4}\epsilon_0|\check{\mathbf{E}}|^2\frac{\omega_p^2}{\check{\omega}^2}, \tag{4.77}$$

which is always positive. In this expression the first term represents the time-average energy stored in the vacuum, while the second term represents the energy stored in the kinetic energy of the electrons. For harmonic excitation, the time-average electron kinetic energy density is

$$\langle w_q \rangle = \frac{1}{4}Nm_e\check{\mathbf{v}} \cdot \check{\mathbf{v}}^*.$$

Substituting $\check{\mathbf{v}}$ from (4.72) with $\nu = 0$ we see that

$$\frac{1}{4}Nm_e\check{\mathbf{v}} \cdot \check{\mathbf{v}}^* = \frac{1}{4}\frac{Nq_e^2}{m_e\check{\omega}^2}|\check{\mathbf{E}}|^2 = \frac{1}{4}\epsilon_0|\check{\mathbf{E}}|^2\frac{\omega_p^2}{\check{\omega}^2},$$

which matches the second term of (4.77).

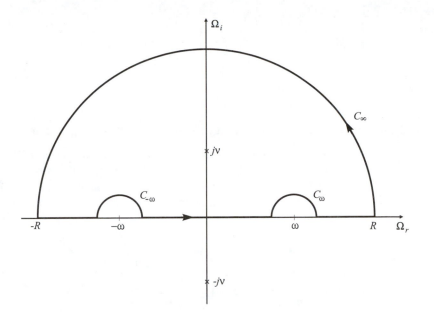

Figure 4.3: Integration contour used in Kronig–Kramers relations to find $\tilde{\epsilon}^{c\prime}$ from $\tilde{\epsilon}^{c\prime\prime}$ for a non-magnetized plasma.

The complex permittivity of a plasma (4.76) obviously obeys the required frequency-symmetry conditions (4.27). It also obeys the Kronig–Kramers relations required for a causal material. From (4.76) we see that the imaginary part of the complex plasma permittivity is

$$\tilde{\epsilon}^{c\prime\prime}(\omega) = -\frac{\epsilon_0 \omega_p^2 \nu}{\omega(\omega^2 + \nu^2)}.$$

Substituting this into (4.37) we have

$$\tilde{\epsilon}^{c\prime}(\omega) - \epsilon_0 = -\frac{2}{\pi} \, \text{P.V.} \int_0^\infty \left[-\frac{\epsilon_0 \omega_p^2 \nu}{\Omega(\Omega^2 + \nu^2)} \right] \frac{\Omega}{\Omega^2 - \omega^2} d\Omega.$$

We can evaluate the principal value integral and thus verify that it produces $\tilde{\epsilon}^{c\prime}$ by using the contour method of § A.1. Because the integrand is even we can extend the domain of integration to $(-\infty, \infty)$ and divide the result by two. Thus

$$\tilde{\epsilon}^{c\prime}(\omega) - \epsilon_0 = \frac{1}{\pi} \, \text{P.V.} \int_{-\infty}^\infty \frac{\epsilon_0 \omega_p^2 \nu}{(\Omega - j\nu)(\Omega + j\nu)} \frac{d\Omega}{(\Omega - \omega)(\Omega + \omega)}.$$

We integrate around the closed contour shown in Figure 4.3. Since the integrand falls off as $1/\Omega^4$ the contribution from C_∞ is zero. The contributions from the semicircles C_ω and $C_{-\omega}$ are given by πj times the residues of the integrand at $\Omega = \omega$ and at $\Omega = -\omega$, respectively, which are identical but of opposite sign. Thus, the semicircle contributions cancel and leave only the contribution from the residue at the upper-half-plane pole $\Omega = j\nu$. Evaluation of the residue gives

$$\tilde{\epsilon}^{c\prime}(\omega) - \epsilon_0 = \frac{1}{\pi} 2\pi j \frac{\epsilon_0 \omega_p^2 \nu}{j\nu + j\nu} \frac{1}{(j\nu - \omega)(j\nu + \omega)} = -\frac{\epsilon_0 \omega_p^2}{\nu^2 + \omega^2}$$

and thus

$$\tilde{\epsilon}^{cl}(\omega) = \epsilon_0 \left(1 - \frac{\omega_p^2}{\nu^2 + \omega^2}\right),$$

which matches (4.76) as expected.

4.6.2 Complex dyadic permittivity of a magnetized plasma

When an electron plasma is exposed to a magnetostatic field, as occurs in the earth's ionosphere, the behavior of the plasma is altered so that the secondary current is no longer aligned with the electric field, requiring the constitutive relationships to be written in terms of a complex dyadic permittivity. If the static field is $\mathbf{B}_0$, the velocity field of the plasma is determined by adding the magnetic component of the Lorentz force to (4.71), giving

$$-q_e[\tilde{\mathbf{E}} + \tilde{\mathbf{v}} \times \mathbf{B}_0] = \tilde{\mathbf{v}}(j\omega m_e + m_e\nu)$$

or equivalently

$$\tilde{\mathbf{v}} - j\frac{q_e}{m_e(\omega - j\nu)}\tilde{\mathbf{v}} \times \mathbf{B}_0 = j\frac{q_e}{m_e(\omega - j\nu)}\tilde{\mathbf{E}}. \tag{4.78}$$

Writing this expression generically as

$$\mathbf{v} + \mathbf{v} \times \mathbf{C} = \mathbf{A}, \tag{4.79}$$

we can solve for $\mathbf{v}$ as follows. Dotting both sides of the equation with $\mathbf{C}$ we quickly establish that $\mathbf{C} \cdot \mathbf{v} = \mathbf{C} \cdot \mathbf{A}$. Crossing both sides of the equation with $\mathbf{C}$, using (B.7), and substituting $\mathbf{C} \cdot \mathbf{A}$ for $\mathbf{C} \cdot \mathbf{v}$, we have

$$\mathbf{v} \times \mathbf{C} = \mathbf{A} \times \mathbf{C} + \mathbf{v}(\mathbf{C} \cdot \mathbf{C}) - \mathbf{C}(\mathbf{A} \cdot \mathbf{C}).$$

Finally, substituting $\mathbf{v} \times \mathbf{C}$ back into (4.79) we obtain

$$\mathbf{v} = \frac{\mathbf{A} - \mathbf{A} \times \mathbf{C} + (\mathbf{A} \cdot \mathbf{C})\mathbf{C}}{1 + \mathbf{C} \cdot \mathbf{C}}. \tag{4.80}$$

Let us first consider a lossless plasma for which $\nu = 0$. We can solve (4.78) for $\tilde{\mathbf{v}}$ by setting

$$\mathbf{C} = -j\frac{\boldsymbol{\omega}_c}{\omega}, \qquad \mathbf{A} = j\frac{\epsilon_0\omega_p^2}{\omega N q_e}\tilde{\mathbf{E}},$$

where

$$\boldsymbol{\omega}_c = \frac{q_e}{m_e}\mathbf{B}_0.$$

Here $\omega_c = q_e B_0/m_e = |\boldsymbol{\omega}_c|$ is called the *electron cyclotron frequency*. Substituting these into (4.80) we have

$$(\omega^2 - \omega_c^2)\tilde{\mathbf{v}} = j\frac{\epsilon_0\omega\omega_p^2}{Nq_e}\tilde{\mathbf{E}} + \frac{\epsilon_0\omega_p^2}{Nq_e}\boldsymbol{\omega}_c \times \tilde{\mathbf{E}} - j\frac{\omega_c}{\omega}\frac{\epsilon_0\omega_p^2}{Nq_e}\boldsymbol{\omega}_c \cdot \tilde{\mathbf{E}}.$$

Since the secondary current produced by the moving electrons is just $\tilde{\mathbf{J}}^s = -Nq_e\tilde{\mathbf{v}}$, we have

$$\tilde{\mathbf{J}}^s = j\omega\left[-\frac{\epsilon_0\omega_p^2}{\omega^2 - \omega_c^2}\tilde{\mathbf{E}} + j\frac{\epsilon_0\omega_p^2}{\omega(\omega^2 - \omega_c^2)}\boldsymbol{\omega}_c \times \tilde{\mathbf{E}} + \frac{\boldsymbol{\omega}_c}{\omega^2}\frac{\epsilon_0\omega_p^2}{\omega^2 - \omega_c^2}\boldsymbol{\omega}_c \cdot \tilde{\mathbf{E}}\right]. \tag{4.81}$$

Now, by the Ampere–Maxwell law we can write for currents in free space

$$\nabla \times \tilde{\mathbf{H}} = \tilde{\mathbf{J}}^i + \tilde{\mathbf{J}}^s + j\omega\epsilon_0\tilde{\mathbf{E}}. \tag{4.82}$$

Considering the plasma to be a material implies that we can describe the gas in terms of a complex permittivity dyadic $\bar{\bar{\epsilon}}^c$ such that the Ampere–Maxwell law is

$$\nabla \times \tilde{\mathbf{H}} = \tilde{\mathbf{J}}^i + j\omega\bar{\bar{\epsilon}}^c \cdot \tilde{\mathbf{E}}.$$

Substituting (4.81) into (4.82), and defining the dyadic $\bar{\bar{\omega}}_c$ so that $\bar{\bar{\omega}}_c \cdot \tilde{\mathbf{E}} = \boldsymbol{\omega}_c \times \tilde{\mathbf{E}}$, we identify the dyadic permittivity

$$\bar{\bar{\epsilon}}^c(\omega) = \left[\epsilon_0 - \epsilon_0 \frac{\omega_p^2}{\omega^2 - \omega_c^2} \right] \bar{\mathbf{I}} + j \frac{\epsilon_0\omega_p^2}{\omega(\omega^2 - \omega_c^2)} \bar{\bar{\omega}}_c + \frac{\epsilon_0\omega_p^2}{\omega^2(\omega^2 - \omega_c^2)} \boldsymbol{\omega}_c\boldsymbol{\omega}_c. \tag{4.83}$$

Note that in rectangular coordinates

$$[\bar{\bar{\omega}}_c] = \begin{bmatrix} 0 & -\omega_{cz} & \omega_{cy} \\ \omega_{cz} & 0 & -\omega_{cx} \\ -\omega_{cy} & \omega_{cx} & 0 \end{bmatrix}. \tag{4.84}$$

To examine the properties of the dyadic permittivity it is useful to write it in matrix form. To do this we must choose a coordinate system. We shall assume that $\mathbf{B}_0$ is aligned along the z-axis such that $\mathbf{B}_0 = \hat{\mathbf{z}}B_0$ and $\boldsymbol{\omega}_c = \hat{\mathbf{z}}\omega_c$. Then (4.84) becomes

$$[\bar{\bar{\omega}}_c] = \begin{bmatrix} 0 & -\omega_c & 0 \\ \omega_c & 0 & 0 \\ 0 & 0 & 0 \end{bmatrix} \tag{4.85}$$

and we can write the permittivity dyadic (4.83) as

$$[\bar{\bar{\epsilon}}(\omega)] = \begin{bmatrix} \epsilon & -j\delta & 0 \\ j\delta & \epsilon & 0 \\ 0 & 0 & \epsilon_z \end{bmatrix} \tag{4.86}$$

where

$$\epsilon = \epsilon_0\left(1 - \frac{\omega_p^2}{\omega^2 - \omega_c^2}\right), \qquad \epsilon_z = \epsilon_0\left(1 - \frac{\omega_p^2}{\omega^2}\right), \qquad \delta = \frac{\epsilon_0\omega_c\omega_p^2}{\omega(\omega^2 - \omega_c^2)}.$$

Note that the form of the permittivity dyadic is that for a lossless *gyrotropic* material (2.33).

Since the plasma is lossless, equation (4.49) shows that the dyadic permittivity must be hermitian. Equation (4.86) confirms this. We also note that since the sign of $\boldsymbol{\omega}_c$ is determined by the sign of $\mathbf{B}_0$, the dyadic permittivity obeys the symmetry relation

$$\tilde{\epsilon}_{ij}^c(\mathbf{B}_0) = \tilde{\epsilon}_{ji}^c(-\mathbf{B}_0) \tag{4.87}$$

as does the permittivity matrix of any material that has anisotropic properties dependent on an externally applied magnetic field [141]. We will find later in this section that the permeability matrix of a magnetized ferrite also obeys such a symmetry condition.

We can let $\omega \to \omega - j\nu$ in (4.81) to obtain the secondary current in a plasma with collisions:

$$\tilde{\mathbf{J}}^s(\mathbf{r}, \omega) = j\omega \left[-\frac{\epsilon_0 \omega_p^2 (\omega - j\nu)}{\omega[(\omega - j\nu)^2 - \omega_c^2]} \tilde{\mathbf{E}}(\mathbf{r}, \omega) + \right.$$

$$+ j \frac{\epsilon_0 \omega_p^2 (\omega - j\nu)}{\omega(\omega - j\nu)[(\omega - j\nu)^2 - \omega_c^2)]} \boldsymbol{\omega}_c \times \tilde{\mathbf{E}}(\mathbf{r}, \omega) +$$

$$\left. + \frac{\boldsymbol{\omega}_c}{(\omega - j\nu)^2} \frac{\epsilon_0 \omega_p^2 (\omega - j\nu)}{\omega[(\omega - j\nu)^2 - \omega_c^2]} \boldsymbol{\omega}_c \cdot \tilde{\mathbf{E}}(\mathbf{r}, \omega) \right].$$

From this we find the dyadic permittivity

$$\bar{\tilde{\boldsymbol{\epsilon}}}^c(\omega) = \left[\epsilon_0 - \frac{\epsilon_0 \omega_p^2 (\omega - j\nu)}{\omega[(\omega - j\nu)^2 - \omega_c^2]} \right] \bar{\mathbf{I}} + j \frac{\epsilon_0 \omega_p^2}{\omega[(\omega - j\nu)^2 - \omega_c^2]} \bar{\boldsymbol{\omega}}_c +$$

$$+ \frac{1}{(\omega - j\nu)} \frac{\epsilon_0 \omega_p^2}{\omega[(\omega - j\nu)^2 - \omega_c^2]} \boldsymbol{\omega}_c \boldsymbol{\omega}_c.$$

Assuming that $\mathbf{B}_0$ is aligned with the z-axis we can use (4.85) to find the components of the dyadic permittivity matrix:

$$\tilde{\epsilon}_{xx}^c(\omega) = \tilde{\epsilon}_{yy}^c(\omega) = \epsilon_0 \left(1 - \frac{\omega_p^2 (\omega - j\nu)}{\omega[(\omega - j\nu)^2 - \omega_c^2]} \right), \tag{4.88}$$

$$\tilde{\epsilon}_{xy}^c(\omega) = -\tilde{\epsilon}_{yx}^c(\omega) = -j\epsilon_0 \frac{\omega_p^2 \omega_c}{\omega[(\omega - j\nu)^2 - \omega_c^2)]}, \tag{4.89}$$

$$\tilde{\epsilon}_{zz}^c(\omega) = \epsilon_0 \left(1 - \frac{\omega_p^2}{\omega(\omega - j\nu)} \right), \tag{4.90}$$

and

$$\tilde{\epsilon}_{zx}^c = \tilde{\epsilon}_{xz}^c = \tilde{\epsilon}_{zy}^c = \tilde{\epsilon}_{yz}^c = 0. \tag{4.91}$$

We see that $[\tilde{\epsilon}^c]$ is not hermitian when $\nu \neq 0$. We expect this since the plasma is lossy when collisions occur. However, we can decompose $[\bar{\tilde{\epsilon}}^c]$ as a sum of two matrices:

$$[\bar{\tilde{\epsilon}}^c] = [\bar{\tilde{\epsilon}}] + \frac{[\bar{\tilde{\sigma}}]}{j\omega},$$

where $[\bar{\tilde{\epsilon}}]$ and $[\bar{\tilde{\sigma}}]$ are hermitian [141]. The details are left as an exercise. We also note that, as in the case of the lossless plasma, the permittivity dyadic obeys the symmetry condition $\tilde{\epsilon}_{ij}^c(\mathbf{B}_0) = \tilde{\epsilon}_{ji}^c(-\mathbf{B}_0)$.

4.6.3 Simple models of dielectrics

We define an isotropic dielectric material (also called an *insulator*) as one that obeys the macroscopic frequency-domain constitutive relationship

$$\tilde{\mathbf{D}}(\mathbf{r}, \omega) = \tilde{\epsilon}(\mathbf{r}, \omega) \tilde{\mathbf{E}}(\mathbf{r}, \omega).$$

Since the polarization vector $\mathbf{P}$ was defined in Chapter 2 as $\mathbf{P}(\mathbf{r}, t) = \mathbf{D}(\mathbf{r}, t) - \epsilon_0 \mathbf{E}(\mathbf{r}, t)$, an isotropic dielectric can also be described through

$$\tilde{\mathbf{P}}(\mathbf{r}, \omega) = (\tilde{\epsilon}(\mathbf{r}, \omega) - \epsilon_0) \tilde{\mathbf{E}}(\mathbf{r}, \omega) = \tilde{\chi}_e(\mathbf{r}, \omega) \epsilon_0 \tilde{\mathbf{E}}(\mathbf{r}, \omega)$$

where $\tilde{\chi}_e$ is the dielectric susceptibility. In this section we shall model a homogeneous dielectric consisting of a single, uniform material type.

We found in Chapter 3 that for a dielectric material immersed in a static electric field, the polarization vector $\mathbf{P}$ can be viewed as a volume density of dipole moments. We choose to retain this view as the fundamental link between microscopic dipole moments and the macroscopic polarization vector. Within the framework of our model we thus describe the polarization through the expression

$$\mathbf{P}(\mathbf{r}, t) = \frac{1}{\Delta V} \sum_{\mathbf{r} - \mathbf{r}_i(t) \in B} \mathbf{p}_i. \tag{4.92}$$

Here $\mathbf{p}_i$ is the dipole moment of the ith elementary microscopic constituent, and we form the macroscopic density function as in § 1.3.1.

We may also write (4.92) as

$$\mathbf{P}(\mathbf{r}, t) = \left[\frac{N_B}{\Delta V} \right] \left[\frac{1}{N_B} \sum_{i=1}^{N_B} \mathbf{p}_i \right] = N(\mathbf{r}, t) \mathbf{p}(\mathbf{r}, t) \tag{4.93}$$

where N_B is the number of constituent particles within ΔV. We identify

$$\mathbf{p}(\mathbf{r}, t) = \frac{1}{N_B} \sum_{i=1}^{N_B} \mathbf{p}_i(\mathbf{r}, t)$$

as the average dipole moment within ΔV, and

$$N(\mathbf{r}, t) = \frac{N_B}{\Delta V}$$

as the dipole moment number density. In this model a dielectric material does not require higher-order multipole moments to describe its behavior. Since we are only interested in homogeneous materials in this section we shall assume that the number density is constant: $N(\mathbf{r}, t) = N$.

To understand how dipole moments arise, we choose to adopt the simple idea that matter consists of atomic particles, each of which has a positively charged nucleus surrounded by a negatively charged electron cloud. Isolated, these particles have no net charge and no net electric dipole moment. However, there are several ways in which individual particles, or aggregates of particles, may take on a dipole moment. When exposed to an external electric field the electron cloud of an individual atom may be displaced, resulting in an *induced* dipole moment which gives rise to *electronic polarization*. When groups of atoms form a molecule, the individual electron clouds may combine to form an asymmetric structure having a *permanent* dipole moment. In some materials these molecules are randomly distributed and no net dipole moment results. However, upon application of an external field the torque acting on the molecules may tend to align them, creating an *induced* dipole moment and *orientation*, or *dipole*, polarization. In other materials, the asymmetric structure of the molecules may be weak until an external field causes the displacement of atoms within each molecule, resulting in an *induced* dipole moment causing *atomic*, or *molecular*, polarization. If a material maintains a permanent polarization without the application of an external field, it is called an *electret* (and is thus similar in behavior to a permanently magnetized magnet).

To describe the constitutive relations, we must establish a link between $\mathbf{P}$ (now describable in microscopic terms) and $\mathbf{E}$. We do this by postulating that the average constituent dipole moment is proportional to the *local electric field strength* $\mathbf{E}'$:

$$\mathbf{p} = \alpha \mathbf{E}', \tag{4.94}$$

where α is called the *polarizability* of the elementary constituent. Each of the polarization effects listed above may have its own polarizability: α_e for electronic polarization, α_a for atomic polarization, and α_d for dipole polarization. The total polarizability is merely the sum $\alpha = \alpha_e + \alpha_a + \alpha_d$.

In a rarefied gas the particles are so far apart that their interaction can be neglected. Here the localized field $\mathbf{E}'$ is the same as the applied field $\mathbf{E}$. In liquids and solids where particles are tightly packed, $\mathbf{E}'$ depends on the manner in which the material is polarized and may differ from $\mathbf{E}$. We therefore proceed to determine a relationship between $\mathbf{E}'$ and $\mathbf{P}$.

The Clausius–Mosotti equation.   We seek the local field at an observation point within a polarized material. Let us first assume that the fields are static. We surround the observation point with an artificial spherical surface of radius a and write the field at the observation point as a superposition of the field $\mathbf{E}$ applied, the field $\mathbf{E}_2$ of the polarized molecules external to the sphere, and the field $\mathbf{E}_3$ of the polarized molecules within the sphere. We take a large enough that we may describe the molecules outside the sphere in terms of the macroscopic dipole moment density $\mathbf{P}$, but small enough to assume that $\mathbf{P}$ is uniform over the surface of the sphere. We also assume that the major contribution to $\mathbf{E}_2$ comes from the dipoles nearest the observation point. We then approximate $\mathbf{E}_2$ using the electrostatic potential produced by the equivalent polarization surface charge on the sphere $\rho_{Ps} = \hat{\mathbf{n}} \cdot \mathbf{P}$ (where $\hat{\mathbf{n}}$ points toward the center of the sphere). Placing the origin of coordinates at the observation point and orienting the z-axis with the polarization $\mathbf{P}$ so that $\mathbf{P} = P_0 \hat{\mathbf{z}}$, we find that $\hat{\mathbf{n}} \cdot \mathbf{P} = -\cos\theta$ and thus the electrostatic potential at any point $\mathbf{r}$ within the sphere is merely

$$\Phi(\mathbf{r}) = -\frac{1}{4\pi\epsilon_0} \oint_S \frac{P_0 \cos\theta'}{|\mathbf{r} - \mathbf{r}'|}\, dS'.$$

This integral has been computed in § 3.2.7 with the result given by (3.103). Hence

$$\Phi(\mathbf{r}) = -\frac{P_0}{3\epsilon_0} r \cos\theta = -\frac{P_0}{3\epsilon_0} z$$

and therefore

$$\mathbf{E}_2 = \frac{\mathbf{P}}{3\epsilon_0}. \tag{4.95}$$

Note that this is uniform and independent of a.

The assumption that the localized field varies spatially as the electrostatic field, even when $\mathbf{P}$ may depend on frequency, is quite good. In Chapter 5 we will find that for a frequency-dependent source (or, equivalently, a time-varying source), the fields very near the source have a spatial dependence nearly identical to that of the electrostatic case.

We now have the seemingly more difficult task of determining the field $\mathbf{E}_3$ produced by the dipoles within the sphere. This would seem difficult since the field produced by dipoles near the observation point should be highly-dependent on the particular dipole arrangement. As mentioned above, there are various mechanisms for polarization, and the distribution of charge near any particular point depends on the molecular arrangement. However, Lorentz showed [115] that for crystalline solids with cubical symmetry, or for a randomly-structured gas, the contribution from dipoles within the sphere is zero. Indeed, it is convenient and reasonable to assume that for most dielectrics the effects of the dipoles immediately surrounding the observation point cancel so that $\mathbf{E}_3 = 0$. This was first suggested by O.F. Mosotti in 1850 [52].

With $\mathbf{E}_2$ approximated as (4.95) and $\mathbf{E}_3$ assumed to be zero, we have the value of the resulting local field:

$$\mathbf{E}'(\mathbf{r}) = \mathbf{E}(\mathbf{r}) + \frac{\mathbf{P}(\mathbf{r})}{3\epsilon_0}. \tag{4.96}$$

This is called the *Mosotti field*. Substituting the Mosotti field into (4.94) and using $\mathbf{P} = N\mathbf{p}$, we obtain

$$\mathbf{P}(\mathbf{r}) = N\alpha\mathbf{E}'(\mathbf{r}) = N\alpha\left(\mathbf{E}(\mathbf{r}) + \frac{\mathbf{P}(\mathbf{r})}{3\epsilon_0}\right).$$

Solving for $\mathbf{P}$ we obtain

$$\mathbf{P}(\mathbf{r}) = \left(\frac{3\epsilon_0 N\alpha}{3\epsilon_0 - N\alpha}\right)\mathbf{E}(\mathbf{r}) = \chi_e \epsilon_0 \mathbf{E}(\mathbf{r}).$$

So the electric susceptibility of a dielectric may be expressed as

$$\chi_e = \frac{3N\alpha}{3\epsilon_0 - N\alpha}. \tag{4.97}$$

Using $\chi_e = \epsilon_r - 1$ we can rewrite (4.97) as

$$\epsilon = \epsilon_0 \epsilon_r = \epsilon_0 \frac{3 + 2N\alpha/\epsilon_0}{3 - N\alpha/\epsilon_0}, \tag{4.98}$$

which we can arrange to obtain

$$\alpha = \alpha_e + \alpha_a + \alpha_d = \frac{3\epsilon_0}{N}\frac{\epsilon_r - 1}{\epsilon_r + 2}.$$

This has been named the *Clausius–Mosotti formula*, after O.F. Mosotti who proposed it in 1850 and R. Clausius who proposed it independently in 1879. When written in terms of the index of refraction n (where $n^2 = \epsilon_r$), it is also known as the *Lorentz–Lorenz formula*, after H. Lorentz and L. Lorenz who proposed it independently for optical materials in 1880. The Clausius–Mosotti formula allows us to determine the dielectric constant from the polarizability and number density of a material. It is reasonably accurate for certain simple gases (with pressures up to 1000 atmospheres) but becomes less reliable for liquids and solids, especially for those with large dielectric constants.

The response of the microscopic structure of matter to an applied field is not instantaneous. When exposed to a rapidly oscillating sinusoidal field, the induced dipole moments may lag in time. This results in a loss mechanism that can be described macroscopically by a complex permittivity. We can modify the Clausius–Mosotti formula by assuming that both the relative permittivity and polarizability are complex numbers, but this will not model the dependence of these parameters on frequency. Instead we shall (in later paragraphs) model the time response of the dipole moments to the applied field.

An interesting application of the Clausius–Mosotti formula is to determine the permittivity of a mixture of dielectrics with different permittivities. Consider the simple case in which many small spheres of permittivity ϵ_2, radius a, and volume V are embedded within a dielectric matrix of permittivity ϵ_1. If we assume that a is much smaller than the wavelength of the electromagnetic field, and that the spheres are sparsely distributed within the matrix, then we may ignore any mutual interaction between the spheres. Since the expression for the permittivity of a uniform dielectric given by (4.98) describes the effect produced by dipoles in free space, we can use the Clausius–Mosotti formula to define an *effective* permittivity ϵ_e for a material consisting of spheres in a background dielectric by replacing ϵ_0 with ϵ_1 to obtain

$$\epsilon_e = \epsilon_1 \frac{3 + 2N\alpha/\epsilon_1}{3 - N\alpha/\epsilon_1}. \tag{4.99}$$

In this expression α is the polarizability of a single dielectric sphere embedded in the background dielectric, and N is the number density of dielectric spheres. To find α we use the static field solution for a dielectric sphere immersed in a field (§ 3.2.10). Remembering that $\mathbf{p} = \alpha \mathbf{E}$ and that for a uniform region of volume V we have $\mathbf{p} = V\mathbf{P}$, we can make the replacements $\epsilon_0 \to \epsilon_1$ and $\epsilon \to \epsilon_2$ in (3.117) to get

$$\alpha = 3\epsilon_1 V \frac{\epsilon_2 - \epsilon_1}{\epsilon_2 + 2\epsilon_1}. \tag{4.100}$$

Defining $f = NV$ as the *fractional volume* occupied by the spheres, we can substitute (4.100) into (4.99) to find that

$$\epsilon_e = \epsilon_1 \frac{1 + 2fy}{1 - fy}$$

where

$$y = \frac{\epsilon_2 - \epsilon_1}{\epsilon_2 + 2\epsilon_1}.$$

This is known as the *Maxwell–Garnett mixing formula*. Rearranging we obtain

$$\frac{\epsilon_e - \epsilon_1}{\epsilon_e + 2\epsilon_1} = f \frac{\epsilon_2 - \epsilon_1}{\epsilon_2 + 2\epsilon_1},$$

which is known as the *Rayleigh mixing formula*. As expected, $\epsilon_e \to \epsilon_1$ as $f \to 0$. Even though as $f \to 1$ the formula also reduces to $\epsilon_e = \epsilon_2$, our initial assumption that $f \ll 1$ (sparsely distributed spheres) is violated and the result is inaccurate for non-spherical inhomogeneities [90]. For a discussion of more accurate mixing formulas, see Ishimaru [90] or Sihvola [175].

The dispersion formula of classical physics.   We may determine the frequency dependence of the permittivity by modeling the time response of induced dipole moments. This was done by H. Lorentz using the simple atomic model we introduced earlier. Consider what happens when a molecule consisting of heavy particles (nuclei) surrounded by clouds of electrons is exposed to a time-harmonic electromagnetic wave. Using the same arguments we made when we studied the interactions of fields with a plasma in § 4.6.1, we assume that each electron experiences a Lorentz force $\mathbf{F}_e = -q_e \mathbf{E}'$. We neglect the magnetic component of the force for nonrelativistic charge velocities, and ignore the motion of the much heavier nuclei in favor of studying the motion of the electron cloud. However, several important distinctions exist between the behavior of charges within a plasma and those within a solid or liquid material. Because of the surrounding polarized matter, any molecule responds to the local field $\mathbf{E}'$ instead of the applied field $\mathbf{E}$. Also, as the electron cloud is displaced by the Lorentz force, the attraction from the positive nuclei provides a restoring force $\mathbf{F}_r$. In the absence of loss the restoring force causes the electron cloud (and thus the induced dipole moment) to oscillate in phase with the applied field. In addition, there will be loss due to radiation by the oscillating molecules and collisions between charges that can be modeled using a "frictional force" $\mathbf{F}_s$ in the same manner as for a mechanical harmonic oscillator.

We can express the restoring and frictional forces by the use of a mechanical analogue. The restoring force acting on each electron is taken to be proportional to the displacement from equilibrium $\mathbf{l}$:

$$\mathbf{F}_r(\mathbf{r}, t) = -m_e \omega_r^2 \mathbf{l}(\mathbf{r}, t),$$

where m_e is the mass of an electron and ω_r is a material constant that depends on the molecular structure. The frictional force is similar to the collisional term in § 4.6.1 in that it is assumed to be proportional to the electron momentum $m_e \mathbf{v}$:

$$\mathbf{F}_s(\mathbf{r}, t) = -2\Gamma m_e \mathbf{v}(\mathbf{r}, t)$$

where Γ is a material constant. With these we can apply Newton's second law to obtain

$$\mathbf{F}(\mathbf{r}, t) = -q_e \mathbf{E}'(\mathbf{r}, t) - m_e \omega_r^2 \mathbf{l}(\mathbf{r}, t) - 2\Gamma m_e \mathbf{v}(\mathbf{r}, t) = m_e \frac{d\mathbf{v}(\mathbf{r}, t)}{dt}.$$

Using $\mathbf{v} = d\mathbf{l}/dt$ we find that the equation of motion for the electron is

$$\frac{d^2 \mathbf{l}(\mathbf{r}, t)}{dt^2} + 2\Gamma \frac{d\mathbf{l}(\mathbf{r}, t)}{dt} + \omega_r^2 \mathbf{l}(\mathbf{r}, t) = -\frac{q_e}{m_e} \mathbf{E}'(\mathbf{r}, t). \tag{4.101}$$

We recognize this differential equation as the damped harmonic equation. When $\mathbf{E}' = 0$ we have the homogeneous solution

$$\mathbf{l}(\mathbf{r}, t) = \mathbf{l}_0(\mathbf{r}) e^{-\Gamma t} \cos\left(t\sqrt{\omega_r^2 - \Gamma^2}\right).$$

Thus the electron position is a damped oscillation. The resonant frequency $\sqrt{\omega_r^2 - \Gamma^2}$ is usually only slightly reduced from ω_r since radiation damping is generally quite low.

Since the dipole moment for an electron displaced from equilibrium by $\mathbf{l}$ is $\mathbf{p} = -q_e \mathbf{l}$, and the polarization density is $\mathbf{P} = N\mathbf{p}$ from (4.93), we can write

$$\mathbf{P}(\mathbf{r}, t) = -Nq_e \mathbf{l}(\mathbf{r}, t).$$

Multiplying (4.101) by $-Nq_e$ and substituting the above expression, we have a differential equation for the polarization:

$$\frac{d^2 \mathbf{P}}{dt^2} + 2\Gamma \frac{d\mathbf{P}}{dt} + \omega_r^2 \mathbf{P} = \frac{Nq_e^2}{m_e} \mathbf{E}'.$$

To obtain a constitutive equation we must relate the polarization to the applied field $\mathbf{E}$. We can accomplish this by relating the local field $\mathbf{E}'$ to the polarization using the Mosotti field (4.90). Substitution gives

$$\frac{d^2 \mathbf{P}}{dt^2} + 2\Gamma \frac{d\mathbf{P}}{dt} + \omega_0^2 \mathbf{P} = \frac{Nq_e^2}{m_e} \mathbf{E} \tag{4.102}$$

where

$$\omega_0 = \sqrt{\omega_r^2 - \frac{Nq_e^2}{3m_e \epsilon_0}}$$

is the resonance frequency of the dipole moments. We see that this frequency is reduced from the resonance frequency of the electron oscillation because of the polarization of the surrounding medium.

We can now obtain a dispersion equation for the electrical susceptibility by taking the Fourier transform of (4.102). We have

$$-\omega^2 \tilde{\mathbf{P}} + j\omega 2\Gamma \tilde{\mathbf{P}} + \omega_0^2 \tilde{\mathbf{P}} = \frac{N q_e^2}{m_e} \tilde{\mathbf{E}}.$$

Thus we obtain the dispersion relation

$$\tilde{\chi}_e(\omega) = \frac{\tilde{\mathbf{P}}}{\epsilon_0 \tilde{\mathbf{E}}} = \frac{\omega_p^2}{\omega_0^2 - \omega^2 + j\omega 2\Gamma}$$

where ω_p is the plasma frequency (4.74). Since $\tilde{\epsilon}_r(\omega) = 1 + \tilde{\chi}_e(\omega)$ we also have

$$\tilde{\epsilon}(\omega) = \epsilon_0 + \epsilon_0 \frac{\omega_p^2}{\omega_0^2 - \omega^2 + j\omega 2\Gamma}. \tag{4.103}$$

If more than one type of oscillating moment contributes to the permittivity, we may extend (4.103) to

$$\tilde{\epsilon}(\omega) = \epsilon_0 + \sum_i \epsilon_0 \frac{\omega_{pi}^2}{\omega_i^2 - \omega^2 + j\omega 2\Gamma_i} \tag{4.104}$$

where $\omega_{pi} = N_i q_e^2 / \epsilon_0 m_i$ is the plasma frequency of the ith resonance component, and ω_i and Γ_i are the oscillation frequency and damping coefficient, respectively, of this component. This expression is the *dispersion formula for classical physics*, so called because it neglects quantum effects. When losses are negligible, (4.104) reduces to the *Sellmeier equation*

$$\tilde{\epsilon}(\omega) = \epsilon_0 + \sum_i \epsilon_0 \frac{\omega_{pi}^2}{\omega_i^2 - \omega^2}. \tag{4.105}$$

Let us now study the frequency behavior of the dispersion relation (4.104). Splitting the permittivity into real and imaginary parts we have

$$\tilde{\epsilon}'(\omega) - \epsilon_0 = \epsilon_0 \sum_i \omega_{pi}^2 \frac{\omega_i^2 - \omega^2}{[\omega_i^2 - \omega^2]^2 + 4\omega^2 \Gamma_i^2},$$

$$\tilde{\epsilon}''(\omega) = -\epsilon_0 \sum_i \omega_{pi}^2 \frac{2\omega \Gamma_i}{[\omega_i^2 - \omega^2]^2 + 4\omega^2 \Gamma_i^2}.$$

As $\omega \to 0$ the permittivity reduces to

$$\epsilon = \epsilon_0 \left(1 + \sum_i \frac{\omega_{pi}^2}{\omega_i^2} \right),$$

which is the static permittivity of the material. As $\omega \to \infty$ the permittivity behaves as

$$\tilde{\epsilon}'(\omega) \to \epsilon_0 \left(1 - \frac{\sum_i \omega_{pi}^2}{\omega^2} \right), \qquad \tilde{\epsilon}''(\omega) \to -\epsilon_0 \frac{2 \sum_i \omega_{pi}^2 \Gamma_i}{\omega^3}.$$

This high frequency behavior is identical to that of a plasma as described by (4.76).

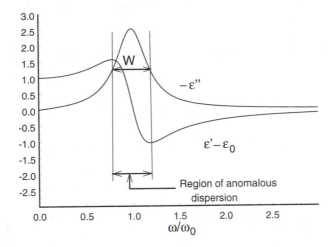

Figure 4.4: Real and imaginary parts of permittivity for a single resonance model of a dielectric with $\Gamma/\omega_0 = 0.2$. Permittivity normalized by dividing by $\epsilon_0(\omega_p/\omega_0)^2$.

The major characteristic of the dispersion relation (4.104) is the presence of one or more *resonances*. Figure 4.4 shows a plot of a single resonance component, where we have normalized the permittivity as

$$(\tilde{\epsilon}'(\omega) - \epsilon_0)/(\epsilon_0\bar{\omega}_p^2) = \frac{1 - \bar{\omega}^2}{[1 - \bar{\omega}^2]^2 + 4\bar{\omega}^2\bar{\Gamma}^2},$$

$$-\tilde{\epsilon}''(\omega)/(\epsilon_0\bar{\omega}_p^2) = \frac{2\bar{\omega}\bar{\Gamma}}{[1 - \bar{\omega}^2]^2 + 4\bar{\omega}^2\bar{\Gamma}^2},$$

with $\bar{\omega} = \omega/\omega_0$, $\bar{\omega}_p = \omega_p/\omega_0$, and $\bar{\Gamma} = \Gamma/\omega_0$. We see a distinct resonance centered at $\omega = \omega_0$. Approaching this resonance through frequencies less than ω_0, we see that $\tilde{\epsilon}'$ increases slowly until peaking at $\omega_{\max} = \omega_0\sqrt{1 - 2\Gamma/\omega_0}$ where it attains a value of

$$\tilde{\epsilon}'_{\max} = \epsilon_0 + \frac{1}{4}\epsilon_0 \frac{\bar{\omega}_p^2}{\bar{\Gamma}(1 - \bar{\Gamma})}.$$

After peaking, $\tilde{\epsilon}'$ undergoes a rapid decrease, passing through $\tilde{\epsilon}' = \epsilon_0$ at $\omega = \omega_0$, and then continuing to decrease until reaching a minimum value of

$$\tilde{\epsilon}'_{\min} = \epsilon_0 - \frac{1}{4}\epsilon_0 \frac{\bar{\omega}_p^2}{\bar{\Gamma}(1 + \bar{\Gamma})}$$

at $\omega_{\min} = \omega_0\sqrt{1 + 2\Gamma/\omega_0}$. As ω continues to increase, $\tilde{\epsilon}'$ again increases slowly toward a final value of $\tilde{\epsilon}' = \epsilon_0$. The regions of slow variation of $\tilde{\epsilon}'$ are called regions of *normal dispersion*, while the region where $\tilde{\epsilon}'$ decreases abruptly is called the region of *anomalous dispersion*. Anomalous dispersion is unusual only in the sense that it occurs over a narrower range of frequencies than normal dispersion.

The imaginary part of the permittivity peaks near the resonant frequency, dropping off monotonically in each direction away from the peak. The width of the curve is an important parameter that we can most easily determine by approximating the behavior of $\tilde{\epsilon}''$ near ω_0. Letting $\Delta\bar{\omega} = (\omega_0 - \omega)/\omega_0$ and using

$$\omega_0^2 - \omega^2 = (\omega_0 - \omega)(\omega_0 + \omega) \approx 2\omega_0^2 \Delta\bar{\omega},$$

we get

$$\tilde{\epsilon}''(\omega) \approx -\frac{1}{2}\epsilon_0\bar{\omega}_p^2 \frac{\bar{\Gamma}}{(\Delta\bar{\omega})^2 + \bar{\Gamma}^2}.$$

This approximation has a maximum value of

$$\tilde{\epsilon}''_{\max} = \tilde{\epsilon}''(\omega_0) = -\frac{1}{2}\epsilon_0\bar{\omega}_p^2 \frac{1}{\bar{\Gamma}}$$

located at $\omega = \omega_0$, and has half-amplitude points located at $\Delta\bar{\omega} = \pm\bar{\Gamma}$. Thus the width of the resonance curve is

$$W = 2\Gamma.$$

Note that for a material characterized by a low-loss resonance ($\Gamma \ll \omega_0$), the location of $\tilde{\epsilon}'_{\max}$ can be approximated as

$$\omega_{\max} = \omega_0\sqrt{1 - 2\Gamma/\omega_0} \approx \omega_0 - \Gamma$$

while $\tilde{\epsilon}'_{\min}$ is located at

$$\omega_{\min} = \omega_0\sqrt{1 + 2\Gamma/\omega_0} \approx \omega_0 + \Gamma.$$

The region of anomalous dispersion thus lies between the half amplitude points of $\tilde{\epsilon}''$: $\omega_0 - \Gamma < \omega < \omega_0 + \Gamma$.

As $\Gamma \to 0$ the resonance curve becomes narrower and taller. Thus, a material characterized by a very low-loss resonance may be modeled very simply using $\tilde{\epsilon}'' = A\delta(\omega - \omega_0)$, where A is a constant to be determined. We can find A by applying the Kronig–Kramers formula (4.37):

$$\tilde{\epsilon}'(\omega) - \epsilon_0 = -\frac{2}{\pi}\,\text{P.V.}\int\limits_0^\infty A\delta(\Omega - \omega_0)\frac{\Omega\,d\Omega}{\Omega^2 - \omega^2} = -\frac{2}{\pi}A\frac{\omega_0}{\omega_0^2 - \omega^2}.$$

Since the material approaches the lossless case, this expression should match the Sellmeier equation (4.105):

$$-\frac{2}{\pi}A\frac{\omega_0}{\omega_0^2 - \omega^2} = \epsilon_0\frac{\omega_p^2}{\omega_0^2 - \omega^2},$$

giving $A = -\pi\epsilon_0\omega_p^2/2\omega_0$. Hence the permittivity of a material characterized by a low-loss resonance may be approximated as

$$\tilde{\epsilon}^c(\omega) = \epsilon_0\left(1 + \frac{\omega_p^2}{\omega_0^2 - \omega^2}\right) - j\epsilon_0\frac{\pi}{2}\frac{\omega_p^2}{\omega_0}\delta(\omega - \omega_0).$$

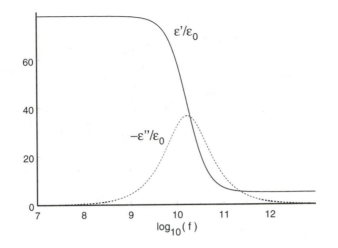

Figure 4.5: Relaxation spectrum for water at 20° C found using Debye equation.

Debye relaxation and the Cole–Cole equation. In solids or liquids consisting of *polar molecules* (those retaining a permanent dipole moment, e.g., water), the resonance effect is replaced by *relaxation*. We can view the molecule as attempting to rotate in response to an applied field within a background medium dominated by the frictional term in (4.101). The rotating molecule experiences many weak collisions which continuously drain off energy, preventing it from accelerating under the force of the applied field. J.W.P. Debye proposed that such materials are described by an exponential damping of their polarization and a complete absence of oscillations. If we neglect the acceleration term in (4.101) we have the equation of motion

$$2\Gamma\frac{d\mathbf{l}(\mathbf{r},t)}{dt} + \omega_r^2 \mathbf{l}(\mathbf{r},t) = -\frac{q_e}{m_e}\mathbf{E}'(\mathbf{r},t),$$

which has homogeneous solution

$$\mathbf{l}(\mathbf{r},t) = \mathbf{l}_0(\mathbf{r})e^{-\frac{\omega_r^2}{2\Gamma}t} = \mathbf{l}_0(\mathbf{r})e^{-t/\tau}$$

where τ is Debye's *relaxation time*.

By neglecting the acceleration term in (4.102) we obtain from (4.103) the dispersion equation, or *relaxation spectrum*

$$\tilde{\epsilon}(\omega) = \epsilon_0 + \epsilon_0 \frac{\omega_p^2}{\omega_0^2 + j\omega 2\Gamma}.$$

Debye proposed a relaxation spectrum a bit more general than this, now called the *Debye equation*:

$$\tilde{\epsilon}(\omega) = \epsilon_\infty + \frac{\epsilon_s - \epsilon_\infty}{1 + j\omega\tau}. \tag{4.106}$$

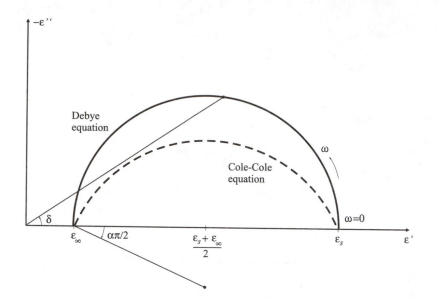

Figure 4.6: Arc plots for Debye and Cole–Cole descriptions of a polar material.

Here ϵ_s is the real static permittivity obtained when $\omega \to 0$, while ϵ_∞ is the real "optical" permittivity describing the high frequency behavior of $\tilde{\epsilon}$. If we split (4.106) into real and imaginary parts we find that

$$\tilde{\epsilon}'(\omega) - \epsilon_\infty = \frac{\epsilon_s - \epsilon_\infty}{1 + \omega^2 \tau^2}, \qquad \tilde{\epsilon}''(\omega) = -\frac{\omega\tau(\epsilon_s - \epsilon_\infty)}{1 + \omega^2 \tau^2}.$$

For a passive material we must have $\tilde{\epsilon}'' < 0$, which requires $\epsilon_s > \epsilon_\infty$. It is straightforward to show that these expressions obey the Kronig–Kramers relationships. The details are left as an exercise.

A plot of the Debye spectrum of water at $T = 20°$ C is shown in Figure 4.5, where we have used $\epsilon_s = 78.3\epsilon_0$, $\epsilon_\infty = 5\epsilon_0$, and $\tau = 9.6 \times 10^{-12}$ s [49]. We see that $\tilde{\epsilon}'$ decreases over the entire frequency range. The frequency dependence of the imaginary part of the permittivity is similar to that found in the resonance model, forming a curve which peaks at the *critical frequency*

$$\omega_{\text{max}} = 1/\tau$$

where it obtains a maximum value of

$$-\tilde{\epsilon}''_{\text{max}} = \frac{\epsilon_s - \epsilon_\infty}{2}.$$

At this point $\tilde{\epsilon}'$ achieves the average value of ϵ_s and ϵ_∞:

$$\epsilon'(\omega_{\text{max}}) = \frac{\epsilon_s + \epsilon_\infty}{2}.$$

Since the frequency label is logarithmic, we see that the peak is far broader than that for the resonance model.

Interestingly, a plot of $-\tilde{\epsilon}''$ versus $\tilde{\epsilon}'$ traces out a semicircle centered along the real axis at $(\epsilon_s + \epsilon_\infty)/2$ and with radius $(\epsilon_s - \epsilon_\infty)/2$. Such a plot, shown in Figure 4.6, was first described by K.S. Cole and R.H. Cole [38] and is thus called a *Cole–Cole diagram* or "arc

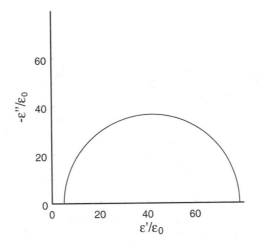

Figure 4.7: Cole–Cole diagram for water at 20° C.

plot." We can think of the vector extending from the origin to a point on the semicircle as a phasor whose phase angle δ is described by the *loss tangent* of the material:

$$\tan \delta = -\frac{\tilde{\epsilon}''}{\tilde{\epsilon}'} = \frac{\omega\tau(\epsilon_s - \epsilon_\infty)}{\epsilon_s + \epsilon_\infty \omega^2 \tau^2}. \tag{4.107}$$

The Cole–Cole plot shows that the maximum value of $-\tilde{\epsilon}''$ is $(\epsilon_s - \epsilon_\infty)/2$ and that $\tilde{\epsilon}' = (\epsilon_s + \epsilon_\infty)/2$ at this point.

A Cole–Cole plot for water, shown in Figure 4.7, displays the typical semicircular nature of the arc plot. However, not all polar materials have a relaxation spectrum that follows the Debye equation as closely as water. Cole and Cole found that for many materials the arc plot traces a circular arc centered *below* the real axis, and that the line through its center makes an angle of $\alpha(\pi/2)$ with the real axis as shown in Figure 4.6. This relaxation spectrum can be described in terms of a modified Debye equation

$$\tilde{\epsilon}(\omega) = \epsilon_\infty + \frac{\epsilon_s - \epsilon_\infty}{1 + (j\omega\tau)^{1-\alpha}},$$

called the *Cole–Cole equation*. A nonzero Cole–Cole parameter α tends to broaden the relaxation spectrum, and results from a spread of relaxation times centered around τ [4]. For water the Cole–Cole parameter is only $\alpha = 0.02$, suggesting that a Debye description is sufficient, but for other materials α may be much higher. For instance, consider a transformer oil with a measured Cole–Cole parameter of $\alpha = 0.23$, along with a measured relaxation time of $\tau = 2.3 \times 10^{-9}$ s, a static permittivity of $\epsilon_s = 5.9\epsilon_0$, and an optical permittivity of $\epsilon_\infty = 2.9\epsilon_0$ [4]. Figure 4.8 shows the Cole–Cole plot calculated using both $\alpha = 0$ and $\alpha = 0.23$, demonstrating a significant divergence from the Debye model. Figure 4.9 shows the relaxation spectrum for the transformer oil calculated with these same two parameters.

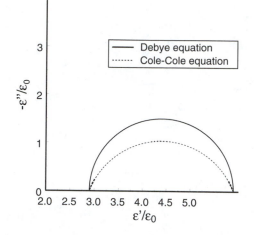

Figure 4.8: Cole–Cole diagram for transformer oil found using Debye equation and Cole–Cole equation with $\alpha = 0.23$.

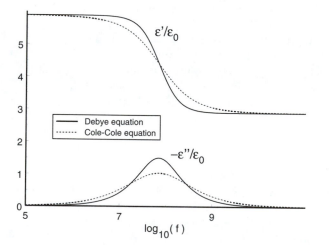

Figure 4.9: Relaxation spectrum for transformer oil found using Debye equation and Cole–Cole equation with $\alpha = 0.23$.

4.6.4 Permittivity and conductivity of a conductor

The free electrons within a conductor may be considered as an electron gas which is free to move under the influence of an applied field. Since the electrons are not bound to the atoms of the conductor, there is no restoring force acting on them. However, there is a damping term associated with electron collisions. We therefore model a conductor as a plasma, but with a very high collision frequency; in a good metallic conductor ν is typically in the range 10^{13}–10^{14} Hz.

We therefore have the conductivity of a conductor from (4.75) as

$$\tilde{\sigma}(\omega) = \frac{\epsilon_0 \omega_p^2 \nu}{\omega^2 + \nu^2}$$

and the permittivity as

$$\tilde{\epsilon}(\omega) = \epsilon_0 \left[1 - \frac{\omega_p^2}{\omega^2 + \nu^2} \right].$$

Since ν is so large, the conductivity is approximately

$$\tilde{\sigma}(\omega) \approx \frac{\epsilon_0 \omega_p^2}{\nu} = \frac{N q_e^2}{m_e \nu}$$

and the permittivity is

$$\tilde{\epsilon}(\omega) \approx \epsilon_0$$

well past microwave frequencies and into the infrared. Hence the dc conductivity is often employed by engineers throughout the communications bands. When approaching the visible spectrum the permittivity and conductivity begin to show a strong frequency dependence. In the violet and ultraviolet frequency ranges the free-charge conductivity becomes proportional to $1/\omega$ and is driven toward zero. However, at these frequencies the resonances of the bound electrons of the metal become important and the permittivity behaves more like that of a dielectric. At these frequencies the permittivity is best described using the resonance formula (4.104).

4.6.5 Permeability dyadic of a ferrite

The magnetic properties of materials are complicated and diverse. The formation of accurate models based on atomic behavior requires an understanding of quantum mechanics, but simple models may be constructed using classical mechanics along with very simple quantum-mechanical assumptions, such as the existence of a spin moment. For an excellent review of the magnetic properties of materials, see Elliott [65].

The magnetic properties of matter ultimately result from atomic currents. In our simple microscopic view these currents arise from the spin and orbital motion of negatively charged electrons. These atomic currents potentially give each atom a *magnetic moment* **m**. In *diamagnetic* materials the orbital and spin moments cancel unless the material is exposed to an external magnetic field, in which case the orbital electron velocity changes to produce a net moment opposite the applied field. In *paramagnetic* materials the spin moments are greater than the orbital moments, leaving the atoms with a net permanent magnetic moment. When exposed to an external magnetic field, these moments align in the same direction as an applied field. In either case, the density of magnetic moments **M** is zero in the absence of an applied field.

In most paramagnetic materials the alignment of the permanent moment of neighboring atoms is random. However, in the subsets of paramagnetic materials known as *ferromagnetic, anti-ferromagnetic,* and *ferrimagnetic* materials, there is a strong coupling between the spin moments of neighboring atoms resulting in either parallel or antiparallel alignment of moments. The most familiar case is the parallel alignment of moments within the domains of ferromagnetic permanent magnets made of iron, nickel, and cobalt. Anti-ferromagnetic materials, such as chromium and manganese, have strongly coupled moments that alternate in direction between small domains, resulting in zero net magnetic moment. Ferrimagnetic materials also have alternating moments, but these are unequal and thus do not cancel completely.

Ferrites form a particularly useful subgroup of ferrimagnetic materials. They were first developed during the 1940s by researchers at the Phillips Laboratories as low-loss magnetic media for supporting electromagnetic waves [65]. Typically, ferrites have conductivities ranging from 10^{-4} to 10^0 S/m (compared to 10^7 for iron), relative permeabilities in the thousands, and dielectric constants in the range 10–15. Their low loss makes them useful for constructing transformer cores and for a variety of microwave applications. Their chemical formula is $XO \cdot Fe_2O_3$, where X is a divalent metal or mixture of metals, such as cadmium, copper, iron, or zinc. When exposed to static magnetic fields, ferrites exhibit gyrotropic magnetic (or *gyromagnetic*) properties and have permeability matrices of the form (2.32). The properties of a wide variety of ferrites are given by von Aulock [204].

To determine the permeability matrix of a ferrite we will model its electrons as simple spinning tops and examine the torque exerted on the magnetic moment by the application of an external field. Each electron has an angular momentum $\mathbf{L}$ and a magnetic dipole moment $\mathbf{m}$, with these two vectors anti-parallel:

$$\mathbf{m}(\mathbf{r}, t) = -\gamma \mathbf{L}(\mathbf{r}, t)$$

where

$$\gamma = \frac{q_e}{m_e} = 1.7592 \times 10^{11} \text{ C/kg}$$

is called the *gyromagnetic ratio.*

Let us first consider a single spinning electron immersed in an applied static magnetic field $\mathbf{B}_0$. Any torque applied to the electron results in a change of angular momentum as given by Newton's second law

$$\mathbf{T}(\mathbf{r}, t) = \frac{d\mathbf{L}(\mathbf{r}, t)}{dt}.$$

We found in (3.179) that a very small loop of current in a magnetic field experiences a torque $\mathbf{m} \times \mathbf{B}$. Thus, when first placed into a static magnetic field $\mathbf{B}_0$ an electron's angular momentum obeys the equation

$$\frac{d\mathbf{L}(\mathbf{r}, t)}{dt} = -\gamma \mathbf{L}(\mathbf{r}, t) \times \mathbf{B}_0(\mathbf{r}) = \boldsymbol{\omega}_0(\mathbf{r}) \times \mathbf{L}(\mathbf{r}, t) \qquad (4.108)$$

where $\boldsymbol{\omega}_0 = \gamma \mathbf{B}_0$. This equation of motion describes the *precession* of the electron spin axis about the direction of the applied field, which is analogous to the precession of a gyroscope [129]. The spin axis rotates at the *Larmor precessional frequency* $\omega_0 = \gamma B_0 = \gamma \mu_0 H_0$.

We can use this to understand what happens when we insert a homogeneous ferrite material into a uniform static magnetic field $\mathbf{B}_0 = \mu_0 \mathbf{H}_0$. The internal field $\mathbf{H}_i$ experienced by any magnetic dipole is not the same as the external field $\mathbf{H}_0$, and need not even be in the same direction. In general we write

$$\mathbf{H}_0(\mathbf{r}, t) - \mathbf{H}_i(\mathbf{r}, t) = \mathbf{H}_d(\mathbf{r}, t)$$

where $\mathbf{H}_d$ is the *demagnetizing field* produced by the magnetic dipole moments of the material. Each electron responds to the internal field by precessing as described above until the precession damps out and the electron moments align with the magnetic field. At this point the ferrite is *saturated*. Because the demagnetizing field depends strongly on the shape of the material we choose to ignore it as a first approximation, and this allows us to concentrate our study on the fundamental atomic properties of the ferrite.

For purposes of understanding its magnetic properties, we view the ferrite as a dense collection of electrons and write

$$\mathbf{M}(\mathbf{r}, t) = N\mathbf{m}(\mathbf{r}, t)$$

where N is the number density of electrons. Since we are assuming the ferrite is homogeneous, we take N to be independent of time and position. Multiplying (4.108) by $-N\gamma$, we obtain an equation describing the evolution of $\mathbf{M}$:

$$\frac{d\mathbf{M}(\mathbf{r}, t)}{dt} = -\gamma \mathbf{M}(\mathbf{r}, t) \times \mathbf{B}_i(\mathbf{r}, t). \tag{4.109}$$

To determine the temporal response of the ferrite we must include a time-dependent component of the applied field. We now let

$$\mathbf{H}_0(\mathbf{r}, t) = \mathbf{H}_i(\mathbf{r}, t) = \mathbf{H}_T(\mathbf{r}, t) + \mathbf{H}_{dc}$$

where $\mathbf{H}_T$ is the time-dependent component superimposed with the uniform static component $\mathbf{H}_{dc}$. Using $\mathbf{B} = \mu_0(\mathbf{H} + \mathbf{M})$ we have from (4.109)

$$\frac{d\mathbf{M}(\mathbf{r}, t)}{dt} = -\gamma\mu_0 \mathbf{M}(\mathbf{r}, t) \times [\mathbf{H}_T(\mathbf{r}, t) + \mathbf{H}_{dc} + \mathbf{M}(\mathbf{r}, t)].$$

With $\mathbf{M} = \mathbf{M}_T(\mathbf{r}, t) + \mathbf{M}_{dc}$ and $\mathbf{M} \times \mathbf{M} = 0$ this becomes

$$\frac{d\mathbf{M}_T(\mathbf{r}, t)}{dt} + \frac{d\mathbf{M}_{dc}}{dt} = -\gamma\mu_0[\mathbf{M}_T(\mathbf{r}, t) \times \mathbf{H}_T(\mathbf{r}, t) + \mathbf{M}_T(\mathbf{r}, t) \times \mathbf{H}_{dc} +$$
$$+ \mathbf{M}_{dc} \times \mathbf{H}_T(\mathbf{r}, t) + \mathbf{M}_{dc} \times \mathbf{H}_{dc}. \tag{4.110}$$

Let us assume that the ferrite is saturated. Then $\mathbf{M}_{dc}$ is aligned with $\mathbf{H}_{dc}$ and their cross product vanishes. Let us further assume that the spectrum of H_T is small compared to H_{dc} at all frequencies: $|\tilde{H}_T(\mathbf{r}, \omega)| \ll H_{dc}$. This small-signal assumption allows us to neglect $\mathbf{M}_T \times \mathbf{H}_T$. Using these and noting that the time derivative of $\mathbf{M}_{dc}$ is zero, we see that (4.110) reduces to

$$\frac{d\mathbf{M}_T(\mathbf{r}, t)}{dt} = -\gamma\mu_0[\mathbf{M}_T(\mathbf{r}, t) \times \mathbf{H}_{dc} + \mathbf{M}_{dc} \times \mathbf{H}_T(\mathbf{r}, t)]. \tag{4.111}$$

To determine the frequency response we write (4.111) in terms of inverse Fourier transforms and invoke the Fourier integral theorem to find that

$$j\omega\tilde{\mathbf{M}}_T(\mathbf{r}, \omega) = -\gamma\mu_0[\tilde{\mathbf{M}}_T(\mathbf{r}, \omega) \times \mathbf{H}_{dc} + \mathbf{M}_{dc} \times \tilde{\mathbf{H}}_T(\mathbf{r}, \omega)].$$

Defining

$$\gamma\mu_0\mathbf{M}_{dc} = \boldsymbol{\omega}_M,$$

where $\omega_M = |\boldsymbol{\omega}_M|$ is the *saturation magnetization frequency*, we find that

$$\tilde{\mathbf{M}}_T + \tilde{\mathbf{M}}_T \times \left[\frac{\boldsymbol{\omega}_0}{j\omega}\right] = \left[-\frac{1}{j\omega}\boldsymbol{\omega}_M \times \tilde{\mathbf{H}}_T\right], \qquad (4.112)$$

where $\boldsymbol{\omega}_0 = \gamma\mu_0\mathbf{H}_{dc}$ with ω_0 now called the *gyromagnetic response frequency*. This has the form $\mathbf{v} + \mathbf{v} \times \mathbf{C} = \mathbf{A}$, which has solution (4.80). Substituting into this expression and remembering that $\boldsymbol{\omega}_0$ is parallel to $\boldsymbol{\omega}_M$, we find that

$$\tilde{\mathbf{M}}_T = \frac{-\frac{1}{j\omega}\boldsymbol{\omega}_M \times \tilde{\mathbf{H}}_T + \frac{1}{\omega^2}\left\{\boldsymbol{\omega}_M[\boldsymbol{\omega}_0 \cdot \tilde{\mathbf{H}}_T] - (\boldsymbol{\omega}_0 \cdot \boldsymbol{\omega}_M)\tilde{\mathbf{H}}_T\right\}}{1 - \frac{\omega_0^2}{\omega^2}}.$$

If we define the dyadic $\bar{\boldsymbol{\omega}}_M$ such that $\bar{\boldsymbol{\omega}}_M \cdot \tilde{\mathbf{H}}_T = \boldsymbol{\omega}_M \times \tilde{\mathbf{H}}_T$, then we identify the dyadic magnetic susceptibility

$$\tilde{\bar{\chi}}_m(\omega) = \frac{j\omega\bar{\boldsymbol{\omega}}_M + \boldsymbol{\omega}_M\boldsymbol{\omega}_0 - \omega_M\omega_0\bar{\mathbf{I}}}{\omega^2 - \omega_0^2} \qquad (4.113)$$

with which we can write $\tilde{\mathbf{M}}(\mathbf{r}, \omega) = \bar{\chi}_m(\omega) \cdot \tilde{\mathbf{H}}(\mathbf{r}, \omega)$. In rectangular coordinates $\bar{\boldsymbol{\omega}}_M$ is represented by

$$[\bar{\boldsymbol{\omega}}_M] = \begin{bmatrix} 0 & -\omega_{Mz} & \omega_{My} \\ \omega_{Mz} & 0 & -\omega_{Mx} \\ -\omega_{My} & \omega_{Mx} & 0 \end{bmatrix}. \qquad (4.114)$$

Finally, using $\tilde{\mathbf{B}} = \mu_0(\tilde{\mathbf{H}} + \tilde{\mathbf{M}}) = \mu_0(\bar{\mathbf{I}} + \tilde{\bar{\chi}}_m) \cdot \tilde{\mathbf{H}} = \tilde{\bar{\mu}} \cdot \tilde{\mathbf{H}}$ we find that

$$\tilde{\bar{\mu}}(\omega) = \mu_0[\bar{\mathbf{I}} + \tilde{\bar{\chi}}_m(\omega)].$$

To examine the properties of the dyadic permeability it is useful to write it in matrix form. To do this we must choose a coordinate system. We shall assume that $\mathbf{H}_{dc}$ is aligned with the z-axis so that $\mathbf{H}_{dc} = \hat{\mathbf{z}}H_{dc}$ and thus $\boldsymbol{\omega}_M = \hat{\mathbf{z}}\omega_M$ and $\boldsymbol{\omega}_0 = \hat{\mathbf{z}}\omega_0$. Then (4.114) becomes

$$[\bar{\boldsymbol{\omega}}_M] = \begin{bmatrix} 0 & -\omega_M & 0 \\ \omega_M & 0 & 0 \\ 0 & 0 & 0 \end{bmatrix}$$

and we can write the susceptibility dyadic (4.113) as

$$[\tilde{\bar{\chi}}_m(\omega)] = \frac{\omega_M}{\omega^2 - \omega_0^2}\begin{bmatrix} -\omega_0 & -j\omega & 0 \\ j\omega & -\omega_0 & 0 \\ 0 & 0 & 0 \end{bmatrix}.$$

The permeability dyadic becomes

$$[\tilde{\bar{\mu}}(\omega)] = \begin{bmatrix} \mu & -j\kappa & 0 \\ j\kappa & \mu & 0 \\ 0 & 0 & \mu_0 \end{bmatrix} \qquad (4.115)$$

where

$$\mu = \mu_0\left(1 - \frac{\omega_0\omega_M}{\omega^2 - \omega_0^2}\right), \qquad (4.116)$$

$$\kappa = \mu_0\frac{\omega\omega_M}{\omega^2 - \omega_0^2}. \qquad (4.117)$$

Because its permeability dyadic is that for a lossless *gyrotropic* material (2.33), we call the ferrite *gyromagnetic*.

Since the ferrite is lossless, the dyadic permeability must be hermitian according to (4.49). The specific form of (4.115) shows this explicitly. We also note that since the sign of $\boldsymbol{\omega}_M$ is determined by that of $\mathbf{H}_{dc}$, the dyadic permittivity obeys the symmetry relation

$$\tilde{\mu}_{ij}(\mathbf{H}_{dc}) = \tilde{\mu}_{ji}(-\mathbf{H}_{dc}),$$

which is the symmetry condition observed for a plasma in (4.87).

A lossy ferrite material can be modeled by adding a damping term to (4.111):

$$\frac{d\mathbf{M}(\mathbf{r},t)}{dt} = -\gamma\mu_0 \left[\mathbf{M}_T(\mathbf{r},t) \times \mathbf{H}_{dc} + \mathbf{M}_{dc} \times \mathbf{H}_T(\mathbf{r},t) \right] + \alpha \frac{\mathbf{M}_{dc}}{M_{dc}} \times \frac{d\mathbf{M}_T(\mathbf{r},t)}{dt},$$

where α is the damping parameter [40, 204]. This term tends to reduce the angle of precession. Fourier transformation gives

$$j\omega\tilde{\mathbf{M}}_T = \boldsymbol{\omega}_0 \times \tilde{\mathbf{M}}_T - \boldsymbol{\omega}_M \times \tilde{\mathbf{H}}_T + \alpha \frac{\boldsymbol{\omega}_M}{\omega_M} \times j\omega\tilde{\mathbf{M}}_T.$$

Remembering that $\boldsymbol{\omega}_0$ and $\boldsymbol{\omega}_M$ are aligned we can write this as

$$\tilde{\mathbf{M}}_T + \tilde{\mathbf{M}}_T \times \left[\frac{\boldsymbol{\omega}_0 \left(1 + j\alpha\frac{\omega}{\omega_0} \right)}{j\omega} \right] = \left[-\frac{1}{j\omega}\boldsymbol{\omega}_M \times \tilde{\mathbf{H}}_T \right].$$

This is identical to (4.112) with

$$\boldsymbol{\omega}_0 \to \boldsymbol{\omega}_0 \left(1 + j\alpha\frac{\omega}{\omega_0} \right).$$

Thus, we merely substitute this into (4.113) to find the susceptibility dyadic for a lossy ferrite:

$$\tilde{\bar{\boldsymbol{\chi}}}_m(\omega) = \frac{j\omega\bar{\boldsymbol{\omega}}_M + \boldsymbol{\omega}_M\omega_0 \left(1 + j\alpha\omega/\omega_0 \right) - \omega_M\omega_0 \left(1 + j\alpha\omega/\omega_0 \right) \bar{\mathbf{I}}}{\omega^2(1 + \alpha^2) - \omega_0^2 - 2j\alpha\omega\omega_0}.$$

Making the same substitution into (4.115) we can write the dyadic permeability matrix as

$$[\tilde{\bar{\boldsymbol{\mu}}}(\omega)] = \begin{bmatrix} \tilde{\mu}_{xx} & \tilde{\mu}_{xy} & 0 \\ \tilde{\mu}_{yx} & \tilde{\mu}_{yy} & 0 \\ 0 & 0 & \mu_0 \end{bmatrix} \tag{4.118}$$

where

$$\tilde{\mu}_{xx} = \tilde{\mu}_{yy} = \mu_0 - \mu_0\omega_M \frac{\omega_0 \left[\omega^2(1 - \alpha^2) - \omega_0^2 \right] + j\omega\alpha \left[\omega^2(1 + \alpha^2) + \omega_0^2 \right]}{\left[\omega^2(1 + \alpha^2) - \omega_0^2 \right]^2 + 4\alpha^2\omega^2\omega_0^2} \tag{4.119}$$

and

$$\tilde{\mu}_{xy} = -\tilde{\mu}_{yx} = \frac{2\mu_0\alpha\omega^2\omega_0\omega_M - j\mu_0\omega\omega_M \left[\omega^2(1 + \alpha^2) - \omega_0^2 \right]}{\left[\omega^2(1 + \alpha^2) - \omega_0^2 \right]^2 + 4\alpha^2\omega^2\omega_0^2}. \tag{4.120}$$

In the case of a lossy ferrite, the hermitian nature of the permeability dyadic is lost.

4.7   Monochromatic fields and the phasor domain

The Fourier transform is very efficient for representing the nearly sinusoidal signals produced by electronic systems such as oscillators. However, we should realize that the elemental term $e^{j\omega t}$ by itself cannot represent any physical quantity; only a continuous superposition of such terms can have physical meaning, because no physical process can be truly monochromatic. All events must have transient periods during which they are established. Even "monochromatic" light appears in bundles called quanta, interpreted as containing finite numbers of oscillations.

Arguments about whether "monochromatic" or "sinusoidal steady-state" fields can actually exist may sound purely academic. After all, a microwave oscillator can create a wave train of 10^{10} oscillations within the first second after being turned on. Such a waveform is surely as close to monochromatic as we would care to measure. But as with all mathematical models of physical systems, we can get into trouble by making non-physical assumptions, in this instance by assuming a physical system has always been in the steady state. Sinusoidal steady-state solutions to Maxwell's equations can lead to troublesome infinities linked to the infinite energy content of each elemental component. For example, an attempt to compute the energy stored within a lossless microwave cavity under steady-state conditions gives an infinite result since the cavity has been building up energy since $t = -\infty$. We handle this by considering time-averaged quantities, but even then must be careful when materials are dispersive (§ 4.5). Nevertheless, the steady-state concept is valuable because of its simplicity and finds widespread application in electromagnetics.

Since the elemental term is complex, we may use its real part, its imaginary part, or some combination of both to represent a monochromatic (or *time-harmonic*) field. We choose the representation

$$\psi(\mathbf{r}, t) = \psi_0(\mathbf{r}) \cos[\breve{\omega} t + \xi(\mathbf{r})], \tag{4.121}$$

where ξ is the temporal phase angle of the sinusoidal function. The Fourier transform is

$$\tilde{\psi}(\mathbf{r}, \omega) = \int_{-\infty}^{\infty} \psi_0(\mathbf{r}) \cos[\breve{\omega} t + \xi(\mathbf{r})] e^{-j\omega t} \, dt. \tag{4.122}$$

Here we run into an immediate problem: the transform in (4.122) does not exist in the ordinary sense since $\cos(\breve{\omega} t + \xi)$ is not absolutely integrable on $(-\infty, \infty)$. We should not be surprised by this: the cosine function cannot describe an actual physical process (it extends in time to $\pm\infty$), so it lacks a classical Fourier transform. One way out of this predicament is to extend the meaning of the Fourier transform as we do in § A.1. Then the monochromatic field (4.121) is viewed as having the generalized transform

$$\tilde{\psi}(\mathbf{r}, \omega) = \psi_0(\mathbf{r})\pi \left[e^{j\xi(\mathbf{r})} \delta(\omega - \breve{\omega}) + e^{-j\xi(\mathbf{r})} \delta(\omega + \breve{\omega}) \right]. \tag{4.123}$$

We can compute the inverse Fourier transform by substituting (4.123) into (4.2):

$$\psi(\mathbf{r}, t) = \frac{1}{2\pi} \int_{-\infty}^{\infty} \psi_0(\mathbf{r})\pi \left[e^{j\xi(\mathbf{r})} \delta(\omega - \breve{\omega}) + e^{-j\xi(\mathbf{r})} \delta(\omega + \breve{\omega}) \right] e^{j\omega t} \, d\omega. \tag{4.124}$$

By our interpretation of the Dirac delta, we see that the decomposition of the cosine function has only two discrete components, located at $\omega = \pm\breve{\omega}$. So we have realized our initial intention of having only a single elemental function present. The sifting property gives

$$\psi(\mathbf{r},t) = \psi_0(\mathbf{r})\frac{e^{j\breve{\omega}t}e^{j\xi(\mathbf{r})} + e^{-j\breve{\omega}t}e^{-j\xi(\mathbf{r})}}{2} = \psi_0(\mathbf{r})\cos[\breve{\omega}t + \xi(\mathbf{r})]$$

as expected.

4.7.1 The time-harmonic EM fields and constitutive relations

The time-harmonic fields are described using the representation (4.121) for each field component. The electric field is

$$\mathbf{E}(\mathbf{r},t) = \sum_{i=1}^{3} \hat{\mathbf{i}}_i |E_i(\mathbf{r})| \cos[\breve{\omega}t + \xi_i^E(\mathbf{r})]$$

for example. Here $|E_i|$ is the complex magnitude of the ith vector component, and ξ_i^E is the phase angle $(-\pi < \xi_i^E \le \pi)$. Similar terminology is used for the remaining fields.

The frequency-domain constitutive relations (4.11)–(4.15) may be written for the time-harmonic fields by employing (4.124). For instance, for an isotropic material where

$$\tilde{\mathbf{D}}(\mathbf{r},\omega) = \tilde{\epsilon}(\mathbf{r},\omega)\tilde{\mathbf{E}}(\mathbf{r},\omega), \qquad \tilde{\mathbf{B}}(\mathbf{r},\omega) = \tilde{\mu}(\mathbf{r},\omega)\tilde{\mathbf{H}}(\mathbf{r},\omega),$$

with

$$\tilde{\epsilon}(\mathbf{r},\omega) = |\tilde{\epsilon}(\mathbf{r},\omega)|e^{\xi^\epsilon(\mathbf{r},\omega)}, \qquad \tilde{\mu}(\mathbf{r},\omega) = |\tilde{\mu}(\mathbf{r},\omega)|e^{\xi^\mu(\mathbf{r},\omega)},$$

we can write

$$\mathbf{D}(\mathbf{r},t) = \sum_{i=1}^{3} \hat{\mathbf{i}}_i |D_i(\mathbf{r})| \cos[\breve{\omega}t + \xi_i^D(\mathbf{r})]$$

$$= \frac{1}{2\pi}\int_{-\infty}^{\infty} \sum_{i=1}^{3} \hat{\mathbf{i}}_i \tilde{\epsilon}(\mathbf{r},\omega)|E_i(\mathbf{r})|\pi \left[e^{j\xi_i^E(\mathbf{r})}\delta(\omega - \breve{\omega}) + e^{-j\xi_i^E(\mathbf{r})}\delta(\omega + \breve{\omega})\right] e^{j\omega t}\,d\omega$$

$$= \frac{1}{2}\sum_{i=1}^{3} \hat{\mathbf{i}}_i |E_i(\mathbf{r})| \left[\tilde{\epsilon}(\mathbf{r},\breve{\omega})e^{j(\breve{\omega}t + j\xi_i^E(\mathbf{r}))} + \tilde{\epsilon}(\mathbf{r},-\breve{\omega})e^{-j(\breve{\omega}t + j\xi_i^E(\mathbf{r}))}\right].$$

Since (4.25) shows that $\tilde{\epsilon}(\mathbf{r},-\breve{\omega}) = \tilde{\epsilon}^*(\mathbf{r},\breve{\omega})$, we have

$$\mathbf{D}(\mathbf{r},t) = \frac{1}{2}\sum_{i=1}^{3} \hat{\mathbf{i}}_i |E_i(\mathbf{r})||\tilde{\epsilon}(\mathbf{r},\breve{\omega})| \left[e^{j(\breve{\omega}t + j\xi_i^E(\mathbf{r}) + j\xi^\epsilon(\mathbf{r},\breve{\omega}))} + e^{-j(\breve{\omega}t + j\xi_i^E(\mathbf{r}) + j\xi^\epsilon(\mathbf{r},\breve{\omega}))}\right]$$

$$= \sum_{i=1}^{3} \hat{\mathbf{i}}_i |\tilde{\epsilon}(\mathbf{r},\breve{\omega})||E_i(\mathbf{r})| \cos[\breve{\omega}t + \xi_i^E(\mathbf{r}) + \xi^\epsilon(\mathbf{r},\breve{\omega})]. \tag{4.125}$$

Similarly

$$\mathbf{B}(\mathbf{r},t) = \sum_{i=1}^{3} \hat{\mathbf{i}}_i |B_i(\mathbf{r})| \cos[\breve{\omega}t + \xi_i^B(\mathbf{r})]$$

$$= \sum_{i=1}^{3} \hat{\mathbf{i}}_i |\tilde{\mu}(\mathbf{r},\breve{\omega})||H_i(\mathbf{r})| \cos[\breve{\omega}t + \xi_i^H(\mathbf{r}) + \xi^\mu(\mathbf{r},\breve{\omega})].$$

4.7.2   The phasor fields and Maxwell's equations

Sinusoidal steady-state computations using the forward and inverse transform formulas are unnecessarily cumbersome. A much more efficient approach is to use the *phasor* concept. If we define the complex function

$$\check{\psi}(\mathbf{r}) = \psi_0(\mathbf{r})e^{j\xi(\mathbf{r})}$$

as the *phasor form* of the monochromatic field $\tilde{\psi}(\mathbf{r}, \omega)$, then the inverse Fourier transform is easily computed by multiplying $\check{\psi}(\mathbf{r})$ by $e^{j\breve{\omega}t}$ and taking the real part. That is,

$$\psi(\mathbf{r}, t) = \mathrm{Re}\left\{\check{\psi}(\mathbf{r})e^{j\breve{\omega}t}\right\} = \psi_0(\mathbf{r})\cos[\breve{\omega}t + \xi(\mathbf{r})]. \qquad (4.126)$$

Using the phasor representation of the fields, we can obtain a set of Maxwell equations relating the phasor components. Let

$$\check{\mathbf{E}}(\mathbf{r}) = \sum_{i=1}^{3}\hat{\mathbf{i}}_i\check{E}_i(\mathbf{r}) = \sum_{i=1}^{3}\hat{\mathbf{i}}_i|E_i(\mathbf{r})|e^{j\xi_i^E(\mathbf{r})}$$

represent the phasor monochromatic electric field, with similar formulas for the other fields. Then

$$\mathbf{E}(\mathbf{r}, t) = \mathrm{Re}\left\{\check{\mathbf{E}}(\mathbf{r})e^{j\breve{\omega}t}\right\} = \sum_{i=1}^{3}\hat{\mathbf{i}}_i|E_i(\mathbf{r})|\cos[\breve{\omega}t + \xi_i^E(\mathbf{r})].$$

Substituting these expressions into Ampere's law (2.2), we have

$$\nabla \times \mathrm{Re}\left\{\check{\mathbf{H}}(\mathbf{r})e^{j\breve{\omega}t}\right\} = \frac{\partial}{\partial t}\mathrm{Re}\left\{\check{\mathbf{D}}(\mathbf{r})e^{j\breve{\omega}t}\right\} + \mathrm{Re}\left\{\check{\mathbf{J}}(\mathbf{r})e^{j\breve{\omega}t}\right\}.$$

Since the real part of a sum of complex variables equals the sum of the real parts, we can write

$$\mathrm{Re}\left\{\nabla \times \check{\mathbf{H}}(\mathbf{r})e^{j\breve{\omega}t} - \check{\mathbf{D}}(\mathbf{r})\frac{\partial}{\partial t}e^{j\breve{\omega}t} - \check{\mathbf{J}}(\mathbf{r})e^{j\breve{\omega}t}\right\} = 0. \qquad (4.127)$$

If we examine for an arbitrary complex function $F = F_r + jF_i$ the quantity

$$\mathrm{Re}\left\{(F_r + jF_i)e^{j\breve{\omega}t}\right\} = \mathrm{Re}\left\{(F_r\cos\breve{\omega}t - F_i\sin\breve{\omega}t) + j(F_r\sin\breve{\omega}t + F_i\cos\breve{\omega}t)\right\},$$

we see that both F_r and F_i must be zero for the expression to vanish for all t. Thus (4.127) requires that

$$\nabla \times \check{\mathbf{H}}(\mathbf{r}) = j\breve{\omega}\check{\mathbf{D}}(\mathbf{r}) + \check{\mathbf{J}}(\mathbf{r}), \qquad (4.128)$$

which is the phasor Ampere's law. Similarly we have

$$\nabla \times \check{\mathbf{E}}(\mathbf{r}) = -j\breve{\omega}\check{\mathbf{B}}(\mathbf{r}), \qquad (4.129)$$
$$\nabla \cdot \check{\mathbf{D}}(\mathbf{r}) = \check{\rho}(\mathbf{r}), \qquad (4.130)$$
$$\nabla \cdot \check{\mathbf{B}}(\mathbf{r}) = 0, \qquad (4.131)$$

and

$$\nabla \cdot \check{\mathbf{J}}(\mathbf{r}) = -j\breve{\omega}\check{\rho}(\mathbf{r}). \qquad (4.132)$$

The constitutive relations may be easily incorporated into the phasor concept. If we use

$$\check{D}_i(\mathbf{r}) = \tilde{\epsilon}(\mathbf{r}, \breve{\omega})\check{E}_i(\mathbf{r}) = |\tilde{\epsilon}(\mathbf{r}, \breve{\omega})|e^{j\xi^\epsilon(\mathbf{r}, \breve{\omega})}|E_i(\mathbf{r})|e^{j\xi_i^E(\mathbf{r})},$$

then forming
$$D_i(\mathbf{r}, t) = \text{Re}\left\{\check{D}_i(\mathbf{r})e^{j\check{\omega}t}\right\}$$
we reproduce (4.125). Thus we may write
$$\check{\mathbf{D}}(\mathbf{r}) = \tilde{\epsilon}(\mathbf{r}, \check{\omega})\check{\mathbf{E}}(\mathbf{r}).$$

Note that we never write $\check{\epsilon}$ or refer to a "phasor permittivity" since the permittivity does not vary sinusoidally in the time domain.

An obvious benefit of the phasor method is that we can manipulate field quantities without involving the sinusoidal time dependence. When our manipulations are complete, we return to the time domain using (4.126).

The phasor Maxwell equations (4.128)–(4.131) are identical in form to the temporal frequency-domain Maxwell equations (4.7)–(4.10), except that $\omega = \check{\omega}$ in the phasor equations. This is sensible, since the phasor fields represent a single component of the complete frequency-domain spectrum of the arbitrary time-varying fields. Thus, if the phasor fields are calculated for some $\check{\omega}$, we can make the replacements
$$\check{\omega} \to \omega, \qquad \check{\mathbf{E}}(\mathbf{r}) \to \tilde{\mathbf{E}}(\mathbf{r}, \omega), \qquad \check{\mathbf{H}}(\mathbf{r}) \to \tilde{\mathbf{H}}(\mathbf{r}, \omega), \qquad \dots,$$
and obtain the general time-domain expressions by performing the inversion (4.2). Similarly, if we evaluate the frequency-domain field $\tilde{\mathbf{E}}(\mathbf{r}, \omega)$ at $\omega = \check{\omega}$, we produce the phasor field $\check{\mathbf{E}}(\mathbf{r}) = \tilde{\mathbf{E}}(\mathbf{r}, \check{\omega})$ for this frequency. That is
$$\text{Re}\left\{\tilde{\mathbf{E}}(\mathbf{r}, \check{\omega})e^{j\check{\omega}t}\right\} = \sum_{i=1}^{3} \hat{\mathbf{i}}_i |\tilde{E}_i(\mathbf{r}, \check{\omega})| \cos\left(\check{\omega}t + \xi^E(\mathbf{r}, \check{\omega})\right).$$

4.7.3 Boundary conditions on the phasor fields

The boundary conditions developed in § 4.3 for the frequency-domain fields may be adapted for use with the phasor fields by selecting $\omega = \check{\omega}$. Let us include the effects of fictitious magnetic sources and write
$$\hat{\mathbf{n}}_{12} \times (\check{\mathbf{H}}_1 - \check{\mathbf{H}}_2) = \check{\mathbf{J}}_s, \tag{4.133}$$
$$\hat{\mathbf{n}}_{12} \times (\check{\mathbf{E}}_1 - \check{\mathbf{E}}_2) = -\check{\mathbf{J}}_{ms}, \tag{4.134}$$
$$\hat{\mathbf{n}}_{12} \cdot (\check{\mathbf{D}}_1 - \check{\mathbf{D}}_2) = \check{\rho}_s, \tag{4.135}$$
$$\hat{\mathbf{n}}_{12} \cdot (\check{\mathbf{B}}_1 - \check{\mathbf{B}}_2) = \check{\rho}_{ms}, \tag{4.136}$$
and
$$\hat{\mathbf{n}}_{12} \cdot (\check{\mathbf{J}}_1 - \check{\mathbf{J}}_2) = -\nabla_s \cdot \check{\mathbf{J}}_s - j\check{\omega}\check{\rho}_s, \tag{4.137}$$
$$\hat{\mathbf{n}}_{12} \cdot (\check{\mathbf{J}}_{m1} - \check{\mathbf{J}}_{m2}) = -\nabla_s \cdot \check{\mathbf{J}}_{ms} - j\check{\omega}\check{\rho}_{ms}, \tag{4.138}$$
where $\hat{\mathbf{n}}_{12}$ points into region 1 from region 2.

4.8 Poynting's theorem for time-harmonic fields

We can specialize Poynting's theorem to time-harmonic form by substituting the time-harmonic field representations. The result depends on whether we use the general form

(2.301), which is valid for dispersive materials, or (2.299). For nondispersive materials (2.299) allows us to interpret the volume integral term as the time rate of change of stored energy. But if the operating frequency lies within the realm of material dispersion and loss, then we can no longer identify an explicit stored energy term.

4.8.1 General form of Poynting's theorem

We begin with (2.301). Substituting the time-harmonic representations we obtain the term

$$
\mathbf{E}(\mathbf{r}, t) \cdot \frac{\partial \mathbf{D}(\mathbf{r}, t)}{\partial t} = \left[\sum_{i=1}^{3} \hat{\mathbf{i}}_i |E_i| \cos[\breve{\omega} t + \xi_i^E] \right] \cdot \frac{\partial}{\partial t} \left[\sum_{i=1}^{3} \hat{\mathbf{i}}_i |D_i| \cos[\breve{\omega} t + \xi_i^D] \right]
$$

$$
= -\breve{\omega} \sum_{i=1}^{3} |E_i||D_i| \cos[\breve{\omega} t + \xi_i^E] \sin[\breve{\omega} t + \xi_i^D].
$$

Since $2 \sin A \cos B \equiv \sin(A + B) + \sin(A - B)$ we have

$$
\mathbf{E}(\mathbf{r}, t) \cdot \frac{\partial}{\partial t} \mathbf{D}(\mathbf{r}, t) = -\frac{1}{2} \sum_{i=1}^{3} \breve{\omega} |E_i||D_i| S_{ii}^{DE}(t),
$$

where

$$
S_{ii}^{DE}(t) = \sin(2\breve{\omega} t + \xi_i^D + \xi_i^E) + \sin(\xi_i^D - \xi_i^E)
$$

describes the temporal dependence of the field product. Separating the current into an impressed term $\mathbf{J}^i$ and a secondary term $\mathbf{J}^c$ (assumed to be the conduction current) as $\mathbf{J} = \mathbf{J}^i + \mathbf{J}^c$ and repeating the above steps with the other terms, we obtain

$$
-\frac{1}{2} \int_V \sum_{i=1}^{3} |J_i^i||E_i| C_{ii}^{J^i E}(t) \, dV = \frac{1}{2} \oint_S \sum_{i,j=1}^{3} |E_i||H_j|(\hat{\mathbf{i}}_i \times \hat{\mathbf{i}}_j) \cdot \hat{\mathbf{n}} C_{ij}^{EH}(t) \, dS +
$$

$$
+\frac{1}{2} \int_V \sum_{i=1}^{3} \left\{ -\breve{\omega} |D_i||E_i| S_{ii}^{DE}(t) - \breve{\omega} |B_i||H_i| S_{ii}^{BH}(t) + |J_i^c||E_i| C_{ii}^{J^c E}(t) \right\} \, dV, \quad (4.139)
$$

where

$$
S_{ii}^{BH}(t) = \sin(2\breve{\omega} t + \xi_i^B + \xi_i^H) + \sin(\xi_i^B - \xi_i^H),
$$
$$
C_{ij}^{EH}(t) = \cos(2\breve{\omega} t + \xi_i^E + \xi_j^H) + \cos(\xi_i^E - \xi_j^H),
$$

and so on.

We see that each power term has two temporal components: one oscillating at frequency $2\breve{\omega}$, and one constant with time. The oscillating component describes power that cycles through the various mechanisms of energy storage, dissipation, and transfer across the boundary. Dissipation may be produced through conduction processes or through polarization and magnetization phase lag, as described by the volume term on the right-hand side of (4.139). Power may also be delivered to the fields either from the sources, as described by the volume term on the left-hand side, or from an active medium, as described by the volume term on the right-hand side. The time-average balance of power supplied to the fields and extracted from the fields throughout each cycle, including that transported across the surface S, is given by the constant terms in (4.139):

$$-\frac{1}{2}\int_V \sum_{i=1}^3 |J_i^i||E_i|\cos(\xi_i^{J^i}-\xi_i^E)\,dV = \frac{1}{2}\int_V \sum_{i=1}^3 \left\{\breve{\omega}|E_i||D_i|\sin(\xi_i^E-\xi_i^D)+\right.$$

$$\left.+\breve{\omega}|B_i||H_i|\sin(\xi_i^H-\xi_i^B)+|J_i^c||E_i|\cos(\xi_i^{J^c}-\xi_i^E)\right\}\,dV +$$

$$+\frac{1}{2}\oint_S \sum_{i,j=1}^3 |E_i||H_j|(\hat{\mathbf{i}}_i\times\hat{\mathbf{i}}_j)\cdot\hat{\mathbf{n}}\cos(\xi_i^E-\xi_j^H)\,dS. \tag{4.140}$$

We associate one mechanism for time-average power loss with the phase lag between applied field and resulting polarization or magnetization. We can see this more clearly if we use the alternative form of the Poynting theorem (2.302) written in terms of the polarization and magnetization vectors. Writing

$$\mathbf{P}(\mathbf{r},t)=\sum_{i=1}^3 |P_i(\mathbf{r})|\cos[\breve{\omega}t+\xi_i^P(\mathbf{r})], \qquad \mathbf{M}(\mathbf{r},t)=\sum_{i=1}^3 |M_i(\mathbf{r})|\cos[\breve{\omega}t+\xi_i^M(\mathbf{r})],$$

and substituting the time-harmonic fields, we see that

$$-\frac{1}{2}\int_V \sum_{i=1}^3 |J_i||E_i|C_{ii}^{JE}(t)\,dV + \frac{\breve{\omega}}{2}\int_V \sum_{i=1}^3 \left[|P_i||E_i|S_{ii}^{PE}(t)+\mu_0|M_i||H_i|S_{ii}^{MH}(t)\right]\,dV$$

$$=-\frac{\breve{\omega}}{2}\int_V \sum_{i=1}^3 \left[\epsilon_0|E_i|^2 S_{ii}^{EE}(t)+\mu_0|H_i|^2 S_{ii}^{HH}(t)\right]\,dV +$$

$$+\frac{1}{2}\oint_S \sum_{i,j=1}^3 |E_i||H_j|(\hat{\mathbf{i}}_i\times\hat{\mathbf{i}}_j)\cdot\hat{\mathbf{n}}C_{ij}^{EH}(t)\,dS. \tag{4.141}$$

Selection of the constant part gives the balance of time-average power:

$$-\frac{1}{2}\int_V \sum_{i=1}^3 |J_i||E_i|\cos(\xi_i^J-\xi_i^E)\,dV$$

$$=\frac{\breve{\omega}}{2}\int_V \sum_{i=1}^3 \left[|E_i||P_i|\sin(\xi_i^E-\xi_i^P)+\mu_0|H_i||M_i|\sin(\xi_i^H-\xi_i^M)\right]\,dV +$$

$$+\frac{1}{2}\oint_S \sum_{i,j=1}^3 |E_i||H_j|(\hat{\mathbf{i}}_i\times\hat{\mathbf{i}}_j)\cdot\hat{\mathbf{n}}\cos(\xi_i^E-\xi_j^H)\,dS. \tag{4.142}$$

Here the power loss associated with the lag in alignment of the electric and magnetic dipoles is easily identified as the volume term on the right-hand side, and is seen to arise through the interaction of the fields with the equivalent sources as described through the phase difference between $\mathbf{E}$ and $\mathbf{P}$ and between $\mathbf{H}$ and $\mathbf{M}$. If these pairs are in phase, then the time-average power balance reduces to that for a dispersionless material, equation (4.146).

4.8.2 Poynting's theorem for nondispersive materials

For nondispersive materials (2.299) is appropriate. We shall carry out the details here so that we may examine the power-balance implications of nondispersive media. We have, substituting the field expressions,

$$-\frac{1}{2}\int_V \sum_{i=1}^{3} |J_i^i||E_i|C_{ii}^{J^i E}(t)\,dV = \frac{1}{2}\int_V \sum_{i=1}^{3} |J_i^c||E_i|C_{ii}^{J^c E}(t)\,dV +$$

$$+\frac{\partial}{\partial t}\int_V \sum_{i=1}^{3} \left\{\frac{1}{4}|D_i||E_i|C_{ii}^{DE}(t) + \frac{1}{4}|B_i||H_i|C_{ii}^{BH}(t)\right\}dV +$$

$$+\frac{1}{2}\oint_S \sum_{i,j=1}^{3} |E_i||H_j|(\hat{\mathbf{i}}_i \times \hat{\mathbf{i}}_j)\cdot\hat{\mathbf{n}}C_{ij}^{EH}(t)\,dS. \tag{4.143}$$

Here we remember that the conductivity relating $\mathbf{E}$ to $\mathbf{J}^c$ must also be nondispersive. Note that the electric and magnetic energy densities $w_e(\mathbf{r},t)$ and $w_m(\mathbf{r},t)$ have the time-average values $\langle w_e(\mathbf{r},t)\rangle$ and $\langle w_m(\mathbf{r},t)\rangle$ given by

$$\langle w_e(\mathbf{r},t)\rangle = \frac{1}{T}\int_{-T/2}^{T/2}\frac{1}{2}\mathbf{E}(\mathbf{r},t)\cdot\mathbf{D}(\mathbf{r},t)\,dt = \frac{1}{4}\sum_{i=1}^{3}|E_i||D_i|\cos(\xi_i^E - \xi_i^D)$$

$$= \frac{1}{4}\,\text{Re}\left\{\check{\mathbf{E}}(\mathbf{r})\cdot\check{\mathbf{D}}^*(\mathbf{r})\right\} \tag{4.144}$$

and

$$\langle w_m(\mathbf{r},t)\rangle = \frac{1}{T}\int_{-T/2}^{T/2}\frac{1}{2}\mathbf{B}(\mathbf{r},t)\cdot\mathbf{H}(\mathbf{r},t)\,dt = \frac{1}{4}\sum_{i=1}^{3}|B_i||H_i|\cos(\xi_i^H - \xi_i^B)$$

$$= \frac{1}{4}\,\text{Re}\left\{\check{\mathbf{H}}(\mathbf{r})\cdot\check{\mathbf{B}}^*(\mathbf{r})\right\}, \tag{4.145}$$

where $T = 2\pi/\check{\omega}$. We have already identified the energy stored in a nondispersive material (§ 4.5.2). If (4.144) is to match with (4.62), the phases of $\check{\mathbf{E}}$ and $\check{\mathbf{D}}$ must match: $\xi_i^E = \xi_i^D$. We must also have $\xi_i^H = \xi_i^B$. Since in a dispersionless material σ must be independent of frequency, from $\check{\mathbf{J}}^c = \sigma\check{\mathbf{E}}$ we also see that $\xi_i^{J^c} = \xi_i^E$.

Upon differentiation the time-average stored energy terms in (4.143) disappear, giving

$$-\frac{1}{2}\int_V \sum_{i=1}^{3} |J_i^i||E_i|C_{ii}^{J^i E}(t)\,dV = \frac{1}{2}\int_V \sum_{i=1}^{3} |J_i^c||E_i|C_{ii}^{EE}(t)\,dV -$$

$$-2\check{\omega}\int_V \sum_{i=1}^{3} \left\{\frac{1}{4}|D_i||E_i|S_{ii}^{EE}(t) + \frac{1}{4}|B_i||H_i|S_{ii}^{BB}(t)\right\}dV +$$

$$+\frac{1}{2}\oint_S \sum_{i,j=1}^{3} |E_i||H_j|(\hat{\mathbf{i}}_i \times \hat{\mathbf{i}}_j)\cdot\hat{\mathbf{n}}C_{ij}^{EH}(t)\,dS.$$

Equating the constant terms, we find the time-average power balance expression

$$-\frac{1}{2}\int_V \sum_{i=1}^{3} |J_i^i||E_i|\cos(\xi_i^{J^i} - \xi_i^E)\,dV = \frac{1}{2}\int_V \sum_{i=1}^{3} |J_i^c||E_i|\,dV +$$

$$+\frac{1}{2}\oint_S \sum_{i,j=1}^{3} |E_i||H_j|(\hat{\mathbf{i}}_i \times \hat{\mathbf{i}}_j)\cdot\hat{\mathbf{n}}\cos(\xi_i^E - \xi_j^H)\,dS. \tag{4.146}$$

This can be written more compactly using phasor notation as

$$\int_V p_J(\mathbf{r})\, dV = \int_V p_\sigma(\mathbf{r})\, dV + \oint_S \mathbf{S}_{av}(\mathbf{r}) \cdot \hat{\mathbf{n}}\, dS \qquad (4.147)$$

where

$$p_J(\mathbf{r}) = -\frac{1}{2}\,\mathrm{Re}\left\{\check{\mathbf{E}}(\mathbf{r}) \cdot \check{\mathbf{J}}^{i*}(\mathbf{r})\right\}$$

is the time-average density of power delivered by the sources to the fields in V,

$$p_\sigma(\mathbf{r}) = \frac{1}{2}\check{\mathbf{E}}(\mathbf{r}) \cdot \check{\mathbf{J}}^{c*}(\mathbf{r})$$

is the time-average density of power transferred to the conducting material as heat, and

$$\mathbf{S}_{av}(\mathbf{r}) \cdot \hat{\mathbf{n}} = \frac{1}{2}\,\mathrm{Re}\left\{\check{\mathbf{E}}(\mathbf{r}) \times \check{\mathbf{H}}^*(\mathbf{r})\right\} \cdot \hat{\mathbf{n}}$$

is the density of time-average power transferred across the boundary surface S. Here

$$\mathbf{S}^c = \check{\mathbf{E}}(\mathbf{r}) \times \check{\mathbf{H}}^*(\mathbf{r})$$

is called the *complex Poynting vector* and $\mathbf{S}_{av}$ is called the *time-average Poynting vector*.

Comparison of (4.146) with (4.140) shows that nondispersive materials cannot manifest the dissipative (or active) properties determined by the term

$$\frac{1}{2}\int_V \sum_{i=1}^{3}\left\{\check{\omega}|E_i||D_i|\sin(\xi_i^E - \xi_i^D) + \check{\omega}|B_i||H_i|\sin(\xi_i^H - \xi_i^B) + |J_i^c||E_i|\cos(\xi_i^{J^c} - \xi_i^E)\right\}\, dV.$$

This term can be used to classify materials as lossless, lossy, or active, as shown next.

4.8.3 Lossless, lossy, and active media

In § 4.5.1 we classified materials based on whether they dissipate (or provide) energy over the period of a transient event. We can provide the same classification based on their steady-state behavior.

We classify a material as *lossless* if the time-average flow of power entering a homogeneous body is zero when there are sources external to the body, but no sources internal to the body. This implies that the mechanisms within the body either do not dissipate power that enters, or that there is a mechanism that creates energy to exactly balance the dissipation. If the time-average power entering is positive, then the material dissipates power and is termed *lossy*. If the time-average power entering is negative, then power must originate from within the body and the material is termed *active*. (Note that the power associated with an active body is not described as arising from sources, but is rather described through the constitutive relations.)

Since materials are generally inhomogeneous we may apply this concept to a vanishingly small volume, thus invoking the point-form of Poynting's theorem. From (4.140) we see that the time-average influx of power density is given by

$$-\nabla \cdot \mathbf{S}_{av}(\mathbf{r}) = p_{in}(\mathbf{r}) = \frac{1}{2}\sum_{i=1}^{3}\left\{\check{\omega}|E_i||D_i|\sin(\xi_i^E - \xi_i^D) + \check{\omega}|B_i||H_i|\sin(\xi_i^H - \xi_i^B) + \right.$$
$$\left. + |J_i^c||E_i|\cos(\xi_i^{J^c} - \xi_i^E)\right\}.$$

Materials are then classified as follows:

$$p_{in}(\mathbf{r}) = 0, \qquad \text{lossless},$$
$$p_{in}(\mathbf{r}) > 0, \qquad \text{lossy},$$
$$p_{in}(\mathbf{r}) \geq 0, \qquad \text{passive},$$
$$p_{in}(\mathbf{r}) < 0, \qquad \text{active}.$$

We see that if $\xi_i^E = \xi_i^D$, $\xi_i^H = \xi_i^B$, and $\mathbf{J}^c = 0$, then the material is lossless. This implies that $(\mathbf{D},\mathbf{E})$ and $(\mathbf{B},\mathbf{H})$ are exactly in phase and there is no conduction current. If the material is isotropic, we may substitute from the constitutive relations (4.21)–(4.23) to obtain

$$p_{in}(\mathbf{r}) = -\frac{\breve{\omega}}{2} \sum_{i=1}^{3} \left\{ |E_i|^2 \left[|\tilde{\epsilon}| \sin(\xi^\epsilon) - \frac{|\tilde{\sigma}|}{\breve{\omega}} \cos(\xi^\sigma) \right] + |\tilde{\mu}||H_i|^2 \sin(\xi^\mu) \right\}. \qquad (4.148)$$

The first two terms can be regarded as resulting from a single complex permittivity (4.26). Then (4.148) simplifies to

$$p_{in}(\mathbf{r}) = -\frac{\breve{\omega}}{2} \sum_{i=1}^{3} \left\{ |\tilde{\epsilon}^c||E_i|^2 \sin(\xi^{\epsilon^c}) + |\tilde{\mu}||H_i|^2 \sin(\xi^\mu) \right\}. \qquad (4.149)$$

Now we can see that a lossless medium, which requires (4.149) to vanish, has $\xi^{\epsilon^c} = \xi^\mu = 0$ (or perhaps the unlikely condition that dissipative and active effects within the electric and magnetic terms exactly cancel). To have $\xi^\mu = 0$ we need $\mathbf{B}$ and $\mathbf{H}$ to be in phase, hence we need $\tilde{\mu}(\mathbf{r}, \omega)$ to be real. To have $\xi^{\epsilon^c} = 0$ we need $\xi^\epsilon = 0$ ($\tilde{\epsilon}(\mathbf{r}, \omega)$ real) and $\tilde{\sigma}(\mathbf{r}, \omega) = 0$ (or perhaps the unlikely condition that the active and dissipative effects of the permittivity and conductivity exactly cancel).

A lossy medium requires (4.149) to be positive. This occurs when $\xi^\mu < 0$ or $\xi^{\epsilon^c} < 0$, meaning that the imaginary part of the permeability or complex permittivity is negative. The complex permittivity has a negative imaginary part if the imaginary part of $\tilde{\epsilon}$ is negative or if the real part of $\tilde{\sigma}$ is positive. Physically, $\xi^\epsilon < 0$ means that $\xi^D < \xi^E$ and thus the phase of the response field $\mathbf{D}$ lags that of the excitation field $\mathbf{E}$. This results from a delay in the polarization alignment of the atoms, and leads to dissipation of power within the material.

An active medium requires (4.149) to be negative. This occurs when $\xi^\mu > 0$ or $\xi^{\epsilon^c} > 0$, meaning that the imaginary part of the permeability or complex permittivity is positive. The complex permittivity has a positive imaginary part if the imaginary part of $\tilde{\epsilon}$ is positive or if the real part of $\tilde{\sigma}$ is negative.

In summary, a passive isotropic medium is lossless when the permittivity and permeability are real and when the conductivity is zero. A passive isotropic medium is lossy when one or more of the following holds: the permittivity is complex with negative imaginary part, the permeability is complex with negative imaginary part, or the conductivity has a positive real part. Finally, a complex permittivity or permeability with positive imaginary part or a conductivity with negative real part indicates an *active* medium.

For anisotropic materials the interpretation of p_{in} is not as simple. Here we find that the permittivity or permeability dyadic may be complex, and yet the material may still be lossless. To determine the condition for a lossless medium, let us recompute p_{in} using the constitutive relations (4.18)–(4.20). With these we have

$$\mathbf{E} \cdot \left[\frac{\partial \mathbf{D}}{\partial t} + \mathbf{J}^c \right] + \mathbf{H} \cdot \frac{\partial \mathbf{B}}{\partial t} = \breve{\omega} \sum_{i,j=1}^{3} |E_i||E_j| \left[-|\tilde{\epsilon}_{ij}| \sin(\breve{\omega}t + \xi_j^E + \xi_{ij}^\epsilon) \cos(\breve{\omega}t + \xi_i^E) + \right.$$

$$+ \frac{|\tilde{\sigma}_{ij}|}{\breve{\omega}} \cos(\breve{\omega}t + \xi_j^E + \xi_{ij}^\sigma) \cos(\breve{\omega}t + \xi_i^E) \Big] +$$

$$+ \breve{\omega} \sum_{i,j=1}^{3} |H_i||H_j| \left[-|\tilde{\mu}_{ij}| \sin(\breve{\omega}t + \xi_j^H + \xi_{ij}^\mu) \cos(\breve{\omega}t + \xi_i^H) \right].$$

Using the angle-sum formulas and discarding the time-varying quantities, we may obtain the time-average input power density:

$$p_{in}(\mathbf{r}) = -\frac{\breve{\omega}}{2} \sum_{i,j=1}^{3} |E_i||E_j| \left[|\tilde{\epsilon}_{ij}| \sin(\xi_j^E - \xi_i^E + \xi_{ij}^\epsilon) - \frac{|\tilde{\sigma}_{ij}|}{\breve{\omega}} \cos(\xi_j^E - \xi_i^E + \xi_{ij}^\sigma) \right] -$$

$$- \frac{\breve{\omega}}{2} \sum_{i,j=1}^{3} |H_i||H_j||\tilde{\mu}_{ij}| \sin(\xi_j^H - \xi_i^H + \xi_{ij}^\mu).$$

The reader can easily verify that the conditions that make this quantity vanish, thus describing a lossless material, are

$$|\tilde{\epsilon}_{ij}| = |\tilde{\epsilon}_{ji}|, \qquad \xi_{ij}^\epsilon = -\xi_{ji}^\epsilon, \qquad\qquad (4.150)$$

$$|\tilde{\sigma}_{ij}| = |\tilde{\sigma}_{ji}|, \qquad \xi_{ij}^\sigma = -\xi_{ji}^\sigma + \pi, \qquad (4.151)$$

$$|\tilde{\mu}_{ij}| = |\tilde{\mu}_{ji}|, \qquad \xi_{ij}^\mu = -\xi_{ji}^\mu. \qquad\qquad (4.152)$$

Note that this requires $\xi_{ii}^\epsilon = \xi_{ii}^\mu = \xi_{ii}^\sigma = 0$.

The condition (4.152) is easily written in dyadic form as

$$\bar{\tilde{\mu}}(\mathbf{r}, \breve{\omega})^\dagger = \bar{\tilde{\mu}}(\mathbf{r}, \breve{\omega}) \qquad\qquad (4.153)$$

where "†" stands for the conjugate-transpose operation. The dyadic permeability $\bar{\tilde{\mu}}$ is hermitian. The set of conditions (4.150)–(4.151) can also be written quite simply using the complex permittivity dyadic (4.24):

$$\bar{\tilde{\epsilon}}^c(\mathbf{r}, \breve{\omega})^\dagger = \bar{\tilde{\epsilon}}^c(\mathbf{r}, \breve{\omega}). \qquad\qquad (4.154)$$

Thus, an anisotropic material is lossless when the both the dyadic permeability and the complex dyadic permittivity are hermitian. Since $\breve{\omega}$ is arbitrary, these results are exactly those obtained in § 4.5.1. Note that in the special case of an isotropic material the conditions (4.153) and (4.154) can only hold if $\tilde{\epsilon}$ and $\tilde{\mu}$ are real and $\tilde{\sigma}$ is zero, agreeing with our earlier conclusions.

4.9 The complex Poynting theorem

An equation having a striking resemblance to Poynting's theorem can be obtained by direct manipulation of the phasor-domain Maxwell equations. The result, although certainly satisfied by the phasor fields, does *not* replace Poynting's theorem as the power-balance equation for time-harmonic fields. We shall be careful to contrast the interpretation of the phasor expression with the actual time-harmonic Poynting theorem.

We begin by dotting both sides of the phasor-domain Faraday's law with $\breve{\mathbf{H}}^*$ to obtain

$$\breve{\mathbf{H}}^* \cdot (\nabla \times \breve{\mathbf{E}}) = -j\breve{\omega}\breve{\mathbf{H}}^* \cdot \breve{\mathbf{B}}.$$

Taking the complex conjugate of the phasor-domain Ampere's law and dotting with $\check{\mathbf{E}}$, we have

$$\check{\mathbf{E}} \cdot (\nabla \times \check{\mathbf{H}}^*) = \check{\mathbf{E}} \cdot \check{\mathbf{J}}^* - j\check{\omega}\check{\mathbf{E}} \cdot \check{\mathbf{D}}^*.$$

We subtract these expressions and use (B.44) to write

$$-\check{\mathbf{E}} \cdot \check{\mathbf{J}}^* = \nabla \cdot (\check{\mathbf{E}} \times \check{\mathbf{H}}^*) - j\check{\omega}[\check{\mathbf{E}} \cdot \check{\mathbf{D}}^* - \check{\mathbf{B}} \cdot \check{\mathbf{H}}^*].$$

Finally, integrating over the volume region V and dividing by two, we have

$$-\frac{1}{2} \int_V \check{\mathbf{E}} \cdot \check{\mathbf{J}}^* \, dV = \frac{1}{2} \oint_S (\check{\mathbf{E}} \times \check{\mathbf{H}}^*) \cdot \mathbf{dS} - 2j\check{\omega} \int_V \left[\frac{1}{4} \check{\mathbf{E}} \cdot \check{\mathbf{D}}^* - \frac{1}{4} \check{\mathbf{B}} \cdot \check{\mathbf{H}}^* \right] dV. \quad (4.155)$$

This is known as the *complex Poynting theorem*, and is an expression that must be obeyed by the phasor fields.

As a power balance theorem, the complex Poynting theorem has meaning only for dispersionless materials. If we let $\mathbf{J} = \mathbf{J}^i + \mathbf{J}^c$ and assume no dispersion, (4.155) becomes

$$-\frac{1}{2} \int_V \check{\mathbf{E}} \cdot \check{\mathbf{J}}^{i*} \, dV = \frac{1}{2} \int_V \check{\mathbf{E}} \cdot \check{\mathbf{J}}^{c*} \, dV + \frac{1}{2} \oint_S (\check{\mathbf{E}} \times \check{\mathbf{H}}^*) \cdot \mathbf{dS} - $$
$$- 2j\omega \int_V [\langle w_e \rangle - \langle w_m \rangle] \, dV \quad (4.156)$$

where $\langle w_e \rangle$ and $\langle w_m \rangle$ are the time-average stored electric and magnetic energy densities as described in (4.62)–(4.63). Selection of the real part now gives

$$-\frac{1}{2} \int_V \operatorname{Re} \left\{ \check{\mathbf{E}} \cdot \check{\mathbf{J}}^{i*} \right\} dV = \frac{1}{2} \int_V \check{\mathbf{E}} \cdot \check{\mathbf{J}}^{c*} \, dV + \frac{1}{2} \oint_S \operatorname{Re} \left\{ \check{\mathbf{E}} \times \check{\mathbf{H}}^* \right\} \cdot \mathbf{dS}, \quad (4.157)$$

which is identical to (4.147). Thus the real part of the complex Poynting theorem gives the balance of time-average power for a dispersionless material.

Selection of the imaginary part of (4.156) gives the balance of imaginary, or *reactive* power:

$$-\frac{1}{2} \int_V \operatorname{Im} \left\{ \check{\mathbf{E}} \cdot \check{\mathbf{J}}^{i*} \right\} dV = \frac{1}{2} \oint_S \operatorname{Im} \left\{ \check{\mathbf{E}} \times \check{\mathbf{H}}^* \right\} \cdot \mathbf{dS} - 2\check{\omega} \int_V [\langle w_e \rangle - \langle w_m \rangle] \, dV. \quad (4.158)$$

In general, the reactive power balance does not have a simple physical interpretation (it is *not* the balance of the oscillating terms in (4.139)). However, an interesting concept can be gleaned from it. If the source current and electric field are in phase, and there is no reactive power leaving S, then the time-average stored electric energy is equal to the time-average stored magnetic energy:

$$\int_V \langle w_e \rangle \, dV = \int_V \langle w_m \rangle \, dV.$$

This is the condition for "resonance." An example is a series RLC circuit with the source current and voltage in phase. Here the stored energy in the capacitor is equal to the stored energy in the inductor and the input impedance (ratio of voltage to current) is real. Such a resonance occurs at only one value of frequency. In more complicated electromagnetic systems resonance may occur at many discrete eigenfrequencies.

4.9.1 Boundary condition for the time-average Poynting vector

In § 2.9.5 we developed a boundary condition for the normal component of the time-domain Poynting vector. For time-harmonic fields we can derive a similar boundary condition using the time-average Poynting vector. Consider a surface S across which the electromagnetic sources and constitutive parameters are discontinuous, as shown in Figure 2.6. Let $\hat{\mathbf{n}}_{12}$ be the unit normal to the surface pointing into region 1 from region 2. If we apply the large-scale form of the complex Poynting theorem (4.155) to the two separate surfaces shown in Figure 2.6, we obtain

$$\frac{1}{2}\int_V \left[\check{\mathbf{E}} \cdot \check{\mathbf{J}}^* - 2j\check{\omega}\left(\frac{1}{4}\check{\mathbf{E}} \cdot \check{\mathbf{D}}^* - \frac{1}{4}\check{\mathbf{B}} \cdot \check{\mathbf{H}}^*\right)\right] dV + \frac{1}{2}\oint_S \mathbf{S}^c \cdot \hat{\mathbf{n}} \, dS$$
$$= \frac{1}{2}\int_{S_{10}} \hat{\mathbf{n}}_{12} \cdot (\mathbf{S}_1^c - \mathbf{S}_2^c) \, dS \qquad (4.159)$$

where $\mathbf{S}^c = \check{\mathbf{E}} \times \check{\mathbf{H}}^*$ is the complex Poynting vector. If, on the other hand, we apply the large-scale form of Poynting's theorem to the entire volume region including the surface of discontinuity, and include the surface current contribution, we have

$$\frac{1}{2}\int_V \left[\check{\mathbf{E}} \cdot \check{\mathbf{J}}^* - 2j\check{\omega}\int_V \left(\frac{1}{4}\check{\mathbf{E}} \cdot \check{\mathbf{D}}^* - \frac{1}{4}\check{\mathbf{B}} \cdot \check{\mathbf{H}}^*\right)\right] dV + \frac{1}{2}\oint_S \mathbf{S}^c \cdot \hat{\mathbf{n}} \, dS$$
$$= -\frac{1}{2}\int_{S_{10}} \check{\mathbf{J}}_s^* \cdot \check{\mathbf{E}} \, dS. \qquad (4.160)$$

If we wish to have the integrals over V and S in (4.159) and (4.160) produce identical results, then we must postulate the two conditions

$$\hat{\mathbf{n}}_{12} \times (\check{\mathbf{E}}_1 - \check{\mathbf{E}}_2) = 0$$

and

$$\hat{\mathbf{n}}_{12} \cdot (\mathbf{S}_1^c - \mathbf{S}_2^c) = -\check{\mathbf{J}}_s^* \cdot \check{\mathbf{E}}. \qquad (4.161)$$

The first condition is merely the continuity of tangential electric field; it allows us to be nonspecific as to which value of $\mathbf{E}$ we use in the second condition. If we take the real part of the second condition we have

$$\hat{\mathbf{n}}_{12} \cdot (\mathbf{S}_{av,1} - \mathbf{S}_{av,2}) = p_{Js}, \qquad (4.162)$$

where $\mathbf{S}_{av} = \frac{1}{2}\text{Re}\{\check{\mathbf{E}} \times \check{\mathbf{H}}^*\}$ is the time-average Poynting power flow density and $p_{Js} = -\frac{1}{2}\text{Re}\{\check{\mathbf{J}}_s^* \cdot \check{\mathbf{E}}\}$ is the time-average density of power delivered by the surface sources. This is the desired boundary condition on the time-average power flow density.

4.10 Fundamental theorems for time-harmonic fields

4.10.1 Uniqueness

If we think of a sinusoidal electromagnetic field as the steady-state culmination of a transient event that has an identifiable starting time, then the conditions for uniqueness established in § 2.2.1 are applicable. However, a true time-harmonic wave, which has existed since $t = -\infty$ and thus has infinite energy, must be interpreted differently.

Our approach is similar to that of § 2.2.1. Consider a simply-connected region of space V bounded by surface S, where both V and S contain only ordinary points. The phasor-domain fields within V are associated with a phasor current distribution $\check{\mathbf{J}}$, which may be internal to V (entirely or in part). We seek conditions under which the phasor electromagnetic fields are uniquely determined. Let the field set $(\check{\mathbf{E}}_1, \check{\mathbf{D}}_1, \check{\mathbf{B}}_1, \check{\mathbf{H}}_1)$ satisfy Maxwell's equations (4.128) and (4.129) associated with the current $\check{\mathbf{J}}$ (along with an appropriate set of constitutive relations), and let $(\check{\mathbf{E}}_2, \check{\mathbf{D}}_2, \check{\mathbf{B}}_2, \check{\mathbf{H}}_2)$ be a second solution. To determine the conditions for uniqueness of the fields, we look for a situation that results in $\check{\mathbf{E}}_1 = \check{\mathbf{E}}_2$, $\check{\mathbf{H}}_1 = \check{\mathbf{H}}_2$, and so on. The electromagnetic fields must obey

$$\nabla \times \check{\mathbf{H}}_1 = j\breve{\omega}\check{\mathbf{D}}_1 + \check{\mathbf{J}},$$
$$\nabla \times \check{\mathbf{E}}_1 = -j\breve{\omega}\check{\mathbf{B}}_1,$$
$$\nabla \times \check{\mathbf{H}}_2 = j\breve{\omega}\check{\mathbf{D}}_2 + \check{\mathbf{J}},$$
$$\nabla \times \check{\mathbf{E}}_2 = -j\breve{\omega}\check{\mathbf{B}}_2.$$

Subtracting these and defining the difference fields $\check{\mathbf{E}}_0 = \check{\mathbf{E}}_1 - \check{\mathbf{E}}_2$, $\check{\mathbf{H}}_0 = \check{\mathbf{H}}_1 - \check{\mathbf{H}}_2$, and so on, we find that

$$\nabla \times \check{\mathbf{H}}_0 = j\breve{\omega}\check{\mathbf{D}}_0, \tag{4.163}$$
$$\nabla \times \check{\mathbf{E}}_0 = -j\breve{\omega}\check{\mathbf{B}}_0. \tag{4.164}$$

Establishing the conditions under which the difference fields vanish throughout V, we shall determine the conditions for uniqueness.

Dotting (4.164) by $\check{\mathbf{H}}_0^*$ and dotting the complex conjugate of (4.163) by $\check{\mathbf{E}}_0$, we have

$$\check{\mathbf{H}}_0^* \cdot \left(\nabla \times \check{\mathbf{E}}_0 \right) = -j\breve{\omega}\check{\mathbf{B}}_0 \cdot \check{\mathbf{H}}_0^*,$$
$$\check{\mathbf{E}}_0 \cdot \left(\nabla \times \check{\mathbf{H}}_0^* \right) = -j\breve{\omega}\check{\mathbf{D}}_0^* \cdot \check{\mathbf{E}}_0.$$

Subtraction yields

$$\check{\mathbf{H}}_0^* \cdot \left(\nabla \times \check{\mathbf{E}}_0 \right) - \check{\mathbf{E}}_0 \cdot \left(\nabla \times \check{\mathbf{H}}_0^* \right) = -j\breve{\omega}\check{\mathbf{B}}_0 \cdot \check{\mathbf{H}}_0^* + j\breve{\omega}\check{\mathbf{D}}_0^* \cdot \check{\mathbf{E}}_0$$

which, by (B.44), can be written as

$$\nabla \cdot \left(\check{\mathbf{E}}_0 \times \check{\mathbf{H}}_0^* \right) = j\breve{\omega} \left[\check{\mathbf{E}}_0 \cdot \check{\mathbf{D}}_0^* - \check{\mathbf{B}}_0 \cdot \check{\mathbf{H}}_0^* \right].$$

Adding this expression to its complex conjugate, integrating over V, and using the divergence theorem, we obtain

$$\text{Re} \oint_S \left[\check{\mathbf{E}}_0 \times \check{\mathbf{H}}_0^* \right] \cdot d\mathbf{S} = -j\frac{\breve{\omega}}{2} \int_V \left[\left(\check{\mathbf{E}}_0^* \cdot \check{\mathbf{D}}_0 - \check{\mathbf{E}}_0 \cdot \check{\mathbf{D}}_0^* \right) + \left(\check{\mathbf{H}}_0^* \cdot \check{\mathbf{B}}_0 - \check{\mathbf{H}}_0 \cdot \check{\mathbf{B}}_0^* \right) \right] dV.$$

Breaking S into two arbitrary portions and using (B.6), we obtain

$$\text{Re} \oint_{S_1} \check{\mathbf{H}}_0^* \cdot (\hat{\mathbf{n}} \times \check{\mathbf{E}}_0) \, dS - \text{Re} \oint_{S_2} \check{\mathbf{E}}_0 \cdot (\hat{\mathbf{n}} \times \check{\mathbf{H}}_0^*) \, dS =$$
$$-j\frac{\breve{\omega}}{2} \int_V \left[\left(\check{\mathbf{E}}_0^* \cdot \check{\mathbf{D}}_0 - \check{\mathbf{E}}_0 \cdot \check{\mathbf{D}}_0^* \right) + \left(\check{\mathbf{H}}_0^* \cdot \check{\mathbf{B}}_0 - \check{\mathbf{H}}_0 \cdot \check{\mathbf{B}}_0^* \right) \right] dV. \tag{4.165}$$

Now if $\hat{\mathbf{n}} \times \mathbf{E}_0 = 0$ or $\hat{\mathbf{n}} \times \mathbf{H}_0 = 0$ over all of S, or some combination of these conditions holds over all of S, then

$$\int_V \left[\left(\check{\mathbf{E}}_0^* \cdot \check{\mathbf{D}}_0 - \check{\mathbf{E}}_0 \cdot \check{\mathbf{D}}_0^* \right) + \left(\check{\mathbf{H}}_0^* \cdot \check{\mathbf{B}}_0 - \check{\mathbf{H}}_0 \cdot \check{\mathbf{B}}_0^* \right) \right] dV = 0. \tag{4.166}$$

This implies a relationship between $\check{\mathbf{E}}_0$, $\check{\mathbf{D}}_0$, $\check{\mathbf{B}}_0$, and $\check{\mathbf{H}}_0$. Since V is arbitrary we see that one possible relationship is simply to have one of each pair $(\check{\mathbf{E}}_0, \check{\mathbf{D}}_0)$ and $(\check{\mathbf{H}}_0, \check{\mathbf{B}}_0)$ equal to zero. Then, by (4.163) and (4.164), $\check{\mathbf{E}}_0 = 0$ implies $\check{\mathbf{B}}_0 = 0$, and $\check{\mathbf{D}}_0 = 0$ implies $\check{\mathbf{H}}_0 = 0$. Thus $\check{\mathbf{E}}_1 = \check{\mathbf{E}}_2$, etc., and the solution is unique throughout V. However, we cannot in general rule out more complicated relationships. The number of possibilities depends on the additional constraints on the relationship between $\check{\mathbf{E}}_0$, $\check{\mathbf{D}}_0$, $\check{\mathbf{B}}_0$, and $\check{\mathbf{H}}_0$ that we must supply to describe the material supporting the field — i.e., the constitutive relationships. For a simple medium described by $\tilde{\mu}(\omega)$ and $\tilde{\epsilon}^c(\omega)$, equation (4.166) becomes

$$\int_V \left(|\check{\mathbf{E}}_0|^2 [\tilde{\epsilon}^c(\check{\omega}) - \tilde{\epsilon}^{c*}(\check{\omega})] + |\check{\mathbf{H}}_0|^2 [\tilde{\mu}(\check{\omega}) - \tilde{\mu}^*(\check{\omega})] \right) dV = 0$$

or

$$\int_V \left[|\check{\mathbf{E}}_0|^2 \tilde{\epsilon}^{c\prime\prime}(\check{\omega}) + |\check{\mathbf{H}}_0|^2 \tilde{\mu}^{\prime\prime}(\check{\omega}) \right] dV = 0.$$

For a lossy medium, $\tilde{\epsilon}^{c\prime\prime} < 0$ and $\tilde{\mu}^{\prime\prime} < 0$ as shown in § 4.5.1. So both terms in the integral must be negative. For the integral to be zero each term must vanish, requiring $\check{\mathbf{E}}_0 = \check{\mathbf{H}}_0 = 0$, and uniqueness is guaranteed.

When establishing more complicated constitutive relations we must be careful to ensure that they lead to a unique solution, and that the condition for uniqueness is understood. In the case above, the assumption $\hat{\mathbf{n}} \times \check{\mathbf{E}}_0 \big|_S = 0$ implies that the tangential components of $\check{\mathbf{E}}_1$ and $\check{\mathbf{E}}_2$ are identical over S — that is, we must give specific values of these quantities on S to ensure uniqueness. A similar statement holds for the condition $\hat{\mathbf{n}} \times \check{\mathbf{H}}_0 \big|_S = 0$.

In summary, the conditions for the fields within a region V containing lossy isotropic materials to be unique are as follows:

1. the sources within V must be specified;

2. the tangential component of the electric field must be specified over all or part of the bounding surface S;

3. the tangential component of the magnetic field must be specified over the remainder of S.

We may question the requirement of a *lossy* medium to demonstrate uniqueness of the phasor fields. Does this mean that within a vacuum the specification of tangential fields is insufficient? Experience shows that the fields in such a region are indeed properly described by the surface fields, and it is just a case of the mathematical model being slightly out of sync with the physics. As long as we recognize that the sinusoidal steady state requires an initial transient period, we know that specification of the tangential fields is sufficient. We must be careful, however, to understand the restrictions of the mathematical model. Any attempt to describe the fields within a lossless cavity, for instance, is fraught with difficulty if true time-harmonic fields are used to model the actual physical fields. A helpful mathematical strategy is to think of free space as the limit of a lossy medium as the loss recedes to zero. Of course this does not represent the physical state of "empty" space. Although even interstellar space may have a few particles for every cubic meter to interact with the electromagnetic field, the density of these particles invalidates our initial macroscopic assumptions.

Another important concern is whether we can extend the uniqueness argument to all of space. If we let S recede to infinity, must we continue to specify the fields over S, or is it sufficient to merely specify the sources within S? Since the boundary fields provide information to the internal region about sources that exist outside S, it is sensible to assume that as $S \to \infty$ there are no sources external to S and thus no need for the boundary fields. This is indeed the case. If all sources are localized, the fields they produce behave in just the right manner for the surface integral in (4.165) to vanish, and thus uniqueness is again guaranteed. Later we will find that the electric and magnetic fields produced by a localized source at great distance have the form of a spherical wave:

$$\check{\mathbf{E}} \sim \check{\mathbf{H}} \sim \frac{e^{-jkr}}{r}.$$

If space is taken to be slightly lossy, then k is complex with negative imaginary part, and thus the fields decrease exponentially with distance from the source. As we argued above, it may not be physically meaningful to assume that space is lossy. Sommerfeld postulated that even for lossless space the surface integral in (4.165) vanishes as $S \to \infty$. This has been verified experimentally, and provides the following restrictions on the free-space fields known as the *Sommerfeld radiation condition*:

$$\lim_{r \to \infty} r \left[\eta_0 \hat{\mathbf{r}} \times \check{\mathbf{H}}(\mathbf{r}) + \check{\mathbf{E}}(\mathbf{r}) \right] = 0, \tag{4.167}$$

$$\lim_{r \to \infty} r \left[\hat{\mathbf{r}} \times \check{\mathbf{E}}(\mathbf{r}) - \eta_0 \check{\mathbf{H}}(\mathbf{r}) \right] = 0, \tag{4.168}$$

where $\eta_0 = (\mu_0/\epsilon_0)^{1/2}$. Later we shall see how these expressions arise from the integral solutions to Maxwell's equations.

4.10.2 Reciprocity revisited

In § 2.9.3 we discussed the basic concept of reciprocity, but were unable to examine its real potential since we had not yet developed the theory of time-harmonic fields. In this section we shall apply the reciprocity concept to time-harmonic sources and fields, and investigate the properties a material must display to be reciprocal.

The general form of the reciprocity theorem. As in § 2.9.3, we consider a closed surface S enclosing a volume V. Sources of an electromagnetic field are located either inside or outside S. Material media may lie within S, and their properties are described in terms of the constitutive relations. To obtain the time-harmonic (phasor) form of the reciprocity theorem we proceed as in § 2.9.3 but begin with the phasor forms of Maxwell's equations. We find

$$\nabla \cdot (\check{\mathbf{E}}_a \times \check{\mathbf{H}}_b - \check{\mathbf{E}}_b \times \check{\mathbf{H}}_a) = j\check{\omega}[\check{\mathbf{H}}_a \cdot \check{\mathbf{B}}_b - \check{\mathbf{H}}_b \cdot \check{\mathbf{B}}_a] - j\check{\omega}[\check{\mathbf{E}}_a \cdot \check{\mathbf{D}}_b - \check{\mathbf{E}}_b \cdot \check{\mathbf{D}}_a] +$$
$$+ [\check{\mathbf{E}}_b \cdot \check{\mathbf{J}}_a - \check{\mathbf{E}}_a \cdot \check{\mathbf{J}}_b - \check{\mathbf{H}}_b \cdot \check{\mathbf{J}}_{ma} + \check{\mathbf{H}}_a \cdot \check{\mathbf{J}}_{mb}], \tag{4.169}$$

where $(\check{\mathbf{E}}_a, \check{\mathbf{D}}_a, \check{\mathbf{B}}_a, \check{\mathbf{H}}_a)$ are the fields produced by the phasor sources $(\check{\mathbf{J}}_a, \check{\mathbf{J}}_{ma})$ and $(\check{\mathbf{E}}_b, \check{\mathbf{D}}_b, \check{\mathbf{B}}_b, \check{\mathbf{H}}_b)$ are the fields produced by an independent set of sources $(\check{\mathbf{J}}_b, \check{\mathbf{J}}_{mb})$.

As in § 2.9.3, we are interested in the case in which the first two terms on the right-hand side of (4.169) are zero. To see the conditions under which this might occur, we substitute the constitutive equations for a bianisotropic medium

$$\check{\mathbf{D}} = \bar{\bar{\xi}} \cdot \check{\mathbf{H}} + \bar{\bar{\epsilon}} \cdot \check{\mathbf{E}},$$
$$\check{\mathbf{B}} = \bar{\bar{\mu}} \cdot \check{\mathbf{H}} + \bar{\bar{\zeta}} \cdot \check{\mathbf{E}},$$

into (4.169), where each of the constitutive parameters is evaluated at $\check{\omega}$. Setting the two terms to zero gives

$$j\check{\omega} \left[\check{\mathbf{H}}_a \cdot \left(\bar{\bar{\mu}} \cdot \check{\mathbf{H}}_b + \bar{\bar{\zeta}} \cdot \check{\mathbf{E}}_b \right) - \check{\mathbf{H}}_b \cdot \left(\bar{\bar{\mu}} \cdot \check{\mathbf{H}}_a + \bar{\bar{\zeta}} \cdot \check{\mathbf{E}}_a \right) \right] -$$

$$-j\check{\omega}\left[\check{\mathbf{E}}_a \cdot \left(\bar{\bar{\boldsymbol{\xi}}}\cdot\check{\mathbf{H}}_b + \bar{\bar{\boldsymbol{\epsilon}}}\cdot\check{\mathbf{E}}_b\right) - \check{\mathbf{E}}_b \cdot \left(\bar{\bar{\boldsymbol{\xi}}}\cdot\check{\mathbf{H}}_a + \bar{\bar{\boldsymbol{\epsilon}}}\cdot\check{\mathbf{E}}_a\right)\right] = 0,$$

which holds if

$$\check{\mathbf{H}}_a \cdot \bar{\bar{\boldsymbol{\mu}}}\cdot\check{\mathbf{H}}_b - \check{\mathbf{H}}_b \cdot \bar{\bar{\boldsymbol{\mu}}}\cdot\check{\mathbf{H}}_a = 0,$$
$$\check{\mathbf{H}}_a \cdot \bar{\bar{\boldsymbol{\zeta}}}\cdot\check{\mathbf{E}}_b + \check{\mathbf{E}}_b \cdot \bar{\bar{\boldsymbol{\xi}}}\cdot\check{\mathbf{H}}_a = 0,$$
$$\check{\mathbf{E}}_a \cdot \bar{\bar{\boldsymbol{\xi}}}\cdot\check{\mathbf{H}}_b + \check{\mathbf{H}}_b \cdot \bar{\bar{\boldsymbol{\zeta}}}\cdot\check{\mathbf{E}}_a = 0,$$
$$\check{\mathbf{E}}_a \cdot \bar{\bar{\boldsymbol{\epsilon}}}\cdot\check{\mathbf{E}}_b - \check{\mathbf{E}}_b \cdot \bar{\bar{\boldsymbol{\epsilon}}}\cdot\check{\mathbf{E}}_a = 0.$$

These in turn hold if

$$\bar{\bar{\boldsymbol{\epsilon}}} = \bar{\bar{\boldsymbol{\epsilon}}}^T, \qquad \bar{\bar{\boldsymbol{\mu}}} = \bar{\bar{\boldsymbol{\mu}}}^T, \qquad \bar{\bar{\boldsymbol{\xi}}} = -\bar{\bar{\boldsymbol{\zeta}}}^T, \qquad \bar{\bar{\boldsymbol{\zeta}}} = -\bar{\bar{\boldsymbol{\xi}}}^T. \qquad (4.170)$$

These are the conditions for a *reciprocal medium*. For example, an anisotropic dielectric is a reciprocal medium if its permittivity dyadic is symmetric. An isotropic medium described by scalar quantities μ and ϵ is certainly reciprocal. In contrast, lossless Gyrotropic media are nonreciprocal since the constitutive parameters obey $\bar{\bar{\boldsymbol{\epsilon}}} = \bar{\bar{\boldsymbol{\epsilon}}}^\dagger$ or $\bar{\bar{\boldsymbol{\mu}}} = \bar{\bar{\boldsymbol{\mu}}}^\dagger$ rather than $\bar{\bar{\boldsymbol{\epsilon}}} = \bar{\bar{\boldsymbol{\epsilon}}}^T$ or $\bar{\bar{\boldsymbol{\mu}}} = \bar{\bar{\boldsymbol{\mu}}}^T$.

For a reciprocal medium (4.169) reduces to

$$\nabla \cdot (\check{\mathbf{E}}_a \times \check{\mathbf{H}}_b - \check{\mathbf{E}}_b \times \check{\mathbf{H}}_a) = \left[\check{\mathbf{E}}_b \cdot \check{\mathbf{J}}_a - \check{\mathbf{E}}_a \cdot \check{\mathbf{J}}_b - \check{\mathbf{H}}_b \cdot \check{\mathbf{J}}_{ma} + \check{\mathbf{H}}_a \cdot \check{\mathbf{J}}_{mb}\right]. \qquad (4.171)$$

At points where the sources are zero, or are conduction currents described entirely by Ohm's law $\check{\mathbf{J}} = \sigma\check{\mathbf{E}}$, we have

$$\nabla \cdot (\check{\mathbf{E}}_a \times \check{\mathbf{H}}_b - \check{\mathbf{E}}_b \times \check{\mathbf{H}}_a) = 0, \qquad (4.172)$$

known as *Lorentz's lemma*. If we integrate (4.171) over V and use the divergence theorem we obtain

$$\oint_S \left[\check{\mathbf{E}}_a \times \check{\mathbf{H}}_b - \check{\mathbf{E}}_b \times \check{\mathbf{H}}_a\right] \cdot d\mathbf{S} = \int_V \left[\check{\mathbf{E}}_b \cdot \check{\mathbf{J}}_a - \check{\mathbf{E}}_a \cdot \check{\mathbf{J}}_b - \check{\mathbf{H}}_b \cdot \check{\mathbf{J}}_{ma} + \check{\mathbf{H}}_a \cdot \check{\mathbf{J}}_{mb}\right] dV.$$
$$(4.173)$$

This is the general form of the *Lorentz reciprocity theorem*, and is valid when V contains reciprocal media as defined in (4.170).

Note that by an identical set of steps we find that the frequency-domain fields obey an identical Lorentz lemma and reciprocity theorem.

The condition for reciprocal systems. The quantity

$$\langle \check{\mathbf{f}}_a, \check{\mathbf{g}}_b \rangle = \int_V \left[\check{\mathbf{E}}_a \cdot \check{\mathbf{J}}_b - \check{\mathbf{H}}_a \cdot \check{\mathbf{J}}_{mb}\right] dV$$

is called the *reaction* between the source fields $\check{\mathbf{g}}$ of set b and the mediating fields $\check{\mathbf{f}}$ of an independent set a. Note that $\check{\mathbf{E}}_a \cdot \check{\mathbf{J}}_b$ is not quite a power density, since the current lacks a complex conjugate. Using this reaction concept, first introduced by Rumsey [161], we can write (4.173) as

$$\langle \check{\mathbf{f}}_b, \check{\mathbf{g}}_a \rangle - \langle \check{\mathbf{f}}_a, \check{\mathbf{g}}_b \rangle = \oint_S \left[\check{\mathbf{E}}_a \times \check{\mathbf{H}}_b - \check{\mathbf{E}}_b \times \check{\mathbf{H}}_a\right] \cdot d\mathbf{S}. \qquad (4.174)$$

We see that if there are no sources within S then

$$\oint_S \left[\check{\mathbf{E}}_a \times \check{\mathbf{H}}_b - \check{\mathbf{E}}_b \times \check{\mathbf{H}}_a \right] \cdot d\mathbf{S} = 0. \tag{4.175}$$

Whenever (4.175) holds we say that the "system" within S is *reciprocal*. Thus, for instance, a region of empty space is a reciprocal system.

A system need not be source-free in order for (4.175) to hold. Suppose the relationship between $\check{\mathbf{E}}$ and $\check{\mathbf{H}}$ on S is given by the *impedance boundary condition*

$$\check{\mathbf{E}}_t = -Z(\hat{\mathbf{n}} \times \check{\mathbf{H}}), \tag{4.176}$$

where $\check{\mathbf{E}}_t$ is the component of $\check{\mathbf{E}}$ tangential to S so that $\hat{\mathbf{n}} \times \mathbf{E} = \hat{\mathbf{n}} \times \mathbf{E}_t$, and the complex *wall impedance* Z may depend on position. By (4.176) we can write

$$(\check{\mathbf{E}}_a \times \check{\mathbf{H}}_b - \check{\mathbf{E}}_b \times \check{\mathbf{H}}_a) \cdot \hat{\mathbf{n}} = \check{\mathbf{H}}_b \cdot (\hat{\mathbf{n}} \times \check{\mathbf{E}}_a) - \check{\mathbf{H}}_a \cdot (\hat{\mathbf{n}} \times \check{\mathbf{E}}_b)$$
$$= -Z\check{\mathbf{H}}_b \cdot [\hat{\mathbf{n}} \times (\hat{\mathbf{n}} \times \check{\mathbf{H}}_a)] + Z\check{\mathbf{H}}_a \cdot [\hat{\mathbf{n}} \times (\hat{\mathbf{n}} \times \check{\mathbf{H}}_b)].$$

Since $\hat{\mathbf{n}} \times (\hat{\mathbf{n}} \times \check{\mathbf{H}}) = \hat{\mathbf{n}}(\hat{\mathbf{n}} \cdot \check{\mathbf{H}}) - \check{\mathbf{H}}$, the right-hand side vanishes. Hence (4.175) still holds even though there are sources within S.

The reaction theorem.   When sources lie within the surface S, and the fields on S obey (4.176), we obtain an important corollary of the Lorentz reciprocity theorem. We have from (4.174) the additional result

$$\langle \check{\mathbf{f}}_a, \check{\mathbf{g}}_b \rangle - \langle \check{\mathbf{f}}_b, \check{\mathbf{g}}_a \rangle = 0.$$

Hence a reciprocal system has

$$\langle \check{\mathbf{f}}_a, \check{\mathbf{g}}_b \rangle = \langle \check{\mathbf{f}}_b, \check{\mathbf{g}}_a \rangle \tag{4.177}$$

(which holds even if there are no sources within S, since then the reactions would be identically zero). This condition for reciprocity is sometimes called the *reaction theorem* and has an important physical meaning which we shall explore below in the form of the Rayleigh–Carson reciprocity theorem. Note that in obtaining this relation we must assume that the medium is reciprocal in order to eliminate the terms in (4.169). Thus, in order for a system to be reciprocal, it must involve *both* a reciprocal medium and a boundary over which (4.176) holds.

It is important to note that the impedance boundary condition (4.176) is widely applicable. If $Z \to 0$, then the boundary condition is that for a PEC: $\hat{\mathbf{n}} \times \check{\mathbf{E}} = 0$. If $Z \to \infty$, a PMC is described: $\hat{\mathbf{n}} \times \check{\mathbf{H}} = 0$. Suppose S represents a sphere of infinite radius. We know from (4.168) that if the sources and material media within S are spatially finite, the fields far removed from these sources are described by the Sommerfeld radiation condition

$$\hat{\mathbf{r}} \times \check{\mathbf{E}} = \eta_0 \check{\mathbf{H}}$$

where $\hat{\mathbf{r}}$ is the radial unit vector of spherical coordinates. This condition is of the type (4.176) since $\hat{\mathbf{r}} = \hat{\mathbf{n}}$ on S, hence the unbounded region that results from S receding to infinity is also reciprocal.

Summary of reciprocity for reciprocal systems.   We can summarize reciprocity as follows. Unbounded space containing sources and materials of finite size is a reciprocal system if the media are reciprocal; a bounded region of space is a reciprocal system only if the materials within are reciprocal and the boundary fields obey (4.176), or if the region is source-free. In each of these cases

$$\oint_S \left[\check{\mathbf{E}}_a \times \check{\mathbf{H}}_b - \check{\mathbf{E}}_b \times \check{\mathbf{H}}_a \right] \cdot d\mathbf{S} = 0 \qquad (4.178)$$

and

$$\langle \check{\mathbf{f}}_a, \check{\mathbf{g}}_b \rangle - \langle \check{\mathbf{f}}_b, \check{\mathbf{g}}_a \rangle = 0. \qquad (4.179)$$

Rayleigh–Carson reciprocity theorem. The physical meaning behind reciprocity can be made clear with a simple example. Consider two electric Hertzian dipoles, each oscillating with frequency $\check{\omega}$ and located within an empty box consisting of PEC walls. These dipoles can be described in terms of volume current density as

$$\check{\mathbf{J}}_a(\mathbf{r}) = \check{\mathbf{I}}_a \delta(\mathbf{r} - \mathbf{r}'_a),$$
$$\check{\mathbf{J}}_b(\mathbf{r}) = \check{\mathbf{I}}_b \delta(\mathbf{r} - \mathbf{r}'_b).$$

Since the fields on the surface obey (4.176) (specifically, $\hat{\mathbf{n}} \times \check{\mathbf{E}} = 0$), and since the medium within the box is empty space (a reciprocal medium), the fields produced by the sources must obey (4.179). We have

$$\int_V \check{\mathbf{E}}_b(\mathbf{r}) \cdot \left[\check{\mathbf{I}}_a \delta(\mathbf{r} - \mathbf{r}'_a) \right] \, dV = \int_V \check{\mathbf{E}}_a(\mathbf{r}) \cdot \left[\check{\mathbf{I}}_b \delta(\mathbf{r} - \mathbf{r}'_b) \right] \, dV,$$

hence

$$\check{\mathbf{I}}_a \cdot \check{\mathbf{E}}_b(\mathbf{r}'_a) = \check{\mathbf{I}}_b \cdot \check{\mathbf{E}}_a(\mathbf{r}'_b). \qquad (4.180)$$

This is the *Rayleigh–Carson reciprocity theorem*. It also holds for two Hertzian dipoles located in unbounded free space, because in that case the Sommerfeld radiation condition satisfies (4.176).

As an important application of this principle, consider a closed PEC body located in free space. Reciprocity holds in the region external to the body since we have $\hat{\mathbf{n}} \times \check{\mathbf{E}} = 0$ at the boundary of the perfect conductor and the Sommerfeld radiation condition on the boundary at infinity. Now let us place dipole a somewhere external to the body, and dipole b adjacent and tangential to the perfectly conducting body. We regard dipole a as the source of an electromagnetic field and dipole b as "sampling" that field. Since the tangential electric field is zero at the surface of the conductor, the reaction between the two dipoles is zero. Now let us switch the roles of the dipoles so that b is regarded as the source and a is regarded as the sampler. By reciprocity the reaction is again zero and thus there is no field produced by b at the position of a. Now the position and orientation of a are arbitrary, so we conclude that an impressed electric source current placed tangentially to a perfectly conducting body produces no field external to the body. This result is used in Chapter 6 to develop a field equivalence principle useful in the study of antennas and scattering.

4.10.3 Duality

A duality principle analogous to that found for time-domain fields in § 2.9.2 may be established for frequency-domain and time-harmonic fields. Consider a closed surface S enclosing a region of space that includes a frequency-domain electric source current $\tilde{\mathbf{J}}$

and a frequency-domain magnetic source current $\tilde{\mathbf{J}}_m$. The fields $(\tilde{\mathbf{E}}_1, \tilde{\mathbf{D}}_1, \tilde{\mathbf{B}}_1, \tilde{\mathbf{H}}_1)$ within the region (which may also contain arbitrary media) are described by

$$\nabla \times \tilde{\mathbf{E}}_1 = -\tilde{\mathbf{J}}_m - j\omega\tilde{\mathbf{B}}_1, \tag{4.181}$$

$$\nabla \times \tilde{\mathbf{H}}_1 = \tilde{\mathbf{J}} + j\omega\tilde{\mathbf{D}}_1, \tag{4.182}$$

$$\nabla \cdot \tilde{\mathbf{D}}_1 = \tilde{\rho}, \tag{4.183}$$

$$\nabla \cdot \tilde{\mathbf{B}}_1 = \tilde{\rho}_m. \tag{4.184}$$

Suppose we have been given a mathematical description of the sources $(\tilde{\mathbf{J}}, \tilde{\mathbf{J}}_m)$ and have solved for the field vectors $(\tilde{\mathbf{E}}_1, \tilde{\mathbf{D}}_1, \tilde{\mathbf{B}}_1, \tilde{\mathbf{H}}_1)$. Of course, we must also have been supplied with a set of boundary values and constitutive relations in order to make the solution unique. We note that if we replace the formula for $\tilde{\mathbf{J}}$ with the formula for $\tilde{\mathbf{J}}_m$ in (4.182) (and $\tilde{\rho}$ with $\tilde{\rho}_m$ in (4.183)) and also replace $\tilde{\mathbf{J}}_m$ with $-\tilde{\mathbf{J}}$ in (4.181) (and $\tilde{\rho}_m$ with $-\tilde{\rho}$ in (4.184)) we get a new problem. However, the symmetry of the equations allows us to specify the solution immediately. The new set of curl equations requires

$$\nabla \times \tilde{\mathbf{E}}_2 = \tilde{\mathbf{J}} - j\omega\tilde{\mathbf{B}}_2, \tag{4.185}$$

$$\nabla \times \tilde{\mathbf{H}}_2 = \tilde{\mathbf{J}}_m + j\omega\tilde{\mathbf{D}}_2. \tag{4.186}$$

If we can resolve the question of how the constitutive parameters must be altered to reflect these replacements, then we can conclude by comparing (4.185) with (4.182) and (4.186) with (4.181) that

$$\tilde{\mathbf{E}}_2 = \tilde{\mathbf{H}}_1, \qquad \tilde{\mathbf{B}}_2 = -\tilde{\mathbf{D}}_1, \qquad \tilde{\mathbf{D}}_2 = \tilde{\mathbf{B}}_1, \qquad \tilde{\mathbf{H}}_2 = -\tilde{\mathbf{E}}_1.$$

The discussion regarding units in § 2.9.2 carries over to the present case. Multiplying Ampere's law by $\eta_0 = (\mu_0/\epsilon_0)^{1/2}$, we have

$$\nabla \times \tilde{\mathbf{E}} = -\tilde{\mathbf{J}}_m - j\omega\tilde{\mathbf{B}}, \qquad \nabla \times (\eta_0\tilde{\mathbf{H}}) = (\eta_0\tilde{\mathbf{J}}) + j\omega(\eta_0\tilde{\mathbf{D}}).$$

Thus if the original problem has solution $(\tilde{\mathbf{E}}_1, \eta_0\tilde{\mathbf{D}}_1, \tilde{\mathbf{B}}_1, \eta_0\tilde{\mathbf{H}}_1)$, then the dual problem with $\tilde{\mathbf{J}}$ replaced by $\tilde{\mathbf{J}}_m/\eta_0$ and $\tilde{\mathbf{J}}_m$ replaced by $-\eta_0\tilde{\mathbf{J}}$ has solution

$$\tilde{\mathbf{E}}_2 = \eta_0\tilde{\mathbf{H}}_1, \tag{4.187}$$

$$\tilde{\mathbf{B}}_2 = -\eta_0\tilde{\mathbf{D}}_1, \tag{4.188}$$

$$\eta_0\tilde{\mathbf{D}}_2 = \tilde{\mathbf{B}}_1, \tag{4.189}$$

$$\eta_0\tilde{\mathbf{H}}_2 = -\tilde{\mathbf{E}}_1. \tag{4.190}$$

As with duality in the time domain, the constitutive parameters for the dual problem must be altered from those of the original problem. For linear anisotropic media we have from (4.13) and (4.14) the constitutive relationships

$$\tilde{\mathbf{D}}_1 = \bar{\tilde{\epsilon}}_1 \cdot \tilde{\mathbf{E}}_1, \tag{4.191}$$

$$\tilde{\mathbf{B}}_1 = \bar{\tilde{\mu}}_1 \cdot \tilde{\mathbf{H}}_1, \tag{4.192}$$

for the original problem, and

$$\tilde{\mathbf{D}}_2 = \bar{\tilde{\epsilon}}_2 \cdot \tilde{\mathbf{E}}_2, \tag{4.193}$$

$$\tilde{\mathbf{B}}_2 = \bar{\tilde{\mu}}_2 \cdot \tilde{\mathbf{H}}_2, \tag{4.194}$$

for the dual problem. Substitution of (4.187)–(4.190) into (4.191) and (4.192) gives

$$\tilde{\mathbf{D}}_2 = \left(\frac{\tilde{\bar{\mu}}_1}{\eta_0^2}\right) \cdot \tilde{\mathbf{E}}_2, \tag{4.195}$$

$$\tilde{\mathbf{B}}_2 = (\eta_0^2 \tilde{\bar{\epsilon}}_1) \cdot \tilde{\mathbf{H}}_2. \tag{4.196}$$

Comparing (4.195) with (4.193) and (4.196) with (4.194), we conclude that

$$\tilde{\bar{\mu}}_2 = \eta_0^2 \tilde{\bar{\epsilon}}_1, \qquad \tilde{\bar{\epsilon}}_2 = \tilde{\bar{\mu}}_1 / \eta_0^2. \tag{4.197}$$

For a linear, isotropic medium specified by $\tilde{\epsilon}$ and $\tilde{\mu}$, the dual problem is obtained by replacing $\tilde{\epsilon}_r$ with $\tilde{\mu}_r$ and $\tilde{\mu}_r$ with $\tilde{\epsilon}_r$. The solution to the dual problem is then

$$\tilde{\mathbf{E}}_2 = \eta_0 \tilde{\mathbf{H}}_1, \qquad \eta_0 \tilde{\mathbf{H}}_2 = -\tilde{\mathbf{E}}_1,$$

as before. The medium in the dual problem must have electric properties numerically equal to the magnetic properties of the medium in the original problem, and magnetic properties numerically equal to the electric properties of the medium in the original problem. Alternatively we may divide Ampere's law by $\eta = (\tilde{\mu}/\tilde{\epsilon})^{1/2}$ instead of η_0. Then the dual problem has $\tilde{\mathbf{J}}$ replaced by $\tilde{\mathbf{J}}_m/\eta$, and $\tilde{\mathbf{J}}_m$ replaced by $-\eta\tilde{\mathbf{J}}$, and the solution is

$$\tilde{\mathbf{E}}_2 = \eta \tilde{\mathbf{H}}_1, \qquad \eta \tilde{\mathbf{H}}_2 = -\tilde{\mathbf{E}}_1. \tag{4.198}$$

There is no need to swap $\tilde{\epsilon}_r$ and $\tilde{\mu}_r$ since information about these parameters is incorporated into the replacement sources.

We may also apply duality to a problem where we have separated the impressed and secondary sources. In a homogeneous, isotropic, conducting medium we may let $\tilde{\mathbf{J}} = \tilde{\mathbf{J}}^i + \tilde{\sigma}\tilde{\mathbf{E}}$. With this the curl equations become

$$\nabla \times \eta\tilde{\mathbf{H}} = \eta\tilde{\mathbf{J}}^i + j\omega\eta\tilde{\epsilon}^c\tilde{\mathbf{E}},$$
$$\nabla \times \tilde{\mathbf{E}} = -\tilde{\mathbf{J}}_m - j\omega\tilde{\mu}\tilde{\mathbf{H}}.$$

The solution to the dual problem is again given by (4.198), except that now $\eta = (\tilde{\mu}/\tilde{\epsilon}^c)^{1/2}$.

As we did near the end of § 2.9.2, we can consider duality in a source-free region. We let S enclose a source-free region of space and, for simplicity, assume that the medium within S is linear, isotropic, and homogeneous. The fields within S are described by

$$\nabla \times \tilde{\mathbf{E}}_1 = -j\omega\tilde{\mu}\tilde{\mathbf{H}}_1,$$
$$\nabla \times \eta\tilde{\mathbf{H}}_1 = j\omega\tilde{\epsilon}\eta\tilde{\mathbf{E}}_1,$$
$$\nabla \cdot \tilde{\epsilon}\tilde{\mathbf{E}}_1 = 0,$$
$$\nabla \cdot \tilde{\mu}\tilde{\mathbf{H}}_1 = 0.$$

The symmetry of the equations is such that the mathematical form of the solution for $\tilde{\mathbf{E}}$ is the same as that for $\eta\tilde{\mathbf{H}}$. Since the fields

$$\tilde{\mathbf{E}}_2 = \eta\tilde{\mathbf{H}}_1, \qquad \tilde{\mathbf{H}}_2 = -\tilde{\mathbf{E}}_1/\eta,$$

also satisfy Maxwell's equations, the dual problem merely involves replacing $\tilde{\mathbf{E}}$ by $\eta\tilde{\mathbf{H}}$ and $\tilde{\mathbf{H}}$ by $-\tilde{\mathbf{E}}/\eta$.

4.11   The wave nature of the time-harmonic EM field

Time-harmonic electromagnetic waves have been studied in great detail. Narrowband waves are widely used for signal transmission, heating, power transfer, and radar. They share many of the properties of more general transient waves, and the discussions of § 2.10.1 are applicable. Here we shall investigate some of the unique properties of time-harmonic waves and introduce such fundamental quantities as wavelength, phase and group velocity, and polarization.

4.11.1   The frequency-domain wave equation

We begin by deriving the frequency-domain wave equation for dispersive bianisotropic materials. A solution to this equation may be viewed as the transform of a general time-dependent field. If one specific frequency is considered the time-harmonic solution is produced.

In § 2.10.2 we derived the time-domain wave equation for bianisotropic materials. There it was necessary to consider only time-independent constitutive parameters. We can overcome this requirement, and thus deal with dispersive materials, by using a Fourier transform approach. We solve a frequency-domain wave equation that includes the frequency dependence of the constitutive parameters, and then use an inverse transform to return to the time domain.

The derivation of the equation parallels that of § 2.10.2. We substitute the frequency-domain constitutive relationships

$$\tilde{\mathbf{D}} = \tilde{\bar{\epsilon}} \cdot \tilde{\mathbf{E}} + \tilde{\bar{\xi}} \cdot \tilde{\mathbf{H}},$$

$$\tilde{\mathbf{B}} = \tilde{\bar{\zeta}} \cdot \tilde{\mathbf{E}} + \tilde{\bar{\mu}} \cdot \tilde{\mathbf{H}},$$

into Maxwell's curl equations (4.7) and (4.8) to get the coupled differential equations

$$\nabla \times \tilde{\mathbf{E}} = -j\omega[\tilde{\bar{\zeta}} \cdot \tilde{\mathbf{E}} + \tilde{\bar{\mu}} \cdot \tilde{\mathbf{H}}] - \tilde{\mathbf{J}}_m,$$

$$\nabla \times \tilde{\mathbf{H}} = j\omega[\tilde{\bar{\epsilon}} \cdot \tilde{\mathbf{E}} + \tilde{\bar{\xi}} \cdot \tilde{\mathbf{H}}] + \tilde{\mathbf{J}},$$

for $\tilde{\mathbf{E}}$ and $\tilde{\mathbf{H}}$. Here we have included magnetic sources $\tilde{\mathbf{J}}_m$ in Faraday's law. Using the dyadic operator $\bar{\nabla}$ defined in (2.308) we can write these equations as

$$\left(\bar{\nabla} + j\omega\tilde{\bar{\zeta}}\right) \cdot \tilde{\mathbf{E}} = -j\omega\tilde{\bar{\mu}} \cdot \tilde{\mathbf{H}} - \tilde{\mathbf{J}}_m, \qquad (4.199)$$

$$\left(\bar{\nabla} - j\omega\tilde{\bar{\xi}}\right) \cdot \tilde{\mathbf{H}} = j\omega\tilde{\bar{\epsilon}} \cdot \tilde{\mathbf{E}} + \tilde{\mathbf{J}}. \qquad (4.200)$$

We can obtain separate equations for $\tilde{\mathbf{E}}$ and $\tilde{\mathbf{H}}$ by defining the inverse dyadics

$$\tilde{\bar{\epsilon}} \cdot \tilde{\bar{\epsilon}}^{-1} = \bar{\mathbf{I}}, \qquad \tilde{\bar{\mu}} \cdot \tilde{\bar{\mu}}^{-1} = \bar{\mathbf{I}}.$$

Using $\tilde{\bar{\mu}}^{-1}$ we can write (4.199) as

$$-j\omega\tilde{\mathbf{H}} = \tilde{\bar{\mu}}^{-1} \cdot \left(\bar{\nabla} + j\omega\tilde{\bar{\zeta}}\right) \cdot \tilde{\mathbf{E}} + \tilde{\bar{\mu}}^{-1} \cdot \tilde{\mathbf{J}}_m.$$

Substituting this into (4.200) we get

$$\left[\left(\bar{\nabla} - j\omega\tilde{\bar{\xi}}\right) \cdot \tilde{\bar{\mu}}^{-1} \cdot \left(\bar{\nabla} + j\omega\tilde{\bar{\zeta}}\right) - \omega^2\tilde{\bar{\epsilon}}\right] \cdot \tilde{\mathbf{E}} = -\left(\bar{\nabla} - j\omega\tilde{\bar{\xi}}\right) \cdot \tilde{\bar{\mu}}^{-1} \cdot \tilde{\mathbf{J}}_m - j\omega\tilde{\mathbf{J}}. \quad (4.201)$$

This is the general frequency-domain wave equation for $\tilde{\mathbf{E}}$. Using $\tilde{\bar{\epsilon}}^{-1}$ we can write (4.200) as

$$j\omega\tilde{\mathbf{E}} = \tilde{\bar{\epsilon}}^{-1} \cdot \left(\bar{\nabla} - j\omega\tilde{\bar{\xi}}\right) \cdot \tilde{\mathbf{H}} - \tilde{\bar{\epsilon}}^{-1} \cdot \tilde{\mathbf{J}}.$$

Substituting this into (4.199) we get

$$\left[\left(\bar{\nabla} + j\omega\tilde{\bar{\zeta}}\right) \cdot \tilde{\bar{\epsilon}}^{-1} \cdot \left(\bar{\nabla} - j\omega\tilde{\bar{\xi}}\right) - \omega^2\tilde{\bar{\mu}}\right] \cdot \tilde{\mathbf{H}} = \left(\bar{\nabla} + j\omega\tilde{\bar{\zeta}}\right) \cdot \tilde{\bar{\epsilon}}^{-1} \cdot \tilde{\mathbf{J}} - j\omega\tilde{\mathbf{J}}_m. \quad (4.202)$$

This is the general frequency-domain wave equation for $\tilde{\mathbf{H}}$.

Wave equation for a homogeneous, lossy, isotropic medium. We may specialize (4.201) and (4.202) to the case of a homogeneous, lossy, isotropic medium by setting $\tilde{\bar{\zeta}} = \tilde{\bar{\xi}} = 0$, $\tilde{\bar{\mu}} = \tilde{\mu}\bar{\mathbf{I}}$, $\tilde{\bar{\epsilon}} = \tilde{\epsilon}\bar{\mathbf{I}}$, and $\tilde{\mathbf{J}} = \tilde{\mathbf{J}}^i + \tilde{\mathbf{J}}^c$:

$$\nabla \times (\nabla \times \tilde{\mathbf{E}}) - \omega^2\tilde{\mu}\tilde{\epsilon}\tilde{\mathbf{E}} = -\nabla \times \tilde{\mathbf{J}}_m - j\omega\tilde{\mu}(\tilde{\mathbf{J}}^i + \tilde{\mathbf{J}}^c), \quad (4.203)$$

$$\nabla \times (\nabla \times \tilde{\mathbf{H}}) - \omega^2\tilde{\mu}\tilde{\epsilon}\tilde{\mathbf{H}} = \nabla \times (\tilde{\mathbf{J}}^i + \tilde{\mathbf{J}}^c) - j\omega\tilde{\epsilon}\tilde{\mathbf{J}}_m. \quad (4.204)$$

Using (B.47) and using Ohm's law $\tilde{\mathbf{J}}^c = \tilde{\sigma}\tilde{\mathbf{E}}$ to describe the secondary current, we get from (4.203)

$$\nabla(\nabla \cdot \tilde{\mathbf{E}}) - \nabla^2\tilde{\mathbf{E}} - \omega^2\tilde{\mu}\tilde{\epsilon}\tilde{\mathbf{E}} = -\nabla \times \tilde{\mathbf{J}}_m - j\omega\tilde{\mu}\tilde{\mathbf{J}}^i - j\omega\tilde{\mu}\tilde{\sigma}\tilde{\mathbf{E}}$$

which, using $\nabla \cdot \tilde{\mathbf{E}} = \tilde{\rho}/\tilde{\epsilon}$, can be simplified to

$$(\nabla^2 + k^2)\tilde{\mathbf{E}} = \nabla \times \tilde{\mathbf{J}}_m + j\omega\tilde{\mu}\tilde{\mathbf{J}}^i + \frac{1}{\tilde{\epsilon}}\nabla\tilde{\rho}. \quad (4.205)$$

This is the *vector Helmholtz equation* for $\tilde{\mathbf{E}}$. Here k is the *complex wavenumber* defined through

$$k^2 = \omega^2\tilde{\mu}\tilde{\epsilon} - j\omega\tilde{\mu}\tilde{\sigma} = \omega^2\tilde{\mu}\left[\tilde{\epsilon} + \frac{\tilde{\sigma}}{j\omega}\right] = \omega^2\tilde{\mu}\tilde{\epsilon}^c \quad (4.206)$$

where $\tilde{\epsilon}^c$ is the complex permittivity (4.26).

By (4.204) we have

$$\nabla(\nabla \cdot \tilde{\mathbf{H}}) - \nabla^2\tilde{\mathbf{H}} - \omega^2\tilde{\mu}\tilde{\epsilon}\tilde{\mathbf{H}} = \nabla \times \tilde{\mathbf{J}}^i + \nabla \times \tilde{\mathbf{J}}^c - j\omega\tilde{\epsilon}\tilde{\mathbf{J}}_m.$$

Using

$$\nabla \times \tilde{\mathbf{J}}^c = \nabla \times (\tilde{\sigma}\tilde{\mathbf{E}}) = \tilde{\sigma}\nabla \times \tilde{\mathbf{E}} = \tilde{\sigma}(-j\omega\tilde{\mathbf{B}} - \tilde{\mathbf{J}}_m)$$

and $\nabla \cdot \tilde{\mathbf{H}} = \tilde{\rho}_m/\tilde{\mu}$ we then get

$$(\nabla^2 + k^2)\tilde{\mathbf{H}} = -\nabla \times \tilde{\mathbf{J}}^i + j\omega\tilde{\epsilon}^c\tilde{\mathbf{J}}_m + \frac{1}{\tilde{\mu}}\nabla\tilde{\rho}_m, \quad (4.207)$$

which is the vector Helmholtz equation for $\tilde{\mathbf{H}}$.

4.11.2 Field relationships and the wave equation for two-dimensional fields

Many important canonical problems are two-dimensional in nature, with the sources and fields invariant along one direction. Two-dimensional fields have a simple structure compared to three-dimensional fields, and this structure often allows a decomposition into even simpler field structures.

Consider a homogeneous region of space characterized by the permittivity $\tilde{\epsilon}$, permeability $\tilde{\mu}$, and conductivity $\tilde{\sigma}$. We assume that all sources and fields are z-invariant, and wish to find the relationship between the various components of the frequency-domain fields in a source-free region. It is useful to define the transverse vector component of an arbitrary vector $\mathbf{A}$ as the component of $\mathbf{A}$ perpendicular to the axis of invariance:

$$\mathbf{A}_t = \mathbf{A} - \hat{\mathbf{z}}(\hat{\mathbf{z}} \cdot \mathbf{A}).$$

For the position vector $\mathbf{r}$, this component is the transverse position vector $\mathbf{r}_t = \boldsymbol{\rho}$. For instance we have

$$\boldsymbol{\rho} = \hat{\mathbf{x}}x + \hat{\mathbf{y}}y, \qquad \boldsymbol{\rho} = \hat{\boldsymbol{\rho}}\rho,$$

in the rectangular and cylindrical coordinate systems, respectively.

Because the region is source-free, the fields $\tilde{\mathbf{E}}$ and $\tilde{\mathbf{H}}$ obey the homogeneous Helmholtz equations

$$(\nabla^2 + k^2) \left\{ \begin{matrix} \tilde{\mathbf{E}} \\ \tilde{\mathbf{H}} \end{matrix} \right\} = 0.$$

Writing the fields in terms of rectangular components, we find that each component must obey a homogeneous scalar Helmholtz equation. In particular, we have for the *axial components* $\tilde{E}_z$ and $\tilde{H}_z$,

$$(\nabla^2 + k^2) \left\{ \begin{matrix} \tilde{E}_z \\ \tilde{H}_z \end{matrix} \right\} = 0.$$

But since the fields are independent of z we may also write

$$(\nabla_t^2 + k^2) \left\{ \begin{matrix} \tilde{E}_z \\ \tilde{H}_z \end{matrix} \right\} = 0 \tag{4.208}$$

where ∇_t^2 is the transverse Laplacian operator

$$\nabla_t^2 = \nabla^2 - \hat{\mathbf{z}}\frac{\partial^2}{\partial z^2}. \tag{4.209}$$

In rectangular coordinates we have

$$\nabla_t^2 = \frac{\partial^2}{\partial x^2} + \frac{\partial^2}{\partial y^2},$$

while in circular cylindrical coordinates

$$\nabla_t^2 = \frac{\partial^2}{\partial \rho^2} + \frac{1}{\rho}\frac{\partial}{\partial \rho} + \frac{1}{\rho^2}\frac{\partial^2}{\partial \phi^2}. \tag{4.210}$$

With our condition on z-independence we can relate the transverse fields $\tilde{\mathbf{E}}_t$ and $\tilde{\mathbf{H}}_t$ to $\tilde{E}_z$ and $\tilde{H}_z$. By Faraday's law we have

$$\nabla \times \tilde{\mathbf{E}}(\boldsymbol{\rho}, \omega) = -j\omega\tilde{\mu}\tilde{\mathbf{H}}(\boldsymbol{\rho}, \omega)$$

and thus

$$\tilde{\mathbf{H}}_t = -\frac{1}{j\omega\tilde{\mu}}\left[\nabla \times \tilde{\mathbf{E}}\right]_t.$$

The transverse portion of the curl is merely

$$\left[\nabla \times \tilde{\mathbf{E}}\right]_t = \hat{\mathbf{x}}\left[\frac{\partial \tilde{E}_z}{\partial y} - \frac{\partial \tilde{E}_y}{\partial z}\right] + \hat{\mathbf{y}}\left[\frac{\partial \tilde{E}_x}{\partial z} - \frac{\partial \tilde{E}_z}{\partial x}\right] = -\hat{\mathbf{z}} \times \left[\hat{\mathbf{x}}\frac{\partial \tilde{E}_z}{\partial x} + \hat{\mathbf{y}}\frac{\partial \tilde{E}_z}{\partial y}\right]$$

since the derivatives with respect to z vanish. The term in brackets is the transverse gradient of $\tilde{E}_z$, where the transverse gradient operator is

$$\nabla_t = \nabla - \hat{\mathbf{z}}\frac{\partial}{\partial z}.$$

In circular cylindrical coordinates this operator becomes

$$\nabla_t = \hat{\boldsymbol{\rho}}\frac{\partial}{\partial \rho} + \hat{\boldsymbol{\phi}}\frac{1}{\rho}\frac{\partial}{\partial \phi}. \tag{4.211}$$

Thus we have

$$\tilde{\mathbf{H}}_t(\boldsymbol{\rho}, \omega) = \frac{1}{j\omega\tilde{\mu}}\hat{\mathbf{z}} \times \nabla_t\tilde{E}_z(\boldsymbol{\rho}, \omega).$$

Similarly, the source-free Ampere's law yields

$$\tilde{\mathbf{E}}_t(\boldsymbol{\rho}, \omega) = -\frac{1}{j\omega\tilde{\epsilon}^c}\hat{\mathbf{z}} \times \nabla_t\tilde{H}_z(\boldsymbol{\rho}, \omega).$$

These results suggest that we can solve a two-dimensional problem by superposition. We first consider the case where $\tilde{E}_z \neq 0$ and $\tilde{H}_z = 0$, called *electric polarization*. This case is also called *TM* or *transverse magnetic* polarization because the magnetic field is transverse to the z-direction (TM_z). We have

$$(\nabla_t^2 + k^2)\tilde{E}_z = 0, \qquad \tilde{\mathbf{H}}_t(\boldsymbol{\rho}, \omega) = \frac{1}{j\omega\tilde{\mu}}\hat{\mathbf{z}} \times \nabla_t\tilde{E}_z(\boldsymbol{\rho}, \omega). \tag{4.212}$$

Once we have solved the Helmholtz equation for $\tilde{E}_z$, the remaining field components follow by simple differentiation. We next consider the case where $\tilde{H}_z \neq 0$ and $\tilde{E}_z = 0$. This is the case of *magnetic polarization*, also called *TE* or *transverse electric* polarization (TE_z). In this case

$$(\nabla_t^2 + k^2)\tilde{H}_z = 0, \qquad \tilde{\mathbf{E}}_t(\boldsymbol{\rho}, \omega) = -\frac{1}{j\omega\tilde{\epsilon}^c}\hat{\mathbf{z}} \times \nabla_t\tilde{H}_z(\boldsymbol{\rho}, \omega). \tag{4.213}$$

A problem involving both $\tilde{E}_z$ and $\tilde{H}_z$ is solved by adding the results for the individual TE_z and TM_z cases.

Note that we can obtain the expression for the TE fields from the expression for the TM fields, and vice versa, using duality. For instance, knowing that the TM fields obey (4.212) we may replace $\tilde{\mathbf{H}}_t$ with $\tilde{\mathbf{E}}_t/\eta$ and $\tilde{E}_z$ with $-\eta\tilde{H}_z$ to obtain

$$\frac{\tilde{\mathbf{E}}_t(\boldsymbol{\rho}, \omega)}{\eta} = \frac{1}{j\omega\tilde{\mu}}\hat{\mathbf{z}} \times \nabla_t[-\eta\tilde{H}_z(\boldsymbol{\rho}, \omega)],$$

which reproduces (4.213).

4.11.3   Plane waves in a homogeneous, isotropic, lossy material

The plane-wave field. In later sections we will solve the frequency-domain wave equation with an arbitrary source distribution. At this point we are more interested in the general behavior of EM waves in the frequency domain, so we seek simple solutions to the homogeneous equation

$$(\nabla^2 + k^2)\tilde{\mathbf{E}}(\mathbf{r},\omega) = 0 \tag{4.214}$$

that governs the fields in source-free regions of space. Here

$$[k(\omega)]^2 = \omega^2 \tilde{\mu}(\omega)\tilde{\epsilon}^c(\omega).$$

Many properties of plane waves are best understood by considering the behavior of a monochromatic field oscillating at a single frequency $\breve{\omega}$. In these cases we merely make the replacements

$$\omega \to \breve{\omega}, \qquad \tilde{\mathbf{E}}(\mathbf{r},\omega) \to \breve{\mathbf{E}}(\mathbf{r}),$$

and apply the rules developed in § 4.7 for the manipulation of phasor fields.

For our first solutions we choose those that demonstrate rectangular symmetry. *Plane waves* have planar spatial phase loci. That is, the spatial surfaces over which the phase of the complex frequency-domain field is constant are planes. Solutions of this type may be obtained using separation of variables in rectangular coordinates. Writing

$$\tilde{\mathbf{E}}(\mathbf{r},\omega) = \hat{\mathbf{x}}\tilde{E}_x(\mathbf{r},\omega) + \hat{\mathbf{y}}\tilde{E}_y(\mathbf{r},\omega) + \hat{\mathbf{z}}\tilde{E}_z(\mathbf{r},\omega)$$

we find that (4.214) reduces to three scalar equations of the form

$$(\nabla^2 + k^2)\tilde{\psi}(\mathbf{r},\omega) = 0$$

where $\tilde{\psi}$ is representative of $\tilde{E}_x$, $\tilde{E}_y$, and $\tilde{E}_z$. This is called the homogeneous scalar Helmholtz equation. Product solutions to this equation are considered in § A.4. In rectangular coordinates

$$\tilde{\psi}(\mathbf{r},\omega) = X(x,\omega)Y(y,\omega)Z(z,\omega)$$

where X, Y, and Z are chosen from the list (A.102). Since the exponentials describe propagating wave functions, we choose

$$\tilde{\psi}(\mathbf{r},\omega) = A(\omega)e^{\pm jk_x(\omega)x}e^{\pm jk_y(\omega)y}e^{\pm jk_z(\omega)z}$$

where A is the *amplitude spectrum* of the plane wave and $k_x^2 + k_y^2 + k_z^2 = k^2$. Using this solution to represent each component of $\tilde{\mathbf{E}}$, we have a propagating-wave solution to the homogeneous vector Helmholtz equation:

$$\tilde{\mathbf{E}}(\mathbf{r},\omega) = \tilde{\mathbf{E}}_0(\omega)e^{\pm jk_x(\omega)x}e^{\pm jk_y(\omega)y}e^{\pm jk_z(\omega)z}, \tag{4.215}$$

where $\mathbf{E}_0(\omega)$ is the vector amplitude spectrum. If we define the *wave vector*

$$\mathbf{k}(\omega) = \hat{\mathbf{x}}k_x(\omega) + \hat{\mathbf{y}}k_y(\omega) + \hat{\mathbf{z}}k_z(\omega),$$

then we can write (4.215) as

$$\tilde{\mathbf{E}}(\mathbf{r},\omega) = \tilde{\mathbf{E}}_0(\omega)e^{-j\mathbf{k}(\omega)\cdot\mathbf{r}}. \tag{4.216}$$

Note that we choose the negative sign in the exponential function and allow the vector components of $\mathbf{k}$ to be either positive or negative as required by the physical nature of a specific problem. Also note that the magnitude of the wave vector is the wavenumber: $|\mathbf{k}| = k$.

We may always write the wave vector as a sum of real and imaginary vector components

$$\mathbf{k} = \mathbf{k}' + j\mathbf{k}'' \tag{4.217}$$

which must obey

$$\mathbf{k} \cdot \mathbf{k} = k^2 = k'^2 - k''^2 + 2j\mathbf{k}' \cdot \mathbf{k}''. \tag{4.218}$$

When the real and imaginary components are collinear, (4.216) describes a *uniform plane wave* with

$$\mathbf{k} = \hat{\mathbf{k}}(k' + jk''). $$

When $\mathbf{k}'$ and $\mathbf{k}''$ have different directions, (4.216) describes a *nonuniform plane wave*. We shall find in § 4.13 that any frequency-domain electromagnetic field in free space may be represented as a continuous superposition of elemental plane-wave components of the type (4.216), but that both uniform and nonuniform terms are required.

The TEM nature of a uniform plane wave. Given the plane-wave solution to the wave equation for the electric field, it is straightforward to find the magnetic field. Substitution of (4.216) into Faraday's law gives

$$\nabla \times \left[\tilde{\mathbf{E}}_0(\omega) e^{-j\mathbf{k}(\omega) \cdot \mathbf{r}} \right] = -j\omega \tilde{\mathbf{B}}(\mathbf{r}, \omega). $$

Computation of the curl is straightforward and easily done in rectangular coordinates. This and similar derivatives often appear when manipulating plane-wave solutions; see the tabulation in Appendix B. By (B.78) we have

$$\tilde{\mathbf{H}} = \frac{\mathbf{k} \times \tilde{\mathbf{E}}}{\omega \tilde{\mu}}. \tag{4.219}$$

Taking the cross product of this expression with $\mathbf{k}$, we also have

$$\mathbf{k} \times \tilde{\mathbf{H}} = \frac{\mathbf{k} \times (\mathbf{k} \times \tilde{\mathbf{E}})}{\omega \tilde{\mu}} = \frac{\mathbf{k}(\mathbf{k} \cdot \tilde{\mathbf{E}}) - \tilde{\mathbf{E}}(\mathbf{k} \cdot \mathbf{k})}{\omega \tilde{\mu}}. \tag{4.220}$$

We can show that $\mathbf{k} \cdot \tilde{\mathbf{E}} = 0$ by examining Gauss' law and employing (B.77):

$$\nabla \cdot \tilde{\mathbf{E}} = -j\mathbf{k} \cdot \tilde{\mathbf{E}} e^{-j\mathbf{k} \cdot \mathbf{r}} = \frac{\tilde{\rho}}{\tilde{\epsilon}} = 0. \tag{4.221}$$

Using this and $\mathbf{k} \cdot \mathbf{k} = k^2 = \omega^2 \tilde{\mu} \tilde{\epsilon}^c$, we obtain from (4.220)

$$\tilde{\mathbf{E}} = -\frac{\mathbf{k} \times \tilde{\mathbf{H}}}{\omega \tilde{\epsilon}^c}. \tag{4.222}$$

Now for a uniform plane wave $\mathbf{k} = \hat{\mathbf{k}}k$, so we can also write (4.219) as

$$\tilde{\mathbf{H}} = \frac{\hat{\mathbf{k}} \times \tilde{\mathbf{E}}}{\eta} = \frac{\hat{\mathbf{k}} \times \tilde{\mathbf{E}}_0}{\eta} e^{-j\mathbf{k} \cdot \mathbf{r}} \tag{4.223}$$

and (4.222) as

$$\tilde{\mathbf{E}} = -\eta \hat{\mathbf{k}} \times \tilde{\mathbf{H}}.$$

Here

$$\eta = \frac{\omega \tilde{\mu}}{k} = \sqrt{\frac{\tilde{\mu}}{\tilde{\epsilon}^c}}$$

is the complex intrinsic impedance of the medium.

Equations (4.223) and (4.221) show that the electric and magnetic fields and the wave vector are mutually orthogonal. The wave is said to be *transverse electromagnetic* or *TEM* to the direction of propagation.

The phase and attenuation constants of a uniform plane wave.   For a uniform plane wave we may write

$$\mathbf{k} = k' \hat{\mathbf{k}} + j k'' \hat{\mathbf{k}} = k \hat{\mathbf{k}} = (\beta - j\alpha) \hat{\mathbf{k}}$$

where $k' = \beta$ and $k'' = -\alpha$. Here α is called the *attenuation constant* and β is the *phase constant*. Since k is defined through (4.206), we have

$$k^2 = (\beta - j\alpha)^2 = \beta^2 - 2j\alpha\beta - \alpha^2 = \omega^2 \tilde{\mu} \tilde{\epsilon}^c = \omega^2 (\tilde{\mu}' + j\tilde{\mu}'')(\tilde{\epsilon}^{c\prime} + j\tilde{\epsilon}^{c\prime\prime}).$$

Equating real and imaginary parts we have

$$\beta^2 - \alpha^2 = \omega^2 [\tilde{\mu}' \tilde{\epsilon}^{c\prime} - \tilde{\mu}'' \tilde{\epsilon}^{c\prime\prime}], \qquad -2\alpha\beta = \omega^2 [\tilde{\mu}'' \tilde{\epsilon}^{c\prime} + \tilde{\mu}' \tilde{\epsilon}^{c\prime\prime}].$$

We assume the material is passive so that $\tilde{\mu}'' \le 0$, $\tilde{\epsilon}^{c\prime\prime} \le 0$. Letting

$$\beta^2 - \alpha^2 = \omega^2 [\tilde{\mu}' \tilde{\epsilon}^{c\prime} - \tilde{\mu}'' \tilde{\epsilon}^{c\prime\prime}] = A, \qquad 2\alpha\beta = \omega^2 [|\tilde{\mu}''| \tilde{\epsilon}^{c\prime} + \tilde{\mu}' |\tilde{\epsilon}^{c\prime\prime}|] = B,$$

we may solve simultaneously to find that

$$\beta^2 = \frac{1}{2} \left[A + \sqrt{A^2 + B^2} \right], \qquad \alpha^2 = \frac{1}{2} \left[-A + \sqrt{A^2 + B^2} \right].$$

Since $A^2 + B^2 = \omega^4 (\tilde{\epsilon}^{c\prime 2} + \tilde{\epsilon}^{c\prime\prime 2})(\tilde{\mu}'^2 + \tilde{\mu}''^2)$, we have

$$\beta = \omega \sqrt{\tilde{\mu}' \tilde{\epsilon}^{c\prime}} \sqrt{\frac{1}{2} \left[\sqrt{\left(1 + \frac{\tilde{\epsilon}^{c\prime\prime 2}}{\tilde{\epsilon}^{c\prime 2}}\right)\left(1 + \frac{\tilde{\mu}''^2}{\tilde{\mu}'^2}\right)} + \left(1 - \frac{\tilde{\mu}''}{\tilde{\mu}'} \frac{\tilde{\epsilon}^{c\prime\prime}}{\tilde{\epsilon}^{c\prime}}\right) \right]}, \qquad (4.224)$$

$$\alpha = \omega \sqrt{\tilde{\mu}' \tilde{\epsilon}^{c\prime}} \sqrt{\frac{1}{2} \left[\sqrt{\left(1 + \frac{\tilde{\epsilon}^{c\prime\prime 2}}{\tilde{\epsilon}^{c\prime 2}}\right)\left(1 + \frac{\tilde{\mu}''^2}{\tilde{\mu}'^2}\right)} - \left(1 - \frac{\tilde{\mu}''}{\tilde{\mu}'} \frac{\tilde{\epsilon}^{c\prime\prime}}{\tilde{\epsilon}^{c\prime}}\right) \right]}, \qquad (4.225)$$

where $\tilde{\epsilon}^c$ and $\tilde{\mu}$ are functions of ω. If $\tilde{\epsilon}(\omega) = \epsilon$, $\tilde{\mu}(\omega) = \mu$, and $\tilde{\sigma}(\omega) = \sigma$ are real and frequency independent, then

$$\alpha = \omega \sqrt{\mu\epsilon} \sqrt{\frac{1}{2} \left[\sqrt{1 + \left(\frac{\sigma}{\omega\epsilon}\right)^2} - 1 \right]}, \qquad (4.226)$$

$$\beta = \omega \sqrt{\mu\epsilon} \sqrt{\frac{1}{2} \left[\sqrt{1 + \left(\frac{\sigma}{\omega\epsilon}\right)^2} + 1 \right]}. \qquad (4.227)$$

These values of α and β are valid for $\omega > 0$. For negative frequencies we must be more careful in evaluating the square root in $k = \omega(\tilde{\mu}\tilde{\epsilon}^c)^{1/2}$. Writing

$$\tilde{\mu}(\omega) = \tilde{\mu}'(\omega) + j\tilde{\mu}''(\omega) = |\tilde{\mu}(\omega)|e^{j\xi^\mu(\omega)},$$
$$\tilde{\epsilon}^c(\omega) = \tilde{\epsilon}^{c\prime}(\omega) + j\tilde{\epsilon}^{c\prime\prime}(\omega) = |\tilde{\epsilon}^c(\omega)|e^{j\xi^\epsilon(\omega)},$$

we have

$$k(\omega) = \beta(\omega) - j\alpha(\omega) = \omega\sqrt{\tilde{\mu}(\omega)\tilde{\epsilon}^c(\omega)} = \omega\sqrt{|\tilde{\mu}(\omega)||\tilde{\epsilon}^c(\omega)|}e^{j\frac{1}{2}[\xi^\mu(\omega)+\xi^\epsilon(\omega)]}.$$

Now for passive materials we must have, by (4.48), $\tilde{\mu}'' < 0$ and $\tilde{\epsilon}^{c\prime\prime} < 0$ for $\omega > 0$. Since we also have $\tilde{\mu}' > 0$ and $\tilde{\epsilon}^{c\prime} > 0$ for $\omega > 0$, we find that $-\pi/2 < \xi^\mu < 0$ and $-\pi/2 < \xi^\epsilon < 0$, and thus $-\pi/2 < (\xi^\mu + \xi^\epsilon)/2 < 0$. Thus we must have $\beta > 0$ and $\alpha > 0$ for $\omega > 0$. For $\omega < 0$ we have by (4.44) and (4.45) that $\tilde{\mu}'' > 0$, $\tilde{\epsilon}^{c\prime\prime} > 0$, $\tilde{\mu}' > 0$, and $\tilde{\epsilon}^{c\prime} > 0$. Thus $\pi/2 > (\xi^\mu + \xi^\epsilon)/2 > 0$, and so $\beta < 0$ and $\alpha > 0$ for $\omega < 0$. In summary, $\alpha(\omega)$ is an even function of frequency and $\beta(\omega)$ is an odd function of frequency:

$$\beta(\omega) = -\beta(-\omega), \qquad \alpha(\omega) = \alpha(-\omega), \tag{4.228}$$

where $\beta(\omega) > 0, \alpha(\omega) > 0$ when $\omega > 0$. From this we find a condition on $\tilde{\mathbf{E}}_0$ in (4.216). Since by (4.47) we must have $\tilde{\mathbf{E}}(\omega) = \tilde{\mathbf{E}}^*(-\omega)$, we see that the uniform plane-wave field obeys

$$\tilde{\mathbf{E}}_0(\omega)e^{[-j\beta(\omega)-\alpha(\omega)]\hat{\mathbf{k}}\cdot\mathbf{r}} = \tilde{\mathbf{E}}_0^*(-\omega)e^{[+j\beta(-\omega)-\alpha(-\omega)]\mathbf{k}\cdot\mathbf{r}}$$

or

$$\tilde{\mathbf{E}}_0(\omega) = \tilde{\mathbf{E}}_0^*(-\omega),$$

since $\beta(-\omega) = -\beta(\omega)$ and $\alpha(-\omega) = \alpha(\omega)$.

Propagation of a uniform plane wave: the group and phase velocities. We have derived the plane-wave solution to the wave equation in the frequency domain, but can discover the wave nature of the solution only by examining its behavior in the time domain. Unfortunately, the explicit form of the time-domain field is highly dependent on the frequency behavior of the constitutive parameters. Even the simplest case in which ϵ, μ, and σ are frequency independent is quite complicated, as we discovered in § 2.10.6. To overcome this difficulty, it is helpful to examine the behavior of a narrowband (but non-monochromatic) signal in a lossy medium with arbitrary constitutive parameters. We will find that the time-domain wave field propagates as a disturbance through the surrounding medium with a velocity determined by the constitutive parameters of the medium. The temporal wave shape does not change as the wave propagates, but the amplitude of the wave attenuates at a rate dependent on the constitutive parameters.

For clarity of presentation we shall assume a linearly polarized plane wave (§ 4.11.4) with

$$\tilde{\mathbf{E}}(\mathbf{r}, \omega) = \hat{\mathbf{e}}\tilde{E}_0(\omega)e^{-j\mathbf{k}(\omega)\cdot\mathbf{r}}. \tag{4.229}$$

Here $\tilde{E}_0(\omega)$ is the spectrum of the temporal dependence of the wave. For the temporal dependence we choose the narrowband signal

$$E_0(t) = E_0 f(t) \cos(\omega_0 t)$$

where $f(t)$ has a narrowband spectrum centered about $\omega = 0$ (and is therefore called a *baseband signal*). An appropriate choice for $f(t)$ is the Gaussian function used in (4.52):

$$f(t) = e^{-a^2 t^2} \leftrightarrow \tilde{F}(\omega) = \sqrt{\frac{\pi}{a^2}}e^{-\frac{\omega^2}{4a^2}},$$

producing

$$E_0(t) = E_0 e^{-a^2 t^2} \cos(\omega_0 t). \tag{4.230}$$

We think of $f(t)$ as *modulating* the single-frequency cosine *carrier wave*, thus providing the *envelope*. By using a large value of a we obtain a narrowband signal whose spectrum is centered about $\pm\omega_0$. Later we shall let $a \to 0$, thereby driving the width of $f(t)$ to infinity and producing a monochromatic waveform.

By (4.1) we have

$$\tilde{E}_0(\omega) = E_0 \frac{1}{2} \left[\tilde{F}(\omega - \omega_0) + \tilde{F}(\omega + \omega_0) \right]$$

where $f(t) \leftrightarrow \tilde{F}(\omega)$. A plot of this spectrum is shown in Figure 4.2. We see that the narrowband signal is centered at $\omega = \pm\omega_0$. Substituting into (4.229) and using $\mathbf{k} = (\beta - j\alpha)\hat{\mathbf{k}}$ for a uniform plane wave, we have the frequency-domain field

$$\tilde{\mathbf{E}}(\mathbf{r}, \omega) = \hat{\mathbf{e}} E_0 \frac{1}{2} \left[\tilde{F}(\omega - \omega_0) e^{-j[\beta(\omega) - j\alpha(\omega)]\hat{\mathbf{k}} \cdot \mathbf{r}} + \tilde{F}(\omega + \omega_0) e^{-j[\beta(\omega) - j\alpha(\omega)]\hat{\mathbf{k}} \cdot \mathbf{r}} \right]. \tag{4.231}$$

The field at any time t and position $\mathbf{r}$ can now be found by inversion:

$$\hat{\mathbf{e}} E(\mathbf{r}, t) = \frac{1}{2\pi} \int_{-\infty}^{\infty} \hat{\mathbf{e}} E_0 \frac{1}{2} \left[\tilde{F}(\omega - \omega_0) e^{-j[\beta(\omega) - j\alpha(\omega)]\hat{\mathbf{k}} \cdot \mathbf{r}} + \right.$$

$$\left. + \tilde{F}(\omega + \omega_0) e^{-j[\beta(\omega) - j\alpha(\omega)]\hat{\mathbf{k}} \cdot \mathbf{r}} \right] e^{j\omega t} \, d\omega. \tag{4.232}$$

We assume that $\beta(\omega)$ and $\alpha(\omega)$ vary slowly within the band occupied by $\tilde{E}_0(\omega)$. With this assumption we can expand β and α near $\omega = \omega_0$ as

$$\beta(\omega) = \beta(\omega_0) + \beta'(\omega_0)(\omega - \omega_0) + \frac{1}{2}\beta''(\omega_0)(\omega - \omega_0)^2 + \cdots,$$

$$\alpha(\omega) = \alpha(\omega_0) + \alpha'(\omega_0)(\omega - \omega_0) + \frac{1}{2}\alpha''(\omega_0)(\omega - \omega_0)^2 + \cdots,$$

where $\beta'(\omega) = d\beta(\omega)/d\omega$, $\beta''(\omega) = d^2\beta(\omega)/d\omega^2$, and so on. In a similar manner we can expand β and α near $\omega = -\omega_0$:

$$\beta(\omega) = \beta(-\omega_0) + \beta'(-\omega_0)(\omega + \omega_0) + \frac{1}{2}\beta''(-\omega_0)(\omega + \omega_0)^2 + \cdots,$$

$$\alpha(\omega) = \alpha(-\omega_0) + \alpha'(-\omega_0)(\omega + \omega_0) + \frac{1}{2}\alpha''(-\omega_0)(\omega + \omega_0)^2 + \cdots.$$

Since we are most interested in the propagation velocity, we need not approximate α with great accuracy, and thus use $\alpha(\omega) \approx \alpha(\pm\omega_0)$ within the narrow band. We must consider β to greater accuracy to uncover the propagating nature of the wave, and thus use

$$\beta(\omega) \approx \beta(\omega_0) + \beta'(\omega_0)(\omega - \omega_0) \tag{4.233}$$

near $\omega = \omega_0$ and

$$\beta(\omega) \approx \beta(-\omega_0) + \beta'(-\omega_0)(\omega + \omega_0) \tag{4.234}$$

near $\omega = -\omega_0$. Substituting these approximations into (4.232) we find

$$\hat{\mathbf{e}} E(\mathbf{r}, t) = \frac{1}{2\pi} \int_{-\infty}^{\infty} \hat{\mathbf{e}} E_0 \frac{1}{2} \left[\tilde{F}(\omega - \omega_0) e^{-j[\beta(\omega_0) + \beta'(\omega_0)(\omega - \omega_0)]\hat{\mathbf{k}} \cdot \mathbf{r}} e^{-[\alpha(\omega_0)]\hat{\mathbf{k}} \cdot \mathbf{r}} + \right.$$

$$\left. + \tilde{F}(\omega + \omega_0) e^{-j[\beta(-\omega_0) + \beta'(-\omega_0)(\omega + \omega_0)]\hat{\mathbf{k}} \cdot \mathbf{r}} e^{-[\alpha(-\omega_0)]\hat{\mathbf{k}} \cdot \mathbf{r}} \right] e^{j\omega t} \, d\omega. \tag{4.235}$$

By (4.228) we know that α is even in ω and β is odd in ω. Since the derivative of an odd function is an even function, we also know that β' is even in ω. We can therefore write (4.235) as

$$\hat{\mathbf{e}}E(\mathbf{r},t) = \hat{\mathbf{e}}E_0 e^{-\alpha(\omega_0)\hat{\mathbf{k}}\cdot\mathbf{r}}\frac{1}{2\pi}\int_{-\infty}^{\infty}\frac{1}{2}\left[\tilde{F}(\omega-\omega_0)e^{-j\beta(\omega_0)\hat{\mathbf{k}}\cdot\mathbf{r}}e^{-j\beta'(\omega_0)(\omega-\omega_0)\hat{\mathbf{k}}\cdot\mathbf{r}}+\right.$$

$$\left.+\ \tilde{F}(\omega+\omega_0)e^{j\beta(\omega_0)\hat{\mathbf{k}}\cdot\mathbf{r}}e^{-j\beta'(\omega_0)(\omega+\omega_0)\hat{\mathbf{k}}\cdot\mathbf{r}}\right]e^{j\omega t}\,d\omega.$$

Multiplying and dividing by $e^{j\omega_0 t}$ and rearranging, we have

$$\hat{\mathbf{e}}E(\mathbf{r},t) = \hat{\mathbf{e}}E_0 e^{-\alpha(\omega_0)\hat{\mathbf{k}}\cdot\mathbf{r}}\frac{1}{2\pi}\int_{-\infty}^{\infty}\frac{1}{2}\left[\tilde{F}(\omega-\omega_0)e^{j\phi}e^{j(\omega-\omega_0)[t-\tau]}+\right.$$

$$\left.+\ \tilde{F}(\omega+\omega_0)e^{-j\phi}e^{j(\omega+\omega_0)[t-\tau]}\right]d\omega$$

where

$$\phi = \omega_0 t - \beta(\omega_0)\hat{\mathbf{k}}\cdot\mathbf{r}, \qquad \tau = \beta'(\omega_0)\hat{\mathbf{k}}\cdot\mathbf{r}.$$

Setting $u = \omega - \omega_0$ in the first term and $u = \omega + \omega_0$ in the second term we have

$$\hat{\mathbf{e}}E(\mathbf{r},t) = \hat{\mathbf{e}}E_0 e^{-\alpha(\omega_0)\hat{\mathbf{k}}\cdot\mathbf{r}}\cos\phi\frac{1}{2\pi}\int_{-\infty}^{\infty}\tilde{F}(u)e^{ju(t-\tau)}\,du.$$

Finally, the time-shifting theorem (Λ.3) gives us the time-domain wave field

$$\hat{\mathbf{e}}E(\mathbf{r},t) = \hat{\mathbf{e}}E_0 e^{-\alpha(\omega_0)\hat{\mathbf{k}}\cdot\mathbf{r}}\cos\left(\omega_0\left[t-\hat{\mathbf{k}}\cdot\mathbf{r}/v_p(\omega_0)\right]\right)f\left(t-\hat{\mathbf{k}}\cdot\mathbf{r}/v_g(\omega_0)\right) \qquad (4.236)$$

where

$$v_g(\omega) = d\omega/d\beta = [d\beta/d\omega]^{-1} \qquad (4.237)$$

is called the *group velocity* and

$$v_p(\omega) = \omega/\beta$$

is called the *phase velocity*.

To interpret (4.236), we note that at any given time t the field is constant over the surface described by

$$\hat{\mathbf{k}}\cdot\mathbf{r} = C \qquad (4.238)$$

where C is some constant. This surface is a plane, as shown in Figure 4.10, with its normal along $\hat{\mathbf{k}}$. It is easy to verify that any point $\mathbf{r}$ on this plane satisfies (4.238). Let $\mathbf{r}_0 = r_0\hat{\mathbf{k}}$ describe the point on the plane with position vector in the direction of $\hat{\mathbf{k}}$, and let $\mathbf{d}$ be a displacement vector from this point to any other point on the plane. Then

$$\hat{\mathbf{k}}\cdot\mathbf{r} = \hat{\mathbf{k}}\cdot(\mathbf{r}_0+\mathbf{d}) = r_0(\hat{\mathbf{k}}\cdot\hat{\mathbf{k}})+\hat{\mathbf{k}}\cdot\mathbf{d}.$$

But $\hat{\mathbf{k}}\cdot\mathbf{d} = 0$, so

$$\hat{\mathbf{k}}\cdot\mathbf{r} = r_0, \qquad (4.239)$$

which is a fixed distance, so (4.238) holds.

Let us identify the plane over which the envelope f takes on a certain value, and follow its motion as time progresses. The value of r_0 associated with this plane must increase with increasing time in such a way that the argument of f remains constant:

$$t - r_0/v_g(\omega_0) = C.$$

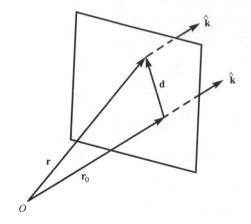

Figure 4.10: Surface of constant $\hat{\mathbf{k}} \cdot \mathbf{r}$.

Differentiation gives

$$\frac{dr_0}{dt} = v_g = \frac{d\omega}{d\beta}. \tag{4.240}$$

So the envelope propagates along $\hat{\mathbf{k}}$ at a rate given by the group velocity v_g. Associated with this propagation is an *attenuation* described by the factor $e^{-\alpha(\omega_0)\hat{\mathbf{k}}\cdot\mathbf{r}}$. This accounts for energy transfer into the lossy medium through Joule heating.

Similarly, we can identify a plane over which the phase of the carrier is constant; this will be parallel to the plane of constant envelope described above. We now set

$$\omega_0 \left[t - \hat{\mathbf{k}} \cdot \mathbf{r}/v_p(\omega_0) \right] = C$$

and differentiate to get

$$\frac{dr_0}{dt} = v_p = \frac{\omega}{\beta}. \tag{4.241}$$

This shows that surfaces of constant carrier phase propagate along $\hat{\mathbf{k}}$ with velocity v_p.

Caution must be exercised in interpreting the two velocities v_g and v_p; in particular, we must be careful not to associate the propagation velocities of energy or information with v_p. Since envelope propagation represents the actual progression of the disturbance, v_g has the recognizable physical meaning of energy velocity. Kraus and Fleisch [105] suggest that we think of a strolling caterpillar: the speed (v_p) of the undulations along the caterpillar's back (representing the carrier wave) may be much faster than the speed (v_g) of the caterpillar's body (representing the envelope of the disturbance).

In fact, v_g is the velocity of energy propagation even for a monochromatic wave (§ 4.11.4). However, for purely monochromatic waves v_g cannot be identified from the time-domain field, whereas v_p can. This leads to some unfortunate misconceptions, especially when v_p exceeds the speed of light. Since v_p is not the velocity of propagation of a physical quantity, but is rather the rate of change of a phase reference point, Einstein's postulate of c as the limiting velocity is not violated.

We can obtain interesting relationships between v_p and v_g by manipulating (4.237) and (4.241). For instance, if we compute

$$\frac{dv_p}{d\omega} = \frac{d}{d\omega}\left(\frac{\omega}{\beta}\right) = \frac{\beta - \omega\frac{d\beta}{d\omega}}{\beta^2}$$

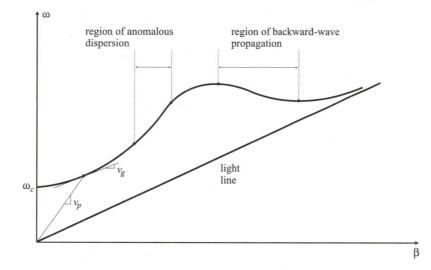

Figure 4.11: An ω–β diagram for a fictitious material.

we find that

$$\frac{v_p}{v_g} = 1 - \beta \frac{dv_p}{d\omega}. \tag{4.242}$$

Hence in frequency ranges where v_p decreases with increasing frequency, we have $v_g < v_p$. These are known as regions of *normal dispersion*. In frequency ranges where v_p increases with increasing frequency, we have $v_g > v_p$. These are known as regions of *anomalous dispersion*. As mentioned in § 4.6.3, the word "anomalous" does not imply that this type of dispersion is unusual.

The propagation of a uniform plane wave through a lossless medium provides a particularly simple example. In a lossless medium we have

$$\beta(\omega) = \omega\sqrt{\mu\epsilon}, \qquad \alpha(\omega) = 0.$$

In this case (4.233) becomes

$$\beta(\omega) = \omega_0\sqrt{\mu\epsilon} + \sqrt{\mu\epsilon}(\omega - \omega_0) = \omega\sqrt{\mu\epsilon}$$

and (4.236) becomes

$$\hat{\mathbf{e}}E(\mathbf{r}, t) = \hat{\mathbf{e}}E_0 \cos\left(\omega_0\left[t - \hat{\mathbf{k}}\cdot\mathbf{r}/v_p(\omega_0)\right]\right) f\left(t - \hat{\mathbf{k}}\cdot\mathbf{r}/v_g(\omega_0)\right).$$

Since the linear approximation to the phase constant β is in this case exact, the wave packet truly propagates without distortion, with a group velocity identical to the phase velocity:

$$v_g = \left[\frac{d}{d\omega}\omega\sqrt{\mu\epsilon}\right]^{-1} = \frac{1}{\sqrt{\mu\epsilon}} = \frac{\omega}{\beta} = v_p.$$

Examples of wave propagation in various media; the ω–β diagram. A plot of ω versus $\beta(\omega)$ can be useful for displaying the dispersive properties of a material. Figure 4.11 shows such an ω–β plot, or *dispersion diagram*, for a fictitious material. The

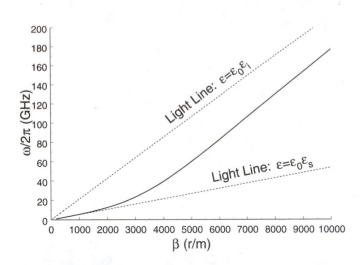

Figure 4.12: Dispersion plot for water computed using the Debye relaxation formula.

slope of the line from the origin to a point (β, ω) is the phase velocity, while the slope of the line tangent to the curve at that point is the group velocity. This plot shows many of the different characteristics of electromagnetic waves (although not necessarily of plane waves). For instance, there may be a minimum frequency ω_c called the *cutoff frequency* at which $\beta = 0$ and below which the wave cannot propagate. This behavior is characteristic of a plane wave propagating in a plasma (as shown below) or of a wave in a hollow pipe waveguide (§ 5.4.3). Over most values of β we have $v_g < v_p$ so the material demonstrates normal dispersion. However, over a small region we do have anomalous dispersion. In another range the slope of the curve is actually negative and thus $v_g < 0$; here the directions of energy and phase front propagation are opposite. Such *backward waves* are encountered in certain guided-wave structures used in microwave oscillators. The ω–β plot also includes the *light line* as a reference curve. For all points on this line $v_g = v_p$; it is generally used to represent propagation within the material under special circumstances, such as when the loss is zero or the material occupies unbounded space. It may also be used to represent propagation within a vacuum.

As an example for which the constitutive parameters depend on frequency, let us consider the relaxation effects of water. By the Debye formula (4.106) we have

$$\tilde{\epsilon}(\omega) = \epsilon_\infty + \frac{\epsilon_s - \epsilon_\infty}{1 + j\omega\tau}.$$

Assuming $\epsilon_\infty = 5\epsilon_0$, $\epsilon_s = 78.3\epsilon_0$, and $\tau = 9.6 \times 10^{-12}$ s [49], we obtain the relaxation spectrum shown in Figure 4.5. If we also assume that $\mu = \mu_0$, we may compute β as a function of ω and construct the ω–β plot. This is shown in Figure 4.12. Since ϵ' varies with frequency, we show both the light line for zero frequency found using $\epsilon_s = 78.3\epsilon_0$, and the light line for infinite frequency found using $\epsilon_i = 5\epsilon_0$. We see that at low values of frequency the dispersion curve follows the low-frequency light line very closely, and thus $v_p \approx v_g \approx c/\sqrt{78.3}$. As the frequency increases, the dispersion curve rises up and

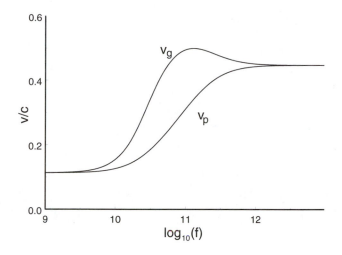

Figure 4.13: Phase and group velocities for water computed using the Debye relaxation formula.

eventually becomes asymptotic with the high-frequency light line. Plots of v_p and v_g shown in Figure 4.13 verify that the velocities start out at $c/\sqrt{78.3}$ for low frequencies, and approach $c/\sqrt{5}$ for high frequencies. Because $v_g > v_p$ at all frequencies, this model of water demonstrates anomalous dispersion.

Another interesting example is that of a non-magnetized plasma. For a collisionless plasma we may set $\nu = 0$ in (4.76) to find

$$k = \begin{cases} \frac{\omega}{c}\sqrt{1 - \frac{\omega_p^2}{\omega^2}}, & \omega > \omega_p, \\ -j\frac{\omega}{c}\sqrt{\frac{\omega_p^2}{\omega^2} - 1}, & \omega < \omega_p. \end{cases}$$

Thus, when $\omega > \omega_p$ we have

$$\tilde{\mathbf{E}}(\mathbf{r}, \omega) = \tilde{E}_0(\omega) e^{-j\beta(\omega)\hat{\mathbf{k}}\cdot\mathbf{r}}$$

and so

$$\beta = \frac{\omega}{c}\sqrt{1 - \frac{\omega_p^2}{\omega^2}}, \qquad \alpha = 0.$$

In this case a plane wave propagates through the plasma without attenuation. However, when $\omega < \omega_p$ we have

$$\tilde{\mathbf{E}}(\mathbf{r}, \omega) = \tilde{E}_0(\omega) e^{-\alpha(\omega)\hat{\mathbf{k}}\cdot\mathbf{r}}$$

with

$$\alpha = \frac{\omega}{c}\sqrt{\frac{\omega_p^2}{\omega^2} - 1}, \qquad \beta = 0,$$

and a plane wave does not propagate, but only attenuates. Such a wave is called an *evanescent wave*. We say that for frequencies below ω_p the wave is *cut off*, and call ω_p the *cutoff frequency*.

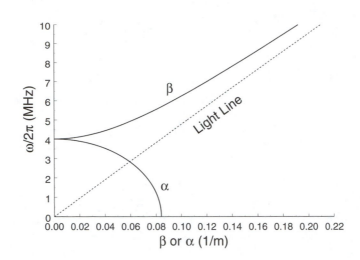

Figure 4.14: Dispersion plot for the ionosphere computed using $N_e = 2 \times 10^{11} m^{-3}$, $\nu = 0$. Light line computed using $\epsilon = \epsilon_0$, $\mu = \mu_0$.

Consider, for instance, a plane wave propagating in the earth's ionosphere. Both the electron density and the collision frequency are highly dependent on such factors as altitude, time of day, and latitude. However, except at the very lowest altitudes, the collision frequency is low enough that the ionosphere may be considered lossless. For instance, at a height of 200 km (the F_1 layer of the ionosphere), as measured for a mid-latitude region, we find that during the day the electron density is approximately $N_e = 2 \times 10^{11}$ m^{-3}, while the collision frequency is only $\nu = 100$ s^{-1} [16]. The attenuation is so small in this case that the ionosphere may be considered essentially lossless above the cutoff frequency (we will develop an approximate formula for the attenuation constant below). Figure 4.14 shows the ω–β diagram for the ionosphere assuming $\nu = 0$, along with the light line $v_p = c$. We see that above the cutoff frequency of $f_p = \omega_p/2\pi = 4.0$ MHz the wave propagates and that $v_g < c$ while $v_p > c$. Below the cutoff frequency the wave does not propagate and the field decays very rapidly because α is large.

A formula for the phase velocity of a plane wave in a lossless plasma is easily derived:

$$v_p = \frac{\omega}{\beta} = \frac{c}{\sqrt{1 - \frac{\omega_p^2}{\omega^2}}} > c.$$

Thus, our observation from the ω–β plot that $v_p > c$ is verified. Similarly, we find that

$$v_g = \left(\frac{d\beta}{d\omega}\right)^{-1} = \left(\frac{1}{c}\sqrt{1 - \frac{\omega_p^2}{\omega^2}} + \frac{1}{c}\frac{\omega_p^2/\omega^2}{\sqrt{1 - \frac{\omega_p^2}{\omega^2}}}\right)^{-1} = c\sqrt{1 - \frac{\omega_p^2}{\omega^2}} < c$$

and our observation that $v_g < c$ is also verified. Interestingly, we find that in this case of an unmagnetized collisionless plasma

$$v_p v_g = c^2.$$

Since $v_p > v_g$, this model of a plasma demonstrates normal dispersion at all frequencies above cutoff.

For the case of a plasma with collisions we retain ν in (4.76) and find that

$$k = \frac{\omega}{c}\sqrt{\left[1 - \frac{\omega_p^2}{\omega^2 + \nu^2}\right] - j\nu \frac{\omega_p^2}{\omega(\omega^2 + \nu^2)}}.$$

When $\nu \neq 0$ a true cutoff effect is not present and the wave may propagate at all frequencies. However, when $\nu \ll \omega_p$ the attenuation for propagating waves of frequency $\omega < \omega_p$ is quite severe, and for all practical purposes the wave is cut off. For waves of frequency $\omega > \omega_p$ there is attenuation. Assuming that $\nu \ll \omega_p$ and that $\nu \ll \omega$, we may approximate the square root with the first two terms of a binomial expansion, and find that to first order

$$\beta = \frac{\omega}{c}\sqrt{1 - \frac{\omega_p^2}{\omega^2}}, \qquad \alpha = \frac{1}{2}\frac{\nu}{c}\frac{\omega_p^2/\omega^2}{\sqrt{1 - \frac{\omega_p^2}{\omega^2}}}.$$

Hence the phase and group velocities above cutoff are essentially those of a lossless plasma, while the attenuation constant is directly proportional to ν.

4.11.4 Monochromatic plane waves in a lossy medium

Many properties of monochromatic plane waves are particularly simple. In fact, certain properties, such as wavelength, only have meaning for monochromatic fields. And since monochromatic or nearly monochromatic waves are employed extensively in radar, communications, and energy transport, it is useful to simplify the results of the preceding section for the special case in which the spectrum of the plane-wave signal consists of a single frequency component. In addition, plane waves of more general time dependence can be viewed as superpositions of individual single-frequency components (through the inverse Fourier transform), and thus we may regard monochromatic waves as building blocks for more complicated plane waves.

We can view the monochromatic field as a specialization of (4.230) for $a \to 0$. This results in $\tilde{F}(\omega) \to \delta(\omega)$, so the linearly-polarized plane wave expression (4.232) reduces to

$$\hat{e}E(\mathbf{r}, t) = \hat{e}E_0 e^{-\alpha(\omega_0)[\hat{\mathbf{k}}\cdot\mathbf{r}]}\cos(\omega_0 t - j\beta(\omega_0)[\hat{\mathbf{k}}\cdot\mathbf{r}]). \tag{4.243}$$

It is convenient to represent monochromatic fields with frequency $\omega = \breve{\omega}$ in phasor form. The phasor form of (4.243) is

$$\breve{\mathbf{E}}(\mathbf{r}) = \hat{e}E_0 e^{-j\beta(\hat{\mathbf{k}}\cdot\mathbf{r})}e^{-\alpha(\hat{\mathbf{k}}\cdot\mathbf{r})} \tag{4.244}$$

where $\beta = \beta(\breve{\omega})$ and $\alpha = \alpha(\breve{\omega})$. We can identify a surface of constant phase as a locus of points obeying

$$\breve{\omega}t - \beta(\hat{\mathbf{k}}\cdot\mathbf{r}) - C_P \tag{4.245}$$

for some constant C_P. This surface is a plane, as shown in Figure 4.10, with its normal in the direction of $\hat{\mathbf{k}}$. It is easy to verify that any point $\mathbf{r}$ on this plane satisfies (4.245). Let $\mathbf{r}_0 = r_0\hat{\mathbf{k}}$ describe the point on the plane with position vector in the $\hat{\mathbf{k}}$ direction, and let $\mathbf{d}$ be a displacement vector from this point to any other point on the plane. Then

$$\hat{\mathbf{k}}\cdot\mathbf{r} = \hat{\mathbf{k}}\cdot(\mathbf{r}_0 + \mathbf{d}) = r_0(\hat{\mathbf{k}}\cdot\hat{\mathbf{k}}) + \hat{\mathbf{k}}\cdot\mathbf{d}.$$

But $\hat{\mathbf{k}} \cdot \mathbf{d} = 0$, so

$$\hat{\mathbf{k}} \cdot \mathbf{r} = r_0, \tag{4.246}$$

which is a spatial constant, hence (4.245) holds for any t. The planar surfaces described by (4.245) are *wavefronts*.

Note that surfaces of constant amplitude are determined by

$$\alpha(\hat{\mathbf{k}} \cdot \mathbf{r}) = C_A$$

where C_A is some constant. As with the phase term, this requires that $\hat{\mathbf{k}} \cdot \mathbf{r} = \text{constant}$, and thus surfaces of constant phase and surfaces of constant amplitude are coplanar. This is a property of uniform plane waves. We shall see later that nonuniform plane waves have planar surfaces that are not parallel.

The cosine term in (4.243) represents a *traveling wave*. As t increases, the argument of the cosine function remains unchanged as long as $\hat{\mathbf{k}} \cdot \mathbf{r}$ increases correspondingly. Thus the planar wavefronts propagate along $\hat{\mathbf{k}}$. As the wavefront progresses, the wave is attenuated because of the factor $e^{-\alpha(\hat{\mathbf{k}} \cdot \mathbf{r})}$. This accounts for energy transferred from the propagating wave to the surrounding medium via Joule heating.

Phase velocity of a uniform plane wave.   The propagation velocity of the progressing wavefront is found by differentiating (4.245) to get

$$\breve{\omega} - \beta \hat{\mathbf{k}} \cdot \frac{d\mathbf{r}}{dt} = 0.$$

By (4.246) we have

$$v_p = \frac{dr_0}{dt} = \frac{\breve{\omega}}{\beta}, \tag{4.247}$$

where the phase velocity v_p represents the propagation speed of the constant-phase surfaces. For the case of a lossy medium with frequency-independent constitutive parameters, (4.227) shows that

$$v_p \leq \frac{1}{\sqrt{\mu\epsilon}},$$

hence the phase velocity in a conducting medium cannot exceed that in a lossless medium with the same parameters μ and ϵ. We cannot draw this conclusion for a medium with frequency-dependent $\tilde{\mu}$ and $\tilde{\epsilon}^c$, since by (4.224) the value of $\breve{\omega}/\beta$ might be greater or less than $1/\sqrt{\tilde{\mu}'\tilde{\epsilon}^{c\prime}}$, depending on the ratios $\tilde{\mu}''/\tilde{\mu}'$ and $\tilde{\epsilon}^{c\prime\prime}/\tilde{\epsilon}^{c\prime}$.

Wavelength of a uniform plane wave.   Another important property of a uniform plane wave is the distance between adjacent wavefronts that produce the same value of the cosine function in (4.243). Note that the field amplitude may not be the same on these two surfaces because of possible attenuation of the wave. Let $\mathbf{r}_1$ and $\mathbf{r}_2$ be points on adjacent wavefronts. We require

$$\beta(\hat{\mathbf{k}} \cdot \mathbf{r}_1) = \beta(\hat{\mathbf{k}} \cdot \mathbf{r}_2) - 2\pi$$

or

$$\lambda = \hat{\mathbf{k}} \cdot (\mathbf{r}_2 - \mathbf{r}_1) = r_{02} - r_{01} = 2\pi/\beta.$$

We call λ the *wavelength*.

Polarization of a uniform plane wave. Plane-wave *polarization* describes the temporal evolution of the vector direction of the electric field, which depends on the manner in which the wave is generated. *Completely polarized* waves are produced by antennas or other equipment; these have a deterministic polarization state which may be described completely by three parameters as discussed below. *Randomly polarized* waves are emitted by some natural sources. *Partially polarized* waves, such as those produced by cosmic radio sources, contain both completely polarized and randomly polarized components. We shall concentrate on the description of completely polarized waves.

The polarization state of a completely polarized monochromatic plane wave propagating in a homogeneous, isotropic region may be described by superposing two simpler plane waves that propagate along the same direction but with different phases and spatially orthogonal electric fields. Without loss of generality we may study propagation along the z-axis and choose the orthogonal field directions to be along $\hat{\mathbf{x}}$ and $\hat{\mathbf{y}}$. So we are interested in the behavior of a wave with electric field

$$\check{\mathbf{E}}(\mathbf{r}) = \hat{\mathbf{x}} E_{x0} e^{j\phi_x} e^{-jkz} + \hat{\mathbf{y}} E_{y0} e^{j\phi_y} e^{-jkz}. \tag{4.248}$$

The time evolution of the direction of $\mathbf{E}$ must be examined in the time domain where we have

$$\mathbf{E}(\mathbf{r}, t) = \mathrm{Re}\left\{ \check{\mathbf{E}} e^{j\omega t} \right\} = \hat{\mathbf{x}} E_{x0} \cos(\omega t - kz + \phi_x) + \hat{\mathbf{y}} E_{y0} \cos(\omega t - kz + \phi_y)$$

and thus, by the identity $\cos(x + y) \equiv \cos x \cos y - \sin x \sin y$,

$$E_x = E_{x0} \left[\cos(\omega t - kz) \cos(\phi_x) - \sin(\omega t - kz) \sin(\phi_x) \right],$$
$$E_y = E_{y0} \left[\cos(\omega t - kz) \cos(\phi_y) - \sin(\omega t - kz) \sin(\phi_y) \right].$$

The tip of the vector $\mathbf{E}$ moves cyclically in the xy-plane with temporal period $T = \omega/2\pi$. Its locus may be found by eliminating the parameter t to obtain a relationship between E_{x0} and E_{y0}. Letting $\delta = \phi_y - \phi_x$ we note that

$$\frac{E_x}{E_{x0}} \sin \phi_y - \frac{E_y}{E_{y0}} \sin \phi_x = \cos(\omega t - kz) \sin \delta,$$
$$\frac{E_x}{E_{x0}} \cos \phi_y - \frac{E_y}{E_{y0}} \cos \phi_x = \sin(\omega t - kz) \sin \delta;$$

squaring these terms we find that

$$\left(\frac{E_x}{E_{x0}} \right)^2 + \left(\frac{E_y}{E_{y0}} \right)^2 - 2 \frac{E_x}{E_{x0}} \frac{E_y}{E_{y0}} \cos \delta = \sin^2 \delta,$$

which is the equation for the ellipse shown in Figure 4.15. By (4.223) the magnetic field of the plane wave is

$$\check{\mathbf{H}} = \frac{\hat{\mathbf{z}} \times \check{\mathbf{E}}}{\eta},$$

hence its tip also traces an ellipse in the xy-plane.

The tip of the electric field vector cycles around the *polarization ellipse* in the xy-plane once every T seconds. The sense of rotation is determined by the sign of δ, and is described by the terms *clockwise/counterclockwise* or *right-hand/left-hand*. There is some disagreement about how to do this. We shall adopt the IEEE definitions (IEEE Standard 145-1983 [189]) and associate with $\delta < 0$ rotation in the *right-hand sense*: if

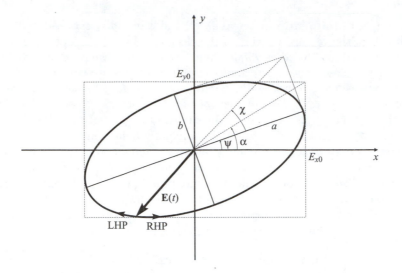

Figure 4.15: Polarization ellipse for a monochromatic plane wave.

the right thumb points in the direction of wave propagation then the fingers curl in the direction of field rotation for increasing time. This is *right-hand polarization* (*RHP*). We associate $\delta > 0$ with *left-hand polarization* (*LHP*).

The polarization ellipse is contained within a rectangle of sides $2E_{x0}$ and $2E_{y0}$, and has its major axis rotated from the x-axis by the *tilt angle* ψ, $0 \leq \psi \leq \pi$. The ratio of E_{y0} to E_{x0} determines an angle α, $0 \leq \alpha \leq \pi/2$:

$$E_{y0}/E_{x0} = \tan \alpha.$$

The shape of the ellipse is determined by the three parameters E_{x0}, E_{y0}, and δ, while the sense of polarization is described by the sign of δ. These may not, however, be the most convenient parameters for describing the polarization of a wave. We can also inscribe the ellipse within a box measuring $2a$ by $2b$, where a and b are the lengths of the semimajor and semiminor axes. Then b/a determines an angle χ, $-\pi/4 \leq \chi \leq \pi/4$, that is analogous to α:

$$\pm b/a = \tan \chi.$$

Here the algebraic sign of χ is used to indicate the sense of polarization: $\chi > 0$ for LHP, $\chi < 0$ for RHP.

The quantities a, b, ψ can also be used to describe the polarization ellipse. When we use the procedure outlined in Born and Wolf [19] to relate the quantities (a, b, ψ) to (E_{x0}, E_{y0}, δ), we find that

$$a^2 + b^2 = E_{x0}^2 + E_{y0}^2,$$

$$\tan 2\psi = (\tan 2\alpha) \cos \delta = \frac{2E_{x0}E_{y0}}{E_{x0}^2 - E_{y0}^2} \cos \delta,$$

$$\sin 2\chi = (\sin 2\alpha) \sin \delta = \frac{2E_{x0}E_{y0}}{E_{x0}^2 + E_{y0}^2} \sin \delta.$$

Alternatively, we can describe the polarization ellipse by the angles ψ and χ and one of the amplitudes E_{x0} or E_{y0}.

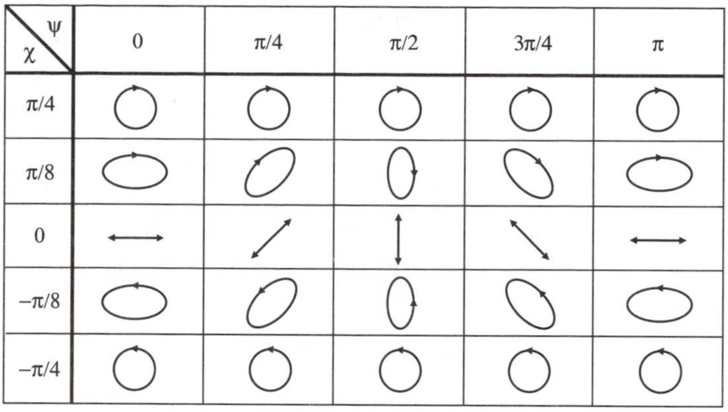

Figure 4.16: Polarization states as a function of tilt angle ψ and ellipse aspect ratio angle χ. Left-hand polarization for $\chi > 0$, right-hand for $\chi < 0$.

Each of these parameter sets is somewhat inconvenient since in each case the units differ among the parameters. In 1852 G. Stokes introduced a system of three independent quantities with identical dimension that can be used to describe plane-wave polarization. Various normalizations of these *Stokes parameters* are employed; when the parameters are chosen to have the dimension of power density we may write them as

$$s_0 = \frac{1}{2\eta} \left[E_{x0}^2 + E_{y0}^2 \right], \tag{4.249}$$

$$s_1 = \frac{1}{2\eta} \left[E_{x0}^2 - E_{y0}^2 \right] = s_0 \cos(2\chi) \cos(2\psi), \tag{4.250}$$

$$s_2 = \frac{1}{\eta} E_{x0} E_{y0} \cos\delta = s_0 \cos(2\chi) \sin(2\psi), \tag{4.251}$$

$$s_3 = \frac{1}{\eta} E_{x0} E_{y0} \sin\delta = s_0 \sin(2\chi). \tag{4.252}$$

Only three of these four parameters are independent since $s_0^2 = s_1^2 + s_2^2 + s_3^2$. Often the Stokes parameters are designated (I, Q, U, V) rather than (s_0, s_1, s_2, s_3).

Figure 4.16 summarizes various polarization states as a function of the angles ψ and χ. Two interesting special cases occur when $\chi = 0$ and $\chi = \pm\pi/4$. The case $\chi = 0$ corresponds to $b = 0$ and thus $\delta = 0$. In this case the electric vector traces out a straight line and we call the polarization *linear*. Here

$$\mathbf{E} = (\hat{\mathbf{x}} E_{x0} + \hat{\mathbf{y}} E_{y0}) \cos(\omega t - kz + \phi_x).$$

When $\psi = 0$ we have $E_{y0} = 0$ and refer to this as *horizontal linear polarization* (HLP); when $\psi = \pi/2$ we have $E_{x0} = 0$ and *vertical linear polarization* (VLP).

The case $\chi = \pm\pi/4$ corresponds to $b = a$ and $\delta = \pm\pi/2$. Thus $E_{x0} = E_{y0}$, and $\mathbf{E}$ traces out a circle regardless of the value of ψ. If $\chi = -\pi/4$ we have right-hand rotation of $\mathbf{E}$ and thus refer to this case as *right-hand circular polarization* (RHCP). If $\chi = \pi/4$ we have *left-hand circular polarization* (LHCP). For these cases

$$\mathbf{E} = E_{x0} \left[\hat{\mathbf{x}} \cos(\omega t - kz) \mp \hat{\mathbf{y}} \sin(\omega t - kz) \right],$$

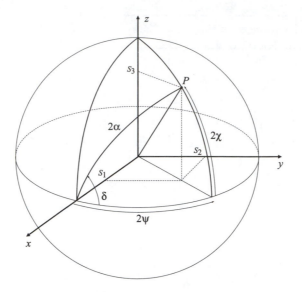

Figure 4.17: Graphical representation of the polarization of a monochromatic plane wave using the Poincaré sphere.

where the upper and lower signs correspond to LHCP and RHCP, respectively. All other values of χ result in the general cases of left-hand or right-hand *elliptical polarization*.

The French mathematician H. Poincaré realized that the Stokes parameters (s_1, s_2, s_3) describe a point on a sphere of radius s_0, and that this *Poincaré sphere* is useful for visualizing the various polarization states. Each state corresponds uniquely to one point on the sphere, and by (4.250)–(4.252) the angles 2χ and 2ψ are the spherical angular coordinates of the point as shown in Figure 4.17. We may therefore map the polarization states shown in Figure 4.16 directly onto the sphere: left- and right-hand polarizations appear in the upper and lower hemispheres, respectively; circular polarization appears at the poles ($2\chi = \pm\pi/2$); linear polarization appears on the equator ($2\chi = 0$), with HLP at $2\psi = 0$ and VLP at $2\psi = \pi$. The angles α and δ also have geometrical interpretations on the Poincaré sphere. The spherical angle of the great-circle route between the point of HLP and a point on the sphere is 2α, while the angle between the great-circle path and the equator is δ.

Uniform plane waves in a good dielectric.   We may base some useful plane-wave approximations on whether the real or imaginary part of $\tilde{\epsilon}^c$ dominates at the frequency of operation. We assume that $\tilde{\mu}(\omega) = \mu$ is independent of frequency and use the notation $\epsilon^c = \tilde{\epsilon}^c(\breve{\omega})$, $\sigma = \tilde{\sigma}(\breve{\omega})$, etc. Remember that

$$\epsilon^c = (\epsilon' + j\epsilon'') + \frac{\sigma}{j\breve{\omega}} = \epsilon' + j\left(\epsilon'' - \frac{\sigma}{\breve{\omega}}\right) = \epsilon^{c\prime} + j\epsilon^{c\prime\prime}.$$

By definition, a "good dielectric" obeys

$$\tan \delta_c = -\frac{\epsilon^{c\prime\prime}}{\epsilon^{c\prime}} = \frac{\sigma}{\breve{\omega}\epsilon'} - \frac{\epsilon''}{\epsilon'} \ll 1. \qquad (4.253)$$

Here $\tan \delta_c$ is the *loss tangent* of the material, as first described in (4.107) for a material without conductivity. For a good dielectric we have

$$k = \beta - j\alpha = \breve\omega\sqrt{\mu\epsilon^c} = \breve\omega\sqrt{\mu\left[\epsilon' + j\epsilon^{c''}\right]} = \breve\omega\sqrt{\mu\epsilon'}\sqrt{1 - j\tan\delta_c},$$

hence

$$k \approx \breve\omega\sqrt{\mu\epsilon'}\left[1 - j\frac{1}{2}\tan\delta_c\right] \tag{4.254}$$

by the binomial approximation for the square root. Therefore

$$\beta \approx \breve\omega\sqrt{\mu\epsilon'} \tag{4.255}$$

and

$$\alpha \approx \frac{\beta}{2}\tan\delta_c = \frac{\sigma}{2}\sqrt{\frac{\mu}{\epsilon'}}\left[1 - \frac{\breve\omega\epsilon''}{\sigma}\right]. \tag{4.256}$$

We conclude that $\alpha \ll \beta$. Using this and the binomial approximation we establish

$$\eta = \frac{\breve\omega\mu}{k} = \frac{\breve\omega\mu}{\beta}\frac{1}{1 - j\alpha/\beta} \approx \frac{\breve\omega\mu}{\beta}\left(1 + j\frac{\alpha}{\beta}\right).$$

Finally,

$$v_p = \frac{\breve\omega}{\beta} \approx \frac{1}{\sqrt{\mu\epsilon'}}$$

and

$$v_g = \left[\frac{d\beta}{d\omega}\right]^{-1} \approx \frac{1}{\sqrt{\mu\epsilon'}}.$$

To first order, the phase constant, phase velocity, and group velocity are the same as those of a lossless medium.

Uniform plane waves in a good conductor. We classify a material as a "good conductor" if

$$\tan\delta_c \approx \frac{\sigma}{\breve\omega\epsilon} \gg 1.$$

In a good conductor the conduction current $\sigma\breve{\mathbf{E}}$ is much greater than the displacement current $j\breve\omega\epsilon'\breve{\mathbf{E}}$, and ϵ'' is usually ignored. Now we may approximate

$$k = \beta - j\alpha = \breve\omega\sqrt{\mu\epsilon'}\sqrt{1 - j\tan\delta_c} \approx \breve\omega\sqrt{\mu\epsilon'}\sqrt{-j\tan\delta_c}.$$

Since $\sqrt{-j} = (1 - j)/\sqrt{2}$ we find that

$$\beta = \alpha \approx \sqrt{\pi f\mu\sigma}. \tag{4.257}$$

Hence

$$v_p = \frac{\breve\omega}{\beta} \approx \sqrt{\frac{2\breve\omega}{\mu\sigma}} = \frac{1}{\sqrt{\mu\epsilon'}}\sqrt{\frac{2}{\tan\delta_c}}.$$

To find v_g we must replace $\breve\omega$ by ω and differentiate, obtaining

$$v_g = \left[\frac{d\beta}{d\omega}\right]^{-1}\Bigg|_{\omega=\breve\omega} \approx \left[\frac{1}{2}\sqrt{\frac{\mu\sigma}{2\breve\omega}}\right]^{-1} = 2\sqrt{\frac{2\breve\omega}{\mu\sigma}} = 2v_p.$$

In a good conductor the group velocity is approximately twice the phase velocity. We could have found this relation from the phase velocity using (4.242). Indeed, noting that

$$\frac{dv_p}{d\omega} = \frac{d}{d\omega}\sqrt{\frac{2\omega}{\mu\sigma}} = \frac{1}{2}\sqrt{\frac{2}{\omega\mu\sigma}}$$

and

$$\beta\frac{dv_p}{d\omega} = \sqrt{\frac{\omega\mu\sigma}{2}}\frac{1}{2}\sqrt{\frac{2}{\omega\mu\sigma}} = \frac{1}{2},$$

we see that

$$\frac{v_p}{v_g} = 1 - \frac{1}{2} = \frac{1}{2}.$$

Note that the phase and group velocities may be only small fractions of the free-space light velocity. For example, in copper ($\sigma = 5.8 \times 10^7$ S/m, $\mu = \mu_0$, $\epsilon = \epsilon_0$) at 1 MHz, we have $v_p = 415$ m/s.

A factor often used to judge the quality of a conductor is the distance required for a propagating uniform plane wave to decrease in amplitude by the factor $1/e$. By (4.244) this distance is given by

$$\delta = \frac{1}{\alpha} = \frac{1}{\sqrt{\pi f \mu \sigma}}. \tag{4.258}$$

We call δ the *skin depth*. A good conductor is characterized by a small skin depth. For example, copper at 1 MHz has $\delta = 0.066$ mm.

Power carried by a uniform plane wave.   Since a plane wavefront is infinite in extent, we usually speak of the *power density* carried by the wave. This is identical to the time-average Poynting flux. Substitution from (4.223) and (4.244) gives

$$\mathbf{S}_{av} = \frac{1}{2}\operatorname{Re}\{\check{\mathbf{E}}\times\check{\mathbf{H}}^*\} = \frac{1}{2}\operatorname{Re}\left\{\check{\mathbf{E}}\times\left(\frac{\hat{\mathbf{k}}\times\check{\mathbf{E}}}{\eta}\right)^*\right\}. \tag{4.259}$$

Expanding the cross products and remembering that $\mathbf{k}\cdot\check{\mathbf{E}} = 0$, we get

$$\mathbf{S}_{av} = \frac{1}{2}\hat{\mathbf{k}}\operatorname{Re}\left\{\frac{|\check{\mathbf{E}}|^2}{\eta^*}\right\} = \hat{\mathbf{k}}\operatorname{Re}\left\{\frac{E_0^2}{2\eta^*}\right\}e^{-2\alpha\hat{\mathbf{k}}\cdot\mathbf{r}}.$$

Hence a uniform plane wave propagating in an isotropic medium carries power in the direction of wavefront propagation.

Velocity of energy transport.   The group velocity (4.237) has an additional interpretation as the velocity of energy transport. If the time-average volume density of energy is given by

$$\langle w_{em}\rangle = \langle w_e\rangle + \langle w_m\rangle$$

and the time-average volume density of energy flow is given by the Poynting flux density

$$\mathbf{S}_{av} = \frac{1}{2}\operatorname{Re}\left\{\check{\mathbf{E}}(\mathbf{r})\times\check{\mathbf{H}}^*(\mathbf{r})\right\} = \frac{1}{4}\left[\check{\mathbf{E}}(\mathbf{r})\times\check{\mathbf{H}}^*(\mathbf{r}) + \check{\mathbf{E}}^*(\mathbf{r})\times\check{\mathbf{H}}(\mathbf{r})\right], \tag{4.260}$$

then the velocity of energy flow, $\mathbf{v}_e$, is defined by

$$\mathbf{S}_{av} = \langle w_{em}\rangle\mathbf{v}_e. \tag{4.261}$$

Let us calculate $\mathbf{v}_e$ for a plane wave propagating in a lossless, source-free medium where $\mathbf{k} = \hat{\mathbf{k}}\omega\sqrt{\mu\epsilon}$. By (4.216) and (4.223) we have

$$\tilde{\mathbf{E}}(\mathbf{r}, \omega) = \tilde{\mathbf{E}}_0(\omega)e^{-j\beta\hat{\mathbf{k}}\cdot\mathbf{r}}, \tag{4.262}$$

$$\tilde{\mathbf{H}}(\mathbf{r}, \omega) = \left(\frac{\hat{\mathbf{k}} \times \tilde{\mathbf{E}}_0(\omega)}{\eta}\right)e^{-j\beta\hat{\mathbf{k}}\cdot\mathbf{r}} = \tilde{\mathbf{H}}_0(\omega)e^{-j\beta\hat{\mathbf{k}}\cdot\mathbf{r}}. \tag{4.263}$$

We can compute the time-average stored energy density using the energy theorem (4.68). In point form we have

$$-\nabla \cdot \left(\tilde{\mathbf{E}}^* \times \frac{\partial\tilde{\mathbf{H}}}{\partial\omega} + \frac{\partial\tilde{\mathbf{E}}}{\partial\omega} \times \tilde{\mathbf{H}}^*\right)\Bigg|_{\omega=\breve{\omega}} = 4j\langle w_{em}\rangle. \tag{4.264}$$

Upon substitution of (4.262) and (4.263) we find that we need to compute the frequency derivatives of $\tilde{\mathbf{E}}$ and $\tilde{\mathbf{H}}$. Using

$$\frac{\partial}{\partial\omega}e^{-j\beta\hat{\mathbf{k}}\cdot\mathbf{r}} = \left(\frac{\partial}{\partial\beta}e^{-j\beta\hat{\mathbf{k}}\cdot\mathbf{r}}\right)\frac{d\beta}{d\omega} = -j\hat{\mathbf{k}}\cdot\mathbf{r}\frac{d\beta}{d\omega}e^{-j\beta\hat{\mathbf{k}}\cdot\mathbf{r}}$$

and remembering that $\mathbf{k} = \hat{\mathbf{k}}\beta$, we have

$$\frac{\partial\tilde{\mathbf{E}}(\mathbf{r}, \omega)}{\partial\omega} = \frac{d\tilde{\mathbf{E}}_0(\omega)}{d\omega}e^{-j\mathbf{k}\cdot\mathbf{r}} + \tilde{\mathbf{E}}_0(\omega)\left(-j\mathbf{r}\cdot\frac{d\mathbf{k}}{d\omega}\right)e^{-j\mathbf{k}\cdot\mathbf{r}},$$

$$\frac{\partial\tilde{\mathbf{H}}(\mathbf{r}, \omega)}{\partial\omega} = \frac{d\tilde{\mathbf{H}}_0(\omega)}{d\omega}e^{-j\mathbf{k}\cdot\mathbf{r}} + \tilde{\mathbf{H}}_0(\omega)\left(-j\mathbf{r}\cdot\frac{d\mathbf{k}}{d\omega}\right)e^{-j\mathbf{k}\cdot\mathbf{r}}.$$

Equation (4.264) becomes

$$-\nabla \cdot \left\{\tilde{\mathbf{E}}_0^*(\omega) \times \frac{d\tilde{\mathbf{H}}_0(\omega)}{d\omega} + \frac{d\tilde{\mathbf{E}}_0(\omega)}{d\omega} \times \tilde{\mathbf{H}}_0^*(\omega)-\right.$$

$$\left.-j\mathbf{r}\cdot\frac{d\mathbf{k}}{d\omega}\left[\tilde{\mathbf{E}}_0^*(\omega) \times \tilde{\mathbf{H}}_0(\omega) + \tilde{\mathbf{E}}_0(\omega) \times \tilde{\mathbf{H}}_0^*(\omega)\right]\right\}\Bigg|_{\omega=\breve{\omega}} = 4j\langle w_{em}\rangle.$$

The first two terms on the left-hand side have zero divergence, since these terms do not depend on $\mathbf{r}$. By the product rule (B.42) we have

$$\left[\tilde{\mathbf{E}}_0^*(\breve{\omega}) \times \tilde{\mathbf{H}}_0(\breve{\omega}) + \tilde{\mathbf{E}}_0(\breve{\omega}) \times \tilde{\mathbf{H}}_0^*(\breve{\omega})\right] \cdot \nabla\left(\mathbf{r}\cdot\frac{d\mathbf{k}}{d\omega}\right)\Bigg|_{\omega=\breve{\omega}} = 4\langle w_{em}\rangle.$$

The gradient term is merely

$$\nabla\left(\mathbf{r}\cdot\frac{d\mathbf{k}}{d\omega}\right)\Bigg|_{\omega=\breve{\omega}} = \nabla\left(x\frac{dk_x}{d\omega} + y\frac{dk_y}{d\omega} + z\frac{dk_z}{d\omega}\right)\Bigg|_{\omega=\breve{\omega}} = \frac{d\mathbf{k}}{d\omega}\Bigg|_{\omega=\breve{\omega}},$$

hence

$$\left[\tilde{\mathbf{E}}_0^*(\breve{\omega}) \times \tilde{\mathbf{H}}_0(\breve{\omega}) + \tilde{\mathbf{E}}_0(\breve{\omega}) \times \tilde{\mathbf{H}}_0^*(\breve{\omega})\right] \cdot \frac{d\mathbf{k}}{d\omega}\Bigg|_{\omega=\breve{\omega}} = 4\langle w_{em}\rangle. \tag{4.265}$$

Finally, the left-hand side of this expression can be written in terms of the time-average Poynting vector. By (4.260) we have

$$\mathbf{S}_{av} = \frac{1}{2}\operatorname{Re}\left\{\breve{\mathbf{E}} \times \breve{\mathbf{H}}^*\right\} = \frac{1}{4}\left[\tilde{\mathbf{E}}_0(\breve{\omega}) \times \tilde{\mathbf{H}}_0^*(\breve{\omega}) + \tilde{\mathbf{E}}_0^*(\breve{\omega}) \times \tilde{\mathbf{H}}_0(\breve{\omega})\right]$$

and thus we can write (4.265) as

$$\mathbf{S}_{av} \cdot \frac{d\mathbf{k}}{d\omega}\bigg|_{\omega=\breve{\omega}} = \langle w_{em} \rangle.$$

Since for a uniform plane wave in an isotropic medium $\mathbf{k}$ and $\mathbf{S}_{av}$ are in the same direction, we have

$$\mathbf{S}_{av} = \hat{\mathbf{k}} \frac{d\omega}{d\beta}\bigg|_{\omega=\breve{\omega}} \langle w_{em} \rangle$$

and the velocity of energy transport for a plane wave of frequency $\breve{\omega}$ is then

$$\mathbf{v}_e = \hat{\mathbf{k}} \frac{d\omega}{d\beta}\bigg|_{\omega=\breve{\omega}}.$$

Thus, for a uniform plane wave in a lossless medium the velocity of energy transport is identical to the group velocity.

Nonuniform plane waves. A nonuniform plane wave has the same form (4.216) as a uniform plane wave, but the vectors $\mathbf{k}'$ and $\mathbf{k}''$ described in (4.217) are not aligned. Thus

$$\breve{\mathbf{E}}(\mathbf{r}) = \mathbf{E}_0 e^{-j\mathbf{k}' \cdot \mathbf{r}} e^{\mathbf{k}'' \cdot \mathbf{r}}.$$

In the time domain this becomes

$$\breve{\mathbf{E}}(\mathbf{r}) = \mathbf{E}_0 e^{\mathbf{k}'' \cdot \mathbf{r}} \cos[\breve{\omega}t - k'(\hat{\mathbf{k}}' \cdot \mathbf{r})]$$

where $\mathbf{k}' = \hat{\mathbf{k}}' k'$. The surfaces of constant phase are planes perpendicular to $\mathbf{k}'$ and propagating in the direction of $\hat{\mathbf{k}}'$. The phase velocity is now

$$v_p = \breve{\omega}/k'$$

and the wavelength is

$$\lambda = 2\pi/k'.$$

In contrast, surfaces of constant amplitude must obey

$$\mathbf{k}'' \cdot \mathbf{r} = C$$

and thus are planes perpendicular to $\mathbf{k}''$.

In a nonuniform plane wave the TEM nature of the fields is lost. This is easily seen by calculating $\breve{\mathbf{H}}$ from (4.219):

$$\breve{\mathbf{H}}(\mathbf{r}) = \frac{\mathbf{k} \times \breve{\mathbf{E}}(\mathbf{r})}{\breve{\omega}\mu} = \frac{\mathbf{k}' \times \breve{\mathbf{E}}(\mathbf{r})}{\breve{\omega}\mu} + j\frac{\mathbf{k}'' \times \breve{\mathbf{E}}(\mathbf{r})}{\breve{\omega}\mu}.$$

Thus, $\breve{\mathbf{H}}$ is no longer perpendicular to the direction of propagation of the phase front. The power carried by the wave also differs from that of the uniform case. The time-average Poynting vector

$$\mathbf{S}_{av} = \frac{1}{2} \operatorname{Re}\left\{ \breve{\mathbf{E}} \times \left(\frac{\mathbf{k} \times \breve{\mathbf{E}}}{\breve{\omega}\mu} \right)^* \right\}$$

can be expanded using the identity (B.7):

$$\mathbf{S}_{av} = \frac{1}{2} \operatorname{Re}\left\{ \frac{1}{\breve{\omega}\mu^*} \left[\mathbf{k}^* \times (\breve{\mathbf{E}} \times \breve{\mathbf{E}}^*) + \breve{\mathbf{E}}^* \times (\mathbf{k}^* \times \breve{\mathbf{E}}) \right] \right\}. \tag{4.266}$$

Since we still have $\mathbf{k} \cdot \mathbf{E} = 0$, we may use the rest of (B.7) to write

$$\check{\mathbf{E}}^* \times (\mathbf{k}^* \times \check{\mathbf{E}}) = \mathbf{k}^*(\check{\mathbf{E}} \cdot \check{\mathbf{E}}^*) + \check{\mathbf{E}}(\mathbf{k} \cdot \check{\mathbf{E}})^* = \mathbf{k}^*(\check{\mathbf{E}} \cdot \check{\mathbf{E}}^*).$$

Substituting this into (4.266), and noting that $\check{\mathbf{E}} \times \check{\mathbf{E}}^*$ is purely imaginary, we find

$$\mathbf{S}_{av} = \frac{1}{2} \mathrm{Re} \left\{ \frac{1}{\check{\omega}\mu^*} \left[j\mathbf{k}^* \times \mathrm{Im}\left\{ \check{\mathbf{E}} \times \check{\mathbf{E}}^* \right\} + \mathbf{k}^* |\check{\mathbf{E}}|^2 \right] \right\}. \tag{4.267}$$

Thus the vector direction of $\mathbf{S}_{av}$ is not generally in the direction of propagation of the plane wavefronts.

Let us examine the special case of nonuniform plane waves propagating in a lossless material. It is intriguing that $\mathbf{k}$ may be complex when k is real, and the implication is important for the plane-wave expansion of complicated fields in free space. By (4.218), real k requires that if $k'' \neq 0$ then

$$\mathbf{k}' \cdot \mathbf{k}'' = 0.$$

Thus, for a nonuniform plane wave in a lossless material the surfaces of constant phase and the surfaces of constant amplitude are orthogonal. To specialize the time-average power to the lossless case we note that μ is purely real and that

$$\mathbf{E} \times \mathbf{E}^* = (\mathbf{E}_0 \times \mathbf{E}_0^*)e^{2\mathbf{k}'' \cdot \mathbf{r}}.$$

Then (4.267) becomes

$$\mathbf{S}_{av} = \frac{1}{2\check{\omega}\mu} e^{2\mathbf{k}'' \cdot \mathbf{r}} \mathrm{Re} \left\{ j(\mathbf{k}' - j\mathbf{k}'') \times \mathrm{Im}\left\{ \mathbf{E}_0 \times \mathbf{E}_0^* \right\} + (\mathbf{k}' - j\mathbf{k}'')|\check{\mathbf{E}}|^2 \right\}$$

or

$$\mathbf{S}_{av} = \frac{1}{2\check{\omega}\mu} e^{2\mathbf{k}'' \cdot \mathbf{r}} \left[\mathbf{k}'' \times \mathrm{Im}\left\{ \mathbf{E}_0 \times \mathbf{E}_0^* \right\} + \mathbf{k}'|\check{\mathbf{E}}|^2 \right].$$

We see that in a lossless medium the direction of energy propagation is perpendicular to the surfaces of constant amplitude (since $\mathbf{k}'' \cdot \mathbf{S}_{av} = 0$), but the direction of energy propagation is not generally in the direction of propagation of the phase planes.

We shall encounter nonuniform plane waves when we study the reflection and refraction of a plane wave from a planar interface in the next section. We shall also find in § 4.13 that nonuniform plane waves are a necessary constituent of the angular spectrum representation of an arbitrary wave field.

4.11.5 Plane waves in layered media

A useful canonical problem in wave propagation involves the reflection of plane waves by planar interfaces between differing material regions. This has many direct applications, from the design of optical coatings and microwave absorbers to the probing of underground oil-bearing rock layers. We shall begin by studying the reflection of a plane wave at a single interface and then extend the results to any number of material layers.

Reflection of a uniform plane wave at a planar material interface. Consider two lossy media separated by the $z = 0$ plane as shown in Figure 4.18. The media are assumed to be isotropic and homogeneous with permeability $\tilde{\mu}(\omega)$ and complex permittivity $\tilde{\epsilon}^c(\omega)$. Both $\tilde{\mu}$ and $\tilde{\epsilon}^c$ may be complex numbers describing magnetic and dielectric loss, respectively. We assume that a linearly-polarized plane-wave field of the form (4.216) is created within region 1 by a process that we shall not study here. We take this field to be the known "incident wave" produced by an impressed source, and wish to compute the total field in regions 1 and 2. Here we shall assume that the incident field is that of a uniform plane wave, and shall extend the analysis to certain types of nonuniform plane waves subsequently.

Since the incident field is uniform, we may write the wave vector associated with this field as

$$\mathbf{k}^i = \hat{\mathbf{k}}^i k^i = \hat{\mathbf{k}}^i (k^{i\prime} + jk^{i\prime\prime})$$

where

$$[k^i(\omega)]^2 = \omega^2 \tilde{\mu}_1(\omega) \tilde{\epsilon}_1^c(\omega).$$

We can assume without loss of generality that $\hat{\mathbf{k}}^i$ lies in the xz-plane and makes an angle θ_i with the interface normal as shown in Figure 4.18. We refer to θ_i as the *incidence angle* of the incident field, and note that it is the angle between the direction of propagation of the planar phase fronts and the normal to the interface. With this we have

$$\mathbf{k}^i = \hat{\mathbf{x}} k_1 \sin\theta_i + \hat{\mathbf{z}} k_1 \cos\theta_i = \hat{\mathbf{x}} k_x^i + \hat{\mathbf{z}} k_z^i.$$

Using $k_1 = \beta_1 - j\alpha_1$ we also have

$$k_x^i = (\beta_1 - j\alpha_1) \sin\theta_i.$$

The term k_z^i is written in a somewhat different form in order to make the result easily applicable to reflections from multiple interfaces. We write

$$k_z^i = (\beta_1 - j\alpha_1) \cos\theta_i = \tau^i e^{-j\gamma^i} = \tau^i \cos\gamma^i - j\tau^i \sin\gamma^i.$$

Thus,

$$\tau^i = \sqrt{\beta_1^2 + \alpha_1^2} \cos\theta_i, \qquad \gamma^i = \tan^{-1}(\alpha_1/\beta_1).$$

We solve for the fields in each region of space directly in the frequency domain. The incident electric field has the form of (4.216),

$$\tilde{\mathbf{E}}^i(\mathbf{r}, \omega) = \tilde{\mathbf{E}}_0^i(\omega) e^{-j\mathbf{k}^i(\omega)\cdot\mathbf{r}}, \tag{4.268}$$

while the magnetic field is found from (4.219) to be

$$\tilde{\mathbf{H}}^i = \frac{\mathbf{k}^i \times \tilde{\mathbf{E}}^i}{\omega \tilde{\mu}_1}. \tag{4.269}$$

The incident field may be decomposed into two orthogonal components, one parallel to the plane of incidence (the plane containing $\hat{\mathbf{k}}$ and the interface normal $\hat{\mathbf{z}}$) and one perpendicular to this plane. We seek unique solutions for the fields in both regions, first for the case in which the incident electric field has only a parallel component, and then for the case in which it has only a perpendicular component. The total field is then determined by superposition of the individual solutions. For perpendicular polarization we have from (4.268) and (4.269)

$$\tilde{\mathbf{E}}_\perp^i = \hat{\mathbf{y}} \tilde{E}_\perp^i e^{-j(k_x^i x + k_z^i z)}, \tag{4.270}$$

$$\tilde{\mathbf{H}}_\perp^i = \frac{-\hat{\mathbf{x}} k_z^i + \hat{\mathbf{z}} k_x^i}{k_1} \frac{\tilde{E}_\perp^i}{\eta_1} e^{-j(k_x^i x + k_z^i z)}, \tag{4.271}$$

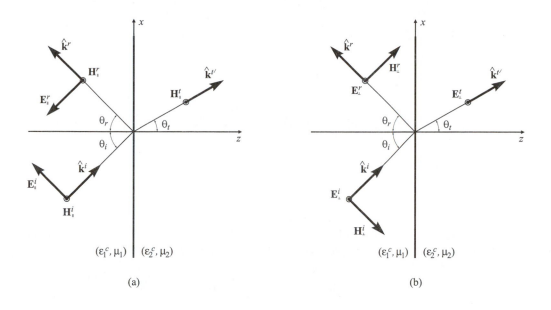

Figure 4.18: Uniform plane wave incident on planar interface between two lossy regions of space. (a) TM polarization, (b) TE polarization.

as shown graphically in Figure 4.18. Here $\eta_1 = (\tilde{\mu}_1/\tilde{\epsilon}_1^c)^{1/2}$ is the intrinsic impedance of medium 1. For parallel polarization, the direction of $\tilde{\mathbf{E}}$ is found by remembering that the wave must be TEM. Thus $\tilde{\mathbf{E}}_\parallel$ is perpendicular to $\mathbf{k}^i$. Since $\tilde{\mathbf{E}}_\parallel$ must also be perpendicular to $\tilde{\mathbf{E}}_\perp$, we have two possible directions for $\tilde{\mathbf{E}}_\parallel$. By convention we choose the one for which $\tilde{\mathbf{H}}$ lies in the same direction as did $\tilde{\mathbf{E}}$ for perpendicular polarization. Thus we have for parallel polarization

$$\tilde{\mathbf{H}}_\parallel^i = \hat{\mathbf{y}}\frac{\tilde{E}_\parallel^i}{\eta_1}e^{-j(k_x^i x + k_z^i z)}, \tag{4.272}$$

$$\tilde{\mathbf{E}}_\parallel^i = \frac{\hat{\mathbf{x}}k_z^i - \hat{\mathbf{z}}k_x^i}{k_1}\tilde{E}_\parallel^i e^{-j(k_x^i x + k_z^i z)}, \tag{4.273}$$

as shown in Figure 4.18. Because $\tilde{\mathbf{E}}$ lies transverse (normal) to the plane of incidence under perpendicular polarization, the field set is often described as *transverse electric* or *TE*. Because $\tilde{\mathbf{H}}$ lies transverse to the plane of incidence under parallel polarization, the fields in that case are *transverse magnetic* or *TM*.

Uniqueness requires that the total field obey the boundary conditions at the planar interface. We hypothesize that the total field within region 1 consists of the incident field superposed with a "reflected" plane-wave field having wave vector $\mathbf{k}^r$, while the field in region 2 consists of a single "transmitted" plane-wave field having wave vector $\mathbf{k}^t$. We cannot at the outset make any assumption regarding whether either of these fields are uniform plane waves. However, we do note that the reflected and transmitted fields cannot have vector components not present in the incident field; extra components would preclude satisfaction of the boundary conditions. Letting $\tilde{E}^r$ be the amplitude of the reflected plane-wave field we may write

$$\tilde{\mathbf{E}}_\perp^r = \hat{\mathbf{y}}\tilde{E}_\perp^r e^{-j(k_x^r x + k_z^r z)}, \qquad \tilde{\mathbf{H}}_\perp^r = \frac{-\hat{\mathbf{x}}k_z^r + \hat{\mathbf{z}}k_x^r}{k_1}\frac{\tilde{E}_\perp^r}{\eta_1}e^{-j(k_x^r x + k_z^r z)},$$

$$\tilde{\mathbf{H}}_{\parallel}^r = \hat{\mathbf{y}} \frac{\tilde{E}_{\parallel}^r}{\eta_1} e^{-j(k_x^r x + k_z^r z)}, \qquad \tilde{\mathbf{E}}_{\parallel}^r = \frac{\hat{\mathbf{x}} k_z^r - \hat{\mathbf{z}} k_x^r}{k_1} \tilde{E}_{\parallel}^r e^{-j(k_x^r x + k_z^r z)},$$

where $(k_x^r)^2 + (k_z^r)^2 = k_1^2$. Similarly, letting $\tilde{E}^t$ be the amplitude of the transmitted field we have

$$\tilde{\mathbf{E}}_{\perp}^t = \hat{\mathbf{y}} \tilde{E}_{\perp}^t e^{-j(k_x^t x + k_z^t z)}, \qquad \tilde{\mathbf{H}}_{\perp}^t = \frac{-\hat{\mathbf{x}} k_z^t + \hat{\mathbf{z}} k_x^t}{k_2} \frac{\tilde{E}_{\perp}^t}{\eta_2} e^{-j(k_x^t x + k_z^t z)},$$

$$\tilde{\mathbf{H}}_{\parallel}^t = \hat{\mathbf{y}} \frac{\tilde{E}_{\parallel}^t}{\eta_2} e^{-j(k_x^t x + k_z^t z)}, \qquad \tilde{\mathbf{E}}_{\parallel}^t = \frac{\hat{\mathbf{x}} k_z^t - \hat{\mathbf{z}} k_x^t}{k_2} \tilde{E}_{\parallel}^t e^{-j(k_x^t x + k_z^t z)},$$

where $(k_x^t)^2 + (k_z^t)^2 = k_2^2$.

The relationships between the field amplitudes $\tilde{E}^i$, $\tilde{E}^r$, $\tilde{E}^t$, and between the components of the reflected and transmitted wave vectors $\mathbf{k}^r$ and $\mathbf{k}^t$, can be found by applying the boundary conditions. The tangential electric and magnetic fields are continuous across the interface at $z = 0$:

$$\hat{\mathbf{z}} \times (\tilde{\mathbf{E}}^i + \tilde{\mathbf{E}}^r)|_{z=0} = \hat{\mathbf{z}} \times \tilde{\mathbf{E}}^t|_{z=0},$$
$$\hat{\mathbf{z}} \times (\tilde{\mathbf{H}}^i + \tilde{\mathbf{H}}^r)|_{z=0} = \hat{\mathbf{z}} \times \tilde{\mathbf{H}}^t|_{z=0}.$$

Substituting the field expressions, we find that for perpendicular polarization the two boundary conditions require

$$\tilde{E}_{\perp}^i e^{-jk_x^i x} + \tilde{E}_{\perp}^r e^{-jk_x^r x} = \tilde{E}_{\perp}^t e^{-jk_x^t x}, \qquad (4.274)$$

$$\frac{k_z^i}{k_1} \frac{\tilde{E}_{\perp}^i}{\eta_1} e^{-jk_x^i x} + \frac{k_z^r}{k_1} \frac{\tilde{E}_{\perp}^r}{\eta_1} e^{-jk_x^r x} = \frac{k_z^t}{k_2} \frac{\tilde{E}_{\perp}^t}{\eta_2} e^{-jk_x^t x}, \qquad (4.275)$$

while for parallel polarization they require

$$\frac{k_z^i}{k_1} \tilde{E}_{\parallel}^i e^{-jk_x^i x} + \frac{k_z^r}{k_1} \tilde{E}_{\parallel}^r e^{-jk_x^r x} = \frac{k_z^t}{k_2} \tilde{E}_{\parallel}^t e^{-jk_x^t x}, \qquad (4.276)$$

$$\frac{\tilde{E}_{\parallel}^i}{\eta_1} e^{-jk_x^i x} + \frac{\tilde{E}_{\parallel}^r}{\eta_1} e^{-jk_x^r x} = \frac{\tilde{E}_{\parallel}^t}{\eta_2} e^{-jk_x^t x}. \qquad (4.277)$$

For the above to hold for all x we must have the exponential terms equal. This requires

$$k_x^i = k_x^r = k_x^t, \qquad (4.278)$$

and also establishes a relation between k_z^i, k_z^r, and k_z^t. Since $(k_x^i)^2 + (k_z^i)^2 = (k_x^r)^2 + (k_z^r)^2 = k_1^2$, we must have $k_z^r = \pm k_z^i$. In order to make the reflected wavefronts propagate away from the interface we select $k_z^r = -k_z^i$. Letting $k_x^i = k_x^r = k_x^t = k_{1x}$ and $k_z^i = -k_z^r = k_{1z}$, we may write the wave vectors in region 1 as

$$\mathbf{k}^i = \hat{\mathbf{x}} k_{1x} + \hat{\mathbf{z}} k_{1z}, \qquad \mathbf{k}^r = \hat{\mathbf{x}} k_{1x} - \hat{\mathbf{z}} k_{1z}.$$

Since $(k_x^t)^2 + (k_z^t)^2 = k_2^2$, letting $k_2 = \beta_2 - j\alpha_2$ we have

$$k_z^t = \sqrt{k_2^2 - k_{1x}^2} = \sqrt{(\beta_2 - j\alpha_2)^2 - (\beta_1 - j\alpha_1)^2 \sin^2 \theta_i} = \tau^t e^{-j\gamma^t}.$$

Squaring out the above relation, we have

$$A - jB = (\tau^t)^2 \cos 2\gamma^t - j(\tau^t)^2 \sin 2\gamma^t$$

where

$$A = \beta_2^2 - \alpha_2^2 - (\beta_1^2 - \alpha_1^2) \sin^2 \theta_i, \qquad B = 2(\beta_2 \alpha_2 - \beta_1 \alpha_1 \sin^2 \theta_i). \qquad (4.279)$$

Thus

$$\tau^t = \left(A^2 + B^2\right)^{1/4}, \qquad \gamma^t = \frac{1}{2} \tan^{-1} \frac{B}{A}. \qquad (4.280)$$

Renaming k_z^t as k_{2z}, we may write the transmitted wave vector as

$$\mathbf{k}^t = \hat{\mathbf{x}} k_{1x} + \hat{\mathbf{z}} k_{2z} = \mathbf{k}_2' + j \mathbf{k}_2''$$

where

$$\mathbf{k}_2' = \hat{\mathbf{x}} \beta_1 \sin \theta_i + \hat{\mathbf{z}} \tau^t \cos \gamma^t, \qquad \mathbf{k}_2'' = -\hat{\mathbf{x}} \alpha_1 \sin \theta_i - \hat{\mathbf{z}} \tau^t \sin \gamma^t.$$

Since the direction of propagation of the transmitted field phase fronts is perpendicular to $\mathbf{k}_2'$, a unit vector in the direction of propagation is

$$\hat{\mathbf{k}}_2' = \frac{\hat{\mathbf{x}} \beta_1 \sin \theta_i + \hat{\mathbf{z}} \tau^t \cos \gamma^t}{\sqrt{\beta_1^2 \sin^2 \theta_i + (\tau^t)^2 \cos^2 \theta_i}}. \qquad (4.281)$$

Similarly, a unit vector perpendicular to planar surfaces of constant amplitude is given by

$$\hat{\mathbf{k}}_2'' = \frac{\hat{\mathbf{x}} \alpha_1 \sin \theta_i + \hat{\mathbf{z}} \tau^t \sin \gamma^t}{\sqrt{\alpha_1^2 \sin^2 \theta_i + (\tau^t)^2 \sin^2 \gamma^t}}. \qquad (4.282)$$

In general $\hat{\mathbf{k}}'$ is not aligned with $\hat{\mathbf{k}}''$ and thus the transmitted field is a nonuniform plane wave.

With these definitions of k_{1x}, k_{1z}, k_{2z}, equations (4.274) and (4.275) can be solved simultaneously and we have

$$\tilde{E}_\perp^r = \tilde{\Gamma}_\perp \tilde{E}_\perp^i, \qquad \tilde{E}_\perp^t = \tilde{T}_\perp \tilde{E}_\perp^i,$$

where

$$\tilde{\Gamma}_\perp = \frac{Z_{2\perp} - Z_{1\perp}}{Z_{2\perp} + Z_{1\perp}}, \qquad \tilde{T}_\perp = 1 + \tilde{\Gamma}_\perp = \frac{2 Z_{2\perp}}{Z_{2\perp} + Z_{1\perp}}, \qquad (4.283)$$

with

$$Z_{1\perp} = \frac{k_1 \eta_1}{k_{1z}}, \qquad Z_{2\perp} = \frac{k_2 \eta_2}{k_{2z}}.$$

Here $\tilde{\Gamma}$ is a frequency-dependent *reflection coefficient* that relates the tangential components of the incident and reflected electric fields, and $\tilde{T}$ is a frequency-dependent *transmission coefficient* that relates the tangential components of the incident and transmitted electric fields. These coefficients are also called the *Fresnel coefficients*.

For the case of parallel polarization we solve (4.276) and (4.277) to find

$$\frac{\tilde{E}_{\parallel,x}^r}{\tilde{E}_{\parallel,x}^i} = \frac{k_x^r}{k_x^i} \frac{\tilde{E}_\parallel^r}{\tilde{E}_\parallel^i} = -\frac{\tilde{E}_\parallel^r}{\tilde{E}_\parallel^i} = \tilde{\Gamma}_\parallel, \qquad \frac{\tilde{E}_{\parallel,x}^t}{\tilde{E}_{\parallel,x}^i} = \frac{(k_z^t / k_2) \tilde{E}_\parallel^t}{(k_z^i / k_1) \tilde{E}_\parallel^i} = \tilde{T}_\parallel.$$

Here

$$\tilde{\Gamma}_\parallel = \frac{Z_{2\parallel} - Z_{1\parallel}}{Z_{2\parallel} + Z_{1\parallel}}, \qquad \tilde{T}_\parallel = 1 + \tilde{\Gamma}_\parallel = \frac{2 Z_{2\parallel}}{Z_{2\parallel} + Z_{1\parallel}}, \qquad (4.284)$$

with

$$Z_{1\parallel} = \frac{k_{1z}\eta_1}{k_1}, \qquad Z_{2\parallel} = \frac{k_{2z}\eta_2}{k_2}.$$

Note that we may also write

$$\tilde{E}_{\parallel}^r = -\tilde{\Gamma}_{\parallel}\tilde{E}_{\parallel}^i, \qquad \tilde{E}_{\parallel}^t = \tilde{T}_{\parallel}\tilde{E}_{\parallel}^i \left(\frac{k_z^i}{k_1}\frac{k_2}{k_z^t}\right).$$

Let us summarize the fields in each region. For perpendicular polarization we have

$$\begin{aligned}
\tilde{\mathbf{E}}_{\perp}^i &= \hat{\mathbf{y}}\tilde{E}_{\perp}^i e^{-j\mathbf{k}^i\cdot\mathbf{r}}, \\
\tilde{\mathbf{E}}_{\perp}^r &= \hat{\mathbf{y}}\tilde{\Gamma}_{\perp}\tilde{E}_{\perp}^i e^{-j\mathbf{k}^r\cdot\mathbf{r}}, \\
\tilde{\mathbf{E}}_{\perp}^t &= \hat{\mathbf{y}}\tilde{T}_{\perp}\tilde{E}_{\perp}^i e^{-j\mathbf{k}^t\cdot\mathbf{r}},
\end{aligned} \tag{4.285}$$

and

$$\tilde{\mathbf{H}}_{\perp}^i = \frac{\mathbf{k}^i \times \tilde{\mathbf{E}}_{\perp}^i}{k_1\eta_1}, \quad \tilde{\mathbf{H}}_{\perp}^r = \frac{\mathbf{k}^r \times \tilde{\mathbf{E}}_{\perp}^r}{k_1\eta_1}, \quad \tilde{\mathbf{H}}_{\perp}^t = \frac{\mathbf{k}^t \times \tilde{\mathbf{E}}_{\perp}^t}{k_2\eta_2}. \tag{4.286}$$

For parallel polarization we have

$$\begin{aligned}
\tilde{\mathbf{E}}_{\parallel}^i &= -\eta_1 \frac{\mathbf{k}^i \times \tilde{\mathbf{H}}_{\parallel}^i}{k_1} e^{-j\mathbf{k}^i\cdot\mathbf{r}}, \\
\tilde{\mathbf{E}}_{\parallel}^r &= -\eta_1 \frac{\mathbf{k}^r \times \tilde{\mathbf{H}}_{\parallel}^r}{k_1} e^{-j\mathbf{k}^r\cdot\mathbf{r}}, \\
\tilde{\mathbf{E}}_{\parallel}^t &= -\eta_2 \frac{\mathbf{k}^t \times \tilde{\mathbf{H}}_{\parallel}^t}{k_2} e^{-j\mathbf{k}^t\cdot\mathbf{r}},
\end{aligned} \tag{4.287}$$

and

$$\begin{aligned}
\tilde{\mathbf{H}}_{\parallel}^i &= \hat{\mathbf{y}}\frac{\tilde{E}_{\parallel}^i}{\eta_1} e^{-j\mathbf{k}^i\cdot\mathbf{r}}, \\
\tilde{\mathbf{H}}_{\parallel}^r &= -\hat{\mathbf{y}}\frac{\tilde{\Gamma}_{\parallel}\tilde{E}_{\parallel}^i}{\eta_1} e^{-j\mathbf{k}^r\cdot\mathbf{r}}, \\
\tilde{\mathbf{H}}_{\parallel}^t &= \hat{\mathbf{y}}\frac{\tilde{T}_{\parallel}\tilde{E}_{\parallel}^i}{\eta_2} \left(\frac{k_z^i}{k_1}\frac{k_2}{k_z^t}\right) e^{-j\mathbf{k}^t\cdot\mathbf{r}}.
\end{aligned} \tag{4.288}$$

The wave vectors are given by

$$\mathbf{k}^i = (\hat{\mathbf{x}}\beta_1\sin\theta_i + \hat{\mathbf{z}}\tau^i\cos\gamma^i) - j(\hat{\mathbf{x}}\alpha_1\sin\theta_i + \hat{\mathbf{z}}\tau^i\sin\gamma^i), \tag{4.289}$$

$$\mathbf{k}^r = (\hat{\mathbf{x}}\beta_1\sin\theta_i - \hat{\mathbf{z}}\tau^i\cos\gamma^i) - j(\hat{\mathbf{x}}\alpha_1\sin\theta_i - \hat{\mathbf{z}}\tau^i\sin\gamma^i), \tag{4.290}$$

$$\mathbf{k}^t = (\hat{\mathbf{x}}\beta_1\sin\theta_i + \hat{\mathbf{z}}\tau^t\cos\gamma^t) - j(\hat{\mathbf{x}}\alpha_1\sin\theta_i + \hat{\mathbf{z}}\tau^t\sin\gamma^t). \tag{4.291}$$

We see that the reflected wave must, like the incident wave, be a uniform plane wave. We define the unsigned *reflection angle* θ_r as the angle between the surface normal and the direction of propagation of the reflected wavefronts (Figure 4.18). Since

$$\mathbf{k}^i \cdot \hat{\mathbf{z}} = k_1\cos\theta_i = -\mathbf{k}^r \cdot \hat{\mathbf{z}} = k_1\cos\theta_r$$

and

$$\mathbf{k}^i \cdot \hat{\mathbf{x}} = k_1\sin\theta_i = \mathbf{k}^r \cdot \hat{\mathbf{x}} = k_1\sin\theta_r$$

we must have

$$\theta_i = \theta_r.$$

This is known as *Snell's law of reflection*. We can similarly define the *transmission angle* to be the angle between the direction of propagation of the transmitted wavefronts and the interface normal. Noting that $\hat{\mathbf{k}}_2' \cdot \hat{\mathbf{z}} = \cos\theta_t$ and $\hat{\mathbf{k}}_2' \cdot \hat{\mathbf{x}} = \sin\theta_t$, we have from (4.281) and (4.282)

$$\cos\theta_t = \frac{\tau^t \cos\gamma^t}{\sqrt{\beta_1^2 \sin^2\theta_i + (\tau^t)^2 \cos^2\gamma^t}}, \tag{4.292}$$

$$\sin\theta_t = \frac{\beta_1 \sin\theta_i}{\sqrt{\beta_1^2 \sin^2\theta_i + (\tau^t)^2 \cos^2\gamma^t}}, \tag{4.293}$$

and thus

$$\theta_t = \tan^{-1}\left(\frac{\beta_1}{\tau^t}\frac{\sin\theta_i}{\cos\gamma^t}\right). \tag{4.294}$$

Depending on the properties of the media, at a certain incidence angle θ_c, called the *critical angle*, the angle of transmission becomes $\pi/2$. Under this condition $\mathbf{k}_2'$ has only an x-component. Thus, surfaces of constant phase propagate parallel to the interface. Later we shall see that for low-loss (or lossless) media, this implies that no time-average power is carried by a monochromatic transmitted wave into the second medium.

We also see that although the transmitted field may be a nonuniform plane wave, its mathematical form is that of the incident plane wave. This allows us to easily generalize the single-interface reflection problem to one involving many layers.

Uniform plane-wave reflection for lossless media. We can specialize the preceding results to the case for which both regions are lossless with $\tilde{\mu} = \mu$ and $\tilde{\epsilon}^c = \epsilon$ real and frequency-independent. By (4.224) we have

$$\beta = \omega\sqrt{\mu\epsilon},$$

while (4.225) gives

$$\alpha = 0.$$

We can easily show that the transmitted wave must be uniform unless the incidence angle exceeds the critical angle. By (4.279) we have

$$A = \beta_2^2 - \beta_1^2 \sin^2\theta_i, \qquad B = 0, \tag{4.295}$$

while (4.280) gives

$$\tau = \left[A^2\right]^{1/4} = \sqrt{|\beta_2^2 - \beta_1^2 \sin^2\theta_i|}$$

and

$$\gamma^t = \frac{1}{2}\tan^{-1}(0).$$

We have several possible choices for γ^t. To choose properly we note that γ^t represents the negative of the phase of the quantity $k_z^t = \sqrt{A}$. If $A > 0$ the phase of the square root is 0. If $A < 0$ the phase of the square root is $-\pi/2$ and thus $\gamma^t = +\pi/2$. Here we choose the plus sign on γ^t to ensure that the transmitted field decays as z increases. We note that if $A = 0$ then $\tau^t = 0$ and from (4.293) we have $\theta_t = \pi/2$. This defines the critical angle, which from (4.295) is

$$\theta_c = \sin^{-1}\left(\frac{\beta_2^2}{\beta_1^2}\right) = \sin^{-1}\left(\frac{\mu_2\epsilon_2}{\mu_1\epsilon_1}\right).$$

Therefore

$$\gamma^t = \begin{cases} 0, & \theta_i < \theta_c, \\ \pi/2, & \theta_i > \theta_c. \end{cases}$$

Using these we can write down the transmitted wave vector from (4.291):

$$\mathbf{k}^t = \mathbf{k}^{t'} + j\mathbf{k}^{t''} = \begin{cases} \hat{\mathbf{x}}\beta_1 \sin\theta_i + \hat{\mathbf{z}}\sqrt{|A|}, & \theta_i < \theta_c, \\ \hat{\mathbf{x}}\beta_1 \sin\theta_i - j\hat{\mathbf{z}}\sqrt{|A|}, & \theta_i > \theta_c. \end{cases} \tag{4.296}$$

By (4.293) we have

$$\sin\theta_t = \frac{\beta_1 \sin\theta_i}{\sqrt{\beta_1^2 \sin^2\theta_i + \beta_2^2 - \beta_1^2 \sin^2\theta_i}} = \frac{\beta_1 \sin\theta_i}{\beta_2}$$

or

$$\beta_2 \sin\theta_t = \beta_1 \sin\theta_i. \tag{4.297}$$

This is known as *Snell's law of refraction*. With this we can write for $\theta_i < \theta_c$

$$A = \beta_2^2 - \beta_1^2 \sin^2\theta_i = \beta_2^2 \cos^2\theta_t.$$

Using this and substituting $\beta_2 \sin\theta_t$ for $\beta_1 \sin\theta_i$, we may rewrite (4.296) for $\theta_i < \theta_c$ as

$$\mathbf{k}^t = \mathbf{k}^{t'} + j\mathbf{k}^{t''} = \hat{\mathbf{x}}\beta_2 \sin\theta_t + \hat{\mathbf{z}}\beta_2 \cos\theta_t. \tag{4.298}$$

Hence the transmitted plane wave is uniform with $\mathbf{k}^{t''} = 0$. When $\theta_i > \theta_c$ we have from (4.296)

$$\mathbf{k}^{t'} = \hat{\mathbf{x}}\beta_1 \sin\theta_i, \qquad \mathbf{k}^{t''} = -\hat{\mathbf{z}}\sqrt{\beta_1^2 \sin^2\theta_i - \beta_2^2}.$$

Since $\mathbf{k}^{t'}$ and $\mathbf{k}^{t''}$ are not collinear, the plane wave is nonuniform. Let us examine the cases $\theta_i < \theta_c$ and $\theta_i > \theta_c$ in greater detail.

Case 1: $\theta_i < \theta_c$. By (4.289)–(4.290) and (4.298) the wave vectors are

$$\mathbf{k}^i = \hat{\mathbf{x}}\beta_1 \sin\theta_i + \hat{\mathbf{z}}\beta_1 \cos\theta_i,$$
$$\mathbf{k}^r = \hat{\mathbf{x}}\beta_1 \sin\theta_i - \hat{\mathbf{z}}\beta_1 \cos\theta_i,$$
$$\mathbf{k}^t = \hat{\mathbf{x}}\beta_2 \sin\theta_t + \hat{\mathbf{z}}\beta_2 \cos\theta_t,$$

and the wave impedances are

$$Z_{1\perp} = \frac{\eta_1}{\cos\theta_i}, \qquad Z_{2\perp} = \frac{\eta_2}{\cos\theta_t},$$
$$Z_{1\|} = \eta_1 \cos\theta_i, \qquad Z_{2\|} = \eta_2 \cos\theta_t.$$

The reflection coefficients are

$$\tilde{\Gamma}_\perp = \frac{\eta_2 \cos\theta_i - \eta_1 \cos\theta_t}{\eta_2 \cos\theta_i + \eta_1 \cos\theta_t}, \qquad \tilde{\Gamma}_\| = \frac{\eta_2 \cos\theta_t - \eta_1 \cos\theta_i}{\eta_2 \cos\theta_t + \eta_1 \cos\theta_i}. \tag{4.299}$$

So the reflection coefficients are purely real, with signs dependent on the constitutive parameters of the media. We can write

$$\tilde{\Gamma}_\perp = \rho_\perp e^{j\phi_\perp}, \qquad \tilde{\Gamma}_\| = \rho_\| e^{j\phi_\|},$$

where ρ and ϕ are real, and where $\phi = 0$ or π.

Under certain conditions the reflection coefficients vanish. For a given set of constitutive parameters we may achieve $\tilde{\Gamma} = 0$ at an incidence angle θ_B, known as the *Brewster* or *polarizing angle*. A wave with an arbitrary combination of perpendicular and parallel polarized components incident at this angle produces a reflected field with a single component. A wave incident with only the appropriate single component produces no reflected field, regardless of its amplitude.

For perpendicular polarization we set $\tilde{\Gamma}_\perp = 0$, requiring

$$\eta_2 \cos\theta_i - \eta_1 \cos\theta_t = 0$$

or equivalently

$$\frac{\mu_2}{\epsilon_2}(1 - \sin^2\theta_i) = \frac{\mu_1}{\epsilon_1}(1 - \sin^2\theta_t).$$

By (4.297) we may put

$$\sin^2\theta_t = \frac{\mu_1\epsilon_1}{\mu_2\epsilon_2}\sin^2\theta_i,$$

resulting in

$$\sin^2\theta_i = \frac{\mu_2}{\epsilon_1}\frac{\epsilon_2\mu_1 - \epsilon_1\mu_2}{\mu_1^2 - \mu_2^2}.$$

The value of θ_i that satisfies this equation must be the Brewster angle, and thus

$$\theta_{B\perp} = \sin^{-1}\sqrt{\frac{\mu_2}{\epsilon_1}\frac{\epsilon_2\mu_1 - \epsilon_1\mu_2}{\mu_1^2 - \mu_2^2}}.$$

When $\mu_1 = \mu_2$ there is no solution to this equation, hence the reflection coefficient cannot vanish. When $\epsilon_1 = \epsilon_2$ we have

$$\theta_{B\perp} = \sin^{-1}\sqrt{\frac{\mu_2}{\mu_1 + \mu_2}} = \tan^{-1}\sqrt{\frac{\mu_2}{\mu_1}}.$$

For parallel polarization we set $\tilde{\Gamma}_\| = 0$ and have

$$\eta_2 \cos\theta_t = \eta_1 \cos\theta_i.$$

Proceeding as above we find that

$$\theta_{B\|} = \sin^{-1}\sqrt{\frac{\epsilon_2}{\mu_1}\frac{\epsilon_1\mu_2 - \epsilon_2\mu_1}{\epsilon_1^2 - \epsilon_2^2}}.$$

This expression has no solution when $\epsilon_1 = \epsilon_2$, and thus the reflection coefficient cannot vanish under this condition. When $\mu_1 = \mu_2$ we have

$$\theta_{B\|} = \sin^{-1}\sqrt{\frac{\epsilon_2}{\epsilon_1 + \epsilon_2}} = \tan^{-1}\sqrt{\frac{\epsilon_2}{\epsilon_1}}.$$

We find that when $\theta_i < \theta_c$ the total field in region 1 behaves as a traveling wave along x, but has characteristics of both a standing wave and a traveling wave along z (Problem 4.7). The traveling-wave component is associated with a Poynting power flux, while the standing-wave component is not. This flux is carried across the boundary into region 2 where the transmitted field consists only of a traveling wave. By (4.161) the normal component of time-average Poynting flux is continuous across the boundary, demonstrating that the time-average power carried by the wave into the interface from region 1 passes out through the interface into region 2 (Problem 4.8).

Case 2: $\theta_i < \theta_c$. The wave vectors are, from (4.289)–(4.290) and (4.296),

$$\mathbf{k}^i = \hat{\mathbf{x}}\beta_1 \sin\theta_i + \hat{\mathbf{z}}\beta_1 \cos\theta_i,$$
$$\mathbf{k}^r = \hat{\mathbf{x}}\beta_1 \sin\theta_i - \hat{\mathbf{z}}\beta_1 \cos\theta_i,$$
$$\mathbf{k}^t = \hat{\mathbf{x}}\beta_1 \sin\theta_i - j\hat{\mathbf{z}}\alpha_c,$$

where

$$\alpha_c = \sqrt{\beta_1^2 \sin^2\theta_i - \beta_2^2}$$

is the *critical angle attenuation constant*. The wave impedances are

$$Z_{1\perp} = \frac{\eta_1}{\cos\theta_i}, \qquad Z_{2\perp} = j\frac{\beta_2\eta_2}{\alpha_c},$$
$$Z_{1\|} = \eta_1 \cos\theta_i, \qquad Z_{2\|} = -j\frac{\alpha_c\eta_2}{\beta_2}.$$

Substituting these into (4.283) and (4.284), we find that the reflection coefficients are the complex quantities

$$\tilde{\Gamma}_\perp = \frac{\beta_2\eta_2 \cos\theta_i + j\eta_1\alpha_c}{\beta_2\eta_2 \cos\theta_i - j\eta_1\alpha_c} = e^{j\phi_\perp},$$
$$\tilde{\Gamma}_\| = -\frac{\beta_2\eta_1 \cos\theta_i + j\eta_2\alpha_c}{\beta_2\eta_1 \cos\theta_i - j\eta_2\alpha_c} = e^{j\phi_\|},$$

where

$$\phi_\perp = 2\tan^{-1}\left(\frac{\eta_1\alpha_c}{\beta_2\eta_2 \cos\theta_i}\right), \qquad \phi_\| = \pi + 2\tan^{-1}\left(\frac{\eta_2\alpha_c}{\beta_2\eta_1 \cos\theta_i}\right).$$

We note with interest that $\rho_\perp = \rho_\| = 1$. So the amplitudes of the reflected waves are identical to those of the incident waves, and we call this the case of *total internal reflection*. The phase of the reflected wave at the interface is changed from that of the incident wave by an amount $\phi_\perp$ or $\phi_\|$. The phase shift incurred by the reflected wave upon total internal reflection is called the *Goos–Hänchen shift*.

In the case of total internal reflection the field in region 1 is a pure standing wave while the field in region 2 decays exponentially in the z-direction and is evanescent (Problem 4.9). Since a standing wave transports no power, there is no Poynting flux into region 2. We find that the evanescent wave also carries no power and thus the boundary condition on power flux at the interface is satisfied (Problem 4.10). We note that for any incident angle except $\theta_i = 0$ (normal incidence) the wave in region 1 does transport power in the x-direction.

Reflection of time-domain uniform plane waves. Solution for the fields reflected and transmitted at an interface shows us the properties of the fields for a certain single excitation frequency and allows us to obtain time-domain fields by Fourier inversion. Under certain conditions it is possible to do the inversion analytically, providing physical insight into the temporal behavior of the fields.

As a simple example, consider a perpendicularly-polarized, uniform plane wave incident from free space at an angle θ_i on the planar surface of a conducting material (Figure 4.18). The material is assumed to have frequency-independent constitutive parameters $\tilde{\mu} = \mu_0$, $\tilde{\epsilon} = \epsilon$, and $\tilde{\sigma} = \sigma$. By (4.285) we have the reflected field

$$\tilde{\mathbf{E}}_\perp^r(\mathbf{r}, \omega) = \hat{\mathbf{y}}\tilde{\Gamma}_\perp(\omega)\tilde{E}_\perp^i(\omega)e^{-j\mathbf{k}^r(\omega)\cdot\mathbf{r}} = \hat{\mathbf{y}}\tilde{E}^r(\omega)e^{-j\omega\frac{\hat{\mathbf{k}}^r\cdot\mathbf{r}}{c}} \qquad (4.300)$$

where $\tilde{E}^r = \tilde{\Gamma}_\perp\tilde{E}_\perp^i$. We can use the time-shifting theorem (A.3) to invert the transform and obtain

$$\mathbf{E}_\perp^r(\mathbf{r}, t) = \mathcal{F}^{-1}\left\{\tilde{\mathbf{E}}_\perp^r(\mathbf{r}, \omega)\right\} = \hat{\mathbf{y}}E^r\left(t - \frac{\hat{\mathbf{k}}^r\cdot\mathbf{r}}{c}\right) \qquad (4.301)$$

where we have by the convolution theorem (12)

$$E^r(t) = \mathcal{F}^{-1}\left\{\tilde{E}^r(\omega)\right\} = \Gamma_\perp(t) * E_\perp(t).$$

Here

$$E_\perp(t) = \mathcal{F}^{-1}\left\{\tilde{E}_\perp^i(\omega)\right\}$$

is the time waveform of the incident plane wave, while

$$\Gamma_\perp(t) = \mathcal{F}^{-1}\left\{\tilde{\Gamma}_\perp(\omega)\right\}$$

is the time-domain reflection coefficient.

By (4.301) the reflected time-domain field propagates along the direction $\hat{\mathbf{k}}^r$ at the speed of light. The time waveform of the field is the convolution of the waveform of the incident field with the time-domain reflection coefficient $\Gamma_\perp(t)$. In the lossless case ($\sigma = 0$), $\Gamma_\perp(t)$ is a δ-function and thus the waveforms of the reflected and incident fields are identical. With the introduction of loss $\Gamma_\perp(t)$ broadens and thus the reflected field waveform becomes a convolution-broadened version of the incident field waveform. To understand the waveform of the reflected field we must compute $\Gamma_\perp(t)$. Note that by choosing the permittivity of region 2 to exceed that of region 1 we preclude total internal reflection.

We can specialize the frequency-domain reflection coefficient (4.283) for our problem by noting that

$$k_{1z} = \beta_1\cos\theta_i, \qquad k_{2z} = \sqrt{k_2^2 - k_{1x}^2} = \omega\sqrt{\mu_0}\sqrt{\epsilon + \frac{\sigma}{j\omega} - \epsilon_0\sin^2\theta_i},$$

and thus

$$Z_{1\perp} = \frac{\eta_0}{\cos\theta_i}, \qquad Z_{2\perp} = \frac{\eta_0}{\sqrt{\epsilon_r + \frac{\sigma}{j\omega\epsilon_0} - \sin^2\theta_i}},$$

where $\epsilon_r = \epsilon/\epsilon_0$ and $\eta_0 = \sqrt{\mu_0/\epsilon_0}$. We thus obtain

$$\tilde{\Gamma}_\perp = \frac{\sqrt{s} - \sqrt{Ds + B}}{\sqrt{s} + \sqrt{Ds + B}} \qquad (4.302)$$

where $s = j\omega$ and

$$D = \frac{\epsilon_r - \sin^2 \theta_i}{\cos^2 \theta_i}, \qquad B = \frac{\sigma}{\epsilon_0 \cos^2 \theta_i}.$$

We can put (4.302) into a better form for inversion. We begin by subtracting $\Gamma_{\perp\infty}$, the high-frequency limit of $\tilde{\Gamma}_\perp$. Noting that

$$\lim_{\omega\to\infty} \tilde{\Gamma}_\perp(\omega) = \Gamma_{\perp\infty} = \frac{1 - \sqrt{D}}{1 + \sqrt{D}},$$

we can form

$$\tilde{\Gamma}_\perp^0(\omega) = \tilde{\Gamma}_\perp(\omega) - \Gamma_{\perp\infty} = \frac{\sqrt{s} - \sqrt{Ds + B}}{\sqrt{s} + \sqrt{Ds + B}} - \frac{1 - \sqrt{D}}{1 + \sqrt{D}}$$

$$= 2\frac{\sqrt{D}}{1 + \sqrt{D}}\left[\frac{\sqrt{s} - \sqrt{s + B/D}}{\sqrt{s} + \sqrt{D}\sqrt{s + D/B}}\right].$$

With a bit of algebra this becomes

$$\tilde{\Gamma}_\perp^0(\omega) = -\frac{2\sqrt{D}}{D - 1}\left(\frac{s}{s + \frac{B}{D-1}}\right)\left(1 - \sqrt{\frac{s + \frac{B}{D}}{s}}\right) - \frac{2B}{\left(1 + \sqrt{D}\right)(D - 1)}\left(\frac{1}{s + \frac{B}{D-1}}\right).$$

Now we can apply (C.12), (C.18), and (C.19) to obtain

$$\Gamma_\perp^0(t) = \mathcal{F}^{-1}\left\{\tilde{\Gamma}_\perp^0(\omega)\right\} = f_1(t) + f_2(t) + f_3(t) \qquad (4.303)$$

where

$$f_1(t) = -\frac{2B}{(1 + \sqrt{D})(D - 1)}e^{-\frac{Bt}{D-1}}U(t),$$

$$f_2(t) = -\frac{B^2}{\sqrt{D}(D - 1)^2}U(t)\int_0^t e^{-\frac{B(t-x)}{D-1}}I\left(\frac{Bx}{2D}\right)dx,$$

$$f_3(t) = \frac{B}{\sqrt{D}(D - 1)}I\left(\frac{Bt}{2D}\right)U(t).$$

Here

$$I(x) = e^{-x}\left[I_0(x) + I_1(x)\right]$$

where $I_0(x)$ and $I_1(x)$ are modified Bessel functions of the first kind. Setting $u = Bx/2D$ we can also write

$$f_2(t) = -\frac{2B\sqrt{D}}{(D - 1)^2}U(t)\int_0^{\frac{Bt}{2D}} e^{-\frac{Bt - 2Du}{D-1}}I(u)\,du.$$

Polynomial approximations for $I(x)$ may be found in Abramowitz and Stegun [1], making the computation of $\Gamma_\perp^0(t)$ straightforward.

The complete time-domain reflection coefficient is

$$\Gamma_\perp(t) = \frac{1 - \sqrt{D}}{1 + \sqrt{D}}\delta(t) + \Gamma_\perp^0(t).$$

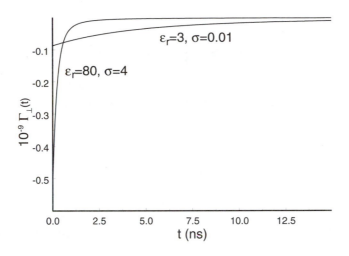

Figure 4.19: Time-domain reflection coefficients.

If $\sigma = 0$ then $\Gamma_\perp^0(t) = 0$ and the reflection coefficient reduces to a single δ-function. Since convolution with this term does not alter wave shape, the reflected field has the same waveform as the incident field.

A plot of $\Gamma_\perp^0(t)$ for normal incidence ($\theta_i = 0^0$) is shown in Figure 4.19. Here two material cases are displayed: $\epsilon_r = 3$, $\sigma = 0.01$ S/m, which is representative of dry water ice, and $\epsilon_r = 80$, $\sigma = 4$ S/m, which is representative of sea water. We see that a pulse waveform experiences more temporal spreading upon reflection from ice than from sea water, but that the amplitude of the dispersive component is less than that for sea water.

Reflection of a nonuniform plane wave from a planar interface. Describing the interaction of a general nonuniform plane wave with a planar interface is problematic because of the non-TEM behavior of the incident wave. We cannot decompose the fields into two mutually orthogonal cases as we did with uniform waves, and thus the analysis is more difficult. However, we found in the last section that when a uniform wave is incident on a planar interface, the transmitted wave, even if nonuniform in nature, takes on the same mathematical form and may be decomposed in the same manner as the incident wave. Thus, we may study the case in which this refracted wave is incident on a successive interface using exactly the same analysis as with a uniform incident wave. This is helpful in the case of multi-layered media, which we shall examine next.

Interaction of a plane wave with multi-layered, planar materials. Consider $N + 1$ regions of space separated by N planar interfaces as shown in Figure 4.20, and assume that a uniform plane wave is incident on the first interface at angle θ_i. Each region is assumed isotropic and homogeneous with a frequency-dependent complex permittivity and permeability. We can easily generalize the previous analysis regarding reflection from a single interface by realizing that in order to satisfy the boundary conditions each

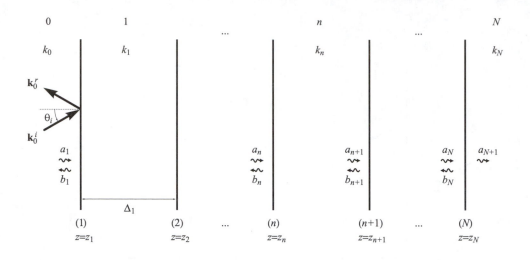

Figure 4.20: Interaction of a uniform plane wave with a multi-layered material.

region, except region N, contains an incident-type wave of the form

$$\tilde{\mathbf{E}}^i(\mathbf{r},\omega) = \tilde{\mathbf{E}}_0^i e^{-j\mathbf{k}^i\cdot\mathbf{r}}$$

and a reflected-type wave of the form

$$\tilde{\mathbf{E}}^r(\mathbf{r},\omega) = \tilde{\mathbf{E}}_0^r e^{-j\mathbf{k}^r\cdot\mathbf{r}}.$$

In region n we may write the wave vectors describing these waves as

$$\mathbf{k}_n^i = \hat{\mathbf{x}}k_{x,n} + \hat{\mathbf{z}}k_{z,n}, \qquad \mathbf{k}_n^r = \hat{\mathbf{x}}k_{x,n} - \hat{\mathbf{z}}k_{z,n},$$

where

$$k_{x,n}^2 + k_{z,n}^2 = k_n^2, \qquad k_n^2 = \omega^2 \tilde{\mu}_n \tilde{\epsilon}_n^c = (\beta_n - j\alpha_n)^2.$$

We note at the outset that, as with the single interface case, the boundary conditions are only satisfied when Snell's law of reflection holds, and thus

$$k_{x,n} = k_{x,0} = k_0 \sin\theta_i \qquad (4.304)$$

where $k_0 = \omega(\tilde{\mu}_0\tilde{\epsilon}_0^c)^{1/2}$ is the wavenumber of the 0th region (not necessarily free space). From this condition we have

$$k_{z,n} = \sqrt{k_n^2 - k_{x,0}^2} = \tau_n e^{-j\gamma_n}$$

where

$$\tau_n = (A_n^2 + B_n^2)^{1/4}, \qquad \gamma_n = \frac{1}{2}\tan^{-1}\left(\frac{B_n}{A_n}\right),$$

and

$$A_n = \beta_n^2 - \alpha_n^2 - (\beta_0^2 - \alpha_0^2)\sin^2\theta_i, \qquad B_n = 2(\beta_n\alpha_n - \beta_0\alpha_0\sin^2\theta_i).$$

Provided that the incident wave is uniform, we can decompose the fields in every region into cases of perpendicular and parallel polarization. This is true even when the waves in certain layers are nonuniform. For the case of perpendicular polarization we can write the electric field in region n, $0 \leq n \leq N-1$, as $\tilde{\mathbf{E}}_{\perp n} = \tilde{\mathbf{E}}^i_{\perp n} + \tilde{\mathbf{E}}^r_{\perp n}$ where

$$\tilde{\mathbf{E}}^i_{\perp n} = \hat{\mathbf{y}} a_{n+1} e^{-jk_{x,n}x} e^{-jk_{z,n}(z-z_{n+1})},$$
$$\tilde{\mathbf{E}}^r_{\perp n} = \hat{\mathbf{y}} b_{n+1} e^{-jk_{x,n}x} e^{+jk_{z,n}(z-z_{n+1})},$$

and the magnetic field as $\tilde{\mathbf{H}}_{\perp n} = \tilde{\mathbf{H}}^i_{\perp n} + \mathbf{H}^r_{\perp n}$ where

$$\tilde{\mathbf{H}}^i_{\perp n} = \frac{-\hat{\mathbf{x}} k_{z,n} + \hat{\mathbf{z}} k_{x,n}}{k_n \eta_n} a_{n+1} e^{-jk_{x,n}x} e^{-jk_{z,n}(z-z_{n+1})},$$

$$\tilde{\mathbf{H}}^r_{\perp n} = \frac{+\hat{\mathbf{x}} k_{z,n} + \hat{\mathbf{z}} k_{x,n}}{k_n \eta_n} b_{n+1} e^{-jk_{x,n}x} e^{+jk_{z,n}(z-z_{n+1})}.$$

When $n = N$ there is no reflected wave; in this region we write

$$\tilde{\mathbf{E}}_{\perp N} = \hat{\mathbf{y}} a_{N+1} e^{-jk_{x,N}x} e^{-jk_{z,N}(z-z_N)},$$

$$\tilde{\mathbf{H}}_{\perp N} = \frac{-\hat{\mathbf{x}} k_{z,N} + \hat{\mathbf{z}} k_{x,N}}{k_N \eta_N} a_{N+1} e^{-jk_{x,N}x} e^{-jk_{z,N}(z-z_N)}.$$

Since a_1 is the known amplitude of the incident wave, there are $2N$ unknown wave amplitudes. We obtain the necessary $2N$ simultaneous equations by applying the boundary conditions at each of the interfaces. At interface n located at $z = z_n$, $1 \leq n \leq N-1$, we have from the continuity of tangential electric field

$$a_n + b_n = a_{n+1} e^{-jk_{z,n}(z_n-z_{n+1})} + b_{n+1} e^{+jk_{z,n}(z_n-z_{n+1})}$$

while from the continuity of magnetic field

$$-a_n \frac{k_{z,n-1}}{k_{n-1}\eta_{n-1}} + b_n \frac{k_{z,n-1}}{k_{n-1}\eta_{n-1}} = -a_{n+1} \frac{k_{z,n}}{k_n \eta_n} e^{-jk_{z,n}(z_n-z_{n+1})} + b_{n+1} \frac{k_{z,n}}{k_n \eta_n} e^{+jk_{z,n}(z_n-z_{n+1})}.$$

Noting that the wave impedance of region n is

$$Z_{\perp n} = \frac{k_n \eta_n}{k_{z,n}}$$

and defining the region n propagation factor as

$$\tilde{P}_n = e^{-jk_{z,n}\Delta_n}$$

where $\Delta_n = z_{n+1} - z_n$, we can write

$$a_n \tilde{P}_n + b_n \tilde{P}_n = a_{n+1} + b_{n+1} \tilde{P}_n^2, \tag{4.305}$$

$$-a_n \tilde{P}_n + b_n \tilde{P}_n = -a_{n+1} \frac{Z_{\perp n-1}}{Z_{\perp n}} + b_{n+1} \frac{Z_{\perp n-1}}{Z_{\perp n}} \tilde{P}_n^2. \tag{4.306}$$

We must still apply the boundary conditions at $z = z_N$. Proceeding as above, we find that (4.305) and (4.306) hold for $n = N$ if we set $b_{N+1} = 0$ and $\tilde{P}_N = 1$.

The $2N$ simultaneous equations (4.305)–(4.306) may be solved using standard matrix methods. However, through a little manipulation we can put the equations into a form easily solved by recursion, providing a very nice physical picture of the multiple reflections that occur within the layered medium. We begin by eliminating b_n by subtracting (4.306) from (4.305):

$$2a_n \tilde{P}_n = a_{n+1} \left[1 + \frac{Z_{\perp n-1}}{Z_{\perp n}} \right] + b_{n+1} \tilde{P}_n^2 \left[1 - \frac{Z_{\perp n-1}}{Z_{\perp n}} \right]. \tag{4.307}$$

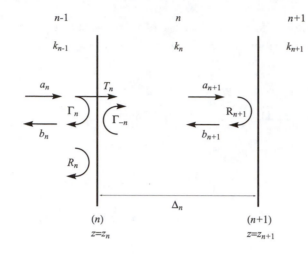

Figure 4.21: Wave flow diagram showing interaction of incident and reflected waves for region n.

Defining

$$\tilde{\Gamma}_n = \frac{Z_{\perp n} - Z_{\perp n-1}}{Z_{\perp n} + Z_{\perp n-1}} \tag{4.308}$$

as the *interfacial reflection coefficient* for interface n (i.e., the reflection coefficient assuming a single interface as in (4.283)), and

$$\tilde{T}_n = \frac{2Z_{\perp n}}{Z_{\perp n} + Z_{\perp n-1}} = 1 + \tilde{\Gamma}_n$$

as the *interfacial transmission coefficient* for interface n, we can write (4.307) as

$$a_{n+1} = a_n \tilde{T}_n \tilde{P}_n + b_{n+1} \tilde{P}_n (-\tilde{\Gamma}_n) \tilde{P}_n.$$

Finally, if we define the *global* reflection coefficient R_n for region n as the ratio of the amplitudes of the reflected and incident waves,

$$\tilde{R}_n = b_n / a_n,$$

we can write

$$a_{n+1} = a_n \tilde{T}_n \tilde{P}_n + a_{n+1} \tilde{R}_{n+1} \tilde{P}_n (-\tilde{\Gamma}_n) \tilde{P}_n. \tag{4.309}$$

For $n = N$ we merely set $R_{N+1} = 0$ to find

$$a_{N+1} = a_N \tilde{T}_N \tilde{P}_N. \tag{4.310}$$

If we choose to eliminate a_{n+1} from (4.305) and (4.306) we find that

$$b_n = a_n \tilde{\Gamma}_n + \tilde{R}_{n+1} \tilde{P}_n (1 - \tilde{\Gamma}_n) a_{n+1}. \tag{4.311}$$

For $n = N$ this reduces to

$$b_N = a_N \tilde{\Gamma}_N. \tag{4.312}$$

Equations (4.309) and (4.311) have nice physical interpretations. Consider Figure 4.21, which shows the wave amplitudes for region n. We may think of the wave incident on interface $n + 1$ with amplitude a_{n+1} as consisting of two terms. The first term is the wave transmitted through interface n (at $z = z_n$). This wave must propagate through a distance Δ_n to reach interface $n+1$ and thus has an amplitude $a_n \tilde{T}_n \tilde{P}_n$. The second term is the reflection at interface n of the wave traveling in the $-z$ direction within region n. The amplitude of the wave before reflection is merely $b_{n+1} \tilde{P}_n$, where the term $\tilde{P}_n$ results from the propagation of the negatively-traveling wave from interface $n + 1$ to interface n. Now, since the interfacial reflection coefficient at interface n for a wave incident from region n is the negative of that for a wave incident from region $n - 1$ (since the wave is traveling in the reverse direction), and since the reflected wave must travel through a distance Δ_n from interface n back to interface $n + 1$, the amplitude of the second term is $b_{n+1} \tilde{P}_n (-\tilde{\Gamma}_n) \tilde{P}_n$. Finally, remembering that $b_{n+1} = \tilde{R}_{n+1} a_{n+1}$, we can write

$$a_{n+1} = a_n \tilde{T}_n \tilde{P}_n + a_{n+1} \tilde{R}_{n+1} \tilde{P}_n (-\tilde{\Gamma}_n) \tilde{P}_n.$$

This equation is exactly the same as (4.309) which was found using the boundary conditions. By similar reasoning, we may say that the wave traveling in the $-z$ direction in region $n - 1$ consists of a term reflected from the interface and a term transmitted through the interface. The amplitude of the reflected term is merely $a_n \tilde{\Gamma}_n$. The amplitude of the transmitted term is found by considering $b_{n+1} = \tilde{R}_{n+1} a_{n+1}$ propagated through a distance Δ_n and then transmitted backwards through interface n. Since the transmission coefficient for a wave going from region n to region $n - 1$ is $1 + (-\tilde{\Gamma}_n)$, the amplitude of the transmitted term is $\tilde{R}_{n+1} \tilde{P}_n (1 - \tilde{\Gamma}_n) a_{n+1}$. Thus we have

$$b_n = \tilde{\Gamma}_n a_n + \tilde{R}_{n+1} \tilde{P}_n (1 - \tilde{\Gamma}_n) a_{n+1},$$

which is identical to (4.311).

We are still left with the task of solving for the various field amplitudes. This can be done using a simple recursive technique. Using $\tilde{T}_n = 1 + \tilde{\Gamma}_n$ we find from (4.309) that

$$a_{n+1} = \frac{(1 + \tilde{\Gamma}_n) \tilde{P}_n}{1 + \tilde{\Gamma}_n \tilde{R}_{n+1} \tilde{P}_n^2} a_n. \qquad (4.313)$$

Substituting this into (4.311) we find

$$b_n = \frac{\tilde{\Gamma}_n + \tilde{R}_{n+1} \tilde{P}_n^2}{1 + \tilde{\Gamma}_n \tilde{R}_{n+1} \tilde{P}_n^2} a_n. \qquad (4.314)$$

Using this expression we find a recursive relationship for the global reflection coefficient:

$$\tilde{R}_n = \frac{b_n}{a_n} = \frac{\tilde{\Gamma}_n + \tilde{R}_{n+1} \tilde{P}_n^2}{1 + \tilde{\Gamma}_n \tilde{R}_{n+1} \tilde{P}_n^2}. \qquad (4.315)$$

The procedure is now as follows. The global reflection coefficient for interface N is, from (4.312),

$$\tilde{R}_N = b_N / a_N = \tilde{\Gamma}_N. \qquad (4.316)$$

This is also obtained from (4.315) with $\tilde{R}_{N+1} = 0$. We next use (4.315) to find $\tilde{R}_{N-1}$:

$$\tilde{R}_{N-1} = \frac{\tilde{\Gamma}_{N-1} + \tilde{R}_N \tilde{P}_{N-1}^2}{1 + \tilde{\Gamma}_{N-1} \tilde{R}_N \tilde{P}_{N-1}^2}.$$

This process is repeated until reaching $\tilde{R}_1$, whereupon all of the global reflection coefficients are known. We then find the amplitudes beginning with a_1, which is the known incident field amplitude. From (4.315) we find $b_1 = a_1 \tilde{R}_1$, and from (4.313) we find

$$a_2 = \frac{(1 + \tilde{\Gamma}_1)\tilde{P}_1}{1 + \tilde{\Gamma}_1 \tilde{R}_2 \tilde{P}_1^2} a_1.$$

This process is repeated until all field amplitudes are known.

We note that the process outlined above holds equally well for parallel polarization as long as we use the parallel wave impedances

$$Z_{\|n} = \frac{k_{z,n}\eta_n}{k_n}$$

when computing the interfacial reflection coefficients. See Problem 4.12.

As a simple example, consider a slab of material of thickness Δ sandwiched between two lossless dielectrics. A time-harmonic uniform plane wave of frequency $\omega = \check{\omega}$ is normally incident onto interface 1, and we wish to compute the amplitude of the wave reflected by interface 1 and determine the conditions under which the reflected wave vanishes. In this case we have $N = 2$, with two interfaces and three regions. By (4.316) we have $R_2 = \Gamma_2$, where $R_2 = \tilde{R}_2(\check{\omega})$, $\Gamma_2 = \tilde{\Gamma}_2(\check{\omega})$, etc. Then by (4.315) we find

$$R_1 = \frac{\Gamma_1 + R_2 P_1^2}{1 + \Gamma_1 R_2 P_1^2} = \frac{\Gamma_1 + \Gamma_2 P_1^2}{1 + \Gamma_1 \Gamma_2 P_1^2}.$$

Hence the reflected wave vanishes when

$$\Gamma_1 + \Gamma_2 P_1^2 = 0.$$

Since the field in region 0 is normally incident we have

$$k_{z,n} = k_n = \beta_n = \check{\omega}\sqrt{\mu_n \epsilon_n}.$$

If we choose $P_1^2 = -1$, then $\Gamma_1 = \Gamma_2$ results in no reflected wave. This requires

$$\frac{Z_1 - Z_0}{Z_1 + Z_0} = \frac{Z_2 - Z_1}{Z_2 + Z_1}.$$

Clearing the denominator we find that $2Z_1^2 = 2Z_0 Z_2$ or

$$Z_1 = \sqrt{Z_0 Z_2}.$$

This condition makes the reflected field vanish if we can ensure that $P_1^2 = -1$. To do this we need

$$e^{-j\beta_1 2\Delta} = -1.$$

The minimum thickness that satisfies this condition is $\beta_1 2\Delta = \pi$. Since $\beta = 2\pi/\lambda$, this is equivalent to

$$\Delta = \lambda/4.$$

A layer of this type is called a *quarter-wave transformer*. Since no wave is reflected from the initial interface, and since all the regions are assumed lossless, all of the power carried by the incident wave in the first region is transferred into the third region. Thus, two regions of differing materials may be "matched" by inserting an appropriate slab between

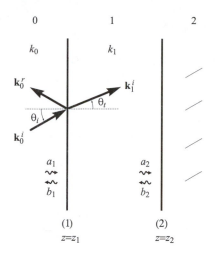

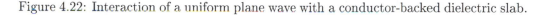

Figure 4.22: Interaction of a uniform plane wave with a conductor-backed dielectric slab.

them. This technique finds use in optical coatings for lenses and for reducing the radar reflectivity of objects.

As a second example, consider a lossless dielectric slab with $\tilde{\epsilon} = \epsilon_1 = \epsilon_{1r}\epsilon_0$, and $\tilde{\mu} = \mu_0$, backed by a perfect conductor and immersed in free space as shown in Figure 4.22. A perpendicularly polarized uniform plane wave is incident on the slab from free space and we wish to find the temporal response of the reflected wave by first calculating the frequency-domain reflected field. Since $\epsilon_0 < \epsilon_1$, total internal reflection cannot occur. Thus the wave vectors in region 1 have real components and can be written as

$$\mathbf{k}_1^i = k_{x,1}\hat{\mathbf{x}} + k_{z,1}\hat{\mathbf{z}}, \qquad \mathbf{k}_1^r = k_{x,1}\hat{\mathbf{x}} - k_{z,1}\hat{\mathbf{z}}.$$

From Snell's law of refraction we know that

$$k_{x,1} = k_0 \sin \theta_i = k_1 \sin \theta_t$$

and so

$$k_{z,1} = \sqrt{k_1^2 - k_{x,1}^2} = \frac{\omega}{c}\sqrt{\epsilon_{1r} - \sin^2 \theta_i} = k_1 \cos \theta_t$$

where θ_t is the transmission angle in region 1. Since region 2 is a perfect conductor we have $\tilde{R}_2 = -1$. By (4.315) we have

$$\tilde{R}_1(\omega) = \frac{\Gamma_1 - \tilde{P}_1^2(\omega)}{1 - \Gamma_1 \tilde{P}_1^2(\omega)}, \tag{4.317}$$

where from (4.308)

$$\Gamma_1 = \frac{Z_1 - Z_0}{Z_1 + Z_0}$$

is not a function of frequency. By the approach we used to obtain (4.300) we write

$$\tilde{\mathbf{E}}_\perp^r(\mathbf{r}, \omega) = \hat{\mathbf{y}} \tilde{R}_1(\omega) \tilde{E}_\perp^i(\omega) e^{-j\mathbf{k}_1^r(\omega) \cdot \mathbf{r}}.$$

So

$$\mathbf{E}_\perp^r(\mathbf{r}, t) = \hat{\mathbf{y}} E^r \left(t - \frac{\hat{\mathbf{k}}_1^r \cdot \mathbf{r}}{c} \right)$$

where by the convolution theorem

$$E^r(t) = R_1(t) * E_\perp^i(t). \tag{4.318}$$

Here

$$E_\perp^i(t) = \mathcal{F}^{-1}\left\{\tilde{E}_\perp^i(\omega)\right\}$$

is the time waveform of the incident plane wave, while

$$R_1(t) = \mathcal{F}^{-1}\left\{\tilde{R}_1(\omega)\right\}$$

is the global time-domain reflection coefficient.

To invert $\tilde{R}_1(\omega)$, we use the binomial expansion $(1-x)^{-1} = 1 + x + x^2 + x^3 + \cdots$ on the denominator of (4.317), giving

$$\tilde{R}_1(\omega) = [\Gamma_1 - \tilde{P}_1^2(\omega)]\left\{1 + [\Gamma_1\tilde{P}_1^2(\omega)] + [\Gamma_1\tilde{P}_1^2(\omega)]^2 + [\Gamma_1\tilde{P}_1^2(\omega)]^3 + \ldots\right\}$$

$$= \Gamma_1 - [1 - \Gamma_1^2]\tilde{P}_1^2(\omega) - [1 - \Gamma_1^2]\Gamma_1\tilde{P}_1^4(\omega) - [1 - \Gamma_1^2]\Gamma_1^2\tilde{P}_1^6(\omega) - \cdots. \tag{4.319}$$

Thus we need the inverse transform of

$$\tilde{P}_1^{2n}(\omega) = e^{-j2nk_{z,1}\Delta_1} = e^{-j2nk_1\Delta_1\cos\theta_t}.$$

Writing $k_1 = \omega/v_1$, where $v_1 = 1/(\mu_0\epsilon_1)^{1/2}$ is the phase velocity of the wave in region 1, and using $1 \leftrightarrow \delta(t)$ along with the time-shifting theorem (A.3) we have

$$\tilde{P}_1^{2n}(\omega) = e^{-j\omega 2n\tau} \leftrightarrow \delta(t - 2n\tau)$$

where $\tau = \Delta_1\cos\theta_t/v_1$. With this the inverse transform of $\tilde{R}_1$ in (4.319) is

$$R_1(t) = \Gamma_1\delta(t) - (1 + \Gamma_1)(1 - \Gamma_1)\delta(t - 2\tau) - (1 + \Gamma_1)(1 - \Gamma_1)\Gamma_1\delta(t - 4\tau) - \cdots$$

and thus from (4.318)

$$E^r(t) = \Gamma_1 E_\perp^i(t) - (1 + \Gamma_1)(1 - \Gamma_1)E_\perp^i(t - 2\tau) - (1 + \Gamma_1)(1 - \Gamma_1)\Gamma_1 E_\perp^i(t - 4\tau) - \cdots.$$

The reflected field consists of time-shifted and amplitude-scaled versions of the incident field waveform. These terms can be interpreted as multiple reflections of the incident wave. Consider Figure 4.23. The first term is the direct reflection from interface 1 and thus has its amplitude multiplied by Γ_1. The next term represents a wave that penetrates the interface (and thus has its amplitude multiplied by the transmission coefficient $1 + \Gamma_1$), propagates to and reflects from the conductor (and thus has its amplitude multiplied by -1), and then propagates back to the interface and passes through in the opposite direction (and thus has its amplitude multiplied by the transmission coefficient for passage from region 1 to region 0, $1 - \Gamma_1$). The time delay between this wave and the initially-reflected wave is given by 2τ, as discussed in detail below. The third term represents a wave that penetrates the interface, reflects from the conductor, returns to and reflects from the interface a second time, again reflects from the conductor, and then passes through the interface in the opposite direction. Its amplitude has an additional multiplicative factor of $-\Gamma_1$ to account for reflection from the interface and an additional factor of -1 to account for the second reflection from the conductor, and is time-delayed by an additional 2τ. Subsequent terms account for additional reflections;

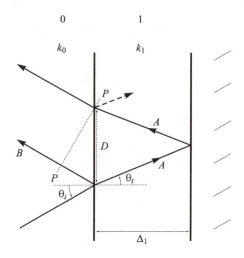

Figure 4.23: Timing diagram for multiple reflections from a conductor-backed dielectric slab.

the nth reflected wave amplitude is multiplied by an additional $(-1)^n$ and $(-\Gamma_1)^n$ and is time-delayed by an additional $2n\tau$.

It is important to understand that the time delay 2τ is *not* just the propagation time for the wave to travel through the slab. To properly describe the timing between the initially-reflected wave and the waves that reflect from the conductor we must consider the field over identical observation planes as shown in Figure 4.23. For example, consider the observation plane designated P-P intersecting the first "exit point" on interface 1. To arrive at this plane the initially-reflected wave takes the path labeled B, arriving at a time

$$\frac{D \sin \theta_i}{v_0}$$

after the time of initial reflection, where $v_0 = c$ is the velocity in region 0. To arrive at this same plane the wave that penetrates the surface takes the path labeled A, arriving at a time

$$\frac{2\Delta_1}{v_1 \cos \theta_t}$$

where v_1 is the wave velocity in region 1 and θ_t is the transmission angle. Noting that $D = 2\Delta_1 \tan \theta_t$, the time delay between the arrival of the two waves at the plane P-P is

$$T = \frac{2\Delta_1}{v_1 \cos \theta_t} - \frac{D \sin \theta_i}{v_0} = \frac{2\Delta_1}{v_1 \cos \theta_t} \left[1 - \frac{\sin \theta_t \sin \theta_i}{v_0/v_1} \right].$$

By Snell's law of refraction (4.297) we can write

$$\frac{v_0}{v_1} = \frac{\sin \theta_i}{\sin \theta_t},$$

which, upon substitution, gives

$$T = 2\frac{\Delta_1 \cos \theta_t}{v_1}.$$

This is exactly the time delay 2τ.

4.11.6   Plane-wave propagation in an anisotropic ferrite medium

Several interesting properties of plane waves, such as Faraday rotation and the existence of stopbands, appear only when the waves propagate through anisotropic media. We shall study the behavior of waves propagating in a magnetized ferrite medium, and note that this behavior is shared by waves propagating in a magnetized plasma, because of the similarity in the dyadic constitutive parameters of the two media.

Consider a uniform ferrite material having scalar permittivity $\tilde{\epsilon} = \epsilon$ and dyadic permeability $\bar{\tilde{\mu}}$. We assume that the ferrite is lossless and magnetized along the z-direction. By (4.115)–(4.117) the permeability of the medium is

$$[\bar{\tilde{\mu}}(\omega)] = \begin{bmatrix} \mu_1 & j\mu_2 & 0 \\ -j\mu_2 & \mu_1 & 0 \\ 0 & 0 & \mu_0 \end{bmatrix}$$

where

$$\mu_1 = \mu_0 \left[1 + \frac{\omega_M \omega_0}{\omega_0^2 - \omega^2} \right], \qquad \mu_2 = \mu_0 \frac{\omega \omega_M}{\omega_0^2 - \omega^2}.$$

The source-free frequency-domain wave equation can be found using (4.201) with $\bar{\tilde{\zeta}} = \bar{\tilde{\xi}} = 0$ and $\bar{\tilde{\epsilon}} = \epsilon \bar{I}$:

$$\left[\bar{\nabla} \cdot \left(\bar{I} \frac{1}{\epsilon} \right) \cdot \bar{\nabla} - \omega^2 \bar{\tilde{\mu}} \right] \cdot \tilde{\mathbf{H}} = 0$$

or, since $\bar{\nabla} \cdot \mathbf{A} = \nabla \times \mathbf{A}$,

$$\frac{1}{\epsilon} \nabla \times \left(\nabla \times \tilde{\mathbf{H}} \right) - \omega^2 \bar{\tilde{\mu}} \cdot \tilde{\mathbf{H}} = 0. \tag{4.320}$$

The simplest solutions to the wave equation for this anisotropic medium are TEM plane waves that propagate along the applied dc magnetic field. We thus seek solutions of the form

$$\tilde{\mathbf{H}}(\mathbf{r}, \omega) = \tilde{\mathbf{H}}_0(\omega) e^{-j\mathbf{k} \cdot \mathbf{r}} \tag{4.321}$$

where $\mathbf{k} = \hat{\mathbf{z}}\beta$ and $\hat{\mathbf{z}} \cdot \tilde{\mathbf{H}}_0 = 0$. We can find β by enforcing (4.320). From (B.78) we find that

$$\nabla \times \tilde{\mathbf{H}} = -j\beta \hat{\mathbf{z}} \times \tilde{\mathbf{H}}_0 e^{-j\beta z}.$$

By Ampere's law we have

$$\tilde{\mathbf{E}} = \frac{\nabla \times \tilde{\mathbf{H}}}{j\omega\epsilon} = -Z_{TEM} \hat{\mathbf{z}} \times \tilde{\mathbf{H}}, \tag{4.322}$$

where

$$Z_{TEM} = \beta/\omega\epsilon$$

is the wave impedance. Note that the wave is indeed TEM. The second curl is found to be

$$\nabla \times \left(\nabla \times \tilde{\mathbf{H}} \right) = -j\beta \nabla \times \left[\hat{\mathbf{z}} \times \tilde{\mathbf{H}}_0 e^{-j\beta z} \right].$$

After an application of (B.43) this becomes

$$\nabla \times \left(\nabla \times \tilde{\mathbf{H}} \right) = -j\beta \left[e^{-j\beta z} \nabla \times (\hat{\mathbf{z}} \times \tilde{\mathbf{H}}_0) - (\hat{\mathbf{z}} \times \tilde{\mathbf{H}}_0) \times \nabla e^{-j\beta z} \right].$$

The first term on the right-hand side is zero, and thus using (B.76) we have

$$\nabla \times \left(\nabla \times \tilde{\mathbf{H}}\right) = \left[-j\beta e^{-j\beta z}\hat{\mathbf{z}} \times (\hat{\mathbf{z}} \times \tilde{\mathbf{H}}_0)\right](-j\beta)$$

or, using (B.7),

$$\nabla \times \left(\nabla \times \tilde{\mathbf{H}}\right) = \beta^2 e^{-j\beta z}\tilde{\mathbf{H}}_0$$

since $\hat{\mathbf{z}} \cdot \tilde{\mathbf{H}}_0 = 0$. With this (4.320) becomes

$$\beta^2 \tilde{\mathbf{H}}_0 = \omega^2 \epsilon \tilde{\bar{\mu}} \cdot \tilde{\mathbf{H}}_0. \tag{4.323}$$

We can solve (4.323) for β by writing the vector equation in component form:

$$\beta^2 H_{0x} = \omega^2 \epsilon \left[\mu_1 H_{0x} + j\mu_2 H_{0y}\right],$$
$$\beta^2 H_{0y} = \omega^2 \epsilon \left[-j\mu_2 H_{0x} + \mu_1 H_{0y}\right].$$

In matrix form these are

$$\begin{bmatrix} \beta^2 - \omega^2 \epsilon \mu_1 & -j\omega^2 \epsilon \mu_2 \\ j\omega^2 \epsilon \mu_2 & \beta^2 - \omega^2 \epsilon \mu_1 \end{bmatrix} \begin{bmatrix} H_{0x} \\ H_{0y} \end{bmatrix} = \begin{bmatrix} 0 \\ 0 \end{bmatrix}, \tag{4.324}$$

and nontrivial solutions occur only if

$$\begin{vmatrix} \beta^2 - \omega^2 \epsilon \mu_1 & -j\omega^2 \epsilon \mu_2 \\ j\omega^2 \epsilon \mu_2 & \beta^2 - \omega^2 \epsilon \mu_1 \end{vmatrix} = 0.$$

Expansion yields the two solutions

$$\beta_\pm = \omega\sqrt{\epsilon\mu_\pm} \tag{4.325}$$

where

$$\mu_\pm = \mu_1 \pm \mu_2 = \mu_0 \left[1 + \frac{\omega_M}{\omega_0 \mp \omega}\right]. \tag{4.326}$$

So the propagation properties of the plane wave are the same as those in a medium with an equivalent scalar permeability given by $\mu_\pm$.

Associated with each of these solutions is a relationship between H_{0x} and H_{0y} that can be found from (4.324). Substituting β_+ into the first equation we have

$$\omega^2 \epsilon \mu_2 H_{0x} - j\omega^2 \epsilon \mu_2 H_{0y} = 0$$

or $H_{0x} = jH_{0y}$. Similarly, substitution of β_- produces $H_{0x} = -jH_{0y}$. Thus, by (4.321) the magnetic field may be expressed as

$$\tilde{\mathbf{H}}(\mathbf{r}, \omega) = H_{0y}[\pm j\hat{\mathbf{x}} + \hat{\mathbf{y}}]e^{-j\beta_\pm z}.$$

By (4.322) we also have the electric field

$$\tilde{\mathbf{E}}(\mathbf{r}, \omega) = Z_{TEM} H_{0y}[\hat{\mathbf{x}} + e^{\mp j\frac{\pi}{2}}\hat{\mathbf{y}}]e^{-j\beta_\pm z}.$$

This field has the form of (4.248). For β_+ we have $\phi_y - \phi_x = -\pi/2$ and thus the wave exhibits RHCP. For β_- we have $\phi_y - \phi_x = \pi/2$ and the wave exhibits LHCP.

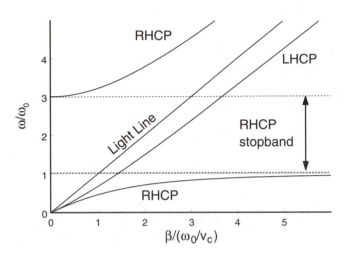

Figure 4.24: Dispersion plot for unmagnetized ferrite with $\omega_M = 2\omega_0$. Light line shows $\omega/\beta = v_c = 1/(\mu_0\epsilon)^{1/2}$.

The dispersion diagram for each polarization case is shown in Figure 4.24, where we have arbitrarily chosen $\omega_M = 2\omega_0$. Here we have combined (4.325) and (4.326) to produce the normalized expression

$$\frac{\beta_\pm}{\omega_0/v_c} = \frac{\omega}{\omega_0}\sqrt{1 + \frac{\omega_M/\omega_0}{1 \mp \omega/\omega_0}}$$

where $v_c = 1/(\mu_0\epsilon)^{1/2}$. Except at low frequencies, an LHCP plane wave passes through the ferrite as if the permeability is close to that of free space. Over all frequencies we have $v_p < v_c$ and $v_g < v_c$. In contrast, an RHCP wave excites the electrons in the ferrite and a resonance occurs at $\omega = \omega_0$. For all frequencies below ω_0 we have $v_p < v_c$ and $v_g < v_c$ and both v_p and v_g reduce to zero as $\omega \to \omega_0$. Because the ferrite is lossless, frequencies between $\omega = \omega_0$ and $\omega = \omega_0 + \omega_M$ result in β being purely imaginary and thus the wave being evanescent. We thus call the frequency range $\omega_0 < \omega < \omega_0 + \omega_M$ a *stopband*; within this band the plane wave cannot transport energy. For frequencies above $\omega_0 + \omega_M$ the RHCP wave propagates as if it is in a medium with permeability less than that of free space. Here we have $v_p > v_c$ and $v_g < v_c$, with $v_p \to v_c$ and $v_g \to v_c$ as $\omega \to \infty$.

Faraday rotation.   The solutions to the wave equation found above do not allow the existence of linearly polarized plane waves. However, by superposing LHCP and RHCP waves we can obtain a wave with the appearance of linear polarization. That is, over any z-plane the electric field vector may be written as $\tilde{\mathbf{E}} = K(E_{x0}\hat{\mathbf{x}} + E_{y0}\hat{\mathbf{y}})$ where E_{x0} and E_{y0} are real (although K may be complex). To see this let us examine

$$\tilde{\mathbf{E}} = \tilde{\mathbf{E}}^+ + \tilde{\mathbf{E}}^- = \frac{E_0}{2}[\hat{\mathbf{x}} - j\hat{\mathbf{y}}]e^{-j\beta_+ z} + \frac{E_0}{2}[\hat{\mathbf{x}} + j\hat{\mathbf{y}}]e^{-j\beta_- z}$$

$$= \frac{E_0}{2} \left[\hat{\mathbf{x}} \left(e^{-j\beta_+ z} + e^{-j\beta_- z} \right) + j\hat{\mathbf{y}} \left(-e^{-j\beta_+ z} + e^{-j\beta_- z} \right) \right]$$

$$= E_0 e^{-j\frac{1}{2}(\beta_+ + \beta_-)z} \left[\hat{\mathbf{x}} \cos \frac{1}{2}(\beta_+ - \beta_-)z + \hat{\mathbf{y}} \sin \frac{1}{2}(\beta_+ - \beta_-)z \right]$$

or

$$\tilde{\mathbf{E}} = E_0 e^{-j\frac{1}{2}(\beta_+ + \beta_-)z} \left[\hat{\mathbf{x}} \cos \theta(z) + \hat{\mathbf{y}} \sin \theta(z) \right]$$

where $\theta(z) = (\beta_+ - \beta_-)z/2$. Because $\beta_+ \neq \beta_-$, the velocities of the two circularly polarized waves differ and the waves superpose to form a linearly polarized wave with a polarization that depends on the observation plane z-value. We may think of the wave as undergoing a phase shift of $(\beta_+ + \beta_-)z/2$ radians as it propagates, while the direction of $\tilde{\mathbf{E}}$ rotates to an angle $\theta(z) = (\beta_+ - \beta_-)z/2$ as the wave propagates. *Faraday rotation* can only occur at frequencies where both the LHCP and RHCP waves propagate, and therefore not within the stopband $\omega_0 < \omega < \omega_0 + \omega_M$.

Faraday rotation is non-reciprocal. That is, if a wave that has undergone a rotation of θ_0 radians by propagating through a distance z_0 is made to propagate an equal distance back in the direction from whence it came, the polarization does not return to its initial state but rather incurs an additional rotation of θ_0. Thus, the polarization angle of the wave when it returns to the starting point is not zero, but $2\theta_0$. This effect is employed in a number of microwave devices including gyrators, isolators, and circulators. The interested reader should see Collin [40], Elliott [67], or Liao [111] for details. We note that for $\omega \gg \omega_M$ we can approximate the rotation angle as

$$\theta(z) = (\beta_+ - \beta_-)z/2 = \frac{1}{2}\omega z \sqrt{\epsilon\mu_0} \left[\sqrt{1 + \frac{\omega_M}{\omega_0 - \omega}} - \sqrt{1 + \frac{\omega_M}{\omega_0 + \omega}} \right] \approx -\frac{1}{2}z\omega_M\sqrt{\epsilon\mu_0},$$

which is independent of frequency. So it is possible to construct Faraday rotation-based ferrite devices that maintain their properties over wide bandwidths.

It is straightforward to extend the above analysis to the case of a lossy ferrite. We find that for typical ferrites the attenuation constant associated with μ_- is small for all frequencies, but the attenuation constant associated with μ_+ is large near the resonant frequency ($\omega \approx \omega_0$) [40]. See Problem 4.16.

4.11.7 Propagation of cylindrical waves

By studying plane waves we have gained insight into the basic behavior of frequency-domain and time-harmonic waves. However, these solutions do not display the fundamental property that waves in space must diverge from their sources. To understand this behavior we shall treat waves having cylindrical and spherical symmetries.

Uniform cylindrical waves. In § 2.10.7 we studied the temporal behavior of cylindrical waves in a homogeneous, lossless medium and found that they diverge from a line source located along the z-axis. Here we shall extend the analysis to lossy media and investigate the behavior of the waves in the frequency domain.

Consider a homogeneous region of space described by the permittivity $\tilde{\epsilon}(\omega)$, permeability $\tilde{\mu}(\omega)$, and conductivity $\tilde{\sigma}(\omega)$. We seek solutions that are invariant over a cylindrical surface: $\tilde{\mathbf{E}}(\mathbf{r}, \omega) = \tilde{\mathbf{E}}(\rho, \omega), \tilde{\mathbf{H}}(\mathbf{r}, \omega) = \tilde{\mathbf{H}}(\rho, \omega)$. Such waves are called *uniform cylindrical waves*. Since the fields are z-independent we may decompose them into TE and TM sets as described in § 4.11.2. For TM polarization we may insert (4.211) into (4.212) to find

$$\tilde{H}_\phi(\rho, \omega) = \frac{1}{j\omega\tilde{\mu}(\omega)} \frac{\partial \tilde{E}_z(\rho, \omega)}{\partial \rho}. \tag{4.327}$$

For TE polarization we have from (4.213)

$$\tilde{E}_\phi(\rho,\omega) = -\frac{1}{j\omega\tilde{\epsilon}^c(\omega)}\frac{\partial\tilde{H}_z(\rho,\omega)}{\partial\rho} \tag{4.328}$$

where $\tilde{\epsilon}^c = \tilde{\epsilon} + \tilde{\sigma}/j\omega$ is the complex permittivity introduced in § 4.4.1. Since $\tilde{\mathbf{E}} = \hat{\phi}\tilde{E}_\phi + \hat{\mathbf{z}}\tilde{E}_z$ and $\tilde{\mathbf{H}} = \hat{\phi}\tilde{H}_\phi + \hat{\mathbf{z}}\tilde{H}_z$, we can always decompose a cylindrical electromagnetic wave into cases of electric and magnetic polarization. In each case the resulting field is TEM$_\rho$ since $\tilde{\mathbf{E}}$, $\tilde{\mathbf{H}}$, and $\hat{\rho}$ are mutually orthogonal.

Wave equations for $\tilde{E}_z$ in the electric polarization case and for $\tilde{H}_z$ in the magnetic polarization case can be derived by substituting (4.210) into (4.208):

$$\left(\frac{\partial^2}{\partial\rho^2} + \frac{1}{\rho}\frac{\partial}{\partial\rho} + k^2\right)\left\{\begin{matrix}\tilde{E}_z \\ \tilde{H}_z\end{matrix}\right\} = 0.$$

Thus the electric field must obey the ordinary differential equation

$$\frac{d^2\tilde{E}_z}{d\rho^2} + \frac{1}{\rho}\frac{d\tilde{E}_z}{d\rho} + k^2\tilde{E}_z = 0. \tag{4.329}$$

This is merely Bessel's equation (A.124). It is a second-order equation with two independent solutions chosen from the list

$$J_0(k\rho), \quad Y_0(k\rho), \quad H_0^{(1)}(k\rho), \quad H_0^{(2)}(k\rho).$$

We find that $J_0(k\rho)$ and $Y_0(k\rho)$ are useful for describing standing waves between boundaries, while $H_0^{(1)}(k\rho)$ and $H_0^{(2)}(k\rho)$ are useful for describing waves propagating in the ρ-direction. Of these, $H_0^{(1)}(k\rho)$ represents waves traveling inward while $H_0^{(2)}(k\rho)$ represents waves traveling outward. At this point we are interested in studying the behavior of outward propagating waves and so we choose

$$\tilde{E}_z(\rho,\omega) = -\frac{j}{4}\tilde{E}_{z0}(\omega)H_0^{(2)}(k\rho). \tag{4.330}$$

As explained in § 2.10.7, $\tilde{E}_{z0}(\omega)$ is the amplitude spectrum of the wave, while the term $-j/4$ is included to make the conversion to the time domain more convenient. By (4.327) we have

$$\tilde{H}_\phi = \frac{1}{j\omega\tilde{\mu}}\frac{\partial\tilde{E}_z}{\partial\rho} = \frac{1}{j\omega\tilde{\mu}}\frac{\partial}{\partial\rho}\left[-\frac{j}{4}\tilde{E}_{z0}H_0^{(2)}(k\rho)\right]. \tag{4.331}$$

Using $dH_0^{(2)}(x)/dx = -H_1^{(2)}(x)$ we find that

$$\tilde{H}_\phi = \frac{1}{Z_{TM}}\frac{\tilde{E}_{z0}}{4}H_1^{(2)}(k\rho) \tag{4.332}$$

where

$$Z_{TM} = \frac{\omega\tilde{\mu}}{k}$$

is called the *TM wave impedance*.

For the case of magnetic polarization, the field $\tilde{H}_z$ must satisfy Bessel's equation (4.329). Thus we choose

$$\tilde{H}_z(\rho,\omega) = -\frac{j}{4}\tilde{H}_{z0}(\omega)H_0^{(2)}(k\rho). \tag{4.333}$$

From (4.328) we find the electric field associated with the wave:

$$\tilde{E}_\phi = -Z_{TE}\frac{\tilde{H}_{z0}}{4}H_1^{(2)}(k\rho),\qquad(4.334)$$

where

$$Z_{TE} = \frac{k}{\omega\tilde{\epsilon}^c}$$

is the *TE wave impedance*.

It is not readily apparent that the terms $H_0^{(2)}(k\rho)$ or $H_1^{(2)}(k\rho)$ describe outward propagating waves. We shall see later that the cylindrical wave may be written as a superposition of plane waves, both uniform and evanescent, propagating in all possible directions. Each of these components does have the expected wave behavior, but it is still not obvious that the sum of such waves is outward propagating.

We saw in § 2.10.7 that when examined in the time domain, a cylindrical wave of the form $H_0^{(2)}(k\rho)$ does indeed propagate outward, and that for lossless media the velocity of propagation of its wavefronts is $v = 1/(\mu\epsilon)^{1/2}$. For time-harmonic fields, the cylindrical wave takes on a familiar behavior when the observation point is sufficiently removed from the source. We may specialize (4.330) to the time-harmonic case by setting $\omega = \check{\omega}$ and using phasors, giving

$$\check{E}_z(\rho) = -\frac{j}{4}\check{E}_{z0}H_0^{(2)}(k\rho).$$

If $|k\rho| \gg 1$ we can use the asymptotic representation (E.62) for the Hankel function

$$H_\nu^{(2)}(z) \sim \sqrt{\frac{2}{\pi z}}e^{-j(z-\pi/4-\nu\pi/2)},\qquad |z| \gg 1,\quad -2\pi < \arg(z) < \pi,$$

to obtain

$$\check{E}_z(\rho) \sim \check{E}_{z0}\frac{e^{-jk\rho}}{\sqrt{8j\pi k\rho}}\qquad(4.335)$$

and

$$\check{H}_\phi(\rho) \sim -\check{E}_{z0}\frac{1}{Z_{TM}}\frac{e^{-jk\rho}}{\sqrt{8j\pi k\rho}}\qquad(4.336)$$

for $|k\rho| \gg 1$. Except for the $\sqrt{\rho}$ term in the denominator, the wave has very much the same form as the plane waves encountered earlier. For the case of magnetic polarization, we can approximate (4.333) and (4.334) to obtain

$$\check{H}_z(\rho) \sim \check{H}_{z0}\frac{e^{-jk\rho}}{\sqrt{8j\pi k\rho}}\qquad(4.337)$$

and

$$\check{E}_\phi(\rho) \sim Z_{TE}\check{H}_{z0}\frac{e^{-jk\rho}}{\sqrt{8j\pi k\rho}}\qquad(4.338)$$

for $|k\rho| \gg 1$.

To interpret the wave nature of the field (4.335) let us substitute $k = \beta - j\alpha$ into the exponential function, where β is the phase constant (4.224) and α is the attenuation constant (4.225). Then

$$\check{E}_z(\rho) \sim \check{E}_{z0}\frac{1}{\sqrt{8j\pi k\rho}}e^{-\alpha\rho}e^{-j\beta\rho}.$$

Assuming $\check{E}_{z0} = |E_{z0}|e^{j\xi^{E}}$, the time-domain representation is found from (4.126):

$$E_z(\rho, t) = \frac{|E_{z0}|}{\sqrt{8\pi k \rho}} e^{-\alpha \rho} \cos[\check{\omega} t - \beta \rho - \pi/4 + \xi^{E}]. \qquad (4.339)$$

We can identify a surface of constant phase as a locus of points obeying

$$\check{\omega} t - \beta \rho - \pi/4 + \xi^{E} = C_P \qquad (4.340)$$

where C_P is some constant. These surfaces are cylinders coaxial with the z-axis, and are called *cylindrical wavefronts*. Note that surfaces of constant amplitude, as determined by

$$\frac{e^{-\alpha \rho}}{\sqrt{\rho}} = C_A$$

where C_A is some constant, are also cylinders.

The cosine term in (4.339) represents a traveling wave. As t is increased the argument of the cosine function remains fixed as long as ρ is increased correspondingly. Hence the cylindrical wavefronts propagate outward as time progresses. As the wavefront travels outward, the field is attenuated because of the factor $e^{-\alpha \rho}$. The velocity of propagation of the phase fronts may be computed by a now familiar technique. Differentiating (4.340) with respect to t we find that

$$\check{\omega} - \beta \frac{d\rho}{dt} = 0,$$

and thus have the phase velocity v_p of the outward expanding phase fronts:

$$v_p = \frac{d\rho}{dt} = \frac{\check{\omega}}{\beta}.$$

Calculation of wavelength also proceeds as before. Examining the two adjacent wavefronts that produce the same value of the cosine function in (4.339), we find $\beta \rho_1 = \beta \rho_2 - 2\pi$ or

$$\lambda = \rho_2 - \rho_1 = 2\pi/\beta.$$

Computation of the power carried by a cylindrical wave is straightforward. Since a cylindrical wavefront is infinite in extent, we usually speak of the *power per unit length* carried by the wave. This is found by integrating the time-average Poynting flux given in (4.157). For electric polarization we find the time-average power flux density using (4.330) and (4.331):

$$\mathbf{S}_{av} = \frac{1}{2} \operatorname{Re}\{\check{E}_z \hat{\mathbf{z}} \times \check{H}_{\phi}^{*} \hat{\boldsymbol{\phi}}\} = \frac{1}{2} \operatorname{Re}\left\{ \hat{\boldsymbol{\rho}} \frac{j}{16 Z_{TM}^{*}} |\check{E}_{z0}|^2 H_0^{(2)}(k\rho) H_1^{(2)*}(k\rho) \right\}. \qquad (4.341)$$

For magnetic polarization we use (4.333) and (4.334):

$$\mathbf{S}_{av} = \frac{1}{2} \operatorname{Re}\{\check{E}_{\phi} \hat{\boldsymbol{\phi}} \times \check{H}_z^{*} \hat{\mathbf{z}}\} = \frac{1}{2} \operatorname{Re}\left\{ -\hat{\boldsymbol{\rho}} \frac{j Z_{TE}}{16} |\check{H}_{z0}|^2 H_0^{(2)*}(k\rho) H_1^{(2)}(k\rho) \right\}.$$

For a lossless medium these expressions can be greatly simplified. By (E.5) we can write

$$j H_0^{(2)}(k\rho) H_1^{(2)*}(k\rho) = j[J_0(k\rho) - jN_0(k\rho)][J_1(k\rho) + jN_1(k\rho)],$$

hence

$$j H_0^{(2)}(k\rho) H_1^{(2)*}(k\rho) = [N_0(k\rho) J_1(k\rho) - J_0(k\rho) N_1(k\rho)] + j[J_0(k\rho) J_1(k\rho) + N_0(k\rho) N_1(k\rho)].$$

Substituting this into (4.341) and remembering that $Z_{TM} = \eta = (\mu/\epsilon)^{1/2}$ is real for lossless media, we have

$$\mathbf{S}_{av} = \hat{\boldsymbol{\rho}} \frac{1}{32\eta} |\check{E}_{z0}|^2 [N_0(k\rho)J_1(k\rho) - J_0(k\rho)N_1(k\rho)].$$

By the Wronskian relation (E.88) we have

$$\mathbf{S}_{av} = \hat{\boldsymbol{\rho}} \frac{|\check{E}_{z0}|^2}{16\pi k\rho\eta}.$$

The power density is inversely proportional to ρ. When we compute the total time-average power per unit length passing through a cylinder of radius ρ, this factor cancels with the ρ-dependence of the surface area to give a result independent of radius:

$$P_{av}/l = \int_0^{2\pi} \mathbf{S}_{av} \cdot \hat{\boldsymbol{\rho}}\rho\, d\phi = \frac{|\check{E}_{z0}|^2}{8k\eta}. \tag{4.342}$$

For a lossless medium there is no mechanism to dissipate the power and so the wave propagates unabated. A similar calculation for the case of magnetic polarization (Problem 4.17) gives

$$\mathbf{S}_{av} = \hat{\boldsymbol{\rho}} \frac{\eta|\check{H}_{z0}|^2}{16\pi k\rho}$$

and

$$P_{av}/l = \frac{\eta|\check{H}_{z0}|^2}{8k}.$$

For a lossy medium the expressions are more difficult to evaluate. In this case we expect the total power passing through a cylinder to depend on the radius of the cylinder, since the fields decay exponentially with distance and thus give up power as they propagate. If we assume that the observation point is far from the z-axis with $|k\rho| \gg 1$, then we can use (4.335) and (4.336) for the electric polarization case to obtain

$$\mathbf{S}_{av} = \frac{1}{2} \operatorname{Re}\{\check{E}_z\hat{\mathbf{z}} \times \check{H}_\phi^*\hat{\boldsymbol{\phi}}\} = \frac{1}{2} \operatorname{Re}\left\{ \hat{\boldsymbol{\rho}} \frac{e^{-2\alpha\rho}}{8\pi\rho|k|Z_{TM}^*} |\check{E}_{z0}|^2 \right\}.$$

Therefore

$$P_{av}/l = \int_0^{2\pi} \mathbf{S}_{av} \cdot \hat{\boldsymbol{\rho}}\rho\, d\phi = \operatorname{Re}\left\{ \frac{1}{Z_{TM}^*} \right\} |\check{E}_{z0}|^2 \frac{e^{-2\alpha\rho}}{8|k|}.$$

We note that for a lossless material $Z_{TM} = \eta$ and $\alpha = 0$, and the expression reduces to (4.342) as expected. Thus for lossy materials the power depends on the radius of the cylinder. In the case of magnetic polarization we use (4.337) and (4.338) to get

$$\mathbf{S}_{av} = \frac{1}{2} \operatorname{Re}\{\check{E}_\phi\hat{\boldsymbol{\phi}} \times \check{H}_z^*\hat{\mathbf{z}}\} = \frac{1}{2} \operatorname{Re}\left\{ \hat{\boldsymbol{\rho}} Z_{TE}^* \frac{e^{-2\alpha\rho}}{8\pi\rho|k|} |\check{H}_{z0}|^2 \right\}$$

and

$$P_{av}/l = \operatorname{Re}\{Z_{TE}^*\} |\check{H}_{z0}|^2 \frac{e^{-2\alpha\rho}}{8|k|}.$$

Example of uniform cylindrical waves: fields of a line source. The simplest example of a uniform cylindrical wave is that produced by an electric or magnetic line source. Consider first an infinite electric line current of amplitude $\tilde{I}(\omega)$ on the z-axis, immersed within a medium of permittivity $\tilde{\epsilon}(\omega)$, permeability $\tilde{\mu}(\omega)$, and conductivity $\tilde{\sigma}(\omega)$. We assume that the current does not vary in the z-direction, and thus the problem is two-dimensional. We can decompose the field produced by the line source into TE and TM cases according to § 4.11.2. It turns out that an electric line source only excites TM fields, as we shall show in § 5.4, and thus we need only $\tilde{E}_z$ to completely describe the fields.

By symmetry the fields are ϕ-independent and thus the wave produced by the line source is a uniform cylindrical wave. Since the wave propagates outward from the line source we have the electric field from (4.330),

$$\tilde{E}_z(\rho,\omega) = -\frac{j}{4}\tilde{E}_{z0}(\omega)H_0^{(2)}(k\rho), \qquad (4.343)$$

and the magnetic field from (4.332),

$$\tilde{H}_\phi(\rho,\omega) = \frac{k}{\omega\tilde{\mu}}\frac{\tilde{E}_{z0}(\omega)}{4}H_1^{(2)}(k\rho).$$

We can find $\tilde{E}_{z0}$ by using Ampere's law:

$$\oint_\Gamma \tilde{\mathbf{H}}\cdot\mathbf{dl} = \int_S \tilde{\mathbf{J}}\cdot\mathbf{dS} + j\omega\int_S \tilde{\mathbf{D}}\cdot\mathbf{dS}.$$

Since $\tilde{\mathbf{J}}$ is the sum of the impressed current $\tilde{I}$ and the secondary conduction current $\tilde{\sigma}\tilde{\mathbf{E}}$, we can also write

$$\oint_\Gamma \tilde{\mathbf{H}}\cdot\mathbf{dl} = \tilde{I} + \int_S (\tilde{\sigma} + j\omega\tilde{\epsilon})\tilde{\mathbf{E}}\cdot\mathbf{dS} = \tilde{I} + j\omega\tilde{\epsilon}^c\int_S \tilde{\mathbf{E}}\cdot\mathbf{dS}.$$

Choosing our path of integration as a circle of radius a in the $z = 0$ plane and substituting for $\tilde{E}_z$ and $\tilde{H}_\phi$, we find that

$$\frac{k}{\omega\tilde{\mu}}\frac{\tilde{E}_{z0}}{4}H_1^{(2)}(ka)2\pi a = \tilde{I} + j\omega\tilde{\epsilon}^c 2\pi\frac{-j\tilde{E}_{z0}}{4}\lim_{\delta\to 0}\int_\delta^a H_0^{(2)}(k\rho)\rho\,d\rho. \qquad (4.344)$$

The limit operation is required because $H_0^{(2)}(k\rho)$ diverges as $\rho \to 0$. By (E.104) the integral is

$$\lim_{\delta\to 0}\int_\delta^a H_0^{(2)}(k\rho)\rho\,d\rho = \frac{a}{k}H_1^{(2)}(ka) - \frac{1}{k}\lim_{\delta\to 0}\delta H_1^{(2)}(k\delta).$$

The limit may be found by using $H_1^{(2)}(x) = J_1(x) - jN_1(x)$ and the small argument approximations (E.50) and (E.53):

$$\lim_{\delta\to 0}\delta H_1^{(2)}(\delta) = \lim_{\delta\to 0}\delta\left[\frac{k\delta}{2} - j\left(-\frac{1}{\pi}\frac{2}{k\delta}\right)\right] = j\frac{2}{\pi k}.$$

Substituting these expressions into (4.344) we obtain

$$\frac{k}{\omega\tilde{\mu}}\frac{\tilde{E}_{z0}}{4}H_1^{(2)}(ka)2\pi a = \tilde{I} + j\omega\tilde{\epsilon}^c 2\pi\frac{-j\tilde{E}_{z0}}{4}\left[\frac{a}{k}H_1^{(2)}(ka) - j\frac{2}{\pi k^2}\right].$$

Using $k^2 = \omega^2 \tilde{\mu} \tilde{\epsilon}^c$ we find that the two Hankel function terms cancel. Solving for $\tilde{E}_{z0}$ we have

$$\tilde{E}_{z0} = -j\omega\tilde{\mu}\tilde{I}$$

and therefore

$$\tilde{E}_z(\rho,\omega) = -\frac{\omega\tilde{\mu}}{4}\tilde{I}(\omega)H_0^{(2)}(k\rho) = -j\omega\tilde{\mu}\tilde{I}(\omega)\tilde{G}(x,y|0,0;\omega). \qquad (4.345)$$

Here $\tilde{G}$ is called the *two-dimensional Green's function* and is given by

$$\tilde{G}(x,y|x',y';\omega) = \frac{1}{4j}H_0^{(2)}\left(k\sqrt{(x-x')^2+(y-y')^2}\right). \qquad (4.346)$$

Green's functions are examined in greater detail in Chapter 5.

It is also possible to determine the field amplitude by evaluating

$$\lim_{a\to 0}\oint_C \tilde{\mathbf{H}}\cdot\mathbf{dl}.$$

This produces an identical result and is a bit simpler since it can be argued that the surface integral of $\tilde{E}_z$ vanishes as $a \to 0$ without having to perform the calculation directly [83, 8].

For a magnetic line source $\tilde{I}_m(\omega)$ aligned along the z-axis we proceed as above, but note that the source only produces TE fields. By (4.333) and (4.334) we have

$$\tilde{H}_z(\rho,\omega) = -\frac{j}{4}\tilde{H}_{z0}(\omega)H_0^{(2)}(k\rho), \qquad \tilde{E}_\phi = -\frac{k}{\omega\tilde{\epsilon}^c}\frac{\tilde{H}_{0z}}{4}H_1^{(2)}(k\rho).$$

We can find $\tilde{H}_{z0}$ by applying Faraday's law

$$\oint_C \tilde{\mathbf{E}}\cdot\mathbf{dl} = -\int_S \tilde{\mathbf{J}}_m\cdot\mathbf{dS} - j\omega\int_S \tilde{\mathbf{B}}\cdot\mathbf{dS}$$

about a circle of radius a in the $z=0$ plane. We have

$$-\frac{k}{\omega\tilde{\epsilon}^c}\frac{\tilde{H}_{z0}}{4}H_1^{(2)}(ka)2\pi a = -\tilde{I}_m - j\omega\tilde{\mu}\left[-\frac{j}{4}\right]\tilde{H}_{z0}2\pi\lim_{\delta\to 0}\int_\delta^a H_0^{(2)}(k\rho)\rho\,d\rho.$$

Proceeding as above we find that

$$\tilde{H}_{z0} = j\omega\tilde{\epsilon}^c\tilde{I}_m$$

hence

$$\tilde{H}_z(\rho,\omega) = -\frac{\omega\tilde{\epsilon}^c}{4}\tilde{I}_m(\omega)H_0^{(2)}(k\rho) = -j\omega\tilde{\epsilon}^c\tilde{I}_m(\omega)\tilde{G}(x,y|0,0;\omega). \qquad (4.347)$$

Note that we could have solved for the magnetic field of a magnetic line current by using the field of an electric line current and the principle of duality. Letting the magnetic current be equal to $-\eta$ times the electric current and using (4.198), we find that

$$\tilde{H}_{z0} = \left(-\frac{1}{\eta}\frac{\tilde{I}_m(\omega)}{\tilde{I}(\omega)}\right)\left(-\frac{1}{\eta}\left[-\frac{\omega\tilde{\mu}}{4}\tilde{I}(\omega)H_0^{(2)}(k\rho)\right]\right) = -\tilde{I}_m(\omega)\frac{\omega\tilde{\epsilon}^c}{4}H_0^{(2)}(k\rho) \qquad (4.348)$$

as in (4.347).

Nonuniform cylindrical waves. When we solve two-dimensional boundary value problems we encounter cylindrical waves that are z-independent but ϕ-dependent. Although such waves propagate outward, they have a more complicated structure than those considered above.

For the case of TM polarization we have, by (4.212),

$$\tilde{H}_\rho = \frac{j}{Z_{TM}k}\frac{1}{\rho}\frac{\partial \tilde{E}_z}{\partial \phi}, \tag{4.349}$$

$$\tilde{H}_\phi = -\frac{j}{Z_{TM}k}\frac{\partial \tilde{E}_z}{\partial \rho}, \tag{4.350}$$

where $Z_{TM} = \omega\tilde{\mu}/k$. For the TE case we have, by (4.213),

$$\tilde{E}_\rho = -\frac{jZ_{TE}}{k}\frac{1}{\rho}\frac{\partial \tilde{H}_z}{\partial \phi}, \tag{4.351}$$

$$\tilde{E}_\phi = \frac{jZ_{TE}}{k}\frac{\partial \tilde{H}_z}{\partial \rho}, \tag{4.352}$$

where $Z_{TE} = k/\omega\tilde{\epsilon}^c$. By (4.208) the wave equations are

$$\left(\frac{\partial^2}{\partial\rho^2} + \frac{1}{\rho}\frac{\partial}{\partial\rho} + \frac{1}{\rho^2}\frac{\partial^2}{\partial\phi^2} + k^2\right)\left\{\begin{matrix}\tilde{E}_z \\ \tilde{H}_z\end{matrix}\right\} = 0.$$

Because this has the form of (A.117) with $\partial/\partial z \to 0$, we have

$$\left\{\begin{matrix}\tilde{E}_z(\rho,\phi,\omega) \\ \tilde{H}_z(\rho,\phi,\omega)\end{matrix}\right\} = P(\rho,\omega)\Phi(\phi,\omega) \tag{4.353}$$

where

$$\Phi(\phi,\omega) = A_\phi(\omega)\sin k_\phi\phi + B_\phi(\omega)\cos k_\phi\phi, \tag{4.354}$$

$$P(\rho) = A_\rho(\omega)B^{(1)}_{k_\phi}(k\rho) + B_\rho(\omega)B^{(2)}_{k_\phi}(k\rho), \tag{4.355}$$

and where $B^{(1)}_\nu(z)$ and $B^{(2)}_\nu(z)$ are any two independent Bessel functions chosen from the set

$$J_\nu(z), \quad N_\nu(z), \quad H^{(1)}_\nu(z), \quad H^{(2)}_\nu(z).$$

In bounded regions we generally use the oscillatory functions $J_\nu(z)$ and $N_\nu(z)$ to represent standing waves. In unbounded regions we generally use $H^{(2)}_\nu(z)$ and $H^{(1)}_\nu(z)$ to represent outward and inward propagating waves, respectively.

Boundary value problems in cylindrical coordinates: scattering by a material cylinder. A variety of problems can be solved using nonuniform cylindrical waves. We shall examine two interesting cases in which an external field is impressed on a two-dimensional object. The impressed field creates secondary sources within or on the object, and these in turn create a secondary field. Our goal is to determine the secondary field by applying appropriate boundary conditions.

As a first example, consider a material cylinder of radius a, complex permittivity $\tilde{\epsilon}^c$, and permeability $\tilde{\mu}$, aligned along the z-axis in free space (Figure 4.25). An incident plane wave propagating in the x-direction is impressed on the cylinder, inducing secondary polarization and conduction currents within the cylinder. These in turn produce

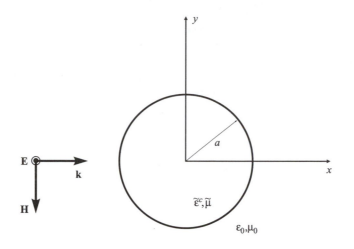

Figure 4.25: TM plane-wave field incident on a material cylinder.

secondary or scattered fields, which are standing waves within the cylinder and outward traveling waves external to the cylinder. Although we have not yet learned how to write the secondary fields in terms of the impressed sources, we can solve for the fields as a boundary value problem. The total field must obey the boundary conditions on tangential components at the interface between the cylinder and surrounding free space. We need not worry about the effect of the secondary sources on the source of the primary field, since by definition impressed sources cannot be influenced by secondary fields.

The scattered field can be found using superposition. When excited by a TM impressed field, the secondary field is also TM. The situation for TE excitation is similar. By decomposing the impressed field into TE and TM components, we may solve for the scattered field in each case and then superpose the results to determine the complete solution.

We first consider the TM case. The impressed electric field may be written as

$$\tilde{\mathbf{E}}^i(\mathbf{r}, \omega) = \hat{\mathbf{z}}\tilde{E}_0(\omega)e^{-jk_0 x} = \hat{\mathbf{z}}\tilde{E}_0(\omega)e^{-jk_0\rho\cos\phi} \tag{4.356}$$

while the magnetic field is, by (4.223),

$$\tilde{\mathbf{H}}^i(\mathbf{r}, \omega) = -\hat{\mathbf{y}}\frac{\tilde{E}_0(\omega)}{\eta_0}e^{-jk_0 x} = -(\hat{\boldsymbol{\rho}}\sin\phi + \hat{\boldsymbol{\phi}}\cos\phi)\frac{\tilde{E}_0(\omega)}{\eta_0}e^{-jk_0\rho\cos\phi}.$$

Here $k_0 = \omega(\mu_0\epsilon_0)^{1/2}$ and $\eta_0 = (\mu_0/\epsilon_0)^{1/2}$. The scattered electric field takes the form of a nonuniform cylindrical wave (4.353). Periodicity in ϕ implies that k_ϕ is an integer, say $k_\phi = n$. Within the cylinder we cannot use any of the functions $N_n(k\rho)$, $H_n^{(2)}(k\rho)$, or $H_n^{(1)}(k\rho)$ to represent the radial dependence of the field, since each is singular at the origin. So we choose $B_n^{(1)}(k\rho) = J_n(k\rho)$ and $B_\rho(\omega) = 0$ in (4.355). Physically, $J_n(k\rho)$ represents the standing wave created by the interaction of outward and inward propagating waves. External to the cylinder we use $H_n^{(2)}(k\rho)$ to represent the radial dependence of the secondary field components: we avoid $N_n(k\rho)$ and $J_n(k\rho)$ since these represent standing waves, and avoid $H_n^{(1)}(k\rho)$ since there are no external secondary sources to create an inward traveling wave.

Any attempt to satisfy the boundary conditions by using a single nonuniform wave fails. This is because the sinusoidal dependence on ϕ of each individual nonuniform wave cannot match the more complicated dependence of the impressed field (4.356). Since the sinusoids are complete, an infinite series of the functions (4.353) can be used to represent the scattered field. So we have internal to the cylinder

$$\tilde{E}_z^s(\mathbf{r}, \omega) = \sum_{n=0}^{\infty} [A_n(\omega) \sin n\phi + B_n(\omega) \cos n\phi] \, J_n(k\rho)$$

where $k = \omega(\tilde{\mu}\tilde{\epsilon}^c)^{1/2}$. External to the cylinder we have free space and thus

$$\tilde{E}_z^s(\mathbf{r}, \omega) = \sum_{n=0}^{\infty} [C_n(\omega) \sin n\phi + D_n(\omega) \cos n\phi] \, H_n^{(2)}(k_0\rho).$$

Equations (4.349) and (4.350) yield the magnetic field internal to the cylinder:

$$\tilde{H}_\rho^s = \sum_{n=0}^{\infty} \frac{jn}{Z_{TM}k} \frac{1}{\rho} [A_n(\omega) \cos n\phi - B_n(\omega) \sin n\phi] \, J_n(k\rho),$$

$$\tilde{H}_\phi^s = -\sum_{n=0}^{\infty} \frac{j}{Z_{TM}} [A_n(\omega) \sin n\phi + B_n(\omega) \cos n\phi] \, J_n'(k\rho),$$

where $Z_{TM} = \omega\tilde{\mu}/k$. Outside the cylinder

$$\tilde{H}_\rho^s = \sum_{n=0}^{\infty} \frac{jn}{\eta_0 k_0} \frac{1}{\rho} [C_n(\omega) \cos n\phi - D_n(\omega) \sin n\phi] \, H_n^{(2)}(k_0\rho),$$

$$\tilde{H}_\phi^s = -\sum_{n=0}^{\infty} \frac{j}{\eta_0} [C_n(\omega) \sin n\phi + D_n(\omega) \cos n\phi] \, H_n^{(2)\prime}(k_0\rho),$$

where $J_n'(z) = dJ_n(z)/dz$ and $H_n^{(2)\prime}(z) = dH_n^{(2)}(z)/dz$.

We have two sets of unknown spectral amplitudes (A_n, B_n) and (C_n, D_n). These can be determined by applying the boundary conditions at the interface. Since the total field outside the cylinder is the sum of the impressed and scattered terms, an application of continuity of the tangential electric field at $\rho = a$ gives us

$$\sum_{n=0}^{\infty} [A_n \sin n\phi + B_n \cos n\phi] \, J_n(ka) =$$

$$\sum_{n=0}^{\infty} [C_n \sin n\phi + D_n \cos n\phi] \, H_n^{(2)}(k_0 a) + \tilde{E}_0 e^{-jk_0 a \cos \phi},$$

which must hold for all $-\pi \leq \phi \leq \pi$. To remove the coefficients from the sum we apply orthogonality. Multiplying both sides by $\sin m\phi$, integrating over $[-\pi, \pi]$, and using the orthogonality conditions (A.129)– (A.131) we obtain

$$\pi A_m J_m(ka) - \pi C_m H_m^{(2)}(k_0 a) = \tilde{E}_0 \int_{-\pi}^{\pi} \sin m\phi e^{-jk_0 a \cos \phi} \, d\phi = 0. \qquad (4.357)$$

Multiplying by $\cos m\phi$ and integrating, we find that

$$2\pi B_m J_m(ka) - 2\pi D_m H_m^{(2)}(k_0 a) = \tilde{E}_0 \epsilon_m \int_{-\pi}^{\pi} \cos m\phi e^{-jk_0 a \cos \phi} \, d\phi$$

$$= 2\pi \tilde{E}_0 \epsilon_m j^{-m} J_m(k_0 a) \qquad (4.358)$$

where ϵ_n is Neumann's number (A.132) and where we have used (E.83) and (E.39) to evaluate the integral.

We must also have continuity of the tangential magnetic field $\tilde{H}_\phi$ at $\rho = a$. Thus

$$-\sum_{n=0}^{\infty} \frac{j}{Z_{TM}} \left[A_n \sin n\phi + B_n \cos n\phi \right] J_n'(ka) =$$

$$-\sum_{n=0}^{\infty} \frac{j}{\eta_0} \left[C_n \sin n\phi + D_n \cos n\phi \right] H_n^{(2)'}(k_0 a) - \cos\phi \frac{\tilde{E}_0}{\eta_0} e^{-jk_0 a \cos \phi}$$

must hold for all $-\pi \le \phi \le \pi$. By orthogonality

$$\pi \frac{j}{Z_{TM}} A_m J_m'(ka) - \pi \frac{j}{\eta_0} C_m H_m^{(2)'}(k_0 a) = \frac{\tilde{E}_0}{\eta_0} \int_{-\pi}^{\pi} \sin m\phi \cos\phi e^{-jk_0 a \cos \phi} \, d\phi = 0 \quad (4.359)$$

and

$$2\pi \frac{j}{Z_{TM}} B_m J_m'(ka) - 2\pi \frac{j}{\eta_0} D_m H_m^{(2)'}(k_0 a) = \epsilon_m \frac{\tilde{E}_0}{\eta_0} \int_{-\pi}^{\pi} \cos m\phi \cos\phi e^{-jk_0 a \cos \phi} \, d\phi.$$

The integral may be computed as

$$\int_{-\pi}^{\pi} \cos m\phi \cos\phi e^{-jk_0 a \cos \phi} \, d\phi = j \frac{d}{d(k_0 a)} \int_{-\pi}^{\pi} \cos m\phi e^{-jk_0 a \cos \phi} \, d\phi = j2\pi j^{-m} J_m'(k_0 a)$$

and thus

$$\frac{1}{Z_{TM}} B_m J_m'(ka) - \frac{1}{\eta_0} D_m H_m^{(2)'}(k_0 a) = \frac{\tilde{E}_0}{\eta_0} \epsilon_m j^{-m} J_m'(k_0 a). \quad (4.360)$$

We now have four equations for the coefficients A_n, B_n, C_n, D_n. We may write (4.357) and (4.359) as

$$\begin{bmatrix} J_m(ka) & -H_m^{(2)}(k_0 a) \\ \frac{\eta_0}{Z_{TM}} J_m'(ka) & -H_m^{(2)'}(k_0 a) \end{bmatrix} \begin{bmatrix} A_m \\ C_m \end{bmatrix} = 0, \quad (4.361)$$

and (4.358) and (4.360) as

$$\begin{bmatrix} J_m(ka) & -H_m^{(2)}(k_0 a) \\ \frac{\eta_0}{Z_{TM}} J_m'(ka) & -H_m^{(2)'}(k_0 a) \end{bmatrix} \begin{bmatrix} B_m \\ D_m \end{bmatrix} = \begin{bmatrix} \tilde{E}_0 \epsilon_m j^{-m} J_m(k_0 a) \\ \tilde{E}_0 \epsilon_m j^{-m} J_m'(k_0 a) \end{bmatrix}. \quad (4.362)$$

Matrix equations (4.361) and (4.362) cannot hold simultaneously unless $A_m = C_m = 0$. Then the solution to (4.362) is

$$B_m = \tilde{E}_0 \epsilon_m j^{-m} \left[\frac{H_m^{(2)}(k_0 a) J_m'(k_0 a) - J_m(k_0 a) H_m^{(2)'}(k_0 a)}{\frac{\eta_0}{Z_{TM}} J_m'(ka) H_m^{(2)}(k_0 a) - H_m^{(2)'}(k_0 a) J_m(ka)} \right], \quad (4.363)$$

$$D_m = -\tilde{E}_0 \epsilon_m j^{-m} \left[\frac{\frac{\eta_0}{Z_{TM}} J_m'(ka) J_m(k_0 a) - J_m'(k_0 a) J_m(ka)}{\frac{\eta_0}{Z_{TM}} J_m'(ka) H_m^{(2)}(k_0 a) - H_m^{(2)'}(k_0 a) J_m(ka)} \right]. \quad (4.364)$$

With these coefficients we can calculate the field inside the cylinder ($\rho \le a$) from

$$\tilde{E}_z(\mathbf{r}, \omega) = \sum_{n=0}^{\infty} B_n(\omega) J_n(k\rho) \cos n\phi,$$

$$\tilde{H}_\rho(\mathbf{r},\omega) = -\sum_{n=0}^{\infty} \frac{jn}{Z_{TM}k}\frac{1}{\rho}B_n(\omega)J_n(k\rho)\sin n\phi,$$

$$\tilde{H}_\phi(\mathbf{r},\omega) = -\sum_{n=0}^{\infty} \frac{j}{Z_{TM}}B_n(\omega)J_n'(k\rho)\cos n\phi,$$

and the field outside the cylinder ($\rho > a$) from

$$\tilde{E}_z(\mathbf{r},\omega) = \tilde{E}_0(\omega)e^{-jk_0\rho\cos\phi} + \sum_{n=0}^{\infty} D_n(\omega)H_n^{(2)}(k_0\rho)\cos n\phi,$$

$$\tilde{H}_\rho(\mathbf{r},\omega) = -\sin\phi\frac{\tilde{E}_0(\omega)}{\eta_0}e^{-jk_0\rho\cos\phi} - \sum_{n=0}^{\infty} \frac{jn}{\eta_0 k_0}\frac{1}{\rho}D_n(\omega)H_n^{(2)}(k_0\rho)\sin n\phi,$$

$$\tilde{H}_\phi(\mathbf{r},\omega) = -\cos\phi\frac{\tilde{E}_0(\omega)}{\eta_0}e^{-jk_0\rho\cos\phi} - \sum_{n=0}^{\infty} \frac{j}{\eta_0}D_n(\omega)H_n^{(2)\prime}(k_0\rho)\cos n\phi.$$

We can easily specialize these results to the case of a perfectly conducting cylinder by allowing $\tilde{\sigma} \to \infty$. Then

$$\frac{\eta_0}{Z_{TM}} = \sqrt{\frac{\mu_0\tilde{\epsilon}^c}{\tilde{\mu}\epsilon_0}} \to \infty$$

and

$$B_n \to 0, \qquad D_n \to -\tilde{E}_0\epsilon_m j^{-m}\frac{J_m(k_0a)}{H_m^{(2)}(k_0a)}.$$

In this case it is convenient to combine the formulas for the impressed and scattered fields when forming the total fields. Since the impressed field is z-independent and obeys the homogeneous Helmholtz equation, we may represent it in terms of nonuniform cylindrical waves:

$$\tilde{E}_z^i = \tilde{E}_0 e^{-jk_0\rho\cos\phi} = \sum_{n=0}^{\infty} [E_n\sin n\phi + F_n\cos n\phi]J_n(k_0\rho),$$

where we have chosen the Bessel function $J_n(k_0\rho)$ since the field is finite at the origin and periodic in ϕ. Applying orthogonality we see immediately that $E_n = 0$ and that

$$\frac{2\pi}{\epsilon_m}F_m J_m(k_0\rho) = \tilde{E}_0\int_{-\pi}^{\pi}\cos m\phi\, e^{-jk_0\rho\cos\phi}\,d\phi = \tilde{E}_0 2\pi j^{-m}J_m(k_0\rho).$$

Thus, $F_n = \tilde{E}_0\epsilon_n j^{-n}$ and

$$\tilde{E}_z^i = \sum_{n=0}^{\infty} \tilde{E}_0\epsilon_n j^{-n}J_n(k_0\rho)\cos n\phi.$$

Adding this impressed field to the scattered field we have the total field outside the cylinder,

$$\tilde{E}_z = \tilde{E}_0\sum_{n=0}^{\infty} \frac{\epsilon_n j^{-n}}{H_n^{(2)}(k_0a)}\left[J_n(k_0\rho)H_n^{(2)}(k_0a) - J_n(k_0a)H_n^{(2)}(k_0\rho)\right]\cos n\phi,$$

while the field within the cylinder vanishes. Then, by (4.350),

$$\tilde{H}_\phi = -\frac{j}{\eta_0}\tilde{E}_0\sum_{n=0}^{\infty} \frac{\epsilon_n j^{-n}}{H_n^{(2)}(k_0a)}\left[J_n'(k_0\rho)H_n^{(2)}(k_0a) - J_n(k_0a)H_n^{(2)\prime}(k_0\rho)\right]\cos n\phi.$$

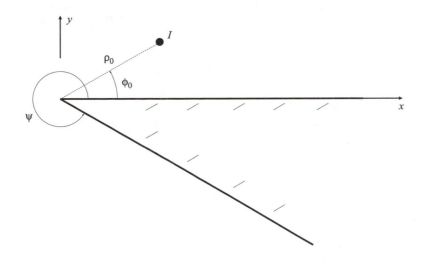

Figure 4.26: Geometry of a perfectly conducting wedge illuminated by a line source.

This in turn gives us the surface current induced on the cylinder. From the boundary condition $\tilde{\mathbf{J}}_s = \hat{\mathbf{n}} \times \tilde{\mathbf{H}}|_{\rho=a} = \hat{\boldsymbol{\rho}} \times [\hat{\boldsymbol{\rho}}\tilde{H}_\rho + \hat{\boldsymbol{\phi}}\tilde{H}_\phi]|_{\rho=a} = \hat{\mathbf{z}}\tilde{H}_\phi|_{\rho=a}$ we have

$$\mathbf{J}_s(\phi,\omega) = -\frac{j}{\eta_0}\hat{\mathbf{z}}\tilde{E}_0 \sum_{n=0}^{\infty} \frac{\epsilon_n j^{-n}}{H_n^{(2)}(k_0 a)} \left[J_n'(k_0 a)H_n^{(2)}(k_0 a) - J_n(k_0 a)H_n^{(2)\prime}(k_0 a) \right] \cos n\phi,$$

and an application of (E.93) gives us

$$\mathbf{J}_s(\phi,\omega) = \hat{\mathbf{z}}\frac{2\tilde{E}_0}{\eta_0 k_0 \pi a} \sum_{n=0}^{\infty} \frac{\epsilon_n j^{-n}}{H_n^{(2)}(k_0 a)} \cos n\phi. \tag{4.365}$$

Computation of the scattered field for a magnetically-polarized impressed field proceeds in the same manner. The impressed electric and magnetic fields are assumed to be

$$\tilde{\mathbf{E}}^i(\mathbf{r},\omega) = \hat{\mathbf{y}}\tilde{E}_0(\omega)e^{-jk_0 x} = (\hat{\boldsymbol{\rho}}\sin\phi + \hat{\boldsymbol{\phi}}\cos\phi)\tilde{E}_0(\omega)e^{-jk_0\rho\cos\phi},$$

$$\tilde{\mathbf{H}}^i(\mathbf{r},\omega) = \hat{\mathbf{z}}\frac{\tilde{E}_0(\omega)}{\eta_0}e^{-jk_0 x} = \hat{\mathbf{z}}\frac{\tilde{E}_0(\omega)}{\eta_0}e^{-jk_0\rho\cos\phi}.$$

For a perfectly conducting cylinder, the total magnetic field is

$$\tilde{H}_z = \frac{\tilde{E}_0}{\eta_0} \sum_{n=0}^{\infty} \frac{\epsilon_n j^{-n}}{H_n^{(2)\prime}(k_0 a)} \left[J_n(k_0\rho)H_n^{(2)\prime}(k_0 a) - J_n'(k_0 a)H_n^{(2)}(k_0\rho) \right] \cos n\phi. \tag{4.366}$$

The details are left as an exercise.

Boundary value problems in cylindrical coordinates: scattering by a perfectly conducting wedge. As a second example, consider a perfectly conducting wedge immersed in free space and illuminated by a line source (Figure 4.26) carrying current $\tilde{I}(\omega)$ and located at (ρ_0, ϕ_0). The current, which is assumed to be z-invariant, induces a secondary current on the surface of the wedge which in turn produces a secondary

(scattered) field. This scattered field, also z-invariant, can be found by solving a boundary value problem. We do this by separating space into the two regions $\rho < \rho_0$ and
$\rho > \rho_0$, $0 < \phi < \psi$. Each of these is source-free, so we can represent the total field using
nonuniform cylindrical waves of the type (4.353). The line source is brought into the
problem by applying the boundary condition on the tangential magnetic field across the
cylindrical surface $\rho = \rho_0$.

Since the impressed electric field has only a z-component, so do the scattered and total
electric fields. We wish to represent the total field $\tilde{E}_z$ in terms of nonuniform cylindrical
waves of the type (4.353). Since the field is not periodic in ϕ, the separation constant
k_ϕ need not be an integer; instead, its value is determined by the positions of the wedge
boundaries. For the region $\rho < \rho_0$ we represent the radial dependence of the field using
the functions J_ν since the field must be finite at the origin. For $\rho > \rho_0$ we use the
outward-propagating wave functions $H_\delta^{(2)}$. Thus

$$\tilde{E}_z(\rho, \phi, \omega) = \begin{cases} \sum_\nu \left[A_\nu \sin \nu\phi + B_\nu \cos \nu\phi \right] J_\nu(k_0\rho), & \rho < \rho_0, \\ \sum_\delta \left[C_\delta \sin \delta\phi + D_\delta \cos \delta\phi \right] H_\delta^{(2)}(k_0\rho), & \rho > \rho_0. \end{cases} \tag{4.367}$$

The coefficients $A_\nu, B_\nu, C_\delta, D_\delta$ and separation constants ν, δ may be found by applying
the boundary conditions on the fields at the surface of the wedge and across the surface
$\rho = \rho_0$. On the wedge face at $\phi = 0$ we must have $\tilde{E}_z = 0$, hence $B_\nu = D_\delta = 0$. On
the wedge face at $\phi = \psi$ we must also have $\tilde{E}_z = 0$, requiring $\sin \nu\psi = \sin \delta\psi = 0$ and
therefore

$$\nu = \delta = \nu_n = n\pi/\psi, \qquad n = 1, 2, \ldots.$$

So

$$\tilde{E}_z = \begin{cases} \sum_{n=0}^\infty A_n \sin \nu_n\phi J_{\nu_n}(k_0\rho), & \rho < \rho_0, \\ \sum_{n=0}^\infty C_n \sin \nu_n\phi H_{\nu_n}^{(2)}(k_0\rho), & \rho > \rho_0. \end{cases} \tag{4.368}$$

The magnetic field can be found from (4.349)–(4.350):

$$\tilde{H}_\rho = \begin{cases} \sum_{n=0}^\infty A_n \frac{j}{\eta_0 k_0} \frac{\nu_n}{\rho} \cos \nu_n\phi J_{\nu_n}(k_0\rho), & \rho < \rho_0, \\ \sum_{n=0}^\infty C_n \frac{j}{\eta_0 k_0} \frac{\nu_n}{\rho} \cos \nu_n\phi H_{\nu_n}^{(2)}(k_0\rho), & \rho > \rho_0, \end{cases} \tag{4.369}$$

$$\tilde{H}_\phi = \begin{cases} -\sum_{n=0}^\infty A_n \frac{j}{\eta_0} \sin \nu_n\phi J'_{\nu_n}(k_0\rho), & \rho < \rho_0, \\ -\sum_{n=0}^\infty C_n \frac{j}{\eta_0} \sin \nu_n\phi H_{\nu_n}^{(2)\prime}(k_0\rho), & \rho > \rho_0. \end{cases} \tag{4.370}$$

The coefficients A_n and C_n are found by applying the boundary conditions at $\rho = \rho_0$.
By continuity of the tangential electric field

$$\sum_{n=0}^\infty A_n \sin \nu_n\phi J_{\nu_n}(k_0\rho_0) = \sum_{n=0}^\infty C_n \sin \nu_n\phi H_{\nu_n}^{(2)}(k_0\rho_0).$$

We now apply orthogonality over the interval $[0, \psi]$. Multiplying by $\sin \nu_m\phi$ and integrating we have

$$\sum_{n=0}^\infty A_n J_{\nu_n}(k_0\rho_0) \int_0^\psi \sin \nu_n\phi \sin \nu_m\phi \, d\phi = \sum_{n=0}^\infty C_n H_{\nu_n}^{(2)}(k_0\rho_0) \int_0^\psi \sin \nu_n\phi \sin \nu_m\phi \, d\phi.$$

Setting $u = \phi\pi/\psi$ we have

$$\int_0^\psi \sin \nu_n\phi \sin \nu_m\phi \, d\phi = \frac{\psi}{\pi} \int_0^\pi \sin nu \sin mu \, du = \frac{\psi}{2}\delta_{mn},$$

thus

$$A_m J_{\nu_m}(k_0 \rho_0) = C_m H^{(2)}_{\nu_m}(k_0 \rho_0). \tag{4.371}$$

The boundary condition $\hat{\mathbf{n}}_{12} \times (\tilde{\mathbf{H}}_1 - \tilde{\mathbf{H}}_2) = \tilde{\mathbf{J}}_s$ requires the surface current at $\rho = \rho_0$. We can write the line current in terms of a surface current density using the δ-function:

$$\tilde{\mathbf{J}}_s = \hat{\mathbf{z}} \tilde{I} \frac{\delta(\phi - \phi_0)}{\rho_0}.$$

This is easily verified as the correct expression since the integral of this density along the circular arc at $\rho = \rho_0$ returns the correct value $\tilde{I}$ for the total current. Thus the boundary condition requires

$$\tilde{H}_\phi(\rho_0^+, \phi, \omega) - \tilde{H}_\phi(\rho_0^-, \phi, \omega) = \tilde{I} \frac{\delta(\phi - \phi_0)}{\rho_0}.$$

By (4.370) we have

$$-\sum_{n=0}^{\infty} C_n \frac{j}{\eta_0} \sin \nu_n \phi H^{(2)\prime}_{\nu_n}(k_0 \rho_0) + \sum_{n=0}^{\infty} A_n \frac{j}{\eta_0} \sin \nu_n \phi J'_{\nu_n}(k_0 \rho_0) = \tilde{I} \frac{\delta(\phi - \phi_0)}{\rho_0}$$

and orthogonality yields

$$-C_m \frac{\psi}{2} \frac{j}{\eta_0} H^{(2)\prime}_{\nu_m}(k_0 \rho_0) + A_m \frac{\psi}{2} \frac{j}{\eta_0} J'_{\nu_m}(k_0 \rho_0) = \tilde{I} \frac{\sin \nu_m \phi_0}{\rho_0}. \tag{4.372}$$

The coefficients A_m and C_m thus obey the matrix equation

$$\begin{bmatrix} J_{\nu_m}(k_0 \rho_0) & -H^{(2)}_{\nu_m}(k_0 \rho_0) \\ J'_{\nu_m}(k_0 \rho_0) & -H^{(2)\prime}_{\nu_m}(k_0 \rho_0) \end{bmatrix} \begin{bmatrix} A_m \\ C_m \end{bmatrix} = \begin{bmatrix} 0 \\ -j2\tilde{I} \frac{\eta_0}{\psi} \frac{\sin \nu_m \phi_0}{\rho_0} \end{bmatrix}$$

and are

$$A_m = \frac{j2\tilde{I} \frac{\eta_0}{\psi} \frac{\sin \nu_m \phi_0}{\rho_0} H^{(2)}_{\nu_m}(k_0 \rho_0)}{H^{(2)\prime}_{\nu_m}(k_0 \rho_0) J_{\nu_m}(k_0 \rho_0) - J'_{\nu_m}(k_0 \rho_0) H^{(2)}_{\nu_m}(k_0 \rho_0)},$$

$$C_m = \frac{j2\tilde{I} \frac{\eta_0}{\psi} \frac{\sin \nu_m \phi_0}{\rho_0} J_{\nu_m}(k_0 \rho_0)}{H^{(2)\prime}_{\nu_m}(k_0 \rho_0) J_{\nu_m}(k_0 \rho_0) - J'_{\nu_m}(k_0 \rho_0) H^{(2)}_{\nu_m}(k_0 \rho_0)}.$$

Using the Wronskian relation (E.93), we replace the denominators in these expressions by $2/(j\pi k_0 \rho_0)$:

$$A_m = -\tilde{I} \frac{\eta_0}{\psi} \pi k_0 \sin \nu_m \phi_0 H^{(2)}_{\nu_m}(k_0 \rho_0),$$

$$C_m = -\tilde{I} \frac{\eta_0}{\psi} \pi k_0 \sin \nu_m \phi_0 J_{\nu_m}(k_0 \rho_0).$$

Hence (4.368) gives

$$\tilde{E}_z(\rho, \phi, \omega) = \begin{cases} -\sum_{n=0}^{\infty} \tilde{I} \frac{\eta_0}{2\psi} \pi k_0 \epsilon_n J_{\nu_n}(k_0 \rho) H^{(2)}_{\nu_n}(k_0 \rho_0) \sin \nu_n \phi \sin \nu_n \phi_0, & \rho < \rho_0, \\ -\sum_{n=0}^{\infty} \tilde{I} \frac{\eta_0}{2\psi} \pi k_0 \epsilon_n H^{(2)}_{\nu_n}(k_0 \rho) J_{\nu_n}(k_0 \rho_0) \sin \nu_n \phi \sin \nu_n \phi_0, & \rho > \rho_0, \end{cases} \tag{4.373}$$

where ϵ_n is Neumann's number (A.132). The magnetic fields can also be found by substituting the coefficients into (4.369) and (4.370).

The fields produced by an impressed plane wave may now be obtained by letting the line source recede to infinity. For large ρ_0 we use the asymptotic form (E.62) and find that

$$\tilde{E}_z(\rho,\phi,\omega) = -\sum_{n=0}^{\infty} \tilde{I}\frac{\eta_0}{2\psi}\pi k_0 \epsilon_n J_{\nu_n}(k_0\rho)\left[\sqrt{\frac{2j}{\pi k_0 \rho_0}}\,j^{\nu_n}e^{-jk_0\rho_0}\right]\sin\nu_n\phi\sin\nu_n\phi_0, \quad \rho < \rho_0.$$

$$(4.374)$$

Since the field of a line source falls off as $\rho_0^{-1/2}$, the amplitude of the impressed field approaches zero as $\rho_0 \to \infty$. We must compensate for the reduction in the impressed field by scaling the amplitude of the current source. To obtain the proper scale factor, we note that the electric field produced at a point ρ by a line source located at ρ_0 may be found from (4.345):

$$\tilde{E}_z = -\tilde{I}\frac{k_0\eta_0}{4}H_0^{(2)}(k_0|\boldsymbol{\rho}-\boldsymbol{\rho}_0|) \approx -\tilde{I}\frac{k_0\eta_0}{4}\sqrt{\frac{2j}{\pi k_0\rho_0}}e^{-jk_0\rho_0}e^{jk_0\rho\cos(\phi-\phi_0)}, \quad k_0\rho_0 \gg 1.$$

But if we write this as

$$\tilde{E}_z \approx \tilde{E}_0 e^{j\mathbf{k}\cdot\boldsymbol{\rho}}$$

then the field looks exactly like that produced by a plane wave with amplitude $\tilde{E}_0$ traveling along the wave vector $\mathbf{k} = -k_0\hat{\mathbf{x}}\cos\phi_0 - k_0\hat{\mathbf{y}}\sin\phi_0$. Solving for $\tilde{I}$ in terms of $\tilde{E}_0$ and substituting it back into (4.374), we get the total electric field scattered from a wedge with an impressed TM plane-wave field:

$$\tilde{E}_z(\rho,\phi,\omega) = \frac{2\pi}{\psi}\tilde{E}_0\sum_{n=0}^{\infty}\epsilon_n j^{\nu_n}J_{\nu_n}(k_0\rho)\sin\nu_n\phi\sin\nu_n\phi_0.$$

Here we interpret the angle ϕ_0 as the incidence angle of the plane wave.

To determine the field produced by an impressed TE plane-wave field, we use a magnetic line source $\tilde{I}_m$ located at ρ_0,ϕ_0 and proceed as above. By analogy with (4.367) we write

$$\tilde{H}_z(\rho,\phi,\omega) = \begin{cases} \sum_{\nu}[A_\nu\sin\nu\phi + B_\nu\cos\nu\phi]J_\nu(k_0\rho), & \rho < \rho_0, \\ \sum_{\delta}[C_\delta\sin\delta\phi + D_\delta\cos\delta\phi]H_\delta^{(2)}(k_0\rho), & \rho > \rho_0. \end{cases}$$

By (4.351) the tangential electric field is

$$\tilde{E}_\rho(\rho,\phi,\omega) = \begin{cases} -\sum_{\nu}[A_\nu\cos\nu\phi - B_\nu\sin\nu\phi]j\frac{Z_{TE}}{k}\frac{1}{\rho}\nu J_\nu(k_0\rho), & \rho < \rho_0, \\ -\sum_{\delta}[C_\delta\cos\delta\phi - D_\delta\sin\delta\phi]j\frac{Z_{TE}}{k}\frac{1}{\rho}\delta H_\delta^{(2)}(k_0\rho), & \rho > \rho_0. \end{cases}$$

Application of the boundary conditions on the tangential electric field at $\phi = 0, \psi$ results in $A_\nu = C_\delta = 0$ and $\nu = \delta = \nu_n = n\pi/\psi$, and thus $\tilde{H}_z$ becomes

$$\tilde{H}_z(\rho,\phi,\omega) = \begin{cases} \sum_{n=0}^{\infty}B_n\cos\nu_n\phi J_{\nu_n}(k_0\rho), & \rho < \rho_0, \\ \sum_{n=0}^{\infty}D_n\cos\nu_n\phi H_{\nu_n}^{(2)}(k_0\rho), & \rho > \rho_0. \end{cases} \qquad (4.375)$$

Application of the boundary conditions on tangential electric and magnetic fields across the magnetic line source then leads directly to

$$\tilde{H}_z(\rho,\phi,\omega) = \begin{cases} -\sum_{n=0}^{\infty}\tilde{I}_m\frac{\eta_0}{2\psi}\pi k_0\epsilon_n J_{\nu_n}(k_0\rho)H_{\nu_n}^{(2)}(k_0\rho_0)\cos\nu_n\phi\cos\nu_n\phi_0, & \rho < \rho_0 \\ -\sum_{n=0}^{\infty}\tilde{I}_m\frac{\eta_0}{2\psi}\pi k_0\epsilon_n H_{\nu_n}^{(2)}(k_0\rho)J_{\nu_n}(k_0\rho_0)\cos\nu_n\phi\cos\nu_n\phi_0, & \rho > \rho_0. \end{cases}$$

$$(4.376)$$

For a plane-wave impressed field this reduces to

$$\tilde{H}_z(\rho, \phi, \omega) = \frac{2\pi}{\psi} \frac{\tilde{E}_0}{\eta_0} \sum_{n=0}^{\infty} \epsilon_n j^{\nu_n} J_{\nu_n}(k_0\rho) \cos \nu_n\phi \cos \nu_n\phi_0.$$

Behavior of current near a sharp edge. In § 3.2.9 we studied the behavior of static charge near a sharp conducting edge by modeling the edge as a wedge. We can follow the same procedure for frequency-domain fields. Assume that the perfectly conducting wedge shown in Figure 4.26 is immersed in a finite, z-independent impressed field of a sort that will not concern us. A current is induced on the surface of the wedge and we wish to study its behavior as we approach the edge.

Because the field is z-independent, we may consider the superposition of TM and TE fields as was done above to solve for the field scattered by a wedge. For TM polarization, if the source is not located near the edge we may write the total field (impressed plus scattered) in terms of nonuniform cylindrical waves. The form of the field that obeys the boundary conditions at $\phi = 0$ and $\phi = \psi$ is given by (4.368):

$$\tilde{E}_z = \sum_{n=0}^{\infty} A_n \sin \nu_n\phi J_{\nu_n}(k_0\rho),$$

where $\nu_n = n\pi/\psi$. Although the A_n depend on the impressed source, the general behavior of the current near the edge is determined by the properties of the Bessel functions. The current on the wedge face at $\phi = 0$ is given by

$$\tilde{\mathbf{J}}_s(\rho, \omega) = \hat{\phi} \times [\hat{\phi}\tilde{H}_\phi + \hat{\rho}\tilde{H}_\rho]|_{\phi=0} = -\hat{\mathbf{z}}\tilde{H}_\rho(\rho, 0, \omega).$$

By (4.349) we have the surface current

$$\tilde{\mathbf{J}}_s(\rho, \omega) = -\hat{\mathbf{z}}\frac{1}{Z_{TM}k_0} \sum_{n=0}^{\infty} A_n \frac{\nu_n}{\rho} J_{\nu_n}(k_0\rho).$$

For $\rho \to 0$ the small-argument approximation (E.51) yields

$$\tilde{\mathbf{J}}_s(\rho, \omega) \approx -\hat{\mathbf{z}}\frac{1}{Z_{TM}k_0} \sum_{n=0}^{\infty} A_n\nu_n \frac{1}{\Gamma(\nu_n + 1)} \left(\frac{k_0}{2}\right)^{\nu_n} \rho^{\nu_n - 1}.$$

The sum is dominated by the smallest power of ρ. Since the $n = 0$ term vanishes we have

$$\tilde{J}_s(\rho, \omega) \sim \rho^{\frac{\pi}{\psi} - 1}, \qquad \rho \to 0.$$

For $\psi < \pi$ the current density, which runs parallel to the edge, is unbounded as $\rho \to 0$. A right-angle wedge ($\psi = 3\pi/2$) carries

$$\tilde{J}_s(\rho, \omega) \sim \rho^{-1/3}.$$

Another important case is that of a half-plane ($\psi = 2\pi$) where

$$\tilde{J}_s(\rho, \omega) \sim \frac{1}{\sqrt{\rho}}. \tag{4.377}$$

This square-root edge singularity dominates the behavior of the current flowing parallel to any flat edge, either straight or with curvature large compared to a wavelength, and is useful for modeling currents on complicated structures.

In the case of TE polarization the magnetic field near the edge is, by (4.375),

$$\tilde{H}_z(\rho, \phi, \omega) = \sum_{n=0}^{\infty} B_n \cos \nu_n \phi J_{\nu_n}(k_0 \rho), \qquad \rho < \rho_0.$$

The current at $\phi = 0$ is

$$\tilde{\mathbf{J}}_s(\rho, \omega) = \hat{\boldsymbol{\phi}} \times \hat{\mathbf{z}} \tilde{H}_z|_{\phi=0} = \hat{\boldsymbol{\rho}} \tilde{H}_z(\rho, 0, \omega)$$

or

$$\tilde{\mathbf{J}}_s(\rho, \omega) = \hat{\boldsymbol{\rho}} \sum_{n=0}^{\infty} B_n J_{\nu_n}(k_0 \rho).$$

For $\rho \to 0$ we use (E.51) to write

$$\tilde{\mathbf{J}}_s(\rho, \omega) = \hat{\boldsymbol{\rho}} \sum_{n=0}^{\infty} B_n \frac{1}{\Gamma(\nu_n + 1)} \left(\frac{k_0}{2} \right)^{\nu_n} \rho^{\nu_n}.$$

The $n = 0$ term gives a constant contribution, so we keep the first two terms to see how the current behaves near $\rho = 0$:

$$\tilde{J}_s \sim b_0 + b_1 \rho^{\frac{\pi}{\psi}}.$$

Here b_0 and b_1 depend on the form of the impressed field. For a thin plate where $\psi = 2\pi$ this becomes

$$\tilde{J}_s \sim b_0 + b_1 \sqrt{\rho}.$$

This is the companion square-root behavior to (4.377). When perpendicular to a sharp edge, the current grows away from the edge as $\rho^{1/2}$. In most cases $b_0 = 0$ since there is no mechanism to store charge along a sharp edge.

4.11.8 Propagation of spherical waves in a conducting medium

We cannot obtain uniform spherical wave solutions to Maxwell's equations. Any field dependent only on r produces the null field external to the source region, as shown in § 4.11.9. Nonuniform spherical waves are in general complicated and most easily handled using potentials. We consider here only the simple problem of fields dependent on r and θ. These waves display the fundamental properties of all spherical waves: they diverge from a localized source and expand with finite velocity.

Consider a homogeneous, source-free region characterized by $\tilde{\epsilon}(\omega)$, $\tilde{\mu}(\omega)$, and $\tilde{\sigma}(\omega)$. We seek wave solutions that are TEM_r in spherical coordinates ($\tilde{H}_r = \tilde{E}_r = 0$) and ϕ-independent. Thus we write

$$\tilde{\mathbf{E}}(\mathbf{r}, \omega) = \hat{\boldsymbol{\theta}} \tilde{E}_\theta(r, \theta, \omega) + \hat{\boldsymbol{\phi}} \tilde{E}_\phi(r, \theta, \omega),$$
$$\tilde{\mathbf{H}}(\mathbf{r}, \omega) = \hat{\boldsymbol{\theta}} \tilde{H}_\theta(r, \theta, \omega) + \hat{\boldsymbol{\phi}} \tilde{H}_\phi(r, \theta, \omega).$$

To determine the behavior of these fields we first examine Faraday's law

$$\nabla \times \tilde{\mathbf{E}}(r, \theta, \omega) = \hat{\mathbf{r}} \frac{1}{r \sin \theta} \frac{\partial}{\partial \theta} [\sin \theta \tilde{E}_\phi(r, \theta, \omega)] - \hat{\boldsymbol{\theta}} \frac{1}{r} \frac{\partial}{\partial r} [r \tilde{E}_\phi(r, \theta, \omega)] + \hat{\boldsymbol{\phi}} \frac{1}{r} \frac{\partial}{\partial r} [r \tilde{E}_\theta(r, \theta, \omega)]$$
$$= -j\omega\tilde{\mu}\tilde{\mathbf{H}}(r, \theta, \omega). \tag{4.378}$$

Since we require $\tilde{H}_r = 0$ we must have

$$\frac{\partial}{\partial \theta}[\sin\theta \tilde{E}_\phi(r,\theta,\omega)] = 0.$$

This implies that either $\tilde{E}_\phi \sim 1/\sin\theta$ or $\tilde{E}_\phi = 0$. We choose $\tilde{E}_\phi = 0$ and investigate whether the resulting fields satisfy the remaining Maxwell equations.

In a source-free, homogeneous region of space we have $\nabla\cdot\tilde{\mathbf{D}} = 0$ and thus also $\nabla\cdot\tilde{\mathbf{E}} = 0$. Since we have only a θ-component of the electric field, this requires

$$\frac{1}{r}\frac{\partial}{\partial\theta}\tilde{E}_\theta(r,\theta,\omega) + \frac{\cot\theta}{r}\tilde{E}_\theta(r,\theta,\omega) = 0.$$

From this we see that when $\tilde{E}_\phi = 0$, the field $\tilde{E}_\theta$ must obey

$$\tilde{E}_\theta(r,\theta,\omega) = \frac{\tilde{f}_E(r,\omega)}{\sin\theta}.$$

By (4.378) there is only a ϕ-component of magnetic field which obeys

$$\tilde{H}_\phi(r,\theta,\omega) = \frac{\tilde{f}_H(r,\omega)}{\sin\theta}$$

where

$$-j\omega\tilde{\mu}\tilde{f}_H(r,\omega) = \frac{1}{r}\frac{\partial}{\partial r}[r\tilde{f}_E(r,\omega)]. \tag{4.379}$$

So the spherical wave is TEM to the r-direction.

We can obtain a wave equation for $\tilde{f}_E$ by taking the curl of (4.378) and substituting from Ampere's law:

$$\nabla\times(\nabla\times\tilde{\mathbf{E}}) = -\hat{\boldsymbol{\theta}}\frac{1}{r}\frac{\partial^2}{\partial r^2}(r\tilde{E}_\theta) = \nabla\times\left(-j\omega\tilde{\mu}\tilde{\mathbf{H}}\right) = -j\omega\tilde{\mu}\left(\tilde{\sigma}\tilde{\mathbf{E}} + j\omega\tilde{\epsilon}\tilde{\mathbf{E}}\right),$$

hence

$$\frac{d^2}{dr^2}[r\tilde{f}_E(r,\omega)] + k^2[r\tilde{f}_E(r,\omega)] = 0. \tag{4.380}$$

Here $k = \omega(\tilde{\mu}\tilde{\epsilon}^c)^{1/2}$ is the complex wavenumber and $\tilde{\epsilon}^c = \tilde{\epsilon} + \tilde{\sigma}/j\omega$ is the complex permittivity. The equation for $\tilde{f}_H$ is identical.

The wave equation (4.380) is merely the second-order harmonic differential equation, with two independent solutions chosen from the list

$$\sin kr, \quad \cos kr, \quad e^{-jkr}, \quad e^{jkr}.$$

We find $\sin kr$ and $\cos kr$ useful for describing standing waves between boundaries, and e^{jkr} and e^{-jkr} useful for describing waves propagating in the r-direction. Of these, e^{jkr} represents waves traveling inward while e^{-jkr} represents waves traveling outward. At this point we choose $r\tilde{f}_E = e^{-jkr}$ and thus

$$\tilde{\mathbf{E}}(r,\theta,\omega) = \hat{\boldsymbol{\theta}}\tilde{E}_0(\omega)\frac{e^{-jkr}}{r\sin\theta}. \tag{4.381}$$

By (4.379) we have

$$\tilde{\mathbf{H}}(r,\theta,\omega) = \hat{\boldsymbol{\phi}}\frac{\tilde{E}_0(\omega)}{Z_{TEM}}\frac{e^{-jkr}}{r\sin\theta} \tag{4.382}$$

where $Z_{TEM} = (\tilde{\mu}/\epsilon^c)^{1/2}$ is the complex wave impedance. Since we can also write

$$\tilde{\mathbf{H}}(r,\theta,\omega) = \frac{\hat{\mathbf{r}} \times \tilde{\mathbf{E}}(r,\theta,\omega)}{Z_{TEM}},$$

the field is TEM to the r-direction, which is the direction of wave propagation as shown below.

The wave nature of the field is easily identified by considering the fields in the phasor domain. Letting $\omega \to \check{\omega}$ and setting $k = \beta - j\alpha$ in the exponential function we find that

$$\check{\mathbf{E}}(r,\theta) = \hat{\boldsymbol{\theta}} \check{E}_0 e^{-\alpha r} \frac{e^{-j\beta r}}{r \sin \theta}$$

where $\check{E}_0 = E_0 e^{j\xi^E}$. The time-domain representation may be found using (4.126):

$$\mathbf{E}(r,\theta,t) = \hat{\boldsymbol{\theta}} E_0 \frac{e^{-\alpha r}}{r \sin \theta} \cos(\check{\omega} t - \beta r + \xi^E). \tag{4.383}$$

We can identify a surface of constant phase as a locus of points obeying

$$\check{\omega} t - \beta r + \xi^E = C_P \tag{4.384}$$

where C_P is some constant. These surfaces, which are spheres centered on the origin, are called *spherical wavefronts*. Note that surfaces of constant amplitude as determined by

$$\frac{e^{-\alpha r}}{r} = C_A,$$

where C_A is some constant, are also spheres.

The cosine term in (4.383) represents a traveling wave with spherical wavefronts that propagate outward as time progresses. Attenuation is caused by the factor $e^{-\alpha r}$. By differentiation we find that the phase velocity is

$$v_p = \check{\omega}/\beta.$$

The wavelength is given by $\lambda = 2\pi/\beta$.

Our solution is not appropriate for unbounded space since the fields have a singularity at $\theta = 0$. To exclude the z-axis we add conducting cones as mentioned on page 105. This results in a biconical structure that can be used as a transmission line or antenna.

To compute the power carried by a spherical wave, we use (4.381) and (4.382) to obtain the time-average Poynting flux

$$\mathbf{S}_{av} = \frac{1}{2} \operatorname{Re}\{\check{E}_\theta \hat{\boldsymbol{\theta}} \times \check{H}_\phi^* \hat{\boldsymbol{\phi}}\} = \frac{1}{2} \hat{\mathbf{r}} \operatorname{Re}\left\{\frac{1}{Z_{TEM}^*}\right\} \frac{E_0^2}{r^2 \sin^2 \theta} e^{-2\alpha r}.$$

The power flux is radial and has density inversely proportional to r^2. The time-average power carried by the wave through a spherical surface at r sandwiched between the cones at θ_1 and θ_2 is

$$P_{av}(r) = \frac{1}{2} \operatorname{Re}\left\{\frac{1}{Z_{TEM}^*}\right\} E_0^2 e^{-2\alpha r} \int_0^{2\pi} d\phi \int_{\theta_1}^{\theta_2} \frac{d\theta}{\sin \theta} = \pi F \operatorname{Re}\left\{\frac{1}{Z_{TEM}^*}\right\} E_0^2 e^{-2\alpha r}$$

where

$$F = \ln\left[\frac{\tan(\theta_2/2)}{\tan(\theta_1/2)}\right]. \tag{4.385}$$

This is independent of r when $\alpha = 0$. For lossy media the power decays exponentially because of Joule heating.

We can write the phasor electric field in terms of the transverse gradient of a scalar potential function $\check{\Phi}$:

$$\mathbf{\check{E}}(r,\theta) = \hat{\boldsymbol{\theta}}\check{E}_0 \frac{e^{-jkr}}{r\sin\theta} = -\nabla_t \check{\Phi}(\theta)$$

where

$$\check{\Phi}(\theta) = -\check{E}_0 e^{-jkr}\ln\left(\tan\frac{\theta}{2}\right).$$

By ∇_t we mean the gradient with the r-component excluded. It is easily verified that

$$\mathbf{\check{E}}(r,\theta) = -\nabla_t\check{\Phi}(\theta) = -\hat{\boldsymbol{\theta}}\check{E}_0 \frac{1}{r}\frac{\partial\check{\Phi}(\theta)}{\partial\theta} = \hat{\boldsymbol{\theta}}\check{E}_0 \frac{e^{-jkr}}{r\sin\theta}.$$

Because $\mathbf{\check{E}}$ and $\check{\Phi}$ are related by the gradient, we can define a unique potential difference between the two cones at any radial position r:

$$\check{V}(r) = -\int_{\theta_1}^{\theta_2}\mathbf{\check{E}}\cdot\mathbf{dl} = \check{\Phi}(\theta_2) - \check{\Phi}(\theta_1) = \check{E}_0 F e^{-jkr},$$

where F is given in (4.385). The existence of a unique voltage difference is a property of all transmission line structures operated in the TEM mode. We can similarly compute the current flowing outward on the cone surfaces. The surface current on the cone at $\theta = \theta_1$ is $\mathbf{\check{J}}_s = \hat{\mathbf{n}}\times\mathbf{\check{H}} = \hat{\boldsymbol{\theta}}\times\hat{\boldsymbol{\phi}}\check{H}_\phi = \hat{\mathbf{r}}\check{H}_\phi$, hence

$$\check{I}(r) = \int_0^{2\pi}\mathbf{\check{J}}_s\cdot\hat{\mathbf{r}}r\sin\theta d\phi = 2\pi\frac{\check{E}_0}{Z_{TEM}}e^{-jkr}.$$

The ratio of voltage to current at any radius r is the *characteristic impedance* of the biconical transmission line (or, equivalently, the *input impedance* of the biconical antenna):

$$Z = \frac{\check{V}(r)}{\check{I}(r)} = \frac{Z_{TEM}}{2\pi}F.$$

If the material between the cones is lossless (and thus $\tilde{\mu}=\mu$ and $\tilde{\epsilon}^c=\epsilon$ are real), this becomes

$$Z = \frac{\eta}{2\pi}F$$

where $\eta = (\mu/\epsilon)^{1/2}$. The frequency independence of this quantity makes biconical antennas (or their approximate representations) useful for broadband applications.

Finally, the time-average power carried by the wave may be found from

$$P_{av}(r) = \frac{1}{2}\mathrm{Re}\left\{\check{V}(r)\check{I}^*(r)\right\} = \pi F \mathrm{Re}\left\{\frac{1}{Z_{TEM}^*}\right\}E_0^2 e^{-2\alpha r}.$$

The complex power relationship $P = VI^*$ is also a property of TEM guided-wave structures.

4.11.9   Nonradiating sources

We showed in § 2.10.9 that not all time-varying sources produce electromagnetic waves. In fact, a subset of localized sources known as *nonradiating sources* produce no field external to the source region. Devaney and Wolf [54] have shown that all nonradiating time-harmonic sources in an unbounded homogeneous medium can be represented in the form

$$\check{\mathbf{J}}^{nr}(\mathbf{r}) = -\nabla \times \left[\nabla \times \check{\mathbf{f}}(\mathbf{r}) \right] + k^2 \check{\mathbf{f}}(\mathbf{r}) \tag{4.386}$$

where $\check{\mathbf{f}}$ is any vector field that is continuous, has partial derivatives up to third order, and vanishes outside some localized region V_s. In fact, $\check{\mathbf{E}}(\mathbf{r}) = j\check{\omega}\mu\check{\mathbf{f}}(\mathbf{r})$ is precisely the phasor electric field produced by $\check{\mathbf{J}}^{nr}(\mathbf{r})$. The reasoning is straightforward. Consider the Helmholtz equation (4.203):

$$\nabla \times (\nabla \times \check{\mathbf{E}}) - k^2 \check{\mathbf{E}} = -j\check{\omega}\mu\check{\mathbf{J}}.$$

By (4.386) we have

$$\left(\nabla \times \nabla \times -k^2 \right) \left[\check{\mathbf{E}} - j\check{\omega}\mu\check{\mathbf{f}} \right] = 0.$$

Since $\check{\mathbf{f}}$ is zero outside the source region it must vanish at infinity. $\check{\mathbf{E}}$ also vanishes at infinity by the radiation condition, and thus the quantity $\check{\mathbf{E}} - j\check{\omega}\mu\check{\mathbf{f}}$ obeys the radiation condition and is a unique solution to the Helmholtz equation throughout all space. Since the Helmholtz equation is homogeneous we have

$$\check{\mathbf{E}} - j\check{\omega}\mu\check{\mathbf{f}} = 0$$

everywhere; since $\check{\mathbf{f}}$ is zero outside the source region, so is $\check{\mathbf{E}}$ (and so is $\check{\mathbf{H}}$).

An interesting special case of nonradiating sources is

$$\check{\mathbf{f}} = \frac{\nabla \check{\Phi}}{k^2}$$

so that

$$\check{\mathbf{J}}^{nr} = -\left(\nabla \times \nabla \times -k^2 \right) \frac{\nabla \check{\Phi}}{k^2} = \nabla \check{\Phi}.$$

Using $\check{\Phi}(\mathbf{r}) = \check{\Phi}(r)$, we see that this source describes the current produced by an oscillating spherical balloon of charge (cf., § 2.10.9). Radially-directed, spherically-symmetric sources cannot produce uniform spherical waves, since these sources are of the nonradiating type.

4.12   Interpretation of the spatial transform

Now that we understand the meaning of a Fourier transform on the time variable, let us consider a single transform involving one of the spatial variables. For a transform over z we shall use the notation

$$\psi^z(x, y, k_z, t) \leftrightarrow \psi(x, y, z, t).$$

Here the spatial frequency transform variable k_z has units of m^{-1}. The forward and inverse transform expressions are

$$\psi^z(x, y, k_z, t) = \int_{-\infty}^{\infty} \psi(x, y, z, t) e^{-jk_z z} \, dz, \qquad (4.387)$$

$$\psi(x, y, z, t) = \frac{1}{2\pi} \int_{-\infty}^{\infty} \psi^z(x, y, k_z, t) e^{jk_z z} \, dk_z, \qquad (4.388)$$

by (A.1) and (A.2).

We interpret (4.388) much as we interpreted the temporal inverse transform (4.2). Any vector component of the electromagnetic field can be decomposed into a continuous superposition of elemental spatial terms $e^{jk_z z}$ with weighting factors $\psi^z(x, y, k_z, t)$. In this case ψ^z is the *spatial frequency spectrum* of ψ. The elemental terms are spatial sinusoids along z with rapidity of variation described by k_z.

As with the temporal transform, ψ^z cannot be arbitrary since ψ must obey a scalar wave equation such as (2.327). For instance, for a source-free region of free space we must have

$$\left(\nabla^2 - \frac{1}{c^2} \frac{\partial}{\partial t^2}\right) \frac{1}{2\pi} \int_{-\infty}^{\infty} \psi^z(x, y, k_z, t) e^{jk_z z} \, dk_z = 0.$$

Decomposing the Laplacian operator as $\nabla^2 = \nabla_t^2 + \partial^2/\partial z^2$ and taking the derivatives into the integrand, we have

$$\frac{1}{2\pi} \int_{-\infty}^{\infty} \left[\left(\nabla_t^2 - k_z^2 - \frac{1}{c^2} \frac{\partial^2}{\partial t^2}\right) \psi^z(x, y, k_z, t)\right] e^{jk_z z} \, dk_z = 0.$$

Hence

$$\left(\nabla_t^2 - k_z^2 - \frac{1}{c^2} \frac{\partial^2}{\partial t^2}\right) \psi^z(x, y, k_z, t) = 0 \qquad (4.389)$$

by the Fourier integral theorem.

The elemental component $e^{jk_z z}$ is spatially sinusoidal and occupies all of space. Because such an element could only be created by a source that spans all of space, it is nonphysical when taken by itself. Nonetheless it is often used to represent more complicated fields. If the elemental spatial term is to be used alone, it is best interpreted physically when combined with a temporal decomposition. That is, we consider a two-dimensional transform, with transforms over both time and space. Then the time-domain representation of the elemental component is

$$\phi(z, t) = \frac{1}{2\pi} \int_{-\infty}^{\infty} e^{jk_z z} e^{j\omega t} \, d\omega. \qquad (4.390)$$

Before attempting to compute this transform, we should note that if the elemental term is to describe an EM field ψ in a source-free region, it must obey the homogeneous scalar wave equation. Substituting (4.390) into the homogeneous wave equation we have

$$\left(\nabla^2 - \frac{1}{c^2} \frac{\partial^2}{\partial t^2}\right) \frac{1}{2\pi} \int_{-\infty}^{\infty} e^{jk_z z} e^{j\omega t} \, d\omega = 0.$$

Differentiation under the integral sign gives

$$\frac{1}{2\pi} \int_{-\infty}^{\infty} \left[\left(-k_z^2 + \frac{\omega^2}{c^2}\right) e^{jk_z z}\right] e^{j\omega t} \, d\omega = 0$$

and thus

$$k_z^2 = \frac{\omega^2}{c^2} = k^2.$$

Substitution of $k_z = k$ into (4.390) gives the time-domain representation of the elemental component

$$\phi(z,t) = \frac{1}{2\pi} \int_{-\infty}^{\infty} e^{j\omega(t+z/c)} \, d\omega.$$

Finally, using the shifting theorem (A.3) along with (A.4), we have

$$\phi(z,t) = \delta\left(t + \frac{z}{c}\right), \tag{4.391}$$

which we recognize as a uniform plane wave propagating in the $-z$-direction with velocity c. There is no variation in the directions transverse to the direction of propagation and the surface describing a constant argument of the δ-function at any time t is a plane perpendicular to the direction of propagation.

We can also consider the elemental spatial component in tandem with a single sinusoidal steady-state elemental component. The phasor representation of the elemental spatial component is

$$\check{\phi}(z) = e^{jk_z z} = e^{jkz}.$$

This elemental term is a time-harmonic plane wave propagating in the $-z$-direction. Indeed, multiplying by $e^{j\check{\omega}t}$ and taking the real part we get

$$\phi(z,t) = \cos(\check{\omega}t + kz),$$

which is the sinusoidal steady-state analogue of (4.391).

Many authors choose to define the temporal and spatial transforms using differing sign conventions. The temporal transform is defined as in (4.1) and (4.2), but the spatial transform is defined through

$$\psi^z(x,y,k_z,t) = \int_{-\infty}^{\infty} \psi(x,y,z,t) e^{jk_z z} \, dz, \tag{4.392}$$

$$\psi(x,y,z,t) = \frac{1}{2\pi} \int_{-\infty}^{\infty} \psi^z(x,y,k_z,t) e^{-jk_z z} \, dk_z. \tag{4.393}$$

This employs a wave traveling in the positive z-direction as the elemental spatial component, which is quite useful for physical interpretation. We shall adopt this notation in § 4.13. The drawback is that we must alter the formulas from standard Fourier transform tables (replacing k by $-k$) to reflect this difference.

In the following sections we shall show how a spatial Fourier decomposition can be used to solve for the electromagnetic fields in a source-free region of space. By employing the spatial transform we may eliminate one or more spatial variables from Maxwell's equations, making the wave equation easier to solve. In the end we must perform an inversion to return to the space domain. This may be difficult or impossible to do analytically, requiring a numerical Fourier inversion.

4.13   Spatial Fourier decomposition of two-dimensional fields

Consider a homogeneous, source-free region characterized by $\tilde{\epsilon}(\omega)$, $\tilde{\mu}(\omega)$, and $\tilde{\sigma}(\omega)$. We seek z-independent solutions to the frequency-domain Maxwell's equations, using the Fourier transform to represent the spatial dependence. By § 4.11.2 a general two-dimensional field may be decomposed into fields TE and TM to the z-direction. In the TM case $\tilde{H}_z = 0$, and $\tilde{E}_z$ obeys the homogeneous scalar Helmholtz equation (4.208). In the TE case $\tilde{E}_z = 0$, and $\tilde{H}_z$ obeys the homogeneous scalar Helmholtz equation. Since each field component obeys the same equation, we let $\tilde{\psi}(x, y, \omega)$ represent either $\tilde{E}_z(x, y, \omega)$ or $\tilde{H}_z(x, y, \omega)$. Then $\tilde{\psi}$ obeys

$$(\nabla_t^2 + k^2)\tilde{\psi}(x, y, \omega) = 0 \tag{4.394}$$

where ∇_t^2 is the transverse Laplacian (4.209) and $k = \omega(\tilde{\mu}\tilde{\epsilon}^c)^{1/2}$ with $\tilde{\epsilon}^c$ the complex permittivity.

We may choose to represent $\tilde{\psi}(x, y, \omega)$ using Fourier transforms over one or both spatial variables. For application to problems in which boundary values or boundary conditions are specified at a constant value of a single variable (e.g., over a plane), one transform suffices. For instance, we may know the values of the field in the $y = 0$ plane (as we will, for example, when we solve the boundary value problems of § 4.13.1). Then we may transform over x and leave the y variable intact so that we may substitute the boundary values.

We adopt (4.392) since the result is more readily interpreted in terms of propagating plane waves. Choosing to transform over x we have

$$\tilde{\psi}^x(k_x, y, \omega) = \int_{-\infty}^{\infty} \tilde{\psi}(x, y, \omega)e^{jk_x x}\, dx, \tag{4.395}$$

$$\tilde{\psi}(x, y, \omega) = \frac{1}{2\pi}\int_{-\infty}^{\infty} \psi^x(k_x, y, \omega)e^{-jk_x x}\, dk_x. \tag{4.396}$$

For convenience in computation or interpretation of the inverse transform, we often regard k_x as a complex variable and perturb the inversion contour into the complex $k_x = k_{xr} + jk_{xi}$ plane. The integral is not altered if the contour is not moved past singularities such as poles or branch points. If the function being transformed has exponential (wave) behavior, then a pole exists in the complex plane; if we move the inversion contour across this pole, the inverse transform does not return the original function. We generally indicate the desire to interpret k_x as complex by indicating that the inversion contour is parallel to the real axis but located in the complex plane at $k_{xi} = \Delta$:

$$\tilde{\psi}(x, y, \omega) = \frac{1}{2\pi}\int_{-\infty+j\Delta}^{\infty+j\Delta} \tilde{\psi}^x(k_x, y, \omega)e^{-jk_x x}\, dk_x. \tag{4.397}$$

Additional perturbations of the contour are allowed provided that the contour is not moved through singularities.

As an example, consider the function

$$u(x) = \begin{cases} 0, & x < 0, \\ e^{-jkx}, & x > 0, \end{cases} \tag{4.398}$$

where $k = k_r + jk_i$ represents a wavenumber. This function has the form of a plane wave propagating in the x-direction and is thus relevant to our studies. If the material through which the wave is propagating is lossy, then $k_i < 0$. The Fourier transform of the function is

$$u^x(k_x) = \int_0^{\infty} e^{-jkx}e^{jk_x x}\, dx = \frac{1}{j(k_x - k)}\left[e^{j(k_{xr}-k_r)x}e^{-(k_{xi}-k_i)x}\right]\Bigg|_0^{\infty}.$$

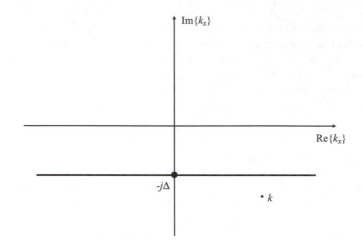

Figure 4.27: Inversion contour for evaluating the spectral integral for a plane wave.

The integral converges if $k_{xi} > k_i$, and the transform is

$$u^x(k_x) = -\frac{1}{j(k_x - k)}.$$

Since $u(x)$ is an exponential function, $u^x(k_x)$ has a pole at $k_x = k$ as anticipated.

To compute the inverse transform we use (4.397):

$$u(x) = \frac{1}{2\pi} \int_{-\infty+j\Delta}^{\infty+j\Delta} \left[-\frac{1}{j(k_x - k)} \right] e^{-jk_x x} \, dk_x. \qquad (4.399)$$

We must be careful to choose Δ in such a way that all values of k_x along the inversion contour lead to a convergent forward Fourier transform. Since we must have $k_{xi} > k_i$, choosing $\Delta > k_i$ ensures proper convergence. This gives the inversion contour shown in Figure 4.27, a special case of which is the real axis. We compute the inversion integral using contour integration as in § A.1. We close the contour in the complex plane and use Cauchy's residue theorem (A.14). For $x > 0$ we take $0 > \Delta > k_i$ and close the contour in the lower half-plane using a semicircular contour C_R of radius R. Then the closed contour integral is equal to $-2\pi j$ times the residue at the pole $k_x = k$. As $R \to \infty$ we find that $k_{xi} \to -\infty$ at all points on the contour C_R. Thus the integrand, which varies as $e^{k_{xi} x}$, vanishes on C_R and there is no contribution to the integral. The inversion integral (4.399) is found from the residue at the pole:

$$u(x) = (-2\pi j) \frac{1}{2\pi} \text{Res}_{k_x = k} \left[-\frac{1}{j(k_x - k)} e^{-jk_x x} \right].$$

Since the residue is merely je^{-jkx} we have $u(x) = e^{-jkx}$. When $x < 0$ we choose $\Delta > 0$ and close the contour along a semicircle C_R of radius R in the upper half-plane. Again we find that on C_R the integrand vanishes as $R \to \infty$, and thus the inversion integral (4.399) is given by $2\pi j$ times the residues of the integrand at any poles within the closed contour. This time, however, there are no poles enclosed and thus $u(x) = 0$. We have recovered the original function (4.398) for both $x > 0$ and $x < 0$. Note that if we had erroneously chosen $\Delta < k_i$ we would not have properly enclosed the pole and would have obtained an incorrect inverse transform.

Now that we know how to represent the Fourier transform pair, let us apply the transform to solve (4.394). Our hope is that by representing $\tilde{\psi}$ in terms of a spatial Fourier integral we will make the equation easier to solve. We have

$$(\nabla_t^2 + k^2)\frac{1}{2\pi} \int\limits_{-\infty+j\Delta}^{\infty+j\Delta} \tilde{\psi}^x(k_x, y, \omega)e^{-jk_x x}\, dk_x = 0.$$

Differentiation under the integral sign with subsequent application of the Fourier integral theorem implies that $\tilde{\psi}$ must obey the second-order harmonic differential equation

$$\left[\frac{d^2}{dy^2} + k_y^2\right]\tilde{\psi}^x(k_x, y, \omega) = 0$$

where we have defined the dependent parameter $k_y = k_{yr} + jk_{yi}$ through $k_x^2 + k_y^2 = k^2$. Two independent solutions to the differential equation are $e^{\mp jk_y y}$ and thus

$$\tilde{\psi}(k_x, y, \omega) = A(k_x, \omega)e^{\mp jk_y y}.$$

Substituting this into the inversion integral, we have the solution to the Helmholtz equation:

$$\tilde{\psi}(x, y, \omega) = \frac{1}{2\pi} \int\limits_{-\infty+j\Delta}^{\infty+j\Delta} A(k_x, \omega)e^{-jk_x x}e^{\mp jk_y y}\, dk_x. \tag{4.400}$$

If we define the wave vector $\mathbf{k} = \hat{\mathbf{x}}k_x \pm \hat{\mathbf{y}}k_y$, we can also write the solution in the form

$$\tilde{\psi}(x, y, \omega) = \frac{1}{2\pi} \int\limits_{-\infty+j\Delta}^{\infty+j\Delta} A(k_x, \omega)e^{-j\mathbf{k}\cdot\boldsymbol{\rho}}\, dk_x \tag{4.401}$$

where $\boldsymbol{\rho} = \hat{\mathbf{x}}x + \hat{\mathbf{y}}y$ is the two-dimensional position vector.

The solution (4.401) has an important physical interpretation. The exponential term looks exactly like a plane wave with its wave vector lying in the xy-plane. For lossy media the plane wave is nonuniform, and the surfaces of constant phase may not be aligned with the surfaces of constant amplitude (see § 4.11.4). For the special case of a lossless medium we have $k_i \to 0$ and can let $\Delta \to 0$ as long as $\Delta > k_i$. As we perform the inverse transform integral over k_x from $-\infty$ to ∞ we will encounter both the condition $k_x^2 > k^2$ and $k_x^2 \leq k^2$. For $k_x^2 \leq k^2$ we have

$$e^{-jk_x x}e^{\mp jk_y y} = e^{-jk_x x}e^{\mp j\sqrt{k^2-k_x^2}\, y}$$

where we choose the upper sign for $y > 0$ and the lower sign for $y < 0$ to ensure that the waves propagate in the $\pm y$-direction, respectively. Thus, in this regime the exponential represents a propagating wave that travels into the half-plane $y > 0$ along a direction which depends on k_x, making an angle ξ with the x-axis as shown in Figure 4.28. For k_x in $[-k, k]$, every possible wave direction is covered, and thus we may think of the inversion integral as constructing the solution to the two-dimensional Helmholtz equation from a continuous superposition of plane waves. The amplitude of each plane wave component is given by $A(k_x, \omega)$, which is often called the *angular spectrum* of the plane waves and

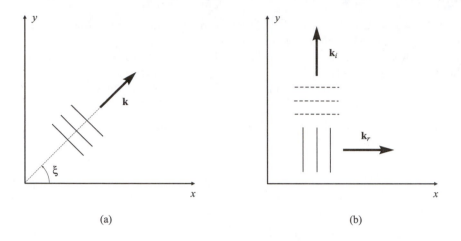

(a)   (b)

Figure 4.28: Propagation behavior of the angular spectrum for (a) $k_x^2 \leq k^2$, (b) $k_x^2 > k^2$.

is determined by the values of the field over the boundaries of the solution region. But this is not the whole picture. The inverse transform integral also requires values of k_x in the intervals $[-\infty, k]$ and $[k, \infty]$. Here we have $k_x^2 > k^2$ and thus

$$e^{-jk_x x}e^{-jk_y y} = e^{-jk_x x}e^{\mp\sqrt{k_x^2-k^2}\,y},$$

where we choose the upper sign for $y > 0$ and the lower sign for $y < 0$ to ensure that the field decays along the y-direction. In these regimes we have an evanescent wave, propagating along x but decaying along y, with surfaces of constant phase and amplitude mutually perpendicular (Figure 4.28). As k_x ranges out to ∞, evanescent waves of all possible decay constants also contribute to the plane-wave superposition.

We may summarize the plane-wave contributions by letting $\mathbf{k} = \hat{\mathbf{x}}k_x + \hat{\mathbf{y}}k_y = \mathbf{k}_r + j\mathbf{k}_i$ where

$$\mathbf{k}_r = \begin{cases} \hat{\mathbf{x}}k_x \pm \hat{\mathbf{y}}\sqrt{k^2 - k_x^2}, & k_x^2 < k^2, \\ \hat{\mathbf{x}}k_x, & k_x^2 > k^2, \end{cases}$$

$$\mathbf{k}_i = \begin{cases} 0, & k_x^2 < k^2, \\ \mp\hat{\mathbf{y}}\sqrt{k_x^2 - k^2}, & k_x^2 > k^2, \end{cases}$$

where the upper sign is used for $y > 0$ and the lower sign for $y < 0$.

In many applications, including the half-plane example considered later, it is useful to write the inversion integral in polar coordinates. Letting

$$k_x = k\cos\xi, \qquad k_y = \pm k\sin\xi,$$

where $\xi = \xi_r + j\xi_i$ is a new complex variable, we have $\mathbf{k} \cdot \boldsymbol{\rho} = kx\cos\xi \pm ky\sin\xi$ and $dk_x = -k\sin\xi\,d\xi$. With this change of variables (4.401) becomes

$$\tilde{\psi}(x,y,\omega) = \frac{k}{2\pi}\int_C A(k\cos\xi,\omega)e^{-jkx\cos\xi}e^{\pm jky\sin\xi}\sin\xi\,d\xi. \qquad (4.402)$$

Since $A(k_x,\omega)$ is a function to be determined, we may introduce a new function

$$f(\xi,\omega) = \frac{k}{2\pi}A(k_x,\omega)\sin\xi$$

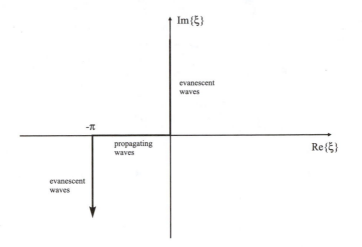

Figure 4.29: Inversion contour for the polar coordinate representation of the inverse Fourier transform.

so that (4.402) becomes

$$\tilde{\psi}(x,y,\omega) = \int_C f(\xi,\omega)e^{-jk\rho\cos(\phi\pm\xi)}\,d\xi \qquad (4.403)$$

where $x = \rho\cos\phi, y = \rho\sin\phi$, and where the upper sign corresponds to $0 < \phi < \pi$ ($y > 0$) while the lower sign corresponds to $\pi < \phi < 2\pi$ ($y < 0$). In these expressions C is a contour in the complex ξ-plane to be determined. Values along this contour must produce identical values of the integrand as did the values of k_x over $[-\infty,\infty]$ in the original inversion integral. By the identities

$$\cos z = \cos(u + jv) = \cos u \cosh v - j \sin u \sinh v,$$
$$\sin z = \sin(u + jv) = \sin u \cosh v + j \cos u \sinh v,$$

we find that the contour shown in Figure 4.29 provides identical values of the integrand (Problem 4.24). The portions of the contour $[0 + j\infty, 0]$ and $[-\pi, -\pi - j\infty]$ together correspond to the regime of evanescent waves ($k < k_x < \infty$ and $-\infty < k_x < k$), while the segment $[0, -\pi]$ along the real axis corresponds to $-k < k_x < k$ and thus describes contributions from propagating plane waves. In this case ξ represents the propagation angle of the waves.

4.13.1 Boundary value problems using the spatial Fourier representation

The field of a line source. As a first example we calculate the Fourier representation of the field of an electric line source. Assume a uniform line current $\tilde{I}(\omega)$ is aligned along the z-axis in a medium characterized by complex permittivity $\tilde{\epsilon}^c(\omega)$ and permeability $\tilde{\mu}(\omega)$. We separate space into two source-free portions, $y > 0$ and $y < 0$, and write the field in each region in terms of an inverse spatial Fourier transform. Then, by applying the boundary conditions in the $y = 0$ plane, we solve for the angular spectrum of the line source.

Since this is a two-dimensional problem we may decompose the fields into TE and TM sets. For an electric line source we need only the TM set, and write E_z as a superposition of plane waves using (4.400). For $y \gtrless 0$ we represent the field in terms of plane waves traveling in the $\pm y$-direction. Thus

$$\tilde{E}_z(x, y, \omega) = \frac{1}{2\pi} \int\limits_{-\infty+j\Delta}^{\infty+j\Delta} A^+(k_x, \omega) e^{-jk_x x} e^{-jk_y y} \, dk_x, \quad y > 0,$$

$$\tilde{E}_z(x, y, \omega) = \frac{1}{2\pi} \int\limits_{-\infty+j\Delta}^{\infty+j\Delta} A^-(k_x, \omega) e^{-jk_x x} e^{+jk_y y} \, dk_x, \quad y < 0.$$

The transverse magnetic field may be found from the axial electric field using (4.212). We find

$$\tilde{H}_x = -\frac{1}{j\omega\tilde{\mu}} \frac{\partial \tilde{E}_z}{\partial y} \tag{4.404}$$

and thus

$$\tilde{H}_x(x, y, \omega) = \frac{1}{2\pi} \int\limits_{-\infty+j\Delta}^{\infty+j\Delta} A^+(k_x, \omega) \left[\frac{k_y}{\omega\tilde{\mu}} \right] e^{-jk_x x} e^{-jk_y y} \, dk_x, \quad y > 0,$$

$$\tilde{H}_x(x, y, \omega) = \frac{1}{2\pi} \int\limits_{-\infty+j\Delta}^{\infty+j\Delta} A^-(k_x, \omega) \left[-\frac{k_y}{\omega\tilde{\mu}} \right] e^{-jk_x x} e^{+jk_y y} \, dk_x, \quad y < 0.$$

To find the spectra $A^\pm(k_x, \omega)$ we apply the boundary conditions at $y = 0$. Since tangential $\tilde{\mathbf{E}}$ is continuous we have, after combining the integrals,

$$\frac{1}{2\pi} \int\limits_{-\infty+j\Delta}^{\infty+j\Delta} \left[A^+(k_x, \omega) - A^-(k_x, \omega) \right] e^{-jk_x x} \, dk_x = 0,$$

and hence by the Fourier integral theorem

$$A^+(k_x, \omega) - A^-(k_x, \omega) = 0. \tag{4.405}$$

We must also apply $\hat{\mathbf{n}}_{12} \times (\tilde{\mathbf{H}}_1 - \tilde{\mathbf{H}}_2) = \tilde{\mathbf{J}}_s$. The line current may be written as a surface current density using the δ-function, giving

$$- \left[\tilde{H}_x(x, 0^+, \omega) - \tilde{H}_x(x, 0^-, \omega) \right] = \tilde{I}(\omega)\delta(x).$$

By (A.4)

$$\delta(x) = \frac{1}{2\pi} \int_{-\infty}^{\infty} e^{-jk_x x} \, dk_x.$$

Then, substituting for the fields and combining the integrands, we have

$$\frac{1}{2\pi} \int\limits_{-\infty+j\Delta}^{\infty+j\Delta} \left[A^+(k_x, \omega) + A^-(k_x, \omega) + \frac{\omega\tilde{\mu}}{k_y} \tilde{I}(\omega) \right] e^{-jk_x x} = 0,$$

hence

$$A^+(k_x, \omega) + A^-(k_x, \omega) = -\frac{\omega\tilde{\mu}}{k_y}\tilde{I}(\omega). \tag{4.406}$$

Solution of (4.405) and (4.406) gives the angular spectra

$$A^+(k_x, \omega) = A^-(k_x, \omega) = -\frac{\omega\tilde{\mu}}{2k_y}\tilde{I}(\omega).$$

Substituting this into the field expressions and combining the cases for $y > 0$ and $y < 0$, we find

$$\tilde{E}_z(x, y, \omega) = -\frac{\omega\tilde{\mu}\tilde{I}(\omega)}{2\pi} \int\limits_{-\infty+j\Delta}^{\infty+j\Delta} \frac{e^{-jk_y|y|}}{2k_y} e^{-jk_x x}\, dk_x = -j\omega\tilde{\mu}\tilde{I}(\omega)\tilde{G}(x, y|0, 0; \omega). \tag{4.407}$$

Here $\tilde{G}$ is the spectral representation of the two-dimensional Green's function first found in § 4.11.7, and is given by

$$\tilde{G}(x, y|x', y'; \omega) = \frac{1}{2\pi j} \int\limits_{-\infty+j\Delta}^{\infty+j\Delta} \frac{e^{-jk_y|y-y'|}}{2k_y} e^{-jk_x(x-x')}\, dk_x. \tag{4.408}$$

By duality we have

$$\tilde{H}_z(x, y, \omega) = -\frac{\omega\tilde{\epsilon}^c\tilde{I}_m(\omega)}{2\pi} \int\limits_{-\infty+j\Delta}^{\infty+j\Delta} \frac{e^{-jk_y|y|}}{2k_y} e^{-jk_x x}\, dk_x = -j\omega\tilde{\epsilon}^c\tilde{I}_m(\omega)G(x, y|0, 0; \omega)$$

$$\tag{4.409}$$

for a magnetic line current $\tilde{I}_m(\omega)$ on the z-axis.

Note that since the earlier expression (4.346) should be equivalent to (4.408), we have the well known identity [33]

$$\frac{1}{\pi} \int\limits_{-\infty+j\Delta}^{\infty+j\Delta} \frac{e^{-jk_y|y|}}{k_y} e^{-jk_x x}\, dk_x = H_0^{(2)}(k\rho).$$

We have not yet specified the contour appropriate for calculating the inverse transform (4.407). We must be careful because the denominator of (4.407) has branch points at $k_y = \sqrt{k^2 - k_x^2} = 0$, or equivalently, $k_x = \pm k = \pm(k_r + jk_i)$. For lossy materials we have $k_i < 0$ and $k_r > 0$, so the branch points appear as in Figure 4.30. We may take the branch cuts outward from these points, and thus choose the inversion contour to lie between the branch points so that the branch cuts are not traversed. This requires $k_i < \Delta < -k_i$. It is natural to choose $\Delta = 0$ and use the real axis as the inversion contour. We must be careful, though, when extending these arguments to the lossless case. If we consider the lossless case to be the limit of the lossy case as $k_i \to 0$, we find that the branch points migrate to the real axis and thus lie on the inversion contour. We can eliminate this problem by realizing that the inversion contour may be perturbed without affecting the value of the integral, as long as it is not made to pass through the branch cuts. If we perturb the inversion contour as shown in Figure 4.30, then as $k_i \to 0$ the branch points do not fall on the contour.

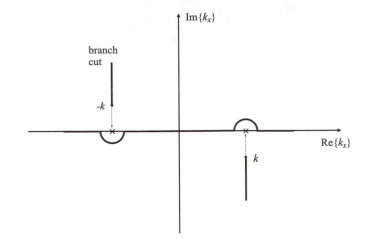

Figure 4.30: Inversion contour in complex k_x-plane for a line source. Dotted arrow shows migration of branch points to real axis as loss goes to zero.

There are many interesting techniques that may be used to compute the inversion integral appearing in (4.407) and in the other expressions we shall obtain in this section. These include direct real-axis integration and closed contour methods that use Cauchy's residue theorem to capture poles of the integrand (which often describe the properties of waves guided by surfaces). Often it is necessary to integrate around the branch cuts in order to meet the conditions for applying the residue theorem. When the observation point is far from the source we may use the method of steepest descents to obtain asymptotic forms for the fields. The interested reader should consult Chew [33], Kong [101], or Sommerfeld [184].

Field of a line source above an interface.   Consider a z-directed electric line current located at $y = h$ within a medium having parameters $\tilde{\mu}_1(\omega)$ and $\tilde{\epsilon}_1^c(\omega)$. The $y = 0$ plane separates this region from a region having parameters $\tilde{\mu}_2(\omega)$ and $\tilde{\epsilon}_2^c(\omega)$. See Figure 4.31. The impressed line current source creates an electromagnetic field that induces secondary polarization and conduction currents in both regions. This current in turn produces a secondary field that adds to the primary field of the line source to satisfy the boundary conditions at the interface. We would like to solve for the secondary field and give its sources an image interpretation.

Since the fields are z-independent we may decompose the fields into sets TE and TM to z. For a z-directed impressed source there is a z-component of $\tilde{\mathbf{E}}$, but no z-component of $\tilde{\mathbf{H}}$; hence the fields are entirely specified by the TM set. The impressed source is unaffected by the secondary field, and we may represent the impressed electric field using (4.407):

$$\tilde{E}_z^i(x,y,\omega) = -\frac{\omega\tilde{\mu}_1\tilde{I}(\omega)}{2\pi} \int\limits_{-\infty+j\Delta}^{\infty+j\Delta} \frac{e^{-jk_{y1}|y-h|}}{2k_{y1}}e^{-jk_x x}\,dk_x, \quad y \geq 0 \qquad (4.410)$$

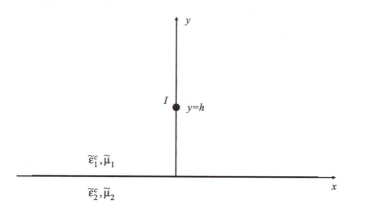

Figure 4.31: Geometry of a z-directed line source above an interface between two material regions.

where $k_{y1} = \sqrt{k_1^2 - k_x^2}$ and $k_1 = \omega(\tilde{\mu}_1\tilde{\epsilon}_1^c)^{1/2}$. From (4.404) we find that

$$
\tilde{H}_x^i = -\frac{1}{j\omega\tilde{\mu}_1}\frac{\partial\tilde{E}_z^i}{\partial y} = \frac{\tilde{I}(\omega)}{2\pi}\int_{-\infty+j\Delta}^{\infty+j\Delta}\frac{e^{jk_{y1}(y-h)}}{2}e^{-jk_x x}\,dk_x, \quad 0 \le y < h.
$$

The scattered field obeys the homogeneous Helmholtz equation for all $y > 0$, and thus may be written using (4.400) as a superposition of upward-traveling waves:

$$
\tilde{E}_{z1}^s(x,y,\omega) = \frac{1}{2\pi}\int_{-\infty+j\Delta}^{\infty+j\Delta} A_1(k_x,\omega)e^{-jk_{y1}y}e^{-jk_x x}\,dk_x,
$$

$$
\tilde{H}_{x1}^s(x,y,\omega) = \frac{1}{2\pi}\int_{-\infty+j\Delta}^{\infty+j\Delta}\frac{k_{y1}}{\omega\tilde{\mu}_1}A_1(k_x,\omega)e^{-jk_{y1}y}e^{-jk_x x}\,dk_x.
$$

Similarly, in region 2 the scattered field may be written as a superposition of downward-traveling waves:

$$
\tilde{E}_{z2}^s(x,y,\omega) = \frac{1}{2\pi}\int_{-\infty+j\Delta}^{\infty+j\Delta} A_2(k_x,\omega)e^{jk_{y2}y}e^{-jk_x x}\,dk_x,
$$

$$
\tilde{H}_{x2}^s(x,y,\omega) = -\frac{1}{2\pi}\int_{-\infty+j\Delta}^{\infty+j\Delta}\frac{k_{y2}}{\omega\tilde{\mu}_2}A_2(k_x,\omega)e^{jk_{y2}y}e^{-jk_x x}\,dk_x,
$$

where $k_{y2} = \sqrt{k_2^2 - k_x^2}$ and $k_2 = \omega(\tilde{\mu}_2\tilde{\epsilon}_2^c)^{1/2}$.

We can solve for the angular spectra A_1 and A_2 by applying the boundary conditions at the interface between the two media. From the continuity of total tangential electric field we find that

$$
\frac{1}{2\pi}\int_{-\infty+j\Delta}^{\infty+j\Delta}\left[-\frac{\omega\tilde{\mu}_1\tilde{I}(\omega)}{2k_{y1}}e^{-jk_{y1}h} + A_1(k_x,\omega) - A_2(k_x,\omega)\right]e^{-jk_x x}\,dk_x = 0,
$$

hence by the Fourier integral theorem

$$A_1(k_x, \omega) - A_2(k_x, \omega) = \frac{\omega \tilde{\mu}_1 \tilde{I}(\omega)}{2k_{y1}} e^{-jk_{y1}h}.$$

The boundary condition on the continuity of $\tilde{H}_x$ yields similarly

$$-\frac{\tilde{I}(\omega)}{2} e^{-jk_{y1}h} = \frac{k_{y1}}{\omega \tilde{\mu}_1} A_1(k_x, \omega) + \frac{k_{y2}}{\omega \tilde{\mu}_2} A_2(k_x, \omega).$$

We obtain

$$A_1(k_x, \omega) = \frac{\omega \tilde{\mu}_1 \tilde{I}(\omega)}{2k_{y1}} R_{TM}(k_x, \omega) e^{-jk_{y1}h},$$

$$A_2(k_x, \omega) = -\frac{\omega \tilde{\mu}_2 \tilde{I}(\omega)}{2k_{y2}} T_{TM}(k_x, \omega) e^{-jk_{y1}h}.$$

Here R_{TM} and $T_{TM} = 1 + R_{TM}$ are reflection and transmission coefficients given by

$$R_{TM}(k_x, \omega) = \frac{\tilde{\mu}_1 k_{y2} - \tilde{\mu}_2 k_{y1}}{\tilde{\mu}_1 k_{y2} + \tilde{\mu}_2 k_{y1}},$$

$$T_{TM}(k_x, \omega) = \frac{2\tilde{\mu}_1 k_{y2}}{\tilde{\mu}_1 k_{y2} + \tilde{\mu}_2 k_{y1}}.$$

These describe the reflection and transmission of each component of the plane-wave spectrum of the impressed field, and thus depend on the parameter k_x. The scattered fields are

$$\tilde{E}_{z1}^s(x, y, \omega) = \frac{\omega \tilde{\mu}_1 \tilde{I}(\omega)}{2\pi} \int\limits_{-\infty+j\Delta}^{\infty+j\Delta} \frac{e^{-jk_{y1}(y+h)}}{2k_{y1}} R_{TM}(k_x, \omega) e^{-jk_x x} \, dk_x, \qquad (4.411)$$

$$\tilde{E}_{z2}^s(x, y, \omega) = -\frac{\omega \tilde{\mu}_2 \tilde{I}(\omega)}{2\pi} \int\limits_{-\infty+j\Delta}^{\infty+j\Delta} \frac{e^{jk_{y2}(y-hk_{y1}/k_{y2})}}{2k_{y2}} T_{TM}(k_x, \omega) e^{-jk_x x} \, dk_x. \quad (4.412)$$

We may now obtain the field produced by an electric line source above a perfect conductor. Letting $\tilde{\sigma}_2 \to \infty$ we have $k_{y2} = \sqrt{k_2^2 - k_x^2} \to \infty$ and

$$R_{TM} \to 1, \qquad T_{TM} \to 2.$$

With these, the scattered fields (4.411) and (4.412) become

$$\tilde{E}_{z1}^s(x, y, \omega) = \frac{\omega \tilde{\mu}_1 \tilde{I}(\omega)}{2\pi} \int\limits_{-\infty+j\Delta}^{\infty+j\Delta} \frac{e^{-jk_{y1}(y+h)}}{2k_{y1}} e^{-jk_x x} \, dk_x, \qquad (4.413)$$

$$\tilde{E}_{z2}^s(x, y, \omega) = 0. \qquad (4.414)$$

Comparing (4.413) to (4.410) we see that the scattered field is exactly the same as that produced by a line source of amplitude $-\tilde{I}(\omega)$ located at $y = -h$. We call this line source the image of the impressed source, and say that the problem of two line sources located

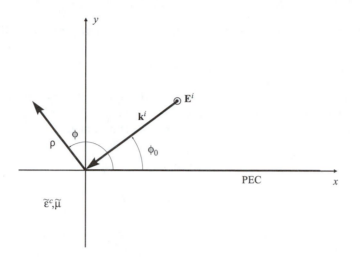

Figure 4.32: Geometry for scattering of a TM plane wave by a conducting half-plane.

symmetrically on the y-axis is equivalent for $y > 0$ to the problem of the line source above a ground plane. The total field is the sum of the impressed and scattered fields:

$$\tilde{E}_z(x, y, \omega) = -\frac{\omega \tilde{\mu}_1 \tilde{I}(\omega)}{2\pi} \int\limits_{-\infty+j\Delta}^{\infty+j\Delta} \frac{e^{-jk_{y1}|y-h|} - e^{-jk_{y1}(y+h)}}{2k_{y1}} e^{-jk_x x} \, dk_x, \quad y \geq 0.$$

We can write this in another form using the Hankel-function representation of the line source (4.345):

$$\tilde{E}_z(x, y, \omega) = -\frac{\omega \tilde{\mu}}{4} \tilde{I}(\omega) H_0^{(2)}(k|\boldsymbol{\rho} - \hat{\mathbf{y}} h|) + \frac{\omega \tilde{\mu}}{4} \tilde{I}(\omega) H_0^{(2)}(k|\boldsymbol{\rho} + \hat{\mathbf{y}} h|)$$

where $|\boldsymbol{\rho} \pm \hat{\mathbf{y}} h| = |\rho \hat{\boldsymbol{\rho}} \pm \hat{\mathbf{y}} h| = \sqrt{x^2 + (y \pm h)^2}$.

Interpreting the general case in terms of images is more difficult. Comparing (4.411) and (4.412) with (4.410), we see that each spectral component of the field in region 1 has the form of an image line source located at $y = -h$ in region 2, but that the amplitude of the line source, $R_{TM}\tilde{I}$, depends on k_x. Similarly, the field in region 2 is composed of spectral components that seem to originate from line sources with amplitudes $-T_{TM}\tilde{I}$ located at $y = hk_{y1}/k_{y2}$ in region 1. In this case the amplitude and position of the image line source producing a spectral component are both dependent on k_x.

The field scattered by a half-plane. Consider a thin planar conductor that occupies the half-plane $y = 0$, $x > 0$. We assume the half-plane lies within a slightly lossy medium having parameters $\tilde{\mu}(\omega)$ and $\tilde{\epsilon}^c(\omega)$, and may consider the case of free space as a lossless limit. The half-plane is illuminated by an impressed uniform plane wave with a z-directed electric field (Figure 4.32). The primary field induces a secondary current on the conductor and this in turn produces a secondary field. The total field must obey the boundary conditions at $y = 0$.

Because the z-directed incident field induces a z-directed secondary current, the fields may be described entirely in terms of a TM set. The impressed plane wave may be written as

$$\tilde{\mathbf{E}}^i(\mathbf{r}, \omega) = \hat{\mathbf{z}} \tilde{E}_0(\omega) e^{jk(x \cos \phi_0 + y \sin \phi_0)}$$

where ϕ_0 is the angle between the incident wave vector and the x-axis. By (4.223) we also have

$$\tilde{\mathbf{H}}^i(\mathbf{r}, \omega) = \frac{\tilde{E}_0(\omega)}{\eta}(\hat{\mathbf{y}} \cos \phi_0 - \hat{\mathbf{x}} \sin \phi_0) e^{jk(x \cos \phi_0 + y \sin \phi_0)}.$$

The scattered fields may be written in terms of the Fourier transform solution to the Helmholtz equation. It is convenient to use the polar coordinate representation (4.403) to develop the necessary equations. Thus, for the scattered electric field we can write

$$\tilde{E}_z^s(x, y, \omega) = \int_C f(\xi, \omega) e^{-jk\rho \cos(\phi \pm \xi)} \, d\xi. \tag{4.415}$$

By (4.404) the x-component of the magnetic field is

$$\tilde{H}_x^s(x, y, \omega) = -\frac{1}{j\omega\tilde{\mu}} \frac{\partial \tilde{E}_z^s}{\partial y} = -\frac{1}{j\omega\tilde{\mu}} \int_C f(\xi, \omega) \frac{\partial}{\partial y} \left(e^{-jkx \cos \xi} e^{\pm jky \sin \xi} \right)$$

$$= -\frac{1}{j\omega\tilde{\mu}} (\pm jk) \int_C f(\xi, \omega) \sin \xi \, e^{-jk\rho \cos(\phi \pm \xi)} \, d\xi.$$

To find the angular spectrum $f(\xi, \omega)$ and ensure uniqueness of solution, we must apply the boundary conditions over the entire $y = 0$ plane. For $x > 0$ where the conductor resides, the total tangential electric field must vanish. Setting the sum of the incident and scattered fields to zero at $\phi = 0$ we have

$$\int_C f(\xi, \omega) e^{-jkx \cos \xi} \, d\xi = -\tilde{E}_0 e^{jkx \cos \phi_0}, \quad x > 0. \tag{4.416}$$

To find the boundary condition for $x < 0$ we note that by symmetry $\tilde{E}_z^s$ is even about $y = 0$ while $\tilde{H}_x^s$, as the y-derivative of $\tilde{E}_z^s$, is odd. Since no current can be induced in the $y = 0$ plane for $x < 0$, the x-directed scattered magnetic field must be continuous and thus equal to zero there. Hence our second condition is

$$\int_C f(\xi, \omega) \sin \xi \, e^{-jkx \cos \xi} \, d\xi = 0, \quad x < 0. \tag{4.417}$$

Now that we have developed the two equations that describe $f(\xi, \omega)$, it is convenient to return to a rectangular-coordinate-based spectral integral to analyze them. Writing $\xi = \cos^{-1}(k_x/k)$ we have

$$\frac{d}{d\xi}(k \cos \xi) = -k \sin \xi = \frac{dk_x}{d\xi}$$

and

$$d\xi = -\frac{dk_x}{k \sin \xi} = -\frac{dk_x}{k\sqrt{1 - \cos^2 \xi}} = -\frac{dk_x}{\sqrt{k^2 - k_x^2}}.$$

Upon substitution of these relations, the inversion contour returns to the real k_x axis (which may then be perturbed by $j\Delta$). Thus, (4.416) and (4.417) may be written as

$$\int_{-\infty+j\Delta}^{\infty+j\Delta} \frac{f\left(\cos^{-1} \frac{k_x}{k}\right)}{\sqrt{k^2 - k_x^2}} e^{-jk_x x} \, dk_x = -\tilde{E}_0 e^{jk_{x0} x}, \quad x > 0, \tag{4.418}$$

$$\int_{-\infty+j\Delta}^{\infty+j\Delta} f\left(\cos^{-1} \frac{k_x}{k}\right) e^{-jk_x x} \, dk_x = 0, \quad x < 0, \tag{4.419}$$

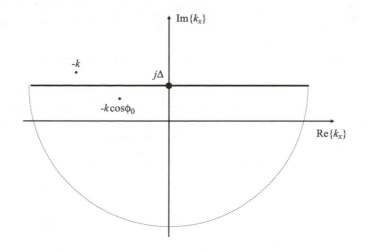

Figure 4.33: Integration contour used to evaluate the function $F(x)$.

where $k_{x0} = k \cos \phi_0$. Equations (4.418) and (4.419) comprise *dual integral equations* for f. We may solve these using an approach called the *Wiener–Hopf technique*.

We begin by considering (4.419). If we close the integration contour in the upper half-plane using a semicircle C_R of radius R where $R \to \infty$, we find that the contribution from the semicircle is

$$\lim_{R \to \infty} \int_{C_R} f \left(\cos^{-1} \frac{k_x}{k} \right) e^{-|x||k_{xi}|} e^{j|x||k_{xr}|} \, dk_x = 0$$

since $x < 0$. This assumes that f does not grow exponentially with R. Thus

$$\oint_C f \left(\cos^{-1} \frac{k_x}{k} \right) e^{-jk_x x} \, dk_x = 0$$

where C now encloses the portion of the upper half-plane $k_{xi} > \Delta$. By Morera's theorem [110], the above relation holds if f is regular (contains no singularities or branch points) in this portion of the upper half-plane. We shall assume this and investigate the other properties of f that follow from (4.418).

In (4.418) we have an integral equated to an exponential function. To understand the implications of the equality it is helpful to write the exponential function as an integral as well. Consider the integral

$$F(x) = \frac{1}{2j\pi} \int_{-\infty+j\Delta}^{\infty+j\Delta} \frac{h(k_x)}{h(-k_{x0})} \frac{1}{k_x + k_{x0}} e^{-jk_x x} \, dk_x.$$

Here $h(k_x)$ is some function regular in the region $k_{xi} < \Delta$, with $h(k_x) \to 0$ as $k_x \to \infty$. If we choose Δ so that $-k_{xi} > \Delta > -k_{xi} \cos \theta_0$ and close the contour with a semicircle in the lower half-plane (Figure 4.33), then the contribution from the semicircle vanishes for large radius and thus, by Cauchy's residue theorem, $F(x) = -e^{jk_{x0} x}$. Using this (4.418) can be written as

$$\int_{-\infty+j\Delta}^{\infty+j\Delta} \left[\frac{f \left(\cos^{-1} \frac{k_x}{k} \right)}{\sqrt{k^2 - k_x^2}} - \frac{\tilde{E}_0}{2j\pi} \frac{h(k_x)}{h(-k_{x0})} \frac{1}{k_x + k_{x0}} \right] e^{-jk_x x} \, dk_x = 0.$$

Setting the integrand to zero and using $\sqrt{k^2 - k_x^2} = \sqrt{k - k_x}\sqrt{k + k_x}$, we have

$$\frac{f\left(\cos^{-1}\frac{k_x}{k}\right)}{\sqrt{k - k_x}}(k_x + k_{x0}) = \frac{\tilde{E}_0}{2j\pi}\sqrt{k + k_x}\,\frac{h(k_x)}{h(-k_{x0})}. \qquad (4.420)$$

The left member has a branch point at $k_x = k$ while the right member has a branch point at $k_x = -k$. If we choose the branch cuts as in Figure 4.30 then since f is regular in the region $k_{xi} > \Delta$ the left side of (4.420) is regular there. Also, since $h(k_x)$ is regular in the region $k_{xi} < \Delta$, the right side is regular there. We assert that since the two sides are equal, both sides must be regular in the entire complex plane. By Liouville's theorem [35] if a function is entire (regular in the entire plane) and bounded, then it must be constant. So

$$\frac{f\left(\cos^{-1}\frac{k_x}{k}\right)}{\sqrt{k - k_x}}(k_x + k_{x0}) = \frac{\tilde{E}_0}{2j\pi}\sqrt{k + k_x}\,\frac{h(k_x)}{h(-k_{x0})} = \text{constant}.$$

We may evaluate the constant by inserting any value of k_x. Using $k_x = -k_{x0}$ on the right we find that

$$\frac{f\left(\cos^{-1}\frac{k_x}{k}\right)}{\sqrt{k - k_x}}(k_x + k_{x0}) = \frac{\tilde{E}_0}{2j\pi}\sqrt{k - k_{x0}}.$$

Substituting $k_x = k\cos\xi$ and $k_{x0} = k\cos\phi_0$ we have

$$f(\xi) = \frac{\tilde{E}_0}{2j\pi}\,\frac{\sqrt{1 - \cos\phi_0}\sqrt{1 - \cos\xi}}{\cos\xi + \cos\phi_0}.$$

Since $\sin(x/2) = \sqrt{(1 - \cos x)/2}$, we may also write

$$f(\xi) = \frac{\tilde{E}_0}{j\pi}\,\frac{\sin\frac{\phi_0}{2}\sin\frac{\xi}{2}}{\cos\xi + \cos\phi_0}.$$

Finally, substituting this into (4.415) we have the spectral representation for the field scattered by a half-plane:

$$\tilde{E}_z^s(\rho, \phi, \omega) = \frac{\tilde{E}_0(\omega)}{j\pi}\int_C \frac{\sin\frac{\phi_0}{2}\sin\frac{\xi}{2}}{\cos\xi + \cos\phi_0}e^{-jk\rho\cos(\phi\pm\xi)}\,d\xi. \qquad (4.421)$$

The scattered field inversion integral in (4.421) may be rewritten in such a way as to separate geometrical optics (plane-wave) terms from diffraction terms. The diffraction terms may be written using standard functions (modified Fresnel integrals) and for large values of ρ appear as cylindrical waves emanating from a line source at the edge of the half-plane. Interested readers should see James [92] for details.

4.14 Periodic fields and Floquet's theorem

In several practical situations EM waves interact with, or are radiated by, structures spatially periodic along one or more directions. Periodic symmetry simplifies field computation, since boundary conditions need only be applied within one period, or *cell*, of the structure. Examples of situations that lead to periodic fields include the guiding of waves in slow-wave structures such as helices and meander lines, the scattering of plane waves from gratings, and the radiation of waves by antenna arrays. In this section we will study the representation of fields with infinite periodicity as spatial Fourier series.

4.14.1 Floquet's theorem

Consider an environment having spatial periodicity along the z-direction. In this environment the frequency-domain field may be represented in terms of a periodic function $\tilde{\psi}_p$ that obeys

$$\tilde{\psi}_p(x,y,z \pm mL, \omega) = \tilde{\psi}_p(x,y,z,\omega)$$

where m is an integer and L is the spatial period. According to *Floquet's theorem*, if $\tilde{\psi}$ represents some vector component of the field, then the field obeys

$$\tilde{\psi}(x,y,z,\omega) = e^{-j\kappa z}\tilde{\psi}_p(x,y,z,\omega). \qquad (4.422)$$

Here $\kappa = \beta - j\alpha$ is a complex wavenumber describing the phase shift and attenuation of the field between the various cells of the environment. The phase shift and attenuation may arise from a wave propagating through a lossy periodic medium (see example below) or may be impressed by a plane wave as it scatters from a periodic surface, or may be produced by the excitation of an antenna array by a distributed terminal voltage. It is also possible to have $\kappa = 0$ as when, for example, a periodic antenna array is driven with all elements in phase.

Because $\tilde{\psi}_p$ is periodic we may expand it in a Fourier series

$$\tilde{\psi}_p(x,y,z,\omega) = \sum_{n=-\infty}^{\infty} \tilde{\psi}_n(x,y,\omega)e^{-j2\pi nz/L}$$

where the $\tilde{\psi}_n$ are found by orthogonality:

$$\tilde{\psi}_n(x,y,\omega) = \frac{1}{L}\int_{-L/2}^{L/2} \tilde{\psi}_p(x,y,z,\omega)e^{j2\pi nz/L}\,dz.$$

Substituting this into (4.422), we have a representation for the field as a Fourier series:

$$\tilde{\psi}(x,y,z,\omega) = \sum_{n=-\infty}^{\infty} \tilde{\psi}_n(x,y,\omega)e^{-j\kappa_n z}$$

where

$$\kappa_n = \beta + 2\pi n/L + j\alpha = \beta_n - j\alpha.$$

We see that within each cell the field consists of a number of constituents called *space harmonics* or *Hartree harmonics*, each with the property of a propagating or evanescent wave. Each has phase velocity

$$v_{pn} = \frac{\omega}{\beta_n} = \frac{\omega}{\beta + 2\pi n/L}.$$

A number of the space harmonics have phase velocities in the $+z$-direction while the remainder have phase velocities in the $-z$-direction, depending on the value of β. However, all of the space harmonics have the same group velocity

$$v_{gn} = \frac{d\omega}{d\beta} = \left(\frac{d\beta_n}{d\omega}\right)^{-1} = \left(\frac{d\beta}{d\omega}\right)^{-1} = v_g.$$

Those space harmonics for which the group and phase velocities are in opposite directions are referred to as *backward waves*, and form the basis of operation of microwave tubes known as "backward wave oscillators."

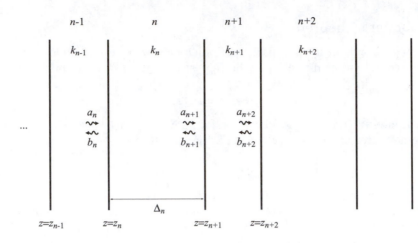

Figure 4.34: Geometry of a periodic stratified medium with each cell consisting of two material layers.

4.14.2 Examples of periodic systems

Plane-wave propagation within a periodically stratified medium. As an example of wave propagation in a periodic structure, let us consider a plane wave propagating within a layered medium consisting of two material layers repeated periodically as shown in Figure 4.34. Each section of two layers is a cell within the periodic medium, and we seek an expression for the propagation constant within the cells, κ.

We developed the necessary tools for studying plane waves within an arbitrary layered medium in § 4.11.5, and can apply them to the case of a periodic medium. In equations (4.305) and (4.306) we have expressions for the wave amplitudes in any region in terms of the amplitudes in the region immediately preceding it. We may write these in matrix form by eliminating one of the variables a_n or b_n from each equation:

$$\begin{bmatrix} T_{11}^{(n)} & T_{12}^{(n)} \\ T_{21}^{(n)} & T_{22}^{(n)} \end{bmatrix} \begin{bmatrix} a_{n+1} \\ b_{n+1} \end{bmatrix} = \begin{bmatrix} a_n \\ b_n \end{bmatrix} \tag{4.423}$$

where

$$T_{11}^{(n)} = \frac{1}{2} \frac{Z_n + Z_{n-1}}{Z_n} \tilde{P}_n^{-1},$$

$$T_{12}^{(n)} = \frac{1}{2} \frac{Z_n - Z_{n-1}}{Z_n} \tilde{P}_n,$$

$$T_{21}^{(n)} = \frac{1}{2} \frac{Z_n - Z_{n-1}}{Z_n} \tilde{P}_n^{-1},$$

$$T_{22}^{(n)} = \frac{1}{2} \frac{Z_n + Z_{n-1}}{Z_n} \tilde{P}_n.$$

Here Z_n represents $Z_{n\perp}$ for perpendicular polarization and $Z_{n\parallel}$ for parallel polarization. The matrix entries are often called *transmission parameters*, and are similar to the parameters used to describe microwave networks, except that in network theory the wave amplitudes are often normalized using the wave impedances. We may use these parameters to describe the cascaded system of two layers:

$$\begin{bmatrix} T_{11}^{(n)} & T_{12}^{(n)} \\ T_{21}^{(n)} & T_{22}^{(n)} \end{bmatrix} \begin{bmatrix} T_{11}^{(n+1)} & T_{12}^{(n+1)} \\ T_{21}^{(n+1)} & T_{22}^{(n+1)} \end{bmatrix} \begin{bmatrix} a_{n+2} \\ b_{n+2} \end{bmatrix} = \begin{bmatrix} a_n \\ b_n \end{bmatrix}.$$

Since for a periodic layered medium the wave amplitudes should obey (4.422), we have

$$\begin{bmatrix} T_{11} & T_{12} \\ T_{21} & T_{22} \end{bmatrix} \begin{bmatrix} a_{n+2} \\ b_{n+2} \end{bmatrix} = \begin{bmatrix} a_n \\ b_n \end{bmatrix} = e^{j\kappa L} \begin{bmatrix} a_{n+2} \\ b_{n+2} \end{bmatrix} \tag{4.424}$$

where $L = \Delta_n + \Delta_{n+1}$ is the period of the structure and

$$\begin{bmatrix} T_{11} & T_{12} \\ T_{21} & T_{22} \end{bmatrix} = \begin{bmatrix} T_{11}^{(n)} & T_{12}^{(n)} \\ T_{21}^{(n)} & T_{22}^{(n)} \end{bmatrix} \begin{bmatrix} T_{11}^{(n+1)} & T_{12}^{(n+1)} \\ T_{21}^{(n+1)} & T_{22}^{(n+1)} \end{bmatrix}.$$

Equation (4.424) is an eigenvalue equation for κ and can be rewritten as

$$\begin{bmatrix} T_{11} - e^{j\kappa L} & T_{12} \\ T_{21} & T_{22} - e^{j\kappa L} \end{bmatrix} \begin{bmatrix} a_{n+2} \\ b_{n+2} \end{bmatrix} = \begin{bmatrix} 0 \\ 0 \end{bmatrix}.$$

This equation only has solutions when the determinant of the matrix vanishes. Expansion of the determinant gives

$$T_{11}T_{22} - T_{12}T_{21} - e^{j\kappa L}(T_{11} + T_{22}) + e^{j2\kappa L} = 0. \tag{4.425}$$

The first two terms are merely

$$T_{11}T_{22} - T_{12}T_{21} = \begin{vmatrix} T_{11} & T_{12} \\ T_{21} & T_{22} \end{vmatrix} = \begin{vmatrix} T_{11}^{(n)} & T_{12}^{(n)} \\ T_{21}^{(n)} & T_{22}^{(n)} \end{vmatrix} \begin{vmatrix} T_{11}^{(n+1)} & T_{12}^{(n+1)} \\ T_{21}^{(n+1)} & T_{22}^{(n+1)} \end{vmatrix}.$$

Since we can show that

$$\begin{vmatrix} T_{11}^{(n)} & T_{12}^{(n)} \\ T_{21}^{(n)} & T_{22}^{(n)} \end{vmatrix} = \frac{Z_{n-1}}{Z_n},$$

we have

$$T_{11}T_{22} - T_{12}T_{21} = \frac{Z_{n-1}}{Z_n} \frac{Z_n}{Z_{n+1}} = 1$$

where we have used $Z_{n-1} = Z_{n+1}$ because of the periodicity of the medium. With this, (4.425) becomes

$$\cos \kappa L = \frac{T_{11} + T_{22}}{2}.$$

Finally, computing the matrix product and simplifying to find $T_{11} + T_{22}$, we have

$$\cos \kappa L = \cos(k_{z,n}\Delta_n) \cos(k_{k,n-1}\Delta_{n-1}) -$$
$$- \frac{1}{2}\left(\frac{Z_{n-1}}{Z_n} + \frac{Z_n}{Z_{n-1}}\right) \sin(k_{z,n}\Delta_n) \sin(k_{z,n-1}\Delta_{n-1}) \tag{4.426}$$

or equivalently

$$\cos \kappa L = \frac{1}{4}\frac{(Z_{n-1} + Z_n)^2}{Z_n Z_{n-1}} \cos(k_{z,n}\Delta_n + k_{z,n-1}\Delta_{n-1}) -$$
$$- \frac{1}{4}\frac{(Z_{n-1} - Z_n)^2}{Z_n Z_{n-1}} \cos(k_{z,n}\Delta_n - k_{z,n-1}\Delta_{n-1}). \tag{4.427}$$

Note that both $\pm\kappa$ satisfy this equation, allowing waves with phase front propagation in both the $\pm z$-directions.

We see in (4.426) that even for lossless materials certain values of ω result in $\cos\kappa L > 1$, causing κL to be imaginary and producing evanescent waves. We refer to the frequency ranges over which $\cos\kappa L > 1$ as *stopbands*, and those over which $\cos\kappa L < 1$ as *passbands*. This terminology is used in filter analysis and, indeed, waves propagating in periodic media experience effects similar to those experienced by signals passing through filters.

Field produced by an infinite array of line sources.   As a second example, consider an infinite number of z-directed line sources within a homogeneous medium of complex permittivity $\tilde{\epsilon}^c(\omega)$ and permeability $\tilde{\mu}(\omega)$, aligned along the x-axis with separation L such that

$$\tilde{\mathbf{J}}(\mathbf{r},\omega) = \sum_{n=-\infty}^{\infty} \hat{\mathbf{z}}\tilde{I}_n\delta(y)\delta(x - nL).$$

The current on each element is allowed to show a progressive phase shift and attenuation. (Such progression may result from a particular method of driving primary currents on successive elements, or, if the currents are secondary, from their excitation by an impressed field such as a plane wave.) Thus we write

$$\tilde{I}_n = \tilde{I}_0 e^{-j\kappa n L} \tag{4.428}$$

where κ is a complex constant.

We may represent the field produced by the source array as a superposition of the fields of individual line sources found earlier. In particular we may use the Hankel function representation (4.345) or the Fourier transform representation (4.407). Using the latter we have

$$\tilde{E}_z(x,y,\omega) = \sum_{n=-\infty}^{\infty} e^{-j\kappa n L}\left[-\frac{\omega\tilde{\mu}\tilde{I}_0(\omega)}{2\pi} \int_{-\infty+j\Delta}^{\infty+j\Delta} \frac{e^{-jk_y|y|}}{2k_y}e^{-jk_x(x-nL)}\,dk_x \right].$$

Interchanging the order of summation and integration we have

$$\tilde{E}_z(x,y,\omega) = -\frac{\omega\tilde{\mu}\tilde{I}_0(\omega)}{2\pi} \int_{-\infty+j\Delta}^{\infty+j\Delta} \frac{e^{-jk_y|y|}}{2k_y}\left[\sum_{n=-\infty}^{\infty} e^{jn(k_x-\kappa)L} \right]e^{-jk_x x}\,dk_x. \tag{4.429}$$

We can rewrite the sum in this expression using Poisson's sum formula [142]

$$\sum_{n=-\infty}^{\infty} f(x - nD) = \frac{1}{D}\sum_{n=-\infty}^{\infty} F(nk_0)e^{jnk_0 x},$$

where $k_0 = 2\pi/D$. Letting $f(x) = \delta(x - x_0)$ in that expression we have

$$\sum_{n=-\infty}^{\infty} \delta\left(x - x_0 - n\frac{2\pi}{L}\right) = \frac{L}{2\pi}\sum_{n=-\infty}^{\infty} e^{jnL(x-x_0)}.$$

Substituting this into (4.429) we have

$$\tilde{E}_z(x,y,\omega) = -\frac{\omega\tilde{\mu}\tilde{I}_0(\omega)}{2\pi} \int_{-\infty+j\Delta}^{\infty+j\Delta} \frac{e^{-jk_y|y|}}{2k_y}\left[\sum_{n=-\infty}^{\infty} \frac{2\pi}{L}\delta\left(k_x - \kappa - n\frac{2\pi}{L}\right) \right]e^{-jk_x x}\,dk_x.$$

Carrying out the integral we replace k_x with $\kappa_n = \kappa + 2n\pi/L$, giving

$$\tilde{E}_z(x, y, \omega) = -\omega\tilde{\mu}\tilde{I}_0(\omega) \sum_{n=-\infty}^{\infty} \frac{e^{-jk_{y,n}|y|}e^{-j\kappa_n x}}{2Lk_{y,n}}$$

$$= -j\omega\tilde{\mu}\tilde{I}_0(\omega)\tilde{G}_\infty(x, y|0, 0, \omega) \qquad (4.430)$$

where $k_{y,n} = \sqrt{k^2 - \kappa_n^2}$, and where

$$\tilde{G}_\infty(x, y|x', y', \omega) = \sum_{n=-\infty}^{\infty} \frac{e^{-jk_{y,n}|y-y'|}e^{-j\kappa_n(x-x')}}{2jLk_{y,n}} \qquad (4.431)$$

is called the *periodic Green's function*.

We may also find the field produced by an infinite array of line sources in terms of the Hankel function representation of a single line source (4.345). Using the current representation (4.428) and summing over the sources, we obtain

$$\tilde{E}_z(\boldsymbol{\rho}, \omega) = -\frac{\omega\tilde{\mu}}{4} \sum_{n=-\infty}^{\infty} \tilde{I}_0(\omega)e^{-j\kappa nL} H_0^{(2)}(k|\boldsymbol{\rho} - \boldsymbol{\rho}_n|) = -j\omega\tilde{\mu}\tilde{I}_0(\omega)\tilde{G}_\infty(x, y|0, 0, \omega)$$

where

$$|\boldsymbol{\rho} - \boldsymbol{\rho}_n| = |\hat{\mathbf{y}}y + \hat{\mathbf{x}}(x - nL)| = \sqrt{y^2 + (x - nL)^2}$$

and where $\tilde{G}_\infty$ is an alternative form of the periodic Green's function

$$\tilde{G}_\infty(x, y|x', y', \omega) = \frac{1}{4j} \sum_{n=-\infty}^{\infty} e^{-j\kappa nL} H_0^{(2)}\left(k\sqrt{(y-y')^2 + (x - nL - x')^2}\right). \qquad (4.432)$$

The periodic Green's functions (4.431) and (4.432) produce identical results, but are each appropriate for certain applications. For example, (4.431) is useful for situations in which boundary conditions at constant values of y are to be applied. Both forms are difficult to compute under certain circumstances, and variants of these forms have been introduced in the literature [203].

4.15 Problems

4.1 Beginning with the Kronig–Kramers formulas (4.35)–(4.36), use the even–odd behavior of the real and imaginary parts of $\tilde{\epsilon}^c$ to derive the alternative relations (4.37)–(4.38).

4.2 Consider the complex permittivity dyadic of a magnetized plasma given by (4.88)–(4.91). Show that we may decompose $[\bar{\bar{\epsilon}}^c]$ as the sum of two matrices

$$[\bar{\bar{\epsilon}}^c] = [\bar{\bar{\epsilon}}] + \frac{[\bar{\bar{\sigma}}]}{j\omega}$$

where $[\bar{\bar{\epsilon}}]$ and $[\bar{\bar{\sigma}}]$ are hermitian.

4.3 Show that the Debye permittivity formulas

$$\tilde{\epsilon}'(\omega) - \epsilon_\infty = \frac{\epsilon_s - \epsilon_\infty}{1 + \omega^2 \tau^2}, \qquad \tilde{\epsilon}''(\omega) = -\frac{\omega\tau(\epsilon_s - \epsilon_\infty)}{1 + \omega^2 \tau^2},$$

obey the Kronig–Kramers relations.

4.4 The frequency-domain duality transformations for the constitutive parameters of an anisotropic medium are given in (4.197). Determine the analogous transformations for the constitutive parameters of a bianisotropic medium.

4.5 Establish the plane-wave identities (B.76)–(B.79) by direct differentiation in rectangular coordinates.

4.6 Assume that sea water has the parameters $\epsilon = 80\epsilon_0$, $\mu = \mu_0$, $\sigma = 4$ S/m, and that these parameters are frequency-independent. Plot the ω–β diagram for a plane wave propagating in this medium and compare to Figure 4.12. Describe the dispersion: is it normal or anomalous? Also plot the phase and group velocities and compare to Figure 4.13. How does the relaxation phenomenon affect the velocity of a wave in this medium?

4.7 Consider a uniform plane wave incident at angle θ_i onto an interface separating two lossless media (Figure 4.18). Assuming perpendicular polarization, write the explicit forms of the total fields in each region under the condition $\theta_i < \theta_c$, where θ_c is the critical angle. Show that the total field in region 1 can be decomposed into a portion that is a pure standing wave in the z-direction and a portion that is a pure traveling wave in the z-direction. Also show that the field in region 2 is a pure traveling wave. Repeat for parallel polarization.

4.8 Consider a uniform plane wave incident at angle θ_i onto an interface separating two lossless media (Figure 4.18). Assuming perpendicular polarization, use the total fields from Problem 4.7 to show that under the condition $\theta_i < \theta_c$ the normal component of the time-average Poynting vector is continuous across the interface. Here θ_c is the critical angle. Repeat for parallel polarization.

4.9 Consider a uniform plane wave incident at angle θ_i onto an interface separating two lossless media (Figure 4.18). Assuming perpendicular polarization, write the explicit forms of the total fields in each region under the condition $\theta_i > \theta_c$, where θ_c is the critical angle. Show that the field in region 1 is a pure standing wave in the z-direction and that the field in region 2 is an evanescent wave. Repeat for parallel polarization.

4.10 Consider a uniform plane wave incident at angle θ_i onto an interface separating two lossless media (Figure 4.18). Assuming perpendicular polarization, use the fields from Problem 4.9 to show that under the condition $\theta_i > \theta_c$ the field in region 1 carries no time-average power in the z-direction, while the field in region 2 carries no time-average power. Here θ_c is the critical angle. Repeat for parallel polarization.

4.11 Consider a uniform plane wave incident at angle θ_i from a lossless material onto a good conductor (Figure 4.18). The conductor has permittivity ϵ_0, permeability μ_0, and conductivity σ. Show that the transmission angle is $\theta_t \approx 0$ and thus the wave in the conductor propagates normal to the interface. Also show that for perpendicular polarization the current per unit width induced by the wave in region 2 is

$$\tilde{\mathbf{K}}(\omega) = \hat{\mathbf{y}}\sigma \tilde{T}_\perp(\omega)\tilde{E}_\perp(\omega)\frac{1-j}{2\beta_2}.$$

and that this is identical to the tangential magnetic field at the surface:

$$\tilde{\mathbf{K}}(\omega) = -\hat{\mathbf{z}} \times \tilde{\mathbf{H}}^t|_{z=0}.$$

If we define the *surface impedance* $Z_s(\omega)$ of the conductor as the ratio of tangential electric and magnetic fields at the interface, show that

$$Z_s(\omega) = \frac{1+j}{\sigma\delta} = R_s(\omega) + jX_s(\omega).$$

Then show that the time-average power flux entering region 2 for a monochromatic wave of frequency $\breve{\omega}$ is simply

$$\mathbf{S}_{av,2} = \hat{\mathbf{z}}\frac{1}{2}(\breve{\mathbf{K}} \cdot \breve{\mathbf{K}}^*)R_s.$$

Note that the since the surface impedance is also the ratio of tangential electric field to induced current per unit width in region 2, it is also called the *internal impedance*.

4.12 Consider a parallel-polarized plane wave obliquely incident from a lossless medium onto a multi-layered material as shown in Figure 4.20. Writing the fields in each region n, $0 \le n \le N-1$, as $\tilde{\mathbf{H}}_{\|n} = \tilde{\mathbf{H}}_{\|n}^i + \tilde{\mathbf{H}}_{\|n}^r$ where

$$\tilde{\mathbf{H}}_{\|n}^i = \hat{\mathbf{y}}a_{n+1}e^{-jk_{x,n}x}e^{-jk_{z,n}(z-z_{n+1})},$$
$$\tilde{\mathbf{H}}_{\|n}^r = -\hat{\mathbf{y}}b_{n+1}e^{-jk_{x,n}x}e^{+jk_{z,n}(z-z_{n+1})},$$

and the field in region N as

$$\tilde{\mathbf{H}}_{\|N} = \hat{\mathbf{y}}a_{N+1}e^{-jk_{x,N}x}e^{-jk_{z,N}(z-z_N)},$$

apply the boundary conditions to solve for the wave amplitudes a_{n+1} and b_n in terms of a global reflection coefficient $\tilde{R}_n$, an interfacial reflection coefficient $\Gamma_{n\|}$, and the wave amplitude a_n. Compare your results to those found for perpendicular polarization (4.313) and (4.314).

4.13 Consider a slab of lossless material with permittivity $\epsilon = \epsilon_r\epsilon_0$ and permeability $\mu = \mu_r\mu_0$ located in free space between the planes $z = z_1$ and $z = z_2$. A right-hand circularly-polarized plane wave is incident on the slab at angle θ_i as shown in Figure 4.22. Determine the conditions (if any) under which the reflected wave is: (a) linearly polarized; (b) right-hand or left-hand circularly polarized; (c) right-hand or left-hand elliptically polarized. Repeat for the transmitted wave.

4.14 Consider a slab of lossless material with permittivity $\epsilon = \epsilon_r\epsilon_0$ and permeability μ_0 located in free space between the planes $z = z_1$ and $z = z_2$. A transient, perpendicularly-polarized plane wave is obliquely incident on the slab as shown in Figure 4.22. If the temporal waveform of the incident wave is $E_\perp^i(t)$, find the transient reflected field in region 0 and the transient transmitted field in region 2 in terms of an infinite superposition of amplitude-scaled, time-shifted versions of the incident wave. Interpret each of the first four terms in the reflected and transmitted fields in terms of multiple reflection within the slab.

4.15 Consider a free-space gap embedded between the planes $z = z_1$ and $z = z_2$ in an infinite, lossless dielectric medium of permittivity $\epsilon_r\epsilon_0$ and permeability μ_0. A perpendicularly-polarized plane wave is incident on the gap at angle $\theta_i > \theta_c$ as shown in

Figure 4.22. Here θ_c is the critical angle for a plane wave incident on the *single* interface between a lossless dielectric of permittivity $\epsilon_r\epsilon_0$ and free space. Apply the boundary conditions and find the fields in each of the three regions. Find the time-average Poynting vector in region 0 at $z = z_1$, in region 1 at $z = z_2$, and in region 2 at $z = z_2$. Is conservation of energy obeyed?

4.16 A uniform ferrite material has scalar permittivity $\tilde{\epsilon} = \epsilon$ and dyadic permeability $\tilde{\bar{\mu}}$. Assume the ferrite is magnetized along the z-direction and has losses so that its permeability dyadic is given by (4.118). Show that the wave equation for a TEM plane wave of the form

$$\tilde{H}(\mathbf{r}, \omega) = \tilde{H}_0(\omega)e^{-jk_z z}$$

is

$$k_z^2 \tilde{H}_0 = \omega^2 \epsilon \tilde{\bar{\mu}} \cdot \tilde{H}_0$$

where $k_z = \beta - j\alpha$. Find explicit formulas for the two solutions $k_{z\pm} = \beta_\pm - j\alpha_\pm$. Show that when the damping parameter $\alpha \ll 1$, near resonance $\alpha_+ \gg \alpha_-$.

4.17 A time-harmonic, TE-polarized, uniform cylindrical wave propagates in a lossy medium. Assuming $|k\rho| \gg 1$, show that the power per unit length passing through a cylinder of radius ρ is given by

$$P_{av}/l = \text{Re}\{Z_{TE}^*\} |\check{H}_{z0}|^2 \frac{e^{-2\alpha\rho}}{8|k|}.$$

If the material is lossless, show that the power per unit length passing through a cylinder is independent of the radius and is given by

$$P_{av}/l = \frac{\eta |\check{H}_{z0}|^2}{8k}.$$

4.18 A TM-polarized plane wave is incident on a cylinder made from a perfect electric conductor such that the current induced on the cylinder is given by (4.365). When the cylinder radius is large compared to the wavelength of the incident wave, we may approximate the current using the principle of *physical optics*. This states that the induced current is zero in the "shadow region" where the cylinder is not directly illuminated by the incident wave. Elsewhere, in the "illuminated region," the induced current is given by

$$\tilde{\mathbf{J}}_s = 2\hat{n} \times \tilde{\mathbf{H}}^i.$$

Plot the current from (4.365) for various values of $k_0 a$ and compare to the current computed from physical optics. How large must $k_0 a$ be for the shadowing effect to be significant?

4.19 The *radar cross section* of a two-dimensional object illuminated by a TM-polarized plane wave is defined by

$$\sigma_{2-D}(\omega, \phi) = \lim_{\rho \to \infty} 2\pi\rho \frac{|\tilde{E}_z^s|^2}{|\tilde{E}_z^i|^2}.$$

This quantity has units of meters and is sometimes called the "scattering width" of the object. Using the asymptotic form of the Hankel function, determine the formula for the radar cross section of a TM-illuminated cylinder made of perfect electric conductor.

Show that when the cylinder radius is small compared to a wavelength the radar cross section may be approximated as

$$\sigma_{2-D}(\omega, \phi) = a \frac{\pi^2}{k_0 a} \frac{1}{\ln^2(0.89 k_0 a)}$$

and is thus independent of the observation angle ϕ.

4.20 A TE-polarized plane wave is incident on a material cylinder with complex permittivity $\tilde{\epsilon}^c(\omega)$ and permeability $\tilde{\mu}(\omega)$, aligned along the z-axis in free space. Apply the boundary conditions on the surface of the cylinder and determine the total field both internal and external to the cylinder. Show that as $\tilde{\sigma} \to \infty$ the magnetic field external to the cylinder reduces to (4.366).

4.21 A TM-polarized plane wave is incident on a PEC cylinder of radius a aligned along the z-axis in free space. The cylinder is coated with a material layer of radius b with complex permittivity $\tilde{\epsilon}^c(\omega)$ and permeability $\tilde{\mu}(\omega)$. Apply the boundary conditions on the surface of the cylinder and across the interface between the material and free space and determine the total field both internal and external to the material layer.

4.22 A PEC cylinder of radius a, aligned along the z-axis in free space, is illuminated by a z-directed electric line source $\tilde{I}(\omega)$ located at (ρ_0, ϕ_0). Expand the fields in the regions $a < \rho < \rho_0$ and $\rho > \rho_0$ in terms of nonuniform cylindrical waves, and apply the boundary conditions at $\rho = a$ and $\rho = \rho_0$ to determine the fields everywhere.

4.23 Repeat Problem 4.22 for the case of a cylinder illuminated by a magnetic line source.

4.24 Assuming

$$f(\xi, \omega) = \frac{k}{2\pi} A(k_x, \omega) \sin \xi,$$

use the relations

$$\cos z = \cos(u + jv) = \cos u \cosh v - j \sin u \sinh v,$$
$$\sin z = \sin(u + jv) = \sin u \cosh v + j \cos u \sinh v,$$

to show that the contour in Figure 4.29 provides identical values of the integrand in

$$\tilde{\psi}(x, y, \omega) = \int_C f(\xi, \omega) e^{-jk\rho \cos(\phi \pm \xi)} \, d\xi$$

as does the contour $[-\infty + j\Delta, \infty + j\Delta]$ in

$$\tilde{\psi}(x, y, \omega) = \frac{1}{2\pi} \int_{-\infty + j\Delta}^{\infty + j\Delta} A(k_x, \omega) e^{-jk_x x} e^{\mp jk_y y} \, dk_x. \qquad (4.433)$$

4.25 Verify (4.409) by writing the TE fields in terms of Fourier transforms and applying boundary conditions.

4.26 Consider a z-directed electric line source $\tilde{I}(\omega)$ located on the y-axis at $y = h$. The region $y < 0$ contains a perfect electric conductor. Write the fields in the regions $0 < y < h$ and $y > h$ in terms of the Fourier transform solution to the homogeneous Helmholtz equation. Note that in the region $0 < y < h$ terms representing waves traveling in *both* the $\pm y$-directions are needed, while in the region $y > h$ only terms traveling in the y-direction are needed. Apply the boundary conditions at $y = 0, h$ to determine the spectral amplitudes. Show that the total field may be decomposed into an impressed term identical to (4.410) and a scattered term identical to (4.413).

4.27 Consider a z-directed magnetic line source $\tilde{I}_m(\omega)$ located on the y-axis at $y = h$. The region $y > 0$ contains a material with parameters $\tilde{\epsilon}_1^c(\omega)$ and $\tilde{\mu}_1(\omega)$, while the region $y < 0$ contains a material with parameters $\tilde{\epsilon}_2^c(\omega)$ and $\tilde{\mu}_2(\omega)$. Using the Fourier transform solution to the Helmholtz equation, write the total field for $y > 0$ as the sum of an impressed field of the magnetic line source and a scattered field, and write the field for $y < 0$ as a scattered field. Apply the boundary conditions at $y = 0$ to determine the spectral amplitudes. Can you interpret the scattered fields in terms of images of the line source?

4.28 Consider a TE-polarized plane wave incident on a PEC half-plane located at $y = 0$, $x > 0$. If the incident magnetic field is given by

$$\tilde{\mathbf{H}}^i(\mathbf{r}, \omega) = \hat{\mathbf{z}} \tilde{H}_0(\omega) e^{jk(x \cos \phi_0 + y \sin \phi_0)},$$

determine the appropriate boundary conditions on the fields at $y = 0$. Solve for the scattered magnetic field using the Fourier transform approach.

4.29 Consider the layered medium of Figure 4.34 with alternating layers of free space and perfect dielectric. The dielectric layer has permittivity $4\epsilon_0$ and thickness Δ while the free space layer has thickness 2Δ. Assuming a normally-incident plane wave, solve for $k_0 \Delta$ in terms of $\kappa \Delta$, and plot k_0 versus κ, identifying the stop and pass bands. This type of $\omega - \beta$ plot for a periodic medium is named a *Brillouin diagram*, after L. Brillouin who investigated energy bands in periodic crystal lattices [23].

4.30 Consider a periodic layered medium as in Figure 4.34, but with each cell consisting of three different layers. Derive an eigenvalue equation similar to (4.427) for the propagation constant.

Chapter 5

Field decompositions and the EM potentials

5.1 Spatial symmetry decompositions

Spatial symmetry can often be exploited to solve electromagnetics problems. For analytic solutions, symmetry can be used to reduce the number of boundary conditions that must be applied. For computer solutions the storage requirements can be reduced. Typical symmetries include rotation about a point or axis, and reflection through a plane, along an axis, or through a point. We shall consider the common case of reflection through a plane. Reflections through the origin and through an axis will be treated in the exercises.

Note that spatial symmetry decompositions may be applied even if the sources and fields possess no spatial symmetry. As long as the boundaries and material media are symmetric, the sources and fields may be decomposed into constituents that individually mimic the symmetry of the environment.

5.1.1 Planar field symmetry

Consider a region of space consisting of linear, isotropic, time-invariant media having material parameters $\epsilon(\mathbf{r})$, $\mu(\mathbf{r})$, and $\sigma(\mathbf{r})$. The electromagnetic fields $(\mathbf{E}, \mathbf{H})$ within this region are related to their impressed sources $(\mathbf{J}^i, \mathbf{J}^i_m)$ and their secondary sources $\mathbf{J}^s = \sigma\mathbf{E}$ through Maxwell's curl equations:

$$\frac{\partial E_z}{\partial y} - \frac{\partial E_y}{\partial z} = -\mu\frac{\partial H_x}{\partial t} - J^i_{mx}, \tag{5.1}$$

$$\frac{\partial E_x}{\partial z} - \frac{\partial E_z}{\partial x} = -\mu\frac{\partial H_y}{\partial t} - J^i_{my}, \tag{5.2}$$

$$\frac{\partial E_y}{\partial x} - \frac{\partial E_x}{\partial y} = -\mu\frac{\partial H_z}{\partial t} - J^i_{mz}, \tag{5.3}$$

$$\frac{\partial H_z}{\partial y} - \frac{\partial H_y}{\partial z} = \epsilon\frac{\partial E_x}{\partial t} + \sigma E_x + J^i_x, \tag{5.4}$$

$$\frac{\partial H_x}{\partial z} - \frac{\partial H_z}{\partial x} = \epsilon\frac{\partial E_y}{\partial t} + \sigma E_y + J^i_y, \tag{5.5}$$

$$\frac{\partial H_y}{\partial x} - \frac{\partial H_x}{\partial y} = \epsilon\frac{\partial E_z}{\partial t} + \sigma E_z + J^i_z. \tag{5.6}$$

We assume the material constants are symmetric about some plane, say $z = 0$. Then

$$\epsilon(x, y, -z) = \epsilon(x, y, z),$$
$$\mu(x, y, -z) = \mu(x, y, z),$$
$$\sigma(x, y, -z) = \sigma(x, y, z).$$

That is, with respect to z the material constants are even functions. We further assume that the boundaries and boundary conditions, which guarantee uniqueness of solution, are also symmetric about the $z = 0$ plane. Then we define two cases of reflection symmetry.

Conditions for even symmetry.   We claim that if the sources obey

$$J_x^i(x, y, z) = J_x^i(x, y, -z), \qquad J_{mx}^i(x, y, z) = -J_{mx}^i(x, y, -z),$$
$$J_y^i(x, y, z) = J_y^i(x, y, -z), \qquad J_{my}^i(x, y, z) = -J_{my}^i(x, y, -z),$$
$$J_z^i(x, y, z) = -J_z^i(x, y, -z), \qquad J_{mz}^i(x, y, z) = J_{mz}^i(x, y, -z),$$

then the fields obey

$$E_x(x, y, z) = E_x(x, y, -z), \qquad H_x(x, y, z) = -H_x(x, y, -z),$$
$$E_y(x, y, z) = E_y(x, y, -z), \qquad H_y(x, y, z) = -H_y(x, y, -z),$$
$$E_z(x, y, z) = -E_z(x, y, -z), \qquad H_z(x, y, z) = H_z(x, y, -z).$$

The electric field shares the symmetry of the electric source: components parallel to the $z = 0$ plane are even in z, and the component perpendicular is odd. The magnetic field shares the symmetry of the magnetic source: components parallel to the $z = 0$ plane are odd in z, and the component perpendicular is even.

We can verify our claim by showing that the symmetric fields and sources obey Maxwell's equations. At an arbitrary point $z = a > 0$ equation (5.1) requires

$$\left.\frac{\partial E_z}{\partial y}\right|_{z=a} - \left.\frac{\partial E_y}{\partial z}\right|_{z=a} = -\mu|_{z=a} \left.\frac{\partial H_x}{\partial t}\right|_{z=a} - J_{mx}^i|_{z=a}.$$

By the assumed symmetry condition on source and material constant we get

$$\left.\frac{\partial E_z}{\partial y}\right|_{z=a} - \left.\frac{\partial E_y}{\partial z}\right|_{z=a} = -\mu|_{z=-a} \left.\frac{\partial H_x}{\partial t}\right|_{z=a} + J_{mx}^i|_{z=-a}.$$

If our claim holds regarding the field behavior, then

$$\left.\frac{\partial E_z}{\partial y}\right|_{z=-a} = -\left.\frac{\partial E_z}{\partial y}\right|_{z=a},$$
$$\left.\frac{\partial E_y}{\partial z}\right|_{z=-a} = -\left.\frac{\partial E_y}{\partial z}\right|_{z=a},$$
$$\left.\frac{\partial H_x}{\partial t}\right|_{z=-a} = -\left.\frac{\partial H_x}{\partial t}\right|_{z=a},$$

and we have

$$-\left.\frac{\partial E_z}{\partial y}\right|_{z=-a} + \left.\frac{\partial E_y}{\partial z}\right|_{z=-a} = \mu|_{z=-a} \left.\frac{\partial H_x}{\partial t}\right|_{z=-a} + J_{mx}^i|_{z=-a}.$$

So this component of Faraday's law is satisfied. With similar reasoning we can show that the symmetric sources and fields satisfy (5.2)–(5.6) as well.

Conditions for odd symmetry. We can also show that if the sources obey

$$J_x^i(x,y,z) = -J_x^i(x,y,-z), \qquad J_{mx}^i(x,y,z) = J_{mx}^i(x,y,-z),$$
$$J_y^i(x,y,z) = -J_y^i(x,y,-z), \qquad J_{my}^i(x,y,z) = J_{my}^i(x,y,-z),$$
$$J_z^i(x,y,z) = J_z^i(x,y,-z), \qquad J_{mz}^i(x,y,z) = -J_{mz}^i(x,y,-z),$$

then the fields obey

$$E_x(x,y,z) = -E_x^i(x,y,-z), \qquad H_x(x,y,z) = H_x(x,y,-z),$$
$$E_y(x,y,z) = -E_y(x,y,-z), \qquad H_y(x,y,z) = H_y(x,y,-z),$$
$$E_z(x,y,z) = E_z(x,y,-z), \qquad H_z(x,y,z) = -H_z(x,y,-z).$$

Again the electric field has the same symmetry as the electric source. However, in this case components parallel to the $z = 0$ plane are odd in z and the component perpendicular is even. Similarly, the magnetic field has the same symmetry as the magnetic source. Here components parallel to the $z = 0$ plane are even in z and the component perpendicular is odd.

Field symmetries and the concept of source images. In the case of odd symmetry the electric field parallel to the $z = 0$ plane is an odd function of z. If we assume that the field is also continuous across this plane, then the electric field tangential to $z = 0$ must vanish: the condition required at the surface of a perfect electric conductor (PEC). We may regard the problem of sources above a perfect conductor in the $z = 0$ plane as *equivalent* to the problem of sources odd about this plane, as long as the sources in both cases are identical for $z > 0$. We refer to the source in the region $z < 0$ as the *image* of the source in the region $z > 0$. Thus the image source $(\mathbf{J}^I, \mathbf{J}_m^I)$ obeys

$$J_x^I(x,y,-z) = -J_x^i(x,y,z), \qquad J_{mx}^I(x,y,-z) = J_{mx}^i(x,y,z),$$
$$J_y^I(x,y,-z) = -J_y^i(x,y,z), \qquad J_{my}^I(x,y,-z) = J_{my}^i(x,y,z),$$
$$J_z^I(x,y,-z) = J_z^i(x,y,z), \qquad J_{mz}^I(x,y,-z) = -J_{mz}^i(x,y,z).$$

That is, parallel components of electric current image in the opposite direction, and the perpendicular component images in the same direction; parallel components of the magnetic current image in the same direction, while the perpendicular component images in the opposite direction.

In the case of even symmetry, the magnetic field parallel to the $z = 0$ plane is odd, and thus the magnetic field tangential to the $z = 0$ plane must be zero. We therefore have an equivalence between the problem of a source above a plane of perfect magnetic conductor (PMC) and the problem of sources even about that plane. In this case we identify image sources that obey

$$J_x^I(x,y,-z) = J_x^i(x,y,z), \qquad J_{mx}^I(x,y,-z) = -J_{mx}^i(x,y,z),$$
$$J_y^I(x,y,-z) = J_y^i(x,y,z), \qquad J_{my}^I(x,y,-z) = -J_{my}^i(x,y,z),$$
$$J_z^I(x,y,-z) = -J_z^i(x,y,z), \qquad J_{mz}^I(x,y,-z) = J_{mz}^i(x,y,z).$$

Parallel components of electric current image in the same direction, and the perpendicular component images in the opposite direction; parallel components of magnetic current image in the opposite direction, and the perpendicular component images in the same direction.

In the case of odd symmetry, we sometimes say that an "electric wall" exists at $z = 0$. The term "magnetic wall" can be used in the case of even symmetry. These terms are particularly common in the description of waveguide fields.

Symmetric field decomposition. Field symmetries may be applied to arbitrary source distributions through a symmetry decomposition of the sources and fields. Consider the general impressed source distributions $(\mathbf{J}^i, \mathbf{J}^i_m)$. The source set

$$J_x^{ie}(x,y,z) = \frac{1}{2}\left[J_x^i(x,y,z) + J_x^i(x,y,-z)\right],$$

$$J_y^{ie}(x,y,z) = \frac{1}{2}\left[J_y^i(x,y,z) + J_y^i(x,y,-z)\right],$$

$$J_z^{ie}(x,y,z) = \frac{1}{2}\left[J_z^i(x,y,z) - J_z^i(x,y,-z)\right],$$

$$J_{mx}^{ie}(x,y,z) = \frac{1}{2}\left[J_{mx}^i(x,y,z) - J_{mx}^i(x,y,-z)\right],$$

$$J_{my}^{ie}(x,y,z) = \frac{1}{2}\left[J_{my}^i(x,y,z) - J_{my}^i(x,y,-z)\right],$$

$$J_{mz}^{ie}(x,y,z) = \frac{1}{2}\left[J_{mz}^i(x,y,z) + J_{mz}^i(x,y,-z)\right],$$

is clearly of even symmetric type while the source set

$$J_x^{io}(x,y,z) = \frac{1}{2}\left[J_x^i(x,y,z) - J_x^i(x,y,-z)\right],$$

$$J_y^{io}(x,y,z) = \frac{1}{2}\left[J_y^i(x,y,z) - J_y^i(x,y,-z)\right],$$

$$J_z^{io}(x,y,z) = \frac{1}{2}\left[J_z^i(x,y,z) + J_z^i(x,y,-z)\right],$$

$$J_{mx}^{io}(x,y,z) = \frac{1}{2}\left[J_{mx}^i(x,y,z) + J_{mx}^i(x,y,-z)\right],$$

$$J_{my}^{io}(x,y,z) = \frac{1}{2}\left[J_{my}^i(x,y,z) + J_{my}^i(x,y,-z)\right],$$

$$J_{mz}^{io}(x,y,z) = \frac{1}{2}\left[J_{mz}^i(x,y,z) - J_{mz}^i(x,y,-z)\right],$$

is of the odd symmetric type. Since $\mathbf{J}^i = \mathbf{J}^{ie} + \mathbf{J}^{io}$ and $\mathbf{J}^i_m = \mathbf{J}^{ie}_m + \mathbf{J}^{io}_m$, we can decompose any source into constituents having, respectively, even and odd symmetry with respect to a plane. The source with even symmetry produces an even field set, while the source with odd symmetry produces an odd field set. The total field is the sum of the fields from each field set.

Planar symmetry for frequency-domain fields. The symmetry conditions introduced above for the time-domain fields also hold for the frequency-domain fields. Because both the conductivity and permittivity must be even functions, we combine their effects and require the complex permittivity to be even. Otherwise the field symmetries and source decompositions are identical.

Example of symmetry decomposition: line source between conducting planes. Consider a z-directed electric line source $\tilde{I}_0$ located at $y = h, x = 0$ between conducting planes at $y = \pm d$, $d > h$. The material between the plates has permeability $\tilde{\mu}(\omega)$ and complex permittivity $\tilde{\epsilon}^c(\omega)$. We decompose the source into one of even symmetric type with line sources $\tilde{I}_0/2$ located at $y = \pm h$, and one of odd symmetric type with a line source $\tilde{I}_0/2$ located at $y = h$ and a line source $-\tilde{I}_0/2$ located at $y = -h$. We solve each of these problems by exploiting the appropriate symmetry, and superpose the results to find the solution to the original problem.

For the even-symmetric case, we begin by using (4.407) to represent the impressed field:

$$\tilde{E}_z^i(x, y, \omega) = -\frac{\omega \tilde{\mu} \frac{\tilde{I}_0(\omega)}{2}}{2\pi} \int_{-\infty+j\Delta}^{\infty+j\Delta} \frac{e^{-jk_y|y-h|} + e^{-jk_y|y+h|}}{2k_y} e^{-jk_x x} \, dk_x.$$

For $y > h$ this becomes

$$\tilde{E}_z^i(x, y, \omega) = -\frac{\omega \tilde{\mu} \frac{\tilde{I}_0(\omega)}{2}}{2\pi} \int_{-\infty+j\Delta}^{\infty+j\Delta} \frac{2 \cos k_y h}{2k_y} e^{-jk_y y} e^{-jk_x x} \, dk_x.$$

The secondary (scattered) field consists of waves propagating in both the $\pm y$-directions:

$$\tilde{E}_z^s(x, y, \omega) = \frac{1}{2\pi} \int_{-\infty+j\Delta}^{\infty+j\Delta} \left[A^+(k_x, \omega) e^{-jk_y y} + A^-(k_x, \omega) e^{jk_y y} \right] e^{-jk_x x} \, dk_x. \tag{5.7}$$

The impressed field is even about $y = 0$. Since the total field $E_z = E_z^i + E_z^s$ must be even in y (E_z is parallel to the plane $y = 0$), the scattered field must also be even. Thus, $A^+ = A^-$ and the total field is for $y > h$

$$\tilde{E}_z(x, y, \omega) = \frac{1}{2\pi} \int_{-\infty+j\Delta}^{\infty+j\Delta} \left[2A^+(k_x, \omega) \cos k_y y - \omega \tilde{\mu} \frac{\tilde{I}_0(\omega)}{2} \frac{2 \cos k_y h}{2k_y} e^{-jk_y y} \right] e^{-jk_x x} \, dk_x.$$

Now the electric field must obey the boundary condition $\tilde{E}_z = 0$ at $y = \pm d$. However, since $\tilde{E}_z$ is even the satisfaction of this condition at $y = d$ automatically implies its satisfaction at $y = -d$. So we set

$$\frac{1}{2\pi} \int_{-\infty+j\Delta}^{\infty+j\Delta} \left[2A^+(k_x, \omega) \cos k_y d - \omega \tilde{\mu} \frac{\tilde{I}_0(\omega)}{2} \frac{2 \cos k_y h}{2k_y} e^{-jk_y d} \right] e^{-jk_x x} \, dk_x = 0$$

and invoke the Fourier integral theorem to get

$$A^+(k_x, \omega) = \omega \tilde{\mu} \frac{\tilde{I}_0(\omega)}{2} \frac{\cos k_y h}{2k_y} \frac{e^{-jk_y d}}{\cos k_y d}.$$

The total field for this case is

$$\tilde{E}_z(x, y, \omega) = -\frac{\omega \tilde{\mu} \frac{\tilde{I}_0(\omega)}{2}}{2\pi} \int_{-\infty+j\Delta}^{\infty+j\Delta} \left[\frac{e^{-jk_y|y-h|} + e^{-jk_y|y+h|}}{2k_y} \right.$$

$$\left. - \frac{2 \cos k_y h}{2k_y} \frac{e^{-jk_y d}}{\cos k_y d} \cos k_y y \right] e^{-jk_x x} \, dk_x.$$

For the odd-symmetric case the impressed field is

$$\tilde{E}_z^i(x, y, \omega) = -\frac{\omega \tilde{\mu} \frac{\tilde{I}_0(\omega)}{2}}{2\pi} \int_{-\infty+j\Delta}^{\infty+j\Delta} \frac{e^{-jk_y|y-h|} - e^{-jk_y|y+h|}}{2k_y} e^{-jk_x x} \, dk_x,$$

which for $y > h$ is

$$\tilde{E}_z^i(x,y,\omega) = -\frac{\omega\tilde{\mu}\frac{\tilde{I}_0(\omega)}{2}}{2\pi} \int\limits_{-\infty+j\Delta}^{\infty+j\Delta} \frac{2j\sin k_y h}{2k_y} e^{-jk_y y} e^{-jk_x x}\, dk_x.$$

The scattered field has the form of (5.7) but must be odd. Thus $A^+ = -A^-$ and the total field for $y > h$ is

$$\tilde{E}_z(x,y,\omega) = \frac{1}{2\pi} \int\limits_{-\infty+j\Delta}^{\infty+j\Delta} \left[2jA^+(k_x,\omega)\sin k_y y - \omega\tilde{\mu}\frac{\tilde{I}_0(\omega)}{2}\frac{2j\sin k_y h}{2k_y} e^{-jk_y y} \right] e^{-jk_x x}\, dk_x.$$

Setting $\tilde{E}_z = 0$ at $z = d$ and solving for A^+ we find that the total field for this case is

$$\tilde{E}_z(x,y,\omega) = -\frac{\omega\tilde{\mu}\frac{\tilde{I}_0(\omega)}{2}}{2\pi} \int\limits_{-\infty+j\Delta}^{\infty+j\Delta} \left[\frac{e^{-jk_y|y-h|} - e^{-jk_y|y+h|}}{2k_y} - \right.$$

$$\left. - \frac{2j\sin k_y h}{2k_y}\frac{e^{-jk_y d}}{\sin k_y d}\sin k_y y \right] e^{-jk_x x}\, dk_x.$$

Adding the fields for the two cases we find that

$$\tilde{E}_z(x,y,\omega) = -\frac{\omega\tilde{\mu}\tilde{I}_0(\omega)}{2\pi} \int\limits_{-\infty+j\Delta}^{\infty+j\Delta} \frac{e^{-jk_y|y-h|}}{2k_y} e^{-jk_x x}\, dk_x +$$

$$+ \frac{\omega\tilde{\mu}\tilde{I}_0(\omega)}{2\pi} \int\limits_{-\infty+j\Delta}^{\infty+j\Delta} \left[\frac{\cos k_y h\cos k_y y}{\cos k_y d} + j\frac{\sin k_y h\sin k_y y}{\sin k_y d} \right] \frac{e^{-jk_y d}}{2k_y} e^{-jk_x x}\, dk_x,$$

$$(5.8)$$

which is a superposition of impressed and scattered fields.

5.2   Solenoidal–lamellar decomposition

We now discuss the decomposition of a general vector field into a *lamellar* component having zero curl and a *solenoidal* component having zero divergence. This is known as a *Helmholtz decomposition*. If $\mathbf{V}$ is any vector field then we wish to write

$$\mathbf{V} = \mathbf{V}_s + \mathbf{V}_l, \tag{5.9}$$

where $\mathbf{V}_s$ and $\mathbf{V}_l$ are the solenoidal and lamellar components of $\mathbf{V}$. Formulas expressing these components in terms of $\mathbf{V}$ are obtained as follows. We first write $\mathbf{V}_s$ in terms of a "vector potential" $\mathbf{A}$ as

$$\mathbf{V}_s = \nabla \times \mathbf{A}. \tag{5.10}$$

This is possible by virtue of (B.49). Similarly, we write $\mathbf{V}_l$ in terms of a "scalar potential" ϕ as

$$\mathbf{V}_l = \nabla\phi. \tag{5.11}$$

To obtain a formula for $\mathbf{V}_l$ we take the divergence of (5.9) and use (5.11) to get

$$\nabla \cdot \mathbf{V} = \nabla \cdot \mathbf{V}_l = \nabla \cdot \nabla\phi = \nabla^2\phi.$$

The result,

$$\nabla^2\phi = \nabla \cdot \mathbf{V},$$

may be regarded as Poisson's equation for the unknown ϕ. This equation is solved in Chapter 3. By (3.61) we have

$$\phi(\mathbf{r}) = -\int_V \frac{\nabla' \cdot \mathbf{V}(\mathbf{r}')}{4\pi R} \, dV',$$

where $R = |\mathbf{r} - \mathbf{r}'|$, and we have

$$\mathbf{V}_l(\mathbf{r}) = -\nabla \int_V \frac{\nabla' \cdot \mathbf{V}(\mathbf{r}')}{4\pi R} \, dV'. \tag{5.12}$$

Similarly, a formula for $\mathbf{V}_s$ can be obtained by taking the curl of (5.9) to get

$$\nabla \times \mathbf{V} = \nabla \times \mathbf{V}_s.$$

Substituting (5.10) we have

$$\nabla \times \mathbf{V} = \nabla \times (\nabla \times \mathbf{A}) = \nabla(\nabla \cdot \mathbf{A}) - \nabla^2 \mathbf{A}.$$

We may choose any value we wish for $\nabla \cdot \mathbf{A}$, since this does not alter $\mathbf{V}_s = \nabla \times \mathbf{A}$. (We discuss such "gauge transformations" in greater detail later in this chapter.) With $\nabla \cdot \mathbf{A} = 0$ we obtain

$$-\nabla \times \mathbf{V} = \nabla^2 \mathbf{A}.$$

This is Poisson's equation for each rectangular component of $\mathbf{A}$; therefore

$$\mathbf{A}(\mathbf{r}) = \int_V \frac{\nabla' \times \mathbf{V}(\mathbf{r}')}{4\pi R} \, dV',$$

and we have

$$\mathbf{V}_s(\mathbf{r}) = \nabla \times \int_V \frac{\nabla' \times \mathbf{V}(\mathbf{r}')}{4\pi R} \, dV'.$$

Summing the results we obtain the Helmholtz decomposition

$$\mathbf{V} = \mathbf{V}_l + \mathbf{V}_s = -\nabla \int_V \frac{\nabla' \cdot \mathbf{V}(\mathbf{r}')}{4\pi R} \, dV' + \nabla \times \int_V \frac{\nabla' \times \mathbf{V}(\mathbf{r}')}{4\pi R} \, dV'. \tag{5.13}$$

Identification of the electromagnetic potentials. Let us write the electromagnetic fields as a general superposition of solenoidal and lamellar components:

$$\mathbf{E} = \nabla \times \mathbf{A}_E + \nabla\phi_E, \tag{5.14}$$
$$\mathbf{B} = \nabla \times \mathbf{A}_B + \nabla\phi_B. \tag{5.15}$$

One possible form of the potentials $\mathbf{A}_E$, $\mathbf{A}_B$, ϕ_E, and ϕ_B appears in (5.13). However, because $\mathbf{E}$ and $\mathbf{B}$ are related by Maxwell's equations, the potentials should be related to the sources. We can determine the explicit relationship by substituting (5.14) and (5.15) into Ampere's and Faraday's laws. It is most convenient to analyze the relationships using superposition of the cases for which $\mathbf{J}_m = 0$ and $\mathbf{J} = 0$.

With $\mathbf{J}_m = 0$ Faraday's law is

$$\nabla \times \mathbf{E} = -\frac{\partial \mathbf{B}}{\partial t}. \tag{5.16}$$

Since $\nabla \times \mathbf{E}$ is solenoidal, $\mathbf{B}$ must be solenoidal and thus $\nabla \phi_B = 0$. This implies that $\phi_B = 0$, which is equivalent to the auxiliary Maxwell equation $\nabla \cdot \mathbf{B} = 0$. Now, substitution of (5.14) and (5.15) into (5.16) gives

$$\nabla \times [\nabla \times \mathbf{A}_E + \nabla \phi_E] = -\frac{\partial}{\partial t} [\nabla \times \mathbf{A}_B].$$

Using $\nabla \times (\nabla \phi_E) = 0$ and combining the terms we get

$$\nabla \times \left[\nabla \times \mathbf{A}_E + \frac{\partial \mathbf{A}_B}{\partial t} \right] = 0,$$

hence

$$\nabla \times \mathbf{A}_E = -\frac{\partial \mathbf{A}_B}{\partial t} + \nabla \xi.$$

Substitution into (5.14) gives

$$\mathbf{E} = -\frac{\partial \mathbf{A}_B}{\partial t} + [\nabla \phi_E + \nabla \xi].$$

Combining the two gradient functions together, we see that we can write both $\mathbf{E}$ and $\mathbf{B}$ in terms of two potentials:

$$\mathbf{E} = -\frac{\partial \mathbf{A}_e}{\partial t} - \nabla \phi_e, \tag{5.17}$$

$$\mathbf{B} = \nabla \times \mathbf{A}_e, \tag{5.18}$$

where the negative sign on the gradient term is introduced by convention.

Gauge transformations and the Coulomb gauge. We pay a price for the simplicity of using only two potentials to represent $\mathbf{E}$ and $\mathbf{B}$. While $\nabla \times \mathbf{A}_e$ is definitely solenoidal, $\mathbf{A}_e$ itself may not be: because of this (5.17) may not be a decomposition into solenoidal and lamellar components. However, a corollary of the Helmholtz theorem states that a vector field is uniquely specified only when *both* its curl and divergence are specified. Here there is an ambiguity in the representation of $\mathbf{E}$ and $\mathbf{B}$; we may remove this ambiguity and define $\mathbf{A}_e$ uniquely by requiring that

$$\nabla \cdot \mathbf{A}_e = 0. \tag{5.19}$$

Then $\mathbf{A}_e$ is solenoidal and the decomposition (5.17) is solenoidal–lamellar. This requirement on $\mathbf{A}_e$ is called the *Coulomb gauge*.

The ambiguity implied by the non-uniqueness of $\nabla \cdot \mathbf{A}_e$ can also be expressed by the observation that a transformation of the type

$$\mathbf{A}_e \rightarrow \mathbf{A}_e + \nabla \Gamma, \tag{5.20}$$

$$\phi_e \rightarrow \phi_e - \frac{\partial \Gamma}{\partial t}, \tag{5.21}$$

leaves the expressions (5.17) and (5.18) unchanged. This is called a *gauge transformation*, and the choice of a certain Γ alters the specification of $\nabla \cdot \mathbf{A}_e$. Thus we may begin with the Coulomb gauge as our baseline, and allow any alteration of $\mathbf{A}_e$ according to (5.20) as long as we augment $\nabla \cdot \mathbf{A}_e$ by $\nabla \cdot \nabla\Gamma = \nabla^2\Gamma$.

Once $\nabla \cdot \mathbf{A}_e$ is specified, the relationship between the potentials and the current $\mathbf{J}$ can be found by substitution of (5.17) and (5.18) into Ampere's law. At this point we assume media that are linear, homogeneous, isotropic, and described by the time-invariant parameters μ, ϵ, and σ. Writing $\mathbf{J} = \mathbf{J}^i + \sigma\mathbf{E}$ we have

$$\frac{1}{\mu}\nabla \times (\nabla \times \mathbf{A}_e) = \mathbf{J}^i - \sigma\frac{\partial \mathbf{A}_e}{\partial t} - \sigma\nabla\phi_e - \epsilon\frac{\partial^2 \mathbf{A}_e}{\partial t^2} - \epsilon\frac{\partial}{\partial t}\nabla\phi_e. \tag{5.22}$$

Taking the divergence of both sides of (5.22) we get

$$0 = \nabla \cdot \mathbf{J}^i - \sigma\frac{\partial}{\partial t}\nabla \cdot \mathbf{A} - \sigma\nabla \cdot \nabla\phi_e - \epsilon\frac{\partial^2}{\partial t^2}\nabla \cdot \mathbf{A}_e - \epsilon\frac{\partial}{\partial t}\nabla \cdot \nabla\phi_e. \tag{5.23}$$

Then, by substitution from the continuity equation and use of (5.19) along with $\nabla \cdot \nabla\phi_e = \nabla^2\phi_e$ we obtain

$$\frac{\partial}{\partial t}\left(\rho^i + \epsilon\nabla^2\phi_e\right) = -\sigma\nabla^2\phi_e.$$

For a lossless medium this reduces to

$$\nabla^2\phi_e = -\rho^i/\epsilon \tag{5.24}$$

and we have

$$\phi_e(\mathbf{r}, t) = \int_V \frac{\rho^i(\mathbf{r}', t)}{4\pi\epsilon R}\, dV'. \tag{5.25}$$

We can obtain an equation for $\mathbf{A}_e$ by expanding the left-hand side of (5.22) to get

$$\nabla(\nabla \cdot \mathbf{A}_e) - \nabla^2\mathbf{A}_e = \mu\mathbf{J}^i - \sigma\mu\frac{\partial \mathbf{A}_e}{\partial t} - \sigma\mu\nabla\phi_e - \mu\epsilon\frac{\partial^2 \mathbf{A}_e}{\partial t^2} - \mu\epsilon\frac{\partial}{\partial t}\nabla\phi_e, \tag{5.26}$$

hence

$$\nabla^2\mathbf{A}_e - \mu\epsilon\frac{\partial^2 \mathbf{A}_e}{\partial t^2} = -\mu\mathbf{J}^i + \sigma\mu\frac{\partial \mathbf{A}_e}{\partial t} + \sigma\mu\nabla\phi_e + \mu\epsilon\frac{\partial}{\partial t}\nabla\phi_e$$

under the Coulomb gauge. For lossless media this becomes

$$\nabla^2\mathbf{A}_e - \mu\epsilon\frac{\partial^2 \mathbf{A}_e}{\partial t^2} = -\mu\mathbf{J}^i + \mu\epsilon\frac{\partial}{\partial t}\nabla\phi_e. \tag{5.27}$$

Observe that the left-hand side of (5.27) is solenoidal (since the Laplacian term came from the curl-curl, and $\nabla \cdot \mathbf{A}_e = 0$), while the right-hand side contains a general vector field $\mathbf{J}^i$ and a lamellar term. We might expect the $\nabla\phi_e$ term to cancel the lamellar portion of $\mathbf{J}^i$, and this does happen [91]. By (5.12) and the continuity equation we can write the lamellar component of the current as

$$\mathbf{J}_l^i(\mathbf{r}, t) = -\nabla\int_V \frac{\nabla' \cdot \mathbf{J}^i(\mathbf{r}', t)}{4\pi R}\, dV' = \frac{\partial}{\partial t}\nabla\int_V \frac{\rho^i(\mathbf{r}', t)}{4\pi R}\, dV' = \epsilon\frac{\partial}{\partial t}\nabla\phi_e.$$

Thus (5.27) becomes

$$\nabla^2\mathbf{A}_e - \mu\epsilon\frac{\partial^2 \mathbf{A}_e}{\partial t^2} = -\mu\mathbf{J}_s^i. \tag{5.28}$$

Therefore the vector potential $\mathbf{A}_e$, which describes the solenoidal portion of both $\mathbf{E}$ and $\mathbf{B}$, is found from just the solenoidal portion of the current. On the other hand, the scalar potential, which describes the lamellar portion of $\mathbf{E}$, is found from ρ^i which arises from $\nabla \cdot \mathbf{J}^i$, the lamellar portion of the current.

From the perspective of field computation, we see that the introduction of potential functions has reoriented the solution process from dealing with two coupled first-order partial differential equations (Maxwell's equations), to two uncoupled second-order equations (the potential equations (5.24) and (5.28)). The decoupling of the equations is often worth the added complexity of dealing with potentials, and, in fact, is the solution technique of choice in such areas as radiation and guided waves. It is worth pausing for a moment to examine the form of these equations. We see that the scalar potential obeys Poisson's equation with the solution (5.25), while the vector potential obeys the wave equation. As a wave, the vector potential must propagate away from the source with finite velocity. However, the solution for the scalar potential (5.25) shows no such behavior. In fact, any change to the charge distribution instantaneously permeates all of space. This apparent violation of Einstein's postulate shows that we must be careful when interpreting the physical meaning of the potentials. Once the computations (5.17) and (5.18) are undertaken, we find that both $\mathbf{E}$ and $\mathbf{B}$ behave as waves, and thus propagate at finite velocity. Mathematically, the conundrum can be resolved by realizing that individually the solenoidal and lamellar components of current must occupy all of space, even if their sum, the actual current $\mathbf{J}^i$, is localized [91].

The Lorentz gauge. A different choice of gauge condition can allow both the vector and scalar potentials to act as waves. In this case $\mathbf{E}$ may be written as a sum of two terms: one purely solenoidal, and the other a superposition of lamellar and solenoidal parts.

Let us examine the effect of choosing the *Lorentz gauge*

$$\nabla \cdot \mathbf{A}_e = -\mu\epsilon\frac{\partial\phi_e}{\partial t} - \mu\sigma\phi_e. \tag{5.29}$$

Substituting this expression into (5.26) we find that the gradient terms cancel, giving

$$\nabla^2\mathbf{A}_e - \mu\sigma\frac{\partial\mathbf{A}_e}{\partial t} - \mu\epsilon\frac{\partial^2\mathbf{A}_e}{\partial t^2} = -\mu\mathbf{J}^i. \tag{5.30}$$

For lossless media

$$\nabla^2\mathbf{A}_e - \mu\epsilon\frac{\partial^2\mathbf{A}_e}{\partial t^2} = -\mu\mathbf{J}^i, \tag{5.31}$$

and (5.23) becomes

$$\nabla^2\phi_e - \mu\epsilon\frac{\partial^2\phi_e}{\partial t^2} = -\frac{\rho^i}{\epsilon}. \tag{5.32}$$

For lossy media we have obtained a second-order differential equation for $\mathbf{A}_e$, but ϕ_e must be found through the somewhat cumbersome relation (5.29). For lossless media the coupled Maxwell equations have been decoupled into two second-order equations, one involving $\mathbf{A}_e$ and one involving ϕ_e. Both (5.31) and (5.32) are wave equations, with $\mathbf{J}^i$ as the source for $\mathbf{A}_e$ and ρ^i as the source for ϕ_e. Thus the expected finite-velocity wave nature of the electromagnetic fields is also manifested in each of the potential functions. The drawback is that, even though we can still use (5.17) and (5.18), the expression for $\mathbf{E}$ is no longer a decomposition into solenoidal and lamellar components. Nevertheless, the choice of the Lorentz gauge is very popular in the study of radiated and guided waves.

The Hertzian potentials. With a little manipulation and the introduction of a new notation, we can maintain the wave nature of the potential functions and still provide a decomposition into purely lamellar and solenoidal components. In this analysis we shall assume lossless media only.

When we chose the Lorentz gauge to remove the arbitrariness of the divergence of the vector potential, we established a relationship between $\mathbf{A}_e$ and ϕ_e. Thus we should be able to write both the electric and magnetic fields in terms of a single potential function. From the Lorentz gauge we can write ϕ_e as

$$\phi_e(\mathbf{r}, t) = -\frac{1}{\mu\epsilon} \int_{-\infty}^{t} \nabla \cdot \mathbf{A}_e(\mathbf{r}, t)\, dt.$$

By (5.17) and (5.18) we can thus write the EM fields as

$$\mathbf{E} = \frac{1}{\mu\epsilon} \nabla \int_{-\infty}^{t} \nabla \cdot \mathbf{A}_e dt - \frac{\partial \mathbf{A}_e}{\partial t}, \tag{5.33}$$

$$\mathbf{B} = \nabla \times \mathbf{A}_e. \tag{5.34}$$

The integro-differential representation of $\mathbf{E}$ in (5.33) is somewhat clumsy in appearance. We can make it easier to manipulate by defining the *Hertzian potential*

$$\mathbf{\Pi}_e = \frac{1}{\mu\epsilon} \int_{-\infty}^{t} \mathbf{A}_e\, dt.$$

In differential form

$$\mathbf{A}_e = \mu\epsilon \frac{\partial \mathbf{\Pi}_e}{dt}. \tag{5.35}$$

With this, (5.33) and (5.34) become

$$\mathbf{E} = \nabla(\nabla \cdot \mathbf{\Pi}_e) - \mu\epsilon \frac{\partial^2}{\partial t^2} \mathbf{\Pi}_e, \tag{5.36}$$

$$\mathbf{B} = \mu\epsilon \nabla \times \frac{\partial \mathbf{\Pi}_e}{\partial t}. \tag{5.37}$$

An equation for $\mathbf{\Pi}_e$ in terms of the source current can be found by substituting (5.35) into (5.31):

$$\mu\epsilon \frac{\partial}{\partial t} \left(\nabla^2 \mathbf{\Pi}_e - \mu\epsilon \frac{\partial^2}{\partial t^2} \mathbf{\Pi}_e \right) = -\mu \mathbf{J}^i.$$

Let us define

$$\mathbf{J}^i = \frac{\partial \mathbf{P}^i}{\partial t}. \tag{5.38}$$

For general impressed current sources (5.38) is just a convenient notation. However, we can conceive of an *impressed polarization current* that is independent of $\mathbf{E}$ and defined through the relation $\mathbf{D} = \epsilon_0 \mathbf{E} + \mathbf{P} + \mathbf{P}^i$. Then (5.38) has a physical interpretation as described in (2.119). We now have

$$\nabla^2 \mathbf{\Pi}_e - \mu\epsilon \frac{\partial^2}{\partial t^2} \mathbf{\Pi}_e = -\frac{1}{\epsilon} \mathbf{P}^i, \tag{5.39}$$

which is a wave equation for $\mathbf{\Pi}_e$. Thus the Hertzian potential has the same wave behavior as the vector potential under the Lorentz gauge.

We can use (5.39) to perform one final simplification of the EM field representation. By the vector identity $\nabla(\nabla \cdot \mathbf{\Pi}) = \nabla \times (\nabla \times \mathbf{\Pi}) + \nabla^2 \mathbf{\Pi}$ we get

$$\nabla(\nabla \cdot \mathbf{\Pi}_e) = \nabla \times (\nabla \times \mathbf{\Pi}_e) - \frac{1}{\epsilon}\mathbf{P}^i + \mu\epsilon\frac{\partial^2}{\partial t^2}\mathbf{\Pi}_e.$$

Substituting this into (5.36) we obtain

$$\mathbf{E} = \nabla \times (\nabla \times \mathbf{\Pi}_e) - \frac{\mathbf{P}^i}{\epsilon}, \tag{5.40}$$

$$\mathbf{B} = \mu\epsilon\nabla \times \frac{\partial \mathbf{\Pi}_e}{\partial t}. \tag{5.41}$$

Let us examine these closely. We know that $\mathbf{B}$ is solenoidal since it is written as the curl of another vector (this is also clear from the auxiliary Maxwell equation $\nabla \cdot \mathbf{B} = 0$). The first term in the expression for $\mathbf{E}$ is also solenoidal. So the lamellar part of $\mathbf{E}$ must be contained within the source term $\mathbf{P}^i$. If we write $\mathbf{P}^i$ in terms of its lamellar and solenoidal components by using

$$\mathbf{J}_s^i = \frac{\partial \mathbf{P}_s^i}{\partial t}, \qquad \mathbf{J}_l^i = \frac{\partial \mathbf{P}_l^i}{\partial t},$$

then (5.40) becomes

$$\mathbf{E} = \left[\nabla \times (\nabla \times \mathbf{\Pi}_e) - \frac{\mathbf{P}_s^i}{\epsilon}\right] - \frac{\mathbf{P}_l^i}{\epsilon}. \tag{5.42}$$

So we have again succeeded in dividing $\mathbf{E}$ into lamellar and solenoidal components.

Potential functions for magnetic current. We can proceed as above to derive the field–potential relationships when $\mathbf{J}^i = 0$ but $\mathbf{J}_m^i \neq 0$. We assume a homogeneous, lossless, isotropic medium with permeability μ and permittivity ϵ, and begin with Faraday's and Ampere's laws

$$\nabla \times \mathbf{E} = -\mathbf{J}_m^i - \frac{\partial \mathbf{B}}{\partial t}, \tag{5.43}$$

$$\nabla \times \mathbf{H} = \frac{\partial \mathbf{D}}{\partial t}. \tag{5.44}$$

We write $\mathbf{H}$ and $\mathbf{D}$ in terms of two potential functions $\mathbf{A}_h$ and ϕ_h as

$$\mathbf{H} = -\frac{\partial \mathbf{A}_h}{\partial t} - \nabla\phi_h,$$

$$\mathbf{D} = -\nabla \times \mathbf{A}_h,$$

and the differential equation for the potentials is found by substitution into (5.43):

$$\nabla \times (\nabla \times \mathbf{A}_h) = \epsilon\mathbf{J}_m^i - \mu\epsilon\frac{\partial^2 \mathbf{A}_h}{\partial t^2} - \mu\epsilon\frac{\partial}{\partial t}\nabla\phi_h. \tag{5.45}$$

Taking the divergence of this equation and substituting from the magnetic continuity equation we obtain

$$\mu\epsilon\frac{\partial^2}{\partial t^2}\nabla \cdot \mathbf{A}_h + \mu\epsilon\frac{\partial}{\partial t}\nabla^2\phi_h = -\epsilon\frac{\partial \rho_m^i}{\partial t}.$$

Under the Lorentz gauge condition

$$\nabla \cdot \mathbf{A}_h = -\mu\epsilon\frac{\partial \phi_h}{\partial t}$$

this reduces to

$$\nabla^2 \phi_h - \mu\epsilon\frac{\partial^2 \phi_h}{\partial t^2} = -\frac{\rho_m^i}{\mu}.$$

Expanding the curl-curl operation in (5.45) we have

$$\nabla(\nabla \cdot \mathbf{A}_h) - \nabla^2 \mathbf{A}_h = \epsilon\mathbf{J}_m^i - \mu\epsilon\frac{\partial^2 \mathbf{A}_h}{\partial t^2} - \mu\epsilon\frac{\partial}{\partial t}\nabla\phi_h,$$

which, upon substitution of the Lorentz gauge condition gives

$$\nabla^2 \mathbf{A}_h - \mu\epsilon\frac{\partial^2 \mathbf{A}_h}{\partial t^2} = -\epsilon\mathbf{J}_m^i. \tag{5.46}$$

We can also derive a Hertzian potential for the case of magnetic current. Letting

$$\mathbf{A}_h = \mu\epsilon\frac{\partial \mathbf{\Pi}_h}{\partial t} \tag{5.47}$$

and employing the Lorentz condition we have

$$\mathbf{D} = -\mu\epsilon\nabla \times \frac{\partial \mathbf{\Pi}_h}{\partial t},$$

$$\mathbf{H} = \nabla(\nabla \cdot \mathbf{\Pi}_h) - \mu\epsilon\frac{\partial^2 \mathbf{\Pi}_h}{\partial t^2}.$$

The wave equation for $\mathbf{\Pi}_h$ is found by substituting (5.47) into (5.46) to give

$$\frac{\partial}{\partial t}\left[\nabla^2\mathbf{\Pi}_h - \mu\epsilon\frac{\partial^2 \mathbf{\Pi}_h}{\partial t^2}\right] = -\frac{1}{\mu}\mathbf{J}_m^i. \tag{5.48}$$

Defining $\mathbf{M}^i$ through

$$\mathbf{J}_m^i = \mu\frac{\partial \mathbf{M}^i}{\partial t},$$

we write the wave equation as

$$\nabla^2\mathbf{\Pi}_h - \mu\epsilon\frac{\partial^2 \mathbf{\Pi}_h}{\partial t^2} = -\mathbf{M}^i.$$

We can think of $\mathbf{M}^i$ as a convenient way of representing $\mathbf{J}_m^i$, or we can conceive of an *impressed magnetization current* that is independent of $\mathbf{H}$ and defined through $\mathbf{B} = \mu_0(\mathbf{H} + \mathbf{M} + \mathbf{M}^i)$. With the help of (5.48) we can also write the fields as

$$\mathbf{H} = \nabla \times (\nabla \times \mathbf{\Pi}_h) - \mathbf{M}_i,$$

$$\mathbf{D} = -\mu\epsilon\nabla \times \frac{\partial \mathbf{\Pi}_h}{\partial t}.$$

Summary of potential relations for lossless media. When both electric and magnetic sources are present, we may superpose the potential representations derived above. We assume a homogeneous, lossless medium with time-invariant parameters μ and ϵ. For the scalar/vector potential representation we have

$$\mathbf{E} = -\frac{\partial \mathbf{A}_e}{\partial t} - \nabla \phi_e - \frac{1}{\epsilon}\nabla \times \mathbf{A}_h, \tag{5.49}$$

$$\mathbf{H} = \frac{1}{\mu}\nabla \times \mathbf{A}_e - \frac{\partial \mathbf{A}_h}{\partial t} - \nabla \phi_h. \tag{5.50}$$

Here the potentials satisfy the wave equations

$$\left(\nabla^2 - \mu\epsilon\frac{\partial^2}{\partial t^2}\right)\left\{\begin{array}{c}\mathbf{A}_e\\ \phi_e\end{array}\right\} = \left\{\begin{array}{c}-\mu\mathbf{J}^i\\ -\frac{\rho^i}{\epsilon}\end{array}\right\}, \tag{5.51}$$

$$\left(\nabla^2 - \mu\epsilon\frac{\partial^2}{\partial t^2}\right)\left\{\begin{array}{c}\mathbf{A}_h\\ \phi_h\end{array}\right\} = \left\{\begin{array}{c}-\epsilon\mathbf{J}_m^i\\ -\frac{\rho_m^i}{\mu}\end{array}\right\},$$

and are linked by the Lorentz conditions

$$\nabla \cdot \mathbf{A}_e = -\mu\epsilon\frac{\partial \phi_e}{\partial t},$$

$$\nabla \cdot \mathbf{A}_h = -\mu\epsilon\frac{\partial \phi_h}{\partial t}.$$

We also have the Hertz potential representation

$$\mathbf{E} = \nabla(\nabla \cdot \mathbf{\Pi}_e) - \mu\epsilon\frac{\partial^2 \mathbf{\Pi}_e}{\partial t^2} - \mu\nabla \times \frac{\partial \mathbf{\Pi}_h}{\partial t}$$

$$= \nabla \times (\nabla \times \mathbf{\Pi}_e) - \frac{\mathbf{P}^i}{\epsilon} - \mu\nabla \times \frac{\partial \mathbf{\Pi}_h}{\partial t}, \tag{5.52}$$

$$\mathbf{H} = \epsilon\nabla \times \frac{\partial \mathbf{\Pi}_e}{\partial t} + \nabla(\nabla \cdot \mathbf{\Pi}_h) - \mu\epsilon\frac{\partial^2 \mathbf{\Pi}_h}{\partial t^2}$$

$$= \epsilon\nabla \times \frac{\partial \mathbf{\Pi}_e}{\partial t} + \nabla \times (\nabla \times \mathbf{\Pi}_h) - \mathbf{M}_i. \tag{5.53}$$

The Hertz potentials satisfy the wave equations

$$\left(\nabla^2 - \mu\epsilon\frac{\partial^2}{\partial t^2}\right)\left\{\begin{array}{c}\mathbf{\Pi}_e\\ \mathbf{\Pi}_h\end{array}\right\} = \left\{\begin{array}{c}-\frac{1}{\epsilon}\mathbf{P}^i\\ -\mathbf{M}^i\end{array}\right\}.$$

Potential functions for the frequency-domain fields. In the frequency domain it is much easier to handle lossy media. Consider a lossy, isotropic, homogeneous medium described by the frequency-dependent parameters $\tilde{\mu}$, $\tilde{\epsilon}$, and $\tilde{\sigma}$. Maxwell's curl equations are

$$\nabla \times \tilde{\mathbf{E}} = -\tilde{\mathbf{J}}_m^i - j\omega\tilde{\mu}\tilde{\mathbf{H}}, \tag{5.54}$$

$$\nabla \times \tilde{\mathbf{H}} = \tilde{\mathbf{J}}^i + j\omega\tilde{\epsilon}^c\tilde{\mathbf{E}}. \tag{5.55}$$

Here we have separated the primary and secondary currents through $\tilde{\mathbf{J}} = \tilde{\mathbf{J}}^i + \tilde{\sigma}\tilde{\mathbf{E}}$, and used the complex permittivity $\tilde{\epsilon}^c = \tilde{\epsilon} + \tilde{\sigma}/j\omega$. As with the time-domain equations we introduce the potential functions using superposition. If $\tilde{\mathbf{J}}_m^i = 0$ and $\tilde{\mathbf{J}}^i \neq 0$ then we may introduce the electric potentials through the relationships

$$\tilde{\mathbf{E}} = -\nabla \tilde{\phi}_e - j\omega \tilde{\mathbf{A}}_e, \tag{5.56}$$

$$\tilde{\mathbf{H}} = \frac{1}{\tilde{\mu}} \nabla \times \tilde{\mathbf{A}}_e. \tag{5.57}$$

Assuming the Lorentz condition

$$\nabla \cdot \tilde{\mathbf{A}}_e = -j\omega \tilde{\mu} \tilde{\epsilon}^c \tilde{\phi}_e,$$

we find that upon substitution of (5.56)–(5.57) into (5.54)–(5.55) the potentials must obey the Helmholtz equation

$$(\nabla^2 + k^2) \left\{ \begin{array}{c} \tilde{\phi}_e \\ \tilde{\mathbf{A}}_e \end{array} \right\} = \left\{ \begin{array}{c} -\tilde{\rho}^i / \tilde{\epsilon}^c \\ -\tilde{\mu} \tilde{\mathbf{J}}^i \end{array} \right\}.$$

If $\tilde{\mathbf{J}}_m^i \neq 0$ and $\tilde{\mathbf{J}}^i = 0$ then we may introduce the magnetic potentials through

$$\tilde{\mathbf{E}} = -\frac{1}{\tilde{\epsilon}^c} \nabla \times \tilde{\mathbf{A}}_h, \tag{5.58}$$

$$\tilde{\mathbf{H}} = -\nabla \tilde{\phi}_h - j\omega \tilde{\mathbf{A}}_h. \tag{5.59}$$

Assuming

$$\nabla \cdot \tilde{\mathbf{A}}_h = -j\omega \tilde{\mu} \tilde{\epsilon}^c \tilde{\phi}_h,$$

we find that upon substitution of (5.58)–(5.59) into (5.54)–(5.55) the potentials must obey

$$(\nabla^2 + k^2) \left\{ \begin{array}{c} \tilde{\phi}_h \\ \tilde{\mathbf{A}}_h \end{array} \right\} = \left\{ \begin{array}{c} -\tilde{\rho}_m^i / \tilde{\mu} \\ -\tilde{\epsilon}^c \tilde{\mathbf{J}}_m^i \end{array} \right\}.$$

When both electric and magnetic sources are present, we use superposition:

$$\tilde{\mathbf{E}} = -\nabla \tilde{\phi}_e - j\omega \tilde{\mathbf{A}}_e - \frac{1}{\tilde{\epsilon}^c} \nabla \times \tilde{\mathbf{A}}_h,$$

$$\tilde{\mathbf{H}} = \frac{1}{\tilde{\mu}} \nabla \times \tilde{\mathbf{A}}_e - \nabla \tilde{\phi}_h - j\omega \tilde{\mathbf{A}}_h.$$

Using the Lorentz conditions we can also write the fields in terms of the vector potentials alone:

$$\tilde{\mathbf{E}} = -\frac{j\omega}{k^2} \nabla (\nabla \cdot \tilde{\mathbf{A}}_e) - j\omega \tilde{\mathbf{A}}_e - \frac{1}{\tilde{\epsilon}^c} \nabla \times \tilde{\mathbf{A}}_h, \tag{5.60}$$

$$\tilde{\mathbf{H}} = \frac{1}{\tilde{\mu}} \nabla \times \tilde{\mathbf{A}}_e - \frac{j\omega}{k^2} \nabla (\nabla \cdot \tilde{\mathbf{A}}_h) - j\omega \tilde{\mathbf{A}}_h. \tag{5.61}$$

We can also define Hertzian potentials for the frequency-domain fields. When $\tilde{\mathbf{J}}_m^i = 0$ and $\tilde{\mathbf{J}}^i \neq 0$ we let

$$\tilde{\mathbf{A}}_e = j\omega \tilde{\mu} \tilde{\epsilon}^c \tilde{\boldsymbol{\Pi}}_e$$

and find

$$\tilde{\mathbf{E}} = \nabla (\nabla \cdot \tilde{\boldsymbol{\Pi}}_e) + k^2 \tilde{\boldsymbol{\Pi}}_e = \nabla \times (\nabla \times \tilde{\boldsymbol{\Pi}}_e) - \frac{\tilde{\mathbf{J}}^i}{j\omega \tilde{\epsilon}^c} \tag{5.62}$$

and

$$\tilde{\mathbf{H}} = j\omega\tilde{\epsilon}^c \nabla \times \tilde{\mathbf{\Pi}}_e. \tag{5.63}$$

Here $\tilde{\mathbf{J}}^i$ can represent either an impressed electric current source or an impressed polarization current source $\tilde{\mathbf{J}}^i = j\omega\tilde{\mathbf{P}}^i$. The electric Hertzian potential obeys

$$(\nabla^2 + k^2)\tilde{\mathbf{\Pi}}_e = -\frac{\tilde{\mathbf{J}}^i}{j\omega\tilde{\epsilon}^c}. \tag{5.64}$$

When $\tilde{\mathbf{J}}_m^i \neq 0$ and $\tilde{\mathbf{J}}^i = 0$ we let

$$\tilde{\mathbf{A}}_h = j\omega\tilde{\mu}\tilde{\epsilon}^c\tilde{\mathbf{\Pi}}_h$$

and find

$$\tilde{\mathbf{E}} = -j\omega\tilde{\mu}\nabla \times \tilde{\mathbf{\Pi}}_h \tag{5.65}$$

and

$$\tilde{\mathbf{H}} = \nabla(\nabla \cdot \tilde{\mathbf{\Pi}}_h) + k^2\tilde{\mathbf{\Pi}}_h = \nabla \times (\nabla \times \tilde{\mathbf{\Pi}}_h) - \frac{\tilde{\mathbf{J}}_m^i}{j\omega\tilde{\mu}}. \tag{5.66}$$

Here $\tilde{\mathbf{J}}_m^i$ can represent either an impressed magnetic current source or an impressed magnetization current source $\tilde{\mathbf{J}}_m^i = j\omega\tilde{\mu}\tilde{\mathbf{M}}^i$. The magnetic Hertzian potential obeys

$$(\nabla^2 + k^2)\tilde{\mathbf{\Pi}}_h = -\frac{\tilde{\mathbf{J}}_m^i}{j\omega\tilde{\mu}}. \tag{5.67}$$

When both electric and magnetic sources are present we have by superposition

$$\tilde{\mathbf{E}} = \nabla(\nabla \cdot \tilde{\mathbf{\Pi}}_e) + k^2\tilde{\mathbf{\Pi}}_e - j\omega\tilde{\mu}\nabla \times \tilde{\mathbf{\Pi}}_h$$

$$= \nabla \times (\nabla \times \tilde{\mathbf{\Pi}}_e) - \frac{\tilde{\mathbf{J}}^i}{j\omega\tilde{\epsilon}^c} - j\omega\tilde{\mu}\nabla \times \tilde{\mathbf{\Pi}}_h$$

and

$$\tilde{\mathbf{H}} = j\omega\tilde{\epsilon}^c\nabla \times \tilde{\mathbf{\Pi}}_e + \nabla(\nabla \cdot \tilde{\mathbf{\Pi}}_h) + k^2\tilde{\mathbf{\Pi}}_h$$

$$= j\omega\tilde{\epsilon}^c\nabla \times \tilde{\mathbf{\Pi}}_e + \nabla \times (\nabla \times \tilde{\mathbf{\Pi}}_h) - \frac{\tilde{\mathbf{J}}_m^i}{j\omega\tilde{\mu}}.$$

5.2.1   Solution for potentials in an unbounded medium: the retarded potentials

Under the Lorentz condition each of the potential functions obeys the wave equation. This equation can be solved using the method of Green's functions to determine the potentials, and the electromagnetic fields can therefore be determined. We now examine the solution for an unbounded medium. Solutions for bounded regions are considered in § 5.2.2.

Consider a linear operator $\mathcal{L}$ that operates on a function of $\mathbf{r}$ and t. If we wish to solve the equation

$$\mathcal{L}\{\psi(\mathbf{r},t)\} = S(\mathbf{r},t), \tag{5.68}$$

we first solve

$$\mathcal{L}\{G(\mathbf{r},t|\mathbf{r}',t')\} = \delta(\mathbf{r} - \mathbf{r}')\delta(t - t')$$

and determine the Green's function G for the operator $\mathcal{L}$. Provided that S resides within V we have

$$\mathcal{L}\left\{\int_V \int_{-\infty}^{\infty} S(\mathbf{r}',t')G(\mathbf{r},t|\mathbf{r}',t')\,dt'\,dV'\right\} = \int_V \int_{-\infty}^{\infty} S(\mathbf{r}',t')\,\mathcal{L}\{G(\mathbf{r},t|\mathbf{r}',t')\}\,dt'\,dV'$$

$$= \int_V \int_{-\infty}^{\infty} S(\mathbf{r}',t')\delta(\mathbf{r}-\mathbf{r}')\delta(t-t')\,dt'\,dV'$$

$$= S(\mathbf{r},t),$$

hence

$$\psi(\mathbf{r},t) = \int_V \int_{-\infty}^{\infty} S(\mathbf{r}',t')G(\mathbf{r},t|\mathbf{r}',t')\,dt'\,dV' \tag{5.69}$$

by comparison with (5.68).

We can also apply this idea in the frequency domain. The solution to

$$\mathcal{L}\{\tilde{\psi}(\mathbf{r},\omega)\} = \tilde{S}(\mathbf{r},\omega) \tag{5.70}$$

is

$$\tilde{\psi}(\mathbf{r},\omega) = \int_V \tilde{S}(\mathbf{r}',\omega)G(\mathbf{r}|\mathbf{r}';\omega)\,dV'$$

where the Green's function G satisfies

$$\mathcal{L}\{G(\mathbf{r}|\mathbf{r}';\omega)\} = \delta(\mathbf{r}-\mathbf{r}').$$

Equation (5.69) is the basic superposition integral that allows us to find the potentials in an infinite, unbounded medium. We note that if the medium is bounded then we must use Green's theorem to include the effects of sources that reside external to the boundaries. These are manifested in terms of the values of the potentials on the boundaries in the same manner as with the static potentials in Chapter 3. In order to determine whether (5.69) is the unique solution to the wave equation, we must also examine the behavior of the fields on the boundary as the boundary recedes to infinity. In the frequency domain we find that an additional "radiation condition" is required to ensure uniqueness.

The retarded potentials in the time domain. Consider an unbounded, homogeneous, lossy, isotropic medium described by parameters μ, ϵ, σ. In the time domain the vector potential $\mathbf{A}_e$ satisfies (5.30). The scalar components of $\mathbf{A}_e$ must obey

$$\nabla^2 A_{e,n}(\mathbf{r},t) - \mu\sigma\frac{\partial A_{e,n}(\mathbf{r},t)}{\partial t} - \mu\epsilon\frac{\partial^2 A_{e,n}(\mathbf{r},t)}{\partial t^2} = -\mu J_n^i(\mathbf{r},t), \qquad n=x,y,z.$$

We may write this in the form

$$\left(\nabla^2 - \frac{2\Omega}{v^2}\frac{\partial}{\partial t} - \frac{1}{v^2}\frac{\partial^2}{\partial t^2}\right)\psi(\mathbf{r},t) = -S(\mathbf{r},t) \tag{5.71}$$

where $\psi = A_{e,n}$, $v^2 = 1/\mu\epsilon$, $\Omega = \sigma/2\epsilon$, and $S = \mu J_n^i$. The solution is

$$\psi(\mathbf{r},t) = \int_V \int_{-\infty}^{\infty} S(\mathbf{r}',t')G(\mathbf{r},t|\mathbf{r}',t')\,dt'\,dV' \tag{5.72}$$

where G satisfies

$$\left(\nabla^2 - \frac{2\Omega}{v^2}\frac{\partial}{\partial t} - \frac{1}{v^2}\frac{\partial^2}{\partial t^2}\right) G(\mathbf{r},t|\mathbf{r}',t') = -\delta(\mathbf{r}-\mathbf{r}')\delta(t-t'). \tag{5.73}$$

In § A.1 we find that

$$G(\mathbf{r},t|\mathbf{r}',t') = e^{-\Omega(t-t')}\frac{\delta(t-t'-R/v)}{4\pi R} +$$

$$+ \frac{\Omega^2}{4\pi v}e^{-\Omega(t-t')}\frac{I_1\left(\Omega\sqrt{(t-t')^2-(R/v)^2}\right)}{\Omega\sqrt{(t-t')^2-(R/v)^2}}, \quad t-t' > \frac{R}{v},$$

where $R = |\mathbf{r}-\mathbf{r}'|$. For lossless media where $\sigma = 0$ this becomes

$$G(\mathbf{r},t|\mathbf{r}',t') = \frac{\delta(t-t'-R/v)}{4\pi R}$$

and thus

$$\psi(\mathbf{r},t) = \int_V \int_{-\infty}^{\infty} S(\mathbf{r}',t')\frac{\delta(t-t'-R/v)}{4\pi R}\,dt'\,dV'$$

$$= \int_V \frac{S(\mathbf{r}',t-R/v)}{4\pi R}\,dV'. \tag{5.74}$$

For lossless media, the scalar potentials and all rectangular components of the vector potentials obey the same wave equation. Thus we have, for instance, the solutions to (5.51):

$$\mathbf{A}_e(\mathbf{r},t) = \frac{\mu}{4\pi}\int_V \frac{\mathbf{J}^i(\mathbf{r}',t-R/v)}{R}\,dV',$$

$$\phi_e(\mathbf{r},t) = \frac{1}{4\pi\epsilon}\int_V \frac{\rho^i(\mathbf{r}',t-R/v)}{R}\,dV'.$$

These are called the *retarded potentials* since their values at time t are determined by the values of the sources at an earlier (or retardation) time $t - R/v$. The retardation time is determined by the propagation velocity v of the potential waves.

The fields are determined by the potentials:

$$\mathbf{E}(\mathbf{r},t) = -\nabla\frac{1}{4\pi\epsilon}\int_V \frac{\rho^i(\mathbf{r}',t-R/v)}{R}\,dV' - \frac{\partial}{\partial t}\frac{\mu}{4\pi}\int_V \frac{\mathbf{J}^i(\mathbf{r}',t-R/v)}{R}\,dV',$$

$$\mathbf{H}(\mathbf{r},t) = \nabla \times \frac{1}{4\pi}\int_V \frac{\mathbf{J}^i(\mathbf{r}',t-R/v)}{R}\,dV'.$$

The derivatives may be brought inside the integrals, but some care must be taken when the observation point $\mathbf{r}$ lies within the source region. In this case the integrals must be performed in a principal value sense by excluding a small volume around the observation point. We discuss this in more detail below for the frequency-domain fields. For details regarding this procedure in the time domain the reader may see Hansen [81].

The retarded potentials in the frequency domain. Consider an unbounded, homogeneous, isotropic medium described by $\tilde{\mu}(\omega)$ and $\tilde{\epsilon}^c(\omega)$. If $\tilde{\psi}(\mathbf{r}, \omega)$ represents a scalar potential or any rectangular component of a vector or Hertzian potential then it must satisfy

$$(\nabla^2 + k^2)\tilde{\psi}(\mathbf{r}, \omega) = -\tilde{S}(\mathbf{r}, \omega) \tag{5.75}$$

where $k = \omega(\tilde{\mu}\tilde{\epsilon}^c)^{1/2}$. This Helmholtz equation has the form of (5.70) and thus

$$\tilde{\psi}(\mathbf{r}, \omega) = \int_V \tilde{S}(\mathbf{r}', \omega) G(\mathbf{r}|\mathbf{r}'; \omega)\, dV'$$

where

$$(\nabla^2 + k^2)G(\mathbf{r}|\mathbf{r}'; \omega) = -\delta(\mathbf{r} - \mathbf{r}'). \tag{5.76}$$

This is equation (A.46) and its solution, as given by (A.49), is

$$G(\mathbf{r}|\mathbf{r}'; \omega) = \frac{e^{-jkR}}{4\pi R}. \tag{5.77}$$

Here we use $v^2 = 1/\tilde{\mu}\tilde{\epsilon}$ and $\Omega = \tilde{\sigma}/2\epsilon$ in (A.47):

$$k = \frac{1}{v}\sqrt{\omega^2 - j2\omega\Omega} = \omega\sqrt{\tilde{\mu}\left(\tilde{\epsilon} - j\frac{\tilde{\sigma}}{\omega}\right)} = \omega\sqrt{\tilde{\mu}\tilde{\epsilon}^c}.$$

The solution to (5.75) is therefore

$$\tilde{\psi}(\mathbf{r}, \omega) = \int_V \tilde{S}(\mathbf{r}', \omega)\frac{e^{-jkR}}{4\pi R}\, dV'. \tag{5.78}$$

When the medium is lossless, the potential must also satisfy the *radiation condition*

$$\lim_{r \to \infty} r\left(\frac{\partial}{\partial r} + jk\right)\tilde{\psi}(\mathbf{r}) = 0 \tag{5.79}$$

to guarantee uniqueness of solution. In § 5.2.2 we shall show how this requirement arises from the solution within a bounded region. For a uniqueness proof for the Helmholtz equation, the reader may consult Chew [33].

We may use (5.78) to find that

$$\tilde{\mathbf{A}}_e(\mathbf{r}, \omega) = \frac{\tilde{\mu}}{4\pi}\int_V \tilde{\mathbf{J}}^i(\mathbf{r}', \omega)\frac{e^{-jkR}}{R}\, dV'. \tag{5.80}$$

Comparison with (5.74) shows that in the frequency domain, time retardation takes the form of a phase shift. Similarly,

$$\tilde{\phi}(\mathbf{r}, \omega) = \frac{1}{4\pi\tilde{\epsilon}^c}\int_V \tilde{\rho}^i(\mathbf{r}', \omega)\frac{e^{-jkR}}{R}\, dV'. \tag{5.81}$$

The electric and magnetic dyadic Green's functions. The frequency-domain electromagnetic fields may be found for electric sources from the electric vector potential using (5.60) and (5.61):

$$\tilde{\mathbf{E}}(\mathbf{r}, \omega) = -j\omega\tilde{\mu}(\omega)\int_V \tilde{\mathbf{J}}^i(\mathbf{r}', \omega)G(\mathbf{r}|\mathbf{r}'; \omega)\, dV' - \frac{j\omega\tilde{\mu}(\omega)}{k^2}\nabla\nabla \cdot \int_V \tilde{\mathbf{J}}^i(\mathbf{r}', \omega)G(\mathbf{r}|\mathbf{r}'; \omega)\, dV',$$

$$\tilde{\mathbf{H}} = \nabla \times \int_V \tilde{\mathbf{J}}^i(\mathbf{r}', \omega)G(\mathbf{r}|\mathbf{r}'; \omega)\, dV'. \tag{5.82}$$

As long as the observation point $\mathbf{r}$ does not lie within the source region we may take the derivatives inside the integrals. Using

$$\nabla \cdot \left[\tilde{\mathbf{J}}^i(\mathbf{r}',\omega) G(\mathbf{r}|\mathbf{r}';\omega) \right] = \tilde{\mathbf{J}}^i(\mathbf{r}',\omega) \cdot \nabla G(\mathbf{r}|\mathbf{r}';\omega) + G(\mathbf{r}|\mathbf{r}';\omega) \nabla \cdot \tilde{\mathbf{J}}(\mathbf{r}',\omega)$$

$$= \nabla G(\mathbf{r}|\mathbf{r}';\omega) \cdot \tilde{\mathbf{J}}^i(\mathbf{r}',\omega)$$

we have

$$\tilde{\mathbf{E}}(\mathbf{r},\omega) = -j\omega\tilde{\mu}(\omega) \int_V \left\{ \tilde{\mathbf{J}}^i(\mathbf{r}',\omega) G(\mathbf{r}|\mathbf{r}';\omega) + \frac{1}{k^2}\nabla\left[\nabla G(\mathbf{r}|\mathbf{r}';\omega) \cdot \tilde{\mathbf{J}}^i(\mathbf{r}',\omega) \right] \right\} dV'.$$

This can be written more compactly as

$$\tilde{\mathbf{E}}(\mathbf{r},\omega) = -j\omega\tilde{\mu}(\omega) \int_V \bar{\mathbf{G}}_e(\mathbf{r}|\mathbf{r}';\omega) \cdot \tilde{\mathbf{J}}^i(\mathbf{r}',\omega)\, dV'$$

where

$$\bar{\mathbf{G}}_e(\mathbf{r}|\mathbf{r}';\omega) = \left[\bar{\mathbf{I}} + \frac{\nabla\nabla}{k^2} \right] G(\mathbf{r}|\mathbf{r}';\omega) \qquad (5.83)$$

is called the *electric dyadic Green's function*. Using

$$\nabla \times [\tilde{\mathbf{J}}^i G] = \nabla G \times \tilde{\mathbf{J}}^i + G\nabla \times \tilde{\mathbf{J}}^i = \nabla G \times \tilde{\mathbf{J}}^i$$

we have for the magnetic field

$$\tilde{\mathbf{H}}(\mathbf{r},\omega) = \int_V \nabla G(\mathbf{r}|\mathbf{r}';\omega) \times \tilde{\mathbf{J}}^i(\mathbf{r}',\omega)\, dV'.$$

Now, using the dyadic identity (B.15) we may show that

$$\tilde{\mathbf{J}}^i \times \nabla G = (\tilde{\mathbf{J}}^i \times \nabla G) \cdot \bar{\mathbf{I}} = (\nabla G \times \bar{\mathbf{I}}) \cdot \tilde{\mathbf{J}}^i.$$

So

$$\tilde{\mathbf{H}}(\mathbf{r},\omega) = -\int_V \bar{\mathbf{G}}_m(\mathbf{r}|\mathbf{r}';\omega) \cdot \tilde{\mathbf{J}}^i(\mathbf{r}',\omega)\, dV'$$

where

$$\bar{\mathbf{G}}_m(\mathbf{r}|\mathbf{r}';\omega) = \nabla G(\mathbf{r}|\mathbf{r}';\omega) \times \bar{\mathbf{I}} \qquad (5.84)$$

is called the *magnetic dyadic Green's function*.

Proceeding similarly for magnetic sources (or using duality) we have

$$\tilde{\mathbf{H}}(\mathbf{r}) = -j\omega\tilde{\epsilon}^c \int_V \bar{\mathbf{G}}_e(\mathbf{r}|\mathbf{r}';\omega) \cdot \tilde{\mathbf{J}}^i_m(\mathbf{r}',\omega)\, dV',$$

$$\tilde{\mathbf{E}}(\mathbf{r}) = \int_V \bar{\mathbf{G}}_m(\mathbf{r}|\mathbf{r}';\omega) \cdot \tilde{\mathbf{J}}^i_m(\mathbf{r}',\omega)\, dV'.$$

When both electric and magnetic sources are present we simply use superposition and add the fields.

When the observation point lies within the source region, we must be much more careful about how we formulate the dyadic Green's functions. In (5.82) we encounter the integral

$$\int_V \tilde{\mathbf{J}}^i(\mathbf{r}',\omega) G(\mathbf{r}|\mathbf{r}';\omega)\, dV'.$$

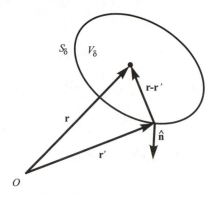

Figure 5.1: Geometry of excluded region used to compute the electric field within a source region.

If $\mathbf{r}$ lies within the source region then G is singular since $R \to 0$ when $\mathbf{r} \to \mathbf{r}'$. However, the integral converges and the potentials exist within the source region. While we run into trouble when we pass both derivatives in the operator $\nabla\nabla\cdot$ through the integral and allow them to operate on G, since differentiation of G increases the order of the singularity, we may safely take one derivative of G.

Even when we allow one derivative on G we must be careful in how we compute the integral. We exclude the point $\mathbf{r}$ by surrounding it with a small volume element V_δ as shown in Figure 5.1 and write

$$\nabla\nabla \cdot \int_V \tilde{\mathbf{J}}^i(\mathbf{r}',\omega)G(\mathbf{r}|\mathbf{r}';\omega)\,dV' =$$

$$\lim_{V_\delta \to 0}\int_{V-V_\delta} \nabla\left[\nabla G(\mathbf{r}|\mathbf{r}';\omega)\cdot\tilde{\mathbf{J}}^i(\mathbf{r}',\omega)\right]\,dV' + \lim_{V_\delta \to 0}\nabla\int_{V_\delta}\nabla G(\mathbf{r}|\mathbf{r}';\omega)\cdot\tilde{\mathbf{J}}^i(\mathbf{r}',\omega)\,dV'.$$

The first integral on the right-hand side is called the *principal value integral* and is usually abbreviated

$$\text{P.V.}\int_V \nabla\left[\nabla G(\mathbf{r}|\mathbf{r}';\omega)\cdot\tilde{\mathbf{J}}^i(\mathbf{r}',\omega)\right]\,dV'.$$

It converges to a value dependent on the shape of the excluded region V_δ, as does the second integral. However, the sum of these two integrals produces a unique result. Using $\nabla G = -\nabla'G$, the identity $\nabla'\cdot(\tilde{\mathbf{J}}G) = \tilde{\mathbf{J}}\cdot\nabla'G + G\nabla'\cdot\tilde{\mathbf{J}}$, and the divergence theorem, we can write

$$-\int_{V_\delta}\nabla'G(\mathbf{r}|\mathbf{r}';\omega)\cdot\tilde{\mathbf{J}}^i(\mathbf{r}',\omega)\,dV' =$$

$$-\oint_{S_\delta}G(\mathbf{r}|\mathbf{r}';\omega)\tilde{\mathbf{J}}^i(\mathbf{r}',\omega)\cdot\hat{\mathbf{n}}'\,dS' + \int_{V_\delta}G(\mathbf{r}|\mathbf{r}';\omega)\nabla'\cdot\tilde{\mathbf{J}}^i(\mathbf{r}',\omega)\,dV'$$

where S_δ is the surface surrounding V_δ. By the continuity equation the second integral on the right-hand side is proportional to the scalar potential produced by the charge within V_δ, and thus vanishes as $V_\delta \to 0$. The first term is proportional to the field at $\mathbf{r}$ produced by surface charge on S_δ, which results in a value proportional to $\mathbf{J}^i$. Thus

$$\lim_{V_\delta \to 0}\nabla\int_{V_\delta}\nabla G(\mathbf{r}|\mathbf{r}';\omega)\cdot\tilde{\mathbf{J}}^i(\mathbf{r}',\omega)\,dV' = -\lim_{V_\delta \to 0}\nabla\oint_{S_\delta}G(\mathbf{r}|\mathbf{r}';\omega)\tilde{\mathbf{J}}^i(\mathbf{r}',\omega)\cdot\hat{\mathbf{n}}'\,dS'$$

$$= -\bar{\mathbf{L}}\cdot\tilde{\mathbf{J}}^i(\mathbf{r},\omega), \tag{5.85}$$

so

$$\nabla\nabla \cdot \int_V \tilde{\mathbf{J}}^i(\mathbf{r}',\omega)G(\mathbf{r}|\mathbf{r}';\omega)\,dV' = \text{P.V.} \int_V \nabla\left[\nabla G(\mathbf{r}|\mathbf{r}';\omega)\cdot\tilde{\mathbf{J}}^i(\mathbf{r}',\omega)\right]dV' - \bar{\mathbf{L}}\cdot\tilde{\mathbf{J}}^i(\mathbf{r},\omega).$$

Here $\bar{\mathbf{L}}$ is usually called the *depolarizing dyadic* [113]. Its value depends on the shape of V_δ, as considered below.

We may now write

$$\tilde{\mathbf{E}}(\mathbf{r},\omega) = -j\omega\tilde{\mu}(\omega)\,\text{P.V.}\int_V \bar{\mathbf{G}}_e(\mathbf{r}|\mathbf{r}';\omega)\cdot\tilde{\mathbf{J}}(\mathbf{r}',\omega)\,dV' - \frac{1}{j\omega\tilde{\epsilon}^c(\omega)}\bar{\mathbf{L}}\cdot\tilde{\mathbf{J}}^i(\mathbf{r},\omega). \qquad (5.86)$$

We may also incorporate both terms into a single dyadic Green's function using the notation

$$\bar{\mathbf{G}}(\mathbf{r}|\mathbf{r}';\omega) = \text{P.V.}\,\bar{\mathbf{G}}_e(\mathbf{r}|\mathbf{r}';\omega) - \frac{1}{k^2}\bar{\mathbf{L}}\delta(\mathbf{r}-\mathbf{r}').$$

Hence when we compute

$$\begin{aligned}
\tilde{\mathbf{E}}(\mathbf{r},\omega) &= -j\omega\tilde{\mu}(\omega)\int_V \bar{\mathbf{G}}(\mathbf{r}|\mathbf{r}';\omega)\cdot\tilde{\mathbf{J}}^i(\mathbf{r}',\omega)\,dV'\\
&= -j\omega\tilde{\mu}(\omega)\int_V\left[\text{P.V.}\,\bar{\mathbf{G}}_e(\mathbf{r}|\mathbf{r}';\omega) - \frac{1}{k^2}\bar{\mathbf{L}}\delta(\mathbf{r}-\mathbf{r}')\right]\cdot\tilde{\mathbf{J}}^i(\mathbf{r}',\omega)\,dV'
\end{aligned}$$

we reproduce (5.86). That is, the symbol P.V. on G_e indicates that a principal value integral must be performed.

Our final task is to compute $\bar{\mathbf{L}}$ from (5.85). When we remove the excluded region from the principal value computation we leave behind a hole in the source region. The contribution to the field at $\mathbf{r}$ by the sources in the excluded region is found from the scalar potential produced by the surface distribution $\hat{\mathbf{n}}\cdot\mathbf{J}^i$. The value of this *correction term* depends on the shape of the excluding volume. However, the correction term always adds to the principal value integral to give the true field at $\mathbf{r}$, regardless of the shape of the volume. So we must always match the shape of the excluded region used to compute the principal value integral with that used to compute the correction term so that the true field is obtained. Note that as $V_\delta \to 0$ the phase factor in the Green's function becomes insignificant, and the values of the current on the surface approach the value at $\mathbf{r}$ (assuming $\mathbf{J}^i$ is continuous at $\mathbf{r}$). Thus we may write

$$\lim_{V_\delta\to 0}\nabla\oint_{S_\delta}\frac{\tilde{\mathbf{J}}^i(\mathbf{r},\omega)\cdot\hat{\mathbf{n}}'}{4\pi|\mathbf{r}-\mathbf{r}'|}\,dS' = \bar{\mathbf{L}}\cdot\tilde{\mathbf{J}}^i(\mathbf{r},\omega).$$

This has the form of a static field integral. For a spherical excluded region we may compute the above quantity quite simply by assuming the current to be uniform throughout V_δ and by aligning the current with the z-axis and placing the center of the sphere at the origin. We then compute the integral at a point $\mathbf{r}$ within the sphere, take the gradient, and allow $\mathbf{r}\to 0$. We thus have for a sphere

$$\lim_{V_\delta\to 0}\nabla\oint_S\frac{\tilde{J}^i\cos\theta'}{4\pi|\mathbf{r}-\mathbf{r}'|}\,dS' = \bar{\mathbf{L}}\cdot[\hat{\mathbf{z}}\tilde{J}^i(\mathbf{r},\omega)].$$

This integral has been computed in § 3.2.7 with the result given by (3.103). Using this we find

$$\lim_{V_\delta\to 0}\left[\nabla\left(\frac{1}{3}\tilde{J}^i z\right)\right]\bigg|_{\mathbf{r}=0} = \hat{\mathbf{z}}\frac{\tilde{J}^i}{3} = \bar{\mathbf{L}}\cdot[\hat{\mathbf{z}}\tilde{J}^i(\mathbf{r},\omega)]$$

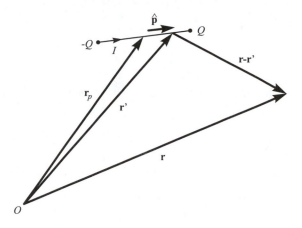

Figure 5.2: Geometry of an electric Hertzian dipole.

and thus

$$\bar{\mathbf{L}} = \frac{1}{3}\bar{\mathbf{I}}.$$

We leave it as an exercise to show that for a cubical excluding volume the depolarizing dyadic is also $\bar{\mathbf{L}} = \bar{\mathbf{I}}/3$. Values for other shapes may be found in Yaghjian [215].

The theory of dyadic Green's functions is well developed and there exist techniques for their construction under a variety of conditions. For an excellent overview the reader may see Tai [192].

Example of field calculation using potentials: the Hertzian dipole. Consider a short line current of length $l \ll \lambda$ at position $\mathbf{r}_p$, oriented along a direction $\hat{\mathbf{p}}$ in a medium with constitutive parameters $\tilde{\mu}(\omega)$, $\tilde{\epsilon}^c(\omega)$, as shown in Figure 5.2. We assume that the frequency-domain current $\tilde{I}(\omega)$ is independent of position, and therefore this *Hertzian dipole* must be terminated by point charges

$$\tilde{Q}(\omega) = \pm\frac{\tilde{I}(\omega)}{j\omega}$$

as required by the continuity equation. The electric vector potential produced by this short current element is

$$\tilde{\mathbf{A}}_e = \frac{\tilde{\mu}}{4\pi} \int_\Gamma \tilde{I}\hat{\mathbf{p}} \frac{e^{-jkR}}{R} \, dl'.$$

At observation points far from the dipole (compared to its length) such that $|\mathbf{r} - \mathbf{r}_p| \gg l$ we may approximate

$$\frac{e^{-jkR}}{R} \approx \frac{e^{-jk|\mathbf{r}-\mathbf{r}_p|}}{|\mathbf{r} - \mathbf{r}_p|}.$$

Then

$$\tilde{\mathbf{A}}_e = \hat{\mathbf{p}}\tilde{\mu}\tilde{I}G(\mathbf{r}|\mathbf{r}_p;\omega) \int_\Gamma dl' = \hat{\mathbf{p}}\tilde{\mu}\tilde{I}lG(\mathbf{r}|\mathbf{r}_p;\omega). \qquad (5.87)$$

Note that we obtain the same answer if we let the current density of the dipole be

$$\tilde{\mathbf{J}} = j\omega\tilde{\mathbf{p}}\delta(\mathbf{r} - \mathbf{r}_p)$$

where $\tilde{\mathbf{p}}$ is the *dipole moment* defined by

$$\tilde{\mathbf{p}} = \tilde{Q}l\hat{\mathbf{p}} = \frac{\tilde{I}l}{j\omega}\hat{\mathbf{p}}.$$

That is, we consider a Hertzian dipole to be a "point source" of electromagnetic radiation. With this notation we have

$$\tilde{\mathbf{A}}_e = \tilde{\mu}\int_V [j\omega\tilde{\mathbf{p}}\delta(\mathbf{r}' - \mathbf{r}_p)]\,G(\mathbf{r}|\mathbf{r}';\omega)\,dV' = j\omega\tilde{\mu}\tilde{\mathbf{p}}G(\mathbf{r}|\mathbf{r}_p;\omega),$$

which is identical to (5.87). The electromagnetic fields are then

$$\tilde{\mathbf{H}}(\mathbf{r},\omega) = j\omega\nabla \times [\tilde{\mathbf{p}}G(\mathbf{r}|\mathbf{r}_p;\omega)], \tag{5.88}$$

$$\tilde{\mathbf{E}}(\mathbf{r},\omega) = \frac{1}{\tilde{\epsilon}^c}\nabla \times \nabla \times [\tilde{\mathbf{p}}G(\mathbf{r}|\mathbf{r}_p;\omega)]. \tag{5.89}$$

Here we have obtained $\tilde{\mathbf{E}}$ from $\tilde{\mathbf{H}}$ outside the source region by applying Ampere's law. By duality we may obtain the fields produced by a magnetic Hertzian dipole of moment

$$\tilde{\mathbf{p}}_m = \frac{\tilde{I}_m l}{j\omega}\hat{\mathbf{p}}$$

located at $\mathbf{r} = \mathbf{r}_p$ as

$$\tilde{\mathbf{E}}(\mathbf{r},\omega) = -j\omega\nabla \times [\tilde{\mathbf{p}}_m G(\mathbf{r}|\mathbf{r}_p;\omega)],$$

$$\tilde{\mathbf{H}}(\mathbf{r},\omega) = \frac{1}{\tilde{\mu}}\nabla \times \nabla \times [\tilde{\mathbf{p}}_m G(\mathbf{r}|\mathbf{r}_p;\omega)].$$

We can learn much about the fields produced by localized sources by considering the simple case of a Hertzian dipole aligned along the z-axis and centered at the origin. Using $\hat{\mathbf{p}} = \hat{\mathbf{z}}$ and $\mathbf{r}_p = 0$ in (5.88) we find that

$$\tilde{\mathbf{H}}(\mathbf{r},\omega) = j\omega\nabla \times \left[\hat{\mathbf{z}}\frac{\tilde{I}}{j\omega}l\frac{e^{-jkr}}{4\pi r}\right] = \hat{\boldsymbol{\phi}}\frac{1}{4\pi}\tilde{I}l\left[\frac{1}{r^2} + j\frac{k}{r}\right]\sin\theta e^{-jkr}. \tag{5.90}$$

By Ampere's law

$$\tilde{\mathbf{E}}(\mathbf{r},\omega) = \frac{1}{j\omega\tilde{\epsilon}^c}\nabla \times \tilde{\mathbf{H}}(\mathbf{r},\omega)$$
$$= \hat{\mathbf{r}}\frac{\eta}{4\pi}\tilde{I}l\left[\frac{2}{r^2} - j\frac{2}{kr^3}\right]\cos\theta e^{-jkr} + \hat{\boldsymbol{\theta}}\frac{\eta}{4\pi}\tilde{I}l\left[j\frac{k}{r} + \frac{1}{r^2} - j\frac{1}{kr^3}\right]\sin\theta e^{-jkr}. $$
$$\tag{5.91}$$

The fields involve various inverse powers of r, with the $1/r$ and $1/r^3$ terms 90° out-of-phase from the $1/r^2$ term. Some terms dominate the field close to the source, while others dominate far away. The terms that dominate near the source[1] are called the *near-zone* or *induction-zone fields*:

$$\tilde{\mathbf{H}}^{NZ}(\mathbf{r},\omega) = \hat{\boldsymbol{\phi}}\frac{\tilde{I}l}{4\pi}\frac{e^{-jkr}}{r^2}\sin\theta,$$

$$\tilde{\mathbf{E}}^{NZ}(\mathbf{r},\omega) = -j\eta\frac{\tilde{I}l}{4\pi}\frac{e^{-jkr}}{kr^3}\left[2\hat{\mathbf{r}}\cos\theta + \hat{\boldsymbol{\theta}}\sin\theta\right].$$

[1]Note that we still require $r \gg l$.

We note that $\tilde{\mathbf{H}}^{NZ}$ and $\tilde{\mathbf{E}}^{NZ}$ are 90° out-of-phase. Also, the electric field has the same spatial dependence as the field of a static electric dipole. The terms that dominate far from the source are called the *far-zone* or *radiation fields*:

$$\tilde{\mathbf{H}}^{FZ}(\mathbf{r}, \omega) = \hat{\boldsymbol{\phi}}\frac{jk\tilde{I}l}{4\pi}\frac{e^{-jkr}}{r}\sin\theta, \tag{5.92}$$

$$\tilde{\mathbf{E}}^{FZ}(\mathbf{r}, \omega) = \hat{\boldsymbol{\theta}}\eta\frac{jk\tilde{I}l}{4\pi}\frac{e^{-jkr}}{r}\sin\theta. \tag{5.93}$$

The far-zone fields are in-phase and in fact form a TEM spherical wave with

$$\tilde{\mathbf{H}}^{FZ} = \frac{\hat{\mathbf{r}} \times \tilde{\mathbf{E}}^{FZ}}{\eta}. \tag{5.94}$$

We speak of the time-average power *radiated* by a time-harmonic source as the integral of the time-average power density over a very large sphere. Thus *radiated power* is the power delivered by the sources to infinity. If the dipole is situated within a lossy medium, all of the time-average power delivered by the sources is dissipated by the medium. If the medium is lossless then all the time-average power is delivered to infinity. Let us compute the power radiated by a time-harmonic Hertzian dipole immersed in a lossless medium. Writing (5.90) and (5.91) in terms of phasors we have the complex Poynting vector

$$\mathbf{S}^c(\mathbf{r}) = \check{\mathbf{E}}(\mathbf{r}) \times \check{\mathbf{H}}^*(\mathbf{r})$$
$$= \hat{\boldsymbol{\theta}}\eta\left(\frac{|\check{I}|l}{4\pi}\right)^2 j\frac{2}{kr^5}\left[k^2r^2 + 1\right]\cos\theta\sin\theta + \hat{\mathbf{r}}\eta\left(\frac{|\check{I}|l}{4\pi}\right)^2\frac{k^2}{r^2}\left[1 - j\frac{1}{k^3r^5}\right]\sin^2\theta.$$

We notice that the θ-component of $\mathbf{S}^c$ is purely imaginary and gives rise to no time-average power flux. This component falls off as $1/r^3$ for large r and produces no net flux through a sphere with radius $r \to \infty$. Additionally, the angular variation $\sin\theta\cos\theta$ integrates to zero over a sphere. In contrast, the r-component has a real part that varies as $1/r^2$ and as $\sin^2\theta$. Hence we find that the total time-average power passing through a sphere expanding to infinity is nonzero:

$$P_{av} = \lim_{r\to\infty}\int_0^{2\pi}\int_0^{\pi}\frac{1}{2}\mathrm{Re}\left\{\hat{\mathbf{r}}\eta\left(\frac{|\check{I}|l}{4\pi}\right)^2\frac{k^2}{r^2}\sin^2\theta\right\} \cdot \hat{\mathbf{r}}r^2\sin\theta\, d\theta\, d\phi$$
$$= \eta\frac{\pi}{3}|\check{I}|^2\left(\frac{l}{\lambda}\right)^2 \tag{5.95}$$

where $\lambda = 2\pi/k$ is the wavelength in the lossless medium. This is the power radiated by the Hertzian dipole. The power is proportional to $|\check{I}|^2$ as it is in a circuit, and thus we may define a *radiation resistance*

$$R_r = \frac{2P_{av}}{|\check{I}|^2} = \eta\frac{2\pi}{3}\left(\frac{l}{\lambda}\right)^2$$

that represents the resistance of a lumped element that would absorb the same power as radiated by the Hertzian dipole when presented with the same current. We also note that the power radiated by a Hertzian dipole (and, in fact, by any source of finite extent) may

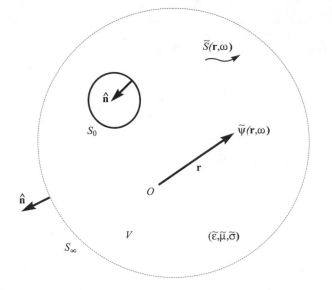

Figure 5.3: Geometry for solution to the frequency-domain Helmholtz equation.

be calculated directly from its far-zone fields. In fact, from (5.94) we have the simple formula for the time-average power density in lossless media

$$\mathbf{S}_{av} = \frac{1}{2}\operatorname{Re}\left\{\check{\mathbf{E}}^{FZ} \times \check{\mathbf{H}}^{FZ*}\right\} = \hat{\mathbf{r}}\frac{1}{2}\frac{|\check{\mathbf{E}}^{FZ}|^2}{\eta}.$$

The dipole field is the first term in a general expansion of the electromagnetic fields in terms of the multipole moments of the sources. Either a Taylor expansion or a spherical-harmonic expansion may be used. The reader may see Papas [141] for details.

5.2.2 Solution for potential functions in a bounded medium

In the previous section we solved for the frequency-domain potential functions in an unbounded region of space. Here we shall extend the solution to a bounded region and identify the physical meaning of the radiation condition (5.79).

Consider a bounded region of space V containing a linear, homogeneous, isotropic medium characterized by $\tilde{\mu}(\omega)$ and $\tilde{\epsilon}^c(\omega)$. As shown in Figure 5.3 we decompose the multiply-connected boundary into a closed "excluding surface" S_0 and a closed "encompassing surface" S_∞ that we shall allow to expand outward to infinity. S_0 may consist of more than one closed surface and is often used to exclude unknown sources from V. We wish to solve the Helmholtz equation (5.75) for $\tilde{\psi}$ within V in terms of the sources within V and the values of $\tilde{\psi}$ on S_0. The actual sources of $\tilde{\psi}$ lie entirely with S_∞ but may lie partly, or entirely, within S_0.

We solve the Helmholtz equation in much the same way that we solved Poisson's equation in § 3.2.4. We begin with Green's second identity, written in terms of the source point (primed) variables and applied to the region V:

$$\int_V [\psi(\mathbf{r}',\omega)\nabla'^2 G(\mathbf{r}|\mathbf{r}';\omega) - G(\mathbf{r}|\mathbf{r}';\omega)\nabla'^2\psi(\mathbf{r}',\omega)]\,dV' =$$

$$\oint_{S_0+S_\infty} \left[\psi(\mathbf{r}',\omega) \frac{\partial G(\mathbf{r}|\mathbf{r}';\omega)}{\partial n'} - G(\mathbf{r}|\mathbf{r}';\omega) \frac{\partial \psi(\mathbf{r}',\omega)}{\partial n'} \right] dS'.$$

We note that $\hat{\mathbf{n}}$ points outward from V, and G is the Green's function (5.77). By inspection, this Green's function obeys the reciprocity condition

$$G(\mathbf{r}|\mathbf{r}';\omega) = G(\mathbf{r}'|\mathbf{r};\omega)$$

and satisfies

$$\nabla^2 G(\mathbf{r}|\mathbf{r}';\omega) = \nabla'^2 G(\mathbf{r}|\mathbf{r}';\omega).$$

Substituting $\nabla'^2\tilde{\psi} = -k^2\tilde{\psi} - \tilde{S}$ from (5.75) and $\nabla'^2 G = -k^2 G - \delta(\mathbf{r} - \mathbf{r}')$ from (5.76) we get

$$\tilde{\psi}(\mathbf{r},\omega) = \int_V \tilde{S}(\mathbf{r}',\omega) G(\mathbf{r}|\mathbf{r}';\omega)\, dV' -$$

$$- \oint_{S_0+S_\infty} \left[\tilde{\psi}(\mathbf{r}',\omega) \frac{\partial G(\mathbf{r}|\mathbf{r}';\omega)}{\partial n'} - G(\mathbf{r}|\mathbf{r}';\omega) \frac{\partial \tilde{\psi}(\mathbf{r}',\omega)}{\partial n'} \right] dS'.$$

Hence $\tilde{\psi}$ within V may be written in terms of the sources within V and the values of $\tilde{\psi}$ and its normal derivative over $S_0 + S_\infty$. The surface contributions account for sources excluded by S_0.

Let us examine the integral over S_∞ more closely. If we let S_∞ recede to infinity, we expect no contribution to the potential at $\mathbf{r}$ from the fields on S_∞. Choosing a sphere centered at the origin, we note that $\hat{\mathbf{n}}' = \hat{\mathbf{r}}'$ and that as $r' \to \infty$

$$G(\mathbf{r}|\mathbf{r}';\omega) = \frac{e^{-jk|\mathbf{r}-\mathbf{r}'|}}{4\pi|\mathbf{r}-\mathbf{r}'|} \approx \frac{e^{-jkr'}}{4\pi r'},$$

$$\frac{\partial G(\mathbf{r}|\mathbf{r}';\omega)}{\partial n'} = \hat{\mathbf{n}}' \cdot \nabla' G(\mathbf{r}|\mathbf{r}';\omega) \approx \frac{\partial}{\partial r'} \frac{e^{-jkr'}}{4\pi r'} = -(1+jkr')\frac{e^{-jkr'}}{4\pi r'}.$$

Substituting these, we find that as $r' \to \infty$

$$\oint_{S_\infty} \left[\tilde{\psi}\frac{\partial G}{\partial n'} - G\frac{\partial\tilde{\psi}}{\partial n'} \right] dS' \approx \int_0^{2\pi} \int_0^\pi \left[-\frac{1+jkr'}{r'^2}\tilde{\psi} - \frac{1}{r'}\frac{\partial\tilde{\psi}}{\partial r'} \right] \frac{e^{-jkr'}}{4\pi} r'^2 \sin\theta'\, d\theta'\, d\phi'$$

$$\approx -\int_0^{2\pi} \int_0^\pi \left[\tilde{\psi} + r'\left(jk\tilde{\psi} + \frac{\partial\tilde{\psi}}{\partial r'} \right) \right] \frac{e^{-jkr'}}{4\pi} \sin\theta'\, d\theta'\, d\phi'.$$

Since this gives the contribution to the field in V from the fields on the surface receding to infinity, we expect that this term should be zero. If the medium has loss, then the exponential term decays and drives the contribution to zero. For a lossless medium the contribution is zero if

$$\lim_{r\to\infty} \tilde{\psi}(\mathbf{r},\omega) = 0, \tag{5.96}$$

$$\lim_{r\to\infty} r \left[jk\tilde{\psi}(\mathbf{r},\omega) + \frac{\partial\tilde{\psi}(\mathbf{r},\omega)}{\partial r} \right] = 0. \tag{5.97}$$

This is called the *radiation condition* for the Helmholtz equation. It is also called the *Sommerfeld radiation condition* after the German physicist A. Sommerfeld. Note that we have not derived this condition: we have merely postulated it. As with all postulates it is subject to experimental verification.

The radiation condition implies that for points far from the source the potentials behave as spherical waves:

$$\tilde{\psi}(\mathbf{r}, \omega) \sim \frac{e^{-jkr}}{r}, \qquad r \to \infty.$$

Substituting this into (5.96) and (5.97) we find that the radiation condition is satisfied.

With $S_\infty \to \infty$ we have

$$\tilde{\psi}(\mathbf{r}, \omega) = \int_V \tilde{S}(\mathbf{r}', \omega) G(\mathbf{r}|\mathbf{r}'; \omega) \, dV' -$$

$$- \oint_{S_0} \left[\tilde{\psi}(\mathbf{r}', \omega) \frac{\partial G(\mathbf{r}|\mathbf{r}'; \omega)}{\partial n'} - G(\mathbf{r}|\mathbf{r}'; \omega) \frac{\partial \tilde{\psi}(\mathbf{r}', \omega)}{\partial n'} \right] dS',$$

which is the expression for the potential within an infinite medium having source-excluding regions. As $S_0 \to 0$ we obtain the expression for the potential in an unbounded medium:

$$\tilde{\psi}(\mathbf{r}, \omega) = \int_V \tilde{S}(\mathbf{r}', \omega) G(\mathbf{r}|\mathbf{r}'; \omega) \, dV',$$

as expected.

The time-domain equation (5.71) may also be solved (at least for the lossless case) in a bounded region of space. The interested reader should see Pauli [143] for details.

5.3   Transverse–longitudinal decomposition

We have seen that when only electric sources are present, the electromagnetic fields in a homogeneous, isotropic region can be represented by a single vector potential $\mathbf{\Pi}_e$. Similarly, when only magnetic sources are present, the fields can be represented by a single vector potential $\mathbf{\Pi}_h$. Hence two vector potentials may be used to represent the field if both electric and magnetic sources are present.

We may also represent the electromagnetic field in a homogeneous, isotropic region using two scalar functions and the sources. This follows naturally from another important field decomposition: a splitting of each field vector into (1) a component along a certain pre-chosen constant direction, and (2) a component transverse to this direction. Depending on the geometry of the sources, it is possible that only one of these components will be present. A special case of this decomposition, the *TE–TM field decomposition*, holds for a source-free region and will be discussed in the next section.

5.3.1   Transverse–longitudinal decomposition in terms of fields

Consider a direction defined by a constant unit vector $\hat{\mathbf{u}}$. We define the *longitudinal component* of $\mathbf{A}$ as $\hat{\mathbf{u}} A_u$ where

$$A_u = \hat{\mathbf{u}} \cdot \mathbf{A},$$

and the *transverse component* of $\mathbf{A}$ as

$$\mathbf{A}_t = \mathbf{A} - \hat{\mathbf{u}} A_u.$$

We may thus decompose any vector into a sum of longitudinal and transverse parts. An important consequence of Maxwell's equations is that the transverse fields may be written entirely in terms of the longitudinal fields and the sources. This holds in both the time and frequency domains; we derive the decomposition in the frequency domain and leave the derivation of the time-domain expressions as exercises. We begin by decomposing the operators in Maxwell's equations into longitudinal and transverse components. We note that

$$\frac{\partial}{\partial u} \equiv \hat{\mathbf{u}} \cdot \nabla$$

and define a *transverse del operator* as

$$\nabla_t \equiv \nabla - \hat{\mathbf{u}}\frac{\partial}{\partial u}.$$

Using these basic definitions, the identities listed in Appendix B may be derived. We shall find it helpful to express the vector curl and Laplacian operations in terms of their longitudinal and transverse components. Using (B.93) and (B.96) we find that the transverse component of the curl is given by

$$
\begin{aligned}
(\nabla \times \mathbf{A})_t &= -\hat{\mathbf{u}} \times \hat{\mathbf{u}} \times (\nabla \times \mathbf{A}) \\
&= -\hat{\mathbf{u}} \times \hat{\mathbf{u}} \times (\nabla_t \times \mathbf{A}_t) - \hat{\mathbf{u}} \times \hat{\mathbf{u}} \times \left(\hat{\mathbf{u}} \times \left[\frac{\partial \mathbf{A}_t}{\partial u} - \nabla_t A_u\right]\right).
\end{aligned}
\tag{5.98}
$$

The first term in the right member is zero by property (B.91). Using (B.7) we can replace the second term by

$$-\hat{\mathbf{u}}\left\{\hat{\mathbf{u}} \cdot \left(\hat{\mathbf{u}} \times \left[\frac{\partial \mathbf{A}_t}{\partial u} - \nabla_t A_u\right]\right)\right\} + (\hat{\mathbf{u}} \cdot \hat{\mathbf{u}})\left(\hat{\mathbf{u}} \times \left[\frac{\partial \mathbf{A}_t}{\partial u} - \nabla_t A_u\right]\right).$$

The first of these terms is zero since

$$\hat{\mathbf{u}} \cdot \left(\hat{\mathbf{u}} \times \left[\frac{\partial \mathbf{A}_t}{\partial u} - \nabla_t A_u\right]\right) = \left[\frac{\partial \mathbf{A}_t}{\partial u} - \nabla_t A_u\right] \cdot (\hat{\mathbf{u}} \times \hat{\mathbf{u}}) = 0,$$

hence

$$(\nabla \times \mathbf{A})_t = \hat{\mathbf{u}} \times \left[\frac{\partial \mathbf{A}_t}{\partial u} - \nabla_t A_u\right].
\tag{5.99}$$

The longitudinal part is then, by property (B.80), merely the difference between the curl and its transverse part, or

$$\hat{\mathbf{u}}\left(\hat{\mathbf{u}} \cdot \nabla \times \mathbf{A}\right) = \nabla_t \times \mathbf{A}_t.
\tag{5.100}$$

A similar set of steps gives the transverse component of the Laplacian as

$$(\nabla^2 \mathbf{A})_t = \left[\nabla_t(\nabla_t \cdot \mathbf{A}_t) + \frac{\partial^2 \mathbf{A}_t}{\partial u^2} - \nabla_t \times \nabla_t \times \mathbf{A}_t\right],
\tag{5.101}$$

and the longitudinal part as

$$\hat{\mathbf{u}}\left(\hat{\mathbf{u}} \cdot \nabla^2 \mathbf{A}\right) = \hat{\mathbf{u}}\nabla^2 A_u.
\tag{5.102}$$

Verification is left as an exercise.

Now we are ready to give a longitudinal–transverse decomposition of the fields in a lossy, homogeneous, isotropic region in terms of the direction $\hat{\mathbf{u}}$. We write Maxwell's equations as

$$\nabla \times \tilde{\mathbf{E}} = -j\omega\tilde{\mu}\tilde{\mathbf{H}}_t - j\omega\tilde{\mu}\hat{\mathbf{u}}\tilde{H}_u - \tilde{\mathbf{J}}_{mt}^i - \hat{\mathbf{u}}\tilde{J}_{mu}^i, \tag{5.103}$$

$$\nabla \times \tilde{\mathbf{H}} = j\omega\tilde{\epsilon}^c\tilde{\mathbf{E}}_t + j\omega\tilde{\epsilon}^c\hat{\mathbf{u}}\tilde{E}_u + \tilde{\mathbf{J}}_t^i + \hat{\mathbf{u}}\tilde{J}_u^i, \tag{5.104}$$

where we have split the right-hand sides into longitudinal and transverse parts. Then, using (5.99) and (5.100), we can equate the transverse and longitudinal parts of each equation to obtain

$$\nabla_t \times \tilde{\mathbf{E}}_t = -j\omega\tilde{\mu}\hat{\mathbf{u}}\tilde{H}_u - \hat{\mathbf{u}}\tilde{J}_{mu}^i, \tag{5.105}$$

$$-\hat{\mathbf{u}} \times \nabla_t \tilde{E}_u + \hat{\mathbf{u}} \times \frac{\partial \tilde{\mathbf{E}}_t}{\partial u} = -j\omega\tilde{\mu}\tilde{\mathbf{H}}_t - \tilde{\mathbf{J}}_{mt}^i, \tag{5.106}$$

$$\nabla_t \times \tilde{\mathbf{H}}_t = j\omega\tilde{\epsilon}^c\hat{\mathbf{u}}\tilde{E}_u + \hat{\mathbf{u}}\tilde{J}_u^i, \tag{5.107}$$

$$-\hat{\mathbf{u}} \times \nabla_t \tilde{H}_u + \hat{\mathbf{u}} \times \frac{\partial \tilde{\mathbf{H}}_t}{\partial u} = j\omega\tilde{\epsilon}^c\tilde{\mathbf{E}}_t + \tilde{\mathbf{J}}_t^i. \tag{5.108}$$

We shall isolate the transverse fields in terms of the longitudinal fields. Forming the cross product of $\hat{\mathbf{u}}$ and the partial derivative of (5.108) with respect to u, we have

$$-\hat{\mathbf{u}} \times \hat{\mathbf{u}} \times \nabla_t\frac{\partial \tilde{H}_u}{\partial u} + \hat{\mathbf{u}} \times \hat{\mathbf{u}} \times \frac{\partial^2 \tilde{\mathbf{H}}_t}{\partial u^2} = j\omega\tilde{\epsilon}^c\hat{\mathbf{u}} \times \frac{\partial \tilde{\mathbf{E}}_t}{\partial u} + \hat{\mathbf{u}} \times \frac{\partial \tilde{\mathbf{J}}_t^i}{\partial u}.$$

Using (B.7) and (B.80) we find that

$$\nabla_t\frac{\partial \tilde{H}_u}{\partial u} - \frac{\partial^2 \tilde{\mathbf{H}}_t}{\partial u^2} = j\omega\tilde{\epsilon}^c\hat{\mathbf{u}} \times \frac{\partial \mathbf{E}_t}{\partial u} + \hat{\mathbf{u}} \times \frac{\partial \tilde{\mathbf{J}}_t^i}{\partial u}. \tag{5.109}$$

Multiplying (5.106) by $j\omega\tilde{\epsilon}^c$ we have

$$-j\omega\tilde{\epsilon}^c\hat{\mathbf{u}} \times \nabla_t \tilde{E}_u + j\omega\tilde{\epsilon}^c\hat{\mathbf{u}} \times \frac{\partial \tilde{\mathbf{E}}_t}{\partial u} = \omega^2\tilde{\mu}\tilde{\epsilon}^c\tilde{\mathbf{H}}_t - j\omega\tilde{\epsilon}^c\tilde{\mathbf{J}}_{mt}^i. \tag{5.110}$$

We now add (5.109) to (5.110) and eliminate $\tilde{\mathbf{E}}_t$ to get

$$\left(\frac{\partial^2}{\partial u^2} + k^2\right)\tilde{\mathbf{H}}_t = \nabla_t\frac{\partial \tilde{H}_u}{\partial u} - j\omega\tilde{\epsilon}^c\hat{\mathbf{u}} \times \nabla_t \tilde{E}_u + j\omega\tilde{\epsilon}^c\tilde{\mathbf{J}}_{mt}^i - \hat{\mathbf{u}} \times \frac{\partial \tilde{\mathbf{J}}_t^i}{\partial u}. \tag{5.111}$$

This one-dimensional Helmholtz equation can be solved to find the transverse magnetic field from the longitudinal components of $\tilde{\mathbf{E}}$ and $\tilde{\mathbf{H}}$. Similar steps lead to a formula for the transverse component of $\tilde{\mathbf{E}}$:

$$\left(\frac{\partial^2}{\partial u^2} + k^2\right)\tilde{\mathbf{E}}_t = \nabla_t\frac{\partial \tilde{E}_u}{\partial u} + j\omega\tilde{\mu}\hat{\mathbf{u}} \times \nabla_t \tilde{H}_u + \hat{\mathbf{u}} \times \frac{\partial \tilde{\mathbf{J}}_{mt}^i}{\partial u} + j\omega\tilde{\mu}\tilde{\mathbf{J}}_t^i. \tag{5.112}$$

We find the longitudinal components from the wave equation for $\tilde{\mathbf{E}}$ and $\tilde{\mathbf{H}}$. Recall that the fields satisfy

$$(\nabla^2 + k^2)\tilde{\mathbf{E}} = \frac{1}{\tilde{\epsilon}^c}\nabla\tilde{\rho}^i + j\omega\tilde{\mu}\tilde{\mathbf{J}}^i + \nabla \times \tilde{\mathbf{J}}_m^i,$$

$$(\nabla^2 + k^2)\tilde{\mathbf{H}} = \frac{1}{\tilde{\mu}}\nabla\tilde{\rho}_m^i + j\omega\tilde{\epsilon}^c\tilde{\mathbf{J}}_m^i - \nabla \times \tilde{\mathbf{J}}^i.$$

Splitting the vectors into longitudinal and transverse parts, and using (5.100) and (5.102), we equate the longitudinal components of the wave equations to obtain

$$\left(\nabla^2 + k^2\right)\tilde{E}_u = \frac{1}{\tilde{\epsilon}^c}\frac{\partial \tilde{\rho}^i}{\partial u} + j\omega\tilde{\mu}\tilde{J}_u^i + \nabla_t \times \tilde{\mathbf{J}}_{mt}^i, \tag{5.113}$$

$$\left(\nabla^2 + k^2\right)\tilde{H}_u = \frac{1}{\tilde{\mu}}\frac{\partial \tilde{\rho}_m^i}{\partial u} + j\omega\tilde{\epsilon}^c \tilde{J}_{mu}^i - \nabla_t \times \tilde{\mathbf{J}}_t^i. \tag{5.114}$$

We note that if $\tilde{\mathbf{J}}_m^i = \tilde{\mathbf{J}}_t^i = 0$, then $\tilde{H}_u = 0$ and the fields are TM to the u-direction; these fields may be determined completely from $\tilde{E}_u$. Similarly, if $\tilde{\mathbf{J}}^i = \tilde{\mathbf{J}}_{mt}^i = 0$, then $\tilde{E}_u = 0$ and the fields are TE to the u-direction; these fields may be determined completely from $\tilde{H}_u$. These properties are used in § 4.11.7, where the fields of electric and magnetic line sources aligned along the z-direction are assumed to be purely TM_z or TE_z, respectively.

5.4 TE–TM decomposition

5.4.1 TE–TM decomposition in terms of fields

A particularly useful field decomposition results if we specialize to a source-free region. With $\tilde{\mathbf{J}}^i = \tilde{\mathbf{J}}_m^i = 0$ in (5.111)–(5.112) we obtain

$$\left(\frac{\partial^2}{\partial u^2} + k^2\right)\tilde{\mathbf{H}}_t = \nabla_t \frac{\partial \tilde{H}_u}{\partial u} - j\omega\tilde{\epsilon}^c \hat{\mathbf{u}} \times \nabla_t \tilde{E}_u, \tag{5.115}$$

$$\left(\frac{\partial^2}{\partial u^2} + k^2\right)\tilde{\mathbf{E}}_t = \nabla_t \frac{\partial \tilde{E}_u}{\partial u} + j\omega\tilde{\mu}\hat{\mathbf{u}} \times \nabla_t \tilde{H}_u. \tag{5.116}$$

Setting the sources to zero in (5.113) and (5.114) we get

$$\left(\nabla^2 + k^2\right)\tilde{E}_u = 0,$$
$$\left(\nabla^2 + k^2\right)\tilde{H}_u = 0.$$

Hence the longitudinal field components are solutions to the homogeneous Helmholtz equation, and the transverse components are specified solely in terms of the longitudinal components. The electromagnetic field is completely specified by the two scalar fields $\tilde{E}_u$ and $\tilde{H}_u$ (and, of course, appropriate boundary values).

We can use superposition to simplify the task of solving (5.115)–(5.116). Since each equation has two forcing terms on the right-hand side, we can solve the equations using one forcing term at a time, and add the results. That is, let $\tilde{\mathbf{E}}_1$ and $\tilde{\mathbf{H}}_1$ be the solutions to (5.115)–(5.116) with $\tilde{E}_u = 0$, and $\tilde{\mathbf{E}}_2$ and $\tilde{\mathbf{H}}_2$ be the solutions with $\tilde{H}_u = 0$. This results in a decomposition

$$\tilde{\mathbf{E}} = \tilde{\mathbf{E}}_1 + \tilde{\mathbf{E}}_2, \tag{5.117}$$
$$\tilde{\mathbf{H}} = \tilde{\mathbf{H}}_1 + \tilde{\mathbf{H}}_2, \tag{5.118}$$

with

$$\tilde{\mathbf{E}}_1 = \tilde{\mathbf{E}}_{1t}, \qquad \tilde{\mathbf{H}}_1 = \tilde{\mathbf{H}}_{1t} + \tilde{H}_{1u}\hat{\mathbf{u}},$$
$$\tilde{\mathbf{H}}_2 = \tilde{\mathbf{H}}_{2t}, \qquad \tilde{\mathbf{E}}_2 = \tilde{\mathbf{E}}_{2t} + \tilde{E}_{2u}\hat{\mathbf{u}}.$$

Because $\tilde{\mathbf{E}}_1$ has no u-component, $\tilde{\mathbf{E}}_1$ and $\tilde{\mathbf{H}}_1$ are termed *transverse electric* (or *TE*) to the u-direction; $\tilde{\mathbf{H}}_2$ has no u-component, and $\tilde{\mathbf{E}}_2$ and $\tilde{\mathbf{H}}_2$ are termed *transverse magnetic* (or *TM*) to the u-direction.[2] We see that in a source-free region any electromagnetic field can be decomposed into a set of two fields that are TE and TM, respectively, to some fixed u-direction. This is useful when solving boundary value (e.g., waveguide and scattering) problems where information about external sources is easily specified using the values of the fields on the boundary of the source-free region. In that case $\tilde{E}_u$ and $\tilde{H}_u$ are determined by solving the homogeneous wave equation in an appropriate coordinate system, and the other field components are found from (5.115)–(5.116). Often the boundary conditions can be satisfied by the TM fields or the TE fields alone. This simplifies the analysis of many types of EM systems.

5.4.2   TE–TM decomposition in terms of Hertzian potentials

We are free to represent $\tilde{\mathbf{E}}$ and $\tilde{\mathbf{H}}$ in terms of scalar fields other than $\tilde{E}_u$ and $\tilde{H}_u$. In doing so, it is helpful to retain the wave nature of the solution so that a meaningful physical interpretation is still possible; we thus use Hertzian potentials since they obey the wave equation.

For the TM case let $\tilde{\mathbf{\Pi}}_h = 0$ and $\tilde{\mathbf{\Pi}}_e = \hat{\mathbf{u}}\tilde{\Pi}_e$. Setting $\tilde{\mathbf{J}}^i = 0$ in (5.64) we have

$$(\nabla^2 + k^2)\tilde{\mathbf{\Pi}}_e = 0.$$

Since $\tilde{\mathbf{\Pi}}_e$ is purely longitudinal, we can use (B.99) to obtain the scalar Helmholtz equation for $\tilde{\Pi}_e$:

$$(\nabla^2 + k^2)\tilde{\Pi}_e = 0. \tag{5.119}$$

Once $\tilde{\Pi}_e$ has been found by solving this wave equation, the fields can be found by using (5.62)–(5.63) with $\tilde{\mathbf{J}}^i = 0$:

$$\tilde{\mathbf{E}} = \nabla \times (\nabla \times \tilde{\mathbf{\Pi}}_e), \tag{5.120}$$

$$\tilde{\mathbf{H}} = j\omega\tilde{\epsilon}^c \nabla \times \tilde{\mathbf{\Pi}}_e. \tag{5.121}$$

We can evaluate $\tilde{\mathbf{E}}$ by noting that $\tilde{\mathbf{\Pi}}_e$ is purely longitudinal. Use of property (B.98) gives

$$\nabla \times \nabla \times \tilde{\mathbf{\Pi}}_e = \nabla_t \frac{\partial \tilde{\Pi}_e}{\partial u} - \hat{\mathbf{u}}\nabla_t^2 \tilde{\Pi}_e.$$

Then, by property (B.97),

$$\nabla \times \nabla \times \tilde{\mathbf{\Pi}}_e = \nabla_t \frac{\partial \tilde{\Pi}_e}{\partial u} - \hat{\mathbf{u}}\left[\nabla^2\tilde{\Pi}_e - \frac{\partial^2\tilde{\Pi}_e}{\partial u^2}\right].$$

By (5.119) then,

$$\tilde{\mathbf{E}} = \nabla_t \frac{\partial \tilde{\Pi}_e}{\partial u} + \hat{\mathbf{u}}\left(\frac{\partial^2}{\partial u^2} + k^2\right)\tilde{\Pi}_e. \tag{5.122}$$

The field $\tilde{\mathbf{H}}$ can be found by noting that $\tilde{\mathbf{\Pi}}_e$ is purely longitudinal. Use of property (B.96) in (5.121) gives

$$\tilde{\mathbf{H}} = -j\omega\tilde{\epsilon}^c\hat{\mathbf{u}} \times \nabla_t\tilde{\Pi}_e. \tag{5.123}$$

---

[2]Some authors prefer to use the terminology *E mode* in place of TM, and *H mode* in place of TE, indicating the presence of a u-directed electric or magnetic field component.

Similar steps can be used to find the TE representation. Substitution of $\tilde{\Pi}_e = 0$ and $\tilde{\Pi}_h = \hat{\mathbf{u}}\tilde{\Pi}_h$ into (5.65)–(5.66) gives the fields

$$\tilde{\mathbf{E}} = j\omega\tilde{\mu}\hat{\mathbf{u}} \times \nabla_t\tilde{\Pi}_h, \tag{5.124}$$

$$\tilde{\mathbf{H}} = \nabla_t\frac{\partial\tilde{\Pi}_h}{\partial u} + \hat{\mathbf{u}}\left(\frac{\partial^2}{\partial u^2} + k^2\right)\tilde{\Pi}_h, \tag{5.125}$$

while $\tilde{\Pi}_h$ must satisfy

$$(\nabla^2 + k^2)\tilde{\Pi}_h = 0. \tag{5.126}$$

Hertzian potential representation of TEM fields. An interesting situation occurs when a field is both TE and TM to a particular direction. Such a field is said to be *transverse electromagnetic* (or *TEM*) to that direction. Unfortunately, with $\tilde{E}_u = \tilde{H}_u = 0$ we cannot use (5.115) or (5.116) to find the transverse field components. It turns out that a single scalar potential function is sufficient to represent the field, and we may use either $\tilde{\Pi}_e$ or $\tilde{\Pi}_h$.

For the TM case, equations (5.122) and (5.123) show that we can represent the electromagnetic fields completely with $\tilde{\Pi}_e$. Unfortunately (5.122) has a longitudinal component, and thus cannot describe a TEM field. But if we require that $\tilde{\Pi}_e$ obey the additional equation

$$\left(\frac{\partial^2}{\partial u^2} + k^2\right)\tilde{\Pi}_e = 0, \tag{5.127}$$

then both $\mathbf{E}$ and $\mathbf{H}$ are transverse to u and thus describe a TEM field. Since $\tilde{\Pi}_e$ must also obey

$$\left(\nabla^2 + k^2\right)\tilde{\Pi}_e = 0,$$

using (B.97) we can write (5.127) as

$$\nabla_t^2\tilde{\Pi}_e = 0.$$

Similarly, for the TE case we found that the EM fields were completely described in (5.124) and (5.125) by $\tilde{\Pi}_h$. In this case $\tilde{\mathbf{H}}$ has a longitudinal component. Thus, if we require

$$\left(\frac{\partial^2}{\partial u^2} + k^2\right)\tilde{\Pi}_h = 0, \tag{5.128}$$

then both $\tilde{\mathbf{E}}$ and $\tilde{\mathbf{H}}$ are purely transverse to u and again describe a TEM field. Equation (5.128) is equivalent to

$$\nabla_t^2\tilde{\Pi}_h = 0.$$

We can therefore describe a TEM field using either $\tilde{\Pi}_e$ or $\tilde{\Pi}_h$, since a TEM field is both TE and TM to the longitudinal direction. If we choose $\tilde{\Pi}_e$ we can use (5.122) and (5.123) to obtain the expressions

$$\tilde{\mathbf{E}} = \nabla_t\frac{\partial\tilde{\Pi}_e}{\partial u}, \tag{5.129}$$

$$\tilde{\mathbf{H}} = -j\omega\tilde{\epsilon}^c\hat{\mathbf{u}} \times \nabla_t\tilde{\Pi}_e, \tag{5.130}$$

where $\tilde{\Pi}_e$ must obey

$$\nabla_t^2\tilde{\Pi}_e = 0, \qquad \left(\frac{\partial^2}{\partial u^2} + k^2\right)\tilde{\Pi}_e = 0. \tag{5.131}$$

If we choose $\tilde{\Pi}_h$ we can use (5.124) and (5.125) to obtain

$$\tilde{\mathbf{E}} = j\omega\tilde{\mu}\hat{\mathbf{u}} \times \nabla_t \tilde{\Pi}_h, \tag{5.132}$$

$$\tilde{\mathbf{H}} = \nabla_t \frac{\partial \tilde{\Pi}_h}{\partial u}, \tag{5.133}$$

where $\tilde{\Pi}_h$ must obey

$$\nabla_t^2 \tilde{\Pi}_h = 0, \qquad \left(\frac{\partial^2}{\partial u^2} + k^2 \right) \tilde{\Pi}_h = 0. \tag{5.134}$$

5.4.3   Application: hollow-pipe waveguides

A classic application of the TE–TM decomposition is to the calculation of waveguide fields. Consider a hollow pipe with PEC walls, aligned along the z-axis. The inside is filled with a homogeneous, isotropic material of permeability $\tilde{\mu}(\omega)$ and complex permittivity $\tilde{\epsilon}^c(\omega)$, and the guide cross-sectional shape is assumed to be independent of z. We assume that a current source exists somewhere within the waveguide, creating waves that either propagate or evanesce away from the source. If the source is confined to the region $-d < z < d$ then each of the regions $z > d$ and $z < -d$ is source-free and we may decompose the fields there into TE and TM sets. Such a waveguide is a good candidate for TE–TM analysis because the TE and TM fields independently satisfy the boundary conditions at the waveguide walls. This is not generally the case for certain other guided-wave structures such as fiber optic cables and microstrip lines.

We may represent the fields either in terms of the longitudinal fields $\tilde{E}_z$ and $\tilde{H}_z$, or in terms of the Hertzian potentials. We choose the Hertzian potentials. For TM fields we choose $\tilde{\boldsymbol{\Pi}}_e = \hat{\mathbf{z}}\tilde{\Pi}_e$, $\tilde{\boldsymbol{\Pi}}_h = 0$; for TE fields we choose $\tilde{\boldsymbol{\Pi}}_h = \hat{\mathbf{z}}\tilde{\Pi}_h$, $\tilde{\boldsymbol{\Pi}}_e = 0$. Both of the potentials must obey the same Helmholtz equation:

$$\left(\nabla^2 + k^2\right)\tilde{\Pi}_z = 0, \tag{5.135}$$

where $\tilde{\Pi}_z$ represents either $\tilde{\Pi}_e$ or $\tilde{\Pi}_h$. We seek a solution to this equation using the separation of variables technique, and assume the product solution

$$\tilde{\Pi}_z(\mathbf{r},\omega) = \tilde{Z}(z,\omega)\tilde{\psi}(\boldsymbol{\rho},\omega),$$

where $\boldsymbol{\rho}$ is the transverse position vector ($\mathbf{r} = \hat{\mathbf{z}}z + \boldsymbol{\rho}$). Substituting the trial solution into (5.135) and writing

$$\nabla^2 = \nabla_t^2 + \frac{\partial^2}{\partial z^2}$$

we find that

$$\frac{1}{\tilde{\psi}(\boldsymbol{\rho},\omega)}\nabla_t^2\tilde{\psi}(\boldsymbol{\rho},\omega) + k^2 = -\frac{1}{Z(z,\omega)}\frac{\partial^2}{\partial z^2}Z(z,\omega).$$

Because the left-hand side of this expression has positional dependence only on $\boldsymbol{\rho}$ while the right-hand side has dependence only on z, we must have both sides equal to a constant, say k_z^2. Then

$$\frac{\partial^2 Z}{\partial z^2} + k_z^2 Z = 0,$$

which is an ordinary differential equation with the solutions

$$Z = e^{\mp jk_z z}.$$

We also have

$$\nabla_t^2 \tilde{\psi}(\boldsymbol{\rho}, \omega) + k_c^2 \tilde{\psi}(\boldsymbol{\rho}, \omega) = 0, \tag{5.136}$$

where $k_c = k^2 - k_z^2$ is called the *cutoff wavenumber*. The solution to this equation depends on the geometry of the waveguide cross-section and whether the field is TE or TM.

The fields may be computed from the Hertzian potentials using $u = z$ in (5.122)–(5.123) and (5.124)–(5.125). Because the fields all contain the common term $e^{\mp jk_z z}$, we define the field quantities $\tilde{\mathbf{e}}$ and $\tilde{\mathbf{h}}$ through

$$\tilde{\mathbf{E}}(\mathbf{r}, \omega) = \tilde{\mathbf{e}}(\boldsymbol{\rho}, \omega) e^{\mp jk_z z}, \qquad \tilde{\mathbf{H}}(\mathbf{r}, \omega) = \tilde{\mathbf{h}}(\boldsymbol{\rho}, \omega) e^{\mp jk_z z}.$$

Then, substituting $\tilde{\Pi}_e = \tilde{\psi}_e e^{\mp jk_z z}$, we have for TM fields

$$\tilde{\mathbf{e}} = \mp jk_z \nabla_t \tilde{\psi}_e + \hat{\mathbf{z}} k_c^2 \tilde{\psi}_e,$$
$$\tilde{\mathbf{h}} = -j\omega\tilde{\epsilon}^c \hat{\mathbf{z}} \times \nabla_t \tilde{\psi}_e.$$

Because we have a simple relationship between the transverse parts of $\tilde{\mathbf{E}}$ and $\tilde{\mathbf{H}}$, we may also write the fields as

$$\tilde{e}_z = k_c^2 \tilde{\psi}_e, \tag{5.137}$$
$$\tilde{\mathbf{e}}_t = \mp jk_z \nabla_t \tilde{\psi}_e, \tag{5.138}$$
$$\tilde{\mathbf{h}}_t = \pm Y_e(\hat{\mathbf{z}} \times \tilde{\mathbf{e}}_t). \tag{5.139}$$

Here

$$Y_e = \frac{\omega\tilde{\epsilon}^c}{k_z}$$

is the complex *TM wave admittance*. For TE fields we have with $\tilde{\Pi}_h = \tilde{\psi}_h e^{\mp jk_z z}$

$$\tilde{\mathbf{e}} = j\omega\tilde{\mu} \hat{\mathbf{z}} \times \nabla_t \tilde{\psi}_h,$$
$$\tilde{\mathbf{h}} = \mp jk_z \nabla_t \tilde{\psi}_h + \hat{\mathbf{z}} k_c^2 \tilde{\psi}_h,$$

or

$$\tilde{h}_z = k_c^2 \tilde{\psi}_h, \tag{5.140}$$
$$\tilde{\mathbf{h}}_t = \mp jk_z \nabla_t \tilde{\psi}_h, \tag{5.141}$$
$$\tilde{\mathbf{e}}_t = \mp Z_h(\hat{\mathbf{z}} \times \tilde{\mathbf{h}}_t). \tag{5.142}$$

Here

$$Z_h = \frac{\omega\tilde{\mu}}{k_z}$$

is the *TM wave impedance*.

Modal solutions for the transverse field dependence. Equation (5.136) describes the transverse behavior of the waveguide fields. When coupled with an appropriate boundary condition, this homogeneous equation has an infinite spectrum of discrete solutions called *eigenmodes* or simply *modes*. Each mode has associated with it a real *eigenvalue* k_c that is dependent on the cross-sectional shape of the waveguide, but independent of frequency and homogeneous material parameters. We number the modes so that $k_c = k_{cn}$ for the nth mode. The amplitude of each modal solution depends on the excitation source within the waveguide.

The appropriate boundary conditions can be found by employing the condition that for both TM and TE fields the tangential component of $\tilde{\mathbf{E}}$ must be zero on the waveguide walls: $\hat{\mathbf{n}} \times \tilde{\mathbf{E}} = 0$, where $\hat{\mathbf{n}}$ is the unit inward normal to the waveguide wall. For TM fields we have $\tilde{E}_z = 0$ and thus

$$\tilde{\psi}_e(\boldsymbol{\rho}, \omega) = 0, \qquad \boldsymbol{\rho} \in \Gamma, \tag{5.143}$$

where Γ is the contour describing the waveguide boundary. For TE fields we have $\hat{\mathbf{n}} \times \tilde{\mathbf{E}}_t = 0$, or

$$\hat{\mathbf{n}} \times (\hat{\mathbf{z}} \times \nabla_t \tilde{\psi}_h) = 0.$$

Using

$$\hat{\mathbf{n}} \times (\hat{\mathbf{z}} \times \nabla_t \tilde{\psi}_h) = \hat{\mathbf{z}}(\hat{\mathbf{n}} \cdot \nabla_t \tilde{\psi}_h) - (\hat{\mathbf{n}} \cdot \hat{\mathbf{z}})\nabla_t \tilde{\psi}_h$$

and noting that $\hat{\mathbf{n}} \cdot \hat{\mathbf{z}} = 0$, we have the boundary condition

$$\hat{\mathbf{n}} \cdot \nabla_t \tilde{\psi}_h(\boldsymbol{\rho}, \omega) = \frac{\partial \tilde{\psi}_h(\boldsymbol{\rho}, \omega)}{\partial n} = 0, \qquad \boldsymbol{\rho} \in \Gamma. \tag{5.144}$$

The wave nature of the waveguide fields. We have seen that all waveguide field components, for both TE and TM modes, vary as $e^{\mp j k_{zn} z}$. Here $k_{zn}^2 = k^2 - k_{cn}^2$ is the *propagation constant* of the nth mode. Letting

$$k_z = \beta - j\alpha$$

we thus have

$$\tilde{\mathbf{E}}, \tilde{\mathbf{H}} \sim e^{\mp j\beta z} e^{\mp \alpha z}.$$

For $z > d$ we choose the minus sign so that we have a wave propagating away from the source; for $z < -d$ we choose the plus sign.

When the guide is filled with a good dielectric we may assume $\tilde{\mu} = \mu$ is real and independent of frequency and use (4.254) to show that

$$k_z = \beta - j\alpha = \sqrt{[\omega^2 \mu\epsilon' - k_c^2] - j\omega^2 \mu\epsilon' \tan \delta_c}$$
$$= \sqrt{\mu\epsilon'}\sqrt{\omega^2 - \omega_c^2}\sqrt{1 - j\frac{\tan \delta_c}{1 - (\omega_c/\omega)^2}}$$

where δ_c is the loss tangent (4.253) and where

$$\omega_c = \frac{k_c}{\sqrt{\mu\epsilon'}}$$

is called the *cutoff frequency*. Under the condition

$$\frac{\tan \delta_c}{1 - (\omega_c/\omega)^2} \ll 1 \tag{5.145}$$

we may approximate the square root using the first two terms of the binomial series to show that

$$\beta - j\alpha \approx \sqrt{\mu\epsilon'}\sqrt{\omega^2 - \omega_c^2}\left[1 - j\frac{1}{2}\frac{\tan \delta_c}{1 - (\omega_c/\omega)^2}\right]. \tag{5.146}$$

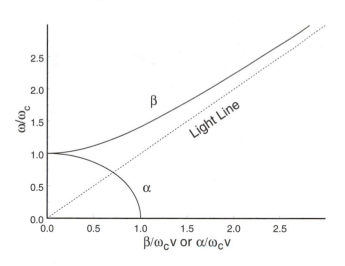

Figure 5.4: Dispersion plot for a hollow-pipe waveguide. Light line computed using $v = 1/\sqrt{\mu\epsilon}$.

Condition (5.145) requires that ω be sufficiently removed from ω_c, either by having $\omega > \omega_c$ or $\omega < \omega_c$. When $\omega > \omega_c$ we say that the frequency is *above cutoff* and find from (5.146) that

$$\beta \approx \omega\sqrt{\mu\epsilon'}\sqrt{1 - \omega_c^2/\omega^2}, \qquad \alpha \approx \frac{\omega^2\mu\epsilon'}{2\beta}\tan\delta_c.$$

Here $\alpha \ll \beta$ and the wave propagates down the waveguide with relatively little loss. When $\omega < \omega_c$ we say that the waveguide is *cut off* or that the frequency is *below cutoff* and find that

$$\alpha \approx \omega\sqrt{\mu\epsilon'}\sqrt{\omega_c^2/\omega^2 - 1}, \qquad \beta \approx \frac{\omega^2\mu\epsilon'}{2\alpha}\tan\delta_c.$$

In this case the wave has a very small phase constant and a very large rate of attenuation. For frequencies near ω_c there is an abrupt but continuous transition between these two types of wave behavior.

When the waveguide is filled with a lossless material having permittivity ϵ and permeability μ, the transition across the cutoff frequency is discontinuous. For $\omega > \omega_c$ we have

$$\beta = \omega\sqrt{\mu\epsilon}\sqrt{1 - \omega_c^2/\omega^2}, \qquad \alpha = 0,$$

and the wave propagates without loss. For $\omega < \omega_c$ we have

$$\alpha = \omega\sqrt{\mu\epsilon}\sqrt{\omega_c^2/\omega^2 - 1}, \qquad \beta = 0,$$

and the wave is evanescent. The dispersion diagram shown in Figure 5.4 clearly shows the abrupt cutoff phenomenon. We can compute the phase and group velocities of the wave above cutoff just as we did for plane waves:

$$v_p = \frac{\omega}{\beta} = \frac{v}{\sqrt{1 - \omega_c^2/\omega^2}},$$

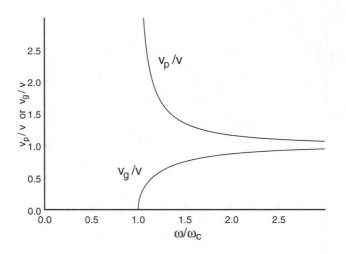

Figure 5.5: Phase and group velocity for a hollow-pipe waveguide.

$$v_g = \frac{d\omega}{d\beta} = v\sqrt{1 - \omega_c^2/\omega^2}, \qquad\qquad (5.147)$$

where $v = 1/\sqrt{\mu\epsilon}$. Note that $v_g v_p = v^2$. We show later that v_g is the velocity of energy transport within a lossless guide. We also see that as $\omega \to \infty$ we have $v_p \to v$ and $v_g \to v$. More interestingly, as $\omega \to \omega_c$ we find that $v_p \to \infty$ and $v_g \to 0$. This is shown graphically in Figure 5.5.

We may also speak of the *guided wavelength* of a monochromatic wave propagating with frequency $\check{\omega}$ in a waveguide. We define this wavelength as

$$\lambda_g = \frac{2\pi}{\beta} = \frac{\lambda}{\sqrt{1 - \omega_c^2/\check{\omega}^2}} = \frac{\lambda}{\sqrt{1 - \lambda^2/\lambda_c^2}}.$$

Here

$$\lambda = \frac{2\pi}{\check{\omega}\sqrt{\mu\epsilon}}, \qquad \lambda_c = \frac{2\pi}{k_c}.$$

Orthogonality of waveguide modes.   The modal fields in a closed-pipe waveguide obey several orthogonality relations. Let $(\check{\mathbf{E}}_n, \check{\mathbf{H}}_n)$ be the time-harmonic electric and magnetic fields of one particular waveguide mode (TE or TM), and let $(\check{\mathbf{E}}_m, \check{\mathbf{H}}_m)$ be the fields of a different mode (TE or TM). One very useful relation states that for a waveguide containing lossless materials

$$\int_{CS} \hat{\mathbf{z}} \cdot (\check{\mathbf{e}}_n \times \check{\mathbf{h}}_m^*) \, dS = 0, \qquad m \neq n, \qquad\qquad (5.148)$$

where CS is the guide cross-section. This is used to establish that the total power carried by a wave is the sum of the powers carried by individual modes (see below).

Other important relationships include the orthogonality of the longitudinal fields,

$$\int_{CS} \check{E}_{zm}\check{E}_{zn}\,dS = 0, \qquad m \neq n, \tag{5.149}$$

$$\int_{CS} \check{H}_{zm}\check{H}_{zn}\,dS = 0, \qquad m \neq n, \tag{5.150}$$

and the orthogonality of transverse fields,

$$\int_{CS} \check{\mathbf{E}}_{tm} \cdot \check{\mathbf{E}}_{tn}\,dS = 0, \qquad m \neq n,$$

$$\int_{CS} \check{\mathbf{H}}_{tm} \cdot \check{\mathbf{H}}_{tn}\,dS = 0, \qquad m \neq n.$$

These may also be combined to give an orthogonality relation for the complete fields:

$$\int_{CS} \check{\mathbf{E}}_{m} \cdot \check{\mathbf{E}}_{n}\,dS = 0, \qquad m \neq n, \tag{5.151}$$

$$\int_{CS} \check{\mathbf{H}}_{m} \cdot \check{\mathbf{H}}_{n}\,dS = 0, \qquad m \neq n. \tag{5.152}$$

For proofs of these relations the reader should see Collin [39].

Power carried by time-harmonic waves in lossless waveguides. The power carried by a time-harmonic wave propagating down a waveguide is defined as the time-average Poynting flux passing through the guide cross-section. Thus we may write

$$P_{av} = \frac{1}{2}\int_{CS} \operatorname{Re}\left\{\check{\mathbf{E}} \times \check{\mathbf{H}}^*\right\} \cdot \hat{z}\,dS.$$

The field within the guide is assumed to be a superposition of all possible waveguide modes. For waves traveling in the $+z$-direction this implies

$$\check{\mathbf{E}} = \sum_m \left(\check{\mathbf{e}}_{tm} + \hat{\mathbf{z}}\check{e}_{zm}\right)e^{-jk_{zm}z}, \qquad \check{\mathbf{H}} = \sum_n \left(\check{\mathbf{h}}_{tn} + \hat{\mathbf{z}}\check{h}_{zn}\right)e^{-jk_{zn}z}.$$

Substituting we have

$$P_{av} = \frac{1}{2}\operatorname{Re}\left\{\int_{CS}\left[\sum_m (\check{\mathbf{e}}_{tm} + \hat{\mathbf{z}}\check{e}_{zm})e^{-jk_{zm}z} \times \sum_n \left(\check{\mathbf{h}}_{tn}^* + \hat{\mathbf{z}}\check{h}_{zn}^*\right)e^{jk_{zn}^*z}\right] \cdot \hat{\mathbf{z}}\,dS\right\}$$

$$= \frac{1}{2}\operatorname{Re}\left\{\sum_m\sum_n e^{-j(k_{zm}-k_{zn}^*)z}\int_{CS}\hat{\mathbf{z}} \cdot \left(\check{\mathbf{e}}_{tm} \times \check{\mathbf{h}}_{tn}^*\right)dS\right\}.$$

By (5.148) we have

$$P_{av} = \frac{1}{2}\operatorname{Re}\left\{\sum_n e^{-j(k_{zn}-k_{zn}^*)z}\int_{CS}\hat{\mathbf{z}} \cdot \left(\check{\mathbf{e}}_{tn} \times \check{\mathbf{h}}_{tn}^*\right)dS\right\}.$$

For modes propagating in a lossless guide $k_{zn} = \beta_{zn}$. For modes that are cut off $k_{zn} = -j\alpha_{zn}$. However, we find below that terms in this series representing modes that are cut off are zero. Thus

$$P_{av} = \sum_n \frac{1}{2}\operatorname{Re}\left\{\int_{CS}\hat{\mathbf{z}} \cdot \left(\check{\mathbf{e}}_{tn} \times \check{\mathbf{h}}_{tn}^*\right)dS\right\} = \sum_n P_{n,av}.$$

Hence for waveguides filled with lossless media the total time-average power flow is given by the superposition of the individual modal powers.

Simple formulas for the individual modal powers in a lossless guide may be obtained by substituting the expressions for the fields. For TM modes we use (5.138) and (5.139) to get

$$P_{av} = \frac{1}{2} \operatorname{Re} \left\{ |k_z|^2 Y_e^* e^{-j(k_z - k_z^*)} \int_{CS} \hat{\mathbf{z}} \cdot (\nabla_t \check{\psi}_e \times [\hat{\mathbf{z}} \times \nabla_t \check{\psi}_e^*]) \, dS \right\}$$

$$= \frac{1}{2} |k_z|^2 \operatorname{Re} \{Y_e^*\} e^{-j(k_z - k_z^*)} \int_{CS} \nabla_t \check{\psi}_e \cdot \nabla_t \check{\psi}_e^* \, dS.$$

Here we have used (B.7) and $\hat{\mathbf{z}} \cdot \nabla_t \check{\psi}_e = 0$. This expression can be simplified by using the two-dimensional version of Green's first identity (B.29):

$$\int_S (\nabla_t a \cdot \nabla_t b + a \nabla_t^2 b) \, dS = \oint_\Gamma a \frac{\partial b}{\partial n} \, dl.$$

Using $a = \check{\psi}_e$ and $b = \check{\psi}_e^*$ and integrating over the waveguide cross-section we have

$$\int_{CS} (\nabla_t \check{\psi}_e \cdot \nabla_t \check{\psi}_e^* + \check{\psi}_e \nabla^2 \check{\psi}_e^*) \, dS = \oint_\Gamma \check{\psi}_e \frac{\partial \check{\psi}_e^*}{\partial n} \, dl.$$

Substituting $\nabla_t^2 \check{\psi}_e^* = -k_c^2 \check{\psi}_e^*$ and remembering that $\check{\psi}_e = 0$ on Γ we reduce this to

$$\int_{CS} \nabla_t \check{\psi}_e \cdot \nabla_t \check{\psi}_e^* \, dS = k_c^2 \int_{CS} \check{\psi}_e \check{\psi}_e^* \, dS. \tag{5.153}$$

Thus the power is

$$P_{av} = \frac{1}{2} \operatorname{Re} \{Y_e^*\} |k_z|^2 k_c^2 e^{-j(k_z - k_z^*)z} \int_{CS} \check{\psi}_e \check{\psi}_e^* \, dS.$$

For modes above cutoff we have $k_z = \beta$ and $Y_e = \omega \epsilon / k_z = \omega \epsilon / \beta$. The power carried by these modes is thus

$$P_{av} = \frac{1}{2} \omega \epsilon \beta k_c^2 \int_{CS} \check{\psi}_e \check{\psi}_e^* \, dS. \tag{5.154}$$

For modes below cutoff we have $k_z = -j\alpha$ and $Y_e = j\omega\epsilon/\alpha$. Thus $\operatorname{Re}\{Y_e^*\} = 0$ and $P_{av} = 0$. For frequencies below cutoff the fields are evanescent and do not carry power in the manner of propagating waves.

For TE modes we may proceed similarly and show that

$$P_{av} = \frac{1}{2} \omega \mu \beta k_c^2 \int_{CS} \check{\psi}_h \check{\psi}_h^* \, dS. \tag{5.155}$$

The details are left as an exercise.

Stored energy in a waveguide and the velocity of energy transport. Consider a source-free section of lossless waveguide bounded on its two ends by the cross-sectional surfaces CS_1 and CS_2. Setting $\check{\mathbf{J}}^i = \check{\mathbf{J}}^c = 0$ in (4.156) we have

$$\frac{1}{2} \oint_S (\check{\mathbf{E}} \times \check{\mathbf{H}}^*) \cdot d\mathbf{S} = 2j\omega \int_V [\langle w_e \rangle - \langle w_m \rangle] \, dV,$$

where V is the region of the guide between CS_1 and CS_2. The right-hand side represents the difference between the total time-average stored electric and magnetic energies. Thus

$$2j\omega\left[\langle W_e\rangle - \langle W_m\rangle\right] =$$
$$\frac{1}{2}\int_{CS_1} -\hat{\mathbf{z}}\cdot(\check{\mathbf{E}}\times\check{\mathbf{H}}^*)\,dS + \frac{1}{2}\int_{CS_2}\hat{\mathbf{z}}\cdot(\check{\mathbf{E}}\times\check{\mathbf{H}}^*)\,dS - \frac{1}{2}\int_{S_{\mathrm{cond}}}(\check{\mathbf{E}}\times\check{\mathbf{H}}^*)\cdot d\mathbf{S},$$

where S_{cond} indicates the conducting walls of the guide and $\hat{\mathbf{n}}$ points into the guide. For a propagating mode the first two terms on the right-hand side cancel since with no loss $\check{\mathbf{E}}\times\check{\mathbf{H}}^*$ is the same on CS_1 and CS_2. The third term is zero since $(\check{\mathbf{E}}\times\check{\mathbf{H}}^*)\cdot\hat{\mathbf{n}} = (\hat{\mathbf{n}}\times\check{\mathbf{E}})\cdot\check{\mathbf{H}}^*$, and $\hat{\mathbf{n}}\times\check{\mathbf{E}} = 0$ on the waveguide walls. Thus we have

$$\langle W_e\rangle = \langle W_m\rangle$$

for any section of a lossless waveguide.

We may compute the time-average stored magnetic energy in a section of lossless waveguide of length l as

$$\langle W_m\rangle = \frac{\mu}{4}\int_0^l\int_{CS}\check{\mathbf{H}}\cdot\check{\mathbf{H}}^*\,dS\,dz.$$

For propagating TM modes we can substitute (5.139) to find

$$\langle W_m\rangle/l = \frac{\mu}{4}(\beta Y_e)^2\int_{CS}(\hat{\mathbf{z}}\times\nabla_t\check{\psi}_e)\cdot(\hat{\mathbf{z}}\times\nabla_t\check{\psi}_e^*)\,dS.$$

Using

$$(\hat{\mathbf{z}}\times\nabla_t\check{\psi}_e)\cdot(\hat{\mathbf{z}}\times\nabla_t\check{\psi}_e^*) = \hat{\mathbf{z}}\cdot\left[\nabla_t\check{\psi}_e^*\times(\hat{\mathbf{z}}\times\nabla_t\check{\psi}_e)\right] = \nabla_t\check{\psi}_e\cdot\nabla_t\check{\psi}_e^*$$

we have

$$\langle W_m\rangle/l = \frac{\mu}{4}(\beta Y_e)^2\int_{CS}\nabla_t\check{\psi}_e\cdot\nabla_t\check{\psi}_e^*\,dS.$$

Finally, using (5.153) we have the stored energy per unit length for a propagating TM mode:

$$\langle W_m\rangle/l = \langle W_e\rangle/l = \frac{\mu}{4}(\omega\epsilon)^2 k_c^2\int_{CS}\check{\psi}_e\check{\psi}_e^*\,dS.$$

Similarly we may show that for a TE mode

$$\langle W_e\rangle/l = \langle W_m\rangle/l = \frac{\epsilon}{4}(\omega\mu)^2 k_c^2\int_{CS}\check{\psi}_h\check{\psi}_h^*\,dS.$$

The details are left as an exercise.

As with plane waves in (4.261) we may describe the velocity of energy transport as the ratio of the Poynting flux density to the total stored energy density:

$$\mathbf{S}_{av} = \langle w_T\rangle\mathbf{v}_e.$$

For TM modes this energy velocity is

$$v_e = \frac{\frac{1}{2}\omega\epsilon\beta k_c^2\check{\psi}_e\check{\psi}_e^*}{2\frac{\mu}{4}(\omega\epsilon)^2 k_c^2\check{\psi}_e\check{\psi}_e^*} = \frac{\beta}{\omega\mu\epsilon} = v\sqrt{1-\omega_c^2/\omega^2},$$

which is identical to the group velocity (5.147). This is also the case for TE modes, for which

$$v_e = \frac{\frac{1}{2}\omega\mu\beta k_c^2\check{\psi}_h\check{\psi}_h^*}{2\frac{\epsilon}{4}(\omega\mu)^2 k_c^2\check{\psi}_h\check{\psi}_h^*} = \frac{\beta}{\omega\mu\epsilon} = v\sqrt{1-\omega_c^2/\omega^2}.$$

Example: fields of a rectangular waveguide. Consider a rectangular waveguide with a cross-section occupying $0 \le x \le a$ and $0 \le y \le b$. The material within the guide is assumed to be a lossless dielectric of permittivity ϵ and permeability μ. We seek the modal fields within the guide.

Both TE and TM fields exist within the guide. In each case we must solve the differential equation

$$\nabla_t^2 \tilde{\psi} + k_c^2 \tilde{\psi} = 0.$$

A product solution in rectangular coordinates may be sought using the separation of variables technique (§ A.4). We find that

$$\tilde{\psi}(x,y,\omega) = [A_x \sin k_x x + B_x \cos k_x x]\,[A_y \sin k_y y + B_y \cos k_y y]$$

where $k_x^2 + k_y^2 = k_c^2$. This solution is easily verified by substitution.

For TM modes the solution is subject to the boundary condition (5.143):

$$\tilde{\psi}_e(\boldsymbol{\rho},\omega) = 0, \qquad \boldsymbol{\rho} \in \Gamma.$$

Applying this at $x = 0$ and $y = 0$ we find $B_x = B_y = 0$. Applying the boundary condition at $x = a$ we then find $\sin k_x a = 0$ and thus

$$k_x = \frac{n\pi}{a}, \qquad n = 1, 2, \ldots.$$

Note that $n = 0$ corresponds to the trivial solution $\tilde{\psi}_e = 0$. Similarly, from the condition at $y = b$ we find that

$$k_y = \frac{m\pi}{b}, \qquad m = 1, 2, \ldots.$$

Thus

$$\tilde{\psi}_e(x,y,\omega) = A_{nm} \sin\left(\frac{n\pi x}{a}\right) \sin\left(\frac{m\pi y}{b}\right).$$

From (5.137)–(5.139) we find that the fields are

$$\tilde{E}_z = k_{c_{nm}}^2 A_{nm}\left[\sin\frac{n\pi x}{a}\sin\frac{m\pi y}{b}\right] e^{\mp jk_z z},$$

$$\tilde{\mathbf{E}}_t = \mp jk_z A_{nm}\left[\hat{\mathbf{x}}\frac{n\pi}{a}\cos\frac{n\pi x}{a}\sin\frac{m\pi y}{b} + \hat{\mathbf{y}}\frac{m\pi}{b}\sin\frac{n\pi x}{a}\cos\frac{m\pi y}{b}\right] e^{\mp jk_z z},$$

$$\tilde{\mathbf{H}}_t = jk_z Y_e A_{nm}\left[\hat{\mathbf{x}}\frac{m\pi}{b}\sin\frac{n\pi x}{a}\cos\frac{m\pi y}{b} - \hat{\mathbf{y}}\frac{n\pi}{a}\cos\frac{n\pi x}{a}\sin\frac{m\pi y}{b}\right] e^{\mp jk_z z}.$$

Here

$$Y_e = \frac{1}{\eta\sqrt{1 - \omega_{c_{nm}}^2/\omega^2}}$$

with $\eta = (\mu\epsilon)^{1/2}$.

Each combination of m, n describes a different field pattern and thus a different mode, designated TM_{nm}. The cutoff wavenumber of the TM_{nm} mode is

$$k_{c_{nm}} = \sqrt{\left(\frac{n\pi}{a}\right)^2 + \left(\frac{m\pi}{b}\right)^2}, \qquad m, n = 1, 2, 3, \ldots$$

and the cutoff frequency is

$$\omega_{c_{nm}} = v\sqrt{\left(\frac{n\pi}{a}\right)^2 + \left(\frac{m\pi}{b}\right)^2}, \qquad m, n = 1, 2, 3, \ldots$$

where $v = 1/(\mu\epsilon)^{1/2}$. Thus the TM_{11} mode has the lowest cutoff frequency of any TM mode. There is a range of frequencies for which this is the only propagating TM mode.

For TE modes the solution is subject to

$$\hat{\mathbf{n}} \cdot \nabla_t \tilde{\psi}_h(\boldsymbol{\rho}, \omega) = \frac{\partial \tilde{\psi}_h(\boldsymbol{\rho}, \omega)}{\partial n} = 0, \qquad \boldsymbol{\rho} \in \Gamma.$$

At $x = 0$ we have

$$\frac{\partial \tilde{\psi}_h}{\partial x} = 0$$

leading to $A_x = 0$. At $y = 0$ we have

$$\frac{\partial \tilde{\psi}_h}{\partial y} = 0$$

leading to $A_y = 0$. At $x = a$ we require $\sin k_x a = 0$ and thus

$$k_x = \frac{n\pi}{a}, \qquad n = 0, 1, 2, \ldots.$$

Similarly, from the condition at $y = b$ we find

$$k_y = \frac{m\pi}{b}, \qquad m = 0, 1, 2, \ldots.$$

The case $n = m = 0$ is not allowed since it produces the trivial solution. Thus

$$\tilde{\psi}_h(x, y, \omega) = B_{nm} \cos\left(\frac{n\pi x}{a}\right) \cos\left(\frac{m\pi y}{b}\right), \qquad m, n = 0, 1, 2, \ldots, \quad m + n > 0.$$

From (5.140)–(5.142) we find that the fields are

$$\tilde{H}_z = k_{c_{nm}}^2 B_{nm} \left[\cos\frac{n\pi x}{a} \cos\frac{m\pi y}{b}\right] e^{\mp jk_z z},$$

$$\tilde{\mathbf{H}}_t = \pm jk_z B_{nm} \left[\hat{\mathbf{x}}\frac{n\pi}{a} \sin\frac{n\pi x}{a} \cos\frac{m\pi y}{b} + \hat{\mathbf{y}}\frac{m\pi}{b} \cos\frac{n\pi x}{a} \sin\frac{m\pi y}{b}\right] e^{\mp jk_z z},$$

$$\tilde{\mathbf{E}}_t = jk_z Z_h B_{nm} \left[\hat{\mathbf{x}}\frac{m\pi}{b} \cos\frac{n\pi x}{a} \sin\frac{m\pi y}{b} - \hat{\mathbf{y}}\frac{n\pi}{a} \sin\frac{n\pi x}{a} \cos\frac{m\pi y}{b}\right] e^{\mp jk_z z}.$$

Here

$$Z_h = \frac{\eta}{\sqrt{1 - \omega_{c_{nm}}^2/\omega^2}}.$$

In this case the modes are designated TE_{nm}. The cutoff wavenumber of the TE_{nm} mode is

$$k_{c_{nm}} = \sqrt{\left(\frac{n\pi}{a}\right)^2 + \left(\frac{m\pi}{b}\right)^2}, \qquad m, n = 0, 1, 2, \ldots, \quad m + n > 0$$

and the cutoff frequency is

$$\omega_{c_{nm}} = v\sqrt{\left(\frac{n\pi}{a}\right)^2 + \left(\frac{m\pi}{b}\right)^2}, \qquad m, n = 0, 1, 2, \ldots, \quad m + n > 0$$

where $v = 1/(\mu\epsilon)^{1/2}$. Modes having the same cutoff frequency are said to be *degenerate*. This is the case with the TE and TM modes. However, the field distributions differ and thus the modes are distinct. Note that we may also have degeneracy among the TE or TM modes. For instance, if $a = b$ then the cutoff frequency of the TE_{nm} mode is identical to that of the TE_{mn} mode. If $a \geq b$ then the TE_{10} mode has the lowest cutoff frequency and is termed the *dominant* mode in a rectangular guide. There is a finite band of frequencies in which this is the only mode propagating (although the bandwidth is small if $a \approx b$.)

Calculation of the time-average power carried by propagating TE and TM modes is left as an exercise.

5.4.4   TE–TM decomposition in spherical coordinates

It is not necessary for the longitudinal direction to be constant to achieve a TE–TM decomposition. It is possible, for instance, to represent the electromagnetic field in terms of components either TE or TM to the radial direction of spherical coordinates. This may be shown using a procedure identical to that used for the longitudinal–transverse decomposition in rectangular coordinates. We carry out the decomposition in the frequency domain and leave the time-domain decomposition as an exercise.

TE–TM decomposition in terms of the radial fields.   Consider a source-free region of space filled with a homogeneous, isotropic material described by parameters $\tilde{\mu}(\omega)$ and $\tilde{\epsilon}^c(\omega)$. We substitute the spherical coordinate representation of the curl into Faraday's and Ampere's laws with source terms $\tilde{\mathbf{J}}$ and $\tilde{\mathbf{J}}_m$ set equal to zero. Equating vector components we have, in particular,

$$\frac{1}{r}\left[\frac{1}{\sin\theta}\frac{\partial \tilde{E}_r}{\partial \phi} - \frac{\partial}{\partial r}(r\tilde{E}_\phi)\right] = -j\omega\tilde{\mu}\tilde{H}_\theta \tag{5.156}$$

and

$$\frac{1}{r}\left[\frac{\partial}{\partial r}(r\tilde{H}_\theta) - \frac{\partial \tilde{H}_r}{\partial \theta}\right] = j\omega\tilde{\epsilon}^c\tilde{E}_\phi. \tag{5.157}$$

We seek to isolate the transverse components of the fields in terms of the radial components. Multiplying (5.156) by $j\omega\tilde{\epsilon}^c r$ we get

$$j\omega\tilde{\epsilon}^c\frac{1}{\sin\theta}\frac{\partial \tilde{E}_r}{\partial \phi} - j\omega\tilde{\epsilon}^c\frac{\partial(r\tilde{E}_\phi)}{\partial r} = k^2 r\tilde{H}_\theta;$$

next, multiplying (5.157) by r and then differentiating with respect to r we get

$$\frac{\partial^2}{\partial r^2}(r\tilde{H}_\theta) - \frac{\partial^2 \tilde{H}_r}{\partial\theta\partial r} = j\omega\tilde{\epsilon}^c\frac{\partial(r\tilde{E}_\phi)}{\partial r}.$$

Subtracting these two equations and rearranging, we obtain

$$\left(\frac{\partial^2}{\partial r^2} + k^2\right)(r\tilde{H}_\theta) = j\omega\tilde{\epsilon}^c\frac{1}{\sin\theta}\frac{\partial \tilde{E}_r}{\partial \phi} + \frac{\partial^2 \tilde{H}_r}{\partial r\partial\theta}.$$

This is a one-dimensional wave equation for the product of r with the transverse field component $\tilde{H}_\theta$. Similarly

$$\left(\frac{\partial^2}{\partial r^2} + k^2\right)(r\tilde{H}_\phi) = -j\omega\tilde{\epsilon}^c\frac{\partial \tilde{E}_r}{\partial \theta} + \frac{1}{\sin\theta}\frac{\partial^2 \tilde{H}_r}{\partial r\partial\phi},$$

and

$$\left(\frac{\partial^2}{\partial r^2} + k^2\right)(r\tilde{E}_\phi) = \frac{1}{\sin\theta}\frac{\partial^2 \tilde{E}_r}{\partial\phi\partial r} + j\omega\tilde{\mu}\frac{\partial \tilde{H}_r}{\partial\theta}, \qquad (5.158)$$

$$\left(\frac{\partial^2}{\partial r^2} + k^2\right)(r\tilde{E}_\theta) = \frac{\partial^2 \tilde{E}_r}{\partial\theta\partial r} + j\omega\tilde{\mu}\frac{1}{\sin\theta}\frac{\partial \tilde{H}_r}{\partial\phi}. \qquad (5.159)$$

Hence we can represent the electromagnetic field in a source-free region in terms of the two scalar quantities $\tilde{E}_r$ and $\tilde{H}_r$. Superposition allows us to solve the TE case with $\tilde{E}_r = 0$ and the TM case with $\tilde{H}_r = 0$, and combine the results for the general expansion of the field.

TE–TM decomposition in terms of potential functions. If we allow the vector potential (or Hertzian potential) to have only an r-component, then the resulting fields are TE or TM to the r-direction. Unfortunately, this scalar component does not satisfy the Helmholtz equation. If we wish to use a potential component that satisfies the Helmholtz equation then we must discard the Lorentz condition and choose a different relationship between the vector and scalar potentials.

1. TM fields. To generate fields TM to r we recall that the electromagnetic fields may be written in terms of electric vector and scalar potentials as

$$\tilde{\mathbf{E}} = -j\omega\tilde{\mathbf{A}}_e - \nabla\tilde{\phi}_e, \qquad (5.160)$$

$$\tilde{\mathbf{B}} = \nabla \times \tilde{\mathbf{A}}_e. \qquad (5.161)$$

In a source-free region we have by Ampere's law

$$\tilde{\mathbf{E}} = \frac{1}{j\omega\tilde{\mu}\tilde{\epsilon}^c}\nabla \times \tilde{\mathbf{B}} = \frac{1}{j\omega\tilde{\mu}\tilde{\epsilon}^c}\nabla \times (\nabla \times \tilde{\mathbf{A}}_e).$$

Here $\tilde{\phi}_e$ and $\tilde{\mathbf{A}}_e$ must satisfy a differential equation that may be derived by examining

$$\nabla \times (\nabla \times \tilde{\mathbf{E}}) = -j\omega\nabla \times \tilde{\mathbf{B}} = -j\omega(j\omega\tilde{\mu}\tilde{\epsilon}^c\tilde{\mathbf{E}}) = k^2\tilde{\mathbf{E}},$$

where $k^2 = \omega^2\tilde{\mu}\tilde{\epsilon}^c$. Substitution from (5.160) gives

$$\nabla \times \left(\nabla \times [-j\omega\tilde{\mathbf{A}}_e - \nabla\tilde{\phi}_e]\right) = k^2[-j\omega\tilde{\mathbf{A}}_e - \nabla\tilde{\phi}_e]$$

or

$$\nabla \times (\nabla \times \tilde{\mathbf{A}}_e) - k^2\tilde{\mathbf{A}}_e = \frac{k^2}{j\omega}\nabla\tilde{\phi}_e. \qquad (5.162)$$

We are still free to specify $\nabla \cdot \tilde{\mathbf{A}}_e$.

At this point let us examine the effect of choosing a vector potential with only an r-component: $\tilde{\mathbf{A}}_e = \hat{\mathbf{r}}\tilde{A}_e$. Since

$$\nabla \times (\hat{\mathbf{r}}\tilde{A}_e) = \frac{\hat{\boldsymbol{\theta}}}{r\sin\theta}\frac{\partial\tilde{A}_e}{\partial\phi} - \frac{\hat{\boldsymbol{\phi}}}{r}\frac{\partial\tilde{A}_e}{\partial\theta} \qquad (5.163)$$

we see that $\mathbf{B} = \nabla \times \tilde{\mathbf{A}}_e$ has no r-component. Since

$$\nabla \times (\nabla \times \tilde{\mathbf{A}}_e) = -\frac{\hat{\mathbf{r}}}{r\sin\theta}\left[\frac{1}{r}\frac{\partial}{\partial\theta}\left(\sin\theta\frac{\partial\tilde{A}_e}{\partial\theta}\right) + \frac{1}{r\sin\theta}\frac{\partial^2 \tilde{A}_e}{\partial\phi^2}\right] + \frac{\hat{\boldsymbol{\theta}}}{r}\frac{\partial^2 \tilde{A}_e}{\partial r\partial\theta} + \frac{\hat{\boldsymbol{\phi}}}{r\sin\theta}\frac{\partial^2 \tilde{A}_e}{\partial r\partial\phi}$$

we see that $\tilde{\mathbf{E}} \sim \nabla \times (\nabla \times \tilde{\mathbf{A}}_e)$ has all three components. This choice of $\tilde{\mathbf{A}}_e$ produces a field TM to the r-direction. We need only choose $\nabla \cdot \tilde{\mathbf{A}}_e$ so that the resulting differential equation is convenient to solve. Substituting the above expressions into (5.162) we find that

$$
-\frac{\hat{\mathbf{r}}}{r\sin\theta}\left[\frac{1}{r}\frac{\partial}{\partial\theta}\left(\sin\theta\frac{\partial\tilde{A}_e}{\partial\theta}\right) + \frac{1}{r\sin\theta}\frac{\partial^2\tilde{A}_e}{\partial\phi^2}\right] + \frac{\hat{\boldsymbol{\theta}}}{r}\frac{\partial^2\tilde{A}_e}{\partial r\partial\theta} + \frac{\hat{\boldsymbol{\phi}}}{r\sin\theta}\frac{\partial^2\tilde{A}_e}{\partial r\partial\phi} - \hat{\mathbf{r}}k^2\tilde{A}_e =
$$
$$
\hat{\mathbf{r}}\frac{k^2}{j\omega}\frac{\partial\tilde{\phi}_e}{\partial r} + \frac{\hat{\boldsymbol{\theta}}}{r}\frac{k^2}{j\omega}\frac{\partial\tilde{\phi}_e}{\partial\theta} + \frac{\hat{\boldsymbol{\phi}}}{r\sin\theta}\frac{k^2}{j\omega}\frac{\partial\tilde{\phi}_e}{\partial\phi}. \tag{5.164}
$$

Since $\nabla \cdot \tilde{\mathbf{A}}_e$ only involves the derivatives of $\tilde{\mathbf{A}}_e$ with respect to r, we may specify $\nabla \cdot \tilde{\mathbf{A}}_e$ indirectly through

$$
\tilde{\phi}_e = \frac{j\omega}{k^2}\frac{\partial\tilde{A}_e}{\partial r}.
$$

With this (5.164) becomes

$$
\frac{1}{r\sin\theta}\left[\frac{1}{r}\frac{\partial}{\partial\theta}\left(\sin\theta\frac{\partial\tilde{A}_e}{\partial\theta}\right) + \frac{1}{r\sin\theta}\frac{\partial^2\tilde{A}_e}{\partial\phi^2}\right] + k^2\tilde{A}_e + \frac{\partial^2\tilde{A}_e}{\partial r^2} = 0.
$$

Using

$$
\frac{1}{r}\frac{\partial}{\partial r}\left[r^2\frac{\partial}{\partial r}\left(\frac{\tilde{A}_e}{r}\right)\right] = \frac{\partial^2\tilde{A}_e}{\partial r^2}
$$

we can write the differential equation as

$$
\frac{1}{r^2}\frac{\partial}{\partial r}\left[r^2\frac{\partial(\tilde{A}_e/r)}{\partial r}\right] + \frac{1}{r^2\sin\theta}\frac{\partial}{\partial\theta}\left[\sin\theta\frac{\partial(\tilde{A}_e/r)}{\partial\theta}\right] + \frac{1}{r^2\sin^2\theta}\frac{\partial^2(\tilde{A}_e/r)}{\partial\phi^2} + k^2\frac{\tilde{A}_e}{r} = 0.
$$

The first three terms of this expression are precisely the Laplacian of $\tilde{A}_e/r$. Thus we have

$$
(\nabla^2 + k^2)\left(\frac{\tilde{A}_e}{r}\right) = 0 \tag{5.165}
$$

and the quantity $\tilde{A}_e/r$ satisfies the homogeneous Helmholtz equation.

The TM fields generated by the vector potential $\tilde{\mathbf{A}}_e = \hat{\mathbf{r}}\tilde{A}_e$ may be found by using (5.160) and (5.161). From (5.160) we have the electric field

$$
\tilde{\mathbf{E}} = -j\omega\tilde{\mathbf{A}}_e - \nabla\tilde{\phi}_e = -j\omega\hat{\mathbf{r}}\tilde{A}_e - \nabla\left(\frac{j\omega}{k^2}\frac{\partial\tilde{A}_e}{\partial r}\right).
$$

Expanding the gradient we have the field components

$$
\tilde{E}_r = \frac{1}{j\omega\tilde{\mu}\tilde{\epsilon}^c}\left(\frac{\partial^2}{\partial r^2} + k^2\right)\tilde{A}_e, \tag{5.166}
$$

$$
\tilde{E}_\theta = \frac{1}{j\omega\tilde{\mu}\tilde{\epsilon}^c}\frac{1}{r}\frac{\partial^2\tilde{A}_e}{\partial r\partial\theta}, \tag{5.167}
$$

$$
\tilde{E}_\phi = \frac{1}{j\omega\tilde{\mu}\tilde{\epsilon}^c}\frac{1}{r\sin\theta}\frac{\partial^2\tilde{A}_e}{\partial r\partial\phi}. \tag{5.168}
$$

The magnetic field components are found using (5.161) and (5.163):

$$\tilde{H}_\theta = \frac{1}{\tilde{\mu}} \frac{1}{r \sin\theta} \frac{\partial \tilde{A}_e}{\partial \phi}, \tag{5.169}$$

$$\tilde{H}_\phi = -\frac{1}{\tilde{\mu}} \frac{1}{r} \frac{\partial \tilde{A}_e}{\partial \theta}. \tag{5.170}$$

2. TE fields. To generate fields TE to r we recall that the electromagnetic fields in a source-free region may be written in terms of magnetic vector and scalar potentials as

$$\tilde{\mathbf{H}} = -j\omega \tilde{\mathbf{A}}_h - \nabla \tilde{\phi}_h, \tag{5.171}$$
$$\tilde{\mathbf{D}} = -\nabla \times \tilde{\mathbf{A}}_h. \tag{5.172}$$

In a source-free region we have from Faraday's law

$$\tilde{\mathbf{H}} = \frac{1}{-j\omega\tilde{\mu}\tilde{\epsilon}^c} \nabla \times \tilde{\mathbf{D}} = \frac{1}{j\omega\tilde{\mu}\tilde{\epsilon}^c} \nabla \times (\nabla \times \tilde{\mathbf{A}}_h).$$

Here $\tilde{\phi}_h$ and $\tilde{\mathbf{A}}_h$ must satisfy a differential equation that may be derived by examining

$$\nabla \times (\nabla \times \tilde{\mathbf{H}}) = j\omega \nabla \times \tilde{\mathbf{D}} = j\omega\tilde{\epsilon}^c(-j\omega\tilde{\mu}\tilde{\mathbf{H}}) = k^2\tilde{\mathbf{H}},$$

where $k^2 = \omega^2 \tilde{\mu}\tilde{\epsilon}^c$. Substitution from (5.171) gives

$$\nabla \times \left(\nabla \times [-j\omega\tilde{\mathbf{A}}_h - \nabla\tilde{\phi}_h] \right) = k^2[-j\omega\tilde{\mathbf{A}}_h - \nabla\tilde{\phi}_h]$$

or

$$\nabla \times (\nabla \times \tilde{\mathbf{A}}_h) - k^2\tilde{\mathbf{A}}_h = \frac{k^2}{j\omega} \nabla\tilde{\phi}_h. \tag{5.173}$$

Choosing $\tilde{\mathbf{A}}_h = \hat{\mathbf{r}}\tilde{A}_h$ and

$$\tilde{\phi}_h = \frac{j\omega}{k^2} \frac{\partial \tilde{A}_h}{\partial r}$$

we find, as with the TM fields,

$$(\nabla^2 + k^2)\left(\frac{\tilde{A}_h}{r} \right) = 0. \tag{5.174}$$

Thus the quantity $\tilde{A}_h/r$ obeys the Helmholtz equation.

We can find the TE fields using (5.171) and (5.172). Substituting we find that

$$\tilde{H}_r = \frac{1}{j\omega\tilde{\mu}\tilde{\epsilon}^c} \left(\frac{\partial^2}{\partial r^2} + k^2 \right) \tilde{A}_h, \tag{5.175}$$

$$\tilde{H}_\theta = \frac{1}{j\omega\tilde{\mu}\tilde{\epsilon}^c} \frac{1}{r} \frac{\partial^2 \tilde{A}_h}{\partial r \partial \theta}, \tag{5.176}$$

$$\tilde{H}_\phi = \frac{1}{j\omega\tilde{\mu}\tilde{\epsilon}^c} \frac{1}{r \sin\theta} \frac{\partial^2 \tilde{A}_h}{\partial r \partial \phi}, \tag{5.177}$$

$$\tilde{E}_\theta = -\frac{1}{\tilde{\epsilon}^c} \frac{1}{r \sin\theta} \frac{\partial \tilde{A}_h}{\partial \phi}, \tag{5.178}$$

$$\tilde{E}_\phi = \frac{1}{\tilde{\epsilon}^c} \frac{1}{r} \frac{\partial \tilde{A}_h}{\partial \theta}. \tag{5.179}$$

Example of spherical TE–TM decomposition: a plane wave. Consider a uniform plane wave propagating in the z-direction in a lossless, homogeneous material of permittivity ϵ and permeability μ, such that its electromagnetic field is

$$\tilde{\mathbf{E}}(\mathbf{r},\omega) = \hat{\mathbf{x}}\tilde{E}_0(\omega)e^{-jkz} = \hat{\mathbf{x}}\tilde{E}_0(\omega)e^{-jkr\cos\theta},$$

$$\tilde{\mathbf{H}}(\mathbf{r},\omega) = \hat{\mathbf{y}}\frac{\tilde{E}_0(\omega)}{\eta}e^{-jkz} = \hat{\mathbf{x}}\frac{\tilde{E}_0(\omega)}{\eta}e^{-jkr\cos\theta}.$$

We wish to represent this field in terms of the superposition of a field TE to r and a field TM to r. We first find the potential functions $\tilde{\mathbf{A}}_e = \hat{\mathbf{r}}\tilde{A}_e$ and $\tilde{\mathbf{A}}_h = \hat{\mathbf{r}}\tilde{A}_h$ that represent the field. Then we may use (5.166)–(5.170) and (5.175)–(5.179) to find the TE and TM representations.

From (5.166) we see that $\tilde{A}_e$ is related to $\tilde{E}_r$, where $\tilde{E}_r$ is given by

$$\tilde{E}_r = \tilde{E}_0\sin\theta\cos\phi e^{-jkr\cos\theta} = \frac{\tilde{E}_0\cos\phi}{jkr}\frac{\partial}{\partial\theta}\left[e^{-jkr\cos\theta}\right].$$

We can separate the r and θ dependences of the exponential function by using the identity (E.101). Since $j_n(-z) = (-1)^n j_n(z) = j^{-2n}j_n(z)$ we have

$$e^{-jkr\cos\theta} = \sum_{n=0}^{\infty} j^{-n}(2n+1)j_n(kr)P_n(\cos\theta).$$

Using

$$\frac{\partial P_n(\cos\theta)}{\partial\theta} = \frac{\partial P_n^0(\cos\theta)}{\partial\theta} = P_n^1(\cos\theta)$$

we thus have

$$\tilde{E}_r = -\frac{j\tilde{E}_0\cos\phi}{kr}\sum_{n=1}^{\infty} j^{-n}(2n+1)j_n(kr)P_n^1(\cos\theta).$$

Here we start the sum at $n=1$ since $P_0^1(x) = 0$. We can now identify the vector potential as

$$\frac{\tilde{A}_e}{r} = \frac{\tilde{E}_0 k}{\omega}\cos\phi\sum_{n=1}^{\infty}\frac{j^{-n}(2n+1)}{n(n+1)}j_n(kr)P_n^1(\cos\theta) \qquad (5.180)$$

since by direct differentiation we have

$$\tilde{E}_r = \frac{1}{j\omega\tilde{\mu}\tilde{\epsilon}^c}\left(\frac{\partial^2}{\partial r^2}+k^2\right)\tilde{A}_e$$

$$= \frac{\tilde{E}_0 k}{j\omega^2\tilde{\mu}\tilde{\epsilon}^c}\cos\phi\sum_{n=1}^{\infty}\frac{j^{-n}(2n+1)}{n(n+1)}P_n^1(\cos\theta)\left(\frac{\partial^2}{\partial r^2}+k^2\right)[rj_n(kr)]$$

$$= -\frac{j\tilde{E}_0\cos\phi}{kr}\sum_{n=1}^{\infty} j^{-n}(2n+1)j_n(kr)P_n^1(\cos\theta),$$

which satisfies (5.166). Here we have used the defining equation of the spherical Bessel functions (E.15) to show that

$$\left(\frac{\partial^2}{\partial r^2}+k^2\right)[rj_n(kr)] = r\frac{\partial^2}{\partial r^2}j_n(kr) + 2\frac{\partial}{\partial r}j_n(kr) + k^2 rj_n(kr)$$

$$= k^2 r\left[\frac{\partial^2}{\partial(kr)^2}+\frac{2}{kr}\frac{\partial}{\partial(kr)}\right]j_n(kr) + k^2 rj_n(kr)$$

$$= -k^2 r\left[1 - \frac{n(n+1)}{(kr)^2}\right]j_n(kr) + k^2 rj_n(kr) = \frac{n(n+1)}{r}j_n(kr).$$

We note immediately that $\tilde{A}_e/r$ satisfies the Helmholtz equation (5.165) since it has the form of the separation of variables solution (D.113).

We may find the vector potential $\tilde{\mathbf{A}}_h = \hat{\mathbf{r}}\tilde{A}_h$ in the same manner. Noting that

$$\tilde{H}_r = \frac{\tilde{E}_0}{\eta}\sin\theta\sin\phi\, e^{-jkr\cos\theta} = \frac{\tilde{E}_0\sin\phi}{\eta jkr}\frac{\partial}{\partial\theta}\left[e^{-jkr\cos\theta}\right]$$

$$= \frac{1}{j\omega\tilde{\mu}\tilde{\epsilon}^c}\left(\frac{\partial^2}{\partial r^2} + k^2\right)\tilde{A}_h,$$

we have the potential

$$\frac{\tilde{A}_h}{r} = \frac{\tilde{E}_0 k}{\eta\omega}\sin\phi\sum_{n=1}^{\infty}\frac{j^{-n}(2n+1)}{n(n+1)}j_n(kr)P_n^1(\cos\theta). \tag{5.181}$$

We may now compute the transverse components of the TM field using (5.167)–(5.170). For convenience, let us define a new function $\hat{J}_n$ by

$$\hat{J}_n(x) = xj_n(x).$$

Then we may write

$$\tilde{E}_r = -\frac{j\tilde{E}_0\cos\phi}{(kr)^2}\sum_{n=1}^{\infty}j^{-n}(2n+1)\hat{J}_n(kr)P_n^1(\cos\theta), \tag{5.182}$$

$$\tilde{E}_\theta = \frac{j\tilde{E}_0}{kr}\sin\theta\cos\phi\sum_{n=1}^{\infty}a_n\hat{J}_n'(kr)P_n^{1\,\prime}(\cos\theta), \tag{5.183}$$

$$\tilde{E}_\phi = \frac{j\tilde{E}_0}{kr\sin\theta}\sin\phi\sum_{n=1}^{\infty}a_n\hat{J}_n'(kr)P_n^1(\cos\theta), \tag{5.184}$$

$$\tilde{H}_\theta = -\frac{\tilde{E}_0}{kr\eta\sin\theta}\sin\phi\sum_{n=1}^{\infty}a_n\hat{J}_n(kr)P_n^1(\cos\theta), \tag{5.185}$$

$$\tilde{H}_\phi = \frac{\tilde{E}_0}{kr\eta}\sin\theta\cos\phi\sum_{n=1}^{\infty}a_n\hat{J}_n(kr)P_n^{1\,\prime}(\cos\theta). \tag{5.186}$$

Here

$$\hat{J}_n'(x) = \frac{d}{dx}\hat{J}_n(x) = \frac{d}{dx}[xj_n(x)] = xj_n'(x) + j_n(x)$$

and

$$a_n = \frac{j^{-n}(2n+1)}{n(n+1)}. \tag{5.187}$$

Similarly, we have the TE fields from (5.176)–(5.179):

$$\tilde{H}_r = -\frac{j\tilde{E}_0\sin\phi}{\eta(kr)^2}\sum_{n=1}^{\infty}j^{-n}(2n+1)\hat{J}_n(kr)P_n^1(\cos\theta), \tag{5.188}$$

$$\tilde{H}_\theta = j\frac{\tilde{E}_0}{\eta kr}\sin\theta\sin\phi\sum_{n=1}^{\infty}a_n\hat{J}_n'(kr)P_n^{1\,\prime}(\cos\theta), \tag{5.189}$$

$$\tilde{H}_\phi = -j\frac{\tilde{E}_0}{\eta kr\sin\theta}\cos\phi\sum_{n=1}^{\infty}a_n\hat{J}_n'(kr)P_n^1(\cos\theta), \tag{5.190}$$

$$\tilde{E}_\theta = -\frac{\tilde{E}_0}{kr \sin \theta} \cos \phi \sum_{n=1}^{\infty} a_n \hat{J}_n(kr) P_n^1(\cos \theta), \qquad (5.191)$$

$$\tilde{E}_\phi = -\frac{\tilde{E}_0}{kr} \sin \theta \sin \phi \sum_{n=1}^{\infty} a_n \hat{J}_n(kr) P_n^{1\prime}(\cos \theta). \qquad (5.192)$$

The total field is then the sum of the TE and TM components.

Example of spherical TE–TM decomposition: scattering by a conducting sphere. Consider a PEC sphere of radius a centered at the origin and imbedded in a homogeneous, isotropic material having parameters $\tilde{\mu}$ and $\tilde{\epsilon}^c$. The sphere is illuminated by a plane wave incident along the z-axis with the fields

$$\tilde{\mathbf{E}}(\mathbf{r}, \omega) = \hat{\mathbf{x}} \tilde{E}_0(\omega) e^{-jkz} = \hat{\mathbf{x}} \tilde{E}_0(\omega) e^{-jkr \cos \theta},$$

$$\tilde{\mathbf{H}}(\mathbf{r}, \omega) = \hat{\mathbf{y}} \frac{\tilde{E}_0(\omega)}{\eta} e^{-jkz} = \hat{\mathbf{x}} \frac{\tilde{E}_0(\omega)}{\eta} e^{-jkr \cos \theta}.$$

We wish to find the field scattered by the sphere.

The boundary condition that determines the scattered field is that the total (incident plus scattered) electric field tangential to the sphere must be zero. We saw in the previous example that the incident electric field may be written as the sum of a field TE to the r-direction and a field TM to the r-direction. Since the region external to the sphere is source-free, we may also represent the scattered field as a sum of TE and TM fields. These may be found from the functions $\tilde{A}_e^s$ and $\tilde{A}_h^s$, which obey the Helmholtz equations (5.165) and (5.174). The general solution to the Helmholtz equation may be found using the separation of variables technique in spherical coordinates, as shown in § A.4, and is given by

$$\left\{ \begin{array}{c} \tilde{A}_e^s / r \\ \tilde{A}_h^s / r \end{array} \right\} = \sum_{n=0}^{\infty} \sum_{m=-n}^{n} C_{nm} Y_{nm}(\theta, \phi) h_n^{(2)}(kr).$$

Here Y_{nm} is the spherical harmonic and we have chosen the spherical Hankel function $h_n^{(2)}$ as the radial dependence since it represents the expected outward-going wave behavior of the scattered field. Since the incident field generated by the potentials (5.180) and (5.181) exactly cancels the field generated by $\tilde{A}_e^s$ and $\tilde{A}_h^s$ on the surface of the sphere, by orthogonality the scattered potential must have ϕ and θ dependencies that match those of the incident field. Thus

$$\frac{\tilde{A}_e^s}{r} = \frac{\tilde{E}_0 k}{\omega} \cos \phi \sum_{n=1}^{\infty} b_n h_n^{(2)}(kr) P_n^1(\cos \theta),$$

$$\frac{\tilde{A}_h^s}{r} = \frac{\tilde{E}_0 k}{\eta \omega} \sin \phi \sum_{n=1}^{\infty} c_n h_n^{(2)}(kr) P_n^1(\cos \theta),$$

where b_n and c_n are constants to be determined by the boundary conditions. By superposition the total field may be computed from the total potentials, which are the sum of the incident and scattered potentials. These are given by

$$\frac{\tilde{A}_e^t}{r} = \frac{\tilde{E}_0 k}{\omega} \cos \phi \sum_{n=1}^{\infty} \left[a_n j_n(kr) + b_n h_n^{(2)}(kr) \right] P_n^1(\cos \theta),$$

$$\frac{\tilde{A}_h^t}{r} = \frac{\tilde{E}_0 k}{\eta \omega} \sin \phi \sum_{n=1}^{\infty} \left[a_n j_n(kr) + c_n h_n^{(2)}(kr) \right] P_n^1(\cos \theta),$$

where a_n is given by (5.187).

The total transverse electric field is found by superposing the TE and TM transverse fields found from the total potentials. We have already computed the transverse incident fields and may easily generalize these results to the total potentials. By (5.183) and (5.191) we have

$$
\tilde{E}_\theta^t(a) = \frac{j\tilde{E}_0}{ka} \sin\theta \cos\phi \sum_{n=1}^\infty \left[a_n \hat{J}_n'(ka) + b_n \hat{H}_n^{(2)'}(ka) \right] P_n^{1'}(\cos\theta) -
$$

$$
- \frac{\tilde{E}_0}{ka\sin\theta} \cos\phi \sum_{n=1}^\infty \left[a_n \hat{J}_n(ka) + c_n \hat{H}_n^{(2)}(ka) \right] P_n^1(\cos\theta) = 0,
$$

where

$$
\hat{H}_n^{(2)}(x) = x h_n^{(2)}(x).
$$

By (5.184) and (5.192) we have

$$
\tilde{E}_\phi^t(a) = \frac{j\tilde{E}_0}{ka\sin\theta} \sin\phi \sum_{n=1}^\infty \left[a_n \hat{J}_n'(ka) + b_n \hat{H}_n^{(2)'}(ka) \right] P_n^1(\cos\theta) -
$$

$$
- \frac{\tilde{E}_0}{ka} \sin\theta \sin\phi \sum_{n=1}^\infty \left[a_n \hat{J}_n(ka) + c_n \hat{H}_n^{(2)}(ka) \right] P_n^{1'}(\cos\theta) = 0.
$$

These two sets of equations are satisfied by the conditions

$$
b_n = -\frac{\hat{J}_n'(ka)}{\hat{H}_n^{(2)'}(ka)} a_n, \qquad c_n = -\frac{\hat{J}_n(ka)}{\hat{H}_n^{(2)}(ka)} a_n.
$$

We can now write the scattered electric fields as

$$
\tilde{\mathbf{E}}_r^s = -j\tilde{E}_0 \cos\phi \sum_{n=1}^\infty b_n \left[\hat{H}_n^{(2)''}(kr) + \hat{H}_n^{(2)}(kr) \right] P_n^1(\cos\theta),
$$

$$
\tilde{\mathbf{E}}_\theta^s = \frac{\tilde{E}_0}{kr} \cos\phi \sum_{n=1}^\infty \left[jb_n \sin\theta \hat{H}_n^{(2)'}(kr) P_n^{1'}(\cos\theta) - c_n \frac{1}{\sin\theta} \hat{H}_n^{(2)}(kr) P_n^1(\cos\theta) \right],
$$

$$
\tilde{\mathbf{E}}_\phi^s = \frac{\tilde{E}_0}{kr} \sin\phi \sum_{n=1}^\infty \left[jb_n \frac{1}{\sin\theta} \hat{H}_n^{(2)'}(kr) P_n^1(\cos\theta) - c_n \sin\theta \hat{H}_n^{(2)}(kr) P_n^{1'}(\cos\theta) \right].
$$

Let us approximate the scattered field for observation points far from the sphere. We may approximate the spherical Hankel functions using (E.68) as

$$
\hat{H}_n^{(2)}(z) = z h_n^{(2)}(z) \approx j^{n+1} e^{-jz}, \qquad \hat{H}_n^{(2)'}(z) \approx j^n e^{-jz}, \qquad \hat{H}_n^{(2)''}(z) \approx -j^{n+1} e^{-jz}.
$$

Substituting these we find that $\tilde{E}_r \to 0$ as expected for the far-zone field, while

$$
\tilde{E}_\theta^s \approx \tilde{E}_0 \frac{e^{-jkr}}{kr} \cos\phi \sum_{n=1}^\infty j^{n+1} \left[b_n \sin\theta P_n^{1'}(\cos\theta) - c_n \frac{1}{\sin\theta} P_n^1(\cos\theta) \right],
$$

$$
\tilde{E}_\phi^s \approx \tilde{E}_0 \frac{e^{-jkr}}{kr} \sin\phi \sum_{n=1}^\infty j^{n+1} \left[b_n \frac{1}{\sin\theta} P_n^1(\cos\theta) - c_n \sin\theta P_n^{1'}(\cos\theta) \right].
$$

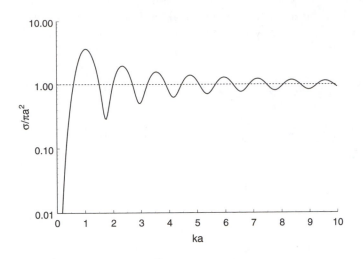

Figure 5.6: Monostatic radar cross-section of a conducting sphere.

From the far-zone fields we can compute the *radar cross-section* (RCS) or *echo area* of the sphere, which is defined by

$$\sigma = \lim_{r \to \infty} \left(4\pi r^2 \frac{|\tilde{\mathbf{E}}^s|^2}{|\tilde{\mathbf{E}}^i|^2} \right). \tag{5.193}$$

Carrying units of m^2, this quantity describes the relative energy density of the scattered field normalized by the distance from the scattering object. Figure 5.6 shows the RCS of a conducting sphere in free space for the *monostatic* case: when the observation direction is aligned with the direction of the incident wave (i.e., $\theta = \pi$), also called the *backscatter* direction. At low frequencies the RCS is proportional to λ^{-4}; this is the range of *Rayleigh scattering*, showing that higher-frequency light scatters more strongly from microscopic particles in the atmosphere (explaining why the sky is blue) [19]. At high frequencies the result approaches that of geometrical optics, and the RCS becomes the interception area of the sphere, πa^2. This is the region of *optical scattering*. Between these two regions lies the *resonance region*, or the region of *Mie scattering*, named for G. Mie who in 1908 published the first rigorous solution for scattering by a sphere (followed soon after by Debye in 1909).

Several interesting phenomena of sphere scattering are best examined in the time domain. We may compute the temporal scattered field by taking the inverse transform of the frequency-domain field. Figure 5.7 shows $E_\theta(t)$ computed in the backscatter direction ($\theta = \pi$) when the incident field waveform $E_0(t)$ is a gaussian pulse and the sphere is in free space. Two distinct features are seen in the scattered field waveform. The first is a sharp pulse almost duplicating the incident field waveform, but of opposite polarity. This is the *specular reflection* produced when the incident field first contacts the sphere and begins to induce a current on the sphere surface. The second feature, called the *creeping wave*, occurs at a time approximately $(2 + \pi)a/c$ seconds after the

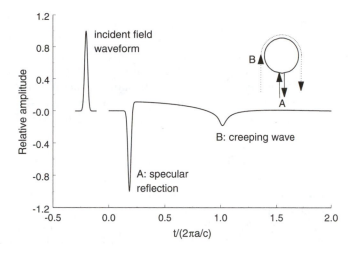

Figure 5.7: Time-domain field back-scattered by a conducting sphere.

specular reflection. This represents the field radiated back along the incident direction by a wave of current excited by the incident field at the tangent point, which travels around the sphere at approximately the speed of light in free space. Although this wave continues to traverse the sphere, its amplitude is reduced so significantly by radiation damping that only a single feature is seen.

5.5   Problems

5.1 Verify that the fields and sources obeying even planar reflection symmetry obey the component Maxwell's equations (5.1)–(5.6). Repeat for fields and sources obeying odd planar reflection symmetry.

5.2 We wish to investigate reflection symmetry through the origin in a homogeneous medium. Under what conditions on magnetic field, magnetic current density, and electric current density are we guaranteed that

$$E_x(x, y, z) = E_x(-x, -y, -z),$$
$$E_y(x, y, z) = E_y(-x, -y, -z),$$
$$E_z(x, y, z) = E_z(-x, -y, -z)?$$

5.3 We wish to investigate reflection symmetry through an axis in a homogeneous medium. Under what conditions on magnetic field, magnetic current density, and electric current density are we guaranteed that

$$E_x(x, y, z) = -E_x(-x, -y, z),$$

$$E_y(x, y, z) = -E_y(-x, -y, z),$$
$$E_z(x, y, z) = E_z(-x, -y, z)?$$

5.4 Consider an electric Hertzian dipole located on the z-axis at $z = h$. Show that if the dipole is parallel to the plane $z = 0$, then adding an oppositely-directed dipole of the same strength at $z = -h$ produces zero electric field tangential to the plane. Also show that if the dipole is z-directed, then adding another z-directed dipole at $z = -h$ produces zero electric field tangential to the $z = 0$ plane. Since the field for $z > 0$ is unaltered in each case if we place a PEC in the $z = 0$ plane, we establish that tangential components of electric current image in the opposite direction while vertical components image in the same direction.

5.5 Consider a z-directed electric line source $\tilde{I}_0$ located at $y = h, x = 0$ between conducting planes at $y = \pm d, d > h$. The material between the plates has permeability $\tilde{\mu}(\omega)$ and complex permittivity $\tilde{\epsilon}^c(\omega)$. Write the impressed and scattered fields in terms of Fourier transforms and apply the boundary conditions at $z = \pm d$ to determine the electric field between the plates. Show that the result is identical to the expression (5.8) obtained using symmetry decomposition, which required the boundary condition to be applied only on the top plate.

5.6 Consider a z-directed electric line source $\tilde{I}_0$ located at $y = h, x = 0$ in free space above a dielectric slab occupying $-d < y < d, d < h$. The slab has permeability μ_0 and permittivity ϵ. Decompose the source into even and odd constituents and solve for the electric field everywhere using the Fourier transform approach. Describe how you would use the even and odd solutions to solve the problem of a dielectric slab located on top of a PEC ground plane.

5.7 Consider an unbounded, homogeneous, isotropic medium described by permeability $\tilde{\mu}(\omega)$ and complex permittivity $\tilde{\epsilon}^c(\omega)$. Assuming there are magnetic sources present, but no electric sources, show that the fields may be written as

$$\tilde{\mathbf{H}}(\mathbf{r}) = -j\omega\tilde{\epsilon}^c \int_V \bar{\mathbf{G}}_e(\mathbf{r}|\mathbf{r}';\omega) \cdot \tilde{\mathbf{J}}_m^i(\mathbf{r}',\omega)\, dV',$$

$$\tilde{\mathbf{E}}(\mathbf{r}) = \int_V \bar{\mathbf{G}}_m(\mathbf{r}|\mathbf{r}';\omega) \cdot \tilde{\mathbf{J}}_m^i(\mathbf{r}',\omega)\, dV',$$

where $\bar{\mathbf{G}}_e$ is given by (5.83) and $\bar{\mathbf{G}}_m$ is given by (5.84).

5.8 Show that for a cubical excluding volume the depolarizing dyadic is $\bar{\mathbf{L}} = \bar{\mathbf{I}}/3$.

5.9 Compute the depolarizing dyadic for a cylindrical excluding volume with height and diameter both $2a$, and with the limit taken as $a \to 0$. Show that $\bar{\mathbf{L}} = 0.293\bar{\mathbf{I}}$.

5.10 Show that the spherical wave function

$$\tilde{\psi}(\mathbf{r}, \omega) = \frac{e^{-jkr}}{4\pi r}$$

obeys the radiation conditions (5.96) and (5.97).

5.11 Verify that the transverse component of the Laplacian of $\mathbf{A}$ is

$$(\nabla^2 \mathbf{A})_t = \left[\nabla_t(\nabla_t \cdot \mathbf{A}_t) + \frac{\partial^2 \mathbf{A}_t}{\partial u^2} - \nabla_t \times \nabla_t \times \mathbf{A}_t\right].$$

Verify that the longitudinal component of the Laplacian of $\mathbf{A}$ is

$$\hat{\mathbf{u}} \left(\hat{\mathbf{u}} \cdot \nabla^2 \mathbf{A} \right) = \hat{\mathbf{u}} \nabla^2 A_u.$$

5.12 Verify the identities (B.82)–(B.93).

5.13 Verify the identities (B.94)–(B.98).

5.14 Derive the formula (5.112) for the transverse component of the electric field.

5.15 The longitudinal/transverse decomposition can be performed beginning with the time-domain Maxwell's equations. Show that for a homogeneous, lossless, isotropic region described by permittivity ϵ and permeability μ the longitudinal fields obey the wave equations

$$\left(\frac{\partial^2}{\partial u^2} - \frac{1}{v^2} \frac{\partial^2}{\partial t^2} \right) \mathbf{H}_t = \nabla_t \frac{\partial H_u}{\partial u} - \epsilon \hat{\mathbf{u}} \times \nabla_t \frac{\partial E_u}{\partial t} + \epsilon \frac{\partial \mathbf{J}_{mt}}{\partial t} - \hat{\mathbf{u}} \times \frac{\partial \mathbf{J}_t}{\partial u},$$

$$\left(\frac{\partial^2}{\partial u^2} - \frac{1}{v^2} \frac{\partial^2}{\partial t^2} \right) \mathbf{E}_t = \nabla_t \frac{\partial E_u}{\partial u} + \mu \hat{\mathbf{u}} \times \nabla_t \frac{\partial H_u}{\partial t} + \hat{\mathbf{u}} \times \frac{\partial \mathbf{J}_{mt}}{\partial u} + \mu \frac{\partial \mathbf{J}_t}{\partial t}.$$

Also show that the transverse fields may be found from the longitudinal fields by solving

$$\left(\nabla^2 - \frac{1}{v^2} \frac{\partial}{\partial t^2} \right) E_u = \frac{1}{\epsilon} \frac{\partial \rho}{\partial u} + \mu \frac{\partial J_u}{\partial t} + \nabla_t \times \mathbf{J}_{mt},$$

$$\left(\nabla^2 - \frac{1}{v^2} \frac{\partial}{\partial t^2} \right) H_u = \frac{1}{\mu} \frac{\partial \rho_m}{\partial u} + \epsilon \frac{\partial J_{mu}}{\partial t} - \nabla_t \times \mathbf{J}_t.$$

Here $v = 1/\sqrt{\mu \epsilon}$.

5.16 Consider a homogeneous, lossless, isotropic region of space described by permittivity ϵ and permeability μ. Beginning with the source-free time-domain Maxwell equations in rectangular coordinates, choose z as the longitudinal direction and show that the TE–TM decomposition is given by

$$\left(\frac{\partial^2}{\partial z^2} - \frac{1}{v^2} \frac{\partial^2}{\partial t^2} \right) E_y = \frac{\partial^2 E_z}{\partial z \partial y} + \mu \frac{\partial^2 H_z}{\partial x \partial t}, \tag{5.194}$$

$$\left(\frac{\partial^2}{\partial z^2} - \frac{1}{v^2} \frac{\partial^2}{\partial t^2} \right) E_x = \frac{\partial^2 E_z}{\partial x \partial z} - \mu \frac{\partial^2 H_z}{\partial y \partial t}, \tag{5.195}$$

$$\left(\frac{\partial^2}{\partial z^2} - \frac{1}{v^2} \frac{\partial^2}{\partial t^2} \right) H_y = -\epsilon \frac{\partial^2 E_z}{\partial x \partial t} + \frac{\partial^2 H_z}{\partial y \partial z}, \tag{5.196}$$

$$\left(\frac{\partial^2}{\partial z^2} - \frac{1}{v^2} \frac{\partial^2}{\partial t^2} \right) H_x = \epsilon \frac{\partial^2 E_z}{\partial y \partial t} + \frac{\partial^2 H_z}{\partial x \partial z}, \tag{5.197}$$

with

$$\left(\nabla^2 - \frac{1}{v^2} \frac{\partial^2}{\partial t^2} \right) E_z = 0, \tag{5.198}$$

$$\left(\nabla^2 - \frac{1}{v^2} \frac{\partial^2}{\partial t^2} \right) H_z = 0. \tag{5.199}$$

Here $v = 1/\sqrt{\mu \epsilon}$.

5.17 Consider the case of TM fields in the time domain. Show that for a homogeneous, isotropic, lossless medium with permittivity ϵ and permeability μ the fields may be derived from a single Hertzian potential $\mathbf{\Pi}_e(\mathbf{r}, t) = \hat{\mathbf{u}}\tilde{\Pi}_e(\mathbf{r}, t)$ that satisfies the wave equation

$$\left(\nabla^2 - \frac{1}{v^2}\frac{\partial^2}{\partial t^2}\right)\Pi_e = 0$$

and that the fields are

$$\mathbf{E} = \nabla_t\frac{\partial\Pi_e}{\partial u} + \hat{\mathbf{u}}\left(\frac{\partial^2}{\partial u^2} - \frac{1}{v^2}\frac{\partial^2}{\partial t^2}\right)\Pi_e, \qquad \mathbf{H} = -\epsilon\hat{\mathbf{u}} \times \nabla_t\frac{\partial\Pi_e}{\partial t}.$$

5.18 Consider the case of TE fields in the time domain. Show that for a homogeneous, isotropic, lossless medium with permittivity ϵ and permeability μ the fields may be derived from a single Hertzian potential $\mathbf{\Pi}_h(\mathbf{r}, t) = \hat{\mathbf{u}}\tilde{\Pi}_h(\mathbf{r}, t)$ that satisfies the wave equation

$$\left(\nabla^2 - \frac{1}{v^2}\frac{\partial^2}{\partial t^2}\right)\Pi_h = 0$$

and that the fields are

$$\mathbf{E} = \mu\hat{\mathbf{u}} \times \nabla_t\frac{\partial\Pi_h}{\partial t}, \qquad \mathbf{H} = \nabla_t\frac{\partial\Pi_h}{\partial u} + \hat{\mathbf{u}}\left(\frac{\partial^2}{\partial u^2} - \frac{1}{v^2}\frac{\partial^2}{\partial t^2}\right)\Pi_h.$$

5.19 Show that in the time domain TEM fields may be written for a homogeneous, isotropic, lossless medium with permittivity ϵ and permeability μ in terms of a Hertzian potential $\mathbf{\Pi}_e = \hat{\mathbf{u}}\Pi_e$ that satisfies

$$\nabla_t^2\Pi_e = 0$$

and that the fields are

$$\mathbf{E} = \nabla_t\frac{\partial\Pi_e}{\partial u}, \qquad \mathbf{H} = -\epsilon\hat{\mathbf{u}} \times \nabla_t\frac{\partial\Pi_e}{\partial t}.$$

5.20 Show that in the time domain TEM fields may be written for a homogeneous, isotropic, lossless medium with permittivity ϵ and permeability μ in terms of a Hertzian potential $\mathbf{\Pi}_h = \hat{\mathbf{u}}\Pi_h$ that satisfies

$$\nabla_t^2\Pi_h = 0$$

and that the fields are

$$\mathbf{E} = \mu\hat{\mathbf{u}} \times \nabla_t\frac{\partial\Pi_h}{\partial t}, \qquad \mathbf{H} = \nabla_t\frac{\partial\Pi_h}{\partial u}.$$

5.21 Consider a TEM plane-wave field of the form

$$\tilde{\mathbf{E}} = \hat{\mathbf{x}}\tilde{E}_0 e^{-jkz}, \qquad \tilde{\mathbf{H}} = \hat{\mathbf{y}}\frac{\tilde{E}_0}{\eta}e^{-jkz},$$

where $k = \omega\sqrt{\mu\epsilon}$ and $\eta = \sqrt{\mu/\epsilon}$. Show that:

(a) $\tilde{\mathbf{E}}$ may be obtained from $\tilde{\mathbf{H}}$ using the equations for a field that is TE_y;

(b) $\tilde{\mathbf{H}}$ may be obtained from $\tilde{\mathbf{E}}$ using the equations for a field that is TM_x;

(c) $\tilde{\mathbf{E}}$ and $\tilde{\mathbf{H}}$ may be obtained from the potential $\tilde{\mathbf{\Pi}}_h = \hat{\mathbf{y}}(\tilde{E}_0/k^2\eta)e^{-jkz}$;

(d) $\tilde{\mathbf{E}}$ and $\tilde{\mathbf{H}}$ may be obtained from the potential $\tilde{\mathbf{\Pi}}_e = \hat{\mathbf{x}}(\tilde{E}_0/k^2)e^{-jkz}$;

(e) $\tilde{\mathbf{E}}$ and $\tilde{\mathbf{H}}$ may be obtained from the potential $\tilde{\mathbf{\Pi}}_e = \hat{\mathbf{z}}(j\tilde{E}_0 x/k)e^{-jkz}$;

(f) $\tilde{\mathbf{E}}$ and $\tilde{\mathbf{H}}$ may be obtained from the potential $\tilde{\mathbf{\Pi}}_h = \hat{\mathbf{z}}(j\tilde{E}_0 y/k\eta)e^{-jkz}$.

5.22 Prove the orthogonality relationships (5.149) and (5.150) for the longitudinal fields in a lossless waveguide. *Hint*: Substitute $a = \check{\psi}_e$ and $b = \check{\psi}_h$ into Green's second identity (B.30) and apply the boundary conditions for TE and TM modes.

5.23 Verify the waveguide orthogonality conditions (5.151)-(5.152) by substituting the field expressions for a rectangular waveguide.

5.24 Show that the time-average power carried by a propagating TE mode in a lossless waveguide is given by

$$P_{av} = \frac{1}{2}\omega\mu\beta k_c^2 \int_{CS} \check{\psi}_h \check{\psi}_h^* \, dS.$$

5.25 Show that the time-average stored energy per unit length for a propagating TE mode in a lossless waveguide is

$$\langle W_e\rangle/l = \langle W_m\rangle/l = \frac{\epsilon}{4}(\omega\mu)^2 k_c^2 \int_{CS} \check{\psi}_h \check{\psi}_h^* \, dS.$$

5.26 Consider a waveguide of circular cross-section aligned on the z-axis and filled with a lossless material having permittivity ϵ and permeability μ. Solve for both the TE and TM fields within the guide. List the first ten modes in order by cutoff frequency.

5.27 Consider a propagating TM mode in a lossless rectangular waveguide. Show that the time-average power carried by the propagating wave is

$$P_{av_{nm}} = \frac{1}{2}\omega\epsilon\beta_{nm}k_{c_{nm}}^2 |A_{nm}|^2 \frac{ab}{4}.$$

5.28 Consider a propagating TE mode in a lossless rectangular waveguide. Show that the time-average power carried by the propagating wave is

$$P_{av_{nm}} = \frac{1}{2}\omega\mu\beta_{nm}k_{c_{nm}}^2 |B_{nm}|^2 \frac{ab}{4}.$$

5.29 Consider a homogeneous, lossless region of space characterized by permeability μ and permittivity ϵ. Beginning with the time-domain Maxwell equations, show that the θ and ϕ components of the electromagnetic fields can be written in terms of the radial components. From this give the TE_r–TM_r field decomposition.

5.30 Consider the formula for the radar cross-section of a PEC sphere (5.193). Show that for the monostatic case the RCS becomes

$$\sigma = \frac{\lambda^2}{4\pi}\left|\sum_{n=1}^{\infty} \frac{(-1)^n(2n+1)}{\hat{H}_n^{(2)\prime}(ka)\hat{H}_n^{(2)}(ka)}\right|^2.$$

5.31 Beginning with the monostatic formula for the RCS of a conducting sphere given in Problem 5.30, use the small-argument approximation to the spherical Hankel functions to show that the RCS is proportional to λ^{-4} when $ka \ll 1$.

5.32 Beginning with the monostatic formula for the RCS of a conducting sphere given in Problem 5.30, use the large-argument approximation to the spherical Hankel functions to show that the RCS approaches the interception area of the sphere, πa^2, as $ka \to \infty$.

5.33 A material sphere of radius a has permittivity ϵ and permeability μ. The sphere is centered at the origin and illuminated by a plane wave traveling in the z-direction with the fields

$$\tilde{\mathbf{E}}(\mathbf{r}, \omega) = \hat{\mathbf{x}}\tilde{E}_0(\omega)e^{-jkz}, \qquad \tilde{\mathbf{H}}(\mathbf{r}, \omega) = \hat{\mathbf{y}}\frac{\tilde{E}_0(\omega)}{\eta}e^{-jkz}.$$

Find the fields internal and external to the sphere.

Chapter 6

Integral solutions of Maxwell's equations

6.1 Vector Kirchoff solution: method of Stratton and Chu

One of the most powerful tools for the analysis of electromagnetics problems is the integral solution to Maxwell's equations formulated by Stratton and Chu [187, 188]. These authors used the vector Green's theorem to solve for $\tilde{\mathbf{E}}$ and $\tilde{\mathbf{H}}$ in much the same way as is done in static fields with the scalar Green's theorem. An alternative approach is to use the Lorentz reciprocity theorem of § 4.10.2, as done by Fradin [74]. The reciprocity approach allows the identification of terms arising from surface discontinuities, which must be added to the result obtained from the other approach [187].

6.1.1 The Stratton–Chu formula

Consider an isotropic, homogeneous medium occupying a bounded region V in space. The medium is described by permeability $\tilde{\mu}(\omega)$, permittivity $\tilde{\epsilon}(\omega)$, and conductivity $\tilde{\sigma}(\omega)$. The region V is bounded by a surface S, which can be multiply-connected so that S is the union of several surfaces $S_1, \ldots, S_N$ as shown in Figure 6.1; these are used to exclude unknown sources and to formulate the *vector Huygens principle*. Impressed electric and magnetic sources may thus reside both inside and outside V.

We wish to solve for the electric and magnetic fields at a point $\mathbf{r}$ within V. To do this we employ the Lorentz reciprocity theorem (4.173), written here using the frequency-domain fields as an integral over primed coordinates:

$$-\oint_S \left[\tilde{\mathbf{E}}_a(\mathbf{r}',\omega) \times \tilde{\mathbf{H}}_b(\mathbf{r}',\omega) - \tilde{\mathbf{E}}_b(\mathbf{r}',\omega) \times \tilde{\mathbf{H}}_a(\mathbf{r}',\omega) \right] \cdot \hat{\mathbf{n}}' dS' =$$

$$\int_V \left[\tilde{\mathbf{E}}_b(\mathbf{r}',\omega) \cdot \tilde{\mathbf{J}}_a(\mathbf{r}',\omega) - \tilde{\mathbf{E}}_a(\mathbf{r}',\omega) \cdot \tilde{\mathbf{J}}_b(\mathbf{r}',\omega) - \right. \tag{6.1}$$

$$\left. \tilde{\mathbf{H}}_b(\mathbf{r}',\omega) \cdot \tilde{\mathbf{J}}_{ma}(\mathbf{r}',\omega) + \tilde{\mathbf{H}}_a(\mathbf{r}',\omega) \cdot \tilde{\mathbf{J}}_{mb}(\mathbf{r}',\omega) \right] dV'. \tag{6.2}$$

Note that the negative sign on the left arises from the definition of $\hat{\mathbf{n}}$ as the *inward* normal to V as shown in Figure 6.1. We place an electric Hertzian dipole at the point $\mathbf{r} = \mathbf{r}_p$ where we wish to compute the field, and set $\tilde{\mathbf{E}}_b = \tilde{\mathbf{E}}_p$ and $\tilde{\mathbf{H}}_b = \tilde{\mathbf{H}}_p$ in the reciprocity theorem, where $\tilde{\mathbf{E}}_p$ and $\tilde{\mathbf{H}}_p$ are the fields produced by the dipole (5.88)–(5.89):

$$\tilde{\mathbf{H}}_p(\mathbf{r},\omega) = j\omega \nabla \times [\tilde{\mathbf{p}} G(\mathbf{r}|\mathbf{r}_p;\omega)], \tag{6.3}$$

$$\tilde{\mathbf{E}}_p(\mathbf{r},\omega) = \frac{1}{\tilde{\epsilon}^c} \nabla \times (\nabla \times [\tilde{\mathbf{p}} G(\mathbf{r}|\mathbf{r}_p;\omega)]). \tag{6.4}$$

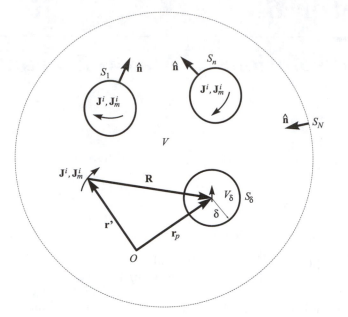

Figure 6.1: Geometry used to derive the Stratton–Chu formula.

We also let $\tilde{\mathbf{E}}_a = \tilde{\mathbf{E}}$ and $\tilde{\mathbf{H}}_a = \tilde{\mathbf{H}}$, where $\tilde{\mathbf{E}}$ and $\tilde{\mathbf{H}}$ are the fields produced by the impressed sources $\tilde{\mathbf{J}}_a = \tilde{\mathbf{J}}^i$ and $\tilde{\mathbf{J}}_{ma} = \tilde{\mathbf{J}}^i_m$ within V that we wish to find at $\mathbf{r} = \mathbf{r}_p$. Since the dipole fields are singular at $\mathbf{r} = \mathbf{r}_p$, we must exclude the point $\mathbf{r}_p$ with a small spherical surface S_δ surrounding the volume V_δ as shown in Figure 6.1. Substituting these fields into (6.2) we obtain

$$-\oint_{S+S_\delta} \left[\tilde{\mathbf{E}} \times \tilde{\mathbf{H}}_p - \tilde{\mathbf{E}}_p \times \tilde{\mathbf{H}} \right] \cdot \hat{\mathbf{n}}' \, dS' = \int_{V-V_\delta} \left[\tilde{\mathbf{E}}_p \cdot \tilde{\mathbf{J}}^i - \tilde{\mathbf{H}}_p \cdot \tilde{\mathbf{J}}^i_m \right] dV'. \qquad (6.5)$$

A useful identity involves the spatially-constant vector $\tilde{\mathbf{p}}$ and the Green's function $G(\mathbf{r}'|\mathbf{r}_p)$:

$$\begin{aligned} \nabla' \times [\nabla' \times (G\tilde{\mathbf{p}})] &= \nabla'[\nabla' \cdot (G\tilde{\mathbf{p}})] - \nabla'^2(G\tilde{\mathbf{p}}) \\ &= \nabla'[\nabla' \cdot (G\tilde{\mathbf{p}})] - \tilde{\mathbf{p}}\nabla'^2 G \\ &= \nabla'(\tilde{\mathbf{p}} \cdot \nabla' G) + \tilde{\mathbf{p}}k^2 G, \end{aligned} \qquad (6.6)$$

where we have used $\nabla'^2 G = -k^2 G$ for $\mathbf{r}' \neq \mathbf{r}_p$.

We begin by computing the terms on the left side of (6.5). We suppress the $\mathbf{r}'$ dependence of the fields and also the dependencies of $G(\mathbf{r}'|\mathbf{r}_p)$. Substituting from (6.3) we have

$$\oint_{S+S_\delta} [\tilde{\mathbf{E}} \times \tilde{\mathbf{H}}_p] \cdot \hat{\mathbf{n}}' \, dS' = j\omega \oint_{S+S_\delta} \left[\tilde{\mathbf{E}} \times \nabla' \times (G\tilde{\mathbf{p}}) \right] \cdot \hat{\mathbf{n}}' \, dS'.$$

Using $\hat{\mathbf{n}}' \cdot [\tilde{\mathbf{E}} \times \nabla' \times (G\tilde{\mathbf{p}})] = \hat{\mathbf{n}}' \cdot [\tilde{\mathbf{E}} \times (\nabla' G \times \tilde{\mathbf{p}})] = (\hat{\mathbf{n}}' \times \tilde{\mathbf{E}}) \cdot (\nabla' G \times \tilde{\mathbf{p}})$ we can write

$$\oint_{S+S_\delta} [\tilde{\mathbf{E}} \times \tilde{\mathbf{H}}_p] \cdot \hat{\mathbf{n}}' \, dS' = j\omega\tilde{\mathbf{p}} \cdot \oint_{S+S_\delta} [\hat{\mathbf{n}}' \times \tilde{\mathbf{E}}] \times \nabla' G \, dS'.$$

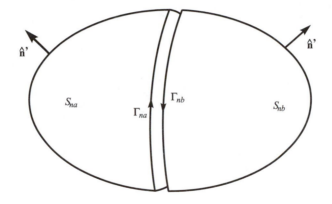

Figure 6.2: Decomposition of surface S_n to isolate surface field discontinuity.

Next we examine

$$\oint_{S+S_\delta} [\tilde{\mathbf{E}}_p \times \tilde{\mathbf{H}}] \cdot \hat{\mathbf{n}}' \, dS' = -\frac{1}{\tilde{\epsilon}^c} \oint_{S+S_\delta} \left[\tilde{\mathbf{H}} \times \nabla' \times \nabla' \times (G\tilde{\mathbf{p}}) \right] \cdot \hat{\mathbf{n}}' \, dS'.$$

Use of (6.6) along with the identity (B.43) gives

$$\oint_{S+S_\delta} [\tilde{\mathbf{E}}_p \times \tilde{\mathbf{H}}] \cdot \hat{\mathbf{n}}' \, dS' = -\frac{1}{\tilde{\epsilon}^c} \oint_{S+S_\delta} \left\{ (\tilde{\mathbf{H}} \times \tilde{\mathbf{p}})k^2 G - \right.$$
$$\left. - \nabla' \times \left[(\tilde{\mathbf{p}} \cdot \nabla'G)\tilde{\mathbf{H}} \right] + (\tilde{\mathbf{p}} \cdot \nabla'G)(\nabla' \times \tilde{\mathbf{H}}) \right\} \cdot \hat{\mathbf{n}}' \, dS'.$$

We would like to use Stokes's theorem on the second term of the right-hand side. Since the theorem is not valid for surfaces on which $\tilde{\mathbf{H}}$ has discontinuities, we break the closed surfaces in Figure 6.1 into open surfaces whose boundary contours isolate the discontinuities as shown in Figure 6.2. Then we may write

$$\oint_{S_n=S_{na}+S_{nb}} \hat{\mathbf{n}}' \cdot \nabla' \times \left[(\tilde{\mathbf{p}} \cdot \nabla'G)\tilde{\mathbf{H}} \right] dS' = \oint_{\Gamma_{na}+\Gamma_{nb}} d\mathbf{l}' \cdot \tilde{\mathbf{H}}(\tilde{\mathbf{p}} \cdot \nabla'G).$$

For surfaces not containing discontinuities of $\tilde{\mathbf{H}}$ the two contour integrals provide equal and opposite contributions and this term vanishes. Thus the left-hand side of (6.5) is

$$-\oint_{S+S_\delta} \left[\tilde{\mathbf{E}} \times \tilde{\mathbf{H}}_p - \tilde{\mathbf{E}}_p \times \tilde{\mathbf{H}} \right] \cdot \hat{\mathbf{n}}' \, dS' =$$
$$-\frac{1}{\tilde{\epsilon}^c} \tilde{\mathbf{p}} \cdot \left\{ \oint_{S+S_\delta} \left[j\omega\tilde{\epsilon}^c(\hat{\mathbf{n}}' \times \tilde{\mathbf{E}}) \times \nabla'G + k^2(\hat{\mathbf{n}}' \times \tilde{\mathbf{H}})G + \hat{\mathbf{n}}' \cdot (\tilde{\mathbf{J}}^i + j\omega\tilde{\epsilon}^c\tilde{\mathbf{E}})\nabla'G \right] dS' \right.$$

where we have substituted $\tilde{\mathbf{J}}^i + j\omega\tilde{\epsilon}^c\tilde{\mathbf{E}}$ for $\nabla' \times \tilde{\mathbf{H}}$ and used $(\tilde{\mathbf{H}} \times \tilde{\mathbf{p}}) \cdot \hat{\mathbf{n}}' = \tilde{\mathbf{p}} \cdot (\hat{\mathbf{n}}' \times \tilde{\mathbf{H}})$.

Now consider the right-hand side of (6.5). Substituting from (6.4) we have

$$\int_{V-V_\delta} \tilde{\mathbf{E}}_p \cdot \tilde{\mathbf{J}}^i \, dV' = \frac{1}{\tilde{\epsilon}^c} \int_{V-V_\delta} \tilde{\mathbf{J}}^i \cdot [\nabla' \times \nabla' \times (\tilde{\mathbf{p}}G)] \, dV'.$$

Using (6.6) and (B.42), we have

$$\int_{V-V_\delta} \tilde{\mathbf{E}}_p \cdot \tilde{\mathbf{J}}^i \, dV' = \frac{1}{\tilde{\epsilon}^c} \int_{V-V_\delta} \left\{ k^2(\tilde{\mathbf{p}} \cdot \tilde{\mathbf{J}}^i)G + \nabla' \cdot [\tilde{\mathbf{J}}^i(\tilde{\mathbf{p}} \cdot \nabla'G)] - (\tilde{\mathbf{p}} \cdot \nabla'G)\nabla' \cdot \tilde{\mathbf{J}}^i \right\} dV'.$$

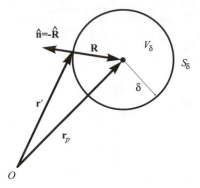

Figure 6.3: Geometry of surface integral used to extract $\mathbf{E}$ at $\mathbf{r}_p$.

Replacing $\nabla' \cdot \tilde{\mathbf{J}}^i$ with $-j\omega\tilde{\rho}^i$ from the continuity equation and using the divergence theorem on the second term on the right-hand side, we then have

$$\int_{V-V_\delta} \tilde{\mathbf{E}}_p \cdot \tilde{\mathbf{J}}^i \, dV' = \frac{1}{\tilde{\epsilon}^c}\tilde{\mathbf{p}} \cdot \left[\int_{V-V_\delta} (k^2\tilde{\mathbf{J}}^iG + j\omega\tilde{\rho}^i\nabla'G)\,dV' - \oint_{S+S_\delta} (\hat{\mathbf{n}}' \cdot \tilde{\mathbf{J}}^i)\nabla'G\,dS' \right].$$

Lastly we examine

$$\int_{V-V_\delta} \tilde{\mathbf{H}}_p \cdot \tilde{\mathbf{J}}_m^i \, dV' = j\omega \int_{V-V_\delta} \tilde{\mathbf{J}}_m^i \cdot \nabla' \times (G\tilde{\mathbf{p}})\,dV'.$$

Use of $\tilde{\mathbf{J}}_m^i \cdot \nabla' \times (G\tilde{\mathbf{p}}) = \tilde{\mathbf{J}}_m^i \cdot (\nabla'G \times \tilde{\mathbf{p}}) = \tilde{\mathbf{p}} \cdot (\tilde{\mathbf{J}}_m^i \times \nabla'G)$ gives

$$\int_{V-V_\delta} \tilde{\mathbf{H}}_p \cdot \tilde{\mathbf{J}}_m^i \, dV' = j\omega\tilde{\mathbf{p}} \cdot \int_{V-V_\delta} \tilde{\mathbf{J}}_m^i \times \nabla'G\,dV'.$$

We now substitute all terms into (6.5) and note that each term involves a dot product with $\tilde{\mathbf{p}}$. Since $\tilde{\mathbf{p}}$ is arbitrary we have

$$-\oint_{S+S_\delta} \left[(\hat{\mathbf{n}}' \times \tilde{\mathbf{E}}) \times \nabla'G + (\hat{\mathbf{n}}' \cdot \tilde{\mathbf{E}})\nabla'G - j\omega\tilde{\mu}(\hat{\mathbf{n}}' \times \tilde{\mathbf{H}})G \right] dS' +$$

$$+\frac{1}{j\omega\tilde{\epsilon}^c} \oint_{\Gamma_a+\Gamma_b} (\mathbf{dl}' \cdot \tilde{\mathbf{H}})\nabla'G = \int_{V-V_\delta} \left[-\tilde{\mathbf{J}}_m^i \times \nabla'G + \frac{\tilde{\rho}^i}{\tilde{\epsilon}^c}\nabla'G - j\omega\tilde{\mu}\tilde{\mathbf{J}}^iG \right] dV'.$$

The electric field may be extracted from the above expression by letting the radius of the excluding volume V_δ recede to zero. We first consider the surface integral over S_δ. Examining Figure 6.3 we see that $R = |\mathbf{r}_p - \mathbf{r}'| = \delta$, $\hat{\mathbf{n}}' = -\hat{\mathbf{R}}$, and

$$\nabla'G(\mathbf{r}'|\mathbf{r}_p) = \frac{d}{dR}\left(\frac{e^{-jkR}}{4\pi R} \right)\nabla'R = \hat{\mathbf{R}}\left(\frac{1+jk\delta}{4\pi\delta^2} \right)e^{-jk\delta} \approx \frac{\hat{\mathbf{R}}}{\delta^2} \quad \text{as } \delta \to 0.$$

Assuming $\tilde{\mathbf{E}}$ is continuous at $\mathbf{r}' = \mathbf{r}_p$ we can write

$$-\lim_{\delta\to 0}\oint_{S_\delta} \left[(\hat{\mathbf{n}}' \times \tilde{\mathbf{E}}) \times \nabla'G + (\hat{\mathbf{n}}' \cdot \tilde{\mathbf{E}})\nabla'G - j\omega\tilde{\mu}(\hat{\mathbf{n}}' \times \tilde{\mathbf{H}})G \right] dS' =$$

$$\lim_{\delta\to 0}\int_{\Omega} \frac{1}{4\pi}\left[(\hat{\mathbf{R}} \times \tilde{\mathbf{E}}) \times \frac{\hat{\mathbf{R}}}{\delta^2} + (\hat{\mathbf{R}} \cdot \tilde{\mathbf{E}})\frac{\hat{\mathbf{R}}}{\delta^2} - j\omega\tilde{\mu}(\hat{\mathbf{R}} \times \tilde{\mathbf{H}})\frac{1}{\delta} \right]\delta^2 \, d\Omega =$$

$$\lim_{\delta\to 0}\int_{\Omega} \frac{1}{4\pi}\left[-(\hat{\mathbf{R}} \cdot \tilde{\mathbf{E}})\hat{\mathbf{R}} + (\hat{\mathbf{R}} \cdot \hat{\mathbf{R}})\tilde{\mathbf{E}} + (\hat{\mathbf{R}} \cdot \tilde{\mathbf{E}})\hat{\mathbf{R}} \right] d\Omega = \tilde{\mathbf{E}}(\mathbf{r}_p).$$

Here we have used $\int_\Omega d\Omega = 4\pi$ for the total solid angle subtending the sphere S_δ. Finally, assuming that the volume sources are continuous, the volume integral over V_δ vanishes and we have

$$\tilde{\mathbf{E}}(\mathbf{r},\omega) = \int_V \left(-\tilde{\mathbf{J}}_m^i \times \nabla'G + \frac{\tilde{\rho}^i}{\tilde{\epsilon}^c}\nabla'G - j\omega\tilde{\mu}\tilde{\mathbf{J}}^iG\right)dV' +$$

$$+ \sum_{n=1}^N \int_{S_n}\left[(\hat{\mathbf{n}}' \times \tilde{\mathbf{E}}) \times \nabla'G + (\hat{\mathbf{n}}' \cdot \tilde{\mathbf{E}})\nabla'G - j\omega\tilde{\mu}(\hat{\mathbf{n}}' \times \tilde{\mathbf{H}})G\right]dS' -$$

$$- \sum_{n=1}^N \frac{1}{j\omega\tilde{\epsilon}^c}\oint_{\Gamma_{na}+\Gamma_{nb}}(\mathbf{dl}' \cdot \tilde{\mathbf{H}})\nabla'G. \tag{6.7}$$

A similar formula for $\tilde{\mathbf{H}}$ can be derived by placing a magnetic dipole of moment $\tilde{\mathbf{p}}_m$ at $\mathbf{r} = \mathbf{r}_p$ and proceeding as above. This leads to

$$\tilde{\mathbf{H}}(\mathbf{r},\omega) = \int_V \left(\tilde{\mathbf{J}}^i \times \nabla'G + \frac{\tilde{\rho}_m^i}{\tilde{\mu}}\nabla'G - j\omega\tilde{\epsilon}^c\tilde{\mathbf{J}}_m^iG\right)dV' +$$

$$+ \sum_{n=1}^N \int_{S_n}\left[(\hat{\mathbf{n}}' \times \tilde{\mathbf{H}}) \times \nabla'G + (\hat{\mathbf{n}}' \cdot \tilde{\mathbf{H}})\nabla'G + j\omega\tilde{\epsilon}^c(\hat{\mathbf{n}}' \times \tilde{\mathbf{E}})G\right]dS' +$$

$$+ \sum_{n=1}^N \frac{1}{j\omega\tilde{\mu}}\oint_{\Gamma_{na}+\Gamma_{nb}}(\mathbf{dl}' \cdot \tilde{\mathbf{E}})\nabla'G. \tag{6.8}$$

We can also obtain this expression by substituting (6.7) into Faraday's law.

6.1.2 The Sommerfeld radiation condition

In § 5.2.2 we found that if the potentials are not to be influenced by effects that are infinitely removed, then they must obey a radiation condition. We can make the same argument about the fields from (6.7) and (6.8). Let us allow one of the excluding surfaces, say S_N, to recede to infinity (enclosing all of the sources as it expands). As $S_N \to \infty$ any contributions from the fields on this surface to the fields at $\mathbf{r}$ should vanish.

Letting S_N be a sphere centered at the origin, we note that $\hat{\mathbf{n}}' = -\hat{\mathbf{r}}'$ and that as $r' \to \infty$

$$G(\mathbf{r}|\mathbf{r}';\omega) = \frac{e^{-jk|\mathbf{r}-\mathbf{r}'|}}{4\pi|\mathbf{r}-\mathbf{r}'|} \approx \frac{e^{-jkr'}}{4\pi r'},$$

$$\nabla'G(\mathbf{r}|\mathbf{r}';\omega) = \hat{\mathbf{R}}\left(\frac{1+jkR}{4\pi R^2}\right)e^{-jkR} \approx -\hat{\mathbf{r}}'\left(\frac{1+jkr'}{r'}\right)\frac{e^{-jkr'}}{4\pi r'}.$$

Substituting these expressions into (6.7) we find that

$$\lim_{S_N\to S_\infty}\oint_{S_N}\left[(\hat{\mathbf{n}}' \times \tilde{\mathbf{E}}) \times \nabla'G + (\hat{\mathbf{n}}' \cdot \tilde{\mathbf{E}})\nabla'G - j\omega\tilde{\mu}(\hat{\mathbf{n}}' \times \tilde{\mathbf{H}})G\right]dS'$$

$$\approx \lim_{r'\to\infty}\int_0^{2\pi}\int_0^\pi \left\{\left[(\hat{\mathbf{r}}' \times \tilde{\mathbf{E}}) \times \hat{\mathbf{r}}' + (\hat{\mathbf{r}}' \cdot \tilde{\mathbf{E}})\hat{\mathbf{r}}'\right]\left(\frac{1+jkr'}{r'}\right)\right.$$

$$\left. + j\omega\tilde{\mu}(\hat{\mathbf{r}}' \times \tilde{\mathbf{H}})\right\}\frac{e^{-jkr'}}{4\pi r'}r'^2\sin\theta'\,d\theta'\,d\phi'$$

$$\approx \lim_{r'\to\infty}\int_0^{2\pi}\int_0^\pi \left\{r'\left[jk\tilde{\mathbf{E}} + j\omega\tilde{\mu}(\hat{\mathbf{r}}' \times \tilde{\mathbf{H}})\right] + \tilde{\mathbf{E}}\right\}\frac{e^{-jkr'}}{4\pi}\sin\theta'\,d\theta'\,d\phi'.$$

Since this gives the contribution to the field in V from the fields on the surface receding to infinity, we expect that this term should be zero. If the medium has loss, then the exponential term decays and drives the contribution to zero. For a lossless medium the contributions are zero if

$$\lim_{r \to \infty} r\tilde{\mathbf{E}}(\mathbf{r}, \omega) < \infty, \tag{6.9}$$

$$\lim_{r \to \infty} r \left[\eta \hat{\mathbf{r}} \times \tilde{\mathbf{H}}(\mathbf{r}, \omega) + \tilde{\mathbf{E}}(\mathbf{r}, \omega) \right] = 0. \tag{6.10}$$

To accompany (6.8) we also have

$$\lim_{r \to \infty} r\tilde{\mathbf{H}}(\mathbf{r}, \omega) < \infty, \tag{6.11}$$

$$\lim_{r \to \infty} r \left[\eta \tilde{\mathbf{H}}(\mathbf{r}, \omega) - \hat{\mathbf{r}} \times \tilde{\mathbf{E}}(\mathbf{r}, \omega) \right] = 0. \tag{6.12}$$

We refer to (6.9) and (6.11) as the *finiteness conditions*, and to (6.10) and (6.12) as the *Sommerfeld radiation condition*, for the electromagnetic field. They show that far from the sources the fields must behave as a wave TEM to the r-direction. We shall see in § 6.2 that the waves are in fact *spherical TEM waves*.

6.1.3   Fields in the excluded region: the extinction theorem

The Stratton–Chu formula provides a solution for the field within the region V, external to the excluded regions. An interesting consequence of this formula, and one that helps us identify the equivalence principle, is that it gives the null result $\tilde{\mathbf{H}} = \tilde{\mathbf{E}} = 0$ when evaluated at points within the excluded regions.

We can show this by considering two cases. In the first case we do *not* exclude the particular region V_m, but do exclude the remaining regions V_n, $n \neq m$. Then the electric field everywhere outside the remaining excluded regions (including at points within V_m) is, by (6.7),

$$\tilde{\mathbf{E}}(\mathbf{r}, \omega) = \int_{V+V_m} \left(-\tilde{\mathbf{J}}_m^i \times \nabla' G + \frac{\tilde{\rho}^i}{\tilde{\epsilon}^c} \nabla' G - j\omega\tilde{\mu}\tilde{\mathbf{J}}^i G \right) dV' +$$
$$+ \sum_{n \neq m} \int_{S_n} \left[(\hat{\mathbf{n}}' \times \tilde{\mathbf{E}}) \times \nabla' G + (\hat{\mathbf{n}}' \cdot \tilde{\mathbf{E}})\nabla' G - j\omega\tilde{\mu}(\hat{\mathbf{n}}' \times \tilde{\mathbf{H}})G \right] dS' -$$
$$- \sum_{n \neq m} \frac{1}{j\omega\tilde{\epsilon}^c} \oint_{\Gamma_{na}+\Gamma_{nb}} (\mathbf{dl}' \cdot \tilde{\mathbf{H}})\nabla' G, \qquad \mathbf{r} \in V + V_m.$$

In the second case we apply the Stratton–Chu formula only to V_m, and exclude all other regions. We incur a sign change on the surface and line integrals compared to the first case because the normal is now directed oppositely. By (6.7) we have

$$\tilde{\mathbf{E}}(\mathbf{r}, \omega) = \int_{V_m} \left(-\tilde{\mathbf{J}}_m^i \times \nabla' G + \frac{\tilde{\rho}^i}{\tilde{\epsilon}^c} \nabla' G - j\omega\tilde{\mu}\tilde{\mathbf{J}}^i G \right) dV' -$$
$$- \int_{S_m} \left[(\hat{\mathbf{n}}' \times \tilde{\mathbf{E}}) \times \nabla' G + (\hat{\mathbf{n}}' \cdot \tilde{\mathbf{E}})\nabla' G - j\omega\tilde{\mu}(\hat{\mathbf{n}}' \times \tilde{\mathbf{H}})G \right] dS' +$$
$$+ \frac{1}{j\omega\tilde{\epsilon}^c} \oint_{\Gamma_{na}+\Gamma_{nb}} (\mathbf{dl}' \cdot \tilde{\mathbf{H}})\nabla' G, \qquad \mathbf{r} \in V_m.$$

Each of the expressions for $\tilde{\mathbf{E}}$ is equally valid for points within V_m. Upon subtraction we get

$$0 = \int_V \left(-\tilde{\mathbf{J}}_m^i \times \nabla'G + \frac{\tilde{\rho}^i}{\tilde{\epsilon}^c} \nabla'G - j\omega\tilde{\mu}\tilde{\mathbf{J}}^i G \right) dV' +$$

$$+ \sum_{n=1}^{N} \int_{S_n} \left[(\hat{\mathbf{n}}' \times \tilde{\mathbf{E}}) \times \nabla'G + (\hat{\mathbf{n}}' \cdot \tilde{\mathbf{E}})\nabla'G - j\omega\tilde{\mu}(\hat{\mathbf{n}}' \times \tilde{\mathbf{H}})G \right] dS' -$$

$$- \sum_{n=1}^{N} \frac{1}{j\omega\tilde{\epsilon}^c} \oint_{\Gamma_{na}+\Gamma_{nb}} (\mathbf{dl}' \cdot \tilde{\mathbf{H}})\nabla'G, \qquad \mathbf{r} \in V_m.$$

This expression is exactly the Stratton–Chu formula (6.7) evaluated at points within the excluded region V_m. The treatment of $\tilde{\mathbf{H}}$ is analogous and is left as an exercise. Since we may repeat this for any excluded region, we find that the Stratton–Chu formula returns the null field when evaluated at points outside V. This is sometimes referred to as the *vector Ewald–Oseen extinction theorem* [90]. We must emphasize that the fields within the excluded regions are *not* generally equal to zero; the Stratton–Chu formula merely returns this result when evaluated there.

6.2 Fields in an unbounded medium

Two special cases of the Stratton–Chu formula are important because of their application to antenna theory. The first is that of sources radiating into an unbounded region. The second involves a bounded region with all sources excluded. We shall consider the former here and the latter in § 6.3.

Assuming that there are no bounding surfaces in (6.7) and (6.8), except for one surface that has been allowed to recede to infinity and therefore provides no surface contribution, we find that the electromagnetic fields in unbounded space are given by

$$\tilde{\mathbf{E}} = \int_V \left(-\tilde{\mathbf{J}}_m^i \times \nabla'G + \frac{\tilde{\rho}^i}{\tilde{\epsilon}^c} \nabla'G - j\omega\tilde{\mu}\tilde{\mathbf{J}}^i G \right) dV',$$

$$\tilde{\mathbf{H}} = \int_V \left(\tilde{\mathbf{J}}^i \times \nabla'G + \frac{\tilde{\rho}_m^i}{\tilde{\mu}} \nabla'G - j\omega\tilde{\epsilon}^c\tilde{\mathbf{J}}_m^i G \right) dV'.$$

We can view the right-hand sides as superpositions of the fields present in the cases where (1) electric sources are present exclusively, and (2) magnetic sources are present exclusively. With $\tilde{\rho}_m^i = 0$ and $\tilde{\mathbf{J}}_m^i = 0$ we find that

$$\tilde{\mathbf{E}} = \int_V \left(\frac{\tilde{\rho}^i}{\tilde{\epsilon}^c} \nabla'G - j\omega\tilde{\mu}\tilde{\mathbf{J}}^i G \right) dV', \tag{6.13}$$

$$\tilde{\mathbf{H}} = \int_V \tilde{\mathbf{J}}^i \times \nabla'G \, dV'. \tag{6.14}$$

Using $\nabla'G = -\nabla G$ we can write

$$\tilde{\mathbf{E}}(\mathbf{r},\omega) = -\nabla \int_V \frac{\tilde{\rho}^i(\mathbf{r}',\omega)}{\tilde{\epsilon}^c(\omega)} G(\mathbf{r}|\mathbf{r}';\omega) \, dV' - j\omega \int_V \tilde{\mu}(\omega)\tilde{\mathbf{J}}^i(\mathbf{r}',\omega)G(\mathbf{r}|\mathbf{r}';\omega) \, dV'$$

$$= -\nabla\tilde{\phi}_e(\mathbf{r},\omega) - j\omega\tilde{\mathbf{A}}_e(\mathbf{r},\omega),$$

where

$$\tilde{\phi}_e(\mathbf{r},\omega) = \int_V \frac{\tilde{\rho}^i(\mathbf{r}',\omega)}{\tilde{\epsilon}^c(\omega)} G(\mathbf{r}|\mathbf{r}';\omega)\,dV',$$

$$\tilde{\mathbf{A}}_e(\mathbf{r},\omega) = \int_V \tilde{\mu}(\omega)\tilde{\mathbf{J}}^i(\mathbf{r}',\omega)G(\mathbf{r}|\mathbf{r}';\omega)\,dV', \tag{6.15}$$

are the electric scalar and vector potential functions introduced in § 5.2. Using $\tilde{\mathbf{J}}^i\times\nabla'G = -\tilde{\mathbf{J}}^i\times\nabla G = \nabla\times(\tilde{\mathbf{J}}^iG)$ we have

$$\tilde{\mathbf{H}}(\mathbf{r},\omega) = \frac{1}{\tilde{\mu}(\omega)}\nabla\times\int_V \tilde{\mu}(\omega)\tilde{\mathbf{J}}^i(\mathbf{r}',\omega)G(\mathbf{r}|\mathbf{r}';\omega)\,dV'$$

$$= \frac{1}{\tilde{\mu}(\omega)}\nabla\times\tilde{\mathbf{A}}_e(\mathbf{r},\omega). \tag{6.16}$$

These expressions for the fields are identical to those of (5.56) and (5.57), and thus the integral formula for the electromagnetic fields produces a result identical to that obtained using potential relations. Similarly, with $\tilde{\rho}^i = 0, \tilde{\mathbf{J}}^i = 0$ we have

$$\tilde{\mathbf{E}} = -\int_V \tilde{\mathbf{J}}_m^i\times\nabla'G\,dV',$$

$$\tilde{\mathbf{H}} = \int_V \left(\frac{\tilde{\rho}_m^i}{\tilde{\mu}}\nabla'G - j\omega\tilde{\epsilon}^c\tilde{\mathbf{J}}_m^iG\right)dV',$$

or

$$\tilde{\mathbf{E}}(\mathbf{r},\omega) = -\frac{1}{\tilde{\epsilon}^c(\omega)}\nabla\times\tilde{\mathbf{A}}_h(\mathbf{r},\omega),$$

$$\tilde{\mathbf{H}}(\mathbf{r},\omega) = -\nabla\tilde{\phi}_h(\mathbf{r},\omega) - j\omega\tilde{\mathbf{A}}_h(\mathbf{r},\omega),$$

where

$$\tilde{\phi}_h(\mathbf{r},\omega) = \int_V \frac{\tilde{\rho}_m^i(\mathbf{r}',\omega)}{\tilde{\mu}(\omega)} G(\mathbf{r}|\mathbf{r}';\omega)\,dV',$$

$$\tilde{\mathbf{A}}_h(\mathbf{r},\omega) = \int_V \tilde{\epsilon}^c(\omega)\tilde{\mathbf{J}}_m^i(\mathbf{r}',\omega)G(\mathbf{r}|\mathbf{r}';\omega)\,dV',$$

are the magnetic scalar and vector potentials introduced in § 5.2.

6.2.1   The far-zone fields produced by sources in unbounded space

Many antennas may be analyzed in terms of electric currents and charges radiating in unbounded space. Since antennas are used to transmit information over great distances, the fields far from the sources are often of most interest. Assume that the sources are contained within a sphere of radius r_s centered at the origin. We define the *far zone* of the sources to consist of all observation points satisfying both $r \gg r_s$ (and thus $r \gg r'$) and $kr \gg 1$. For points in the far zone we may approximate the unit vector $\hat{\mathbf{R}}$ directed from the sources to the observation point by the unit vector $\hat{\mathbf{r}}$ directed from the origin to the observation point. We may also approximate

$$\nabla'G = \frac{d}{dR}\left(\frac{e^{-jkR}}{4\pi R}\right)\nabla'R = \hat{\mathbf{R}}\left(\frac{1+jkR}{R}\right)\frac{e^{-jkR}}{4\pi R} \approx \hat{\mathbf{r}}jk\frac{e^{-jkR}}{4\pi R} = \hat{\mathbf{r}}jkG. \tag{6.17}$$

Using this we can obtain expressions for $\tilde{\mathbf{E}}$ and $\tilde{\mathbf{H}}$ in the far zone of the sources. The approximation (6.17) leads directly to

$$\tilde{\rho}^i \nabla' G \approx \left[j \frac{\nabla' \cdot \tilde{\mathbf{J}}^i}{\omega} \right] (\hat{\mathbf{r}} j k G) = -\frac{k}{\omega} \hat{\mathbf{r}} \left[\nabla' \cdot (G \tilde{\mathbf{J}}^i) - \tilde{\mathbf{J}}^i \cdot \nabla' G \right].$$

Substituting this into (6.13), again using (6.17) and also using the divergence theorem, we have

$$\tilde{\mathbf{E}}(\mathbf{r}, \omega) \approx -\int_V j \omega \tilde{\mu} \left[\tilde{\mathbf{J}}^i - \hat{\mathbf{r}} (\hat{\mathbf{r}} \cdot \tilde{\mathbf{J}}^i) \right] G \, dV' + \hat{\mathbf{r}} \frac{k}{\omega \tilde{\epsilon}^c} \oint_S (\hat{\mathbf{n}}' \cdot \tilde{\mathbf{J}}^i) G \, dS',$$

where the surface S surrounds the volume V that contains the impressed sources. If we let this volume slightly exceed that needed to contain the sources, then we do not change the value of the volume integral above; however, the surface integral vanishes since $\hat{\mathbf{n}}' \cdot \tilde{\mathbf{J}}^i = 0$ everywhere on the surface. Using $\hat{\mathbf{r}} \times (\hat{\mathbf{r}} \times \tilde{\mathbf{J}}^i) = \hat{\mathbf{r}} (\hat{\mathbf{r}} \cdot \tilde{\mathbf{J}}^i) - \tilde{\mathbf{J}}^i$ we then obtain the far-zone expression

$$\tilde{\mathbf{E}}(\mathbf{r}, \omega) \approx j \omega \hat{\mathbf{r}} \times \left[\hat{\mathbf{r}} \times \int_V \tilde{\mu}(\omega) \tilde{\mathbf{J}}^i(\mathbf{r}', \omega) G(\mathbf{r}|\mathbf{r}'; \omega) \, dV' \right]$$

$$= j \omega \hat{\mathbf{r}} \times \left[\hat{\mathbf{r}} \times \tilde{\mathbf{A}}_e(\mathbf{r}, \omega) \right],$$

where $\tilde{\mathbf{A}}_e$ is the electric vector potential. The far-zone electric field has no r-component, and it is often convenient to write

$$\tilde{\mathbf{E}}(\mathbf{r}, \omega) \approx -j \omega \tilde{\mathbf{A}}_{eT}(\mathbf{r}, \omega) \tag{6.18}$$

where $\tilde{\mathbf{A}}_{eT}$ is the vector component of $\tilde{\mathbf{A}}_e$ transverse to the r-direction:

$$\tilde{\mathbf{A}}_{eT} = -\hat{\mathbf{r}} \times \left[\hat{\mathbf{r}} \times \tilde{\mathbf{A}}_e \right] = \tilde{\mathbf{A}}_e - \hat{\mathbf{r}} (\hat{\mathbf{r}} \cdot \tilde{\mathbf{A}}_e) = \hat{\boldsymbol{\theta}} \tilde{A}_{e\theta} + \hat{\boldsymbol{\phi}} \tilde{A}_{e\phi}.$$

We can approximate the magnetic field in a similar fashion. Noting that $\tilde{\mathbf{J}}^i \times \nabla' G = \tilde{\mathbf{J}}^i \times (j k \hat{\mathbf{r}} G)$ we have

$$\tilde{\mathbf{H}}(\mathbf{r}, \omega) \approx -j \frac{k}{\tilde{\mu}(\omega)} \hat{\mathbf{r}} \times \int_V \tilde{\mu}(\omega) \tilde{\mathbf{J}}^i(\mathbf{r}', \omega) G(\mathbf{r}|\mathbf{r}', \omega) \, dV'$$

$$\approx -\frac{1}{\eta} j \omega \hat{\mathbf{r}} \times \tilde{\mathbf{A}}_e(\mathbf{r}, \omega).$$

With this we have

$$\tilde{\mathbf{E}}(\mathbf{r}, \omega) = -\eta \hat{\mathbf{r}} \times \tilde{\mathbf{H}}(\mathbf{r}, \omega), \qquad \tilde{\mathbf{H}}(\mathbf{r}, \omega) = \frac{\hat{\mathbf{r}} \times \tilde{\mathbf{E}}(\mathbf{r}, \omega)}{\eta},$$

in the far zone.

To simplify the computations involved, we often choose to approximate the vector potential in the far zone. Noting that

$$R = \sqrt{(\mathbf{r} - \mathbf{r}') \cdot (\mathbf{r} - \mathbf{r}')} = \sqrt{r^2 + r'^2 - 2(\mathbf{r} \cdot \mathbf{r}')}$$

and remembering that $r \gg r'$ for $\mathbf{r}$ in the far zone, we can use the leading terms of a binomial expansion of the square root to get

$$R = r\sqrt{1 - \frac{2(\hat{\mathbf{r}} \cdot \mathbf{r}')}{r} + \left(\frac{r'}{r}\right)^2} \approx r\sqrt{1 - \frac{2(\hat{\mathbf{r}} \cdot \mathbf{r}')}{r}} \approx r\left[1 - \frac{\hat{\mathbf{r}} \cdot \mathbf{r}'}{r}\right]$$

$$\approx r - \hat{\mathbf{r}} \cdot \mathbf{r}'. \tag{6.19}$$

Thus the Green's function may be approximated as

$$G(\mathbf{r}|\mathbf{r}';\omega) \approx \frac{e^{-jkr}}{4\pi r} e^{jk\hat{\mathbf{r}}\cdot\mathbf{r}'}. \tag{6.20}$$

Here we have kept the approximation (6.19) intact in the phase of G but have used $1/R \approx 1/r$ in the amplitude of G. We must keep a more accurate approximation for the phase since $k(\hat{\mathbf{r}} \cdot \mathbf{r}')$ may be an appreciable fraction of a radian. We thus have the far-zone approximation for the vector potential

$$\tilde{\mathbf{A}}_e(\mathbf{r},\omega) \approx \tilde{\mu}(\omega)\frac{e^{-jkr}}{4\pi r}\int_V \tilde{\mathbf{J}}^i(\mathbf{r}',\omega)e^{jk\hat{\mathbf{r}}\cdot\mathbf{r}'}\, dV',$$

which we may use in computing (6.18).

Let us summarize the expressions for computing the far-zone fields:

$$\tilde{\mathbf{E}}(\mathbf{r},\omega) = -jw\left[\hat{\boldsymbol{\theta}}\tilde{A}_{e\theta}(\mathbf{r},\omega) + \hat{\boldsymbol{\phi}}\tilde{A}_{e\phi}(\mathbf{r},\omega)\right], \tag{6.21}$$

$$\tilde{\mathbf{H}}(\mathbf{r},\omega) = \frac{\hat{\mathbf{r}} \times \tilde{\mathbf{E}}(\mathbf{r},\omega)}{\eta}, \tag{6.22}$$

$$\tilde{\mathbf{A}}_e(\mathbf{r},\omega) = \frac{e^{-jkr}}{4\pi r}\tilde{\mu}(\omega)\tilde{\mathbf{a}}_e(\theta,\phi,\omega), \tag{6.23}$$

$$\tilde{\mathbf{a}}_e(\theta,\phi,\omega) = \int_V \tilde{\mathbf{J}}^i(\mathbf{r}',\omega)e^{jk\hat{\mathbf{r}}\cdot\mathbf{r}'}\, dV'. \tag{6.24}$$

Here $\tilde{\mathbf{a}}_e$ is called the *directional weighting function*. This function is independent of r and describes the angular variation, or *pattern*, of the fields.

In the far zone $\tilde{\mathbf{E}}, \tilde{\mathbf{H}}, \hat{\mathbf{r}}$ are mutually orthogonal. Because of this, and because the fields vary as e^{-jkr}/r, the electromagnetic field in the far zone takes the form of a spherical TEM wave, which is consistent with the Sommerfeld radiation condition.

Power radiated by time-harmonic sources in unbounded space. In § 5.2.1 we defined the power radiated by a time-harmonic source in unbounded space as the total time-average power passing through a sphere of very large radius. We found that for a Hertzian dipole the radiated power could be computed from the far-zone fields through

$$P_{av} = \lim_{r\to\infty}\int_0^{2\pi}\int_0^{\pi} \mathbf{S}_{av}\cdot\hat{\mathbf{r}}r^2\sin\theta\, d\theta\, d\phi$$

where

$$\mathbf{S}_{av} = \frac{1}{2}\mathrm{Re}\left\{\check{\mathbf{E}} \times \check{\mathbf{H}}^*\right\}$$

is the time-average Poynting vector. By superposition this holds for any localized source. Assuming a lossless medium and using phasor notation to describe the time-harmonic

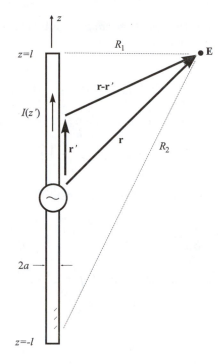

Figure 6.4: Dipole antenna in a lossless unbounded medium.

fields we have, by (6.22),

$$\mathbf{S}_{av} = \frac{1}{2} \operatorname{Re} \left\{ \frac{\breve{\mathbf{E}} \times (\hat{\mathbf{r}} \times \breve{\mathbf{E}}^*)}{\eta} \right\} = \hat{\mathbf{r}} \frac{\breve{\mathbf{E}} \cdot \breve{\mathbf{E}}^*}{2\eta}.$$

Substituting from (6.21), we can also write $\mathbf{S}_{av}$ in terms of the directional weighting function as

$$\mathbf{S}_{av} = \hat{\mathbf{r}} \frac{\breve{\omega}^2}{2\eta} \left(\breve{A}_{e\theta} \breve{A}_{e\theta}^* + \breve{A}_{e\phi} \breve{A}_{e\phi}^* \right) = \hat{\mathbf{r}} \frac{k^2 \eta}{(4\pi r)^2} \left(\frac{1}{2} \breve{a}_{e\theta} \breve{a}_{e\theta}^* + \frac{1}{2} \breve{a}_{e\phi} \breve{a}_{e\phi}^* \right). \tag{6.25}$$

We note that $\mathbf{S}_{av}$ describes the variation of the power density with θ, ϕ, and is thus sometimes used as a descriptor of the *power pattern* of the sources.

Example of a current source radiating into an unbounded medium: the dipole antenna. A common type of antenna consists of a thin wire of length $2l$ and radius a, fed at the center by a voltage generator as shown in Figure 6.4. The generator induces an impressed current on the surface of the wire which in turn radiates an electromagnetic wave. For very thin wires ($a \ll \lambda$, $a \ll l$) embedded in a lossless medium, the current may be accurately approximated using a standing-wave distribution:

$$\tilde{\mathbf{J}}^i(\mathbf{r}, \omega) = \hat{\mathbf{z}} \tilde{I}(\omega) \sin\left[k(l - |z|)\right] \delta(x)\delta(y). \tag{6.26}$$

We may compute the field produced by the dipole antenna by first finding the vector potential from (6.15) and then calculating the magnetic field from (6.16). The electric field may then be found by the use of Ampere's law.

We assume a lossless medium with parameters μ, ϵ. Substituting the current expression into (6.15) and integrating over x and y we find that

$$\tilde{\mathbf{A}}_e(\mathbf{r}, \omega) = \hat{\mathbf{z}} \frac{\mu \tilde{I}}{4\pi} \int_{-l}^{l} \sin k(l - |z'|) \frac{e^{-jkR}}{R} dz' \tag{6.27}$$

where $R = \sqrt{(z - z')^2 + \rho^2}$ and $\rho^2 = x^2 + y^2$. Using (6.16) we have

$$\tilde{\mathbf{H}} = \nabla \times \frac{1}{\mu} \tilde{\mathbf{A}}_e = -\hat{\phi} \frac{1}{\mu} \frac{\partial \tilde{A}_{ez}}{\partial \rho}.$$

Writing the sine function in (6.27) in terms of exponentials, we then have

$$\tilde{H}_\phi = j \frac{\tilde{I}}{8\pi} \left[e^{jkl} \int_{-l}^{0} \frac{\partial}{\partial \rho} \frac{e^{-jk(R-z')}}{R} dz' - e^{-jkl} \int_{-l}^{0} \frac{\partial}{\partial \rho} \frac{e^{-jk(R+z')}}{R} dz' + \right.$$
$$\left. + e^{jkl} \int_{0}^{l} \frac{\partial}{\partial \rho} \frac{e^{-jk(R+z')}}{R} dz' - e^{-jkl} \int_{0}^{l} \frac{\partial}{\partial \rho} \frac{e^{-jk(R-z')}}{R} dz' \right].$$

Noting that

$$\frac{\partial}{\partial \rho} \frac{e^{-jk(R \pm z')}}{R} = \pm \rho \frac{\partial}{\partial z'} \frac{e^{-jk(R \pm z')}}{R [R \mp (z - z')]} = -\rho \frac{1 + jkR}{R^3} e^{-jk(R \pm z')}$$

we can write

$$\tilde{H}_\phi = j \frac{\tilde{I} \rho}{8\pi} \left[-e^{jkl} \frac{e^{-jk(R-z')}}{R [R + (z - z')]} \bigg|_{-l}^{0} - e^{-jkl} \frac{e^{-jk(R+z')}}{R [R - (z - z')]} \bigg|_{-l}^{0} + \right.$$
$$\left. + e^{jkl} \frac{e^{-jk(R+z')}}{R [R - (z - z')]} \bigg|_{0}^{l} + e^{-jkl} \frac{e^{-jk(R-z')}}{R [R + (z - z')]} \bigg|_{0}^{l} \right].$$

Collecting terms and simplifying we get

$$\tilde{H}_\phi(\mathbf{r}, \omega) = j \frac{\tilde{I}(\omega)}{4\pi \rho} \left[e^{-jkR_1} + e^{-jkR_2} - (2 \cos kl) e^{-jkr} \right] \tag{6.28}$$

where $R_1 = \sqrt{\rho^2 + (z - l)^2}$ and $R_2 = \sqrt{\rho^2 + (z + l)^2}$. For points external to the dipole the source current is zero and thus

$$\tilde{\mathbf{E}}(\mathbf{r}, \omega) = \frac{1}{j\omega\epsilon} \nabla \times \tilde{\mathbf{H}}(\mathbf{r}, \omega) = \frac{1}{j\omega\epsilon} \left\{ -\hat{\rho} \frac{\partial}{\partial z} \tilde{H}_\phi(\mathbf{r}, \omega) + \hat{\mathbf{z}} \frac{1}{\rho} \frac{\partial}{\partial \rho} [\rho \tilde{H}_\phi(\mathbf{r}, \omega)] \right\}.$$

Performing the derivatives we have

$$\tilde{E}_\rho(\mathbf{r}, \omega) = j \frac{\eta \tilde{I}(\omega)}{4\pi} \left[\frac{z - l}{\rho} \frac{e^{-jkR_1}}{R_1} + \frac{z + l}{\rho} \frac{e^{-jkR_2}}{R_2} - \frac{z}{\rho} (2 \cos kl) \frac{e^{-jkr}}{r} \right], \tag{6.29}$$

$$\tilde{E}_z(\mathbf{r}, \omega) = -j \frac{\eta \tilde{I}(\omega)}{4\pi} \left[\frac{e^{-jkR_1}}{R_1} + \frac{e^{-jkR_2}}{R_2} - (2 \cos kl) \frac{e^{-jkr}}{r} \right]. \tag{6.30}$$

The work of specializing these expressions for points in the far zone is left as an exercise. Instead, we shall use the general far-zone expressions (6.21)–(6.24). Substituting (6.26) into (6.24) and carrying out the x and y integrals we have the directional weighting function

$$\tilde{\mathbf{a}}_e(\theta, \phi, \omega) = \int_{-l}^{l} \hat{\mathbf{z}} \tilde{I}(\omega) \sin k(l - |z'|) e^{jkz' \cos \theta} \, dz'.$$

Writing the sine functions in terms of exponentials we have

$$\tilde{\mathbf{a}}_e(\theta, \phi, \omega) = \frac{\hat{\mathbf{z}} \tilde{I}(\omega)}{2j} \left[e^{jkl} \int_{0}^{l} e^{jkz'(\cos \theta - 1)} \, dz' - e^{-jkl} \int_{0}^{l} e^{jkz'(\cos \theta + 1)} \, dz' + \right.$$

$$\left. + e^{jkl} \int_{-l}^{0} e^{jkz'(\cos \theta + 1)} - e^{-jkl} \int_{-l}^{0} e^{jkz'(\cos \theta - 1)} \right].$$

Carrying out the integrals and simplifying, we obtain

$$\tilde{\mathbf{a}}_e(\theta, \phi, \omega) = \hat{\mathbf{z}} \frac{2\tilde{I}(\omega)}{k} \frac{F(\theta, kl)}{\sin \theta}$$

where

$$F(\theta, kl) = \frac{\cos(kl \cos \theta) - \cos kl}{\sin \theta}$$

is called the *radiation function*. Using $\hat{\mathbf{z}} = \hat{\mathbf{r}} \cos \theta - \hat{\boldsymbol{\theta}} \sin \theta$ we find that

$$\tilde{a}_{e\theta}(\theta, \phi, \omega) = -\frac{2\tilde{I}(\omega)}{k} F(\theta, kl), \qquad \tilde{a}_{e\phi}(\theta, \phi, \omega) = 0.$$

Thus we have from (6.23) and (6.21) the electric field

$$\tilde{\mathbf{E}}(\mathbf{r}, \omega) = \hat{\boldsymbol{\theta}} \frac{j\eta \tilde{I}(\omega)}{2\pi} \frac{e^{-jkr}}{r} F(\theta, kl) \tag{6.31}$$

and from (6.22) the magnetic field

$$\tilde{\mathbf{H}}(\mathbf{r}, \omega) = \hat{\boldsymbol{\phi}} \frac{j\tilde{I}(\omega)}{2\pi} \frac{e^{-jkr}}{r} F(\theta, kl). \tag{6.32}$$

We see that the radiation function contains all of the angular dependence of the field and thus describes the pattern of the dipole. When the dipole is short compared to a wavelength we may approximate the radiation function as

$$F(\theta, kl \ll 1) \approx \frac{1 - \frac{1}{2}(kl \cos \theta)^2 - 1 + \frac{1}{2}(kl)^2}{\sin \theta} = \frac{1}{2}(kl)^2 \sin \theta. \tag{6.33}$$

So a short dipole antenna has the same pattern as a Hertzian dipole, whose far-zone electric field is (5.93).

We may also calculate the radiated power for time-harmonic fields. The time-average Poynting vector for the far-zone fields is, from (6.25),

$$\mathbf{S}_{av} = \hat{\mathbf{r}} \eta \frac{|\check{I}|^2}{8\pi^2 r^2} F^2(\theta, kl),$$

and thus the radiated power is

$$P_{av} = \eta \frac{|\check{I}|^2}{4\pi} \int_{0}^{\pi} F^2(\theta, kl) \sin \theta \, d\theta.$$

This expression cannot be computed in closed form. For a short dipole we may use (6.33) to approximate the power, but the result is somewhat misleading since the current on a short dipole is much smaller than $\tilde{I}$. A better measure of the strength of the current is its value at the center, or *feedpoint*, of the dipole. This *input current* is by (6.26) merely $\tilde{I}_0(\omega) = \tilde{I}(\omega)\sin(kl)$. Using this we find

$$P_{av} \approx \eta \frac{|\check{I}_0|^2}{4\pi} \frac{1}{4}(kl)^2 \int_0^\pi \sin^3\theta \, d\theta = \eta \frac{\pi}{3}|\check{I}_0|^2 \left(\frac{l}{\lambda}\right)^2.$$

This is exactly $1/4$ of the power radiated by a Hertzian dipole of the same length and current amplitude (5.95). The factor of $1/4$ comes from the difference between the current of the dipole antenna, which is zero at each end, and the current on the Hertzian dipole, which is constant across the length of the antenna. It is more common to use a dipole antenna that is a half wavelength long ($2l = \lambda/2$), since it is then nearly resonant. With this we have through numerical integration the free-space radiated power

$$P_{av} = \eta_0 \frac{|\check{I}_0|^2}{4\pi} \int_0^\pi \frac{\cos^2\left(\frac{\pi}{2}\cos\theta\right)}{\sin\theta} \, d\theta = 36.6|\check{I}_0|^2$$

and the radiation resistance

$$R_r = \frac{2P_{av}}{|\check{I}(z=0)|^2} = \frac{2P_{av}}{|\check{I}_0|^2} = 73.2 \; \Omega.$$

6.3 Fields in a bounded, source-free region

In § 6.2 we considered the first important special case of the Stratton–Chu formula: sources in an unbounded medium. We now consider the second important special case of a bounded, source-free region. This case has important applications to the study of microwave antennas and, in its scalar form, to the study of the diffraction of light.

6.3.1 The vector Huygens principle

We may derive the formula for a bounded, source-free region of space by specializing the general Stratton–Chu formulas. We assume that all sources of the fields are within the excluded regions and thus set the sources to zero within V. From (6.7)–(6.8) we have

$$\tilde{\mathbf{E}}(\mathbf{r}, \omega) = \sum_{n=1}^{N} \int_{S_n} \left[(\hat{\mathbf{n}}' \times \tilde{\mathbf{E}}) \times \nabla'G + (\hat{\mathbf{n}}' \cdot \tilde{\mathbf{E}})\nabla'G - j\omega\tilde{\mu}(\hat{\mathbf{n}}' \times \tilde{\mathbf{H}})G\right] dS' -$$

$$- \sum_{n=1}^{N} \frac{1}{j\omega\tilde{\epsilon}^c} \oint_{\Gamma_{na}+\Gamma_{nb}} (\mathbf{dl}' \cdot \tilde{\mathbf{H}})\nabla'G, \tag{6.34}$$

and

$$\tilde{\mathbf{H}}(\mathbf{r}, \omega) = \sum_{n=1}^{N} \int_{S_n} \left[(\hat{\mathbf{n}}' \times \tilde{\mathbf{H}}) \times \nabla'G + (\hat{\mathbf{n}}' \cdot \tilde{\mathbf{H}})\nabla'G + j\omega\tilde{\epsilon}^c(\hat{\mathbf{n}}' \times \tilde{\mathbf{E}})G\right] dS' +$$

$$+ \sum_{n=1}^{N} \frac{1}{j\omega\tilde{\mu}} \oint_{\Gamma_{na}+\Gamma_{nb}} (\mathbf{dl'} \cdot \tilde{\mathbf{E}}) \nabla' G. \tag{6.35}$$

This is known as the *vector Huygens principle* after the Dutch physicist C. Huygens, who formulated his "secondary source concept" to explain the propagation of light. According to his idea, published in *Traité de la lumière* in 1690, points on a propagating wavefront are secondary sources of spherical waves that add together in just the right way to produce the field on any successive wavefront. We can interpret (6.34) and (6.35) in much the same way. The field at each point within V, where there are no sources, can be imagined to arise from spherical waves emanated from every point on the surface bounding V. The amplitudes of these waves are determined by the values of the fields on the boundaries. Thus, we may consider the boundary fields to be equivalent to secondary sources of the fields within V. We will expand on this concept below by introducing the concept of equivalence and identifying the specific form of the secondary sources.

6.3.2 The Franz formula

The vector Huygens principle as derived above requires secondary sources for the fields within V that involve both the tangential and normal components of the fields on the bounding surface. Since only tangential components are required to guarantee uniqueness within V, we seek an expression involving only $\hat{\mathbf{n}} \times \tilde{\mathbf{H}}$ and $\hat{\mathbf{n}} \times \tilde{\mathbf{E}}$. Physically, the normal component of the field is equivalent to a secondary charge source on the surface while the tangential component is equivalent to a secondary current source. Since charge and current are related by the continuity equation, specification of the normal component is superfluous.

To derive a version of the vector Huygens principle that omits the normal fields we take the curl of (6.35) to get

$$\nabla \times \tilde{\mathbf{H}}(\mathbf{r}, \omega) = \sum_{n=1}^{N} \nabla \times \oint_{S_n} (\hat{\mathbf{n}}' \times \tilde{\mathbf{H}}) \times \nabla' G \, dS' + \sum_{n=1}^{N} \oint_{S_n} \nabla \times \left[(\hat{\mathbf{n}}' \cdot \tilde{\mathbf{H}}) \nabla' G \right] dS' +$$
$$+ \sum_{n=1}^{N} \nabla \times \oint_{S_n} j\omega\tilde{\epsilon}^c (\hat{\mathbf{n}}' \times \tilde{\mathbf{E}}) G \, dS' + \sum_{n=1}^{N} \frac{1}{j\omega\tilde{\mu}} \oint_{\Gamma_{na}+\Gamma_{nb}} \nabla \times \left[(\mathbf{dl'} \cdot \tilde{\mathbf{E}}) \nabla' G \right] dS'. \tag{6.36}$$

Now, using $\nabla' G = -\nabla G$ and employing the vector identity (B.43) we can show that

$$\nabla \times [f(\mathbf{r}') \nabla' G(\mathbf{r}|\mathbf{r}')] = -f(\mathbf{r}') \{ \nabla \times [\nabla G(\mathbf{r}|\mathbf{r}')] \} + [\nabla G(\mathbf{r}|\mathbf{r}')] \times \nabla f(\mathbf{r}') = 0,$$

since $\nabla \times \nabla G = 0$ and $\nabla f(\mathbf{r}') = 0$. This implies that the second and fourth terms of (6.36) are zero. The first term can be modified using

$$\nabla \times \left\{ \left[\hat{\mathbf{n}}' \times \tilde{\mathbf{H}}(\mathbf{r}') \right] G(\mathbf{r}|\mathbf{r}') \right\} = G(\mathbf{r}|\mathbf{r}') \nabla \times \left[\hat{\mathbf{n}}' \times \tilde{\mathbf{H}}(\mathbf{r}') \right] - \left[\hat{\mathbf{n}}' \times \tilde{\mathbf{H}}(\mathbf{r}') \right] \times \nabla G(\mathbf{r}|\mathbf{r}')$$
$$= \left[\hat{\mathbf{n}}' \times \tilde{\mathbf{H}}(\mathbf{r}') \right] \times \nabla' G(\mathbf{r}|\mathbf{r}'),$$

giving

$$\nabla \times \tilde{\mathbf{H}}(\mathbf{r}, \omega) = \sum_{n=1}^{N} \nabla \times \oint_{S_n} \nabla \times \left[(\hat{\mathbf{n}}' \times \tilde{\mathbf{H}}) G \right] dS' + \sum_{n=1}^{N} \nabla \times \oint_{S_n} j\omega\tilde{\epsilon}^c (\hat{\mathbf{n}}' \times \tilde{\mathbf{E}}) G \, dS'.$$

Finally, using Ampere's law $\nabla \times \tilde{\mathbf{H}} = j\omega\tilde{\epsilon}^c\tilde{\mathbf{E}}$ in the source free region V, and taking the curl in the first term outside the integral, we have

$$\tilde{\mathbf{E}}(\mathbf{r},\omega) = \sum_{n=1}^{N} \nabla \times \nabla \times \oint_{S_n} \frac{1}{j\omega\tilde{\epsilon}^c}(\hat{\mathbf{n}}' \times \tilde{\mathbf{H}})G\,dS' + \sum_{n=1}^{N} \nabla \times \oint_{S_n} (\hat{\mathbf{n}}' \times \tilde{\mathbf{E}})G\,dS'. \qquad (6.37)$$

Similarly

$$\tilde{\mathbf{H}}(\mathbf{r},\omega) = -\sum_{n=1}^{N} \nabla \times \nabla \times \oint_{S_n} \frac{1}{j\omega\tilde{\mu}}(\hat{\mathbf{n}}' \times \tilde{\mathbf{E}})G\,dS' + \sum_{n=1}^{N} \nabla \times \oint_{S_n} (\hat{\mathbf{n}}' \times \tilde{\mathbf{H}})G\,dS'. \qquad (6.38)$$

These expressions together constitute the *Franz formula* for the vector Huygens principle [192].

6.3.3   Love's equivalence principle

Love's equivalence principle allows us to identify the equivalent Huygens sources for the fields within a bounded, source-free region V. It then allows us to replace a problem in the bounded region with an "equivalent" problem in unbounded space where the source-excluding surfaces are replaced by equivalent sources. The field produced by both the real and the equivalent sources gives a field in V identical to that of the original problem. This is particularly useful since we know how to compute the fields within an unbounded region by employing potential functions.

We identify the equivalent sources by considering the electric and magnetic Hertzian potentials produced by electric and magnetic current sources. Consider an impressed electric surface current $\tilde{\mathbf{J}}_s^{eq}$ and a magnetic surface current $\tilde{\mathbf{J}}_{ms}^{eq}$ flowing on the closed surface S in a homogeneous, isotropic medium with permeability $\tilde{\mu}(\omega)$ and complex permittivity $\tilde{\epsilon}^c(\omega)$. These sources produce

$$\tilde{\boldsymbol{\Pi}}_e(\mathbf{r},\omega) = \oint_S \frac{\tilde{\mathbf{J}}_s^{eq}(\mathbf{r}',\omega)}{j\omega\tilde{\epsilon}^c(\omega)}G(\mathbf{r}|\mathbf{r}';\omega)\,dS', \qquad (6.39)$$

$$\tilde{\boldsymbol{\Pi}}_h(\mathbf{r},\omega) = \oint_S \frac{\tilde{\mathbf{J}}_{ms}^{eq}(\mathbf{r}',\omega)}{j\omega\tilde{\mu}(\omega)}G(\mathbf{r}|\mathbf{r}';\omega)\,dS', \qquad (6.40)$$

which in turn can be used to find

$$\tilde{\mathbf{E}} = \nabla \times (\nabla \times \tilde{\boldsymbol{\Pi}}_e) - j\omega\tilde{\mu}\nabla \times \tilde{\boldsymbol{\Pi}}_h,$$
$$\tilde{\mathbf{H}} = j\omega\tilde{\epsilon}^c\nabla \times \tilde{\boldsymbol{\Pi}}_e + \nabla \times (\nabla \times \tilde{\boldsymbol{\Pi}}_h).$$

Upon substitution we find that

$$\tilde{\mathbf{E}}(\mathbf{r},\omega) = \nabla \times \nabla \times \oint_S \frac{1}{j\omega\tilde{\epsilon}^c}\left[\tilde{\mathbf{J}}_s^{eq}G\right]dS' + \nabla \times \oint_S [-\tilde{\mathbf{J}}_{ms}^{eq}]G\,dS',$$
$$\tilde{\mathbf{H}}(\mathbf{r},\omega) = -\nabla \times \nabla \times \oint_S \frac{1}{j\omega\tilde{\mu}}\left[-\tilde{\mathbf{J}}_{ms}^{eq}G\right]dS' + \nabla \times \oint_S \tilde{\mathbf{J}}_s^{eq}G\,dS'.$$

These are identical to the Franz equations (6.37) and (6.38) if we identify

$$\tilde{\mathbf{J}}_s^{eq} = \hat{\mathbf{n}} \times \tilde{\mathbf{H}}, \qquad \tilde{\mathbf{J}}_{ms}^{eq} = -\hat{\mathbf{n}} \times \tilde{\mathbf{E}}. \qquad (6.41)$$

These are the equivalent source densities for the Huygens principle.

We now state *Love's equivalence principle* [39]. Consider the fields within a homogeneous, source-free region V with parameters $(\tilde{\epsilon}^c, \tilde{\mu})$ bounded by a surface S. We know how to compute the fields using the Franz formula and the surface fields. Now consider a second problem in which the same surface S exists in an unbounded medium with identical parameters. If the surface carries the equivalent sources (6.41) then the electromagnetic fields within V calculated using the Hertzian potentials (6.39) and (6.40) are identical to those of the first problem, while the fields calculated outside V are zero. We see that this must be true since the Franz formulas and the field/potential formulas are identical, and the Franz formula (since it was derived from the Stratton–Chu formula) gives the null field outside V. The two problems are *equivalent* in the sense that they produce identical fields within V.

The fields produced by the equivalent sources obey the appropriate boundary conditions across S. From (2.194) and (2.195) we have the boundary conditions

$$\hat{\mathbf{n}} \times (\tilde{\mathbf{H}}_1 - \tilde{\mathbf{H}}_2) = \tilde{\mathbf{J}}_s,$$
$$\hat{\mathbf{n}} \times (\tilde{\mathbf{E}}_1 - \tilde{\mathbf{E}}_2) = -\tilde{\mathbf{J}}_{ms}.$$

Here $\hat{\mathbf{n}}$ points inward to V, $(\tilde{\mathbf{E}}_1, \tilde{\mathbf{H}}_1)$ are the fields within V, and $(\tilde{\mathbf{E}}_2, \tilde{\mathbf{H}}_2)$ are the fields within the excluded region. If the fields produced by the equivalent sources within the excluded region are zero, then the fields must obey

$$\hat{\mathbf{n}} \times \tilde{\mathbf{H}}_1 = \tilde{\mathbf{J}}_s^{eq},$$
$$\hat{\mathbf{n}} \times \tilde{\mathbf{E}}_1 = -\tilde{\mathbf{J}}_{ms}^{eq},$$

which is true by the definition of $(\tilde{\mathbf{J}}_s^{eq}, \tilde{\mathbf{J}}_{sm}^{eq})$.

Note that we can extend the equivalence principle to the case where the media are different internal to V than external to V. See Chen [29].

With the equivalent sources identified we may compute the electromagnetic field in V using standard techniques. Specifically, we may use the Hertzian potentials as shown above or, since the Hertzian potentials are a simple remapping of the vector potentials, we may use (5.60) and (5.61) to write

$$\tilde{\mathbf{E}} = -j\frac{\omega}{k^2}\left[\nabla(\nabla \cdot \tilde{\mathbf{A}}_e) + k^2\tilde{\mathbf{A}}_e\right] - \frac{1}{\tilde{\epsilon}^c}\nabla \times \tilde{\mathbf{A}}_h,$$
$$\tilde{\mathbf{H}} = -j\frac{\omega}{k^2}\left[\nabla(\nabla \cdot \tilde{\mathbf{A}}_h) + k^2\tilde{\mathbf{A}}_h\right] + \frac{1}{\tilde{\mu}}\nabla \times \tilde{\mathbf{A}}_e,$$

where

$$\tilde{\mathbf{A}}_e(\mathbf{r}, \omega) = \oint_S \tilde{\mu}(\omega)\tilde{\mathbf{J}}_s^{eq}(\mathbf{r}', \omega)G(\mathbf{r}|\mathbf{r}'; \omega)\, dS' \tag{6.42}$$

$$= \oint_S \tilde{\mu}(\omega)[\hat{\mathbf{n}}' \times \tilde{\mathbf{H}}(\mathbf{r}', \omega)]G(\mathbf{r}|\mathbf{r}'; \omega)\, dS', \tag{6.43}$$

$$\tilde{\mathbf{A}}_h(\mathbf{r}, \omega) = \oint_S \tilde{\epsilon}^c(\omega)\tilde{\mathbf{J}}_{ms}^{eq}(\mathbf{r}', \omega)G(\mathbf{r}|\mathbf{r}'; \omega)\, dS' \tag{6.44}$$

$$= \oint_S \tilde{\epsilon}^c(\omega)[-\hat{\mathbf{n}}' \times \tilde{\mathbf{E}}(\mathbf{r}', \omega)]G(\mathbf{r}|\mathbf{r}'; \omega)\, dS'. \tag{6.45}$$

At points where the source is zero we can write the fields in the alternative form

$$\tilde{\mathbf{E}} = -j\frac{\omega}{k^2}\nabla \times \nabla \times \tilde{\mathbf{A}}_e - \frac{1}{\tilde{\epsilon}^c}\nabla \times \tilde{\mathbf{A}}_h, \tag{6.46}$$

$$\tilde{\mathbf{H}} = -j\frac{\omega}{k^2}\nabla \times \nabla \times \tilde{\mathbf{A}}_h + \frac{1}{\tilde{\mu}}\nabla \times \tilde{\mathbf{A}}. \tag{6.47}$$

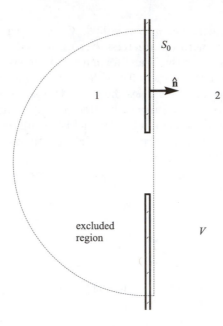

Figure 6.5: Geometry for problem of an aperture in a perfectly conducting ground screen illuminated by an impressed source.

By superposition, if there are volume sources within V we merely add the fields due to these sources as computed from the potential functions.

6.3.4   The Schelkunoff equivalence principle

With Love's equivalence principle we create an equivalent problem by replacing an excluded region by equivalent electric and magnetic sources. These require knowledge of both the tangential electric and magnetic fields over the bounding surface. However, the uniqueness theorem says that only one of either the tangential electric or the tangential magnetic fields need be specified to make the fields within V unique. Thus we may wonder whether it is possible to formulate an equivalent problem that involves only tangential $\tilde{\mathbf{E}}$ or tangential $\tilde{\mathbf{H}}$. It is indeed possible, as shown by Schelkunoff [39, 169].

When we use the equivalent sources to form the equivalent problem, we know that they produce a null field within the excluded region. Thus we may form a different equivalent problem by filling the excluded region with a perfect conductor, and keeping the same equivalent sources. The boundary conditions across S are not changed, and thus by the uniqueness theorem the fields within V are not altered. However, the manner in which we must compute the fields within V is changed. We can no longer use formulas for the fields produced by sources in free space, but must use formulas for fields produced by sources in the vicinity of a conducting body. In general this can be difficult since it requires the formation of a new Green's function that satisfies the boundary condition over the conducting body (which could possess a peculiar shape). Fortunately, we showed in § 4.10.2 that an electric source adjacent and tangential to a perfect electric conductor produces no field, hence we need not consider the equivalent electric sources ($\hat{\mathbf{n}} \times \tilde{\mathbf{H}}$) when computing the fields in V. Thus, in our new equivalent problem we need the single tangential field $-\hat{\mathbf{n}} \times \tilde{\mathbf{E}}$. This is the *Schelkunoff equivalence principle*.

There is one situation in which it is relatively easy to use the Schelkunoff equivalence. Consider a perfectly conducting ground screen with an aperture in it, as shown in Figure 6.5. We assume that the aperture has been illuminated in some way by an electromagnetic wave produced by sources in region 1 so that there are both fields within the aperture and electric current flowing on the region-2 side of the screen due to diffraction from the edges of the aperture. We wish to compute the fields in region 2. We can create an equivalent problem by placing a planar surface S_0 adjacent to the screen, but slightly offset into region 2, and then closing the surface at infinity so that all of the screen plus region 1 is excluded. Then we replace region 1 with homogeneous space and place on S_0 the equivalent currents $\tilde{\mathbf{J}}_s^{eq} = \hat{\mathbf{n}} \times \tilde{\mathbf{H}}$, $\tilde{\mathbf{J}}_{ms}^{eq} = -\hat{\mathbf{n}} \times \tilde{\mathbf{E}}$, where $\tilde{\mathbf{H}}$ and $\tilde{\mathbf{E}}$ are the fields on S_0 in the original problem. We note that over the portion of S_0 adjacent to the screen $\tilde{\mathbf{J}}_{ms}^{eq} = 0$ since $\hat{\mathbf{n}} \times \tilde{\mathbf{E}} = 0$, but that $\tilde{\mathbf{J}}_s^{eq} \neq 0$. From the equivalent currents we can compute the fields in region 2 using the potential functions. However, it is often difficult to determine $\tilde{\mathbf{J}}_s^{eq}$ over the conducting surface. If we apply Schelkunoff's equivalence, we can formulate a second equivalent problem in which we place into region 1 a perfect conductor. Then we have the equivalent source currents $\tilde{\mathbf{J}}_s^{eq}$ and $\tilde{\mathbf{J}}_{ms}^{eq}$ adjacent and tangential to a perfect conductor. By the image theorem of § 5.1.1 we can replace this problem by yet another equivalent problem in which the conductor is replaced by the images of $\tilde{\mathbf{J}}_s^{eq}$ and $\tilde{\mathbf{J}}_{ms}^{eq}$ in homogeneous space. Since the image of the tangential electric current $\tilde{\mathbf{J}}_s^{eq}$ is oppositely directed, the fields of the electric current and its image cancel. Since the image of the magnetic current is in the same direction as $\tilde{\mathbf{J}}_{ms}^{eq}$, the fields produced by the magnetic current and its image add. We also note that $\tilde{\mathbf{J}}_{ms}^{eq}$ is nonzero only over the aperture (since $\hat{\mathbf{n}} \times \tilde{\mathbf{E}} = 0$ on the screen), and thus the field in region 1 can be found from

$$\tilde{\mathbf{E}}(\mathbf{r}, \omega) = -\frac{1}{\tilde{\epsilon}^c(\omega)} \nabla \times \tilde{\mathbf{A}}_h(\mathbf{r}, \omega),$$

where

$$\tilde{\mathbf{A}}_h(\mathbf{r}, \omega) = \int_{S_0} \tilde{\epsilon}^c(\omega)[-2\hat{\mathbf{n}}' \times \tilde{\mathbf{E}}_{ap}(\mathbf{r}', \omega)] G(\mathbf{r}|\mathbf{r}'; \omega) \, dS'$$

and $\tilde{\mathbf{E}}_{ap}$ is the electric field in the aperture in the original problem. We shall present an example in the next section.

6.3.5   Far-zone fields produced by equivalent sources

The equivalence principle is useful for analyzing antennas with complicated source distributions. The sources may be excluded using a surface S, and then a knowledge of the fields over S (found, for example, by estimation or measurement) can be used to compute the fields external to the antenna. Here we describe how to compute these fields in the far zone.

Given that $\tilde{\mathbf{J}}_s^{eq} = \hat{\mathbf{n}} \times \tilde{\mathbf{H}}$ and $\tilde{\mathbf{J}}_{ms}^{eq} = -\hat{\mathbf{n}} \times \tilde{\mathbf{E}}$ are the equivalent sources on S, we may compute the fields using the potentials (6.43) and (6.45). Using (6.20) these can be approximated in the far zone $(r \gg r', kr \gg 1)$ as

$$\tilde{\mathbf{A}}_e(\mathbf{r}, \omega) = \tilde{\mu}(\omega) \frac{e^{-jkr}}{4\pi r} \tilde{\mathbf{a}}_e(\theta, \phi, \omega),$$

$$\tilde{\mathbf{A}}_h(\mathbf{r}, \omega) = \tilde{\epsilon}^c(\omega) \frac{e^{-jkr}}{4\pi r} \tilde{\mathbf{a}}_h(\theta, \phi, \omega), \tag{6.48}$$

where

$$\tilde{\mathbf{a}}_e(\theta, \phi, \omega) = \oint_S \tilde{\mathbf{J}}_s^{eq}(\mathbf{r}', \omega) e^{jk\hat{\mathbf{r}} \cdot \mathbf{r}'} \, dS',$$

$$\tilde{a}_h(\theta,\phi,\omega) = \oint_S \tilde{\mathbf{J}}_{sm}^{eq}(\mathbf{r}',\omega)e^{jk\hat{\mathbf{r}}\cdot\mathbf{r}'}\,dS', \qquad (6.49)$$

are the directional weighting functions.

To compute the fields from the potentials we must apply the curl operator. So we must evaluate

$$\nabla \times \left[\frac{e^{-jkr}}{r}\mathbf{V}(\theta,\phi)\right] = \frac{e^{-jkr}}{r}\nabla \times \mathbf{V}(\theta,\phi) + \nabla\left(\frac{e^{-jkr}}{r}\right) \times \mathbf{V}(\theta,\phi).$$

The curl of $\mathbf{V}$ is proportional to $1/r$ in spherical coordinates, hence the first term on the right is proportional to $1/r^2$. Since we are interested in the far-zone fields, this term can be discarded in favor of $1/r$-type terms. Using

$$\nabla\left(\frac{e^{-jkr}}{r}\right) = -\hat{\mathbf{r}}\left(\frac{1+jkr}{r}\right)\frac{e^{-jkr}}{r} \approx -\hat{\mathbf{r}}jk\frac{e^{-jkr}}{r}, \qquad kr \gg 1,$$

we have

$$\nabla \times \left[\frac{e^{-jkr}}{r}\mathbf{V}(\theta,\phi)\right] \approx -jk\hat{\mathbf{r}} \times \left[\frac{e^{-jkr}}{r}\mathbf{V}(\theta,\phi)\right].$$

Using this approximation we also establish

$$\nabla \times \nabla \times \left[\frac{e^{-jkr}}{r}\mathbf{V}(\theta,\phi)\right] \approx -k^2\hat{\mathbf{r}} \times \hat{\mathbf{r}} \times \left[\frac{e^{-jkr}}{r}\mathbf{V}(\theta,\phi)\right] = k^2\frac{e^{-jkr}}{r}\mathbf{V}_T(\theta,\phi)$$

where $\mathbf{V}_T = \mathbf{V} - \hat{\mathbf{r}}(\hat{\mathbf{r}}\cdot\mathbf{V})$ is the vector component of $\mathbf{V}$ transverse to the r-direction.

With these formulas we can approximate (6.46) and (6.47) as

$$\tilde{\mathbf{E}}(\mathbf{r},\omega) = -j\omega\tilde{\mathbf{A}}_{eT}(\mathbf{r},\omega) + \frac{jk}{\tilde{\epsilon}^c(\omega)}\hat{\mathbf{r}} \times \tilde{\mathbf{A}}_h(\mathbf{r},\omega), \qquad (6.50)$$

$$\tilde{\mathbf{H}}(\mathbf{r},\omega) = -j\omega\tilde{\mathbf{A}}_{hT}(\mathbf{r},\omega) - \frac{jk}{\tilde{\mu}(\omega)}\hat{\mathbf{r}} \times \tilde{\mathbf{A}}_e(\mathbf{r},\omega).$$

Note that

$$\hat{\mathbf{r}} \times \tilde{\mathbf{E}} = -j\omega\hat{\mathbf{r}} \times \tilde{\mathbf{A}}_{eT} + \frac{jk}{\tilde{\epsilon}^c}\hat{\mathbf{r}} \times \hat{\mathbf{r}} \times \tilde{\mathbf{A}}_h.$$

Since $\hat{\mathbf{r}} \times \tilde{\mathbf{A}}_{eT} = \hat{\mathbf{r}} \times \tilde{\mathbf{A}}_e$ and $\hat{\mathbf{r}} \times \hat{\mathbf{r}} \times \tilde{\mathbf{A}}_h = -\tilde{\mathbf{A}}_{hT}$, we have

$$\hat{\mathbf{r}} \times \tilde{\mathbf{E}} = \eta\left[-j\omega\tilde{\mathbf{A}}_{hT} - \frac{jk}{\tilde{\mu}}\hat{\mathbf{r}} \times \tilde{\mathbf{A}}_e\right] = \eta\tilde{\mathbf{H}}.$$

Thus

$$\tilde{\mathbf{H}} = \frac{\hat{\mathbf{r}} \times \tilde{\mathbf{E}}}{\eta}$$

and the electromagnetic field in the far zone is a TEM spherical wave, as expected.

Example of fields produced by equivalent sources: an aperture antenna. As an example of calculating the fields in a bounded region from equivalent sources, let us find the far-zone field in free space produced by a rectangular waveguide opening into a perfectly-conducting ground screen of infinite extent as shown in Figure 6.6. For simplicity assume the waveguide propagates a pure TE_{10} mode, and that all higher-order

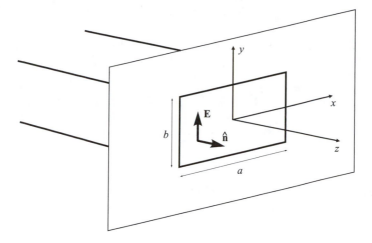

Figure 6.6: Aperture antenna consisting of a rectangular waveguide opening into a conducting ground screen of infinite extent.

modes excited when the guided wave is reflected at the aperture may be ignored. Thus the electric field in the aperture S_0 is

$$\tilde{\mathbf{E}}_a(x, y) = \hat{\mathbf{y}} E_0 \cos\left(\frac{\pi}{a} x\right).$$

We may compute the far-zone field using the Schelkunoff equivalence principle of § 6.3.4. We exclude the region $z < 0^+$ using a planar surface S which we close at infinity. We then fill the region $z < 0$ with a perfect conductor. By the image theory the equivalent electric sources on S cancel while the equivalent magnetic sources double. Since the only nonzero magnetic sources are on S_0 (since $\hat{\mathbf{n}} \times \tilde{\mathbf{E}} = 0$ on the screen), we have the equivalent problem of the source

$$\tilde{\mathbf{J}}_{ms}^{eq} = -2\hat{\mathbf{n}} \times \tilde{\mathbf{E}}_a = 2\hat{\mathbf{x}} E_0 \cos\left(\frac{\pi}{a} x\right)$$

on S_0 in free space, where the equivalence holds for $z > 0$.

We may find the far-zone field created by this equivalent current by first computing the directional weighting function (6.49). Since

$$\hat{\mathbf{r}} \cdot \mathbf{r}' = \hat{\mathbf{r}} \cdot (x'\hat{\mathbf{x}} + y'\hat{\mathbf{y}}) = x' \sin\theta\cos\phi + y'\sin\theta\sin\phi,$$

we find that

$$\tilde{\mathbf{a}}_h(\theta, \phi, \omega) = \int_{-b/2}^{b/2} \int_{-a/2}^{a/2} \hat{\mathbf{x}} 2 E_0 \cos\left(\frac{\pi}{a} x'\right) e^{jkx'\sin\theta\cos\phi} e^{jky'\sin\theta\sin\phi}\, dx'\, dy'$$

$$= \hat{\mathbf{x}} 4\pi E_0 ab \frac{\cos\pi X}{\pi^2 - 4(\pi X)^2} \frac{\sin\pi Y}{\pi Y}$$

where

$$X = \frac{a}{\lambda} \sin\theta\cos\phi, \qquad Y = \frac{b}{\lambda}\sin\theta\sin\phi.$$

Here λ is the free-space wavelength. By (6.50) the electric field is

$$\tilde{\mathbf{E}} = \frac{jk_0}{\epsilon_0} \hat{\mathbf{r}} \times \tilde{\mathbf{A}}_h$$

where $\tilde{\mathbf{A}}_h$ is given in (6.48). Using

$$\hat{\mathbf{r}} \times \hat{\mathbf{x}} = \hat{\boldsymbol{\phi}} \cos \theta \cos \phi + \hat{\boldsymbol{\theta}} \sin \phi$$

we find that

$$\tilde{\mathbf{E}} = jk_0 abE_0 \frac{e^{-jkr}}{r} \left(\hat{\boldsymbol{\theta}} \sin \phi + \hat{\boldsymbol{\phi}} \cos \theta \cos \phi \right) \frac{\cos(\pi X)}{\pi^2 - 4(\pi X)^2} \frac{\sin(\pi Y)}{\pi Y}.$$

The magnetic field is merely $\tilde{\mathbf{H}} = (\hat{\mathbf{r}} \times \tilde{\mathbf{E}})/\eta$.

6.4 Problems

6.1 Beginning with the Lorentz reciprocity theorem, derive (6.8).

6.2 Obtain (6.8) by substitution of (6.7) into Faraday's law.

6.3 Show that (6.8) returns the null result when evaluated within the excluded regions.

6.4 Show that under the condition $kr \gg 1$ the formula for the magnetic field of a dipole antenna (6.28) reduces to (6.32), while the formulas for the electric fields (6.29) and (6.30) reduce to (6.31).

6.5 Consider the dipole antenna shown in Figure 6.4. Instead of a standing-wave current distribution, assume the antenna carries a *traveling-wave current distribution*

$$\tilde{\mathbf{J}}^i(\mathbf{r}, \omega) = \hat{\mathbf{z}}\tilde{I}(\omega)e^{-jk|z|}\delta(x)\delta(y), \qquad -l \le z \le l.$$

Find the electric and magnetic fields at all points away from the current distribution. Specialize the result for $kr \gg 1$.

6.6 A circular loop of thin wire has radius a and lies in the $z = 0$ plane in free space. A current is induced on the wire with the density

$$\tilde{\mathbf{J}}(\mathbf{r}, \omega) = \hat{\boldsymbol{\phi}}\tilde{I}(\omega) \cos\left[k_0 a(\pi - |\phi|)\right] \delta(r - a)\frac{\delta(\theta - \pi/2)}{r}, \qquad |\phi| \le \pi.$$

Compute the far-zone fields produced by this loop antenna. Specialize your results for the electrically-small case of $k_0 a \ll 1$. Compute the time-average power radiated by, and the radiation resistance of, the electrically-small loop.

6.7 Consider a plane wave with the fields

$$\tilde{\mathbf{E}} = \tilde{E}_0\hat{\mathbf{x}}e^{-jkz}, \qquad \tilde{\mathbf{H}} = \frac{\tilde{E}_0}{\eta}\hat{\mathbf{y}}e^{-jkz},$$

normally incident from $z < 0$ on a square aperture of side a in a PEC ground screen at $z = 0$. Assume that the field in the aperture is identical to the field of the plane wave with the screen absent (this is called the *Kirchhoff approximation*). Compute the far-zone electromagnetic fields for $z > 0$.

6.8 Consider a coaxial cable of inner radius a and outer radius b, opening into a PEC ground plane at $z = 0$. Assume that only the TEM wave exists in the line and that no higher-order modes are created when the wave reflects from the aperture. Compute the far-zone electric and magnetic fields of this aperture antenna.

Appendix A

Mathematical appendix

A.1   The Fourier transform

The Fourier transform permits us to decompose a complicated field structure into elemental components. This can simplify the computation of fields and provide physical insight into their spatiotemporal behavior. In this section we review the properties of the transform and demonstrate its usefulness in solving field equations.

One-dimensional case

Let f be a function of a single variable x. The Fourier transform of $f(x)$ is the function $F(k)$ defined by the integral

$$\mathcal{F}\{f(x)\} = F(k) = \int_{-\infty}^{\infty} f(x)e^{-jkx}\,dx. \tag{A.1}$$

Note that x and the corresponding transform variable k must have reciprocal units: if x is time in seconds, then k is a *temporal frequency* in radians per second; if x is a length in meters, then k is a *spatial frequency* in radians per meter. We sometimes refer to $F(k)$ as the *frequency spectrum* of $f(x)$.

Not every function has a Fourier transform. The existence of (A.1) can be guaranteed by a set of sufficient conditions such as the following:

1. f is absolutely integrable: $\int_{-\infty}^{\infty} |f(x)|\,dx < \infty$;

2. f has no infinite discontinuities;

3. f has at most finitely many discontinuities and finitely many extrema in any finite interval (a, b).

While such rigor is certainly of mathematical value, it may be of less ultimate use to the engineer than the following heuristic observation offered by Bracewell [22]: *a good mathematical model of a physical process should be Fourier transformable.* That is, if the Fourier transform of a mathematical model does not exist, the model cannot precisely describe a physical process.

The usefulness of the transform hinges on our ability to recover f through the inverse transform:

$$\mathcal{F}^{-1}\{F(k)\} = f(x) = \frac{1}{2\pi} \int_{-\infty}^{\infty} F(k)\,e^{jkx}\,dk. \tag{A.2}$$

When this is possible we write

$$f(x) \leftrightarrow F(k)$$

and say that $f(x)$ and $F(k)$ form a Fourier transform pair. The *Fourier integral theorem* states that

$$\mathcal{F}\mathcal{F}^{-1}\{f(x)\} = \mathcal{F}^{-1}\mathcal{F}\{f(x)\} = f(x),$$

except at points of discontinuity of f. At a jump discontinuity the inversion formula returns the average value of the one-sided limits $f(x^+)$ and $f(x^-)$ of $f(x)$. At points of continuity the forward and inverse transforms are unique.

Transform theorems and properties. We now review some basic facts pertaining to the Fourier transform. Let $f(x) \leftrightarrow F(k) = R(k) + jX(k)$, and $g(x) \leftrightarrow G(k)$.

1. *Linearity.* $\alpha f(x) + \beta g(x) \leftrightarrow \alpha F(k) + \beta G(k)$ if α and β are arbitrary constants. This follows directly from the linearity of the transform integral, and makes the transform useful for solving linear differential equations (e.g., Maxwell's equations).

2. *Symmetry.* The property $F(x) \leftrightarrow 2\pi f(-k)$ is helpful when interpreting transform tables in which transforms are listed only in the forward direction.

3. *Conjugate function.* We have $f^*(x) \leftrightarrow F^*(-k)$.

4. *Real function.* If f is real, then $F(-k) = F^*(k)$. Also,

$$R(k) = \int_{-\infty}^{\infty} f(x)\cos kx \, dx, \quad X(k) = -\int_{-\infty}^{\infty} f(x)\sin kx \, dx,$$

   and

$$f(x) = \frac{1}{\pi}\,\mathrm{Re}\int_{0}^{\infty} F(k)e^{jkx}\, dk.$$

   A real function is completely determined by its positive frequency spectrum. It is obviously advantageous to know this when planning to collect spectral data.

5. *Real function with reflection symmetry.* If f is real and even, then $X(k) \equiv 0$ and

$$R(k) = 2\int_{0}^{\infty} f(x)\cos kx \, dx, \quad f(x) = \frac{1}{\pi}\int_{0}^{\infty} R(k)\cos kx \, dk.$$

   If f is real and odd, then $R(k) \equiv 0$ and

$$X(k) = -2\int_{0}^{\infty} f(x)\sin kx \, dx, \quad f(x) = -\frac{1}{\pi}\int_{0}^{\infty} X(k)\sin kx \, dk.$$

   (Recall that f is even if $f(-x) = f(x)$ for all x. Similarly f is odd if $f(-x) = -f(x)$ for all x.)

6. *Causal function.* Recall that f is causal if $f(x) = 0$ for $x < 0$.

   (a) If f is real and causal, then

$$X(k) = -\frac{2}{\pi}\int_{0}^{\infty}\int_{0}^{\infty} R(k')\cos k'x \sin kx \, dk' \, dx,$$

$$R(k) = -\frac{2}{\pi}\int_{0}^{\infty}\int_{0}^{\infty} X(k')\sin k'x \cos kx \, dk' \, dx.$$

(b) If f is real and causal, and $f(0)$ is finite, then $R(k)$ and $X(k)$ are related by the *Hilbert transforms*

$$X(k) = -\frac{1}{\pi} \, \text{P.V.} \int_{-\infty}^{\infty} \frac{R(k)}{k - k'} \, dk', \quad R(k) = \frac{1}{\pi} \, \text{P.V.} \int_{-\infty}^{\infty} \frac{X(k)}{k - k'} \, dk'.$$

(c) If f is causal and has finite energy, it is not possible to have $F(k) = 0$ for $k_1 < k < k_2$. That is, the transform of a causal function cannot vanish over an interval.

A causal function is completely determined by the real or imaginary part of its spectrum. As with item 4, this is helpful when performing calculations or measurements in the frequency domain. If the function is not band-limited however, truncation of integrals will give erroneous results.

7. *Time-limited vs. band-limited functions.* Assume $t_2 > t_1$. If $f(t) = 0$ for both $t < t_1$ and $t > t_2$, then it is not possible to have $F(k) = 0$ for both $k < k_1$ and $k > k_2$ where $k_2 > k_1$. That is, a time-limited signal cannot be band-limited. Similarly, a band-limited signal cannot be time-limited.

8. *Null function. If the forward or inverse transform of a function is identically zero, then the function is identically zero.* This important consequence of the Fourier integral theorem is useful when solving homogeneous partial differential equations in the frequency domain.

9. *Space or time shift.* For any fixed x_0,

$$f(x - x_0) \leftrightarrow F(k)e^{-jkx_0}. \tag{A.3}$$

A temporal or spatial shift affects only the phase of the transform, not the magnitude.

10. *Frequency shift.* For any fixed k_0,

$$f(x)e^{jk_0x} \leftrightarrow F(k - k_0).$$

Note that if $f \leftrightarrow F$ where f is real, then frequency-shifting F causes f to become complex — again, this is important if F has been obtained experimentally or through computation in the frequency domain.

11. *Similarity.* We have

$$f(\alpha x) \leftrightarrow \frac{1}{|\alpha|} F\left(\frac{k}{\alpha}\right),$$

where α is any real constant. "Reciprocal spreading" is exhibited by the Fourier transform pair; dilation in space or time results in compression in frequency, and vice versa.

12. *Convolution.* We have

$$\int_{-\infty}^{\infty} f_1(x')f_2(x - x') \, dx' \leftrightarrow F_1(k)F_2(k)$$

and

$$f_1(x)f_2(x) \leftrightarrow \frac{1}{2\pi} \int_{-\infty}^{\infty} F_1(k')F_2(k - k') \, dk'.$$

The first of these is particularly useful when a problem has been solved in the frequency domain and the solution is found to be a product of two or more functions of k.

13. *Parseval's identity.* We have

$$\int_{-\infty}^{\infty} |f(x)|^2 \, dx = \frac{1}{2\pi} \int_{-\infty}^{\infty} |F(k)|^2 \, dk.$$

Computations of energy in the time and frequency domains always give the same result.

14. *Differentiation.* We have

$$\frac{d^n f(x)}{dx^n} \leftrightarrow (jk)^n F(k) \quad \text{and} \quad (-jx)^n f(x) \leftrightarrow \frac{d^n F(k)}{dk^n}.$$

The Fourier transform can convert a differential equation in the x domain into an algebraic equation in the k domain, and vice versa.

15. *Integration.* We have

$$\int_{-\infty}^{x} f(u) \, du \leftrightarrow \pi F(k)\delta(k) + \frac{F(k)}{jk}$$

where $\delta(k)$ is the Dirac delta or unit impulse.

Generalized Fourier transforms and distributions. It is worth noting that many useful functions are not Fourier transformable in the sense given above. An example is the signum function

$$\text{sgn}(x) = \begin{cases} -1, & x < 0, \\ 1, & x > 0. \end{cases}$$

Although this function lacks a Fourier transform in the usual sense, for practical purposes it may still be safely associated with what is known as a *generalized Fourier transform*. A treatment of this notion would be out of place here; however, the reader should certainly be prepared to encounter an entry such as

$$\text{sgn}(x) \leftrightarrow 2/jk$$

in a standard Fourier transform table. Other functions can be regarded as possessing transforms when *generalized functions* are permitted into the discussion. An important example of a generalized function is the Dirac delta $\delta(x)$, which has enormous value in describing distributions that are very thin, such as the charge layers often found on conductor surfaces. We shall not delve into the intricacies of distribution theory. However, we can hardly avoid dealing with generalized functions; to see this we need look no further than the simple function $\cos k_0 x$ with its transform pair

$$\cos k_0 x \leftrightarrow \pi[\delta(k + k_0) + \delta(k - k_0)].$$

The reader of this book must therefore know the standard facts about $\delta(x)$: that it acquires meaning only as part of an integrand, and that it satisfies the *sifting property*

$$\int_{-\infty}^{\infty} \delta(x - x_0)f(x) \, dx = f(x_0)$$

for any continuous function f. With $f(x) = 1$ we obtain the familiar relation

$$\int_{-\infty}^{\infty} \delta(x)\, dx = 1.$$

With $f(x) = e^{-jkx}$ we obtain

$$\int_{-\infty}^{\infty} \delta(x)e^{-jkx}\, dx = 1,$$

thus

$$\delta(x) \leftrightarrow 1.$$

It follows that

$$\frac{1}{2\pi} \int_{-\infty}^{\infty} e^{jkx}\, dk = \delta(x). \tag{A.4}$$

Useful transform pairs.   Some of the more common Fourier transforms that arise in the study of electromagnetics are given in Appendix C. These often involve the simple functions defined here:

1. *Unit step function*

$$U(x) = \begin{cases} 1, & x < 0, \\ 0, & x > 0. \end{cases} \tag{A.5}$$

2. *Signum function*

$$\text{sgn}(x) = \begin{cases} -1, & x < 0, \\ 1, & x > 0. \end{cases} \tag{A.6}$$

3. *Rectangular pulse function*

$$\text{rect}(x) = \begin{cases} 1, & |x| < 1, \\ 0, & |x| > 1. \end{cases} \tag{A.7}$$

4. *Triangular pulse function*

$$\Lambda(x) = \begin{cases} 1 - |x|, & |x| < 1, \\ 0, & |x| > 1. \end{cases} \tag{A.8}$$

5. *Sinc function*

$$\text{sinc}(x) = \frac{\sin x}{x}. \tag{A.9}$$

Transforms of multi-variable functions

Fourier transformations can be performed over multiple variables by successive applications of (A.1). For example, the two-dimensional Fourier transform over x_1 and x_2 of the function $f(x_1, x_2, x_3, \ldots, x_N)$ is the quantity $F(k_{x_1}, k_{x_2}, x_3, \ldots, x_N)$ given by

$$\int_{-\infty}^{\infty} \left[\int_{-\infty}^{\infty} f(x_1, x_2, x_3, \ldots, x_N)\, e^{-jk_{x_1} x_1}\, dx_1 \right] e^{-jk_{x_2} x_2}\, dx_2$$

$$= \int_{-\infty}^{\infty} \int_{-\infty}^{\infty} f(x_1, x_2, x_3, \ldots, x_N)\, e^{-jk_{x_1} x_1} e^{-jk_{x_2} x_2}\, dx_1\, dx_2.$$

The two-dimensional inverse transform is computed by multiple application of (A.2), recovering $f(x_1, x_2, x_3, \ldots, x_N)$ through the operation

$$\frac{1}{(2\pi)^2} \int_{-\infty}^{\infty} \int_{-\infty}^{\infty} F(k_{x_1}, k_{x_2}, x_3, \ldots, x_N)\, e^{jk_{x_1} x_1} e^{jk_{x_2} x_2}\, dk_{x_1}\, dk_{x_2}.$$

Higher-dimensional transforms and inversions are done analogously.

Transforms of separable functions.   If we are able to write

$$f(x_1, x_2, x_3, \ldots, x_N) = f_1(x_1, x_3, \ldots, x_N) f_2(x_2, x_3, \ldots, x_N),$$

then successive transforms on the variables x_1 and x_2 result in

$$f(x_1, x_2, x_3, \ldots, x_N) \leftrightarrow F_1(k_{x_1}, x_3, \ldots, x_N) F_2(k_{x_2}, x_3, \ldots, x_N).$$

In this case a multi-variable transform can be obtained with the help of a table of one-dimensional transforms. If, for instance,

$$f(x, y, z) = \delta(x - x')\delta(y - y')\delta(z - z'),$$

then we obtain

$$F(k_x, k_y, k_z) = e^{-jk_x x'} e^{-jk_y y'} e^{-jk_z z'}$$

by three applications of (A.1).

   A more compact notation for multi-dimensional functions and transforms makes use of the vector notation $\mathbf{k} = \hat{\mathbf{x}} k_x + \hat{\mathbf{y}} k_y + \hat{\mathbf{z}} k_z$ and $\mathbf{r} = \hat{\mathbf{x}} x + \hat{\mathbf{y}} y + \hat{\mathbf{z}} z$ where $\mathbf{r}$ is the position vector. In the example above, for instance, we could have written

$$\delta(x - x')\delta(y - y')\delta(z - z') = \delta(\mathbf{r} - \mathbf{r}'),$$

and

$$F(\mathbf{k}) = \int_{-\infty}^{\infty} \int_{-\infty}^{\infty} \int_{-\infty}^{\infty} \delta(\mathbf{r} - \mathbf{r}') e^{-j\mathbf{k}\cdot\mathbf{r}}\, dx\, dy\, dz = e^{-j\mathbf{k}\cdot\mathbf{r}'}.$$

Fourier–Bessel transform.   If x_1 and x_2 have the same dimensions, it may be convenient to recast the two-dimensional Fourier transform in polar coordinates. Let $x_1 = \rho\cos\phi$, $k_{x_1} = p\cos\theta$, $x_2 = \rho\sin\phi$, and $k_{x_2} = p\sin\theta$, where p and ρ are defined on $(0, \infty)$ and ϕ and θ are defined on $(-\pi, \pi)$. Then

$$F(p, \theta, x_3, \ldots, x_N) = \int_{-\pi}^{\pi} \int_{0}^{\infty} f(\rho, \phi, x_3, \ldots, x_N)\, e^{-jp\rho\cos(\phi-\theta)} \rho\, d\rho\, d\phi. \qquad (A.10)$$

If f is independent of ϕ (due to rotational symmetry about an axis transverse to x_1 and x_2), then the ϕ integral can be computed using the identity

$$J_0(x) = \frac{1}{2\pi} \int_{-\pi}^{\pi} e^{-jx\cos(\phi-\theta)}\, d\phi.$$

Thus (A.10) becomes

$$F(p, x_3, \ldots, x_N) = 2\pi \int_{0}^{\infty} f(\rho, x_3, \ldots, x_N) J_0(\rho p)\, \rho\, d\rho, \qquad (A.11)$$

showing that F is independent of the angular variable θ. Expression (A.11) is termed the *Fourier–Bessel transform* of f. The reader can easily verify that f can be recovered from F through

$$f(\rho, x_3, \ldots, x_N) = \int_0^\infty F(p, x_3, \ldots, x_N) J_0(\rho p) \, p \, dp,$$

the inverse Fourier–Bessel transform.

A review of complex contour integration

Some powerful techniques for the evaluation of integrals rest on complex variable theory. In particular, the computation of the Fourier inversion integral is often aided by these techniques. We therefore provide a brief review of this material. For a fuller discussion the reader may refer to one of many widely available textbooks on complex analysis.

We shall denote by $f(z)$ a complex valued function of a complex variable z. That is,

$$f(z) = u(x, y) + jv(x, y),$$

where the real and imaginary parts $u(x, y)$ and $v(x, y)$ of f are each functions of the real and imaginary parts x and y of z:

$$z = x + jy = \operatorname{Re}(z) + j \operatorname{Im}(z).$$

Here $j = \sqrt{-1}$, as is mostly standard in the electrical engineering literature.

Limits, differentiation, and analyticity. Let $w = f(z)$, and let $z_0 = x_0 + jy_0$ and $w_0 = u_0 + jv_0$ be points in the complex z and w planes, respectively. We say that w_0 is the limit of $f(z)$ as z approaches z_0, and write

$$\lim_{z \to z_0} f(z) = w_0,$$

if and only if both $u(x, y) \to u_0$ and $v(x, y) \to v_0$ as $x \to x_0$ and $y \to y_0$ independently. The derivative of $f(z)$ at a point $z = z_0$ is defined by the limit

$$f'(z_0) = \lim_{z \to z_0} \frac{f(z) - f(z_0)}{z - z_0},$$

if it exists. Existence requires that the derivative be independent of direction of approach; that is, $f'(z_0)$ cannot depend on the manner in which $z \to z_0$ in the complex plane. (This turns out to be a much stronger condition than simply requiring that the functions u and v be differentiable with respect to the variables x and y.) We say that $f(z)$ is *analytic* at z_0 if it is differentiable at z_0 and at all points in some neighborhood of z_0.

If $f(z)$ is not analytic at z_0 but every neighborhood of z_0 contains a point at which $f(z)$ is analytic, then z_0 is called a *singular point of* $f(z)$.

Laurent expansions and residues. Although Taylor series can be used to expand complex functions around points of analyticity, we must often expand functions around points z_0 at or near which the functions fail to be analytic. For this we use the *Laurent* *expansion,* a generalization of the Taylor expansion involving both positive and negative powers of $z - z_0$:

$$f(z) = \sum_{n=-\infty}^{\infty} a_n(z-z_0)^n = \sum_{n=1}^{\infty} \frac{a_{-n}}{(z-z_0)^n} + \sum_{n=0}^{\infty} a_n(z-z_0)^n.$$

The numbers a_n are the coefficients of the Laurent expansion of $f(z)$ at point $z = z_0$. The first series on the right is the *principal part* of the Laurent expansion, and the second series is the *regular part.* The regular part is an ordinary power series, hence it converges in some disk $|z - z_0| < R$ where $R \geq 0$. Putting $\zeta = 1/(z - z_0)$, the principal part becomes $\sum_{n=1}^{\infty} a_{-n}\zeta^n$; this power series converges for $|\zeta| < \rho$ where $\rho \geq 0$, hence the principal part converges for $|z - z_0| > 1/\rho \triangleq r$. When $r < R$, the Laurent expansion converges in the annulus $r < |z - z_0| < R$; when $r > R$, it diverges everywhere in the complex plane.

The function $f(z)$ has an *isolated singularity* at point z_0 if $f(z)$ is not analytic at z_0 but is analytic in the "punctured disk" $0 < |z - z_0| < R$ for some $R > 0$. Isolated singularities are classified by reference to the Laurent expansion. Three types can arise:

1. *Removable singularity.* The point z_0 is a removable singularity of $f(z)$ if the principal part of the Laurent expansion of $f(z)$ about z_0 is identically zero (i.e., if $a_n = 0$ for $n = -1, -2, -3, \ldots$).

2. *Pole of order k.* The point z_0 is a pole of order k if the principal part of the Laurent expansion about z_0 contains only finitely many terms that form a polynomial of degree k in $(z - z_0)^{-1}$. A pole of order 1 is called a *simple pole.*

3. *Essential singularity.* The point z_0 is an essential singularity of $f(z)$ if the principal part of the Laurent expansion of $f(z)$ about z_0 contains infinitely many terms (i.e., if $a_{-n} \neq 0$ for infinitely many n).

The coefficient a_{-1} in the Laurent expansion of $f(z)$ about an isolated singular point z_0 is the *residue of $f(z)$* at z_0. It can be shown that

$$a_{-1} = \frac{1}{2\pi j} \oint_\Gamma f(z)\, dz \tag{A.12}$$

where Γ is any simple closed curve oriented counterclockwise and containing in its interior z_0 and no other singularity of $f(z)$. Particularly useful to us is the formula for evaluation of residues at pole singularities. If $f(z)$ has a pole of order k at $z = z_0$, then the residue of $f(z)$ at z_0 is given by

$$a_{-1} = \frac{1}{(k-1)!} \lim_{z \to z_0} \frac{d^{k-1}}{dz^{k-1}}[(z-z_0)^k f(z)]. \tag{A.13}$$

Cauchy–Goursat and residue theorems.   It can be shown that if $f(z)$ is analytic at all points on and within a simple closed contour C, then

$$\oint_C f(z)\, dz = 0.$$

This central result is known as the *Cauchy–Goursat theorem.* We shall not offer a proof, but shall proceed instead to derive a useful consequence known as the *residue theorem.*

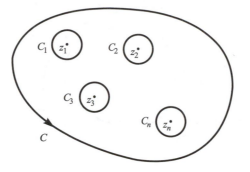

Figure A.1: Derivation of the residue theorem.

Figure A.1 depicts a simple closed curve C enclosing n isolated singularities of a function $f(z)$. We assume that $f(z)$ is analytic on and elsewhere within C. Around each singular point z_k we have drawn a circle C_k so small that it encloses no singular point other than z_k; taken together, the C_k $(k = 1, \ldots, n)$ and C form the boundary of a region in which $f(z)$ is everywhere analytic. By the Cauchy–Goursat theorem

$$\int_C f(z)\,dz + \sum_{k=1}^{n} \int_{C_k} f(z)\,dz = 0.$$

Hence

$$\frac{1}{2\pi j} \int_C f(z)\,dz = \sum_{k=1}^{n} \frac{1}{2\pi j} \int_{C_k} f(z)\,dz,$$

where now the integrations are all performed in a counterclockwise sense. By (A.12)

$$\int_C f(z)\,dz = 2\pi j \sum_{k=1}^{n} r_k \tag{A.14}$$

where $r_1, \ldots, r_n$ are the residues of $f(z)$ at the singularities within C.

Contour deformation. Suppose f is analytic in a region D and Γ is a simple closed curve in D. If Γ can be continuously deformed to another simple closed curve Γ' without passing out of D, then

$$\int_{\Gamma'} f(z)\,dz = \int_{\Gamma} f(z)\,dz. \tag{A.15}$$

To see this, consider Figure A.2 where we have introduced another set of curves $\pm\gamma$; these new curves are assumed parallel and infinitesimally close to each other. Let C be the composite curve consisting of Γ, $+\gamma$, $-\Gamma'$, and $-\gamma$, in that order. Since f is analytic on and within C, we have

$$\int_C f(z)\,dz = \int_{\Gamma} f(z)\,dz + \int_{+\gamma} f(z)\,dz + \int_{-\Gamma'} f(z)\,dz + \int_{-\gamma} f(z)\,dz = 0.$$

But $\int_{-\Gamma'} f(z)\,dz = -\int_{\Gamma'} f(z)\,dz$ and $\int_{-\gamma} f(z)\,dz = -\int_{+\gamma} f(z)\,dz$, hence (A.15) follows. The contour deformation principle often permits us to replace an integration contour by one that is more convenient.

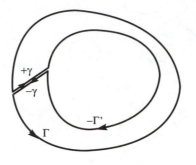

Figure A.2: Derivation of the contour deformation principle.

Principal value integrals.   We must occasionally carry out integrations of the form

$$I = \int_{-\infty}^{\infty} f(x)\, dx$$

where $f(x)$ has a finite number of singularities x_k ($k = 1, \ldots, n$) along the real axis. Such singularities in the integrand force us to interpret I as an improper integral. With just one singularity present at point x_1, for instance, we define

$$\int_{-\infty}^{\infty} f(x)\, dx = \lim_{\varepsilon \to 0} \int_{-\infty}^{x_1 - \varepsilon} f(x)\, dx + \lim_{\eta \to 0} \int_{x_1 + \eta}^{\infty} f(x)\, dx$$

provided that both limits exist. When both limits do not exist, we may still be able to obtain a well-defined result by computing

$$\lim_{\varepsilon \to 0} \left(\int_{-\infty}^{x_1 - \varepsilon} f(x)\, dx + \int_{x_1 + \varepsilon}^{\infty} f(x)\, dx \right)$$

(i.e., by taking $\eta = \varepsilon$ so that the limits are "symmetric"). This quantity is called the *Cauchy principal value* of I and is denoted

$$\text{P.V.} \int_{-\infty}^{\infty} f(x)\, dx.$$

More generally, we have

$$\text{P.V.} \int_{-\infty}^{\infty} f(x)\, dx = \lim_{\varepsilon \to 0} \left(\int_{-\infty}^{x_1 - \varepsilon} f(x)\, dx + \int_{x_1 + \varepsilon}^{x_2 - \varepsilon} f(x)\, dx + \right.$$

$$\left. + \cdots + \int_{x_{n-1} + \varepsilon}^{x_n - \varepsilon} f(x)\, dx + \int_{x_n + \varepsilon}^{\infty} f(x)\, dx \right)$$

for n singularities $x_1 < \cdots < x_n$.

In a large class of problems $f(z)$ (i.e., $f(x)$ with x replaced by the complex variable z) is analytic everywhere except for the presence of finitely many simple poles. Some of these may lie on the real axis (at points $x_1 < \cdots < x_n$, say), and some may not. Consider now the integration contour C shown in Figure A.3. We choose R so large and ε so small that C encloses all the poles of f that lie in the upper half of the complex

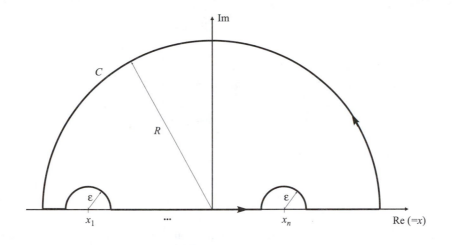

Figure A.3: Complex plane technique for evaluating a principal value integral.

plane. In many problems of interest the integral of f around the large semicircle tends to zero as $R \to \infty$ and the integrals around the small semicircles are well-behaved as $\varepsilon \to 0$. It may then be shown that

$$\text{P.V.} \int_{-\infty}^{\infty} f(x)\,dx = \pi j \sum_{k=1}^{n} r_k + 2\pi j \sum_{\text{UHP}} r_k$$

where r_k is the residue at the kth simple pole. The first sum on the right accounts for the contributions of those poles that lie on the real axis; note that it is associated with a factor πj instead of $2\pi j$, since these terms arose from integrals over semicircles rather than over full circles. The second sum, of course, is extended only over those poles that reside in the upper half-plane.

Fourier transform solution of the 1-D wave equation

Successive applications of the Fourier transform can reduce a partial differential equation to an ordinary differential equation, and finally to an algebraic equation. After the algebraic equation is solved by standard techniques, Fourier inversion can yield a solution to the original partial differential equation. We illustrate this by solving the one-dimensional inhomogeneous wave equation

$$\left(\frac{\partial^2}{\partial z^2} - \frac{1}{c^2} \frac{\partial^2}{\partial t^2} \right) \psi(x,y,z,t) = S(x,y,z,t), \tag{A.16}$$

where the field ψ is the desired unknown and S is the known source term. For uniqueness of solution we must specify ψ and $\partial \psi / \partial z$ over some $z = \text{constant}$ plane. Assume that

$$\psi(x,y,z,t)\Big|_{z=0} = f(x,y,t), \tag{A.17}$$

$$\frac{\partial}{\partial z} \psi(x,y,z,t)\Big|_{z=0} = g(x,y,t). \tag{A.18}$$

We begin by positing inverse temporal Fourier transform relationships for ψ and S:

$$\psi(x,y,z,t) = \frac{1}{2\pi} \int_{-\infty}^{\infty} \tilde{\psi}(x,y,z,\omega) e^{j\omega t}\,d\omega,$$

$$S(x, y, z, t) = \frac{1}{2\pi} \int_{-\infty}^{\infty} \tilde{S}(x, y, z, \omega) e^{j\omega t} \, d\omega.$$

Substituting into (A.16), passing the derivatives through the integral, calculating the derivatives, and combining the inverse transforms, we obtain

$$\frac{1}{2\pi} \int_{-\infty}^{\infty} \left[\left(\frac{\partial^2}{\partial z^2} + k^2 \right) \tilde{\psi}(x, y, z, \omega) - \tilde{S}(x, y, z, \omega) \right] e^{j\omega t} \, d\omega = 0$$

where $k = \omega/c$. By the Fourier integral theorem

$$\left(\frac{\partial^2}{\partial z^2} + k^2 \right) \tilde{\psi}(x, y, z, \omega) - \tilde{S}(x, y, z, \omega) = 0. \tag{A.19}$$

We have thus converted a partial differential equation into an ordinary differential equation. A spatial transform on z will now convert the ordinary differential equation into an algebraic equation. We write

$$\tilde{\psi}(x, y, z, \omega) = \frac{1}{2\pi} \int_{-\infty}^{\infty} \tilde{\psi}^z(x, y, k_z, \omega) e^{jk_z z} \, dk_z,$$

$$\tilde{S}(x, y, z, \omega) = \frac{1}{2\pi} \int_{-\infty}^{\infty} \tilde{S}^z(x, y, k_z, \omega) e^{jk_z z} \, dk_z,$$

in (A.19), pass the derivatives through the integral sign, compute the derivatives, and set the integrand to zero to get

$$(k^2 - k_z^2)\tilde{\psi}^z(x, y, k_z, \omega) - \tilde{S}^z(x, y, k_z, \omega) = 0;$$

hence

$$\tilde{\psi}^z(x, y, k_z, \omega) = -\frac{\tilde{S}^z(x, y, k_z, \omega)}{(k_z - k)(k_z + k)}. \tag{A.20}$$

The price we pay for such an easy solution is that we must now perform a two-dimensional Fourier inversion to obtain $\psi(x, y, z, t)$ from $\tilde{\psi}^z(x, y, k_z, \omega)$. It turns out to be easiest to perform the spatial inverse transform first, so let us examine

$$\tilde{\psi}(x, y, z, \omega) = \frac{1}{2\pi} \int_{-\infty}^{\infty} \tilde{\psi}^z(x, y, k_z, \omega) e^{jk_z z} \, dk_z.$$

By (A.20) we have

$$\tilde{\psi}(x, y, z, \omega) = \frac{1}{2\pi} \int_{-\infty}^{\infty} [\tilde{S}^z(x, y, k_z, \omega)] \left[\frac{-1}{(k_z - k)(k_z + k)} \right] e^{jk_z z} \, dk_z,$$

where the integrand involves a product of two functions. With

$$\tilde{g}^z(k_z, \omega) = \frac{-1}{(k_z - k)(k_z + k)},$$

the convolution theorem gives

$$\tilde{\psi}(x, y, z, \omega) = \int_{-\infty}^{\infty} \tilde{S}(x, y, \zeta, \omega) \tilde{g}(z - \zeta, \omega) \, d\zeta \tag{A.21}$$

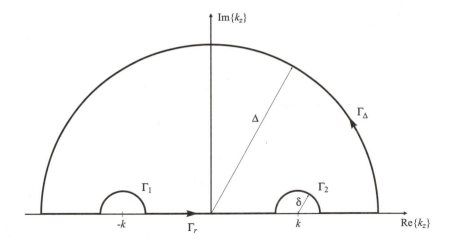

Figure A.4: Contour used to compute inverse transform in solution of the 1-D wave equation.

where

$$\tilde{g}(z,\omega) = \frac{1}{2\pi}\int_{-\infty}^{\infty}\tilde{g}^z(k_z,\omega)e^{jk_zz}\,dk_z = \frac{1}{2\pi}\int_{-\infty}^{\infty}\frac{-1}{(k_z-k)(k_z+k)}e^{jk_zz}\,dk_z.$$

To compute this integral we use complex plane techniques. The domain of integration extends along the real k_z-axis in the complex k_z-plane; because of the poles at $k_z = \pm k$, we must treat the integral as a principal value integral. Denoting

$$I(k_z) = \frac{-e^{jk_zz}}{2\pi(k_z-k)(k_z+k)},$$

we have

$$\int_{-\infty}^{\infty}I(k_z)\,dk_z = \lim\int_{\Gamma_r}I(k_z)\,dk_z$$

$$= \lim\int_{-\Delta}^{-k-\delta}I(k_z)\,dk_z + \lim\int_{-k+\delta}^{k-\delta}I(k_z)\,dk_z + \lim\int_{k+\delta}^{\Delta}I(k_z)\,dk_z$$

where the limits take $\delta \to 0$ and $\Delta \to \infty$. Our k_z-plane contour takes detours around the poles using semicircles of radius δ, and is closed using a semicircle of radius Δ (Figure A.4). Note that if $z > 0$, we must close the contour in the upper half-plane.

By Cauchy's integral theorem

$$\int_{\Gamma_r}I(k_z)\,dk_z + \int_{\Gamma_1}I(k_z)\,dk_z + \int_{\Gamma_2}I(k_z)\,dk_z + \int_{\Gamma_\Delta}I(k_z)\,dk_z = 0.$$

Thus

$$\int_{-\infty}^{\infty}I(k_z)\,dk_z = -\lim_{\delta\to 0}\int_{\Gamma_1}I(k_z)\,dk_z - \lim_{\delta\to 0}\int_{\Gamma_2}I(k_z)\,dk_z - \lim_{\Delta\to\infty}\int_{\Gamma_\Delta}I(k_z)\,dk_z.$$

The contribution from the semicircle of radius Δ can be computed by writing k_z in polar coordinates as $k_z = \Delta e^{j\theta}$:

$$\lim_{\Delta\to\infty}\int_{\Gamma_\Delta}I(k_z)\,dk_z = \frac{1}{2\pi}\lim_{\Delta\to\infty}\int_0^{\pi}\frac{-e^{jz\Delta e^{j\theta}}}{(\Delta e^{j\theta}-k)(\Delta e^{j\theta}+k)}j\Delta e^{j\theta}\,d\theta.$$

Using Euler's identity we can write

$$\lim_{\Delta \to \infty} \int_{\Gamma_\Delta} I(k_z)\, dk_z = \frac{1}{2\pi} \lim_{\Delta \to \infty} \int_0^\pi \frac{-e^{-\Delta z \sin \theta} e^{j\Delta z \cos \theta}}{\Delta^2 e^{2j\theta}} j\Delta e^{j\theta}\, d\theta.$$

Thus, as long as $z > 0$ the integrand will decay exponentially as $\Delta \to \infty$, and

$$\lim_{\Delta \to \infty} \int_{\Gamma_\Delta} I(k_z)\, dk_z \to 0.$$

Similarly, $\int_{\Gamma_\Delta} I(k_z)\, dk_z \to 0$ when $z < 0$ if we close the semicircle in the lower half-plane. Thus,

$$\int_{-\infty}^{\infty} I(k_z)\, dk_z = -\lim_{\delta \to 0} \int_{\Gamma_1} I(k_z)\, dk_z - \lim_{\delta \to 0} \int_{\Gamma_2} I(k_z)\, dk_z. \qquad (A.22)$$

The integrals around the poles can also be computed by writing k_z in polar coordinates. Writing $k_z = -k + \delta e^{j\theta}$ we find

$$\lim_{\delta \to 0} \int_{\Gamma_1} I(k_z)\, dk_z = \frac{1}{2\pi} \lim_{\delta \to 0} \int_\pi^0 \frac{-e^{jz(-k+\delta e^{j\theta})} j\delta e^{j\theta}}{(-k + \delta e^{j\theta} - k)(-k + \delta e^{j\theta} + k)}\, d\theta$$

$$= \frac{1}{2\pi} \int_0^\pi \frac{e^{-jkz}}{-2k} j\, d\theta = -\frac{j}{4k} e^{-jkz}.$$

Similarly, using $k_z = k + \delta e^{j\theta}$, we obtain

$$\lim_{\delta \to 0} \int_{\Gamma_2} I(k_z)\, dk_z = \frac{j}{4k} e^{jkz}.$$

Substituting these into (A.22) we have

$$\tilde{g}(z,\omega) = \frac{j}{4k} e^{-jkz} - \frac{j}{4k} e^{jkz} = \frac{1}{2k} \sin kz, \qquad (A.23)$$

valid for $z > 0$. For $z < 0$, we close in the lower half-plane instead and get

$$\tilde{g}(z,\omega) = -\frac{1}{2k} \sin kz. \qquad (A.24)$$

Substituting (A.23) and (A.24) into (A.21) we obtain

$$\tilde{\psi}(x,y,z,\omega) = \int_{-\infty}^{z} \tilde{S}(x,y,\zeta,\omega) \frac{\sin k(z-\zeta)}{2k}\, d\zeta - \frac{1}{2k} \int_z^\infty \tilde{S}(x,y,\zeta,\omega) \frac{\sin k(z-\zeta)}{2k}\, d\zeta$$

where we have been careful to separate the two cases considered above. To make things a bit easier when we apply the boundary conditions, let us rewrite the above expression. Splitting the domain of integration we write

$$\tilde{\psi}(x,y,z,\omega) = \int_{-\infty}^{0} \tilde{S}(x,y,\zeta,\omega) \frac{\sin k(z-\zeta)}{2k}\, d\zeta + \int_0^z \tilde{S}(x,y,\zeta,\omega) \frac{\sin k(z-\zeta)}{k}\, d\zeta -$$

$$- \int_0^\infty \tilde{S}(x,y,\zeta,\omega) \frac{\sin k(z-\zeta)}{2k}\, d\zeta.$$

Expansion of the trigonometric functions then gives

$$\tilde{\psi}(x,y,z,\omega) = \int_0^z \tilde{S}(x,y,\zeta,\omega)\frac{\sin k(z-\zeta)}{k}\,d\zeta +$$

$$+ \frac{\sin kz}{2k}\int_{-\infty}^0 \tilde{S}(x,y,\zeta,\omega)\cos k\zeta\,d\zeta - \frac{\cos kz}{2k}\int_{-\infty}^0 \tilde{S}(x,y,\zeta,\omega)\sin k\zeta\,d\zeta -$$

$$- \frac{\sin kz}{2k}\int_0^\infty \tilde{S}(x,y,\zeta,\omega)\cos k\zeta\,d\zeta + \frac{\cos kz}{2k}\int_0^\infty \tilde{S}(x,y,\zeta,\omega)\sin k\zeta\,d\zeta.$$

The last four integrals are independent of z, so we can represent them with functions constant in z. Finally, rewriting the trigonometric functions as exponentials we have

$$\tilde{\psi}(x,y,z,\omega) = \int_0^z \tilde{S}(x,y,\zeta,\omega)\frac{\sin k(z-\zeta)}{k}\,d\zeta + \tilde{A}(x,y,\omega)e^{-jkz} + \tilde{B}(x,y,\omega)e^{jkz}.$$

$$(A.25)$$

This formula for $\tilde{\psi}$ was found as a solution to the inhomogeneous ordinary differential equation (A.19). Hence, to obtain the complete solution we should add any possible solutions of the homogeneous differential equation. Since these are exponentials, (A.25) in fact represents the complete solution, where $\tilde{A}$ and $\tilde{B}$ are considered unknown and can be found using the boundary conditions.

If we are interested in the frequency-domain solution to the wave equation, then we are done. However, since our boundary conditions (A.17) and (A.18) pertain to the time domain, we must temporally inverse transform before we can apply them. Writing the sine function in (A.25) in terms of exponentials, we can express the time-domain solution as

$$\psi(x,y,z,t) = \int_0^z \mathcal{F}^{-1}\left\{\frac{c}{2}\frac{\tilde{S}(x,y,\zeta,\omega)}{j\omega}e^{j\frac{\omega}{c}(z-\zeta)} - \frac{c}{2}\frac{\tilde{S}(x,y,\zeta,\omega)}{j\omega}e^{-j\frac{\omega}{c}(z-\zeta)}\right\}\,d\zeta +$$

$$+ \mathcal{F}^{-1}\left\{\tilde{A}(x,y,\omega)e^{-j\frac{\omega}{c}z}\right\} + \mathcal{F}^{-1}\left\{\tilde{B}(x,y,\omega)e^{j\frac{\omega}{c}z}\right\}. \qquad (A.26)$$

A combination of the Fourier integration and time-shifting theorems gives the general identity

$$\mathcal{F}^{-1}\left\{\frac{\tilde{S}(x,y,\zeta,\omega)}{j\omega}e^{-j\omega t_0}\right\} = \int_{-\infty}^{t-t_0} S(x,y,\zeta,\tau)\,d\tau, \qquad (A.27)$$

where we have assumed that $\tilde{S}(x,y,\zeta,0) = 0$. Using this in (A.26) along with the time-shifting theorem we obtain

$$\psi(x,y,z,t) = \frac{c}{2}\int_0^z \left\{\int_{-\infty}^{t-\frac{\zeta-z}{c}} S(x,y,\zeta,\tau)\,d\tau - \int_{-\infty}^{t-\frac{z-\zeta}{c}} S(x,y,\zeta,\tau)\,d\tau\right\}\,d\zeta +$$

$$+ a\left(x,y,t-\frac{z}{c}\right) + b\left(x,y,t+\frac{z}{c}\right),$$

or

$$\psi(x,y,z,t) = \frac{c}{2}\int_0^z \int_{t-\frac{z-\zeta}{c}}^{t+\frac{z-\zeta}{c}} S(x,y,\zeta,\tau)\,d\tau\,d\zeta + a\left(x,y,t-\frac{z}{c}\right) + b\left(x,y,t+\frac{z}{c}\right) \quad (A.28)$$

where

$$a(x,y,t) = \mathcal{F}^{-1}[\tilde{A}(x,y,\omega)], \qquad b(x,y,t) = \mathcal{F}^{-1}[\tilde{B}(x,y,\omega)].$$

To calculate $a(x, y, t)$ and $b(x, y, t)$, we must use the boundary conditions (A.17) and (A.18). To apply (A.17), we put $z = 0$ into (A.28) to give

$$a(x, y, t) + b(x, y, t) = f(x, y, t). \tag{A.29}$$

Using (A.18) is a bit more complicated since we must compute $\partial \psi / \partial z$, and z is a parameter in the limits of the integral describing ψ. To compute the derivative we apply Leibnitz' rule for differentiation:

$$\frac{d}{d\alpha} \int_{\phi(\alpha)}^{\theta(\alpha)} f(x, \alpha)\, dx = \left(\frac{d\theta}{d\alpha} \right) f\left(\theta(\alpha), \alpha \right) - \left(\frac{d\phi}{d\alpha} \right) f\left(\phi(\alpha), \alpha \right) + \int_{\phi(\alpha)}^{\theta(\alpha)} \frac{\partial f}{\partial \alpha}\, dx. \tag{A.30}$$

Using this on the integral term in (A.28) we have

$$\frac{\partial}{\partial z} \left[\frac{c}{2} \int_0^z \left(\int_{t - \frac{z - \zeta}{c}}^{t + \frac{z - \zeta}{c}} S(x, y, \zeta, \tau)\, d\tau \right) d\zeta \right] = \frac{c}{2} \int_0^z \frac{\partial}{\partial z} \left(\int_{t - \frac{z - \zeta}{c}}^{t + \frac{z - \zeta}{c}} S(x, y, \zeta, \tau)\, d\tau \right) d\zeta,$$

which is zero at $z = 0$. Thus

$$\left. \frac{\partial \psi}{\partial z} \right|_{z=0} = g(x, y, t) = -\frac{1}{c} a'(x, y, t) + \frac{1}{c} b'(x, y, t)$$

where $a' = \partial a / \partial t$ and $b' = \partial b / \partial t$. Integration gives

$$-a(x, y, t) + b(x, y, t) = c \int_{-\infty}^t g(x, y, \tau)\, d\tau. \tag{A.31}$$

Equations (A.29) and (A.31) represent two algebraic equations in the two unknown functions a and b. The solutions are

$$2a(x, y, t) = f(x, y, t) - c \int_{-\infty}^t g(x, y, \tau)\, d\tau,$$

$$2b(x, y, t) = f(x, y, t) + c \int_{-\infty}^t g(x, y, \tau)\, d\tau.$$

Finally, substitution of these into (A.28) gives us the solution to the inhomogeneous wave equation

$$\psi(x, y, z, t) = \frac{c}{2} \int_0^z \int_{t - \frac{z - \zeta}{c}}^{t + \frac{z - \zeta}{c}} S(x, y, \zeta, \tau)\, d\tau\, d\zeta + \frac{1}{2} \left[f\left(x, y, t - \frac{z}{c} \right) + f\left(x, y, t + \frac{z}{c} \right) \right] +$$

$$+ \frac{c}{2} \int_{t - \frac{z}{c}}^{t + \frac{z}{c}} g(x, y, \tau)\, d\tau. \tag{A.32}$$

This is known as the *D'Alembert solution*. The terms $f(x, y, t \mp z/c)$ contribute to ψ as waves propagating away from the plane $z = 0$ in the $\pm z$-directions, respectively. The integral over the forcing term S is seen to accumulate values of S over a time interval determined by $z - \zeta$.

The boundary conditions could have been applied while still in the temporal frequency domain (but not the spatial frequency domain, since the spatial position z is lost). But to do this, we would need the boundary conditions to be in the temporal frequency domain. This is easily accomplished by transforming them to give

$$\left. \tilde{\psi}(x, y, z, \omega) \right|_{z=0} = \tilde{f}(x, y, \omega),$$

$$\left. \frac{\partial}{\partial z} \tilde{\psi}(x, y, z, \omega) \right|_{z=0} = \tilde{g}(x, y, \omega).$$

Applying these to (A.25) (and again using Leibnitz' rule) we have

$$\tilde{A}(x, y, \omega) + \tilde{B}(x, y, \omega) = \tilde{f}(x, y, \omega),$$
$$-jk\tilde{A}(x, y, \omega) + jk\tilde{B}(x, y, \omega) = \tilde{g}(x, y, \omega),$$

hence

$$2\tilde{A}(x, y, \omega) = \tilde{f}(x, y, \omega) - c\frac{\tilde{g}(x, y, \omega)}{j\omega},$$
$$2\tilde{B}(x, y, \omega) = \tilde{f}(x, y, \omega) + c\frac{\tilde{g}(x, y, \omega)}{j\omega}.$$

Finally, substituting these back into (A.25) and expanding the sine function we obtain the frequency-domain solution that obeys the given boundary conditions:

$$\tilde{\psi}(x, y, z, \omega) = \frac{c}{2} \int_0^z \left[\frac{\tilde{S}(x, y, \zeta, \omega)e^{j\frac{\omega}{c}(z-\zeta)}}{j\omega} - \frac{\tilde{S}(x, y, \zeta, \omega)e^{-j\frac{\omega}{c}(z-\zeta)}}{j\omega} \right] d\zeta +$$
$$+ \frac{1}{2} \left[\tilde{f}(x, y, \omega)e^{j\frac{\omega}{c}z} + \tilde{f}(x, y, \omega)e^{-j\frac{\omega}{c}z} \right] +$$
$$+ \frac{c}{2} \left[\frac{\tilde{g}(x, y, \omega)e^{j\frac{\omega}{c}z}}{j\omega} - \frac{\tilde{g}(x, y, \omega)e^{-j\frac{\omega}{c}z}}{j\omega} \right].$$

This is easily inverted using (A.27) to give (A.32).

Fourier transform solution of the 1-D homogeneous wave equation for dissipative media

Wave propagation in dissipative media can be studied using the one-dimensional wave equation

$$\left(\frac{\partial^2}{\partial z^2} - \frac{2\Omega}{v^2}\frac{\partial}{\partial t} - \frac{1}{v^2}\frac{\partial^2}{\partial t^2} \right) \psi(x, y, z, t) = S(x, y, z, t). \tag{A.33}$$

This equation is nearly identical to the wave equation for lossless media studied in the previous section, except for the addition of the $\partial \psi / \partial t$ term. This extra term will lead to important physical consequences regarding the behavior of the wave solutions.

We shall solve (A.33) using the Fourier transform approach of the previous section, but to keep the solution simple we shall only consider the homogeneous problem. We begin by writing ψ in terms of its inverse temporal Fourier transform:

$$\psi(x, y, z, t) = \frac{1}{2\pi} \int_{-\infty}^{\infty} \tilde{\psi}(x, y, z, \omega)e^{j\omega t} \, d\omega.$$

Substituting this into the homogeneous version of (A.33) and taking the time derivatives, we obtain

$$\frac{1}{2\pi} \int_{-\infty}^{\infty} \left[(j\omega)^2 + 2\Omega(j\omega) - v^2\frac{\partial^2}{\partial z^2} \right] \tilde{\psi}(x, y, z, \omega)e^{j\omega t} \, d\omega = 0.$$

The Fourier integral theorem leads to

$$\frac{\partial^2 \tilde{\psi}(x, y, z, \omega)}{\partial z^2} - \kappa^2 \tilde{\psi}(x, y, z, \omega) = 0 \tag{A.34}$$

where

$$\kappa = \frac{1}{v}\sqrt{p^2 + 2\Omega p}$$

with $p = j\omega$.

We can solve the homogeneous ordinary differential equation (A.34) by inspection:

$$\tilde{\psi}(x,y,z,\omega) = \tilde{A}(x,y,\omega)e^{-\kappa z} + \tilde{B}(x,y,\omega)e^{\kappa z}. \tag{A.35}$$

Here $\tilde{A}$ and $\tilde{B}$ are frequency-domain coefficients to be determined. We can either specify these coefficients directly, or solve for them by applying specific boundary conditions. We examine each possibility below.

Solution to wave equation by direct application of boundary conditions. The solution to the wave equation (A.33) will be unique if we specify functions $f(x,y,t)$ and $g(x,y,t)$ such that

$$\psi(x,y,z,t)\Big|_{z=0} = f(x,y,t),$$

$$\frac{\partial}{\partial z}\psi(x,y,z,t)\Big|_{z=0} = g(x,y,t). \tag{A.36}$$

Assuming the Fourier transform pairs $f(x,y,t) \leftrightarrow \tilde{f}(x,y,\omega)$ and $g(x,y,t) \leftrightarrow \tilde{g}(x,y,\omega)$, we can apply the boundary conditions (A.36) in the frequency domain:

$$\tilde{\psi}(x,y,z,\omega)\Big|_{z=0} = \tilde{f}(x,y,\omega),$$

$$\frac{\partial}{\partial z}\tilde{\psi}(x,y,z,\omega)\Big|_{z=0} = \tilde{g}(x,y,\omega).$$

From these we find

$$\tilde{A} + \tilde{B} = \tilde{f}, \qquad -\kappa\tilde{A} + \kappa\tilde{B} = \tilde{g},$$

or

$$\tilde{A} = \frac{1}{2}\left[\tilde{f} - \frac{\tilde{g}}{\kappa}\right], \qquad \tilde{B} = \frac{1}{2}\left[\tilde{f} + \frac{\tilde{g}}{\kappa}\right].$$

Substitution into (A.35) gives

$$\tilde{\psi}(x,y,z,\omega) = \tilde{f}(x,y,\omega)\cosh\kappa z + \tilde{g}(x,y,\omega)\frac{\sinh\kappa z}{\kappa}$$

$$= \tilde{f}(x,y,\omega)\frac{\partial}{\partial z}\tilde{Q}(x,y,z,\omega) + \tilde{g}(x,y,\omega)\tilde{Q}(x,y,z,\omega)$$

$$= \tilde{\psi}_1(x,y,z,\omega) + \tilde{\psi}_2(x,y,z,\omega)$$

where $\tilde{Q} = \sinh\kappa z/\kappa$. Assuming that $Q(x,y,z,t) \leftrightarrow \tilde{Q}(x,y,z,\omega)$, we can employ the convolution theorem to immediately write down $\psi(x,y,z,t)$:

$$\psi(x,y,z,t) = f(x,y,t) * \frac{\partial}{\partial z}Q(x,y,z,t) + g(x,y,z,t) * Q(x,y,z,t)$$

$$= \psi_1(x,y,z,t) + \psi_2(x,y,z,t). \tag{A.37}$$

To find ψ we must first compute the inverse transform of $\tilde{Q}$. Here we resort to a tabulated result [26]:

$$\frac{\sinh\left[a\sqrt{p+\lambda}\sqrt{p+\mu}\right]}{\sqrt{p+\lambda}\sqrt{p+\mu}} \leftrightarrow \frac{1}{2}e^{-\frac{1}{2}(\mu+\lambda)t}J_0\left(\frac{1}{2}(\lambda-\mu)\sqrt{a^2-t^2}\right), \quad -a < t < a.$$

Here a is a positive, finite real quantity, and λ and μ are finite complex quantities. Outside the range $|t| < a$ the time-domain function is zero.

Letting $a = z/v$, $\mu = 0$, and $\lambda = 2\Omega$ in the above expression, we find

$$Q(x, y, z, t) = \frac{v}{2} e^{-\Omega t} J_0\left(\frac{\Omega}{v}\sqrt{z^2 - v^2 t^2}\right) [U(t + z/v) - U(t - z/v)] \qquad (A.38)$$

where $U(x)$ is the unit step function (A.5). From (A.37) we see that

$$\psi_2(x, y, z, t) = \int_{-\infty}^{\infty} g(x, y, t - \tau) Q(x, y, z, \tau)\, d\tau = \int_{-z/v}^{z/v} g(x, y, t - \tau) Q(x, y, z, \tau)\, d\tau.$$

Using the change of variables $u = t - \tau$ and substituting (A.38), we then have

$$\psi_2(x, y, z, t) = \frac{v}{2} e^{-\Omega t} \int_{t-\frac{z}{v}}^{t+\frac{z}{v}} g(x, y, u) e^{\Omega u} J_0\left(\frac{\Omega}{v}\sqrt{z^2 - (t-u)^2 v^2}\right) du. \qquad (A.39)$$

To find ψ_1 we must compute $\partial Q/\partial z$. Using the product rule we have

$$\frac{\partial Q(x, y, z, t)}{\partial z} = \frac{v}{2} e^{-\Omega t} J_0\left(\frac{\Omega}{v}\sqrt{z^2 - v^2 t^2}\right) \frac{\partial}{\partial z}[U(t + z/v) - U(t - z/v)] +$$

$$+ \frac{v}{2} e^{-\Omega t}[U(t + z/v) - U(t - z/v)]\frac{\partial}{\partial z} J_0\left(\frac{\Omega}{v}\sqrt{z^2 - v^2 t^2}\right).$$

Next, using $dU(x)/dx = \delta(x)$ and remembering that $J_0'(x) = -J_1(x)$ and $J_0(0) = 1$, we can write

$$\frac{\partial Q(x, y, z, t)}{\partial z} = \frac{1}{2} e^{-\Omega t}[\delta(t + z/v) + \delta(t - z/v)] -$$

$$- \frac{z\Omega^2}{2v} e^{-\Omega t} \frac{J_1\left(\frac{\Omega}{v}\sqrt{z^2 - v^2 t^2}\right)}{\frac{\Omega}{v}\sqrt{z^2 - v^2 t^2}}[U(t + z/v) - U(t - z/v)].$$

Convolving this expression with $f(x, y, t)$ we obtain

$$\psi_1(x, y, z, t) = \frac{1}{2} e^{-\frac{\Omega}{v} z} f\left(x, y, t - \frac{z}{v}\right) + \frac{1}{2} e^{\frac{\Omega}{v} z} f\left(x, y, t + \frac{z}{v}\right) -$$

$$- \frac{z\Omega^2}{2v} e^{-\Omega t} \int_{t-\frac{z}{v}}^{t+\frac{z}{v}} f(x, y, u) e^{\Omega u} \frac{J_1\left(\frac{\Omega}{v}\sqrt{z^2 - (t-u)^2 v^2}\right)}{\frac{\Omega}{v}\sqrt{z^2 - (t-u)^2 v^2}} du. \qquad (A.40)$$

Finally, adding (A.40) and (A.39), we obtain

$$\psi(x, y, z, t) = \frac{1}{2} e^{-\frac{\Omega}{v} z} f\left(x, y, t - \frac{z}{v}\right) + \frac{1}{2} e^{\frac{\Omega}{v} z} f\left(x, y, t + \frac{z}{v}\right) -$$

$$- \frac{z\Omega^2}{2v} e^{-\Omega t} \int_{t-\frac{z}{v}}^{t+\frac{z}{v}} f(x, y, u) e^{\Omega u} \frac{J_1\left(\frac{\Omega}{v}\sqrt{z^2 - (t-u)^2 v^2}\right)}{\frac{\Omega}{v}\sqrt{z^2 - (t-u)^2 v^2}} du +$$

$$+ \frac{v}{2} e^{-\Omega t} \int_{t-\frac{z}{v}}^{t+\frac{z}{v}} g(x, y, u) e^{\Omega u} J_0\left(\frac{\Omega}{v}\sqrt{z^2 - (t-u)^2 v^2}\right) du. \qquad (A.41)$$

Note that when $\Omega = 0$ this reduces to

$$\psi(x, y, z, t) = \frac{1}{2} f\left(x, y, t - \frac{z}{v}\right) + \frac{1}{2} f\left(x, y, t + \frac{z}{v}\right) + \frac{v}{2} \int_{t-\frac{z}{v}}^{t+\frac{z}{v}} g(x, y, u)\, du,$$

which matches (A.32) for the homogeneous case where $S = 0$.

Solution to wave equation by specification of wave amplitudes. An alternative to direct specification of boundary conditions is specification of the amplitude functions $\tilde{A}(x,y,\omega)$ and $\tilde{B}(x,y,\omega)$ or their inverse transforms $A(x,y,t)$ and $B(x,y,t)$. If we specify the time-domain functions we can write $\psi(x,y,z,t)$ as the inverse transform of (A.35). For example, a wave traveling in the $+z$-direction behaves as

$$\psi(x,y,z,t) = A(x,y,t) * F^+(x,y,z,t) \tag{A.42}$$

where

$$F^+(x,y,z,t) \leftrightarrow e^{-\kappa z} = e^{-\frac{z}{v}\sqrt{p^2+2\Omega p}}.$$

We can find F^+ using the following Fourier transform pair [26]):

$$e^{-\frac{x}{v}\sqrt{(p+\rho)^2-\sigma^2}} \leftrightarrow e^{-\frac{\rho}{v}x}\delta(t-x/v) + \frac{\sigma x}{v}e^{-\rho t}\frac{I_1\left(\sigma\sqrt{t^2-(x/v)^2}\right)}{\sqrt{t^2-(x/v)^2}}, \quad \frac{x}{v} < t. \tag{A.43}$$

Here x is real and positive and $I_1(x)$ is the modified Bessel function of the first kind and order 1. Outside the range $x/v < t$ the time-domain function is zero. Letting $\rho = \Omega$ and $\sigma = \Omega$ we find

$$F^+(x,y,z,t) = \frac{\Omega^2 z}{v}e^{-\Omega t}\frac{I_1(\Omega\sqrt{t^2-(z/v)^2})}{\Omega\sqrt{t^2-(z/v)^2}}U(t-z/v) + e^{-\frac{\Omega}{v}z}\delta(t-z/v). \tag{A.44}$$

Note that F^+ is a real functions of time, as expected.

Substituting (A.44) into (A.42) and writing the convolution in integral form we have

$$\psi(x,y,z,t) = \int_{z/v}^{\infty} A(x,y,t-\tau)\left[\frac{\Omega^2 z}{v}e^{-\Omega\tau}\frac{I_1(\Omega\sqrt{\tau^2-(z/v)^2})}{\Omega\sqrt{\tau^2-(z/v)^2}}\right]d\tau +$$

$$+ e^{-\frac{\Omega}{v}z}A\left(x,y,t-\frac{z}{v}\right), \quad z > 0. \tag{A.45}$$

The 3-D Green's function for waves in dissipative media

To understand the fields produced by bounded sources within a dissipative medium we may wish to investigate solutions to the wave equation in three dimensions. The Green's function approach requires the solution to

$$\left(\nabla^2 - \frac{2\Omega}{v^2}\frac{\partial}{\partial t} - \frac{1}{v^2}\frac{\partial^2}{\partial t^2}\right)G(\mathbf{r}|\mathbf{r}';t) = -\delta(t)\delta(\mathbf{r}-\mathbf{r}')$$

$$= -\delta(t)\delta(x-x')\delta(y-y')\delta(z-z').$$

That is, we are interested in the impulse response of a point source located at $\mathbf{r} = \mathbf{r}'$. We begin by substituting the inverse temporal Fourier transform relations

$$G(\mathbf{r}|\mathbf{r}';t) = \frac{1}{2\pi}\int_{-\infty}^{\infty}\tilde{G}(\mathbf{r}|\mathbf{r}';\omega)e^{j\omega t}\,d\omega,$$

$$\delta(t) = \frac{1}{2\pi}\int_{-\infty}^{\infty}e^{j\omega t}\,d\omega,$$

obtaining

$$\frac{1}{2\pi}\int_{-\infty}^{\infty}\left[\left(\nabla^2 - j\omega\frac{2\Omega}{v^2} - \frac{1}{v^2}(j\omega)^2\right)\tilde{G}(\mathbf{r}|\mathbf{r}';\omega) + \delta(\mathbf{r}-\mathbf{r}')\right]e^{j\omega t}d\omega = 0.$$

By the Fourier integral theorem we have

$$(\nabla^2 + k^2)\tilde{G}(\mathbf{r}|\mathbf{r}';\omega) = -\delta(\mathbf{r} - \mathbf{r}'). \tag{A.46}$$

This is known as the *Helmholtz equation.* Here

$$k = \frac{1}{v}\sqrt{\omega^2 - j2\omega\Omega} \tag{A.47}$$

is called the *wavenumber.*

To solve the Helmholtz equation we write $\tilde{G}$ in terms of a 3-dimensional inverse Fourier transform. Substitution of

$$\tilde{G}(\mathbf{r}|\mathbf{r}';\omega) = \frac{1}{(2\pi)^3}\int_{-\infty}^{\infty}\tilde{G}^r(\mathbf{k}|\mathbf{r}';\omega)e^{j\mathbf{k}\cdot\mathbf{r}}\,d^3k,$$

$$\delta(\mathbf{r} - \mathbf{r}') = \frac{1}{(2\pi)^3}\int_{-\infty}^{\infty}e^{j\mathbf{k}\cdot(\mathbf{r}-\mathbf{r}')}\,d^3k,$$

into (A.46) gives

$$\frac{1}{(2\pi)^3}\int_{-\infty}^{\infty}\left[\nabla^2\left(\tilde{G}^r(\mathbf{k}|\mathbf{r}';\omega)e^{j\mathbf{k}\cdot\mathbf{r}}\right) + k^2\tilde{G}^r(\mathbf{k}|\mathbf{r}';\omega)e^{j\mathbf{k}\cdot\mathbf{r}} + e^{j\mathbf{k}\cdot(\mathbf{r}-\mathbf{r}')}\right]d^3k = 0.$$

Here

$$\mathbf{k} = \hat{\mathbf{x}}k_x + \hat{\mathbf{y}}k_y + \hat{\mathbf{z}}k_z$$

with $|\mathbf{k}|^2 = k_x^2 + k_y^2 + k_z^2 = K^2$. Carrying out the derivatives and invoking the Fourier integral theorem we have

$$(K^2 - k^2)\tilde{G}^r(\mathbf{k}|\mathbf{r}';\omega) = e^{-j\mathbf{k}\cdot\mathbf{r}'}.$$

Solving for $\tilde{G}$ and substituting it into the inverse transform relation we have

$$\tilde{G}(\mathbf{r}|\mathbf{r}';\omega) = \frac{1}{(2\pi)^3}\int_{-\infty}^{\infty}\frac{e^{j\mathbf{k}\cdot(\mathbf{r}-\mathbf{r}')}}{(K-k)(K+k)}\,d^3k. \tag{A.48}$$

To compute the inverse transform integral in (A.48) we write the 3-D transform variable in spherical coordinates:

$$\mathbf{k}\cdot(\mathbf{r} - \mathbf{r}') = KR\cos\theta, \qquad d^3k = K^2\sin\theta\,dK\,d\theta\,d\phi,$$

where $R = |\mathbf{r} - \mathbf{r}'|$ and θ is the angle between $\mathbf{k}$ and $\mathbf{r} - \mathbf{r}'$. Hence (A.48) becomes

$$\tilde{G}(\mathbf{r}|\mathbf{r}';\omega) = \frac{1}{(2\pi)^3}\int_0^{\infty}\frac{K^2\,dK}{(K-k)(K+k)}\int_0^{2\pi}d\phi\int_0^{\pi}e^{jKR\cos\theta}\sin\theta\,d\theta$$

$$= \frac{2}{(2\pi)^2 R}\int_0^{\infty}\frac{K\sin(KR)}{(K-k)(K+k)}\,dK,$$

or, equivalently,

$$\tilde{G}(\mathbf{r}|\mathbf{r}';\omega) = \frac{1}{2jR(2\pi)^2}\int_{-\infty}^{\infty}\frac{e^{jKR}}{(K-k)(K+k)}K\,dK -$$

$$- \frac{1}{2jR(2\pi)^2}\int_{-\infty}^{\infty}\frac{e^{-jkR}}{(K-k)(K+k)}K\,dK.$$

We can compute the integrals over K using the complex plane technique. We consider K to be a complex variable, and note that for dissipative media we have $k = k_r + jk_i$, where $k_r > 0$ and $k_i < 0$. Thus the integrand has poles at $K = \pm k$. For the integral involving e^{+jKR} we close the contour in the upper half-plane using a semicircle of radius Δ and use Cauchy's residue theorem. Then at all points on the semicircle the integrand decays exponentially as $\Delta \to \infty$, and there is no contribution to the integral from this part of the contour. The real-line integral is thus equal to $2\pi j$ times the residue at $K = -k$:

$$\int_{-\infty}^{\infty} \frac{e^{jKR}}{(K-k)(K+k)} K \, dK = 2\pi j \frac{e^{-jkR}}{-2k}(-k).$$

For the term involving e^{-jKR} we close in the lower half-plane and again the contribution from the infinite semicircle vanishes. In this case our contour is clockwise and so the real line integral is $-2\pi j$ times the residue at $K = k$:

$$\int_{-\infty}^{\infty} \frac{e^{-jKR}}{(K-k)(K+k)} K \, dK = -2\pi j \frac{e^{-jkR}}{2k} k.$$

Thus

$$\tilde{G}(\mathbf{r}|\mathbf{r}';\omega) = \frac{e^{-jkR}}{4\pi R}. \tag{A.49}$$

Note that if $\Omega = 0$ then this reduces to

$$\tilde{G}(\mathbf{r}|\mathbf{r}';\omega) = \frac{e^{-j\omega R/v}}{4\pi R}. \tag{A.50}$$

Our last step is to find the temporal Green's function. Let $p = j\omega$. Then we can write

$$\tilde{G}(\mathbf{r}|\mathbf{r}';\omega) = \frac{e^{\kappa R}}{4\pi R}$$

where

$$\kappa = -jk = \frac{1}{v}\sqrt{p^2 + 2\Omega p}.$$

We may find the inverse transform using (A.43). Letting $x = R$, $\rho = \Omega$, and $\sigma = \Omega$ we find

$$G(\mathbf{r}|\mathbf{r}';t) = e^{-\frac{\Omega}{v}R}\frac{\delta(t - R/v)}{4\pi R} + \frac{\Omega^2}{4\pi v}e^{-\Omega t}\frac{I_1\left(\Omega\sqrt{t^2 - (R/v)^2}\right)}{\Omega\sqrt{t^2 - (R/v)^2}} U\left(t - \frac{R}{v}\right).$$

We note that in the case of no dissipation where $\Omega = 0$ this reduces to

$$G(\mathbf{r}|\mathbf{r}';t) = \frac{\delta(t - R/v)}{4\pi R}$$

which is the inverse transform of (A.50).

Fourier transform representation of the static Green's function

In the study of static fields, we shall be interested in the solution to the partial differential equation

$$\nabla^2 G(\mathbf{r}|\mathbf{r}') = -\delta(\mathbf{r} - \mathbf{r}') = -\delta(x - x')\delta(y - y')\delta(z - z'). \tag{A.51}$$

Here $G(\mathbf{r}|\mathbf{r}')$, called the "static Green's function," represents the potential at point $\mathbf{r}$ produced by a unit point source at point $\mathbf{r}'$.

In Chapter 3 we find that $G(\mathbf{r}|\mathbf{r}') = 1/4\pi|\mathbf{r} - \mathbf{r}'|$. In a variety of problems it is also useful to have G written in terms of an inverse Fourier transform over the variables x and y. Letting G^r form a three-dimensional Fourier transform pair with G, we can write

$$G(\mathbf{r}|\mathbf{r}') = \frac{1}{(2\pi)^3} \int_{-\infty}^{\infty} G^r(k_x, k_y, k_z|\mathbf{r}')e^{jk_x x}e^{jk_y y}e^{jk_z z}\, dk_x\, dk_y\, dk_z.$$

Substitution into (A.51) along with the inverse transformation representation for the delta function (A.4) gives

$$\frac{1}{(2\pi)^3}\nabla^2 \int_{-\infty}^{\infty} G^r(k_x, k_y, k_z|\mathbf{r}')e^{jk_x x}e^{jk_y y}e^{jk_z z}\, dk_x\, dk_y\, dk_z$$
$$= -\frac{1}{(2\pi)^3} \int_{-\infty}^{\infty} e^{jk_x(x-x')}e^{jk_y(y-y')}e^{jk_z(z-z')}\, dk_x\, dk_y\, dk_z.$$

We then combine the integrands and move the Laplacian operator through the integral to obtain

$$\frac{1}{(2\pi)^3} \int_{-\infty}^{\infty} \left[\nabla^2\left(G^r(\mathbf{k}|\mathbf{r}')e^{j\mathbf{k}\cdot\mathbf{r}}\right) + e^{j\mathbf{k}\cdot(\mathbf{r}-\mathbf{r}')}\right] d^3k = 0,$$

where $\mathbf{k} = \hat{\mathbf{x}}k_x + \hat{\mathbf{y}}k_y + \hat{\mathbf{z}}k_z$. Carrying out the derivatives,

$$\frac{1}{(2\pi)^3} \int_{-\infty}^{\infty} \left[\left(-k_x^2 - k_y^2 - k_z^2\right)G^r(\mathbf{k}|\mathbf{r}') + e^{-j\mathbf{k}\cdot\mathbf{r}'}\right] e^{j\mathbf{k}\cdot\mathbf{r}}\, d^3k = 0.$$

Letting $k_x^2 + k_y^2 = k_\rho^2$ and invoking the Fourier integral theorem we get the algebraic equation

$$\left(-k_\rho^2 - k_z^2\right)G^r(\mathbf{k}|\mathbf{r}') + e^{-j\mathbf{k}\cdot\mathbf{r}'} = 0,$$

which we can easily solve for G^r:

$$G^r(\mathbf{k}|\mathbf{r}') = \frac{e^{-j\mathbf{k}\cdot\mathbf{r}'}}{k_\rho^2 + k_z^2}. \tag{A.52}$$

Equation (A.52) gives us a 3-D transform representation for the Green's function. Since we desire the 2-D representation, we shall have to perform the inverse transform over k_z. Writing

$$G^{xy}(k_x, k_y, z|\mathbf{r}') = \frac{1}{2\pi} \int_{-\infty}^{\infty} G^r(k_x, k_y, k_z|\mathbf{r}')e^{jk_z z}\, dk_z$$

we have

$$G^{xy}(k_x, k_y, z|\mathbf{r}') = \frac{1}{2\pi} \int_{-\infty}^{\infty} \frac{e^{-jk_x x'}e^{-jk_y y'}e^{jk_z(z-z')}}{k_\rho^2 + k_z^2}\, dk_z. \tag{A.53}$$

To compute this integral, we let k_z be a complex variable and consider a closed contour in the complex plane, consisting of a semicircle and the real axis. As previously discussed, we compute the principal value integral as the semicircle radius $\Delta \to \infty$, and find that the contribution along the semicircle reduces to zero. Hence we can use Cauchy's residue theorem (A.14) to obtain the real-line integral:

$$G^{xy}(k_x, k_y, z|\mathbf{r}') = 2\pi j\, \mathrm{res}\left\{\frac{1}{2\pi}\frac{e^{-jk_x x'}e^{-jk_y y'}e^{jk_z(z-z')}}{k_\rho^2 + k_z^2}\right\}.$$

Here res$\{f(k_z)\}$ denotes the residues of the function $f(k_z)$. The integrand in (A.53) has poles of order 1 at $k_z = \pm jk_\rho$, $k_\rho \geq 0$. If $z - z' > 0$ we close in the upper half-plane and enclose only the pole at $k_z = jk_\rho$. Computing the residue using (A.13), we obtain

$$G^{xy}(k_x, k_y, z|\mathbf{r}') = j\frac{e^{-jk_x x'}e^{-jk_y y'}e^{-k_\rho(z-z')}}{2jk_\rho}, \qquad z > z'.$$

Since $z > z'$ this function decays for increasing z, as expected physically. For $z - z' < 0$ we close in the lower half-plane, enclosing the pole at $k_z = -jk_\rho$ and incurring an additional negative sign since our contour is now clockwise. Evaluating the residue we have

$$G^{xy}(k_x, k_y, z|\mathbf{r}') = -j\frac{e^{-jk_x x'}e^{-jk_y y'}e^{k_\rho(z-z')}}{-2jk_\rho}, \qquad z < z'.$$

We can combine both cases $z > z'$ and $z < z'$ by using the absolute value function:

$$G^{xy}(k_x, k_y, z|\mathbf{r}') = \frac{e^{-jk_x x'}e^{-jk_y y'}e^{-k_\rho|z-z'|}}{2k_\rho}. \tag{A.54}$$

Finally, we substitute (A.54) into the inverse transform formula. This gives the Green's function representation

$$G(\mathbf{r}|\mathbf{r}') = \frac{1}{4\pi|\mathbf{r}-\mathbf{r}'|} = \frac{1}{(2\pi)^2}\int_{-\infty}^{\infty}\frac{e^{-k_\rho|z-z'|}}{2k_\rho}e^{j\mathbf{k}_\rho\cdot(\mathbf{r}-\mathbf{r}')}\,d^2k_\rho, \tag{A.55}$$

where $\mathbf{k}_\rho = \hat{x}k_x + \hat{y}k_y$, $k_\rho = |\mathbf{k}_\rho|$, and $d^2k_\rho = dk_x\,dk_y$.

On occasion we may wish to represent the solution of the homogeneous (Laplace) equation

$$\nabla^2\psi(\mathbf{r}) = 0$$

in terms of a 2-D Fourier transform. In this case we represent ψ as a 2-D inverse transform and substitute to obtain

$$\frac{1}{(2\pi)^2}\int_{-\infty}^{\infty}\nabla^2\left(\psi^{xy}(k_x, k_y, z)e^{jk_x x}e^{jk_y y}\right)\,dk_x\,dk_y = 0.$$

Carrying out the derivatives and invoking the Fourier integral theorem we find that

$$\left(\frac{\partial^2}{\partial z^2} - k_\rho^2\right)\psi^{xy}(k_x, k_y, z) = 0.$$

Hence

$$\psi^{xy}(k_x, k_y, z) = Ae^{k_\rho z} + Be^{-k_\rho z}$$

where A and B are constants with respect to z. Inverse transformation gives

$$\psi(\mathbf{r}) = \frac{1}{(2\pi)^2}\int_{-\infty}^{\infty}\left[A(\mathbf{k}_\rho)e^{k_\rho z} + B(\mathbf{k}_\rho)e^{-k_\rho z}\right]e^{j\mathbf{k}_\rho\cdot\mathbf{r}}\,d^2k_\rho. \tag{A.56}$$

A.2   Vector transport theorems

We are often interested in the time rate of change of some field integrated over a moving volume or surface. Such a derivative may be used to describe the transport of a physical quantity (e.g., charge, momentum, energy) through space. Many of the relevant theorems are derived in this section. The results find application in the development of the large-scale forms of Maxwell equations, the continuity equation, and the Poynting theorem.

Partial, total, and material derivatives

The key to understanding transport theorems lies in the difference between the various means of time-differentiating a field. Consider a scalar field $T(\mathbf{r}, t)$ (which could represent one component of a vector or dyadic field). If we fix our position within the field and examine how the field varies with time, we describe the *partial derivative* of T. However, this may not be the most useful means of measuring the time rate of change of a field. For instance, in mechanics we might be interested in the rate at which water cools as it sinks to the bottom of a container. In this case, T could represent temperature. We could create a "depth profile" at any given time (i.e., measure $T(\mathbf{r}, t_0)$ for some fixed t_0) by taking simultaneous data from a series of temperature probes at varying depths. We could also create a temporal profile at any given depth (i.e., measure $T(\mathbf{r}_0, t)$ for some fixed $\mathbf{r}_0$) by taking continuous data from a probe fixed at that depth. But neither of these would describe how an individual sinking water particle "experiences" a change in temperature over time.

Instead, we could use a probe that descends along with a particular water packet (i.e., volume element), measuring the time rate of temperature change of that element. This rate of change is called the *convective* or *material derivative*, since it corresponds to a situation in which a physical material quantity is followed as the derivative is calculated. We anticipate that this quantity will depend on (1) the time rate of change of T at each fixed point that the particle passes, and (2) the spatial rate of change of T as well as the rapidity with which the packet of interest is swept through that space gradient. The faster the packet descends, or the faster the temperature cools with depth, the larger the material derivative should be.

To compute the material derivative we describe the position of a water packet by the vector

$$\mathbf{r}(t) = \hat{\mathbf{x}}x(t) + \hat{\mathbf{y}}y(t) + \hat{\mathbf{z}}z(t).$$

Because no two packets can occupy the same place at the same time, the specification of $\mathbf{r}(0) = \mathbf{r}_0$ uniquely describes (or "tags") a particular packet. The time rate of change of $\mathbf{r}$ with $\mathbf{r}_0$ held constant (the material derivative of the position vector) is thus the velocity field $\mathbf{u}(\mathbf{r}, t)$ of the fluid:

$$\left(\frac{d\mathbf{r}}{dt}\right)_{\mathbf{r}_0} = \frac{D\mathbf{r}}{Dt} = \mathbf{u}. \tag{A.57}$$

Here we use the "big D" notation to denote the material derivative, thereby avoiding confusion with the partial and total derivatives described below.

To describe the time rate of change of the temperature of a particular water packet, we only need to hold $\mathbf{r}_0$ constant while we examine the change. If we write the temperature as

$$T(\mathbf{r}, t) = T(\mathbf{r}(\mathbf{r_0}, t), t) = T[x(\mathbf{r_0}, t), y(\mathbf{r_0}, t), z(\mathbf{r_0}, t), t],$$

then we can use the chain rule to find the time rate of change of T with $\mathbf{r_0}$ held constant:

$$\frac{DT}{Dt} = \left(\frac{dT}{dt}\right)_{\mathbf{r_0}}$$

$$= \left(\frac{\partial T}{\partial x}\right)\left(\frac{dx}{dt}\right)_{\mathbf{r_0}} + \left(\frac{\partial T}{\partial y}\right)\left(\frac{dy}{dt}\right)_{\mathbf{r_0}} + \left(\frac{\partial T}{\partial z}\right)\left(\frac{dz}{dt}\right)_{\mathbf{r_0}} + \frac{\partial T}{\partial t}.$$

We recognize the partial derivatives of the coordinates as the components of the material velocity (A.57), and thus can write

$$\frac{DT}{Dt} = \frac{\partial T}{\partial t} + u_x\frac{\partial T}{\partial x} + u_y\frac{\partial T}{\partial y} + u_z\frac{\partial T}{\partial z} = \frac{\partial T}{\partial t} + \mathbf{u} \cdot \nabla T.$$

As expected, the material derivative depends on both the local time rate of change and the spatial rate of change of temperature.

Suppose next that our probe is motorized and can travel about in the sinking water. If the probe sinks faster than the surrounding water, the time rate of change (measured by the probe) should exceed the material derivative. Let the probe position and velocity be

$$\mathbf{r}(t) = \hat{\mathbf{x}}x(t) + \hat{\mathbf{y}}y(t) + \hat{\mathbf{z}}z(t), \qquad \mathbf{v}(\mathbf{r}, t) = \hat{\mathbf{x}}\frac{dx(t)}{dt} + \hat{\mathbf{y}}\frac{dy(t)}{dt} + \hat{\mathbf{z}}\frac{dz(t)}{dt}.$$

We can use the chain rule to determine the time rate of change of the temperature observed by the probe, but in this case we do *not* constrain the velocity components to represent the moving fluid. Thus, we merely obtain

$$\frac{dT}{dt} = \frac{\partial T}{\partial x}\frac{dx}{dt} + \frac{\partial T}{\partial y}\frac{dy}{dt} + \frac{\partial T}{\partial z}\frac{dz}{dt} + \frac{\partial T}{\partial t}$$

$$= \frac{\partial T}{\partial t} + \mathbf{v} \cdot \nabla T.$$

This is called the *total derivative* of the temperature field.

In summary, the time rate of change of a scalar field T seen by an observer moving with arbitrary velocity $\mathbf{v}$ is given by the total derivative

$$\frac{dT}{dt} = \frac{\partial T}{\partial t} + \mathbf{v} \cdot \nabla T. \tag{A.58}$$

If the velocity of the observer happens to match the velocity $\mathbf{u}$ of a moving substance, the time rate of change is the material derivative

$$\frac{DT}{Dt} = \frac{\partial T}{\partial t} + \mathbf{u} \cdot \nabla T. \tag{A.59}$$

We can obtain the material derivative of a vector field $\mathbf{F}$ by component-wise application of (A.59):

$$\frac{D\mathbf{F}}{Dt} = \frac{D}{Dt}[\hat{\mathbf{x}}F_x + \hat{\mathbf{y}}F_y + \hat{\mathbf{z}}F_z]$$

$$= \hat{\mathbf{x}}\frac{\partial F_x}{\partial t} + \hat{\mathbf{y}}\frac{\partial F_y}{\partial t} + \hat{\mathbf{z}}\frac{\partial F_z}{\partial t} + \hat{\mathbf{x}}[\mathbf{u} \cdot (\nabla F_x)] + \hat{\mathbf{y}}[\mathbf{u} \cdot (\nabla F_y)] + \hat{\mathbf{z}}[\mathbf{u} \cdot (\nabla F_z)].$$

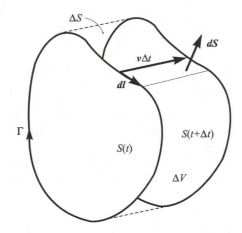

Figure A.5: Derivation of the Helmholtz transport theorem.

Using the notation

$$\mathbf{u} \cdot \nabla = u_x \frac{\partial}{\partial x} + u_y \frac{\partial}{\partial y} + u_z \frac{\partial}{\partial z}$$

we can write

$$\frac{D\mathbf{F}}{Dt} = \frac{\partial \mathbf{F}}{\partial t} + (\mathbf{u} \cdot \nabla)\mathbf{F}. \tag{A.60}$$

This is the material derivative of a vector field $\mathbf{F}$ when $\mathbf{u}$ describes the motion of a physical material. Similarly, the total derivative of a vector field is

$$\frac{d\mathbf{F}}{dt} = \frac{\partial \mathbf{F}}{\partial t} + (\mathbf{v} \cdot \nabla)\mathbf{F}$$

where $\mathbf{v}$ is arbitrary.

The Helmholtz and Reynolds transport theorems

We choose the intuitive approach taken by Tai [190] and Whitaker [214]. Consider an open surface $S(t)$ moving through space and possibly deforming as it moves. The velocity of the points comprising the surface is given by the vector field $\mathbf{v}(\mathbf{r}, t)$. We are interested in computing the time derivative of the flux of a vector field $\mathbf{F}(\mathbf{r}, t)$ through $S(t)$:

$$\psi(t) = \frac{d}{dt} \int_{S(t)} \mathbf{F}(\mathbf{r}, t) \cdot d\mathbf{S}$$

$$= \lim_{\Delta t \to 0} \frac{\int_{S(t+\Delta t)} \mathbf{F}(\mathbf{r}, t + \Delta t) \cdot d\mathbf{S} - \int_{S(t)} \mathbf{F}(\mathbf{r}, t) \cdot d\mathbf{S}}{\Delta t}. \tag{A.61}$$

Here $S(t+\Delta t) = S_2$ is found by extending each point on $S(t) = S_1$ through a displacement $\mathbf{v}\Delta t$, as shown in Figure A.5. Substituting the Taylor expansion

$$\mathbf{F}(\mathbf{r}, t + \Delta t) = \mathbf{F}(\mathbf{r}, t) + \frac{\partial \mathbf{F}(\mathbf{r}, t)}{\partial t} \Delta t + \cdots$$

into (A.61), we find that only the first two terms give non-zero contributions to the integral and

$$\psi(t) = \int_{S(t)} \frac{\partial \mathbf{F}(\mathbf{r}, t)}{\partial t} \cdot \mathbf{dS} + \lim_{\Delta t \to 0} \frac{\int_{S_2} \mathbf{F}(\mathbf{r}, t) \cdot \mathbf{dS} - \int_{S_1} \mathbf{F}(\mathbf{r}, t) \cdot \mathbf{dS}}{\Delta t}. \tag{A.62}$$

The second term on the right can be evaluated with the help of Figure A.5. As the surface moves through a displacement $\mathbf{v}\Delta t$ it sweeps out a volume region ΔV that is bounded on the back by S_1, on the front by S_2, and on the side by a surface $S_3 = \Delta S$. We can thus compute the two surface integrals in (A.62) as the difference between contributions from the surface enclosing ΔV and the side surface ΔS (remembering that the normal to S_1 in (A.62) points *into* ΔV). Thus

$$\psi(t) = \int_{S(t)} \frac{\partial \mathbf{F}(\mathbf{r}, t)}{\partial t} \cdot \mathbf{dS} + \lim_{\Delta t \to 0} \frac{\oint_{S_1 + S_2 + \Delta S} \mathbf{F}(\mathbf{r}, t) \cdot \mathbf{dS} - \int_{\Delta S} \mathbf{F}(\mathbf{r}, t) \cdot \mathbf{dS}_3}{\Delta t}$$

$$= \int_{S(t)} \frac{\partial \mathbf{F}(\mathbf{r}, t)}{\partial t} \cdot \mathbf{dS} + \lim_{\Delta t \to 0} \frac{\int_{\Delta V} \nabla \cdot \mathbf{F}(\mathbf{r}, t) \, dV_3 - \int_{\Delta S} \mathbf{F}(\mathbf{r}, t) \cdot \mathbf{dS}_3}{\Delta t}$$

by the divergence theorem. To compute the integrals over ΔS and ΔV we note from Figure A.5 that the incremental surface and volume elements are just

$$\mathbf{dS}_3 = \mathbf{dl} \times (\mathbf{v}\Delta t), \qquad dV_3 = (\mathbf{v}\Delta t) \cdot \mathbf{dS}.$$

Then, since $\mathbf{F} \cdot [\mathbf{dl} \times (\mathbf{v}\Delta t)] = \Delta t (\mathbf{v} \times \mathbf{F}) \cdot \mathbf{dl}$, we have

$$\psi(t) = \int_{S(t)} \frac{\partial \mathbf{F}(\mathbf{r}, t)}{\partial t} \cdot \mathbf{dS} + \lim_{\Delta t \to 0} \frac{\Delta t \int_{S(t)} [\mathbf{v} \nabla \cdot \mathbf{F}(\mathbf{r}, t)] \cdot \mathbf{dS}}{\Delta t} - \lim_{\Delta t \to 0} \frac{\Delta t \oint_{\Gamma} [\mathbf{v} \times \mathbf{F}(\mathbf{r}, t)] \cdot \mathbf{dl}}{\Delta t}.$$

Taking the limit and using Stokes's theorem on the last integral we have finally

$$\frac{d}{dt} \int_{S(t)} \mathbf{F} \cdot \mathbf{dS} = \int_{S(t)} \left[\frac{\partial \mathbf{F}}{\partial t} + \mathbf{v}\nabla \cdot \mathbf{F} - \nabla \times (\mathbf{v} \times \mathbf{F}) \right] \cdot \mathbf{dS}, \tag{A.63}$$

which is the *Helmholtz transport theorem* [190, 43].

In case the surface corresponds to a moving physical material, we may wish to write the Helmholtz transport theorem in terms of the material derivative. We can set $\mathbf{v} = \mathbf{u}$ and use

$$\nabla \times (\mathbf{u} \times \mathbf{F}) = \mathbf{u}(\nabla \cdot \mathbf{F}) - \mathbf{F}(\nabla \cdot \mathbf{u}) + (\mathbf{F} \cdot \nabla)\mathbf{u} - (\mathbf{u} \cdot \nabla)\mathbf{F}$$

and (A.60) to obtain

$$\frac{d}{dt} \int_{S(t)} \mathbf{F} \cdot \mathbf{dS} = \int_{S(t)} \left[\frac{D\mathbf{F}}{Dt} + \mathbf{F}(\nabla \cdot \mathbf{u}) - (\mathbf{F} \cdot \nabla)\mathbf{u} \right] \cdot \mathbf{dS}.$$

If $S(t)$ in (A.63) is closed, enclosing a volume region $V(t)$, then

$$\oint_{S(t)} [\nabla \times (\mathbf{v} \times \mathbf{F})] \cdot \mathbf{dS} = \int_{V(t)} \nabla \cdot [\nabla \times (\mathbf{v} \times \mathbf{F})] \, dV = 0$$

by the divergence theorem and (B.49). In this case the Helmholtz transport theorem becomes

$$\frac{d}{dt} \oint_{S(t)} \mathbf{F} \cdot \mathbf{dS} = \oint_{S(t)} \left[\frac{\partial \mathbf{F}}{\partial t} + \mathbf{v}\nabla \cdot \mathbf{F} \right] \cdot \mathbf{dS}. \tag{A.64}$$

We now come to an essential tool that we employ throughout the book. Using the divergence theorem we can rewrite (A.64) as

$$\frac{d}{dt} \int_{V(t)} \nabla \cdot \mathbf{F} \, dV = \int_{V(t)} \nabla \cdot \frac{\partial \mathbf{F}}{\partial t} \, dV + \oint_{S(t)} (\nabla \cdot \mathbf{F}) \mathbf{v} \cdot \mathbf{dS}.$$

Replacing $\nabla \cdot \mathbf{F}$ by the scalar field ρ we have

$$\frac{d}{dt} \int_{V(t)} \rho \, dV = \int_{V(t)} \frac{\partial \rho}{\partial t} \, dV + \oint_{S(t)} \rho \mathbf{v} \cdot \mathbf{dS}. \tag{A.65}$$

In this *general form* of the transport theorem $\mathbf{v}$ is an arbitrary velocity. In most applications $\mathbf{v} = \mathbf{u}$ describes the motion of a material substance; then

$$\frac{D}{Dt} \int_{V(t)} \rho \, dV = \int_{V(t)} \frac{\partial \rho}{\partial t} \, dV + \oint_{S(t)} \rho \mathbf{u} \cdot \mathbf{dS}, \tag{A.66}$$

which is the *Reynolds transport theorem* [214]. The D/Dt notation implies that $V(t)$ retains exactly the same material elements as it moves and deforms to follow the material substance.

We may rewrite the Reynolds transport theorem in various forms. By the divergence theorem we have

$$\frac{d}{dt} \int_{V(t)} \rho \, dV = \int_{V(t)} \left[\frac{\partial \rho}{\partial t} + \nabla \cdot (\rho \mathbf{v}) \right] dV.$$

Setting $\mathbf{v} = \mathbf{u}$, using (B.42), and using (A.59) for the material derivative of ρ, we obtain

$$\frac{D}{Dt} \int_{V(t)} \rho \, dV = \int_{V(t)} \left[\frac{D\rho}{Dt} + \rho \nabla \cdot \mathbf{u} \right] dV. \tag{A.67}$$

We may also generate a vector form of the general transport theorem by taking ρ in (A.65) to be a component of a vector. Assembling all of the components we have

$$\frac{d}{dt} \int_{V(t)} \mathbf{A} \, dV = \int_{V(t)} \frac{\partial \mathbf{A}}{\partial t} \, dV + \oint_{S(t)} \mathbf{A}(\mathbf{v} \cdot \hat{\mathbf{n}}) \, dS. \tag{A.68}$$

A.3   Dyadic analysis

Dyadic analysis was introduced in the late nineteenth century by Gibbs to generalize vector analysis to problems in which the components of vectors are related in a linear manner. It has now been widely supplanted by tensor theory, but maintains a foothold in engineering where the transformation properties of tensors are not paramount (except, of course, in considerations such as those involving special relativity). Terms such as "tensor permittivity" and "dyadic permittivity" are often used interchangeably.

Component form representation. We wish to write one vector field $\mathbf{A}(\mathbf{r}, t)$ as a linear function of another vector field $\mathbf{B}(\mathbf{r}, t)$:

$$\mathbf{A} = f(\mathbf{B}).$$

By this we mean that each component of $\mathbf{A}$ is a linear combination of the components of $\mathbf{B}$:

$$A_1(\mathbf{r}, t) = a_{11'}\, B_{1'}(\mathbf{r}, t) + a_{12'}\, B_{2'}(\mathbf{r}, t) + a_{13'}\, B_{3'}(\mathbf{r}, t),$$
$$A_2(\mathbf{r}, t) = a_{21'}\, B_{1'}(\mathbf{r}, t) + a_{22'}\, B_{2'}(\mathbf{r}, t) + a_{23'}\, B_{3'}(\mathbf{r}, t),$$
$$A_3(\mathbf{r}, t) = a_{31'}\, B_{1'}(\mathbf{r}, t) + a_{32'}\, B_{2'}(\mathbf{r}, t) + a_{33'}\, B_{3'}(\mathbf{r}, t).$$

Here the $a_{ij'}$ may depend on space and time (or frequency). The prime on the second index indicates that $\mathbf{A}$ and $\mathbf{B}$ may be expressed in distinct coordinate frames $(\hat{\mathbf{i}}_1, \hat{\mathbf{i}}_2, \hat{\mathbf{i}}_3)$ and $(\hat{\mathbf{i}}_{1'}, \hat{\mathbf{i}}_{2'}, \hat{\mathbf{i}}_{3'})$, respectively. We have

$$A_1 = \left(a_{11'}\hat{\mathbf{i}}_{1'} + a_{12'}\hat{\mathbf{i}}_{2'} + a_{13'}\hat{\mathbf{i}}_{3'} \right) \cdot \left(\hat{\mathbf{i}}_{1'}B_{1'} + \hat{\mathbf{i}}_{2'}B_{2'} + \hat{\mathbf{i}}_{3'}B_{3'} \right),$$
$$A_2 = \left(a_{21'}\hat{\mathbf{i}}_{1'} + a_{22'}\hat{\mathbf{i}}_{2'} + a_{23'}\hat{\mathbf{i}}_{3'} \right) \cdot \left(\hat{\mathbf{i}}_{1'}B_{1'} + \hat{\mathbf{i}}_{2'}B_{2'} + \hat{\mathbf{i}}_{3'}B_{3'} \right),$$
$$A_3 = \left(a_{31'}\hat{\mathbf{i}}_{1'} + a_{32'}\hat{\mathbf{i}}_{2'} + a_{33'}\hat{\mathbf{i}}_{3'} \right) \cdot \left(\hat{\mathbf{i}}_{1'}B_{1'} + \hat{\mathbf{i}}_{2'}B_{2'} + \hat{\mathbf{i}}_{3'}B_{3'} \right),$$

and since $\mathbf{B} = \hat{\mathbf{i}}_{1'}B_{1'} + \hat{\mathbf{i}}_{2'}B_{2'} + \hat{\mathbf{i}}_{3'}B_{3'}$ we can write

$$\mathbf{A} = \hat{\mathbf{i}}_1(\mathbf{a}_1' \cdot \mathbf{B}) + \hat{\mathbf{i}}_2(\mathbf{a}_2' \cdot \mathbf{B}) + \hat{\mathbf{i}}_3(\mathbf{a}_3' \cdot \mathbf{B})$$

where

$$\mathbf{a}_1' = a_{11'}\hat{\mathbf{i}}_{1'} + a_{12'}\hat{\mathbf{i}}_{2'} + a_{13'}\hat{\mathbf{i}}_{3'},$$
$$\mathbf{a}_2' = a_{21'}\hat{\mathbf{i}}_{1'} + a_{22'}\hat{\mathbf{i}}_{2'} + a_{23'}\hat{\mathbf{i}}_{3'},$$
$$\mathbf{a}_3' = a_{31'}\hat{\mathbf{i}}_{1'} + a_{32'}\hat{\mathbf{i}}_{2'} + a_{33'}\hat{\mathbf{i}}_{3'}.$$

In shorthand notation

$$\mathbf{A} = \bar{\mathbf{a}} \cdot \mathbf{B} \qquad\qquad (A.69)$$

where

$$\bar{\mathbf{a}} = \hat{\mathbf{i}}_1\mathbf{a}_1' + \hat{\mathbf{i}}_2\mathbf{a}_2' + \hat{\mathbf{i}}_3\mathbf{a}_3'. \qquad\qquad (A.70)$$

Written out, the quantity $\bar{\mathbf{a}}$ looks like

$$\begin{aligned}
\bar{\mathbf{a}} = \; & a_{11'}(\hat{\mathbf{i}}_1\hat{\mathbf{i}}_{1'}) + a_{12'}(\hat{\mathbf{i}}_1\hat{\mathbf{i}}_{2'}) + a_{13'}(\hat{\mathbf{i}}_1\hat{\mathbf{i}}_{3'}) + \\
+ \; & a_{21'}(\hat{\mathbf{i}}_2\hat{\mathbf{i}}_{1'}) + a_{22'}(\hat{\mathbf{i}}_2\hat{\mathbf{i}}_{2'}) + a_{23'}(\hat{\mathbf{i}}_2\hat{\mathbf{i}}_{3'}) + \\
+ \; & a_{31'}(\hat{\mathbf{i}}_3\hat{\mathbf{i}}_{1'}) + a_{32'}(\hat{\mathbf{i}}_3\hat{\mathbf{i}}_{2'}) + a_{33'}(\hat{\mathbf{i}}_3\hat{\mathbf{i}}_{3'}).
\end{aligned}$$

Terms such as $\hat{\mathbf{i}}_1\hat{\mathbf{i}}_{1'}$ are called *dyads*, while sums of dyads such as $\bar{\mathbf{a}}$ are called *dyadics*. The components $a_{ij'}$ of $\bar{\mathbf{a}}$ may be conveniently placed into an array:

$$[\bar{\mathbf{a}}] = \begin{bmatrix} a_{11'} & a_{12'} & a_{13'} \\ a_{21'} & a_{22'} & a_{23'} \\ a_{31'} & a_{32'} & a_{33'} \end{bmatrix}.$$

Writing

$$[\mathbf{A}] = \begin{bmatrix} A_1 \\ A_2 \\ A_3 \end{bmatrix}, \qquad [\mathbf{B}] = \begin{bmatrix} B_{1'} \\ B_{2'} \\ B_{3'} \end{bmatrix},$$

we see that $\mathbf{A} = \bar{\mathbf{a}} \cdot \mathbf{B}$ can be written as

$$[\mathbf{A}] = [\bar{\mathbf{a}}]\,[\mathbf{B}] = \begin{bmatrix} a_{11'} & a_{12'} & a_{13'} \\ a_{21'} & a_{22'} & a_{23'} \\ a_{31'} & a_{32'} & a_{33'} \end{bmatrix} \begin{bmatrix} B_{1'} \\ B_{2'} \\ B_{3'} \end{bmatrix}.$$

Note carefully that in (A.69) $\bar{\mathbf{a}}$ operates on $\mathbf{B}$ from the left. A reorganization of the components of $\bar{\mathbf{a}}$ allows us to write

$$\bar{\mathbf{a}} = \mathbf{a}_1 \hat{\mathbf{i}}_{1'} + \mathbf{a}_2 \hat{\mathbf{i}}_{2'} + \mathbf{a}_3 \hat{\mathbf{i}}_{3'} \tag{A.71}$$

where

$$\begin{aligned} \mathbf{a}_1 &= a_{11'}\,\hat{\mathbf{i}}_1 + a_{21'}\,\hat{\mathbf{i}}_2 + a_{31'}\,\hat{\mathbf{i}}_3, \\ \mathbf{a}_2 &= a_{12'}\,\hat{\mathbf{i}}_1 + a_{22'}\,\hat{\mathbf{i}}_2 + a_{32'}\,\hat{\mathbf{i}}_3, \\ \mathbf{a}_3 &= a_{13'}\,\hat{\mathbf{i}}_1 + a_{23'}\,\hat{\mathbf{i}}_2 + a_{33'}\,\hat{\mathbf{i}}_3. \end{aligned}$$

We may now consider using $\bar{\mathbf{a}}$ to operate on a vector $\mathbf{C} = \hat{\mathbf{i}}_1 C_1 + \hat{\mathbf{i}}_2 C_2 + \hat{\mathbf{i}}_3 C_3$ from the right:

$$\mathbf{C} \cdot \bar{\mathbf{a}} = (\mathbf{C} \cdot \mathbf{a}_1)\hat{\mathbf{i}}_{1'} + (\mathbf{C} \cdot \mathbf{a}_2)\hat{\mathbf{i}}_{2'} + (\mathbf{C} \cdot \mathbf{a}_3)\hat{\mathbf{i}}_{3'}.$$

In matrix form $\mathbf{C} \cdot \bar{\mathbf{a}}$ is

$$[\bar{\mathbf{a}}]^T\,[\mathbf{C}] = \begin{bmatrix} a_{11'} & a_{21'} & a_{31'} \\ a_{12'} & a_{22'} & a_{32'} \\ a_{13'} & a_{23'} & a_{33'} \end{bmatrix} \begin{bmatrix} C_1 \\ C_2 \\ C_3 \end{bmatrix}$$

where the superscript "T" denotes the matrix transpose operation. That is,

$$\mathbf{C} \cdot \bar{\mathbf{a}} = \bar{\mathbf{a}}^T \cdot \mathbf{C}$$

where $\bar{\mathbf{a}}^T$ is the transpose of $\bar{\mathbf{a}}$.

If the primed and unprimed frames coincide, then

$$\begin{aligned} \bar{\mathbf{a}} = \; &a_{11}(\hat{\mathbf{i}}_1\hat{\mathbf{i}}_1) + a_{12}(\hat{\mathbf{i}}_1\hat{\mathbf{i}}_2) + a_{13}(\hat{\mathbf{i}}_1\hat{\mathbf{i}}_3) + \\ &+ a_{21}(\hat{\mathbf{i}}_2\hat{\mathbf{i}}_1) + a_{22}(\hat{\mathbf{i}}_2\hat{\mathbf{i}}_2) + a_{23}(\hat{\mathbf{i}}_2\hat{\mathbf{i}}_3) + \\ &+ a_{31}(\hat{\mathbf{i}}_3\hat{\mathbf{i}}_1) + a_{32}(\hat{\mathbf{i}}_3\hat{\mathbf{i}}_2) + a_{33}(\hat{\mathbf{i}}_3\hat{\mathbf{i}}_3). \end{aligned}$$

In this case we may compare the results of $\bar{\mathbf{a}} \cdot \mathbf{B}$ and $\mathbf{B} \cdot \bar{\mathbf{a}}$ for a given vector $\mathbf{B} = \hat{\mathbf{i}}_1 B_1 + \hat{\mathbf{i}}_2 B_2 + \hat{\mathbf{i}}_3 B_3$. We leave it to the reader to verify that in general

$$\mathbf{B} \cdot \bar{\mathbf{a}} \neq \bar{\mathbf{a}} \cdot \mathbf{B}.$$

Vector form representation. We can express dyadics in coordinate-free fashion if we expand the concept of a dyad to permit entities such as $\mathbf{AB}$. Here $\mathbf{A}$ and $\mathbf{B}$ are called the *antecedent* and *consequent*, respectively. The operation rules

$$(\mathbf{AB}) \cdot \mathbf{C} = \mathbf{A}(\mathbf{B} \cdot \mathbf{C}), \qquad \mathbf{C} \cdot (\mathbf{AB}) = (\mathbf{C} \cdot \mathbf{A})\mathbf{B},$$

define the anterior and posterior products of $\mathbf{AB}$ with a vector $\mathbf{C}$, and give results consistent with our prior component notation. Sums of dyads such as $\mathbf{AB} + \mathbf{CD}$ are called *dyadic polynomials*, or dyadics. The simple dyadic

$$\mathbf{AB} = (A_1\hat{\mathbf{i}}_1 + A_2\hat{\mathbf{i}}_2 + A_3\hat{\mathbf{i}}_3)(B_{1'}\hat{\mathbf{i}}_{1'} + B_{2'}\hat{\mathbf{i}}_{2'} + B_{3'}\hat{\mathbf{i}}_{3'})$$

can be represented in component form using

$$\mathbf{AB} = \hat{\mathbf{i}}_1 \mathbf{a}_1' + \hat{\mathbf{i}}_2 \mathbf{a}_2' + \hat{\mathbf{i}}_3 \mathbf{a}_3'$$

where

$$\mathbf{a}_1' = A_1 B_{1'} \, \hat{\mathbf{i}}_{1'} + A_1 B_{2'} \, \hat{\mathbf{i}}_{2'} + A_1 B_{3'} \, \hat{\mathbf{i}}_{3'},$$
$$\mathbf{a}_2' = A_2 B_{1'} \, \hat{\mathbf{i}}_{1'} + A_2 B_{2'} \, \hat{\mathbf{i}}_{2'} + A_2 B_{3'} \, \hat{\mathbf{i}}_{3'},$$
$$\mathbf{a}_3' = A_3 B_{1'} \, \hat{\mathbf{i}}_{1'} + A_3 B_{2'} \, \hat{\mathbf{i}}_{2'} + A_3 B_{3'} \, \hat{\mathbf{i}}_{3'},$$

or using

$$\mathbf{AB} = \mathbf{a}_1 \hat{\mathbf{i}}_{1'} + \mathbf{a}_2 \hat{\mathbf{i}}_{2'} + \mathbf{a}_3 \hat{\mathbf{i}}_{3'}$$

where

$$\mathbf{a}_1 = \hat{\mathbf{i}}_1 \, A_1 B_{1'} + \hat{\mathbf{i}}_2 \, A_2 B_{1'} + \hat{\mathbf{i}}_3 \, A_3 B_{1'},$$
$$\mathbf{a}_2 = \hat{\mathbf{i}}_1 \, A_1 B_{2'} + \hat{\mathbf{i}}_2 \, A_2 B_{2'} + \hat{\mathbf{i}}_3 \, A_3 B_{2'},$$
$$\mathbf{a}_3 = \hat{\mathbf{i}}_1 \, A_1 B_{3'} + \hat{\mathbf{i}}_2 \, A_2 B_{3'} + \hat{\mathbf{i}}_3 \, A_3 B_{3'}.$$

Note that if we write $\bar{\mathbf{a}} = \mathbf{AB}$ then $a_{ij} = A_i B_{j'}$.

A simple dyad $\mathbf{AB}$ by itself cannot represent a general dyadic $\bar{\mathbf{a}}$; only six independent quantities are available in $\mathbf{AB}$ (the three components of $\mathbf{A}$ and the three components of $\mathbf{B}$), while an arbitrary dyadic has nine independent components. However, it can be shown that any dyadic can be written as a sum of three dyads:

$$\bar{\mathbf{a}} = \mathbf{AB} + \mathbf{CD} + \mathbf{EF}.$$

This is called a *vector representation* of $\bar{\mathbf{a}}$. If $\mathbf{V}$ is a vector, the distributive laws

$$\bar{\mathbf{a}} \cdot \mathbf{V} = (\mathbf{AB} + \mathbf{CD} + \mathbf{EF}) \cdot \mathbf{V} = \mathbf{A}(\mathbf{B} \cdot \mathbf{V}) + \mathbf{C}(\mathbf{D} \cdot \mathbf{V}) + \mathbf{E}(\mathbf{F} \cdot \mathbf{V}),$$
$$\mathbf{V} \cdot \bar{\mathbf{a}} = \mathbf{V} \cdot (\mathbf{AB} + \mathbf{CD} + \mathbf{EF}) = (\mathbf{V} \cdot \mathbf{A})\mathbf{B} + (\mathbf{V} \cdot \mathbf{C})\mathbf{D} + (\mathbf{V} \cdot \mathbf{E})\mathbf{F},$$

apply.

Dyadic algebra and calculus.   The cross product of a vector with a dyadic produces another dyadic. If $\bar{\mathbf{a}} = \mathbf{AB} + \mathbf{CD} + \mathbf{EF}$ then by definition

$$\bar{\mathbf{a}} \times \mathbf{V} = \mathbf{A}(\mathbf{B} \times \mathbf{V}) + \mathbf{C}(\mathbf{D} \times \mathbf{V}) + \mathbf{E}(\mathbf{F} \times \mathbf{V}),$$
$$\mathbf{V} \times \bar{\mathbf{a}} = (\mathbf{V} \times \mathbf{A})\mathbf{B} + (\mathbf{V} \times \mathbf{C})\mathbf{D} + (\mathbf{V} \times \mathbf{E})\mathbf{F}.$$

The corresponding component forms are

$$\bar{\mathbf{a}} \times \mathbf{V} = \hat{\mathbf{i}}_1 (\mathbf{a}_1' \times \mathbf{V}) + \hat{\mathbf{i}}_2 (\mathbf{a}_2' \times \mathbf{V}) + \hat{\mathbf{i}}_3 (\mathbf{a}_3' \times \mathbf{V}),$$
$$\mathbf{V} \times \bar{\mathbf{a}} = (\mathbf{V} \times \mathbf{a}_1)\hat{\mathbf{i}}_{1'} + (\mathbf{V} \times \mathbf{a}_2)\hat{\mathbf{i}}_{2'} + (\mathbf{V} \times \mathbf{a}_3)\hat{\mathbf{i}}_{3'},$$

where we have used (A.70) and (A.71), respectively. Interactions between dyads or dyadics may also be defined. The dot product of two dyads $\mathbf{AB}$ and $\mathbf{CD}$ is a dyad given by

$$(\mathbf{AB}) \cdot (\mathbf{CD}) = \mathbf{A}(\mathbf{B} \cdot \mathbf{C})\mathbf{D} = (\mathbf{B} \cdot \mathbf{C})(\mathbf{AD}).$$

The dot product of two dyadics can be found by applying the distributive property.

If α is a scalar, then the product $\alpha\bar{\mathbf{a}}$ is a dyadic with components equal to α times the components of $\bar{\mathbf{a}}$. Dyadic addition may be accomplished by adding individual dyadic components as long as the dyadics are expressed in the same coordinate system. Subtraction is accomplished by adding the negative of a dyadic, which is defined through scalar multiplication by -1.

Some useful dyadic identities appear in Appendix B. Many more can be found in Van Bladel [202].

The various vector derivatives may also be extended to dyadics. Computations are easiest in rectangular coordinates, since $\hat{\mathbf{i}}_1 = \hat{\mathbf{x}}$, $\hat{\mathbf{i}}_2 = \hat{\mathbf{y}}$, and $\hat{\mathbf{i}}_3 = \hat{\mathbf{z}}$ are constant with position. The dyadic

$$\bar{\mathbf{a}} = \mathbf{a}_x\hat{\mathbf{x}} + \mathbf{a}_y\hat{\mathbf{y}} + \mathbf{a}_z\hat{\mathbf{z}}$$

has divergence

$$\nabla \cdot \bar{\mathbf{a}} = (\nabla \cdot \mathbf{a}_x)\hat{\mathbf{x}} + (\nabla \cdot \mathbf{a}_y)\hat{\mathbf{y}} + (\nabla \cdot \mathbf{a}_z)\hat{\mathbf{z}},$$

and curl

$$\nabla \times \bar{\mathbf{a}} = (\nabla \times \mathbf{a}_x)\hat{\mathbf{x}} + (\nabla \times \mathbf{a}_y)\hat{\mathbf{y}} + (\nabla \times \mathbf{a}_z)\hat{\mathbf{z}}.$$

Note that the divergence of a dyadic is a vector while the curl of a dyadic is a dyadic. The gradient of a vector $\mathbf{a} = a_x\hat{\mathbf{x}} + a_y\hat{\mathbf{y}} + a_z\hat{\mathbf{z}}$ is

$$\nabla\mathbf{a} = (\nabla a_x)\hat{\mathbf{x}} + (\nabla a_y)\hat{\mathbf{y}} + (\nabla a_z)\hat{\mathbf{z}},$$

a dyadic quantity.

The dyadic derivatives may be expressed in coordinate-free notation by using the vector representation. The dyadic $\mathbf{AB}$ has divergence

$$\nabla \cdot (\mathbf{AB}) = (\nabla \cdot \mathbf{A})\mathbf{B} + \mathbf{A} \cdot (\nabla\mathbf{B})$$

and curl

$$\nabla \times (\mathbf{AB}) = (\nabla \times \mathbf{A})\mathbf{B} - \mathbf{A} \times (\nabla\mathbf{B}).$$

The Laplacian of a dyadic is a dyadic given by

$$\nabla^2\bar{\mathbf{a}} = \nabla(\nabla \cdot \bar{\mathbf{a}}) - \nabla \times (\nabla \times \bar{\mathbf{a}}).$$

The divergence theorem for dyadics is

$$\int_V \nabla \cdot \bar{\mathbf{a}}\, dV = \oint_S \hat{\mathbf{n}} \cdot \bar{\mathbf{a}}\, dS.$$

Some of the other common differential and integral identities for dyadics can be found in Van Bladel [202] and Tai [192].

Special dyadics. We say that $\bar{\mathbf{a}}$ is *symmetric* if

$$\mathbf{B} \cdot \bar{\mathbf{a}} = \bar{\mathbf{a}} \cdot \mathbf{B}$$

for any vector $\mathbf{B}$. This requires $\bar{\mathbf{a}}^T = \bar{\mathbf{a}}$, i.e., $a_{ij'} = a_{ji'}$. We say that $\bar{\mathbf{a}}$ is *antisymmetric* if

$$\mathbf{B} \cdot \bar{\mathbf{a}} = -\bar{\mathbf{a}} \cdot \mathbf{B}$$

for any $\mathbf{B}$. In this case $\bar{\mathbf{a}}^T = -\bar{\mathbf{a}}$. That is, $a_{ij'} = -a_{ji'}$ and $a_{ii'} = 0$. A symmetric dyadic has only six independent components while an antisymmetric dyadic has only three. The reader can verify that any dyadic can be decomposed into symmetric and antisymmetric parts as

$$\bar{\mathbf{a}} = \frac{1}{2}\left(\bar{\mathbf{a}} + \bar{\mathbf{a}}^T\right) + \frac{1}{2}\left(\bar{\mathbf{a}} - \bar{\mathbf{a}}^T\right).$$

A simple example of a symmetric dyadic is the *unit dyadic* $\bar{\mathbf{I}}$ defined by

$$\bar{\mathbf{I}} = \hat{\mathbf{i}}_1\hat{\mathbf{i}}_1 + \hat{\mathbf{i}}_2\hat{\mathbf{i}}_2 + \hat{\mathbf{i}}_3\hat{\mathbf{i}}_3.$$

This quantity often arises in the manipulation of dyadic equations, and satisfies

$$\mathbf{A} \cdot \bar{\mathbf{I}} = \bar{\mathbf{I}} \cdot \mathbf{A} = \mathbf{A}$$

for any vector $\mathbf{A}$. In matrix form $\bar{\mathbf{I}}$ is the identity matrix:

$$[\bar{\mathbf{I}}] = \begin{bmatrix} 1 & 0 & 0 \\ 0 & 1 & 0 \\ 0 & 0 & 1 \end{bmatrix}.$$

The components of a dyadic may be complex. We say that $\bar{\mathbf{a}}$ is *hermitian* if

$$\mathbf{B} \cdot \bar{\mathbf{a}} = \bar{\mathbf{a}}^* \cdot \mathbf{B} \tag{A.72}$$

holds for any $\mathbf{B}$. This requires that $\bar{\mathbf{a}}^* = \bar{\mathbf{a}}^T$. Taking the transpose we can write

$$\bar{\mathbf{a}} = (\bar{\mathbf{a}}^*)^T = \bar{\mathbf{a}}^\dagger$$

where "$\dagger$" stands for the conjugate-transpose operation. We say that $\bar{\mathbf{a}}$ is *anti-hermitian* if

$$\mathbf{B} \cdot \bar{\mathbf{a}} = -\bar{\mathbf{a}}^* \cdot \mathbf{B} \tag{A.73}$$

for arbitrary $\mathbf{B}$. In this case $\bar{\mathbf{a}}^* = -\bar{\mathbf{a}}^T$. Any complex dyadic can be decomposed into hermitian and anti-hermitian parts:

$$\bar{\mathbf{a}} = \frac{1}{2}\left(\bar{\mathbf{a}}^H + \bar{\mathbf{a}}^A\right) \tag{A.74}$$

where

$$\bar{\mathbf{a}}^H = \bar{\mathbf{a}} + \bar{\mathbf{a}}^\dagger, \qquad \bar{\mathbf{a}}^A = \bar{\mathbf{a}} - \bar{\mathbf{a}}^\dagger. \tag{A.75}$$

A dyadic identity important in the study of material parameters is

$$\mathbf{B} \cdot \bar{\mathbf{a}}^* \cdot \mathbf{B}^* = \mathbf{B}^* \cdot \bar{\mathbf{a}}^\dagger \cdot \mathbf{B}. \tag{A.76}$$

We show this by decomposing $\bar{\mathbf{a}}$ according to (A.74), giving

$$\mathbf{B} \cdot \bar{\mathbf{a}}^* \cdot \mathbf{B}^* = \frac{1}{2}\left(\left[\mathbf{B}^* \cdot \bar{\mathbf{a}}^H\right]^* + \left[\mathbf{B}^* \cdot \bar{\mathbf{a}}^A\right]^*\right) \cdot \mathbf{B}^*$$

where we have used $(\mathbf{B} \cdot \bar{\mathbf{a}})^* = (\mathbf{B}^* \cdot \bar{\mathbf{a}}^*)$. Applying (A.72) and (A.73) we obtain

$$\mathbf{B} \cdot \bar{\mathbf{a}}^* \cdot \mathbf{B}^* = \frac{1}{2}\left(\left[\bar{\mathbf{a}}^{H*} \cdot \mathbf{B}^*\right]^* - \left[\bar{\mathbf{a}}^{A*} \cdot \mathbf{B}^*\right]^*\right) \cdot \mathbf{B}^*$$

$$= \mathbf{B}^* \cdot \frac{1}{2}\left(\left[\bar{\mathbf{a}}^H \cdot \mathbf{B}\right] - \left[\bar{\mathbf{a}}^A \cdot \mathbf{B}\right]\right)$$

$$= \mathbf{B}^* \cdot \left(\frac{1}{2}\left[\bar{\mathbf{a}}^H - \bar{\mathbf{a}}^A\right] \cdot \mathbf{B}\right).$$

Since the term in brackets is $\bar{\mathbf{a}}^H - \bar{\mathbf{a}}^A = 2\bar{\mathbf{a}}^\dagger$ by (A.75), the identity is proved.

A.4 Boundary value problems

Many physical phenomena may be described mathematically as the solutions to *boundary value problems*. The desired physical quantity (usually called a "field") in a certain region of space is found by solving one or more partial differential equations subject to certain conditions over the boundary surface. The boundary conditions may specify the values of the field, some manipulated version of the field (such as the normal derivative), or a relationship between fields in adjoining regions. If the field varies with time as well as space, initial or final values of the field must also be specified. Particularly important is whether a boundary value problem is *well-posed* and therefore has a unique solution which depends continuously on the data supplied. This depends on the forms of the differential equation and boundary conditions. The well-posedness of Maxwell's equations is discussed in § 2.2.

The importance of boundary value problems has led to an array of techniques, both analytical and numerical, for solving them. Many problems (such as boundary value problems involving Laplace's equation) may be solved in several different ways. Uniqueness permits an engineer to focus attention on which technique will yield the most efficient solution. In this section we concentrate on the separation of variables technique, which is widely applied in the solution of Maxwell's equations. We first discuss eigenvalue problems and then give an overview of separation of variables. Finally we consider a number of example problems in each of the three common coordinate systems.

Sturm–Liouville problems and eigenvalues

The partial differential equations of electromagnetics can often be reduced to ordinary differential equations. In some cases symmetry permits us to reduce the number of dimensions by inspection; in other cases, we may employ an integral transform (e.g., the Fourier transform) or separation of variables. The resulting ordinary differential equations may be viewed as particular cases of the *Sturm–Liouville differential equation*

$$\frac{d}{dx}\left[p(x)\frac{d\psi(x)}{dx}\right] + q(x)\psi(x) + \lambda\sigma(x)\psi(x) = 0, \qquad x \in [a, b]. \tag{A.77}$$

In linear operator notation

$$\mathcal{L}\left[\psi(x)\right] = -\lambda\sigma(x)\psi(x), \tag{A.78}$$

where $\mathcal{L}$ is the linear Sturm–Liouville operator

$$\mathcal{L} = \left(\frac{d}{dx}\left[p(x)\frac{d}{dx}\right] + q(x)\right).$$

Obviously $\psi(x) = 0$ satisfies (A.78). However, for certain values of λ dependent on p, q, σ, and the boundary conditions we impose, (A.78) has non-trivial solutions. Each λ that satisfies (A.78) is an *eigenvalue* of $\mathcal{L}$, and any non-trivial solution associated with that eigenvalue is an *eigenfunction*. Taken together, the eigenvalues of an operator form its *eigenvalue spectrum*.

We shall restrict ourselves to the case in which $\mathcal{L}$ is *self-adjoint*. Assume p, q, and σ are real and continuous on $[a, b]$. It is straightforward to show that for any two functions $u(x)$ and $v(x)$ Lagrange's identity

$$u\,\mathcal{L}[v] - v\,\mathcal{L}[u] = \frac{d}{dx}\left[p\left(u\frac{dv}{dx} - v\frac{du}{dx}\right)\right] \tag{A.79}$$

holds. Integration gives Green's formula

$$\int_a^b \left(u\,\mathcal{L}[v] - v\,\mathcal{L}[u]\right)\,dx = p\left(u\frac{dv}{dx} - v\frac{du}{dx}\right)\Big|_a^b.$$

The operator $\mathcal{L}$ is self-adjoint if its associated boundary conditions are such that

$$p\left(u\frac{dv}{dx} - v\frac{du}{dx}\right)\Big|_a^b = 0. \tag{A.80}$$

Possible sets of conditions include the *homogeneous* boundary conditions

$$\alpha_1\psi(a) + \beta_1\psi'(a) = 0, \quad \alpha_2\psi(b) + \beta_2\psi'(b) = 0, \tag{A.81}$$

and the *periodic* boundary conditions

$$\psi(a) = \psi(b), \quad p(a)\psi'(a) = p(b)\psi'(b). \tag{A.82}$$

By imposing one of these sets on (A.78) we obtain a *Sturm–Liouville problem*.

The self-adjoint Sturm–Liouville operator has some nice properties. Each eigenvalue is real, and the eigenvalues form a denumerable set with no cluster point. Moreover, eigenfunctions corresponding to distinct eigenvalues are orthogonal, and the eigenfunctions form a complete set. Hence we can expand any sufficiently smooth function in terms of the eigenfunctions of a problem. We discuss this further below.

A *regular* Sturm–Liouville problem involves a self-adjoint operator $\mathcal{L}$ with $p(x) > 0$ and $\sigma(x) > 0$ everywhere, and the homogeneous boundary conditions (A.81). If p or σ vanishes at an endpoint of $[a, b]$, or an endpoint is at infinity, the problem is *singular*. The harmonic differential equation can form the basis of regular problems, while problems involving Bessel's and Legendre's equations are singular. Regular Sturm–Liouville problems have additional properties. There are infinitely many eigenvalues. There is a smallest eigenvalue but no largest eigenvalue, and the eigenvalues can be ordered as $\lambda_0 < \lambda_1 < \cdots < \lambda_n \cdots$. Associated with each λ_n is a unique (to an arbitrary multiplicative constant) eigenfunction ψ_n that has exactly n zeros in (a, b).

If a problem is singular because $p = 0$ at an endpoint, we can also satisfy (A.80) by demanding that ψ be bounded at that endpoint (a *singularity condition*) and that any regular Sturm–Liouville boundary condition hold at the other endpoint. This is the case for Bessel's and Legendre's equations discussed below.

Orthogonality of the eigenfunctions. Let $\mathcal{L}$ be self-adjoint, and let ψ_m and ψ_n be eigenfunctions associated with λ_m and λ_n, respectively. Then by (A.80) we have

$$\int_a^b \left(\psi_m(x)\,\mathcal{L}[\psi_n(x)] - \psi_n(x)\,\mathcal{L}[\psi_m(x)]\right)\,dx = 0.$$

But $\mathcal{L}[\psi_n(x)] = -\lambda_n\sigma(x)\psi_n(x)$ and $\mathcal{L}[\psi_m(x)] = -\lambda_m\sigma(x)\psi_m(x)$. Hence

$$(\lambda_m - \lambda_n)\int_a^b \psi_m(x)\psi_n(x)\sigma(x)\,dx = 0,$$

and $\lambda_m \neq \lambda_n$ implies that

$$\int_a^b \psi_m(x)\psi_n(x)\sigma(x)\,dx = 0. \tag{A.83}$$

We say that ψ_m and ψ_n are orthogonal with respect to the weight function $\sigma(x)$.

Eigenfunction expansion of an arbitrary function. If $\mathcal{L}$ is self-adjoint, then its eigenfunctions form a *complete set*. This means that any piecewise smooth function may be represented as a weighted series of eigenfunctions. Specifically, if f and f' are piecewise continuous on $[a, b]$, then f may be represented as the *generalized Fourier series*

$$f(x) = \sum_{n=0}^{\infty} c_n \psi_n(x). \tag{A.84}$$

Convergence of the series is uniform and gives, at any point of (a, b), the average value $[f(x^+) + f(x^-)]/2$ of the one-sided limits $f(x^+)$ and $f(x^-)$ of $f(x)$. The c_n can be found using orthogonality condition (A.83): multiply (A.84) by $\psi_m \sigma$ and integrate to obtain

$$\int_a^b f(x)\psi_m(x)\sigma(x)\,dx = \sum_{n=0}^{\infty} c_n \int_a^b \psi_n(x)\psi_m(x)\sigma(x)\,dx,$$

hence

$$c_n = \frac{\int_a^b f(x)\psi_n(x)\sigma(x)\,dx}{\int_a^b \psi_n^2(x)\sigma(x)\,dx}. \tag{A.85}$$

These coefficients ensure that the series *converges in mean* to f; i.e., the mean-square error

$$\int_a^b \left| f(x) - \sum_{n=0}^{\infty} c_n \psi_n(x) \right|^2 \sigma(x)\,dx$$

is minimized. Truncation to finitely-many terms generally results in oscillations (*Gibb's phenomena*) near points of discontinuity of f. The c_n are easier to compute if the ψ_n are orthonormal with

$$\int_a^b \psi_n^2(x)\sigma(x)\,dx = 1$$

for each n.

Uniqueness of the eigenfunctions. If both ψ_1 and ψ_2 are associated with the same eigenvalue λ, then

$$\mathcal{L}[\psi_1(x)] + \lambda\sigma(x)\psi_1(x) = 0, \quad \mathcal{L}[\psi_2(x)] + \lambda\sigma(x)\psi_2(x) = 0,$$

hence

$$\psi_1(x)\,\mathcal{L}[\psi_2(x)] - \psi_2(x)\,\mathcal{L}[\psi_1(x)] = 0.$$

By (A.79) we have

$$\frac{d}{dx}\left[p(x)\left(\psi_1(x)\frac{d\psi_2(x)}{dx} - \psi_2(x)\frac{d\psi_1(x)}{dx} \right) \right] = 0$$

or

$$p(x)\left(\psi_1(x)\frac{d\psi_2(x)}{dx} - \psi_2(x)\frac{d\psi_1(x)}{dx} \right) = C$$

where C is constant. Either of (A.81) implies $C = 0$, hence

$$\frac{d}{dx}\left(\frac{\psi_2(x)}{\psi_1(x)} \right) = 0$$

so that $\psi_1(x) = K\psi_2(x)$ for some constant K. So under homogeneous boundary conditions, every eigenvalue is associated with a unique eigenfunction.

This is false for the periodic boundary conditions (A.82). Eigenfunction expansion then becomes difficult, as we can no longer assume eigenfunction orthogonality. However, the Gram–Schmidt algorithm may be used to construct orthogonal eigenfunctions. We refer the interested reader to Haberman [79].

The harmonic differential equation. The ordinary differential equation

$$\frac{d^2\psi(x)}{dx^2} = -k^2\psi(x) \tag{A.86}$$

is Sturm–Liouville with $p \equiv 1$, $q \equiv 0$, $\sigma \equiv 1$, and $\lambda = k^2$. Suppose we take $[a, b] = [0, L]$ and adopt the homogeneous boundary conditions

$$\psi(0) = 0 \quad \text{and} \quad \psi(L) = 0. \tag{A.87}$$

Since $p(x) > 0$ and $\sigma(x) > 0$ on $[0, L]$, equations (A.86) and (A.87) form a regular Sturm–Liouville problem. Thus we should have an infinite number of discrete eigenvalues. A power series technique yields the two independent solutions

$$\psi_a(x) = A_a \sin kx, \quad \psi_b(x) = A_b \cos kx,$$

to (A.86); hence by linearity the most general solution is

$$\psi(x) = A_a \sin kx + A_b \cos kx. \tag{A.88}$$

The condition at $x = 0$ gives $A_a \sin 0 + A_b \cos 0 = 0$, hence $A_b = 0$. The other condition then requires

$$A_a \sin kL = 0. \tag{A.89}$$

Since $A_a = 0$ would give $\psi \equiv 0$, we satisfy (A.89) by choosing $k = k_n = n\pi/L$ for $n = 1, 2, \ldots$. Because $\lambda = k^2$, the eigenvalues are

$$\lambda_n = (n\pi/L)^2$$

with corresponding eigenfunctions

$$\psi_n(x) = \sin k_n x.$$

Note that $\lambda = 0$ is not an eigenvalue; eigenfunctions are nontrivial by definition, and $\sin(0\pi x/L) \equiv 0$. Likewise, the differential equation associated with $\lambda = 0$ can be solved easily, but only its trivial solution can fit homogeneous boundary conditions: with $k = 0$, (A.86) becomes $d^2\psi(x)/dx^2 = 0$, giving $\psi(x) = ax + b$; this can satisfy (A.87) only with $a = b = 0$.

These "eigensolutions" obey the properties outlined earlier. In particular the ψ_n are orthogonal,

$$\int_0^L \sin\left(\frac{n\pi x}{L}\right) \sin\left(\frac{m\pi x}{L}\right) dx = \frac{L}{2}\delta_{mn},$$

and the eigenfunction expansion of a piecewise continuous function f is given by

$$f(x) = \sum_{n=1}^{\infty} c_n \sin\left(\frac{n\pi x}{L}\right)$$

where, with $\sigma(x) = 1$ in (A.85), we have

$$c_n = \frac{\int_0^L f(x) \sin\left(\frac{n\pi x}{L}\right) dx}{\int_0^L \sin^2\left(\frac{n\pi x}{L}\right) dx} = \frac{2}{L} \int_0^L f(x) \sin\left(\frac{n\pi x}{L}\right) dx.$$

Hence we recover the standard Fourier sine series for $f(x)$.

With little extra effort we can examine the eigenfunctions resulting from enforcement of the periodic boundary conditions

$$\psi(0) = \psi(L) \quad \text{and} \quad \psi'(0) = \psi'(L).$$

The general solution (A.88) still holds, so we have the choices $\psi(x) = \sin kx$ and $\psi(x) = \cos kx$. Evidently both

$$\psi(x) = \sin\left(\frac{2n\pi x}{L}\right) \quad \text{and} \quad \psi(x) = \cos\left(\frac{2n\pi x}{L}\right)$$

satisfy the boundary conditions for $n = 1, 2, \ldots$. Thus each eigenvalue $(2n\pi/L)^2$ is associated with two eigenfunctions.

Bessel's differential equation. Bessel's equation

$$\frac{d}{dx}\left(x\frac{d\psi(x)}{dx}\right) + \left(k^2 x - \frac{\nu^2}{x}\right)\psi(x) = 0 \tag{A.90}$$

occurs when problems are solved in circular-cylindrical coordinates. Comparison with (A.77) shows that $\lambda = k^2$, $p(x) = x$, $q(x) = -\nu^2/x$, and $\sigma(x) = x$. We take $[a, b] = [0, L]$ along with the boundary conditions

$$\psi(L) = 0 \quad \text{and} \quad |\psi(0)| < \infty. \tag{A.91}$$

Although the resulting Sturm–Liouville problem is singular, the specified conditions (A.91) maintain satisfaction of (A.80). The eigenfunctions are orthogonal because (A.80) is satisfied by having $\psi(L) = 0$ and $p(x)\, d\psi(x)/dx \to 0$ as $x \to 0$.

As a second-order ordinary differential equation, (A.90) has two solutions denoted by

$$J_\nu(kx) \quad \text{and} \quad N_\nu(kx),$$

and termed *Bessel functions*. Their properties are summarized in Appendix E.1. The function $J_\nu(x)$, the Bessel function of the first kind and order ν, is well-behaved in $[0, L]$. The function $N_\nu(x)$, the Bessel function of the second kind and order ν, is unbounded at $x = 0$; hence it is excluded as an eigenfunction of the Sturm–Liouville problem.

The condition at $x = L$ shows that the eigenvalues are defined by

$$J_\nu(kL) = 0.$$

We denote the mth root of $J_\nu(x) = 0$ by $p_{\nu m}$. Then

$$k_{\nu m} = \sqrt{\lambda_{\nu m}} = p_{\nu m}/L.$$

The infinitely many eigenvalues are ordered as $\lambda_{\nu 1} < \lambda_{\nu 2} < \ldots$. Associated with eigenvalue $\lambda_{\nu m}$ is a single eigenfunction $J_\nu(\sqrt{\lambda_{\nu m}} x)$. The orthogonality relation is

$$\int_0^L J_\nu\left(\frac{p_{\nu m}}{L} x\right) J_\nu\left(\frac{p_{\nu n}}{L} x\right) x\, dx = 0, \quad m \neq n.$$

Since the eigenfunctions are also complete, we can expand any piecewise continuous function f in a *Fourier–Bessel series*

$$f(x) = \sum_{m=1}^{\infty} c_m J_\nu \left(p_{\nu m} \frac{x}{L} \right), \quad 0 \le x \le L, \ \nu > -1.$$

By (A.85) and (E.22) we have

$$c_m = \frac{2}{L^2 J_{\nu+1}^2 (p_{\nu m})} \int_0^L f(x) J_\nu \left(p_{\nu m} \frac{x}{L} \right) x \, dx.$$

The associated Legendre equation.   Legendre's equation occurs when problems are solved in spherical coordinates. It is often written in one of two forms. Letting θ be the polar angle of spherical coordinates ($0 \le \theta \le \pi$), the equation is

$$\frac{d}{d\theta} \left(\sin\theta \frac{d\psi(\theta)}{d\theta} \right) + \left(\lambda \sin\theta - \frac{m^2}{\sin\theta} \right) \psi(\theta) = 0.$$

This is Sturm–Liouville with $p(\theta) = \sin\theta$, $\sigma(\theta) = \sin\theta$, and $q(\theta) = -m^2/\sin\theta$. The boundary conditions

$$|\psi(0)| < \infty \quad \text{and} \quad |\psi(\pi)| < \infty$$

define a singular problem: the conditions are not homogeneous, $p(\theta) = 0$ at both endpoints, and $q(\theta) < 0$. Despite this, the Legendre problem does share properties of a regular Sturm–Liouville problem — including eigenfunction orthogonality and completeness.

Using $x = \cos\theta$, we can put Legendre's equation into its other common form

$$\frac{d}{dx} \left([1 - x^2] \frac{d\psi(x)}{dx} \right) + \left(\lambda - \frac{m^2}{1 - x^2} \right) \psi(x) = 0, \tag{A.92}$$

where $-1 \le x \le 1$. It is found that ψ is bounded at $x = \pm 1$ only if

$$\lambda = n(n+1)$$

where $n \ge m$ is an integer. These λ are the eigenvalues of the Sturm–Liouville problem, and the corresponding $\psi_n(x)$ are the eigenfunctions.

As a second-order partial differential equation, (A.92) has two solutions known as *associated Legendre functions*. The solution bounded at both $x = \pm 1$ is the associated Legendre function of the first kind, denoted $P_n^m(x)$. The second solution, unbounded at $x = \pm 1$, is the associated Legendre function of the second kind $Q_n^m(x)$. Appendix E.2 tabulates some properties of these functions.

For fixed m, each λ_{mn} is associated with a single eigenfunction $P_n^m(x)$. Since $P_n^m(x)$ is bounded at $x = \pm 1$, and since $p(\pm 1) = 0$, the eigenfunctions obey Lagrange's identity (A.79), hence are orthogonal on $[-1, 1]$ with respect to the weight function $\sigma(x) = 1$. Evaluation of the orthogonality integral leads to

$$\int_{-1}^{1} P_l^m(x) P_n^m(x) \, dx = \delta_{ln} \frac{2}{2n+1} \frac{(n+m)!}{(n-m)!} \tag{A.93}$$

or equivalently

$$\int_0^\pi P_l^m(\cos\theta) P_n^m(\cos\theta) \sin\theta \, d\theta = \delta_{ln} \frac{2}{2n+1} \frac{(n+m)!}{(n-m)!}.$$

For $m = 0$, $P_n^m(x)$ is a polynomial of degree n. Each such *Legendre polynomial*, denoted $P_n(x)$, is given by

$$P_n(x) = \frac{1}{2^n n!} \frac{d^n (x^2 - 1)^n}{dx^n}.$$

It turns out that

$$P_n^m(x) = (-1)^m (1 - x^2)^{m/2} \frac{d^m P_n(x)}{dx^m},$$

giving $P_n^m(x) = 0$ for $m > n$.

Because the Legendre polynomials form a complete set in the interval $[-1, 1]$, we may expand any sufficiently smooth function in a *Fourier–Legendre series*

$$f(x) = \sum_{n=0}^{\infty} c_n P_n(x).$$

Convergence in mean is guaranteed if

$$c_n = \frac{2n + 1}{2} \int_{-1}^{1} f(x) P_n(x) \, dx,$$

found using (A.85) along with (A.93).

In practice, the associated Legendre functions appear along with exponential functions in the solutions to spherical boundary value problems. The combined functions are known as *spherical harmonics*, and form solutions to two-dimensional Sturm–Liouville problems. We consider these next.

Higher-dimensional SL problems: Helmholtz's equation. Replacing d/dx by ∇, we generalize the Sturm–Liouville equation to higher dimensions:

$$\nabla \cdot [p(\mathbf{r}) \nabla \psi(\mathbf{r})] + q(\mathbf{r}) \psi(\mathbf{r}) + \lambda \sigma(\mathbf{r}) \psi(\mathbf{r}) = 0,$$

where q, p, σ, ψ are real functions. Of particular interest is the case $q(\mathbf{r}) = 0$, $p(\mathbf{r}) = \sigma(\mathbf{r}) = 1$, giving the Helmholtz equation

$$\nabla^2 \psi(\mathbf{r}) + \lambda \psi(\mathbf{r}) = 0. \tag{A.94}$$

In most boundary value problems, ψ or its normal derivative is specified on the surface of a bounded region. We obtain a three-dimensional analogue to the regular Sturm–Liouville problem by assuming the homogeneous boundary conditions

$$\alpha \psi(\mathbf{r}) + \beta \hat{\mathbf{n}} \cdot \nabla \psi(\mathbf{r}) = 0 \tag{A.95}$$

on the closed surface, where $\hat{\mathbf{n}}$ is the outward unit normal.

The problem consisting of (A.94) and (A.95) has properties analogous to those of the regular one-dimensional Sturm–Liouville problem. All eigenvalues are real. There are infinitely many eigenvalues. There is a smallest eigenvalue but no largest eigenvalue. However, associated with an eigenvalue there may be many eigenfunctions $\psi_\lambda(\mathbf{r})$. The eigenfunctions are orthogonal with

$$\int_V \psi_{\lambda_1}(\mathbf{r}) \psi_{\lambda_2}(\mathbf{r}) \, dV = 0, \qquad \lambda_1 \neq \lambda_2.$$

They are also complete and can be used to represent any piecewise smooth function $f(\mathbf{r})$ according to

$$f(\mathbf{r}) = \sum_{\lambda} a_{\lambda}\psi_{\lambda}(\mathbf{r}),$$

which converges in mean when

$$a_{\lambda_m} = \frac{\int_V f(\mathbf{r})\psi_{\lambda_m}(\mathbf{r})\,dV}{\int_V \psi_{\lambda_m}^2(\mathbf{r})\,dV}.$$

These properties are shared by the two-dimensional eigenvalue problem involving an open surface S with boundary contour Γ.

Spherical harmonics.   We now inspect solutions to the two-dimensional eigenvalue problem

$$\nabla^2 Y(\theta,\phi) + \frac{\lambda}{a^2}Y(\theta,\phi) = 0$$

over the surface of a sphere of radius a. Since the sphere has no boundary contour, we demand that $Y(\theta,\phi)$ be bounded in θ and periodic in ϕ. In the next section we shall apply separation of variables and show that

$$Y_{nm}(\theta,\phi) = \sqrt{\frac{2n+1}{4\pi}\frac{(n-m)!}{(n+m)!}}\,P_n^m(\cos\theta)e^{jm\phi}$$

where $\lambda = n(n+1)$. Note that Q_n^m does not appear as it is not bounded at $\theta = 0, \pi$. The functions Y_{nm} are called *spherical harmonics* (sometimes *zonal* or *tesseral* harmonics, depending on the values of n and m). As expressed above they are in orthonormal form, because the orthogonality relationships for the exponential and associated Legendre functions yield

$$\int_{-\pi}^{\pi}\int_0^{\pi} Y_{n'm'}^*(\theta,\phi)Y_{nm}(\theta,\phi)\sin\theta\,d\theta\,d\phi = \delta_{n'n}\delta_{m'm}. \tag{A.96}$$

As solutions to the Sturm–Liouville problem, these functions form a complete set on the surface of a sphere. Hence they can be used to represent any piecewise smooth function $f(\theta,\phi)$ as

$$f(\theta,\phi) = \sum_{n=0}^{\infty}\sum_{m=-n}^{n} a_{nm}Y_{nm}(\theta,\phi),$$

where

$$a_{nm} = \int_{-\pi}^{\pi}\int_0^{\pi} f(\theta,\phi)Y_{nm}^*(\theta,\phi)\sin\theta\,d\theta\,d\phi$$

by (A.96). The summation index m ranges from $-n$ to n because $P_n^m = 0$ for $m > n$. For negative index we can use

$$Y_{n,-m}(\theta,\phi) = (-1)^m Y_{nm}^*(\theta,\phi).$$

Some properties of the spherical harmonics are tabulated in Appendix E.3.

Separation of variables

We now consider a technique that finds widespread application in solving boundary value problems, applying as it does to many important partial differential equations such as Laplace's equation, the diffusion equation, and the scalar and vector wave equations. These equations are related to the scalar Helmholtz equation

$$\nabla^2 \psi(\mathbf{r}) + k^2 \psi(\mathbf{r}) = 0 \qquad (A.97)$$

where k is a complex constant. If k is real and we supply the appropriate boundary conditions, we have the higher-dimensional Sturm–Liouville problem with $\lambda = k^2$. We shall not pursue the extension of Sturm–Liouville theory to complex values of k.

Laplace's equation is Helmholtz's equation with $k = 0$. With $\lambda = k^2 = 0$ it might appear that Laplace's equation does not involve eigenvalues; however, separation of variables does lead us to lower-dimensional eigenvalue problems to which our previous methods apply. Solutions to the scalar or vector wave equations usually begin with Fourier transformation on the time variable, or with an initial separation of the time variable to reach a Helmholtz form.

The separation of variables idea is simple. We seek a solution to (A.97) in the form of a product of functions each of a single variable. If ψ depends on all three spatial dimensions, then we seek a solution of the type

$$\psi(u, v, w) = U(u)V(v)W(w),$$

where u, v, and w are the coordinate variables used to describe the problem. If ψ depends on only two coordinates, we may seek a product solution involving two functions each dependent on a single coordinate; alternatively, may use the three-variable solution and choose constants so that the result shows no variation with one coordinate. The Helmholtz equation is considered *separable* if it can be reduced to a set of independent ordinary differential equations, each involving a single coordinate variable. The ordinary differential equations, generally of second order, can be solved by conventional techniques resulting in solutions of the form

$$U(u) = A_u U_A(u, k_u, k_v, k_w) + B_u U_B(u, k_u, k_v, k_w),$$
$$V(v) = A_v V_A(v, k_u, k_v, k_w) + B_v V_B(v, k_u, k_v, k_w),$$
$$W(w) = A_w W_A(w, k_u, k_v, k_w) + B_w W_B(w, k_u, k_v, k_w).$$

The constants k_u, k_v, k_w are called *separation constants* and are found, along with the amplitude constants A, B, by applying boundary conditions appropriate for a given problem. At least one separation constant depends on (or equals) k, so only two are independent. In many cases k_u, k_v, and k_w become the discrete eigenvalues of the respective differential equations, and correspond to eigenfunctions $U(u, k_u, k_v, k_w)$, $V(v, k_u, k_v, k_w)$, and $W(w, k_u, k_v, k_w)$. In other cases the separation constants form a continuous spectrum of values, often when a Fourier transform solution is employed.

The Helmholtz equation can be separated in eleven different orthogonal coordinate systems [134]. Undoubtedly the most important of these are the rectangular, circular-cylindrical, and spherical systems, and we shall consider each in detail. We do note, however, that separability in a certain coordinate system does not imply that all problems expressed in that coordinate system can be easily handled using the resulting solutions. Only when the geometry and boundary conditions are simple do the solutions lend themselves to easy application; often other solution techniques are more appropriate.

Although rigorous conditions can be set forth to guarantee solvability by separation of variables [119], we prefer the following, more heuristic list:

1. Use a coordinate system that allows the given partial differential equation to separate into ordinary differential equations.

2. The problem's boundaries must be such that those boundaries not at infinity coincide with a single level surface of the coordinate system.

3. Use superposition to reduce the problem to one involving a single nonhomogeneous boundary condition. Then:

   (a) Solve the resulting Sturm–Liouville problem in one or two dimensions, with homogeneous boundary conditions on all boundaries. Then use a discrete eigenvalue expansion (Fourier series) and eigenfunction orthogonality to satisfy the remaining nonhomogeneous condition.

   (b) If a Sturm–Liouville problem cannot be formulated with the homogeneous boundary conditions (because, for instance, one boundary is at infinity), use a Fourier integral (continuous expansion) to satisfy the remaining nonhomogeneous condition.

If a Sturm–Liouville problem cannot be formulated, discovering the form of the integral transform to use can be difficult. In these cases other approaches, such as conformal mapping, may prove easier.

Solutions in rectangular coordinates. In rectangular coordinates the Helmholtz equation is

$$\frac{\partial^2 \psi(x,y,z)}{\partial x^2} + \frac{\partial^2 \psi(x,y,z)}{\partial y^2} + \frac{\partial^2 \psi(x,y,z)}{\partial z^2} + k^2 \psi(x,y,z) = 0. \qquad (A.98)$$

We seek a solution of the form $\psi(x,y,z) = X(x)Y(y)Z(z)$; substitution into (A.98) followed by division through by $X(x)Y(y)Z(z)$ gives

$$\frac{1}{X(x)} \frac{d^2 X(x)}{dx^2} + \frac{1}{Y(y)} \frac{d^2 Y(y)}{dy^2} + \frac{1}{Z(z)} \frac{d^2 Z(z)}{dz^2} = -k^2. \qquad (A.99)$$

At this point we require the *separation argument*. The left-hand side of (A.99) is a sum of three functions, each involving a single independent variable, whereas the right-hand side is constant. But the only functions of independent variables that always sum to a constant are themselves constants. Thus we may equate each term on the left to a different constant:

$$\frac{1}{X(x)} \frac{d^2 X(x)}{dx^2} = -k_x^2,$$

$$\frac{1}{Y(y)} \frac{d^2 Y(y)}{dy^2} = -k_y^2, \qquad (A.100)$$

$$\frac{1}{Z(z)} \frac{d^2 Z(z)}{dz^2} = -k_z^2,$$

provided that

$$k_x^2 + k_y^2 + k_z^2 = k^2.$$

The negative signs in (A.100) have been introduced for convenience.

Let us discuss the general solutions of equations (A.100). If $k_x = 0$, the two independent solutions for $X(x)$ are

$$X(x) = a_x x \quad \text{and} \quad X(x) = b_x$$

where a_x and b_x are constants. If $k_x \neq 0$, solutions may be chosen from the list of functions

$$e^{-jk_x x}, \quad e^{jk_x x}, \quad \sin k_x x, \quad \cos k_x x,$$

any two of which are independent. Because

$$\sin x = (e^{jx} - e^{-jx})/2j \quad \text{and} \quad \cos x = (e^{jx} + e^{-jx})/2, \quad (A.101)$$

the six possible solutions for $k_x \neq 0$ are

$$X(x) = \begin{cases} A_x e^{jk_x x} + B_x e^{-jk_x x}, \\ A_x \sin k_x x + B_x \cos k_x x, \\ A_x \sin k_x x + B_x e^{-jk_x x}, \\ A_x e^{jk_x x} + B_x \sin k_x x, \\ A_x e^{jk_x x} + B_x \cos k_x x, \\ A_x e^{-jk_x x} + B_x \cos k_x x. \end{cases} \quad (A.102)$$

We may base our choice on convenience (e.g., the boundary conditions may be amenable to one particular form) or on the desired behavior of the solution (e.g., standing waves vs. traveling waves). If k is complex, then so may be k_x, k_y, or k_z; observe that with imaginary arguments the complex exponentials are actually real exponentials, and the trigonometric functions are actually hyperbolic functions.

The solutions for $Y(y)$ and $Z(z)$ are identical to those for $X(x)$. We can write, for instance,

$$X(x) = \begin{cases} A_x e^{jk_x x} + B_x e^{-jk_x x}, & k_x \neq 0, \\ a_x x + b_x, & k_x = 0, \end{cases} \quad (A.103)$$

$$Y(y) = \begin{cases} A_y e^{jk_y y} + B_y e^{-jk_y y}, & k_y \neq 0, \\ a_y y + b_y, & k_y = 0, \end{cases} \quad (A.104)$$

$$Z(z) = \begin{cases} A_z e^{jk_z z} + B_z e^{-jk_z z}, & k_z \neq 0, \\ a_z z + b_z, & k_z = 0. \end{cases} \quad (A.105)$$

Examples. Let us begin by solving the simple equation

$$\nabla^2 V(x) = 0.$$

Since V depends only on x we can use (A.103)–(A.105) with $k_y = k_z = 0$ and $a_y = a_z = 0$. Moreover $k_x = 0$ because $k_x^2 + k_y^2 + k_z^2 = k^2 = 0$ for Laplace's equation. The general solution is therefore

$$V(x) = a_x x + b_x.$$

Boundary conditions must be specified to determine a_x and b_x; for instance, the conditions $V(0) = 0$ and $V(L) = V_0$ yield $V(x) = V_0 x/L$.

Next let us solve

$$\nabla^2 \psi(x, y) = 0.$$

We produce a lack of z-dependence in ψ by letting $k_z = 0$ and choosing $a_z = 0$. Moreover, $k_x^2 = -k_y^2$ since Laplace's equation requires $k = 0$. This leads to three possibilities. If $k_x = k_y = 0$, we have the product solution

$$\psi(x,y) = (a_x x + b_x)(a_y y + b_y). \tag{A.106}$$

If k_y is real and nonzero, then

$$\psi(x,y) = (A_x e^{-k_y x} + B_x e^{k_y x})(A_y e^{jk_y y} + B_y e^{-jk_y y}). \tag{A.107}$$

Using the relations

$$\sinh u = (e^u - e^{-u})/2 \quad \text{and} \quad \cosh u = (e^u + e^{-u})/2 \tag{A.108}$$

along with (A.101), we can rewrite (A.107) as

$$\psi(x,y) = (A_x \sinh k_y x + B_x \cosh k_y x)(A_y \sin k_y y + B_y \cos k_y y). \tag{A.109}$$

(We can reuse the constant names A_x, B_x, A_y, B_y, since the constants are unknown at this point.) If k_x is real and nonzero we have

$$\psi(x,y) = (A_x \sin k_x x + B_x \cos k_x x)(A_y \sinh k_x y + B_y \cosh k_x y). \tag{A.110}$$

Consider the problem consisting of Laplace's equation

$$\nabla^2 V(x,y) = 0 \tag{A.111}$$

holding in the region $0 < x < L_1, 0 < y < L_2, -\infty < z < \infty$, together with the boundary conditions

$$V(0,y) = V_1, \quad V(L_1,y) = V_2, \quad V(x,0) = V_3, \quad V(x,L_2) = V_4.$$

The solution $V(x,y)$ represents the potential within a conducting tube with each wall held at a different potential. Superposition applies: since Laplace's equation is linear we can write the solution as the sum of solutions to four different sub-problems. Each sub-problem has homogeneous boundary conditions on one independent variable and inhomogeneous conditions on the other, giving a Sturm–Liouville problem in one of the variables. For instance, let us examine the solutions found above in relation to the sub-problem consisting of Laplace's equation (A.111) in the region $0 < x < L_1, 0 < y < L_2$, $-\infty < z < \infty$, subject to the conditions

$$V(0,y) = V(L_1,y) = V(x,0) = 0, \quad V(x,L_2) = V_4 \neq 0.$$

First we try (A.106). The boundary condition at $x = 0$ gives

$$V(0,y) = (a_x(0) + b_x)(a_y y + b_y) = 0,$$

which holds for all $y \in (0, L_2)$ only if $b_x = 0$. The condition at $x = L_1$,

$$V(L_1,y) = a_x L_1 (a_y y + b_y) = 0,$$

then requires $a_x = 0$. But $a_x = b_x = 0$ gives $V(x,y) = 0$, and the condition at $y = L_2$ cannot be satisfied; clearly (A.106) was inappropriate. Next we examine (A.109). The condition at $x = 0$ gives

$$V(0,y) = (A_x \sinh 0 + B_x \cosh 0)(A_y \sin k_y y + B_y \cos k_y y) = 0,$$

hence $B_x = 0$. The condition at $x = L_1$ implies

$$V(L_1, y) = [A_x \sinh(k_y L_1)](A_y \sin k_y y + B_y \cos k_y y) = 0.$$

This can hold if either $A_x = 0$ or $k_y = 0$, but the case $k_y = 0$ ($= k_x$) was already considered. Thus $A_x = 0$ and the trivial solution reappears. Our last candidate is (A.110). The condition at $x = 0$ requires

$$V(0, y) = (A_x \sin 0 + B_x \cos 0)(A_y \sinh k_x y + B_y \cosh k_x y) = 0,$$

which implies $B_x = 0$. Next we have

$$V(L_1, y) = [A_x \sin(k_x L_1)](A_y \sinh k_y y + B_y \cosh k_y y) = 0.$$

We avoid $A_x = 0$ by setting $\sin(k_x L_1) = 0$ so that $k_{x_n} = n\pi/L_1$ for $n = 1, 2, \ldots$. (Here $n = 0$ is omitted because it would produce a trivial solution.) These are eigenvalues corresponding to the eigenfunctions $X_n(x) = \sin(k_{x_n} x)$, and were found in § A.4 for the harmonic equation. At this point we have a family of solutions

$$V_n(x, y) = \sin(k_{x_n} x)[A_{y_n} \sinh(k_{x_n} y) + B_{y_n} \cosh(k_{x_n} y)], \quad n = 1, 2, \ldots.$$

The subscript n on the left identifies V_n as the eigensolution associated with eigenvalue k_{x_n}. It remains to satisfy boundary conditions at $y = 0, L_2$. At $y = 0$ we have

$$V_n(x, 0) = \sin(k_{x_n} x)[A_{y_n} \sinh 0 + B_{y_n} \cosh 0] = 0,$$

hence $B_{y_n} = 0$ and

$$V_n(x, y) = A_{y_n} \sin(k_{x_n} x) \sinh(k_{x_n} y), \quad n = 1, 2, \ldots. \tag{A.112}$$

It is clear that no single eigensolution (A.112) can satisfy the one remaining boundary condition. However, we are guaranteed that a series of solutions can represent the constant potential on $y = L_2$; recall that as a solution to a regular Sturm–Liouville problem, the trigonometric functions are complete (hence they could represent any well-behaved function on the interval $0 \le x \le L_1$). In fact, the resulting series is a Fourier sine series for the constant potential at $y = L_2$. So let

$$V(x, y) = \sum_{n=1}^{\infty} V_n(x, y) = \sum_{n=1}^{\infty} A_{y_n} \sin(k_{x_n} x) \sinh(k_{x_n} y).$$

The remaining boundary condition requires

$$V(x, L_2) = \sum_{n=1}^{\infty} A_{y_n} \sin(k_{x_n} x) \sinh(k_{x_n} L_2) = V_4.$$

The constants A_{y_n} can be found using orthogonality; multiplying through by $\sin(k_{x_m} x)$ and integrating, we have

$$\sum_{n=1}^{\infty} A_{y_n} \sinh(k_{x_n} L_2) \int_0^{L_1} \sin\left(\frac{m\pi x}{L_1}\right) \sin\left(\frac{n\pi x}{L_1}\right) dx = V_4 \int_0^{L_1} \sin\left(\frac{m\pi x}{L_1}\right) dx.$$

The integral on the left equals $\delta_{mn} L_1/2$ where δ_{mn} is the Kronecker delta given by

$$\delta_{mn} = \begin{cases} 1, & m = n, \\ 0, & n \neq m. \end{cases}$$

After evaluating the integral on the right we obtain

$$\sum_{n=1}^{\infty} A_{y_n} \delta_{mn} \sinh(k_{x_n} L_2) = \frac{2V_4(1 - \cos m\pi)}{m\pi},$$

hence

$$A_{y_m} = \frac{2V_4(1 - \cos m\pi)}{m\pi \sinh(k_{x_m} L_2)}.$$

The final solution for this sub-problem is therefore

$$V(x, y) = \sum_{n=1}^{\infty} \frac{2V_4(1 - \cos n\pi)}{n\pi \sinh\left(\frac{n\pi L_2}{L_1}\right)} \sin\left(\frac{n\pi x}{L_1}\right) \sinh\left(\frac{n\pi y}{L_1}\right).$$

The remaining three sub-problems are left for the reader.

Let us again consider (A.111), this time for

$$0 \le x \le L_1, \quad 0 \le y < \infty, \quad -\infty < z < \infty,$$

and subject to

$$V(0, y) = V(L_1, y) = 0, \quad V(x, 0) = V_0.$$

Let us try the solution form that worked in the previous example:

$$V(x, y) = [A_x \sin(k_x x) + B_x \cos(k_x x)][A_y \sinh(k_x y) + B_y \cosh(k_x y)].$$

The boundary conditions at $x = 0, L_1$ are the same as before so we have

$$V_n(x, y) = \sin(k_{x_n} x)[A_{y_n} \sinh(k_{x_n} y) + B_{y_n} \cosh(k_{x_n} y)], \quad n = 1, 2, \dots.$$

To find A_{y_n} and B_{y_n} we note that V cannot grow without bound as $y \to \infty$. Individually the hyperbolic functions grow exponentially. However, using (A.108) we see that $B_{y_n} = -A_{y_n}$ gives

$$V_n(x, y) = A_{y_n} \sin(k_{x_n} x)e^{-k_{x_n} y}$$

where A_{y_n} is a new unknown constant. (Of course, we could have chosen this exponential dependence at the beginning.) Lastly, we can impose the boundary condition at $y = 0$ on the infinite series of eigenfunctions

$$V(x, y) = \sum_{n=1}^{\infty} A_{y_n} \sin(k_{x_n} x)e^{-k_{x_n} y}$$

to find A_{y_n}. The result is

$$V(x, y) = \sum_{n=1}^{\infty} \frac{2V_0}{\pi n}(1 - \cos n\pi) \sin(k_{x_n} x)e^{-k_{x_n} y}.$$

As in the previous example, the solution is a discrete superposition of eigenfunctions.

The problem consisting of (A.111) holding for

$$0 \le x \le L_1, \quad 0 \le y < \infty, \quad -\infty < z < \infty,$$

along with

$$V(0, y) = 0, \quad V(L_1, y) = V_0 e^{-ay}, \quad V(x, 0) = 0,$$

requires a continuous superposition of eigenfunctions to satisfy the boundary conditions. Let us try

$$V(x, y) = [A_x \sinh k_y x + B_x \cosh k_y x][A_y \sin k_y y + B_y \cos k_y y].$$

The conditions at $x = 0$ and $y = 0$ require that $B_x = B_y = 0$. Thus

$$V_{k_y}(x, y) = A \sinh k_y x \sin k_y y.$$

A single function of this form cannot satisfy the remaining condition at $x = L_1$. So we form a continuous superposition

$$V(x, y) = \int_0^\infty A(k_y) \sinh k_y x \sin k_y y \, dk_y. \tag{A.113}$$

By the condition at $x = L_1$

$$\int_0^\infty A(k_y) \sinh(k_y L_1) \sin k_y y \, dk_y = V_0 e^{-ay}. \tag{A.114}$$

We can find the amplitude function $A(k_y)$ by using the orthogonality property

$$\delta(y - y') = \frac{2}{\pi} \int_0^\infty \sin xy \sin xy' \, dx. \tag{A.115}$$

Multiplying both sides of (A.114) by $\sin k_y' y$ and integrating, we have

$$\int_0^\infty A(k_y) \sinh(k_y L_1) \left[\int_0^\infty \sin k_y y \sin k_y' y \, dy \right] dk_y = \int_0^\infty V_0 e^{-ay} \sin k_y' y \, dy.$$

We can evaluate the term in brackets using (A.115) to obtain

$$\int_0^\infty A(k_y) \sinh(k_y L_1) \frac{\pi}{2} \delta(k_y - k_y') \, dk_y = \int_0^\infty V_0 e^{-ay} \sin k_y' y \, dy,$$

hence

$$\frac{\pi}{2} A(k_y') \sinh(k_y' L_1) = V_0 \int_0^\infty e^{-ay} \sin k_y' y \, dy.$$

We then evaluate the integral on the right, solve for $A(k_y)$, and substitute into (A.113) to obtain

$$V(x, y) = \frac{2V_0}{\pi} \int_0^\infty \frac{k_y}{a^2 + k_y^2} \frac{\sinh(k_y x)}{\sinh(k_y L_1)} \sin k_y y \, dk_y.$$

Note that our application of the orthogonality property is merely a calculation of the inverse Fourier sine transform. Thus we could have found the amplitude coefficient by reference to a table of transforms.

We can use the Fourier transform solution even when the domain is infinite in more than one dimension. Suppose we solve (A.111) in the region

$$0 \le x < \infty, \quad 0 \le y < \infty, \quad -\infty < z < \infty,$$

subject to

$$V(0, y) = V_0 e^{-ay}, \quad V(x, 0) = 0.$$

Because of the condition at $y = 0$ let us use

$$V(x,y) = (A_x e^{-k_y x} + B_x e^{k_y x})(A_y \sin k_y y + B_y \cos k_y y).$$

The solution form

$$V_{k_y}(x,y) = B(k_y) e^{-k_y x} \sin k_y y$$

satisfies the finiteness condition and the homogeneous condition at $y = 0$. The remaining condition can be satisfied by a continuous superposition of solutions:

$$V(x,y) = \int_0^\infty B(k_y) e^{-k_y x} \sin k_y y \, dk_y.$$

We must have

$$V_0 e^{-ay} = \int_0^\infty B(k_y) \sin k_y y \, dk_y.$$

Use of the orthogonality relationship (A.115) yields the amplitude spectrum $B(k_y)$, and we find that

$$V(x,y) = \frac{2}{\pi} \int_0^\infty e^{-k_y x} \frac{k_y}{a^2 + k_y^2} \sin k_y y \, dk_y. \tag{A.116}$$

As a final example in rectangular coordinates let us consider a problem in which ψ depends on all three variables:

$$\nabla^2 \psi(x,y,z) + k^2 \psi(x,y,z) = 0$$

for

$$0 \le x \le L_1, \quad 0 \le y \le L_2, \quad 0 \le z \le L_3,$$

subject to

$$\psi(0,y,z) = \psi(L_1,y,z) = 0,$$
$$\psi(x,0,z) = \psi(x,L_2,z) = 0,$$
$$\psi(x,y,0) = \psi(x,y,L_3) = 0.$$

Here $k \ne 0$ is a constant. This is a three-dimensional eigenvalue problem as described in § A.4, where $\lambda = k^2$ are the eigenvalues and the closed surface is a rectangular box. Physically, the wave function ψ represents the so-called *eigenvalue* or *normal mode* solutions for the "TM modes" of a rectangular cavity. Since $k_x^2 + k_y^2 + k_z^2 = k^2$, we might have one or two separation constants equal to zero, but not all three. We find, however, that the only solution with a zero separation constant that can fit the boundary conditions is the trivial solution. In light of the boundary conditions and because we expect standing waves in the box, we take

$$\psi(x,y,z) = [A_x \sin(k_x x) + B_x \cos(k_x x)] \cdot$$
$$\cdot \, [A_y \sin(k_y y) + B_y \cos(k_y y)] \cdot$$
$$\cdot \, [A_z \sin(k_z z) + B_z \cos(k_z z)].$$

The conditions $\psi(0,y,z) = \psi(x,0,z) = \psi(x,y,0) = 0$ give $B_x = B_y = B_z = 0$. The conditions at $x = L_1$, $y = L_2$, and $z = L_3$ require the separation constants to assume the discrete values $k_x = k_{x_m} = m\pi/L_1$, $k_y = k_{y_n} = n\pi/L_2$, and $k_z = k_{z_p} = p\pi/L_3$, where $k_{x_m}^2 + k_{y_n}^2 + k_{z_p}^2 = k_{mnp}^2$ and $m,n,p = 1,2,\ldots$. Associated with each of these eigenvalues is an eigenfunction of a one-dimensional Sturm–Liouville problem. For the three-dimensional problem, an eigenfunction

$$\psi_{mnp}(x, y, z) = A_{mnp} \sin(k_{x_m} x) \sin(k_{y_n} y) \sin(k_{z_p} z)$$

is associated with each three-dimensional eigenvalue k_{mnp}. Each choice of m, n, p produces a discrete cavity resonance frequency at which the boundary conditions can be satisfied. Depending on the values of $L_{1,2,3}$, we may have more than one eigenfunction associated with an eigenvalue. For example, if $L_1 = L_2 = L_3 = L$ then $k_{121} = k_{211} = k_{112} = \sqrt{6}\pi/L$. However, the eigenfunctions associated with this single eigenvalue are all different:

$$\psi_{121} = \sin(k_{x_1} x) \sin(k_{y_2} y) \sin(k_{z_1} z),$$
$$\psi_{211} = \sin(k_{x_2} x) \sin(k_{y_1} y) \sin(k_{z_1} z),$$
$$\psi_{112} = \sin(k_{x_1} x) \sin(k_{y_1} y) \sin(k_{z_2} z).$$

When more than one cavity mode corresponds to a given resonant frequency, we call the modes *degenerate*. By completeness, we can represent any well-behaved function as

$$f(x, y, z) = \sum_{m,n,p} A_{mnp} \sin(k_{x_m} x) \sin(k_{y_n} y) \sin(k_{z_p} z).$$

The A_{mnp} are found using orthogonality. When such expansions are used to solve problems involving objects (such as excitation probes) inside the cavity, they are termed *normal mode expansions* of the cavity field.

Solutions in cylindrical coordinates. In cylindrical coordinates the Helmholtz equation is

$$\frac{1}{\rho} \frac{\partial}{\partial \rho} \left(\rho \frac{\partial \psi(\rho, \phi, z)}{\partial \rho} \right) + \frac{1}{\rho^2} \frac{\partial^2 \psi(\rho, \phi, z)}{\partial \phi^2} + \frac{\partial^2 \psi(\rho, \phi, z)}{\partial z^2} + k^2 \psi(\rho, \phi, z) = 0. \quad (A.117)$$

With $\psi(\rho, \phi, z) = P(\rho)\Phi(\phi)Z(z)$ we obtain

$$\frac{1}{\rho} \frac{\partial}{\partial \rho} \left(\rho \frac{\partial (P\Phi Z)}{\partial \rho} \right) + \frac{1}{\rho^2} \frac{\partial^2 (P\Phi Z)}{\partial \phi^2} + \frac{\partial^2 (P\Phi Z)}{\partial z^2} + k^2 (P\Phi Z) = 0;$$

carrying out the ρ derivatives and dividing through by $P\Phi Z$ we have

$$-\frac{1}{Z} \frac{d^2 Z}{dz^2} = k^2 + \frac{1}{\rho^2 \Phi} \frac{d^2 \Phi}{d\phi^2} + \frac{1}{\rho P} \frac{dP}{d\rho} + \frac{1}{P} \frac{d^2 P}{d\rho^2}.$$

The left side depends on z while the right side depends on ρ and ϕ, hence both must equal the same constant k_z^2:

$$-\frac{1}{Z} \frac{d^2 Z}{dz^2} = k_z^2, \quad (A.118)$$

$$k^2 + \frac{1}{\rho^2 \Phi} \frac{d^2 \Phi}{d\phi^2} + \frac{1}{\rho P} \frac{dP}{d\rho} + \frac{1}{P} \frac{d^2 P}{d\rho^2} = k_z^2. \quad (A.119)$$

We have separated the z-dependence from the dependence on the other variables. For the harmonic equation (A.118),

$$Z(z) = \begin{cases} A_z \sin k_z z + B_z \cos k_z z, & k_z \neq 0, \\ a_z z + b_z, & k_z = 0. \end{cases} \quad (A.120)$$

Of course we could use exponentials or a combination of exponentials and trigonometric functions instead. Rearranging (A.119) and multiplying through by ρ^2, we obtain

$$-\frac{1}{\Phi}\frac{d^2\Phi}{d\phi^2} = \left(k^2 - k_z^2\right)\rho^2 + \frac{\rho}{P}\frac{dP}{d\rho} + \frac{\rho^2}{P}\frac{d^2P}{d\rho^2}.$$

The left and right sides depend only on ϕ and ρ, respectively; both must equal some constant k_ϕ^2:

$$-\frac{1}{\Phi}\frac{d^2\Phi}{d\phi^2} = k_\phi^2, \qquad (A.121)$$

$$\left(k^2 - k_z^2\right)\rho^2 + \frac{\rho}{P}\frac{dP}{d\rho} + \frac{\rho^2}{P}\frac{d^2P}{d\rho^2} = k_\phi^2. \qquad (A.122)$$

The variables ρ and ϕ are thus separated, and harmonic equation (A.121) has solutions

$$\Phi(\phi) = \begin{cases} A_\phi \sin k_\phi\phi + B_\phi \cos k_\phi\phi, & k_\phi \neq 0, \\ a_\phi\phi + b_\phi, & k_\phi = 0. \end{cases} \qquad (A.123)$$

Equation (A.122) is a bit more involved. In rearranged form it is

$$\frac{d^2P}{d\rho^2} + \frac{1}{\rho}\frac{dP}{d\rho} + \left(k_c^2 - \frac{k_\phi^2}{\rho^2}\right)P = 0 \qquad (A.124)$$

where

$$k_c^2 = k^2 - k_z^2.$$

The solution depends on whether any of k_z, k_ϕ, or k_c are zero. If $k_c = k_\phi = 0$, then

$$\frac{d^2P}{d\rho^2} + \frac{1}{\rho}\frac{dP}{d\rho} = 0$$

so that

$$P(\rho) = a_\rho \ln \rho + b_\rho.$$

If $k_c = 0$ but $k_\phi \neq 0$, we have

$$\frac{d^2P}{d\rho^2} + \frac{1}{\rho}\frac{dP}{d\rho} - \frac{k_\phi^2}{\rho^2}P = 0$$

so that

$$P(\rho) = a_\rho\rho^{-k_\phi} + b_\rho\rho^{k_\phi}. \qquad (A.125)$$

This includes the case $k = k_z = 0$ (Laplace's equation). If $k_c \neq 0$ then (A.124) is Bessel's differential equation. For noninteger k_ϕ the two independent solutions are denoted $J_{k_\phi}(z)$ and $J_{-k_\phi}(z)$, where $J_\nu(z)$ is the ordinary Bessel function of the first kind of order ν. For k_ϕ an integer n, $J_n(z)$ and $J_{-n}(z)$ are not independent and a second independent solution denoted $N_n(z)$ must be introduced. This is the ordinary Bessel function of the second kind, order n. As it is also independent when the order is noninteger, $J_\nu(z)$ and $N_\nu(z)$ are often chosen as solutions whether ν is integer or not. Linear combinations of these independent solutions may be used to produce new independent solutions. The functions

$H_\nu^{(1)}(z)$ and $H_\nu^{(2)}(z)$ are the *Hankel functions* of the first and second kind of order ν, and are related to the Bessel functions by

$$H_\nu^{(1)}(z) = J_\nu(z) + jN_\nu(z),$$
$$H_\nu^{(2)}(z) = J_\nu(z) - jN_\nu(z).$$

The argument z can be complex (as can ν, but this shall not concern us). When z is imaginary we introduce two new functions $I_\nu(z)$ and $K_\nu(z)$, defined for integer order by

$$I_n(z) = j^{-n} J_n(jz),$$
$$K_n(z) = \frac{\pi}{2} j^{n+1} H_n^{(1)}(jz).$$

Expressions for noninteger order are given in Appendix E.1.

Bessel functions cannot be expressed in terms of simple, standard functions. However, a series solution to (A.124) produces many useful relationships between Bessel functions of differing order and argument. The *recursion relations* for Bessel functions serve to connect functions of various orders and their derivatives. See Appendix E.1.

Of the six possible solutions to (A.124),

$$R(\rho) = \begin{cases} A_\rho J_\nu(k_c\rho) + B_\rho N_\nu(k_c\rho), \\ A_\rho J_\nu(k_c\rho) + B_\rho H_\nu^{(1)}(k_c\rho), \\ A_\rho J_\nu(k_c\rho) + B_\rho H_\nu^{(2)}(k_c\rho), \\ A_\rho N_\nu(k_c\rho) + B_\rho H_\nu^{(1)}(k_c\rho), \\ A_\rho N_\nu(k_c\rho) + B_\rho H_\nu^{(2)}(k_c\rho), \\ A_\rho H_\nu^{(1)}(k_c\rho) + B_\rho H_\nu^{(2)}(k_c\rho), \end{cases}$$

which do we choose? Again, we are motivated by convenience and the physical nature of the problem. If the argument is real or imaginary, we often consider large or small argument behavior. For x real and large,

$$J_\nu(x) \to \sqrt{\frac{2}{\pi x}} \cos\left(x - \frac{\pi}{4} - \nu\frac{\pi}{2}\right),$$

$$N_\nu(x) \to \sqrt{\frac{2}{\pi x}} \sin\left(x - \frac{\pi}{4} - \nu\frac{\pi}{2}\right),$$

$$H_\nu^{(1)}(x) \to \sqrt{\frac{2}{\pi x}} e^{j\left(x - \frac{\pi}{4} - \nu\frac{\pi}{2}\right)},$$

$$H_\nu^{(2)}(x) \to \sqrt{\frac{2}{\pi x}} e^{-j\left(x - \frac{\pi}{4} - \nu\frac{\pi}{2}\right)},$$

$$I_\nu(x) \to \sqrt{\frac{1}{2\pi x}} e^x,$$

$$K_\nu(x) \to \sqrt{\frac{\pi}{2x}} e^{-x},$$

while for x real and small,

$$J_0(x) \to 1,$$

$$N_0(x) \to \frac{2}{\pi} \left(\ln x + 0.5772157 - \ln 2\right),$$

$$J_\nu(x) \to \frac{1}{\nu!}\left(\frac{x}{2}\right)^\nu,$$

$$N_\nu(x) \to -\frac{(\nu-1)!}{\pi}\left(\frac{2}{x}\right)^\nu.$$

Because $J_\nu(x)$ and $N_\nu(x)$ oscillate for large argument, they can represent standing waves along the radial direction. However, $N_\nu(x)$ is unbounded for small x and is inappropriate for regions containing the origin. The Hankel functions become complex exponentials for large argument, hence represent traveling waves. Finally, $K_\nu(x)$ is unbounded for small x and cannot be used for regions containing the origin, while $I_\nu(x)$ increases exponentially for large x and cannot be used for unbounded regions.

Examples.   Consider the boundary value problem for Laplace's equation

$$\nabla^2 V(\rho,\phi) = 0 \tag{A.126}$$

in the region

$$0 \le \rho \le \infty, \quad 0 \le \phi \le \phi_0, \quad -\infty < z < \infty,$$

where the boundary conditions are

$$V(\rho,0) = 0, \quad V(\rho,\phi_0) = V_0.$$

Since there is no z-dependence we let $k_z = 0$ in (A.120) and choose $a_z = 0$. Then $k_c^2 = k^2 - k_z^2 = 0$ since $k = 0$. There are two possible solutions, depending on whether k_ϕ is zero. First let us try $k_\phi \ne 0$. Using (A.123) and (A.125) we have

$$V(\rho,\phi) = [A_\phi \sin(k_\phi\phi) + B_\phi \cos(k_\phi\phi)][a_\rho \rho^{-k_\phi} + b_\rho \rho^{k_\phi}]. \tag{A.127}$$

Assuming $k_\phi > 0$ we must have $b_\rho = 0$ to keep the solution finite. The condition $V(\rho,0) = 0$ requires $B_\phi = 0$. Thus

$$V(\rho,\phi) = A_\phi \sin(k_\phi\phi)\rho^{-k_\phi}.$$

Our final boundary condition requires

$$V(\rho,\phi_0) = V_0 = A_\phi \sin(k_\phi\phi_0)\rho^{-k_\phi}.$$

Because this cannot hold for all ρ, we must resort to $k_\phi = 0$ and

$$V(\rho,\phi) = (a_\phi\phi + b_\phi)(a_\rho \ln\rho + b_\rho). \tag{A.128}$$

Proper behavior as $\rho \to \infty$ dictates that $a_\rho = 0$. $V(\rho,0) = 0$ requires $b_\phi = 0$. Thus $V(\rho,\phi) = V(\phi) = b_\phi\phi$. The constant b_ϕ is found from the remaining boundary condition: $V(\phi_0) = V_0 = b_\phi\phi_0$ so that $b_\phi = V_0/\phi_0$. The final solution is

$$V(\phi) = V_0\phi/\phi_0.$$

It is worthwhile to specialize this to $\phi_0 = \pi/2$ and compare with the solution to the same problem found earlier using rectangular coordinates. With $a = 0$ in (A.116) we have

$$V(x,y) = \frac{2}{\pi}\int_0^\infty e^{-k_y x}\frac{\sin k_y y}{k_y}\,dk_y.$$

Despite its much more complicated form, this must be the same solution by uniqueness. Next let us solve (A.126) subject to the "split cylinder" conditions

$$V(a, \phi) = \begin{cases} V_0, & 0 < \phi < \pi, \\ 0, & -\pi < \phi < 0. \end{cases}$$

Because there is no z-dependence we choose $k_z = a_z = 0$ and have $k_c^2 = k^2 - k_z^2 = 0$. Since $k_\phi = 0$ would violate the boundary conditions at $\rho = a$, we use

$$V(\rho, \phi) = (a_\rho \rho^{-k_\phi} + b_\rho \rho^{k_\phi})(A_\phi \sin k_\phi \phi + B_\phi \cos k_\phi \phi).$$

The potential must be single-valued in ϕ: $V(\rho, \phi + 2n\pi) = V(\rho, \phi)$. This is only possible if k_ϕ is an integer, say $k_\phi = m$. Then

$$V_m(\rho, \phi) = \begin{cases} (A_m \sin m\phi + B_m \cos m\phi)\rho^m, & \rho < a, \\ (C_m \sin m\phi + D_m \cos m\phi)\rho^{-m}, & \rho > a. \end{cases}$$

On physical grounds we have discarded ρ^{-m} for $\rho < a$ and ρ^m for $\rho > a$. To satisfy the boundary conditions at $\rho = a$ we must use an infinite series of the complete set of eigensolutions. For $\rho < a$ the boundary condition requires

$$B_0 + \sum_{m=1}^{\infty} (A_m \sin m\phi + B_m \cos m\phi)a^m = \begin{cases} V_0, & 0 < \phi < \pi, \\ 0, & -\pi < \phi < 0. \end{cases}$$

Application of the orthogonality relations

$$\int_{-\pi}^{\pi} \cos m\phi \cos n\phi \, d\phi = \frac{2\pi}{\epsilon_n} \delta_{mn}, \quad m, n = 0, 1, 2, \ldots, \tag{A.129}$$

$$\int_{-\pi}^{\pi} \sin m\phi \sin n\phi \, d\phi = \pi \delta_{mn}, \quad m, n = 1, 2, \ldots, \tag{A.130}$$

$$\int_{-\pi}^{\pi} \cos m\phi \sin n\phi \, d\phi = 0, \quad m, n = 0, 1, 2, \ldots, \tag{A.131}$$

where

$$\epsilon_n = \begin{cases} 1, & n = 0, \\ 2, & n > 0, \end{cases} \tag{A.132}$$

is Neumann's number, produces appropriate values for the constants A_m and B_m. The full solution is

$$V(\rho, \phi) = \begin{cases} \dfrac{V_0}{2} + \dfrac{V_0}{\pi} \sum_{n=1}^{\infty} \dfrac{[1 - (-1)^n]}{n} \left(\dfrac{\rho}{a}\right)^n \sin n\phi, & \rho < a, \\ \dfrac{V_0}{2} + \dfrac{V_0}{\pi} \sum_{n=1}^{\infty} \dfrac{[1 - (-1)^n]}{n} \left(\dfrac{a}{\rho}\right)^n \sin n\phi, & \rho > a. \end{cases}$$

The boundary value problem

$$\nabla^2 V(\rho, \phi, z) = 0, \quad 0 \le \rho \le a, \, -\pi \le \phi \le \pi, \, 0 \le z \le h,$$
$$V(\rho, \phi, 0) = 0, \quad 0 \le \rho \le a, \, -\pi \le \phi \le \pi,$$
$$V(a, \phi, z) = 0, \quad -\pi \le \phi \le \pi, \, 0 \le z \le h,$$
$$V(\rho, \phi, h) = V_0, \quad 0 \le \rho \le a, \, -\pi \le \phi \le \pi,$$

describes the potential within a grounded "canister" with top at potential V_0. Symmetry precludes ϕ-dependence, hence $k_\phi = a_\phi = 0$. Since $k = 0$ (Laplace's equation) we also have $k_c^2 = k^2 - k_z^2 = -k_z^2$. Thus we have either k_z real and $k_c = jk_z$, or k_c real and $k_z = jk_c$. With k_z real we have

$$V(\rho, z) = [A_z \sin k_z z + B_z \cos k_z z][A_\rho K_0(k_z \rho) + B_\rho I_0(k_z \rho)]; \tag{A.133}$$

with k_c real we have

$$V(\rho, z) = [A_z \sinh k_c z + B_z \cosh k_c z][A_\rho J_0(k_c \rho) + B_\rho N_0(k_c \rho)]. \tag{A.134}$$

The functions K_0 and I_0 are inappropriate for use in this problem, and we proceed to (A.134). Since N_0 is unbounded for small argument, we need $B_\rho = 0$. The condition $V(\rho, 0) = 0$ gives $B_z = 0$, thus

$$V(\rho, z) = A_z \sinh(k_c z) J_0(k_c \rho).$$

The oscillatory nature of J_0 means that we can satisfy the condition at $\rho = a$:

$$V(a, z) = A_z \sinh(k_c z) J_0(k_c a) = 0 \quad \text{for} \quad 0 \le z < h$$

if $J_0(k_c a) = 0$. Letting p_{0m} denote the mth root of $J_0(x) = 0$ for $m = 1, 2, \ldots$, we have $k_{c_m} = p_{0m}/a$. Because we cannot satisfy the boundary condition at $z = h$ with a single eigensolution, we use the superposition

$$V(\rho, z) = \sum_{m=1}^{\infty} A_m \sinh\left(\frac{p_{0m} z}{a}\right) J_0\left(\frac{p_{0m} \rho}{a}\right).$$

We require

$$V(\rho, h) = \sum_{m=1}^{\infty} A_m \sinh\left(\frac{p_{0m} h}{a}\right) J_0\left(\frac{p_{0m} \rho}{a}\right) = V_0, \tag{A.135}$$

where the A_m can be evaluated by orthogonality of the functions $J_0(p_{0m}\rho/a)$. If $p_{\nu m}$ is the mth root of $J_\nu(x) = 0$, then

$$\int_0^a J_\nu\left(\frac{p_{\nu m}\rho}{a}\right) J_\nu\left(\frac{p_{\nu n}\rho}{a}\right) \rho\, d\rho = \delta_{mn}\frac{a^2}{2}J_\nu'^2(p_{\nu n}) = \delta_{mn}\frac{a^2}{2}J_{\nu+1}^2(p_{\nu n}) \tag{A.136}$$

where $J_\nu'(x) = dJ_\nu(x)/dx$. Multiplying (A.135) by $\rho J_0(p_{0n}\rho/a)$ and integrating, we have

$$A_n \sinh\left(\frac{p_{0n} h}{a}\right)\frac{a^2}{2}J_0'^2(p_{0n}a) = \int_0^a V_0 J_0\left(\frac{p_{0n}\rho}{a}\right)\rho\, d\rho.$$

Use of (E.105),

$$\int x^{n+1} J_n(x)\, dx = x^{n+1} J_{n+1}(x) + C,$$

allows us to evaluate

$$\int_0^a J_0\left(\frac{p_{0n}\rho}{a}\right)\rho\, d\rho = \frac{a^2}{p_{0n}}J_1(p_{0n}).$$

With this we finish calculating A_m and have

$$V(\rho, z) = 2V_0 \sum_{m=1}^{\infty} \frac{\sinh(\frac{p_{0m}}{a}z) J_0(\frac{p_{0m}}{a}\rho)}{p_{0m}\sinh(\frac{p_{0m}}{a}h) J_1(p_{0m})}$$

as the desired solution.

Finally, let us assume that $k > 0$ and solve

$$\nabla^2 \psi(\rho, \phi, z) + k^2 \psi(\rho, \phi, z) = 0$$

where $0 \leq \rho \leq a$, $-\pi \leq \phi \leq \pi$, and $-\infty < z < \infty$, subject to the condition

$$\hat{\mathbf{n}} \cdot \nabla \psi(\rho, \phi, z) \Big|_{\rho=a} = \frac{\partial \psi(\rho, \phi, z)}{\partial \rho} \Big|_{\rho=a} = 0$$

for $-\pi \leq \phi \leq \pi$ and $-\infty < z < \infty$. The solution to this problem leads to the transverse-electric (TE$_z$) fields in a lossless circular waveguide, where ψ represents the z-component of the magnetic field. Although there is symmetry with respect to ϕ, we seek ϕ-dependent solutions; the resulting complete eigenmode solution will permit us to expand any well-behaved function within the waveguide in terms of a normal mode (eigenfunction) series. In this problem none of the constants k, k_z, or k_ϕ equal zero, except as a special case. However, the field must be single-valued in ϕ and thus k_ϕ must be an integer m. We consider our possible choices for $P(\rho)$, $Z(z)$, and $\Phi(\phi)$. Since $k_c^2 = k^2 - k_z^2$ and $k^2 > 0$ is arbitrary, we must consider various possibilities for the signs of k_c^2 and k_z^2. We can rule out $k_c^2 < 0$ based on consideration of the behavior of the functions I_m and K_m. We also need not consider $k_c < 0$, since this gives the same solution as $k_c > 0$. We are then left with two possible cases. Writing $k_z^2 = k^2 - k_c^2$, we see that either $k > k_c$ and $k_z^2 > 0$, or $k < k_c$ and $k_z^2 < 0$. For $k_z^2 > 0$ we write

$$\psi(\rho, \phi, z) = [A_z e^{-jk_z z} + B_z e^{jk_z z}][A_\phi \sin m\phi + B_\phi \cos m\phi] J_m(k_c \rho).$$

Here the terms involving $e^{\mp jk_z z}$ represent waves propagating in the $\pm z$ directions. The boundary condition at $\rho = a$ requires

$$J'_m(k_c a) = 0$$

where $J'_m(x) = dJ_m(x)/dx$. Denoting the nth zero of $J'_m(x)$ by p'_{mn} we have $k_c = k_{cm} = p'_{mn}/a$. This gives the eigensolutions

$$\psi_m = [A_{zm} e^{-jk_z z} + B_{zm} e^{jk_z z}][A_{\phi m} \sin m\phi + B_{\phi m} \cos m\phi] k_c J_m \left(\frac{p'_{mn} \rho}{a} \right).$$

The undetermined constants $A_{zm}, B_{zm}, A_{\rho m}, B_{\rho m}$ could be evaluated when the individual eigensolutions are used to represent a function in terms of a modal expansion. For the case $k_z^2 < 0$ we again choose complex exponentials in z; however, $k_z = -j\alpha$ gives $e^{\mp jk_z z} = e^{\mp \alpha z}$ and attenuation along z. The reader can verify that the eigensolutions are

$$\psi_m = [A_{zm} e^{-\alpha z} + B_{zm} e^{\alpha z}][A_{\phi m} \sin m\phi + B_{\phi m} \cos m\phi] k_c J_m \left(\frac{p'_{mn} \rho}{a} \right)$$

where now $k_c^2 = k^2 + \alpha^2$.

We have used Bessel function completeness in the examples above. This property is a consequence of the Sturm–Liouville problem first studied in § A.4. We often use Fourier–Bessel series to express functions over finite intervals. Over infinite intervals we use the Fourier–Bessel transform.

The Fourier–Bessel series can be generalized to Bessel functions of noninteger order, and to the derivatives of Bessel functions. Let $f(\rho)$ be well-behaved over the interval $[0, a]$. Then the series

$$f(\rho) = \sum_{m=1}^{\infty} C_m J_\nu \left(p_{\nu m} \frac{\rho}{a} \right), \quad 0 \leq \rho \leq a, \; \nu > -1$$

converges, and the constants are

$$C_m = \frac{2}{a^2 J_{\nu+1}^2(p_{\nu m})} \int_0^a f(\rho) J_\nu \left(p_{\nu m} \frac{\rho}{a}\right) \rho \, d\rho$$

by (A.136). Here $p_{\nu m}$ is the mth root of $J_\nu(x)$. An alternative form of the series uses $p'_{\nu m}$, the roots of $J'_\nu(x)$, and is given by

$$f(\rho) = \sum_{m=1}^\infty D_m J_\nu \left(p'_{\nu m} \frac{\rho}{a}\right), \quad 0 \le \rho \le a, \ \nu > -1.$$

In this case the expansion coefficients are found using the orthogonality relationship

$$\int_0^a J_\nu \left(\frac{p'_{\nu m}}{a} \rho\right) J_\nu \left(\frac{p'_{\nu n}}{a} \rho\right) \rho \, d\rho = \delta_{mn} \frac{a^2}{2} \left(1 - \frac{\nu^2}{p'^2_{\nu m}}\right) J_\nu^2(p'_{\nu m}),$$

and are

$$D_m = \frac{2}{a^2 \left(1 - \frac{\nu^2}{p'^2_{\nu m}} J_\nu^2(p'_{\nu m})\right)} \int_0^a f(\rho) J_\nu \left(\frac{p'_{\nu m}}{a} \rho\right) \rho \, d\rho.$$

Solutions in spherical coordinates. If into Helmholtz's equation

$$\frac{1}{r^2} \frac{\partial}{\partial r} \left(r^2 \frac{\partial \psi(r,\theta,\phi)}{\partial r}\right) + \frac{1}{r^2 \sin\theta} \frac{\partial}{\partial \theta} \left(\sin\theta \frac{\partial \psi(r,\theta,\phi)}{\partial \theta}\right) +$$

$$+ \frac{1}{r^2 \sin^2\theta} \frac{\partial^2 \psi(r,\theta,\phi)}{\partial \phi^2} + k^2 \psi(r,\theta,\phi) = 0$$

we put $\psi(r,\theta,\phi) = R(r)\Theta(\theta)\Phi(\phi)$ and multiply through by $r^2 \sin^2\theta/\psi(r,\theta,\phi)$, we obtain

$$\frac{\sin^2\theta}{R(r)} \frac{d}{dr} \left(r^2 \frac{dR(r)}{dr}\right) + \frac{\sin\theta}{\Theta(\theta)} \frac{d}{d\theta} \left(\sin\theta \frac{d\Theta(\theta)}{d\theta}\right) + k^2 r^2 \sin^2\theta = -\frac{1}{\Phi(\phi)} \frac{d^2\Phi(\phi)}{d\phi^2}.$$

Since the right side depends only on ϕ while the left side depends only on r and θ, both sides must equal some constant μ^2:

$$\frac{\sin^2\theta}{R(r)} \frac{d}{dr} \left(r^2 \frac{dR(r)}{dr}\right) + \frac{\sin\theta}{\Theta(\theta)} \frac{d}{d\theta} \left(\sin\theta \frac{d\Theta(\theta)}{d\theta}\right) + k^2 r^2 \sin^2\theta = \mu^2, \quad \text{(A.137)}$$

$$\frac{d^2\Phi(\phi)}{d\phi^2} + \mu^2 \Phi(\phi) = 0. \quad \text{(A.138)}$$

We have thus separated off the ϕ-dependence. Harmonic ordinary differential equation (A.138) has solutions

$$\Phi(\phi) = \begin{cases} A_\phi \sin\mu\phi + B_\phi \cos\mu\phi, & \mu \neq 0, \\ a_\phi \phi + b_\phi, & \mu = 0. \end{cases}$$

(We could have used complex exponentials to describe $\Phi(\phi)$, or some combination of exponentials and trigonometric functions, but it is conventional to use only trigonometric functions.) Rearranging (A.137) and dividing through by $\sin^2\theta$ we have

$$\frac{1}{R(r)} \frac{d}{dr} \left(r^2 \frac{dR(r)}{dr}\right) + k^2 r^2 = -\frac{1}{\sin\theta\,\Theta(\theta)} \frac{d}{d\theta} \left(\sin\theta \frac{d\Theta(\theta)}{d\theta}\right) + \frac{\mu^2}{\sin^2\theta}.$$

We introduce a new constant k_θ^2 to separate r from θ:

$$\frac{1}{R(r)}\frac{d}{dr}\left(r^2\frac{dR(r)}{dr}\right) + k^2 r^2 = k_\theta^2, \qquad (A.139)$$

$$-\frac{1}{\sin\theta\,\Theta(\theta)}\frac{d}{d\theta}\left(\sin\theta\frac{d\Theta(\theta)}{d\theta}\right) + \frac{\mu^2}{\sin^2\theta} = k_\theta^2. \qquad (A.140)$$

Equation (A.140),

$$\frac{1}{\sin\theta}\frac{d}{d\theta}\left(\sin\theta\frac{d\Theta(\theta)}{d\theta}\right) + \left(k_\theta^2 - \frac{\mu^2}{\sin^2\theta}\right)\Theta(\theta) = 0,$$

can be put into a standard form by letting

$$\eta = \cos\theta \qquad (A.141)$$

and $k_\theta^2 = \nu(\nu+1)$ where ν is a parameter:

$$(1-\eta^2)\frac{d^2\Theta(\eta)}{d\eta^2} - 2\eta\frac{d\Theta(\eta)}{d\eta} + \left[\nu(\nu+1) - \frac{\mu^2}{1-\eta^2}\right]\Theta(\eta) = 0, \quad -1 \le \eta \le 1.$$

This is the *associated Legendre equation*. It has two independent solutions called *associated Legendre functions of the first* and *second kinds*, denoted $P_\nu^\mu(\eta)$ and $Q_\nu^\mu(\eta)$, respectively. In these functions, all three quantities μ, ν, η may be arbitrary complex constants as long as $\nu + \mu \ne -1, -2, \ldots$. But (A.141) shows that η is real in our discussion; μ will generally be real also, and will be an integer whenever $\Phi(\phi)$ is single-valued. The choice of ν is somewhat more complicated. The function $P_\nu^\mu(\eta)$ diverges at $\eta = \pm 1$ unless ν is an integer, while $Q_\nu^\mu(\eta)$ diverges at $\eta = \pm 1$ regardless of whether ν is an integer. In § A.4 we required that $P_\nu^\mu(\eta)$ be bounded on $[-1, 1]$ to have a Sturm–Liouville problem with suitable orthogonality properties. By (A.141) we must exclude $Q_\nu^\mu(\eta)$ for problems containing the z-axis, and restrict ν to be an integer n in $P_\nu^\mu(\eta)$ for such problems. In case the z-axis is excluded, we choose $\nu = n$ whenever possible, because the finite sums $P_n^m(\eta)$ and $Q_n^m(\eta)$ are much easier to manipulate than $P_\nu^\mu(\eta)$ and $Q_\nu^\mu(\eta)$. In many problems we must count on completeness of the Legendre polynomials $P_n(\eta) = P_n^0(\eta)$ or spherical harmonics $Y_{mn}(\theta, \phi)$ in order to satisfy the boundary conditions. In this book we shall consider only those boundary value problems that can be solved using integer values of ν and μ, hence choose

$$\Theta(\theta) = A_\theta P_n^m(\cos\theta) + B_\theta Q_n^m(\cos\theta). \qquad (A.142)$$

Single-valuedness in $\Phi(\phi)$ is a consequence of having $\mu = m$, and $\phi = $ constant boundary surfaces are thereby disallowed.

The associated Legendre functions have many important properties. For instance,

$$P_n^m(\eta) = \begin{cases} 0, & m > n, \\ (-1)^m \dfrac{(1-\eta^2)^{m/2}}{2^n n!}\dfrac{d^{n+m}(\eta^2-1)^n}{d\eta^{n+m}}, & m \le n. \end{cases} \qquad (A.143)$$

The case $m = 0$ receives particular attention because it corresponds to azimuthal invariance (ϕ-independence). We define $P_n^0(\eta) = P_n(\eta)$ where $P_n(\eta)$ is the Legendre polynomial of order n. From (A.143), we see that[1]

$$P_n(\eta) = \frac{1}{2^n n!} \frac{d^n (\eta^2 - 1)^n}{d\eta^n}$$

is a polynomial of degree n, and that

$$P_n^m(\eta) = (-1)^m (1 - \eta^2)^{m/2} \frac{d^m}{d\eta^m} P_n(\eta).$$

Both the associated Legendre functions and the Legendre polynomials obey orthogonality relations and many recursion formulas.

In problems where the z-axis is included, the product $\Theta(\theta)\Phi(\phi)$ is sometimes defined as the spherical harmonic

$$Y_{nm}(\theta, \phi) = \sqrt{\frac{2n + 1}{4\pi} \frac{(n - m)!}{(n + m)!}} \, P_n^m(\cos\theta) e^{jm\theta}.$$

These functions, which are complete over the surface of a sphere, were treated earlier in this section.

Remembering that $k_r^2 = \nu(\nu + 1)$, the r-dependent equation (A.139) becomes

$$\frac{1}{r^2} \frac{d}{dr} \left(r^2 \frac{dR(r)}{dr} \right) + \left(k^2 + \frac{n(n + 1)}{r^2} \right) R(r) = 0. \qquad \text{(A.144)}$$

When $k = 0$ we have

$$\frac{d^2 R(r)}{dr^2} + \frac{2}{r} \frac{dR(r)}{dr} - \frac{n(n + 1)}{r^2} R(r) = 0$$

so that

$$R(r) = A_r r^n + B_r r^{-(n+1)}.$$

When $k \neq 0$, the substitution $\bar{R}(r) = \sqrt{kr} R(r)$ puts (A.144) into the form

$$r^2 \frac{d^2 \bar{R}(r)}{dr^2} + r \frac{d\bar{R}(r)}{dr} + \left[k^2 r^2 - \left(n + \frac{1}{2} \right)^2 \right] \bar{R}(r) = 0,$$

which we recognize as Bessel's equation of half-integer order. Thus

$$R(r) = \frac{\bar{R}(r)}{\sqrt{kr}} = \frac{Z_{n+\frac{1}{2}}(kr)}{\sqrt{kr}}.$$

For convenience we define the *spherical Bessel functions*

$$j_n(z) = \sqrt{\frac{\pi}{2z}} J_{n+\frac{1}{2}}(z),$$

$$n_n(z) = \sqrt{\frac{\pi}{2z}} N_{n+\frac{1}{2}}(z) = (-1)^{n+1} \sqrt{\frac{\pi}{2z}} J_{-(n+\frac{1}{2})}(z),$$

$$h_n^{(1)}(z) = \sqrt{\frac{\pi}{2z}} H_{n+\frac{1}{2}}^{(1)}(z) = j_n(z) + j n_n(z),$$

$$h_n^{(2)}(z) = \sqrt{\frac{\pi}{2z}} H_{n+\frac{1}{2}}^{(2)}(z) = j_n(z) - j n_n(z).$$

---

[1]Care must be taken when consulting tables of Legendre functions and their properties. In particular, one must be on the lookout for possible disparities regarding the factor $(-1)^m$ (cf., [76, 1, 109, 8] vs. [5, 187]). Similar care is needed with $Q_n^m(x)$.

These can be written as finite sums involving trigonometric functions and inverse powers of z. We have, for instance,

$$j_0(z) = \frac{\sin z}{z},$$

$$n_0(z) = -\frac{\cos z}{z},$$

$$j_1(z) = \frac{\sin z}{z^2} - \frac{\cos z}{z},$$

$$n_1(z) = -\frac{\cos z}{z^2} - \frac{\sin z}{z}.$$

We can now write $R(r)$ as a linear combination of any two of the spherical Bessel functions $j_n, n_n, h_n^{(1)}, h_n^{(2)}$:

$$R(r) = \begin{cases} A_r j_n(kr) + B_r n_n(kr), \\ A_r j_n(kr) + B_r h_n^{(1)}(kr), \\ A_r j_n(kr) + B_r h_n^{(2)}(kr), \\ A_r n_n(kr) + B_r h_n^{(1)}(kr), \\ A_r n_n(kr) + B_r h_n^{(2)}(kr), \\ A_r h_n^{(1)}(kr) + B_r h_n^{(2)}(kr). \end{cases} \tag{A.145}$$

Imaginary arguments produce *modified spherical Bessel functions*; the interested reader is referred to Gradsteyn [76] or Abramowitz [1].

Examples. The problem

$$\nabla^2 V(r, \theta, \phi) = 0, \quad \theta_0 \le \theta \le \pi/2, \ 0 \le r < \infty, \ -\pi \le \phi \le \pi,$$
$$V(r, \theta_0, \phi) = V_0, \quad -\pi \le \phi \le \pi, \ 0 \le r < \infty,$$
$$V(r, \pi/2, \phi) = 0, \quad -\pi \le \phi \le \pi, \ 0 \le r < \infty,$$

gives the potential field between a cone and the $z = 0$ plane. Azimuthal symmetry prompts us to choose $\mu = a_\phi = 0$. Since $k = 0$ we have

$$R(r) = A_r r^n + B_r r^{-(n+1)}. \tag{A.146}$$

Noting that positive and negative powers of r are unbounded for large and small r, respectively, we take $n = B_r = 0$. Hence the solution depends only on θ:

$$V(r, \theta, \phi) = V(\theta) = A_\theta P_0^0(\cos\theta) + B_\theta Q_0^0(\cos\theta).$$

We must retain Q_0^0 since the solution region does not contain the z-axis. Using

$$P_0^0(\cos\theta) = 1 \quad \text{and} \quad Q_0^0(\cos\theta) = \ln\cot(\theta/2)$$

(cf., Appendix E.2), we have

$$V(\theta) = A_\theta + B_\theta \ln\cot(\theta/2).$$

A straightforward application of the boundary conditions gives $A_\theta = 0$ and $B_\theta = V_0/\ln\cot(\theta_0/2)$, hence

$$V(\theta) = V_0 \frac{\ln\cot(\theta/2)}{\ln\cot(\theta_0/2)}.$$

Next we solve the boundary value problem

$$\nabla^2 V(r, \theta, \phi) = 0,$$
$$V(a, \theta, \phi) = -V_0, \quad \pi/2 \leq \theta < \pi, \ -\pi \leq \phi \leq \pi,$$
$$V(a, \theta, \phi) = +V_0, \quad 0 < \theta \leq \pi/2, \ -\pi \leq \phi \leq \pi,$$

for both $r > a$ and $r < a$. This yields the potential field of a conducting sphere split into top and bottom hemispheres and held at a potential difference of $2V_0$. Azimuthal symmetry gives $\mu = 0$. The two possible solutions for $\Theta(\theta)$ are

$$\Theta(\theta) = \begin{cases} A_\theta + B_\theta \ln \cot(\theta/2), & n = 0, \\ A_\theta P_n(\cos \theta), & n \neq 0, \end{cases}$$

where we have discarded $Q_0^0(\cos \theta)$ because the region of interest contains the z-axis. The $n = 0$ solution cannot match the boundary conditions; neither can a single term of the type $A_\theta P_n(\cos \theta)$, but a series of these latter terms can. We use

$$V(r, \theta) = \sum_{n=0}^{\infty} V_n(r, \theta) = \sum_{n=0}^{\infty} [A_r r^n + B_r r^{-(n+1)}] P_n(\cos \theta). \quad (A.147)$$

The terms $r^{-(n+1)}$ and r^n are not allowed, respectively, for $r < a$ and $r > a$. For $r < a$ then,

$$V(r, \theta) = \sum_{n=0}^{\infty} A_n r^n P_n(\cos \theta).$$

Letting $V_0(\theta)$ be the potential on the surface of the split sphere, we impose the boundary condition:

$$V(a, \theta) = V_0(\theta) = \sum_{n=0}^{\infty} A_n a^n P_n(\cos \theta), \quad 0 \leq \theta \leq \pi.$$

This is a Fourier–Legendre expansion of $V_0(\theta)$. The A_n are evaluated by orthogonality. Multiplying by $P_m(\cos \theta) \sin \theta$ and integrating from $\theta = 0$ to π, we obtain

$$\sum_{n=0}^{\infty} A_n a^n \int_0^\pi P_n(\cos \theta) P_m(\cos \theta) \sin \theta \, d\theta = \int_0^\pi V_0(\theta) P_m(\cos \theta) \sin \theta \, d\theta.$$

Using orthogonality relationship (A.93) and the given $V_0(\theta)$ we have

$$A_m a^m \frac{2}{2m+1} = V_0 \int_0^{\pi/2} P_m(\cos \theta) \sin \theta \, d\theta - V_0 \int_{\pi/2}^\pi P_m(\cos \theta) \sin \theta \, d\theta.$$

The substitution $\eta = \cos \theta$ gives

$$A_m a^m \frac{2}{2m+1} = V_0 \int_0^1 P_m(\eta) \, d\eta - V_0 \int_{-1}^0 P_m(\eta) \, d\eta$$
$$= V_0 \int_0^1 P_m(\eta) \, d\eta - V_0 \int_0^1 P_m(-\eta) \, d\eta;$$

then $P_m(-\eta) = (-1)^m P_m(\eta)$ gives

$$A_m = a^{-m} \frac{2m+1}{2} V_0 [1 - (-1)^m] \int_0^1 P_m(\eta) \, d\eta.$$

Because $A_m = 0$ for m even, we can put $m = 2n + 1$ ($n = 0, 1, 2, \ldots$) and have

$$A_{2n+1} = \frac{(4n+3)V_0}{a^{2n+1}} \int_0^1 P_{2n+1}(\eta)\, d\eta = \frac{V_0(-1)^n}{a^{2n+1}} \frac{4n+3}{2n+2} \frac{(2n!)}{(2^n n!)^2}$$

by (E.176). Hence

$$V(r, \theta) = \sum_{n=0}^{\infty} V_0(-1)^n \frac{4n+3}{2n+2} \frac{(2n!)}{(2^n n!)^2} \left(\frac{r}{a}\right)^{2n+1} P_{2n+1}(\cos\theta)$$

for $r < a$. The case $r > a$ is left to the reader.

Finally, consider

$$\nabla^2 \psi(x, y, z) + k^2 \psi(x, y, z) = 0, \quad 0 \le r \le a, \, 0 \le \theta \le \pi, \, -\pi \le \phi \le \pi,$$
$$\psi(a, \theta, \phi) = 0, \quad 0 \le \theta \le \pi, \, -\pi \le \phi \le \pi,$$

where $k \ne 0$ is constant. This is a three-dimensional eigenvalue problem. Wave function ψ represents the solutions for the electromagnetic field within a spherical cavity for modes TE to r. Despite the prevailing symmetry, we choose solutions that vary with both θ and ϕ. We are motivated by a desire to solve problems involving cavity excitation, and eigenmode completeness will enable us to represent any piecewise continuous function within the cavity. We employ spherical harmonics because the boundary surface is a sphere. These exclude $Q_m^n(\cos\theta)$, which is appropriate since our problem contains the z-axis. Since $k \ne 0$ we must choose a radial dependence from (A.145). Small-argument behavior rules out n_n, $h_n^{(1)}$, and $h_n^{(2)}$, leaving us with

$$\psi(r, \theta, \phi) = A_{mn} j_n(kr) Y_{nm}(\theta, \phi)$$

or, equivalently,

$$\psi(r, \theta, \phi) = A_{mn} j_n(kr) P_n^m(\cos\theta) e^{jm\phi}.$$

The eigenvalues $\lambda = k^2$ are found by applying the condition at $r = a$:

$$\psi(a, \theta, \phi) = A_{mn} j_n(ka) Y_{nm}(\theta, \phi) = 0,$$

requiring $j_n(ka) = 0$. Denoting the qth root of $j_n(x) = 0$ by α_{nq}, we have $k_{nq} = \alpha_{nq}/a$ and corresponding eigenfunctions

$$\psi_{mnq}(r, \theta, \phi) = A_{mnq} j_n(k_{nq} r) Y_{nm}(\theta, \phi).$$

The eigenvalues are proportional to the resonant frequencies of the cavity and the eigenfunctions can be used to find the modal field distributions. Since the eigenvalues are independent of m, we may have several eigenfunctions ψ_{mnq} associated with each k_{mnq}. The only limitation is that we must keep $m \le n$ to have $P_m^n(\cos\theta)$ nonzero. This is another instance of mode degeneracy. There are $2n$ degenerate modes associated with each resonant frequency (one for each of $e^{\pm jn\phi}$). By completeness we can expand any piecewise continuous function within or on the sphere as a series

$$f(r, \theta, \phi) = \sum_{m,n,q} A_{mnq} j_n(k_{nq} r) Y_{nm}(\theta, \phi).$$

Appendix B

Useful identities

Algebraic identities for vectors and dyadics

$$\mathbf{A} + \mathbf{B} = \mathbf{B} + \mathbf{A} \tag{B.1}$$

$$\mathbf{A} \cdot \mathbf{B} = \mathbf{B} \cdot \mathbf{A} \tag{B.2}$$

$$\mathbf{A} \times \mathbf{B} = -\mathbf{B} \times \mathbf{A} \tag{B.3}$$

$$\mathbf{A} \cdot (\mathbf{B} + \mathbf{C}) = \mathbf{A} \cdot \mathbf{B} + \mathbf{A} \cdot \mathbf{C} \tag{B.4}$$

$$\mathbf{A} \times (\mathbf{B} + \mathbf{C}) = \mathbf{A} \times \mathbf{B} + \mathbf{A} \times \mathbf{C} \tag{B.5}$$

$$\mathbf{A} \cdot (\mathbf{B} \times \mathbf{C}) = \mathbf{B} \cdot (\mathbf{C} \times \mathbf{A}) = \mathbf{C} \cdot (\mathbf{A} \times \mathbf{B}) \tag{B.6}$$

$$\mathbf{A} \times (\mathbf{B} \times \mathbf{C}) = \mathbf{B}(\mathbf{A} \cdot \mathbf{C}) - \mathbf{C}(\mathbf{A} \cdot \mathbf{B}) = \mathbf{B} \times (\mathbf{A} \times \mathbf{C}) + \mathbf{C} \times (\mathbf{B} \times \mathbf{A}) \tag{B.7}$$

$$(\mathbf{A} \times \mathbf{B}) \cdot (\mathbf{C} \times \mathbf{D}) = \mathbf{A} \cdot [\mathbf{B} \times (\mathbf{C} \times \mathbf{D})] = (\mathbf{B} \cdot \mathbf{D})(\mathbf{A} \cdot \mathbf{C}) - (\mathbf{B} \cdot \mathbf{C})(\mathbf{A} \cdot \mathbf{D}) \tag{B.8}$$

$$(\mathbf{A} \times \mathbf{B}) \times (\mathbf{C} \times \mathbf{D}) = \mathbf{C}[\mathbf{A} \cdot (\mathbf{B} \times \mathbf{D})] - \mathbf{D}[\mathbf{A} \cdot (\mathbf{B} \times \mathbf{C})] \tag{B.9}$$

$$\mathbf{A} \times [\mathbf{B} \times (\mathbf{C} \times \mathbf{D})] = (\mathbf{B} \cdot \mathbf{D})(\mathbf{A} \times \mathbf{C}) - (\mathbf{B} \cdot \mathbf{C})(\mathbf{A} \times \mathbf{D}) \tag{B.10}$$

$$\mathbf{A} \cdot (\bar{\mathbf{c}} \cdot \mathbf{B}) = (\mathbf{A} \cdot \bar{\mathbf{c}}) \cdot \mathbf{B} \tag{B.11}$$

$$\mathbf{A} \times (\bar{\mathbf{c}} \times \mathbf{B}) = (\mathbf{A} \times \bar{\mathbf{c}}) \times \mathbf{B} \tag{B.12}$$

$$\mathbf{C} \cdot (\bar{\mathbf{a}} \cdot \bar{\mathbf{b}}) = (\mathbf{C} \cdot \bar{\mathbf{a}}) \cdot \bar{\mathbf{b}} \tag{B.13}$$

$$(\bar{\mathbf{a}} \cdot \bar{\mathbf{b}}) \cdot \mathbf{C} = \bar{\mathbf{a}} \cdot (\bar{\mathbf{b}} \cdot \mathbf{C}) \tag{B.14}$$

$$\mathbf{A} \cdot (\mathbf{B} \times \bar{\mathbf{c}}) = -\mathbf{B} \cdot (\mathbf{A} \times \bar{\mathbf{c}}) = (\mathbf{A} \times \mathbf{B}) \cdot \bar{\mathbf{c}} \tag{B.15}$$

$$\mathbf{A} \times (\mathbf{B} \times \bar{\mathbf{c}}) = \mathbf{B} \cdot (\mathbf{A} \times \bar{\mathbf{c}}) - \bar{\mathbf{c}}(\mathbf{A} \cdot \mathbf{B}) \tag{B.16}$$

$$\mathbf{A} \cdot \bar{\mathbf{I}} = \bar{\mathbf{I}} \cdot \mathbf{A} = \mathbf{A} \tag{B.17}$$

Integral theorems

Note: S bounds V, Γ bounds S, $\hat{\mathbf{n}}$ is normal to S at $\mathbf{r}$, $\hat{\mathbf{l}}$ and $\hat{\mathbf{m}}$ are tangential to S at $\mathbf{r}$, $\hat{\mathbf{l}}$ is tangential to the contour Γ, $\hat{\mathbf{m}} \times \hat{\mathbf{l}} = \hat{\mathbf{n}}$, $\mathbf{dl} = \hat{\mathbf{l}} \, dl$, and $\mathbf{dS} = \hat{\mathbf{n}} \, dS$.

Divergence theorem

$$\int_V \nabla \cdot \mathbf{A} \, dV = \oint_S \mathbf{A} \cdot \mathbf{dS} \tag{B.18}$$

$$\int_V \nabla \cdot \bar{\mathbf{a}}\, dV = \oint_S \hat{\mathbf{n}} \cdot \bar{\mathbf{a}}\, dS \qquad (\text{B.19})$$

$$\int_S \nabla_s \cdot \mathbf{A}\, dS = \oint_\Gamma \hat{\mathbf{m}} \cdot \mathbf{A}\, dl \qquad (\text{B.20})$$

Gradient theorem

$$\int_V \nabla a\, dV = \oint_S a\, \mathbf{dS} \qquad (\text{B.21})$$

$$\int_V \nabla \mathbf{A}\, dV = \oint_S \hat{\mathbf{n}}\mathbf{A}\, dS \qquad (\text{B.22})$$

$$\int_V \nabla_s a\, dS = \oint_\Gamma \hat{\mathbf{m}}a\, dl \qquad (\text{B.23})$$

Curl theorem

$$\int_V (\nabla \times \mathbf{A})\, dV = -\oint_S \mathbf{A} \times \mathbf{dS} \qquad (\text{B.24})$$

$$\int_V (\nabla \times \bar{\mathbf{a}})\, dV = \oint_S \hat{\mathbf{n}} \times \bar{\mathbf{a}}\, dS \qquad (\text{B.25})$$

$$\int_S \nabla_s \times \mathbf{A}\, dS = \oint_\Gamma \hat{\mathbf{m}} \times \mathbf{A}\, dl \qquad (\text{B.26})$$

Stokes's theorem

$$\int_S (\nabla \times \mathbf{A}) \cdot \mathbf{dS} = \oint_\Gamma \mathbf{A} \cdot \mathbf{dl} \qquad (\text{B.27})$$

$$\int_S \hat{\mathbf{n}} \cdot (\nabla \times \bar{\mathbf{a}})\, dS = \oint_\Gamma \mathbf{dl} \cdot \bar{\mathbf{a}} \qquad (\text{B.28})$$

Green's first identity for scalar fields

$$\int_V (\nabla a \cdot \nabla b + a\nabla^2 b)\, dV = \oint_S a\frac{\partial b}{\partial n}\, dS \qquad (\text{B.29})$$

Green's second identity for scalar fields (Green's theorem)

$$\int_V (a\nabla^2 b - b\nabla^2 a)\, dV = \oint_S \left(a\frac{\partial b}{\partial n} - b\frac{\partial a}{\partial n} \right)\, dS \qquad (\text{B.30})$$

Green's first identity for vector fields

$$\int_V \{(\nabla \times \mathbf{A}) \cdot (\nabla \times \mathbf{B}) - \mathbf{A} \cdot [\nabla \times (\nabla \times \mathbf{B})]\}\, dV =$$

$$\int_V \nabla \cdot [\mathbf{A} \times (\nabla \times \mathbf{B})]\, dV = \oint_S [\mathbf{A} \times (\nabla \times \mathbf{B})] \cdot \mathbf{dS} \qquad (\text{B.31})$$

Green's second identity for vector fields

$$\int_V \{\mathbf{B} \cdot [\nabla \times (\nabla \times \mathbf{A})] - \mathbf{A} \cdot [\nabla \times (\nabla \times \mathbf{B})]\}\, dV =$$

$$\oint_S [\mathbf{A} \times (\nabla \times \mathbf{B}) - \mathbf{B} \times (\nabla \times \mathbf{A})] \cdot \mathbf{dS} \qquad (\text{B.32})$$

Helmholtz theorem

$$\mathbf{A}(\mathbf{r}) = -\nabla \left[\int_V \frac{\nabla' \cdot \mathbf{A}(\mathbf{r}')}{4\pi|\mathbf{r} - \mathbf{r}'|} \, dV' - \oint_S \frac{\mathbf{A}(\mathbf{r}') \cdot \hat{\mathbf{n}}'}{4\pi|\mathbf{r} - \mathbf{r}'|} \, dS' \right] +$$

$$+ \nabla \times \left[\int_V \frac{\nabla' \times \mathbf{A}(\mathbf{r}')}{4\pi|\mathbf{r} - \mathbf{r}'|} \, dV' + \oint_S \frac{\mathbf{A}(\mathbf{r}') \times \hat{\mathbf{n}}'}{4\pi|\mathbf{r} - \mathbf{r}'|} \, dS' \right] \qquad \text{(B.33)}$$

Miscellaneous identities

$$\oint_S d\mathbf{S} = 0 \qquad \text{(B.34)}$$

$$\int_S \hat{\mathbf{n}} \times (\nabla a) \, dS = \oint_\Gamma a \, d\mathbf{l} \qquad \text{(B.35)}$$

$$\int_S (\nabla a \times \nabla b) \cdot d\mathbf{S} = \int_\Gamma a \nabla b \cdot d\mathbf{l} = -\int_\Gamma b \nabla a \cdot d\mathbf{l} \qquad \text{(B.36)}$$

$$\oint d\mathbf{l} \, \mathbf{A} = \int_S \hat{\mathbf{n}} \times (\nabla \mathbf{A}) \, dS \qquad \text{(B.37)}$$

Derivative identities

$$\nabla(a + b) = \nabla a + \nabla b \qquad \text{(B.38)}$$
$$\nabla \cdot (\mathbf{A} + \mathbf{B}) = \nabla \cdot \mathbf{A} + \nabla \cdot \mathbf{B} \qquad \text{(B.39)}$$
$$\nabla \times (\mathbf{A} + \mathbf{B}) = \nabla \times \mathbf{A} + \nabla \times \mathbf{B} \qquad \text{(B.40)}$$
$$\nabla(ab) = a\nabla b + b\nabla a \qquad \text{(B.41)}$$
$$\nabla \cdot (a\mathbf{B}) = a\nabla \cdot \mathbf{B} + \mathbf{B} \cdot \nabla a \qquad \text{(B.42)}$$
$$\nabla \times (a\mathbf{B}) = a\nabla \times \mathbf{B} - \mathbf{B} \times \nabla a \qquad \text{(B.43)}$$
$$\nabla \cdot (\mathbf{A} \times \mathbf{B}) = \mathbf{B} \cdot \nabla \times \mathbf{A} - \mathbf{A} \cdot \nabla \times \mathbf{B} \qquad \text{(B.44)}$$
$$\nabla \times (\mathbf{A} \times \mathbf{B}) = \mathbf{A}(\nabla \cdot \mathbf{B}) - \mathbf{B}(\nabla \cdot \mathbf{A}) + (\mathbf{B} \cdot \nabla)\mathbf{A} - (\mathbf{A} \cdot \nabla)\mathbf{B} \qquad \text{(B.45)}$$
$$\nabla(\mathbf{A} \cdot \mathbf{B}) = \mathbf{A} \times (\nabla \times \mathbf{B}) + \mathbf{B} \times (\nabla \times \mathbf{A}) + (\mathbf{A} \cdot \nabla)\mathbf{B} + (\mathbf{B} \cdot \nabla)\mathbf{A} \qquad \text{(B.46)}$$
$$\nabla \times (\nabla \times \mathbf{A}) = \nabla(\nabla \cdot \mathbf{A}) - \nabla^2 \mathbf{A} \qquad \text{(B.47)}$$
$$\nabla \cdot (\nabla a) = \nabla^2 a \qquad \text{(B.48)}$$
$$\nabla \cdot (\nabla \times \mathbf{A}) = 0 \qquad \text{(B.49)}$$
$$\nabla \times (\nabla a) = 0 \qquad \text{(B.50)}$$
$$\nabla \times (a\nabla b) = \nabla a \times \nabla b \qquad \text{(B.51)}$$
$$\nabla^2(ab) = a\nabla^2 b + 2(\nabla a) \cdot (\nabla b) + b\nabla^2 a \qquad \text{(B.52)}$$
$$\nabla^2(a\mathbf{B}) = a\nabla^2 \mathbf{B} + \mathbf{B}\nabla^2 a + 2(\nabla a \cdot \nabla)\mathbf{B} \qquad \text{(B.53)}$$
$$\nabla^2 \bar{\mathbf{a}} = \nabla(\nabla \cdot \bar{\mathbf{a}}) - \nabla \times (\nabla \times \bar{\mathbf{a}}) \qquad \text{(B.54)}$$
$$\nabla \cdot (\mathbf{AB}) = (\nabla \cdot \mathbf{A})\mathbf{B} + \mathbf{A} \cdot (\nabla \mathbf{B}) = (\nabla \cdot \mathbf{A})\mathbf{B} + (\mathbf{A} \cdot \nabla)\mathbf{B} \qquad \text{(B.55)}$$
$$\nabla \times (\mathbf{AB}) = (\nabla \times \mathbf{A})\mathbf{B} - \mathbf{A} \times (\nabla \mathbf{B}) \qquad \text{(B.56)}$$
$$\nabla \cdot (\nabla \times \bar{\mathbf{a}}) = 0 \qquad \text{(B.57)}$$

$$\nabla \times (\nabla \mathbf{A}) = 0 \tag{B.58}$$

$$\nabla(\mathbf{A} \times \mathbf{B}) = (\nabla \mathbf{A}) \times \mathbf{B} - (\nabla \mathbf{B}) \times \mathbf{A} \tag{B.59}$$

$$\nabla(a\mathbf{B}) = (\nabla a)\mathbf{B} + a(\nabla \mathbf{B}) \tag{B.60}$$

$$\nabla \cdot (a\bar{\mathbf{b}}) = (\nabla a) \cdot \bar{\mathbf{b}} + a(\nabla \cdot \bar{\mathbf{b}}) \tag{B.61}$$

$$\nabla \times (a\bar{\mathbf{b}}) = (\nabla a) \times \bar{\mathbf{b}} + a(\nabla \times \bar{\mathbf{b}}) \tag{B.62}$$

$$\nabla \cdot (a\bar{\mathbf{I}}) = \nabla a \tag{B.63}$$

$$\nabla \times (a\bar{\mathbf{I}}) = \nabla a \times \bar{\mathbf{I}} \tag{B.64}$$

Identities involving the displacement vector

Note: $\mathbf{R} = \mathbf{r} - \mathbf{r}'$, $R = |\mathbf{R}|$, $\hat{\mathbf{R}} = \mathbf{R}/R$, $f'(x) = df(x)/dx$.

$$\nabla f(R) = -\nabla' f(R) = \hat{\mathbf{R}} f'(R) \tag{B.65}$$

$$\nabla R = \hat{\mathbf{R}} \tag{B.66}$$

$$\nabla \left(\frac{1}{R}\right) = -\frac{\hat{\mathbf{R}}}{R^2} \tag{B.67}$$

$$\nabla \left(\frac{e^{-jkR}}{R}\right) = -\hat{\mathbf{R}}\left(\frac{1}{R} + jk\right)\frac{e^{-jkR}}{R} \tag{B.68}$$

$$\nabla \cdot \left[f(R)\hat{\mathbf{R}}\right] = -\nabla' \cdot \left[f(R)\hat{\mathbf{R}}\right] = 2\frac{f(R)}{R} + f'(R) \tag{B.69}$$

$$\nabla \cdot \mathbf{R} = 3 \tag{B.70}$$

$$\nabla \cdot \hat{\mathbf{R}} = \frac{2}{R} \tag{B.71}$$

$$\nabla \cdot \left(\hat{\mathbf{R}}\frac{e^{-jkR}}{R}\right) = \left(\frac{1}{R} - jk\right)\frac{e^{-jkR}}{R} \tag{B.72}$$

$$\nabla \times \left[f(R)\hat{\mathbf{R}}\right] = 0 \tag{B.73}$$

$$\nabla^2 \left(\frac{1}{R}\right) = -4\pi\delta(\mathbf{R}) \tag{B.74}$$

$$(\nabla^2 + k^2)\frac{e^{-jkR}}{R} = -4\pi\delta(\mathbf{R}) \tag{B.75}$$

Identities involving the plane-wave function

Note: $\mathbf{E}$ is a constant vector, $k = |\mathbf{k}|$.

$$\nabla \left(e^{-j\mathbf{k}\cdot\mathbf{r}}\right) = -j\mathbf{k}e^{-j\mathbf{k}\cdot\mathbf{r}} \tag{B.76}$$

$$\nabla \cdot \left(\mathbf{E}e^{-j\mathbf{k}\cdot\mathbf{r}}\right) = -j\mathbf{k} \cdot \mathbf{E}e^{-j\mathbf{k}\cdot\mathbf{r}} \tag{B.77}$$

$$\nabla \times \left(\mathbf{E}e^{-j\mathbf{k}\cdot\mathbf{r}}\right) = -j\mathbf{k} \times \mathbf{E}e^{-j\mathbf{k}\cdot\mathbf{r}} \tag{B.78}$$

$$\nabla^2 \left(\mathbf{E}e^{-j\mathbf{k}\cdot\mathbf{r}}\right) = -k^2\mathbf{E}e^{-j\mathbf{k}\cdot\mathbf{r}} \tag{B.79}$$

Identities involving the transverse/longitudinal decomposition

Note: $\hat{\mathbf{u}}$ is a constant unit vector, $A_u \equiv \hat{\mathbf{u}} \cdot \mathbf{A}$, $\partial/\partial u \equiv \hat{\mathbf{u}} \cdot \nabla$, $\mathbf{A}_t \equiv \mathbf{A} - \hat{\mathbf{u}}A_u$, $\nabla_t \equiv \nabla - \hat{\mathbf{u}}\partial/\partial u$.

$$\mathbf{A} = \mathbf{A}_t + \hat{\mathbf{u}}A_u \tag{B.80}$$

$$\nabla = \nabla_t + \hat{\mathbf{u}}\frac{\partial}{\partial u} \tag{B.81}$$

$$\hat{\mathbf{u}} \cdot \mathbf{A}_t = 0 \tag{B.82}$$

$$(\hat{\mathbf{u}} \cdot \nabla_t)\,\phi = 0 \tag{B.83}$$

$$\nabla_t\phi = \nabla\phi - \hat{\mathbf{u}}\frac{\partial\phi}{\partial u} \tag{B.84}$$

$$\hat{\mathbf{u}} \cdot (\nabla\phi) = (\hat{\mathbf{u}} \cdot \nabla)\phi = \frac{\partial\phi}{\partial u} \tag{B.85}$$

$$\hat{\mathbf{u}} \cdot (\nabla_t\phi) = 0 \tag{B.86}$$

$$\nabla_t \cdot (\hat{\mathbf{u}}\phi) = 0 \tag{B.87}$$

$$\nabla_t \times (\hat{\mathbf{u}}\phi) = -\hat{\mathbf{u}} \times \nabla_t\phi \tag{B.88}$$

$$\nabla_t \times (\hat{\mathbf{u}} \times \mathbf{A}) = \hat{\mathbf{u}}\nabla_t \cdot \mathbf{A}_t \tag{B.89}$$

$$\hat{\mathbf{u}} \times (\nabla_t \times \mathbf{A}) = \nabla_t A_u \tag{B.90}$$

$$\hat{\mathbf{u}} \times (\nabla_t \times \mathbf{A}_t) = 0 \tag{B.91}$$

$$\hat{\mathbf{u}} \cdot (\hat{\mathbf{u}} \times \mathbf{A}) = 0 \tag{B.92}$$

$$\hat{\mathbf{u}} \times (\hat{\mathbf{u}} \times \mathbf{A}) = -\mathbf{A}_t \tag{B.93}$$

$$\nabla\phi = \nabla_t\phi + \hat{\mathbf{u}}\frac{\partial\phi}{\partial u} \tag{B.94}$$

$$\nabla \cdot \mathbf{A} = \nabla_t \cdot \mathbf{A}_t + \frac{\partial A_u}{\partial u} \tag{B.95}$$

$$\nabla \times \mathbf{A} = \nabla_t \times \mathbf{A}_t + \hat{\mathbf{u}} \times \left[\frac{\partial \mathbf{A}_t}{\partial u} - \nabla_t A_u\right] \tag{B.96}$$

$$\nabla^2\phi = \nabla_t^2\phi + \frac{\partial^2\phi}{\partial u^2} \tag{B.97}$$

$$\nabla \times \nabla \times \mathbf{A} = \left[\nabla_t \times \nabla_t \times \mathbf{A}_t - \frac{\partial^2 \mathbf{A}_t}{\partial u^2} + \nabla_t\frac{\partial A_u}{\partial u}\right] + \hat{\mathbf{u}}\left[\frac{\partial}{\partial u}(\nabla_t \cdot \mathbf{A}_t) - \nabla_t^2 A_u\right] \tag{B.98}$$

$$\nabla^2\mathbf{A} = \left[\nabla_t(\nabla_t \cdot \mathbf{A}_t) + \frac{\partial^2 \mathbf{A}_t}{\partial u^2} - \nabla_t \times \nabla_t \times \mathbf{A}_t\right] + \hat{\mathbf{u}}\nabla^2 A_u \tag{B.99}$$

Appendix C

Some Fourier transform pairs

Note:

$$G(k) = \int_{-\infty}^{\infty} g(x)e^{-jkx}\, dx, \quad g(x) = \frac{1}{2\pi}\int_{-\infty}^{\infty} G(k)e^{jkx}\, dk, \quad g(x) \leftrightarrow G(k).$$

$$\text{rect}(x) \leftrightarrow 2\,\text{sinc}\,k \tag{C.1}$$

$$\Lambda(x) \leftrightarrow \text{sinc}^2\frac{k}{2} \tag{C.2}$$

$$\text{sgn}(x) \leftrightarrow \frac{2}{jk} \tag{C.3}$$

$$e^{jk_0 x} \leftrightarrow 2\pi\delta(k - k_0) \tag{C.4}$$

$$\delta(x) \leftrightarrow 1 \tag{C.5}$$

$$1 \leftrightarrow 2\pi\delta(k) \tag{C.6}$$

$$\frac{d^n\delta(x)}{dx^n} \leftrightarrow (jk)^n \tag{C.7}$$

$$x^n \leftrightarrow 2\pi j^n\frac{d^n\delta(k)}{dk^n} \tag{C.8}$$

$$U(x) \leftrightarrow \pi\delta(k) + \frac{1}{jk} \tag{C.9}$$

$$\sum_{n=-\infty}^{\infty} \delta\left(t - n\frac{2\pi}{k_0}\right) \leftrightarrow k_0 \sum_{n=-\infty}^{\infty} \delta(k - nk_0) \tag{C.10}$$

$$e^{-ax^2} \leftrightarrow \sqrt{\frac{\pi}{a}}e^{-\frac{k^2}{4a}} \tag{C.11}$$

$$e^{-ax}U(x) \leftrightarrow \frac{1}{a + jk} \tag{C.12}$$

$$e^{-a|x|} \leftrightarrow \frac{2a}{a^2 + k^2} \tag{C.13}$$

$$e^{-ax} \cos bx \, U(x) \leftrightarrow \frac{a + jk}{(a + jk)^2 + b^2} \tag{C.14}$$

$$e^{-ax} \sin bx \, U(x) \leftrightarrow \frac{b}{(a + jk)^2 + b^2} \tag{C.15}$$

$$\cos k_0 x \leftrightarrow \pi[\delta(k + k_0) + \delta(k - k_0)] \tag{C.16}$$

$$\sin k_0 x \leftrightarrow j\pi[\delta(k + k_0) - \delta(k - k_0)] \tag{C.17}$$

$$\frac{1}{2}be^{-\frac{1}{2}bx}\left[I_0\left(\frac{1}{2}bx\right) + I_1\left(\frac{1}{2}bx\right)\right]U(x) \leftrightarrow \sqrt{\frac{jk + b}{jk} - 1} \tag{C.18}$$

$$g(x) - ae^{-ax}\int_{-\infty}^{x} e^{au}g(u)\,du \leftrightarrow \frac{jk}{jk + a}G(k) \tag{C.19}$$

Appendix D

Coordinate systems

Rectangular coordinate system

Coordinate variables

$$u = x, \quad -\infty < x < \infty \tag{D.1}$$

$$v = y, \quad -\infty < y < \infty \tag{D.2}$$

$$w = z, \quad -\infty < z < \infty \tag{D.3}$$

Vector algebra

$$\mathbf{A} = \hat{\mathbf{x}}A_x + \hat{\mathbf{y}}A_y + \hat{\mathbf{z}}A_z \tag{D.4}$$

$$\mathbf{A} \cdot \mathbf{B} = A_x B_x + A_y B_y + A_z B_z \tag{D.5}$$

$$\mathbf{A} \times \mathbf{B} = \begin{vmatrix} \hat{\mathbf{x}} & \hat{\mathbf{y}} & \hat{\mathbf{z}} \\ A_x & A_y & A_z \\ B_x & B_y & B_z \end{vmatrix} \tag{D.6}$$

Dyadic representation

$$\begin{aligned} \bar{\mathbf{a}} = {} & \hat{\mathbf{x}}a_{xx}\hat{\mathbf{x}} + \hat{\mathbf{x}}a_{xy}\hat{\mathbf{y}} + \hat{\mathbf{x}}a_{xz}\hat{\mathbf{z}} + \\ & + \hat{\mathbf{y}}a_{yx}\hat{\mathbf{x}} + \hat{\mathbf{y}}a_{yy}\hat{\mathbf{y}} + \hat{\mathbf{y}}a_{yz}\hat{\mathbf{z}} + \\ & + \hat{\mathbf{z}}a_{zx}\hat{\mathbf{x}} + \hat{\mathbf{z}}a_{zy}\hat{\mathbf{y}} + \hat{\mathbf{z}}a_{zz}\hat{\mathbf{z}} \end{aligned} \tag{D.7}$$

$$\bar{\mathbf{a}} = \hat{\mathbf{x}}\mathbf{a}'_x + \hat{\mathbf{y}}\mathbf{a}'_y + \hat{\mathbf{z}}\mathbf{a}'_z = \mathbf{a}_x\hat{\mathbf{x}} + \mathbf{a}_y\hat{\mathbf{y}} + \mathbf{a}_z\hat{\mathbf{z}} \tag{D.8}$$

$$\mathbf{a}'_x = a_{xx}\hat{\mathbf{x}} + a_{xy}\hat{\mathbf{y}} + a_{xz}\hat{\mathbf{z}} \tag{D.9}$$

$$\mathbf{a}'_y = a_{yx}\hat{\mathbf{x}} + a_{yy}\hat{\mathbf{y}} + a_{yz}\hat{\mathbf{z}} \tag{D.10}$$

$$\mathbf{a}'_z = a_{zx}\hat{\mathbf{x}} + a_{zy}\hat{\mathbf{y}} + a_{zz}\hat{\mathbf{z}} \tag{D.11}$$

$$\mathbf{a}_x = a_{xx}\hat{\mathbf{x}} + a_{yx}\hat{\mathbf{y}} + a_{zx}\hat{\mathbf{z}} \tag{D.12}$$

$$\mathbf{a}_y = a_{xy}\hat{\mathbf{x}} + a_{yy}\hat{\mathbf{y}} + a_{zy}\hat{\mathbf{z}} \tag{D.13}$$

$$\mathbf{a}_z = a_{xz}\hat{\mathbf{x}} + a_{yz}\hat{\mathbf{y}} + a_{zz}\hat{\mathbf{z}} \tag{D.14}$$

Differential operations

$$\mathbf{dl} = \hat{\mathbf{x}}\, dx + \hat{\mathbf{y}}\, dy + \hat{\mathbf{z}}\, dz \tag{D.15}$$

$$dV = dx\, dy\, dz \tag{D.16}$$

$$dS_x = dy\, dz \tag{D.17}$$
$$dS_y = dx\, dz \tag{D.18}$$
$$dS_z = dx\, dy \tag{D.19}$$

$$\nabla f = \hat{\mathbf{x}}\frac{\partial f}{\partial x} + \hat{\mathbf{y}}\frac{\partial f}{\partial y} + \hat{\mathbf{z}}\frac{\partial f}{\partial z} \tag{D.20}$$

$$\nabla \cdot \mathbf{F} = \frac{\partial F_x}{\partial x} + \frac{\partial F_y}{\partial y} + \frac{\partial F_z}{\partial z} \tag{D.21}$$

$$\nabla \times \mathbf{F} = \begin{vmatrix} \hat{\mathbf{x}} & \hat{\mathbf{y}} & \hat{\mathbf{z}} \\ \frac{\partial}{\partial x} & \frac{\partial}{\partial y} & \frac{\partial}{\partial z} \\ F_x & F_y & F_z \end{vmatrix} \tag{D.22}$$

$$\nabla^2 f = \frac{\partial^2 f}{\partial x^2} + \frac{\partial^2 f}{\partial y^2} + \frac{\partial^2 f}{\partial z^2} \tag{D.23}$$

$$\nabla^2 \mathbf{F} = \hat{\mathbf{x}}\nabla^2 F_x + \hat{\mathbf{y}}\nabla^2 F_y + \hat{\mathbf{z}}\nabla^2 F_z \tag{D.24}$$

Separation of the Helmholtz equation

$$\frac{\partial^2 \psi(x,y,z)}{\partial x^2} + \frac{\partial^2 \psi(x,y,z)}{\partial y^2} + \frac{\partial^2 \psi(x,y,z)}{\partial z^2} + k^2 \psi(x,y,z) = 0 \tag{D.25}$$

$$\psi(x,y,z) = X(x)Y(y)Z(z) \tag{D.26}$$

$$k_x^2 + k_y^2 + k_z^2 = k^2 \tag{D.27}$$

$$\frac{d^2 X(x)}{dx^2} + k_x^2 X(x) = 0 \tag{D.28}$$

$$\frac{d^2 Y(y)}{dy^2} + k_y^2 Y(y) = 0 \tag{D.29}$$

$$\frac{d^2 Z(z)}{dz^2} + k_z^2 Z(z) = 0 \tag{D.30}$$

$$X(x) = \begin{cases} A_x F_1(k_x x) + B_x F_2(k_x x), & k_x \neq 0, \\ a_x x + b_x, & k_x = 0. \end{cases} \tag{D.31}$$

$$Y(y) = \begin{cases} A_y F_1(k_y y) + B_y F_2(k_y y), & k_y \neq 0, \\ a_y y + b_y, & k_y = 0. \end{cases} \tag{D.32}$$

$$Z(z) = \begin{cases} A_z F_1(k_z z) + B_z F_2(k_z z), & k_z \neq 0, \\ a_z z + b_z, & k_z = 0. \end{cases} \tag{D.33}$$

$$F_1(\xi), F_2(\xi) = \begin{cases} e^{j\xi} \\ e^{-j\xi} \\ \sin(\xi) \\ \cos(\xi) \end{cases} \tag{D.34}$$

Cylindrical coordinate system

Coordinate variables

$$u = \rho, \quad 0 \leq \rho < \infty \tag{D.35}$$
$$v = \phi, \quad -\pi \leq \phi \leq \pi \tag{D.36}$$
$$w = z, \quad -\infty < z < \infty \tag{D.37}$$

$$x = \rho \cos \phi \tag{D.38}$$
$$y = \rho \sin \phi \tag{D.39}$$
$$z = z \tag{D.40}$$

$$\rho = \sqrt{x^2 + y^2} \tag{D.41}$$
$$\phi = \tan^{-1} \frac{y}{x} \tag{D.42}$$
$$z = z \tag{D.43}$$

Vector algebra

$$\hat{\boldsymbol{\rho}} = \hat{\mathbf{x}} \cos \phi + \hat{\mathbf{y}} \sin \phi \tag{D.44}$$
$$\hat{\boldsymbol{\phi}} = -\hat{\mathbf{x}} \sin \phi + \hat{\mathbf{y}} \cos \phi \tag{D.45}$$
$$\hat{\mathbf{z}} = \hat{\mathbf{z}} \tag{D.46}$$

$$\mathbf{A} = \hat{\boldsymbol{\rho}} A_\rho + \hat{\boldsymbol{\phi}} A_\phi + \hat{\mathbf{z}} A_z \tag{D.47}$$

$$\mathbf{A} \cdot \mathbf{B} = A_\rho B_\rho + A_\phi B_\phi + A_z B_z \tag{D.48}$$

$$\mathbf{A} \times \mathbf{B} = \begin{vmatrix} \hat{\boldsymbol{\rho}} & \hat{\boldsymbol{\phi}} & \hat{\mathbf{z}} \\ A_\rho & A_\phi & A_z \\ B_\rho & B_\phi & B_z \end{vmatrix} \tag{D.49}$$

Dyadic representation

$$\bar{\mathbf{a}} = \hat{\boldsymbol{\rho}}a_{\rho\rho}\hat{\boldsymbol{\rho}} + \hat{\boldsymbol{\rho}}a_{\rho\phi}\hat{\boldsymbol{\phi}} + \hat{\boldsymbol{\rho}}a_{\rho z}\hat{\mathbf{z}} +$$
$$+ \hat{\boldsymbol{\phi}}a_{\phi\rho}\hat{\boldsymbol{\rho}} + \hat{\boldsymbol{\phi}}a_{\phi\phi}\hat{\boldsymbol{\phi}} + \hat{\boldsymbol{\phi}}a_{\phi z}\hat{\mathbf{z}} +$$
$$+ \hat{\mathbf{z}}a_{z\rho}\hat{\boldsymbol{\rho}} + \hat{\mathbf{z}}a_{z\phi}\hat{\boldsymbol{\phi}} + \hat{\mathbf{z}}a_{zz}\hat{\mathbf{z}} \tag{D.50}$$

$$\bar{\mathbf{a}} = \hat{\boldsymbol{\rho}}\mathbf{a}'_\rho + \hat{\boldsymbol{\phi}}\mathbf{a}'_\phi + \hat{\mathbf{z}}\mathbf{a}'_z = \mathbf{a}_\rho\hat{\boldsymbol{\rho}} + \mathbf{a}_\phi\hat{\boldsymbol{\phi}} + \mathbf{a}_z\hat{\mathbf{z}} \tag{D.51}$$

$$\mathbf{a}'_\rho = a_{\rho\rho}\hat{\boldsymbol{\rho}} + a_{\rho\phi}\hat{\boldsymbol{\phi}} + a_{\rho z}\hat{\mathbf{z}} \tag{D.52}$$
$$\mathbf{a}'_\phi = a_{\phi\rho}\hat{\boldsymbol{\rho}} + a_{\phi\phi}\hat{\boldsymbol{\phi}} + a_{\phi z}\hat{\mathbf{z}} \tag{D.53}$$
$$\mathbf{a}'_z = a_{z\rho}\hat{\boldsymbol{\rho}} + a_{z\phi}\hat{\boldsymbol{\phi}} + a_{zz}\hat{\mathbf{z}} \tag{D.54}$$

$$\mathbf{a}_\rho = a_{\rho\rho}\hat{\boldsymbol{\rho}} + a_{\phi\rho}\hat{\boldsymbol{\phi}} + a_{z\rho}\hat{\mathbf{z}} \tag{D.55}$$
$$\mathbf{a}_\phi = a_{\rho\phi}\hat{\boldsymbol{\rho}} + a_{\phi\phi}\hat{\boldsymbol{\phi}} + a_{z\phi}\hat{\mathbf{z}} \tag{D.56}$$
$$\mathbf{a}_z = a_{\rho z}\hat{\boldsymbol{\rho}} + a_{\phi z}\hat{\boldsymbol{\phi}} + a_{zz}\hat{\mathbf{z}} \tag{D.57}$$

Differential operations

$$\mathbf{dl} = \hat{\boldsymbol{\rho}}\,d\rho + \hat{\boldsymbol{\phi}}\rho\,d\phi + \hat{\mathbf{z}}\,dz \tag{D.58}$$

$$dV = \rho\,d\rho\,d\phi\,dz \tag{D.59}$$

$$dS_\rho = \rho\,d\phi\,dz, \tag{D.60}$$
$$dS_\phi = d\rho\,dz, \tag{D.61}$$
$$dS_z = \rho\,d\rho\,d\phi \tag{D.62}$$

$$\nabla f = \hat{\boldsymbol{\rho}}\frac{\partial f}{\partial\rho} + \hat{\boldsymbol{\phi}}\frac{1}{\rho}\frac{\partial f}{\partial\phi} + \hat{\mathbf{z}}\frac{\partial f}{\partial z} \tag{D.63}$$

$$\nabla\cdot\mathbf{F} = \frac{1}{\rho}\frac{\partial}{\partial\rho}(\rho F_\rho) + \frac{1}{\rho}\frac{\partial F_\phi}{\partial\phi} + \frac{\partial F_z}{\partial z} \tag{D.64}$$

$$\nabla\times\mathbf{F} = \frac{1}{\rho}\begin{vmatrix} \hat{\boldsymbol{\rho}} & \rho\hat{\boldsymbol{\phi}} & \hat{\mathbf{z}} \\ \frac{\partial}{\partial\rho} & \frac{\partial}{\partial\phi} & \frac{\partial}{\partial z} \\ F_\rho & \rho F_\phi & F_z \end{vmatrix} \tag{D.65}$$

$$\nabla^2 f = \frac{1}{\rho}\frac{\partial}{\partial\rho}\left(\rho\frac{\partial f}{\partial\rho}\right) + \frac{1}{\rho^2}\frac{\partial^2 f}{\partial\phi^2} + \frac{\partial^2 f}{\partial z^2} \tag{D.66}$$

$$\nabla^2\mathbf{F} = \hat{\boldsymbol{\rho}}\left(\nabla^2 F_\rho - \frac{2}{\rho^2}\frac{\partial F_\phi}{\partial\phi} - \frac{F_\rho}{\rho^2}\right) + \hat{\boldsymbol{\phi}}\left(\nabla^2 F_\phi + \frac{2}{\rho^2}\frac{\partial F_\rho}{\partial\phi} - \frac{F_\phi}{\rho^2}\right) + \hat{\mathbf{z}}\nabla^2 F_z \tag{D.67}$$

Separation of the Helmholtz equation

$$\frac{1}{\rho}\frac{\partial}{\partial\rho}\left(\rho\frac{\partial\psi(\rho,\phi,z)}{\partial\rho}\right) + \frac{1}{\rho^2}\frac{\partial^2\psi(\rho,\phi,z)}{\partial\phi^2} + \frac{\partial^2\psi(\rho,\phi,z)}{\partial z^2} + k^2\psi(\rho,\phi,z) = 0 \qquad \text{(D.68)}$$

$$\psi(\rho,\phi,z) = P(\rho)\Phi(\phi)Z(z) \qquad \text{(D.69)}$$

$$k_c^2 = k^2 - k_z^2 \qquad \text{(D.70)}$$

$$\frac{d^2 P(\rho)}{d\rho^2} + \frac{1}{\rho}\frac{dP(\rho)}{d\rho} + \left(k_c^2 - \frac{k_\phi^2}{\rho^2}\right)P(\rho) = 0 \qquad \text{(D.71)}$$

$$\frac{\partial^2\Phi(\phi)}{\partial\phi^2} + k_\phi^2\Phi(\phi) = 0 \qquad \text{(D.72)}$$

$$\frac{d^2 Z(z)}{dz^2} + k_z^2 Z(z) = 0 \qquad \text{(D.73)}$$

$$Z(z) = \begin{cases} A_z F_1(k_z z) + B_z F_2(k_z z), & k_z \neq 0, \\ a_z z + b_z, & k_z = 0. \end{cases} \qquad \text{(D.74)}$$

$$\Phi(\phi) = \begin{cases} A_\phi F_1(k_\phi\phi) + B_\phi F_2(k_\phi\phi), & k_\phi \neq 0, \\ a_\phi\phi + b_\phi, & k_\phi = 0. \end{cases} \qquad \text{(D.75)}$$

$$P(\rho) = \begin{cases} a_\rho\ln\rho + b_\rho, & k_c = k_\phi = 0, \\ a_\rho\rho^{-k_\phi} + b_\rho\rho^{k_\phi}, & k_c = 0 \text{ and } k_\phi \neq 0, \\ A_\rho G_1(k_c\rho) + B_\rho G_2(k_c\rho), & \text{otherwise.} \end{cases} \qquad \text{(D.76)}$$

$$F_1(\xi), F_2(\xi) = \begin{cases} e^{j\xi} \\ e^{-j\xi} \\ \sin(\xi) \\ \cos(\xi) \end{cases} \qquad \text{(D.77)}$$

$$G_1(\xi), G_2(\xi) = \begin{cases} J_{k_\phi}(\xi) \\ N_{k_\phi}(\xi) \\ H_{k_\phi}^{(1)}(\xi) \\ H_{k_\phi}^{(2)}(\xi) \end{cases} \qquad \text{(D.78)}$$

Spherical coordinate system

Coordinate variables

$$u = r, \quad 0 \leq r < \infty \qquad \text{(D.79)}$$
$$v = \theta, \quad 0 \leq \theta \leq \pi \qquad \text{(D.80)}$$
$$w = \phi, \quad -\pi \leq \phi \leq \pi \qquad \text{(D.81)}$$

$$x = r \sin\theta \cos\phi \tag{D.82}$$
$$y = r \sin\theta \sin\phi \tag{D.83}$$
$$z = r \cos\theta \tag{D.84}$$

$$r = \sqrt{x^2 + y^2 + z^2} \tag{D.85}$$
$$\theta = \tan^{-1} \frac{\sqrt{x^2 + y^2}}{z} \tag{D.86}$$
$$\phi = \tan^{-1} \frac{y}{x} \tag{D.87}$$

Vector algebra

$$\hat{\mathbf{r}} = \hat{\mathbf{x}} \sin\theta \cos\phi + \hat{\mathbf{y}} \sin\theta \sin\phi + \hat{\mathbf{z}} \cos\theta \tag{D.88}$$
$$\hat{\boldsymbol{\theta}} = \hat{\mathbf{x}} \cos\theta \cos\phi + \hat{\mathbf{y}} \cos\theta \sin\phi - \hat{\mathbf{z}} \sin\theta \tag{D.89}$$
$$\hat{\boldsymbol{\phi}} = -\hat{\mathbf{x}} \sin\phi + \hat{\mathbf{y}} \cos\phi \tag{D.90}$$

$$\mathbf{A} = \hat{\mathbf{r}} A_r + \hat{\boldsymbol{\theta}} A_\theta + \hat{\boldsymbol{\phi}} A_\phi \tag{D.91}$$

$$\mathbf{A} \cdot \mathbf{B} = A_r B_r + A_\theta B_\theta + A_\phi B_\phi \tag{D.92}$$

$$\mathbf{A} \times \mathbf{B} = \begin{vmatrix} \hat{\mathbf{r}} & \hat{\boldsymbol{\theta}} & \hat{\boldsymbol{\phi}} \\ A_r & A_\theta & A_\phi \\ B_r & B_\theta & B_\phi \end{vmatrix} \tag{D.93}$$

Dyadic representation

$$\begin{aligned} \bar{\mathbf{a}} = {}& \hat{\mathbf{r}} a_{rr} \hat{\mathbf{r}} + \hat{\mathbf{r}} a_{r\theta} \hat{\boldsymbol{\theta}} + \hat{\mathbf{r}} a_{r\phi} \hat{\boldsymbol{\phi}} + \\ &+ \hat{\boldsymbol{\theta}} a_{\theta r} \hat{\mathbf{r}} + \hat{\boldsymbol{\theta}} a_{\theta\theta} \hat{\boldsymbol{\theta}} + \hat{\boldsymbol{\theta}} a_{\theta\phi} \hat{\boldsymbol{\phi}} + \\ &+ \hat{\boldsymbol{\phi}} a_{\phi r} \hat{\mathbf{r}} + \hat{\boldsymbol{\phi}} a_{\phi\theta} \hat{\boldsymbol{\theta}} + \hat{\boldsymbol{\phi}} a_{\phi\phi} \hat{\boldsymbol{\phi}} \end{aligned} \tag{D.94}$$

$$\bar{\mathbf{a}} = \hat{\mathbf{r}} \mathbf{a}'_r + \hat{\boldsymbol{\theta}} \mathbf{a}'_\theta + \hat{\boldsymbol{\phi}} \mathbf{a}'_\phi = \mathbf{a}_r \hat{\mathbf{r}} + \mathbf{a}_\theta \hat{\boldsymbol{\theta}} + \mathbf{a}_\phi \hat{\boldsymbol{\phi}} \tag{D.95}$$

$$\mathbf{a}'_r = a_{rr} \hat{\mathbf{r}} + a_{r\theta} \hat{\boldsymbol{\theta}} + a_{r\phi} \hat{\boldsymbol{\phi}} \tag{D.96}$$
$$\mathbf{a}'_\theta = a_{\theta r} \hat{\mathbf{r}} + a_{\theta\theta} \hat{\boldsymbol{\theta}} + a_{\theta\phi} \hat{\boldsymbol{\phi}} \tag{D.97}$$
$$\mathbf{a}'_\phi = a_{\phi r} \hat{\mathbf{r}} + a_{\phi\theta} \hat{\boldsymbol{\theta}} + a_{\phi\phi} \hat{\boldsymbol{\phi}} \tag{D.98}$$

$$\mathbf{a}_r = a_{rr} \hat{\mathbf{r}} + a_{\theta r} \hat{\boldsymbol{\theta}} + a_{\phi r} \hat{\boldsymbol{\phi}} \tag{D.99}$$
$$\mathbf{a}_\theta = a_{r\theta} \hat{\mathbf{r}} + a_{\theta\theta} \hat{\boldsymbol{\theta}} + a_{\phi\theta} \hat{\boldsymbol{\phi}} \tag{D.100}$$
$$\mathbf{a}_\phi = a_{r\phi} \hat{\mathbf{r}} + a_{\theta\phi} \hat{\boldsymbol{\theta}} + a_{\phi\phi} \hat{\boldsymbol{\phi}} \tag{D.101}$$

Differential operations

$$\mathbf{dl} = \hat{\mathbf{r}}\,dr + \hat{\boldsymbol{\theta}}r\,d\theta + \hat{\boldsymbol{\phi}}r\sin\theta\,d\phi \tag{D.102}$$

$$dV = r^2\sin\theta\,dr\,d\theta\,d\phi \tag{D.103}$$

$$dS_r = r^2\sin\theta\,d\theta\,d\phi \tag{D.104}$$
$$dS_\theta = r\sin\theta\,dr\,d\phi \tag{D.105}$$
$$dS_\phi = r\,dr\,d\theta \tag{D.106}$$

$$\nabla f = \hat{\mathbf{r}}\frac{\partial f}{\partial r} + \hat{\boldsymbol{\theta}}\frac{1}{r}\frac{\partial f}{\partial \theta} + \hat{\boldsymbol{\phi}}\frac{1}{r\sin\theta}\frac{\partial f}{\partial \phi} \tag{D.107}$$

$$\nabla \cdot \mathbf{F} = \frac{1}{r^2}\frac{\partial}{\partial r}\left(r^2 F_r\right) + \frac{1}{r\sin\theta}\frac{\partial}{\partial \theta}\left(\sin\theta F_\theta\right) + \frac{1}{r\sin\theta}\frac{\partial F_\phi}{\partial \phi} \tag{D.108}$$

$$\nabla \times \mathbf{F} = \frac{1}{r^2\sin\theta}\begin{vmatrix} \hat{\mathbf{r}} & r\hat{\boldsymbol{\theta}} & r\sin\theta\hat{\boldsymbol{\phi}} \\ \frac{\partial}{\partial r} & \frac{\partial}{\partial \theta} & \frac{\partial}{\partial \phi} \\ F_r & rF_\theta & r\sin\theta F_\phi \end{vmatrix} \tag{D.109}$$

$$\nabla^2 f = \frac{1}{r^2}\frac{\partial}{\partial r}\left(r^2\frac{\partial f}{\partial r}\right) + \frac{1}{r^2\sin\theta}\frac{\partial}{\partial \theta}\left(\sin\theta\frac{\partial f}{\partial \theta}\right) + \frac{1}{r^2\sin^2\theta}\frac{\partial^2 f}{\partial \phi^2} \tag{D.110}$$

$$\nabla^2 \mathbf{F} = \hat{\mathbf{r}}\left[\nabla^2 F_r - \frac{2}{r^2}\left(F_r + \frac{\cos\theta}{\sin\theta}F_\theta + \frac{1}{\sin\theta}\frac{\partial F_\phi}{\partial \phi} + \frac{\partial F_\theta}{\partial \theta}\right)\right] +$$
$$+ \hat{\boldsymbol{\theta}}\left[\nabla^2 F_\theta - \frac{1}{r^2}\left(\frac{1}{\sin^2\theta}F_\theta - 2\frac{\partial F_r}{\partial \theta} + 2\frac{\cos\theta}{\sin^2\theta}\frac{\partial F_\phi}{\partial \phi}\right)\right] +$$
$$+ \hat{\boldsymbol{\phi}}\left[\nabla^2 F_\phi - \frac{1}{r^2}\left(\frac{1}{\sin^2\theta}F_\phi - 2\frac{1}{\sin\theta}\frac{\partial F_r}{\partial \phi} - 2\frac{\cos\theta}{\sin^2\theta}\frac{\partial F_\theta}{\partial \phi}\right)\right] \tag{D.111}$$

Separation of the Helmholtz equation

$$\frac{1}{r^2}\frac{\partial}{\partial r}\left(r^2\frac{\partial \psi(r,\theta,\phi)}{\partial r}\right) + \frac{1}{r^2\sin\theta}\frac{\partial}{\partial \theta}\left(\sin\theta\frac{\partial \psi(r,\theta,\phi)}{\partial \theta}\right) +$$
$$+ \frac{1}{r^2\sin^2\theta}\frac{\partial^2 \psi(r,\theta,\phi)}{\partial \phi^2} + k^2\psi(r,\theta,\phi) = 0 \tag{D.112}$$

$$\psi(r,\theta,\phi) = R(r)\Theta(\theta)\Phi(\phi) \tag{D.113}$$

$$\eta = \cos\theta \tag{D.114}$$

$$\frac{1}{R(r)}\frac{d}{dr}\left(r^2\frac{dR(r)}{dr}\right) + k^2 r^2 = n(n+1) \tag{D.115}$$

$$(1-\eta^2)\frac{d^2\Theta(\eta)}{d\eta^2} - 2\eta\frac{d\Theta(\eta)}{d\eta} + \left[n(n+1) - \frac{\mu^2}{1-\eta^2}\right]\Theta(\eta) = 0, \quad -1 \leq \eta \leq 1 \tag{D.116}$$

$$\frac{d^2\Phi(\phi)}{d\phi^2} + \mu^2\Phi(\phi) = 0 \tag{D.117}$$

$$\Phi(\phi) = \begin{cases} A_\phi \sin(\mu\phi) + B_\phi \cos(\mu\phi), & \mu \neq 0, \\ a_\phi \phi + b_\phi, & \mu = 0. \end{cases} \tag{D.118}$$

$$\Theta(\theta) = A_\theta P_n^\mu(\cos\theta) + B_\theta Q_n^\mu(\cos\theta) \tag{D.119}$$

$$R(r) = \begin{cases} R(r) = A_r r^n + B_r r^{-(n+1)}, & k = 0, \\ A_r F_1(kr) + B_r F_2(kr), & \text{otherwise.} \end{cases} \tag{D.120}$$

$$F_1(\xi), F_2(\xi) = \begin{cases} j_n(\xi) \\ n_n(\xi) \\ h_n^{(1)}(\xi) \\ h_n^{(2)}(\xi) \end{cases} \tag{D.121}$$

Appendix E

Properties of special functions

E.1 Bessel functions

Notation

z = complex number; ν, x = real numbers; n = integer

$J_\nu(z)$ = ordinary Bessel function of the first kind

$N_\nu(z)$ = ordinary Bessel function of the second kind

$I_\nu(z)$ = modified Bessel function of the first kind

$K_\nu(z)$ = modified Bessel function of the second kind

$H_\nu^{(1)}$ = Hankel function of the first kind

$H_\nu^{(2)}$ = Hankel function of the second kind

$j_n(z)$ = ordinary spherical Bessel function of the first kind

$n_n(z)$ = ordinary spherical Bessel function of the second kind

$h_n^{(1)}(z)$ = spherical Hankel function of the first kind

$h_n^{(2)}(z)$ = spherical Hankel function of the second kind

$f'(z) = df(z)/dz$ = derivative with respect to argument

Differential equations

$$\frac{d^2 Z_\nu(z)}{dz^2} + \frac{1}{z}\frac{dZ_\nu(z)}{dz} + \left(1 - \frac{\nu^2}{z^2}\right) Z_\nu(z) = 0 \tag{E.1}$$

$$Z_\nu(z) = \begin{cases} J_\nu(z) \\ N_\nu(z) \\ H_\nu^{(1)}(z) \\ H_\nu^{(2)}(z) \end{cases} \tag{E.2}$$

$$N_\nu(z) = \frac{\cos(\nu\pi)J_\nu(z) - J_{-\nu}(z)}{\sin(\nu\pi)}, \quad \nu \neq n, \quad |\arg(z)| < \pi \tag{E.3}$$

$$H_\nu^{(1)}(z) = J_\nu(z) + jN_\nu(z) \tag{E.4}$$
$$H_\nu^{(2)}(z) = J_\nu(z) - jN_\nu(z) \tag{E.5}$$

$$\frac{d^2 \bar{Z}_\nu(x)}{dz^2} + \frac{1}{z}\frac{d\bar{Z}_\nu(z)}{dz} - \left(1 + \frac{\nu^2}{z^2}\right)\bar{Z}_\nu = 0 \tag{E.6}$$

$$\bar{Z}_\nu(z) = \begin{cases} I_\nu(z) \\ K_\nu(z) \end{cases} \tag{E.7}$$

$$L(z) = \begin{cases} I_\nu(z) \\ e^{j\nu\pi}K_\nu(z) \end{cases} \tag{E.8}$$

$$I_\nu(z) = e^{-j\nu\pi/2}J_\nu(ze^{j\pi/2}), \quad -\pi < \arg(z) \le \frac{\pi}{2} \tag{E.9}$$

$$I_\nu(z) = e^{j3\nu\pi/2}J_\nu(ze^{-j3\pi/2}), \quad \frac{\pi}{2} < \arg(z) \le \pi \tag{E.10}$$

$$K_\nu(z) = \frac{j\pi}{2}e^{j\nu\pi/2}H_\nu^{(1)}(ze^{j\pi/2}), \quad -\pi < \arg(z) \le \frac{\pi}{2} \tag{E.11}$$

$$K_\nu(z) = -\frac{j\pi}{2}e^{-j\nu\pi/2}H_\nu^{(2)}(ze^{-j\pi/2}), \quad -\frac{\pi}{2} < \arg(z) \le \pi \tag{E.12}$$

$$I_n(x) = j^{-n}J_n(jx) \tag{E.13}$$

$$K_n(x) = \frac{\pi}{2}j^{n+1}H_n^{(1)}(jx) \tag{E.14}$$

$$\frac{d^2 z_n(z)}{dz^2} + \frac{2}{z}\frac{dz_n(z)}{dz} + \left[1 - \frac{n(n+1)}{z^2}\right]z_n(z) = 0, \quad n = 0, \pm 1, \pm 2, \ldots \tag{E.15}$$

$$z_n(z) = \begin{cases} j_n(z) \\ n_n(z) \\ h_n^{(1)}(z) \\ h_n^{(2)}(z) \end{cases} \tag{E.16}$$

$$j_n(z) = \sqrt{\frac{\pi}{2z}}J_{n+\frac{1}{2}}(z) \tag{E.17}$$

$$n_n(z) = \sqrt{\frac{\pi}{2z}}N_{n+\frac{1}{2}}(z) \tag{E.18}$$

$$h_n^{(1)}(z) = \sqrt{\frac{\pi}{2z}}H_{n+\frac{1}{2}}^{(1)}(z) = j_n(z) + jn_n(z) \tag{E.19}$$

$$h_n^{(2)}(z) = \sqrt{\frac{\pi}{2z}}H_{n+\frac{1}{2}}^{(2)}(z) = j_n(z) - jn_n(z) \tag{E.20}$$

$$n_n(z) = (-1)^{n+1}j_{-(n+1)}(z) \tag{E.21}$$

Orthogonality relationships

$$\int_0^a J_\nu\left(\frac{p_{\nu m}}{a}\rho\right) J_\nu\left(\frac{p_{\nu n}}{a}\rho\right) \rho\, d\rho = \delta_{mn}\frac{a^2}{2}J_{\nu+1}^2(p_{\nu n}) = \delta_{mn}\frac{a^2}{2}\left[J_\nu'(p_{\nu n})\right]^2, \quad \nu > -1 \tag{E.22}$$

$$\int_0^a J_\nu\left(\frac{p'_{\nu m}}{a}\rho\right) J_\nu\left(\frac{p'_{\nu n}}{a}\rho\right)\rho\,d\rho = \delta_{mn}\frac{a^2}{2}\left(1-\frac{\nu^2}{p'^2_{\nu m}}\right)J_\nu^2(p'_{\nu m}),\quad \nu>-1 \qquad \text{(E.23)}$$

$$\int_0^\infty J_\nu(\alpha x)J_\nu(\beta x)x\,dx = \frac{1}{\alpha}\delta(\alpha-\beta) \qquad \text{(E.24)}$$

$$\int_0^a j_l\left(\frac{\alpha_{lm}}{a}r\right) j_l\left(\frac{\alpha_{ln}}{a}r\right) r^2\,dr = \delta_{mn}\frac{a^3}{2}j_{n+1}^2(\alpha_{ln}a) \qquad \text{(E.25)}$$

$$\int_{-\infty}^\infty j_m(x)j_n(x)\,dx = \delta_{mn}\frac{\pi}{2n+1},\quad m,n\geq 0 \qquad \text{(E.26)}$$

$$J_m(p_{mn}) = 0 \qquad \text{(E.27)}$$
$$J'_m(p'_{mn}) = 0 \qquad \text{(E.28)}$$
$$j_m(\alpha_{mn}) = 0 \qquad \text{(E.29)}$$
$$j'_m(\alpha'_{mn}) = 0 \qquad \text{(E.30)}$$

Specific examples

$$j_0(z) = \frac{\sin z}{z} \qquad \text{(E.31)}$$

$$n_0(z) = -\frac{\cos z}{z} \qquad \text{(E.32)}$$

$$h_0^{(1)}(z) = -\frac{j}{z}e^{jz} \qquad \text{(E.33)}$$

$$h_0^{(2)}(z) = \frac{j}{z}e^{-jz} \qquad \text{(E.34)}$$

$$j_1(z) = \frac{\sin z}{z^2} - \frac{\cos z}{z} \qquad \text{(E.35)}$$

$$n_1(z) = -\frac{\cos z}{z^2} - \frac{\sin z}{z} \qquad \text{(E.36)}$$

$$j_2(z) = \left(\frac{3}{z^3}-\frac{1}{z}\right)\sin z - \frac{3}{z^2}\cos z \qquad \text{(E.37)}$$

$$n_2(z) = \left(-\frac{3}{z^3}+\frac{1}{z}\right)\cos z - \frac{3}{z^2}\sin z \qquad \text{(E.38)}$$

Functional relationships

$$J_n(-z) = (-1)^n J_n(z) \qquad \text{(E.39)}$$
$$I_n(-z) = (-1)^n I_n(z) \qquad \text{(E.40)}$$
$$j_n(-z) = (-1)^n j_n(z) \qquad \text{(E.41)}$$
$$n_n(-z) = (-1)^{n+1} n_n(z) \qquad \text{(E.42)}$$

$$J_{-n}(z) = (-1)^n J_n(z) \qquad \text{(E.43)}$$

$$N_{-n}(z) = (-1)^n N_n(z) \tag{E.44}$$

$$I_{-n}(z) = I_n(z) \tag{E.45}$$

$$K_{-n}(z) = K_n(z) \tag{E.46}$$

$$j_{-n}(z) = (-1)^n n_{n-1}(z), \quad n > 0 \tag{E.47}$$

Power series

$$J_n(z) = \sum_{k=0}^{\infty} (-1)^k \frac{(z/2)^{n+2k}}{k!(n+k)!} \tag{E.48}$$

$$I_n(z) = \sum_{k=0}^{\infty} \frac{(z/2)^{n+2k}}{k!(n+k)!} \tag{E.49}$$

Small argument approximations $|z| \ll 1$.

$$J_n(z) \approx \frac{1}{n!}\left(\frac{z}{2}\right)^n \tag{E.50}$$

$$J_\nu(z) \approx \frac{1}{\Gamma(\nu+1)}\left(\frac{z}{2}\right)^\nu \tag{E.51}$$

$$N_0(z) \approx \frac{2}{\pi}(\ln z + 0.5772157 - \ln 2) \tag{E.52}$$

$$N_n(z) \approx -\frac{(n-1)!}{\pi}\left(\frac{2}{z}\right)^n, \quad n > 0 \tag{E.53}$$

$$N_\nu(z) \approx -\frac{\Gamma(\nu)}{\pi}\left(\frac{2}{z}\right)^\nu, \quad \nu > 0 \tag{E.54}$$

$$I_n(z) \approx \frac{1}{n!}\left(\frac{z}{2}\right)^n \tag{E.55}$$

$$I_\nu(z) \approx \frac{1}{\Gamma(\nu+1)}\left(\frac{z}{2}\right)^\nu \tag{E.56}$$

$$j_n(z) \approx \frac{2^n n!}{(2n+1)!} z^n \tag{E.57}$$

$$n_n(z) \approx -\frac{(2n)!}{2^n n!} z^{-(n+1)} \tag{E.58}$$

Large argument approximations $|z| \gg 1$.

$$J_\nu(z) \approx \sqrt{\frac{2}{\pi z}}\cos\left(z - \frac{\pi}{4} - \frac{\nu\pi}{2}\right), \quad |\arg(z)| < \pi \tag{E.59}$$

$$N_\nu(z) \approx \sqrt{\frac{2}{\pi z}}\sin\left(z - \frac{\pi}{4} - \frac{\nu\pi}{2}\right), \quad |\arg(z)| < \pi \tag{E.60}$$

$$H_\nu^{(1)}(z) \approx \sqrt{\frac{2}{\pi z}}e^{j\left(z-\frac{\pi}{4}-\frac{\nu\pi}{2}\right)}, \quad -\pi < \arg(z) < 2\pi \tag{E.61}$$

$$H_\nu^{(2)}(z) \approx \sqrt{\frac{2}{\pi z}}e^{-j\left(z-\frac{\pi}{4}-\frac{\nu\pi}{2}\right)}, \quad -2\pi < \arg(z) < \pi \tag{E.62}$$

$$I_\nu(z) \approx \sqrt{\frac{1}{2\pi z}}e^z, \quad |\arg(z)| < \frac{\pi}{2} \tag{E.63}$$

$$K_\nu(z) \approx \sqrt{\frac{\pi}{2z}} e^{-z}, \quad |\arg(z)| < \frac{3\pi}{2} \tag{E.64}$$

$$j_n(z) \approx \frac{1}{z} \sin\left(z - \frac{n\pi}{2}\right), \quad |\arg(z)| < \pi \tag{E.65}$$

$$n_n(z) \approx -\frac{1}{z} \cos\left(z - \frac{n\pi}{2}\right), \quad |\arg(z)| < \pi \tag{E.66}$$

$$h_n^{(1)}(z) \approx (-j)^{n+1} \frac{e^{jz}}{z}, \quad -\pi < \arg(z) < 2\pi \tag{E.67}$$

$$h_n^{(2)}(z) \approx j^{n+1} \frac{e^{-jz}}{z}, \quad -2\pi < \arg(z) < \pi \tag{E.68}$$

Recursion relationships

$$z Z_{\nu-1}(z) + z Z_{\nu+1}(z) = 2\nu Z_\nu(z) \tag{E.69}$$

$$Z_{\nu-1}(z) - Z_{\nu+1}(z) = 2 Z_\nu'(z) \tag{E.70}$$

$$z Z_\nu'(z) + \nu Z_\nu(z) = z Z_{\nu-1}(z) \tag{E.71}$$

$$z Z_\nu'(z) - \nu Z_\nu(z) = -z Z_{\nu+1}(z) \tag{E.72}$$

$$z L_{\nu-1}(z) - z L_{\nu+1}(z) = 2\nu L_\nu(z) \tag{E.73}$$

$$L_{\nu-1}(z) + L_{\nu+1}(z) = 2 L_\nu'(z) \tag{E.74}$$

$$z L_\nu'(z) + \nu L_\nu(z) = z L_{\nu-1}(z) \tag{E.75}$$

$$z L_\nu'(z) - \nu L_\nu(z) = z L_{\nu+1}(z) \tag{E.76}$$

$$z z_{n-1}(z) + z z_{n+1}(z) = (2n+1) z_n(z) \tag{E.77}$$

$$n z_{n-1}(z) - (n+1) z_{n+1}(z) = (2n+1) z_n'(z) \tag{E.78}$$

$$z z_n'(z) + (n+1) z_n(z) = z z_{n-1}(z) \tag{E.79}$$

$$-z z_n'(z) + n z_n(z) = z z_{n+1}(z) \tag{E.80}$$

Integral representations

$$J_n(z) = \frac{1}{2\pi} \int_{-\pi}^{\pi} e^{-jn\theta + jz\sin\theta}\, d\theta \tag{E.81}$$

$$J_n(z) = \frac{1}{\pi} \int_0^{\pi} \cos(n\theta - z\sin\theta)\, d\theta \tag{E.82}$$

$$J_n(z) = \frac{1}{2\pi} j^{-n} \int_{-\pi}^{\pi} e^{jz\cos\theta} \cos(n\theta)\, d\theta \tag{E.83}$$

$$I_n(z) = \frac{1}{\pi} \int_0^{\pi} e^{z\cos\theta} \cos(n\theta)\, d\theta \tag{E.84}$$

$$K_n(z) = \int_0^{\infty} e^{-z\cosh(t)} \cosh(nt)\, dt, \quad |\arg(z)| < \frac{\pi}{2} \tag{E.85}$$

$$j_n(z) = \frac{z^n}{2^{n+1} n!} \int_0^{\pi} \cos(z\cos\theta) \sin^{2n+1}\theta\, d\theta \tag{E.86}$$

$$j_n(z) = \frac{(-j)^n}{2} \int_0^{\pi} e^{jz\cos\theta} P_n(\cos\theta) \sin\theta\, d\theta \tag{E.87}$$

Wronskians and cross products

$$J_\nu(z)N_{\nu+1}(z) - J_{\nu+1}(z)N_\nu(z) = -\frac{2}{\pi z} \tag{E.88}$$

$$H_\nu^{(2)}(z)H_{\nu+1}^{(1)}(z) - H_\nu^{(1)}(z)H_{\nu+1}^{(2)}(z) = \frac{4}{j\pi z} \tag{E.89}$$

$$I_\nu(z)K_{\nu+1}(z) + I_{\nu+1}(z)K_\nu(z) = \frac{1}{z} \tag{E.90}$$

$$I_\nu(z)K_\nu'(z) - I_\nu'(z)K_\nu(z) = -\frac{1}{z} \tag{E.91}$$

$$J_\nu(z)H_\nu^{(1)'}(z) - J_\nu'(z)H_\nu^{(1)}(z) = \frac{2j}{\pi z} \tag{E.92}$$

$$J_\nu(z)H_\nu^{(2)'}(z) - J_\nu'(z)H_\nu^{(2)}(z) = -\frac{2j}{\pi z} \tag{E.93}$$

$$H_\nu^{(1)}(z)H_\nu^{(2)'}(z) - H_\nu^{(1)'}(z)H_\nu^{(2)}(z) = -\frac{4j}{\pi z} \tag{E.94}$$

$$j_n(z)n_{n-1}(z) - j_{n-1}(z)n_n(z) = \frac{1}{z^2} \tag{E.95}$$

$$j_{n+1}(z)n_{n-1}(z) - j_{n-1}(z)n_{n+1}(z) = \frac{2n+1}{z^3} \tag{E.96}$$

$$j_n(z)n_n'(z) - j_n'(z)n_n(z) = \frac{1}{z^2} \tag{E.97}$$

$$h_n^{(1)}(z)h_n^{(2)'}(z) - h_n^{(1)'}(z)h_n^{(2)}(z) = -\frac{2j}{z^2} \tag{E.98}$$

Summation formulas

R, r, ρ, ϕ, ψ as shown.

$R = \sqrt{r^2 + \rho^2 - 2r\rho\cos\phi}$.

$$e^{j\nu\psi}Z_\nu(zR) = \sum_{k=-\infty}^{\infty} J_k(z\rho)Z_{\nu+k}(zr)e^{jk\phi}, \quad \rho < r, \quad 0 < \psi < \frac{\pi}{2} \tag{E.99}$$

$$e^{jn\psi}J_n(zR) = \sum_{k=-\infty}^{\infty} J_k(z\rho)J_{n+k}(zr)e^{jk\phi} \tag{E.100}$$

$$e^{jz\rho\cos\phi} = \sum_{k=0}^{\infty} j^k(2k+1)j_k(z\rho)P_k(\cos\phi) \tag{E.101}$$

For $\rho < r$ and $0 < \psi < \pi/2$,

$$\frac{e^{jzR}}{R} = \frac{j\pi}{2\sqrt{r\rho}}\sum_{k=0}^{\infty}(2k+1)J_{k+\frac{1}{2}}(z\rho)H_{k+\frac{1}{2}}^{(1)}(zr)P_k(\cos\phi) \tag{E.102}$$

$$\frac{e^{-jzR}}{R} = -\frac{j\pi}{2\sqrt{r\rho}}\sum_{k=0}^{\infty}(2k+1)J_{k+\frac{1}{2}}(z\rho)H_{k+\frac{1}{2}}^{(2)}(zr)P_k(\cos\phi) \tag{E.103}$$

Integrals

$$\int x^{\nu+1} J_\nu(x)\, dx = x^{\nu+1} J_{\nu+1}(x) + C \tag{E.104}$$

$$\int Z_\nu(ax) Z_\nu(bx) x\, dx = x \frac{[b Z_\nu(ax) Z_{\nu-1}(bx) - a Z_{\nu-1}(ax) Z_\nu(bx)]}{a^2 - b^2} + C, \quad a \neq b \tag{E.105}$$

$$\int x Z_\nu^2(ax)\, dx = \frac{x^2}{2} \left[Z_\nu^2(ax) - Z_{\nu-1}(ax) Z_{\nu+1}(ax) \right] + C \tag{E.106}$$

$$\int_0^\infty J_\nu(ax)\, dx = \frac{1}{a}, \quad \nu > -1, \quad a > 0 \tag{E.107}$$

Fourier–Bessel expansion of a function

$$f(\rho) = \sum_{m=1}^\infty a_m J_\nu \left(p_{\nu m} \frac{\rho}{a} \right), \quad 0 \leq \rho \leq a, \quad \nu > -1 \tag{E.108}$$

$$u_m = \frac{2}{a^2 J_{\nu+1}^2(p_{\nu m})} \int_0^a f(\rho) J_\nu \left(p_{\nu m} \frac{\rho}{a} \right) \rho\, d\rho \tag{E.109}$$

$$f(\rho) = \sum_{m=1}^\infty b_m J_\nu \left(p'_{\nu m} \frac{\rho}{a} \right), \quad 0 \leq \rho \leq a, \quad \nu > -1 \tag{E.110}$$

$$b_m = \frac{2}{a^2 \left(1 - \frac{\nu^2}{p'^2_{\nu m}} J_\nu^2(p'_{\nu m}) \right)} \int_0^a f(\rho) J_\nu \left(\frac{p'_{\nu m}}{a} \rho \right) \rho\, d\rho \tag{E.111}$$

Series of Bessel functions

$$e^{jz \cos \phi} = \sum_{k=-\infty}^\infty j^k J_k(z) e^{jk\phi} \tag{E.112}$$

$$e^{jz \cos \phi} = J_0(z) + 2 \sum_{k=1}^\infty j^k J_k(z) \cos \phi \tag{E.113}$$

$$\sin z = 2 \sum_{k=0}^\infty (-1)^k J_{2k+1}(z) \tag{E.114}$$

$$\cos z = J_0(z) + 2 \sum_{k=1}^\infty (-1)^k J_{2k}(z) \tag{E.115}$$

E.2 Legendre functions

Notation

x, y, θ = real numbers; l, m, n = integers;

$P_n^m(\cos \theta)$ = associated Legendre function of the first kind

$Q_n^m(\cos\theta)$ = associated Legendre function of the second kind

$P_n(\cos\theta) = P_n^0(\cos\theta)$ = Legendre polynomial

$Q_n(\cos\theta) = Q_n^0(\cos\theta)$ = Legendre function of the second kind

Differential equation $x = \cos\theta$.

$$(1 - x^2)\frac{d^2 R_n^m(x)}{dx^2} - 2x\frac{dR_n^m(x)}{dx} + \left[n(n+1) - \frac{m^2}{1-x^2}\right] R_n^m(x) = 0, \quad -1 \le x \le 1 \quad \text{(E.116)}$$

$$R_n^m(x) = \begin{cases} P_n^m(x) \\ Q_n^m(x) \end{cases} \tag{E.117}$$

Orthogonality relationships

$$\int_{-1}^{1} P_l^m(x)P_n^m(x)\,dx = \delta_{ln}\frac{2}{2n+1}\frac{(n+m)!}{(n-m)!} \tag{E.118}$$

$$\int_0^\pi P_l^m(\cos\theta)P_n^m(\cos\theta)\sin\theta\,d\theta = \delta_{ln}\frac{2}{2n+1}\frac{(n+m)!}{(n-m)!} \tag{E.119}$$

$$\int_{-1}^{1} \frac{P_n^m(x)P_n^k(x)}{1-x^2}\,dx = \delta_{mk}\frac{1}{m}\frac{(n+m)!}{(n-m)!} \tag{E.120}$$

$$\int_0^\pi \frac{P_n^m(\cos\theta)P_n^k(\cos\theta)}{\sin\theta}\,d\theta = \delta_{mk}\frac{1}{m}\frac{(n+m)!}{(n-m)!} \tag{E.121}$$

$$\int_{-1}^{1} P_l(x)P_n(x)\,dx = \delta_{ln}\frac{2}{2n+1} \tag{E.122}$$

$$\int_0^\pi P_l(\cos\theta)P_n(\cos\theta)\sin\theta\,d\theta = \delta_{ln}\frac{2}{2n+1} \tag{E.123}$$

Specific examples

$$P_0(x) = 1 \tag{E.124}$$

$$P_1(x) = x = \cos(\theta) \tag{E.125}$$

$$P_2(x) = \frac{1}{2}(3x^2 - 1) = \frac{1}{4}(3\cos 2\theta + 1) \tag{E.126}$$

$$P_3(x) = \frac{1}{2}(5x^3 - 3x) = \frac{1}{8}(5\cos 3\theta + 3\cos\theta) \tag{E.127}$$

$$P_4(x) = \frac{1}{8}(35x^4 - 30x^2 + 3) = \frac{1}{64}(35\cos 4\theta + 20\cos 2\theta + 9) \tag{E.128}$$

$$P_5(x) = \frac{1}{8}(63x^5 - 70x^3 + 15x) = \frac{1}{128}(63\cos 5\theta + 35\cos 3\theta + 30\cos\theta) \tag{E.129}$$

$$Q_0(x) = \frac{1}{2}\ln\left(\frac{1+x}{1-x}\right) = \ln\left(\cot\frac{\theta}{2}\right) \tag{E.130}$$

$$Q_1(x) = \frac{x}{2}\ln\left(\frac{1+x}{1-x}\right) - 1 = \cos\theta\ln\left(\cot\frac{\theta}{2}\right) - 1 \tag{E.131}$$

$$Q_2(x) = \frac{1}{4}(3x^2 - 1)\ln\left(\frac{1+x}{1-x}\right) - \frac{3}{2}x \tag{E.132}$$

$$Q_3(x) = \frac{1}{4}(5x^3 - 3x)\ln\left(\frac{1+x}{1-x}\right) - \frac{5}{2}x^2 + \frac{2}{3} \tag{E.133}$$

$$Q_4(x) = \frac{1}{16}(35x^4 - 30x^2 + 3)\ln\left(\frac{1+x}{1-x}\right) - \frac{35}{8}x^3 + \frac{55}{24}x \tag{E.134}$$

$$P_1^1(x) = -(1-x^2)^{1/2} = -\sin\theta \tag{E.135}$$

$$P_2^1(x) = -3x(1-x^2)^{1/2} = -3\cos\theta\sin\theta \tag{E.136}$$

$$P_2^2(x) = 3(1-x^2) = 3\sin^2\theta \tag{E.137}$$

$$P_3^1(x) = -\frac{3}{2}(5x^2 - 1)(1-x^2)^{1/2} = -\frac{3}{2}(5\cos^2\theta - 1)\sin\theta \tag{E.138}$$

$$P_3^2(x) = 15x(1-x^2) = 15\cos\theta\sin^2\theta \tag{E.139}$$

$$P_3^3(x) = -15(1-x^2)^{3/2} = -15\sin^3\theta \tag{E.140}$$

$$P_4^1(x) = -\frac{5}{2}(7x^3 - 3x)(1-x^2)^{1/2} = -\frac{5}{2}(7\cos^3\theta - 3\cos\theta)\sin\theta \tag{E.141}$$

$$P_4^2(x) = \frac{15}{2}(7x^2 - 1)(1-x^2) = \frac{15}{2}(7\cos^2\theta - 1)\sin^2\theta \tag{E.142}$$

$$P_4^3(x) = -105x(1-x^2)^{3/2} = -105\cos\theta\sin^3\theta \tag{E.143}$$

$$P_4^4(x) = 105(1-x^2)^2 = 105\sin^4\theta \tag{E.144}$$

Functional relationships

$$P_n^m(x) = \begin{cases} 0, & m > n, \\ (-1)^m \frac{(1-x^2)^{m/2}}{2^n n!} \frac{d^{n+m}(x^2-1)^n}{dx^{n+m}}, & m \le n. \end{cases} \tag{E.145}$$

$$P_n(x) = \frac{1}{2^n n!} \frac{d^n(x^2-1)^n}{dx^n} \tag{E.146}$$

$$R_n^m(x) = (-1)^m (1-x^2)^{m/2} \frac{d^m R_n(x)}{dx^m} \tag{E.147}$$

$$P_n^{-m}(x) = (-1)^m \frac{(n-m)!}{(n+m)!} P_n^m(x) \tag{E.148}$$

$$P_n(-x) = (-1)^n P_n(x) \tag{E.149}$$

$$Q_n(-x) = (-1)^{n+1} Q_n(x) \tag{E.150}$$

$$P_n^m(-x) = (-1)^{n+m} P_n^m(x) \tag{E.151}$$

$$Q_n^m(-x) = (-1)^{n+m+1} Q_n^m(x) \tag{E.152}$$

$$P_n^m(1) = \begin{cases} 1, & m = 0, \\ 0, & m > 0. \end{cases} \tag{E.153}$$

$$|P_n(x)| \le P_n(1) = 1 \tag{E.154}$$

$$P_n(0) = \frac{\Gamma\left(\frac{n}{2} + \frac{1}{2}\right)}{\sqrt{\pi}\,\Gamma\left(\frac{n}{2} + 1\right)} \cos\frac{n\pi}{2} \tag{E.155}$$

$$P_n^{-m}(x) = (-1)^m \frac{(n-m)!}{(n+m)!} P_n^m(x) \tag{E.156}$$

Power series

$$P_n(x) = \sum_{k=0}^{n} \frac{(-1)^k (n+k)!}{(n-k)!(k!)^2 2^{k+1}} \left[(1-x)^k + (-1)^n (1+x)^k\right] \tag{E.157}$$

Recursion relationships

$$(n+1-m)R_{n+1}^m(x) + (n+m)R_{n-1}^m(x) = (2n+1)xR_n^m(x) \tag{E.158}$$

$$(1-x^2)R_n^{m\prime}(x) = (n+1)xR_n^m(x) - (n-m+1)R_{n+1}^m(x) \tag{E.159}$$

$$(2n+1)xR_n(x) = (n+1)R_{n+1}(x) + nR_{n-1}(x) \tag{E.160}$$

$$(x^2-1)R_n'(x) = (n+1)[R_{n+1}(x) - xR_n(x)] \tag{E.161}$$

$$R_{n+1}'(x) - R_{n-1}'(x) = (2n+1)R_n(x) \tag{E.162}$$

Integral representations

$$P_n(\cos\theta) = \frac{\sqrt{2}}{\pi} \int_0^\pi \frac{\sin\left(n+\frac{1}{2}\right)u}{\sqrt{\cos\theta - \cos u}}\, du \tag{E.163}$$

$$P_n(x) = \frac{1}{\pi} \int_0^\pi \left[x + (x^2-1)^{1/2}\cos\theta\right]^n d\theta \tag{E.164}$$

Addition formula

$$P_n(\cos\gamma) = P_n(\cos\theta)P_n(\cos\theta') +$$
$$+ 2\sum_{m=1}^{n} \frac{(n-m)!}{(n+m)!} P_n^m(\cos\theta)P_n^m(\cos\theta')\cos m(\phi-\phi'), \tag{E.165}$$

$$\cos\gamma = \cos\theta\cos\theta' + \sin\theta\sin\theta'\cos(\phi-\phi') \tag{E.166}$$

Summations

$$\frac{1}{|\mathbf{r}-\mathbf{r}'|} = \frac{1}{\sqrt{r^2 + r'^2 - 2rr'\cos\gamma}} = \sum_{n=0}^{\infty} \frac{r_<^n}{r_>^{n+1}} P_n(\cos\gamma) \tag{E.167}$$

$$\cos\gamma = \cos\theta\cos\theta' + \sin\theta\sin\theta'\cos(\phi-\phi') \tag{E.168}$$

$$r_< = \min\{|\mathbf{r}|, |\mathbf{r}'|\}, \quad r_> = \max\{|\mathbf{r}|, |\mathbf{r}'|\} \tag{E.169}$$

Integrals

$$\int P_n(x)\, dx = \frac{P_{n+1}(x) - P_{n-1}(x)}{2n+1} + C \tag{E.170}$$

$$\int_{-1}^{1} x^m P_n(x)\, dx = 0, \quad m < n \tag{E.171}$$

$$\int_{-1}^{1} x^n P_n(x)\, dx = \frac{2^{n+1}(n!)^2}{(2n+1)!} \tag{E.172}$$

$$\int_{-1}^{1} x^{2k} P_{2n}(x)\, dx = \frac{2^{2n+1}(2k)!(k+n)!}{(2k+2n+1)!(k-n)!} \tag{E.173}$$

$$\int_{-1}^{1} \frac{P_n(x)}{\sqrt{1-x}}\, dx = \frac{2\sqrt{2}}{2n+1} \tag{E.174}$$

$$\int_{-1}^{1} \frac{P_{2n}(x)}{\sqrt{1-x^2}}\, dx = \left[\frac{\Gamma\left(n+\frac{1}{2}\right)}{n!}\right]^2 \tag{E.175}$$

$$\int_{0}^{1} P_{2n+1}(x)\, dx = (-1)^n \frac{(2n)!}{2n+2} \frac{1}{(2^n n!)^2} \tag{E.176}$$

Fourier–Legendre series expansion of a function

$$f(x) = \sum_{n=0}^{\infty} a_n P_n(x), \quad -1 \le x \le 1 \tag{E.177}$$

$$a_n = \frac{2n+1}{2} \int_{-1}^{1} f(x) P_n(x)\, dx \tag{E.178}$$

E.3   Spherical harmonics

Notation

θ, ϕ = real numbers; m, n = integers

$Y_{nm}(\theta, \phi)$ = spherical harmonic function

Differential equation

$$\frac{1}{\sin\theta} \frac{\partial}{\partial\theta}\left(\sin\theta \frac{\partial Y(\theta,\phi)}{\partial\theta}\right) + \frac{1}{\sin^2\theta} \frac{\partial^2 Y(\theta,\phi)}{\partial\phi^2} + \frac{1}{a^2}\lambda Y(\theta,\phi) = 0 \tag{E.179}$$

$$\lambda = a^2 n(n+1) \tag{E.180}$$

$$Y_{nm}(\theta,\phi) = \sqrt{\frac{2n+1}{4\pi}\frac{(n-m)!}{(n+m)!}} P_n^m(\cos\theta) e^{jm\theta} \tag{E.181}$$

Orthogonality relationships

$$\int_{-\pi}^{\pi} \int_{0}^{\pi} Y_{n'm'}^{*}(\theta, \phi) Y_{nm}(\theta, \phi) \sin \theta \, d\theta \, d\phi = \delta_{n'n} \delta_{m'm} \tag{E.182}$$

$$\sum_{n=0}^{\infty} \sum_{m=-n}^{n} Y_{nm}^{*}(\theta', \phi') Y_{nm}(\theta, \phi) = \delta(\phi - \phi') \delta(\cos \theta - \cos \theta') \tag{E.183}$$

Specific examples

$$Y_{00}(\theta, \phi) = \sqrt{\frac{1}{4\pi}} \tag{E.184}$$

$$Y_{10}(\theta, \phi) = \sqrt{\frac{3}{4\pi}} \cos \theta \tag{E.185}$$

$$Y_{11}(\theta, \phi) = -\sqrt{\frac{3}{8\pi}} \sin \theta e^{j\phi} \tag{E.186}$$

$$Y_{20}(\theta, \phi) = \sqrt{\frac{5}{4\pi}} \left(\frac{3}{2} \cos^{2} \theta - \frac{1}{2} \right) \tag{E.187}$$

$$Y_{21}(\theta, \phi) = -\sqrt{\frac{15}{8\pi}} \sin \theta \cos \theta e^{j\phi} \tag{E.188}$$

$$Y_{22}(\theta, \phi) = \sqrt{\frac{15}{32\pi}} \sin^{2} \theta e^{2j\phi} \tag{E.189}$$

$$Y_{30}(\theta, \phi) = \sqrt{\frac{7}{4\pi}} \left(\frac{5}{2} \cos^{3} \theta - \frac{3}{2} \cos \theta \right) \tag{E.190}$$

$$Y_{31}(\theta, \phi) = -\sqrt{\frac{21}{64\pi}} \sin \theta \left(5 \cos^{2} \theta - 1 \right) e^{j\phi} \tag{E.191}$$

$$Y_{32}(\theta, \phi) = \sqrt{\frac{105}{32\pi}} \sin^{2} \theta \cos \theta e^{2j\phi} \tag{E.192}$$

$$Y_{33}(\theta, \phi) = -\sqrt{\frac{35}{64\pi}} \sin^{3} \theta e^{3j\phi} \tag{E.193}$$

Functional relationships

$$Y_{n0}(\theta, \phi) = \sqrt{\frac{2n+1}{4\pi}} P_{n}(\cos \theta) \tag{E.194}$$

$$Y_{n,-m}(\theta, \phi) = (-1)^{m} Y_{nm}^{*}(\theta, \phi) \tag{E.195}$$

Addition formulas

$$P_{n}(\cos \gamma) = \frac{4\pi}{2n+1} \sum_{m=-n}^{n} Y_{nm}(\theta, \phi) Y_{nm}^{*}(\theta', \phi') \tag{E.196}$$

$$P_{n}(\cos \gamma) = P_{n}(\cos \theta) P_{n}(\cos \theta') +$$
$$+ \sum_{m=-n}^{n} \frac{(n-m)!}{(n+m)!} P_{n}^{m}(\cos \theta) P_{n}^{m}(\cos \theta') \cos \left[m(\phi - \phi') \right] \tag{E.197}$$

$$\cos\gamma = \cos\theta\cos\theta' + \sin\theta\sin\theta'\cos(\phi-\phi') \qquad \text{(E.198)}$$

Series

$$\sum_{m=-n}^{n} |Y_{nm}(\theta,\phi)|^2 = \frac{2n+1}{4\pi} \qquad \text{(E.199)}$$

$$\frac{1}{|\mathbf{r}-\mathbf{r}'|} = \frac{1}{\sqrt{r^2 + r'^2 - 2rr'\cos\gamma}}$$

$$= 4\pi \sum_{n=0}^{\infty} \sum_{m=-n}^{n} \frac{1}{2n+1} \frac{r_<^n}{r_>^{n+1}} Y_{nm}^*(\theta',\phi') Y_{nm}(\theta,\phi), \qquad \text{(E.200)}$$

$$r_< = \min\left\{|\mathbf{r}|, |\mathbf{r}'|\right\}, \quad r_> = \max\left\{|\mathbf{r}|, |\mathbf{r}'|\right\} \qquad \text{(E.201)}$$

Series expansion of a function

$$f(\theta,\phi) = \sum_{n=0}^{\infty} \sum_{m=-n}^{n} a_{nm} Y_{nm}(\theta,\phi) \qquad \text{(E.202)}$$

$$a_{nm} = \int_{-\pi}^{\pi} \int_{0}^{\pi} f(\theta,\phi) Y_{nm}^*(\theta,\phi) \sin\theta \, d\theta \, d\phi \qquad \text{(E.203)}$$

References

[1] Abramowitz, M., and Stegun, I., *Handbook of Mathematical Functions*, Dover Publications, New York, 1965.

[2] Adair, R., *Concepts in Physics*, Academic Press, New York, 1969.

[3] Anderson, J., and Ryon, J., *Electromagnetic Radiation in Accelerated Systems*, Physical Review, vol. 181, no. 5, pp. 1765–1775, May 1969.

[4] Anderson, J. C., *Dielectrics*, Reinhold Publishing Co., New York, 1964.

[5] Arfken, G., *Mathematical Methods for Physicists*, Academic Press, New York, 1970.

[6] Aris, R., *Vectors, Tensors, and the Basic Equations of Fluid Mechanics*, Dover Publications, New York, 1962.

[7] Atiyah, M., and Hitchen, N., *The Geometry and Dynamics of Magnetic Monopoles*, Princeton University Press, Princeton, NJ, 1988.

[8] Balanis, C., *Advanced Engineering Electromagnetics*, Wiley, New York, 1989.

[9] Becker, R., *Electromagnetic Fields and Interactions*, Dover Publications, New York, 1982.

[10] Beiser, A., *Concepts of Modern Physics*, McGraw–Hill, New York, 1981.

[11] Bergmann, P. G., *Basic Theories of Physics: Mechanics and Electrodynamics*, Prentice–Hall, New York, 1949.

[12] Blanpied, W., *Physics: Its Structure and Evolution*, Blaisdell Publishing Co., Waltham, MA, 1969.

[13] Boffi, L., *Electrodynamics of Moving Media*, ScD Thesis, Massachusetts Institute of Technology, Cambridge, 1957.

[14] Bohm, D., *The Special Theory of Relativity*, Routledge, London, 1996.

[15] Bohn, E. V., *Introduction to Electromagnetic Fields and Waves*, Addison–Wesley, Reading, MA, 1968.

[16] Boithias, L., *Radio Wave Propagation*, McGraw–Hill, New York, 1987.

[17] Borisenko, A., and Tarapov, I., *Vector and Tensor Analysis with Applications*, Dover Publications, New York, 1968.

[18] Born, M., *Einstein's Theory of Relativity*, Dover Publications, New York, 1965.

[19] Born, M., and Wolf, E., *Principles of Optics*, Pergamon Press, Oxford, 1980.

[20] Bowman, F., *An Introduction to Bessel Functions*, Dover Publications, New York, 1958.

[21] Bowman, J. J., Senior, T. B. A., and Uslenghi, P. L. E., *Electromagnetic and Acoustic Scattering by Simple Shapes*, North-Holland Publishing Co., Amsterdam, 1969.

[22] Bracewell, R., *The Fourier Transform and its Applications*, McGraw–Hill, New York, 1986.

[23] Brillouin, L., *Wave Propagation in Periodic Structures; Electric Filters and Crystal Lattices*, Dover Publications, New York, 1953.

[24] Bronshtein, I. N., and Semendyayev, K. A., *Handbook of Mathematics*, Springer–Verlag, Berlin, 1997.

[25] Byron, F., and Fuller, R., *Mathematics of Classical and Quantum Physics*, Dover Publications, New York, 1970.

[26] Campbell, G. A., and Foster, R. M., *Fourier Integrals for Practical Applications*, D. Van Nostrand Co., New York, 1948.

[27] Censor, D., *Application-Oriented Relativistic Electrodynamics*, in Progress in Electromagnetics Research 4, Elsevier Science Publishing, New York, 1991.

[28] Chen, H. C., *Theory of Electromagnetic Waves: a Coordinate Free Approach*, McGraw–Hill, New York, 1983.

[29] Chen, K.-M., *A Mathematical Formulation of the Equivalence Principle*, IEEE Transactions on Microwave Theory and Techniques, vol. 37, no. 10, pp. 1576–1580, 1989.

[30] Chen, Q., Katsurai, M., and Aoyagi, P., *An FDTD Formulation for Dispersive Media Using a Current Density*, IEEE Transactions on Antennas and Propagation, vol. 46, no. 10, pp. 1739–1746, October 1988.

[31] Cheng, D. K., *Field and Wave Electromagnetics*, Addison–Wesley, Reading, MA, 1983.

[32] Cheston, W. B., *Theory of Electric and Magnetic Fields*, Wiley, New York, 1964.

[33] Chew, W. C., *Waves and Fields in Inhomogeneous Media*, Van Nostrand Reinhold, New York, 1990.

[34] Christy, R., and Pytte, A., *The Structure of Matter*, W.A. Benjamin, New York, 1965.

[35] Churchill, R. V., Brown, J. W., and Verhey, R. F., *Complex Variables and Applications*, McGraw–Hill, New York, 1974.

[36] Clemmow, P. C., *The Plane Wave Spectrum Representation of Electromagnetic Fields*, IEEE Press, Piscatawaty, NJ, 1996.

[37] Cole, G., *Fluid Dynamics*, Wiley, New York, 1962.

[38] Cole, K. S., and Cole, R. H., *Dispersion and Absorption in Dielectrics*, Journal of Chemical Physics, no. 9, vol. 341, 1941.

[39] Collin, R., *Field Theory of Guided Waves*, IEEE Press, New York, 1991.

[40] Collin, R., *Foundations for Microwave Engineering*, 2nd ed., McGraw–Hill, New York, 1992.

[41] Collin, R., and Zucker, F., *Antenna Theory*, McGraw–Hill, New York, 1969.

[42] Corson, D., *Electromagnetic Induction in Moving Systems*, American Journal of Physics, vol. 24, pp. 126–130, 1956.

[43] Costen, R., and Adamson, D., *Three-dimensional Derivation of the Electrodynamic Jump Conditions and Momentum-Energy Laws at a Moving Boundary*, Proceedings of the IEEE, vol. 53, no. 5, pp. 1181–1196, September 1965.

[44] Cowling, T. G., *Magnetohydrodynamics*, Interscience Publishers, New York, 1957.

[45] Craigie, N., ed., *Theory and Detection of Magnetic Monopoles in Gauge Theories*, World Scientific, Singapore, 1986.

[46] *CRC Handbook of Chemistry and Physics*, 77th edition, CRC Press, Boca Raton, FL, 1996.

[47] *CRC Standard Mathematical Tables*, 24th Edition, Beyer, W.H., and Selby, S.M., eds., CRC Press, Cleveland, Ohio, 1976.

[48] Cullwick, E., *Electromagnetism and Relativity*, Longmans, Green, and Co., London, 1957.

[49] Daniel, V., *Dielectric Relaxation*, Academic Press, London, UK, 1967.

[50] Danielson, D., *Vectors and Tensors in Engineering and Physics*, Addison–Wesley, Reading, MA, 1997.

[51] Davies, K., *Ionospheric Radio Propagation*, National Bureau of Standards Monograph 80, Washington, DC, 1965.

[52] Debye, P., *Polar Molecules*, Dover Publications, New York, 1945.

[53] Dettman, J., *Mathematical Methods in Physics and Engineering*, Dover Publications, New York, 1988.

[54] Devaney, A. J., and Wolf, E., *Radiating and Nonradiating Classical Current Distributions and the Fields They Generate*, Physical Review D, vol. 8, no. 4, pp. 1044–1047, 15 August 1973.

[55] Di Francia, G. T., *Electromagnetic Waves*, Interscience Publishers, New York, 1953.

[56] Dorf, R., ed., *Electrical Engineering Handbook*, CRC Press, Boca Raton, FL, 1993.

[57] Doughty, N., *Lagrangian Interaction*, Addison–Wesley, Reading, MA, 1990.

[58] Dudley, D., *Mathematical Foundations for Electromagnetic Theory*, IEEE Press, New York, 1994.

[59] Duffin, W. J., *Advanced Electricity and Magnetism*, McGraw–Hill, 1968.

[60] Duffy, D., *Advanced Engineering Mathematics*, CRC Press, Boca Raton, FL, 1998.

[61] Durney, C. H., and Johnson, C. C., *Introduction to Modern Electromagnetics*, McGraw–Hill, New York, 1969.

[62] Einstein, A., *Relativity*, Prometheus Books, Amherst, NJ, 1995.

[63] Einstein, A., and Infeld, L., *The Evolution of Physics*, Simon and Schuster, New York, 1938.

[64] Eisenhart, L., *Separable Systems of Stäckel*, Ann. Math., vol. 35, p. 284, 1934.

[65] Elliott, R., *Electromagnetics*, IEEE Press, New York, 1993.

[66] Elliott, R., *Antenna Theory and Design*, Prentice–Hall, Englewood Cliffs, NJ, 1981.

[67] Elliott, R. S., *An Introduction to Guided Waves and Microwave Circuits*, Prentice–Hall, Englewood Cliffs, NJ, 1993.

[68] Elmore, W. C., and Heald, M. A., *Physics of Waves*, Dover Publications, New York, 1985.

[69] Eyges, L., *The Classical Electromagnetic Field*, Dover Publications, New York, 1980.

[70] Fano, R., Chu, L., and Adler, R., *Electromagnetic Fields, Energy, and Forces*, Wiley, New York, 1960.

[71] Feather, N., *Electricity and Matter*, Edinburgh University Press, Edinburgh, 1968.

[72] Felsen, L. B., and Marcuvitz, N., *Radiation and Scattering of Waves*, IEEE Press, Piscataway, NJ, 1994.

[73] Feynman, R., Leighton, R., and Sands, M., *The Feynman Lectures on Physics, Vol. 2*, Addison–Wesley, Reading, MA, 1964.

[74] Fradin, A. Z., *Microwave Antennas*, Pergamon Press, Oxford, UK, 1961.

[75] Goodman, J., *Introduction to Fourier Optics*, McGraw–Hill, New York, 1968.

[76] Gradshteyn, I., and Ryzhik, I., *Table of Integrals, Series, and Products*, Academic Press, Boston, 1994.

[77] Greenwood, D. T., *Classical Dynamics*, Prentice–Hall, Englewood Cliffs, NJ, 1977.

[78] Guenther, R. B., and Lee, J. W., *Partial Differential Equations of Mathematical Physics and Integral Equations*, Dover Publications, New York, 1996.

[79] Haberman, R., *Elementary Applied Partial Differential Equations*, Prentice–Hall, Upper Saddle River, NJ, 1998.

[80] Halliday, D., and Resnick, R., *Fundamentals of Physics*, Wiley, New York, 1981.

[81] Hansen, T. B., and Yaghjian, A. D., *Plane-Wave Theory of Time-Domain Fields*, IEEE Press, New York, 1999.

[82] Harrington, R. F., *Field Computation by Moment Methods*, IEEE Press, New York, 1993.

[83] Harrington, R., *Time Harmonic Electromagnetic Fields*, McGraw–Hill, New York, 1961.

[84] Haus, H. A., and Melcher, J. R., *Electromagnetic Fields and Energy*, Prentice–Hall, Englewood Cliffs, NJ, 1989.

[85] Hertz, H., *Electric Waves*, MacMillan, New York, 1900.

[86] Hesse, M. B., *Forces and Fields*, Greenwood Press, Westport, CT, 1970.

[87] Hlawiczka, P., *Gyrotropic Waveguides*, Academic Press, London, UK, 1981.

[88] Hunsberger, F., Luebbers, R., and Kunz, K., *Finite-Difference Time-Domain Analysis of Gyrotropic Media — I: Magnetized Plasma*, IEEE Transactions on Antennas and Propagation, vol. 40, no. 12, pp. 1489–1495, December 1992.

[89] Ingarden, R., and Jamiolkowski, A., *Classical Electrodynamics*, Elsevier, Amsterdam, 1985.

[90] Ishimaru, A., *Electromagnetic Wave Propagation, Radiation, and Scattering*, Prentice–Hall, Englewood Cliffs, NJ, 1991.

[91] Jackson, J., *Classical Electrodynamics*, Wiley, New York, 1975.

[92] James, G. L., *Geometrical Theory of Diffraction for Electromagnetic Waves*, Peter Peregrinus Ltd., London, UK, 1986.

[93] Jeffrey, A., *Complex Analysis and Applications*, CRC Press, Boca Raton, FL, 1992.

[94] Jiang, B., Wu, J., and Povinelli, L. A., *The Origin of Spurious Solutions in Computational Electromagnetics*, Journal of Computational Physics, vol. 125, pp. 104–123, 1996.

[95] Johnk, C. T. A., *Engineering Electromagnetic Fields and Waves*, Wiley, New York, 1988.

[96] Johnson, C. C., *Field and Wave Electrodynamics*, McGraw–Hill, New York, 1965.

[97] Jones, D., *The Theory of Electromagnetism*, Pergamon Press, New York, 1964.

[98] Joos, G., *Theoretical Physics*, Dover Publications, New York, 1986.

[99] King, R., and Prasad, S., *Fundamental Electromagnetic Theory and Applications*, Prentice–Hall, Englewood Cliffs, NJ, 1986.

[100] King, R. W. P., and T. T. Wu, *The Scattering and Diffraction of Waves*, Harvard University Press, Cambridge, MA, 1959.

[101] Kong, J., *Electromagnetic Wave Theory*, Wiley, New York, 1990.

[102] Kong, J., *Theorems of Bianisotropic Media*, Proceedings of the IEEE, vol. 60, no. 9, pp. 1036–1046, 1972.

[103] Koshlyakov, N. S., *Differential Equations of Mathematical Physics*, Interscience Publishers, New York, 1964.

[104] Kraus, J. D., *Elecromagnetics*, McGraw–Hill, New York, 1992.

[105] Kraus, J. D., and Fleisch, D. A., *Electromagnetics with Applications*, WCB McGraw–Hill, Boston, 1999.

[106] Landau, L., and Lifshitz, E., *The Classical Theory of Fields*, Addison–Wesley, Cambridge, MA, 1951.

[107] Landau, L. D., Lifshitz, E. M., and Pitaevskii, L. P., *Electrodynamics of Theoretical Physics*, 2nd Ed., Butterworth–Heinemann, Oxford, UK, 1984.

[108] Langmuir, R. V., *Electromagnetic Fields and Waves*, McGraw–Hill, New York, 1961.

[109] Lebedev, N., *Special Functions and Their Applications*, Dover Publications, New York, 1972.

[110] LePage, W., *Complex Variables and the Laplace Transform for Engineers*, Dover Publications, New York, 1961.

[111] Liao, S. Y., *Microwave Devices and Circuits*, 3rd ed., Prentice–Hall, Englewood Cliffs, NJ, 1990.

[112] Lindell, I., Sihvola, A., Tretyakov, S., and Vitanen, A., *Electromagnetic Waves in Chiral and Bi-isotropic Media*, Artech House, Boston, 1994.

[113] Lindell, I., *Methods for Electromagnetic Field Analysis*, IEEE Press, New York, 1992.

[114] Lindsay, R., and Margenau, H., *Foundations of Physics*, Dover Publications, New York, 1957.

[115] Lorentz, H. A., *Theory of Electrons*, Dover Publications, New York, 1952.

[116] Lorrain, P., and Corson, D. R., *Electromagnetic Fields and Waves*, W.H. Freeman and Company, San Francisco, 1970.

[117] Lucas, J., and Hodgson, P., *Spacetime and Electromagnetism*, Clarendon Press, Oxford, 1990.

[118] Luebbers, R., and Hunsberger, F., *FDTD for Nth-Order Dispersive Media*, IEEE Transactions on Antennas and Propagation, vol. 40, no. 11, pp. 1297–1301, November 1992.

[119] MacCluer, C., *Boundary Value Problems and Orthogonal Expansions, Physical Problems from a Sobolev Viewpoint*, IEEE Press, New York, 1994.

[120] Marion, J. B., *Classical Electromagnetic Radiation*, Academic Press, New York, 1965.

[121] Marsden, J., *Basic Complex Analysis*, W.H. Freeman and Company, San Francisco, 1973.

[122] Marsden, J., and Tromba, A., *Vector Calculus*, W.H. Freeman and Company, San Francisco, 1976.

[123] Mason, M., and Weaver, W., *The Electromagnetic Field*, University of Chicago Press, Chicago, 1929.

[124] Mathews, J., and Walker, R. L., *Mathematical Methods of Physics*, W.A. Benjamin, New York, 1964.

[125] Maxwell, J., *A Dynamical Theory of the Electromagnetic Field*, Royal Society Transactions, vol. CLV, reprinted in Simpson, T., *Maxwell on the Electromagnetic Field*, Rutgers University Press, New Brunswick, NJ, 1997.

[126] Maxwell, J. C., *A Treatise on Electricity and Magnetism*, Two volumes, Dover Publications, New York, 1954.

[127] McQuistan, R. B., *Scalar and Vector Fields: A Physical Interpretation*, Wiley, New York, 1965.

[128] Mentzer, J. R., *Scattering and Diffraction of Radio Waves*, Pergamon Press, New York, 1955.

[129] Meriam, J. L., *Dynamics*, Wiley, New York, 1978.

[130] Miller, A., *Albert Einstein's Special Theory of Relativity*, Springer–Verlag, New York, 1998.

[131] Mills, D., *Nonlinear Optics*, Springer–Verlag, Berlin, 1998.

[132] Mo, C., *Theory of Electrodynamics in Media in Noninertial Frames and Applications*, Journal of Mathematical Physics, vol. 11, no. 8, pp. 2589–2610, August 1970.

[133] Moon, P., and Spencer, D. E., *Foundations of Electrodynamics*, Boston Technical Publishers, Boston, 1965.

[134] Moon, P., and Spencer, D., *Field Theory Handbook*, Springer–Verlag, New York, 1971.

[135] Moon, P., and Spencer, D., *Vectors*, D. Van Nostrand Company, Princeton, NJ, 1965.

[136] Morse, P., and Feshbach, H., *Methods of Theoretical Physics*, McGraw–Hill, New York, 1953.

[137] Nagel, E., *The Structure of Science*, Harcourt, Brace and World, New York, 1961.

[138] Norwood, J., *Twentieth Century Physics*, Prentice–Hall, Englewood Cliffs, NJ, 1976.

[139] O'Dell, T., *The Electrodynamics of Magneto-Electric Media*, North Holland Publishing, Amsterdam, 1970.

[140] Panofsky, W. K. H., and Philips, M., *Classical Electricity and Magnetism*, Addison–Wesley, Reading, MA, 1962.

[141] Papas, C. H., *Theory of Electromagnetic Wave Propagation*, Dover Publications, New York, 1988.

[142] Papoulis, A., *The Fourier Integral and Its Applications*, McGraw–Hill, New York, 1962.

[143] Pauli, W., *Pauli Lectures on Physics: Volume I. Electrodynamics*, MIT Press, Cambridge, MA, 1973.

[144] Pauli, W., *Theory of Relativity*, Dover Publications, New York, 1981.

[145] Penfield, P., and Haus, H., *Electrodynamics of Moving Media*, MIT Press, Cambridge, MA, 1967.

[146] Peterson, A. F., Ray, S. L., and Mittra, R., *Computational Methods for Electromagnetics*, IEEE Press, New York, 1998.

[147] Pierce, J. R., *Almost All About Waves*, MIT Press, Cambridge, MA, 1974.

[148] Pierce, J. R., *Theory and Design of Electron Beams*, Van Nostrand, New York, 1954.

[149] Plonsey, R., and Collin, R. E., *Principles and Applications of Electromagnetic Fields*, McGraw–Hill, 1961.

[150] Podolsky, B., and Kunz, K., *Fundamentals of Electrodynamics*, Marcel Dekker, New York, 1969.

[151] Post, E., *Formal Structure of Electromagnetics*, Dover Publications, Mineola, NY, 1997.

[152] Purcell, E., *Electricity and Magnetism (Berkeley Physics Course, Vol. 2)*, McGraw-Hill, New York, 1985.

[153] Ramo, S., Whinnery, J., and Van Duzer, T., *Fields and Waves in Communication Electronics*, Wiley, New York, 1965.

[154] Reid, C., *Hilbert*, Springer–Verlag, New York, 1996.

[155] Resnick, R., and Halliday, D., *Physics (Part I)*, Wiley, New York, 1966.

[156] Robinson, P., *Fourier and Laplace Transforms*, Dover Publications, New York, 1968.

[157] Rojansky, V., *Electromagnetic Fields and Waves*, Prentice–Hall, Englewood Cliffs, NJ, 1971.

[158] Rosensweig, R. E., *Ferrohydrodynamics*, Dover Publications, Mineola, NY, 1997.

[159] Rosser, W. V. G., *Classical Electromagnetism via Relativity*, Plenum Press, New York, 1968.

[160] Ruck, G. T., Barrick, D. E., Stuart, W. D., and Krichbaum, C. K., *Radar Cross Section Handbook*, Plenum Press, New York, 1970.

[161] Rumsey, V. H., *Reaction Concept in Electromagnetic Theory*, Physical Review, vol. 94, pp. 1483–1491, June 1954.

[162] Sadiku, M. N. O., *Numerical Techniques in Electromagnetics*, CRC Press, Boca Raton, FL, 1992.

[163] Sagan, H., *Boundary and Eigenvalue Problems in Mathematical Physics*, Dover Publications, New York, 1989.

[164] Scaife, B. K. P., Scaife, W. G. S., Bennett, R. G., and Calderwood, J. H., *Complex Permittivity: Theory and Measurement*, English Universities Press Ltd, London, UK, 1971.

[165] Scanlon, P., Henriksen, R., and Allen, J., *Approaches to Electromagnetic Induction*, American Journal of Physics, vol. 37, no. 7, pp. 698–708, July 1969.

[166] Schelkunoff, S., *Electromagnetic Fields*, Blaisdell, New York, 1963.

[167] Schelkunoff, S., *On Teaching the Undergraduate Electromagnetic Theory*, IEEE Transactions on Education, vol. E-15, no. 1, pp. 15–25, February 1972.

[168] Schelkunoff, S. A., *Electromagnetic Waves*, D. Van Nostrand, Princeton, NJ, 1943.

[169] Schelkunoff, S. A., *Some Equivalence Theorems of Electromagnetics and Their Applications to Radiation Problems*, Bell System Technical Journal, vol. 15, pp. 92–112, 1936.

[170] Schwartz, M., *Principles of Electrodynamics*, Dover Publications, New York, 1972.

[171] Schwartz, W., *Intermediate Electromagnetic Theory*, Robert E. Krieger Publishing Company, Huntington, NY, 1973.

[172] Schwinger, J., DeRaad, L., Milton, K., and Tsai, W., *Classical Electrodynamics*, Perseus Books, Reading, MA, 1998.

[173] Shen, L. C., and Kong, J. A., *Applied Electromagnetism*, PWS Publishing Company, Boston, 1995.

[174] Shen, Y., *The Principles of Nonlinear Optics*, Wiley, New York, 1984.

[175] Sihvola, A., *Electromagnetic Mixing Formulas and Applications*, Institution of Electrical Engineers, London, UK, 1999.

[176] Silver, S., ed., *Microwave Antenna Theory and Design*, Dover Publications, New York, 1965.

[177] Silverman, M., *Waves and Grains*, Princeton University Press, Princeton, NJ, 1998.

[178] Simpson, T., *Maxwell on the Electromagnetic Field*, Rutgers University Press, New Brunswick, NJ, 1997.

[179] Slater, J., and Frank N., *Electromagnetism*, Dover Publications, New York, 1969.

[180] Smith, G. S., *An Introduction to Classical Electromagnetic Radiation*, Cambridge University Press, Cambridge, UK, 1997.

[181] Smythe, W. R., *Static and Dynamic Electricity*, McGraw–Hill, New York, 1968.

[182] Sobolev, S. L., *Partial Differential Equations of Mathematical Physics*, Dover Publications, New York, 1989.

[183] Someda, C., *Electromagnetic Waves*, Chapman and Hall, London, 1998.

[184] Sommerfeld, A., *Partial Differential Equations in Physics*, Academic Press, New York, 1949.

[185] Sommerfeld, A., *Electrodynamics*, Academic Press, New York, 1952.

[186] Staelin, D. H., Morgenthaler, A. W., and Kong, J. A., *Electromagnetic Waves*, Prentice–Hall, Englewood Cliffs, NJ, 1994.

[187] Stratton, J., *Electromagnetic Theory*, McGraw–Hill, New York, 1941.

[188] Stratton, J. A., and Chu, L. J., *Diffraction Theory of Electromagnetic Waves*, Physical Review, vol. 56, pp. 99–107, 1939.

[189] Stutzman, W. L., *Polarization in Electromagnetic Systems*, Artech House, Boston, 1993.

[190] Tai, C.-T., *Generalized Vector and Dyadic Analysis*, IEEE Press, New York, 1997.

[191] Tai, C.-T., *Some Essential Formulas in Dyadic Analysis and Their Applications*, Radio Science, vol. 22, no. 7, pp. 1283–1288, 1987.

[192] Tai, C.-T., *Dyadic Green's Functions in Electromagnetic Theory*, 2nd Ed., IEEE Press, New York, 1994.

[193] Tai, C.-T., *A Study of the Electrodynamics of Moving Media*, Proceedings of the IEEE, pp. 685–689, June 1964.

[194] Tai, C.-T., *On the Presentation of Maxwell's Theory*, Proceedings of the IEEE, vol. 60, no. 8, pp. 936–945, August 1972.

[195] Tanenbaum, B. S., *Plasma Physics*, McGraw–Hill, New York, 1967.

[196] Tellegen, B., *The Gyrator, a New Electric Network Element*, Philips Research Reports, vol. 3, pp. 81–101, April 1948.

[197] Tiersten, H., *A Development of the Equations of Electromagnetism in Material Continua*, Springer–Verlag, New York, 1990.

[198] Tolstoy, I., *Wave Propagation*, McGraw–Hill, New York, 1973.

[199] Truesdell, C., and Toupin, R., *The Classical Field Theories* in *Encyclopedia of Physics*, S. Flugge, Ed., vol. 3, Part 1, Springer–Verlag, Berlin, 1960.

[200] U.S. Army Corps of Engineers, *Electromagnetic Pulse (EMP) and TEMPEST Protection for Facilities*, Pamphlet No. 1110-3-2, December 1990.

[201] Van Bladel, J., *Electromagnetic Fields in the Presence of Rotating Bodies*, Proceedings of the IEEE, vol. 64, no. 3, pp. 301–318, March 1976.

[202] Van Bladel, J., *Electromagnetic Fields*, Hemisphere Publishing, New York, 1985.

[203] Veysoglu, M. E., Ha, Y., Shin, R. T., and Kong, J. A., *Polarimetric Passive Remote Sensing of Periodic Surfaces*, Journal of Electromagnetic Waves and Applications, vol. 5, no. 3, pp. 267–280, 1991.

[204] von Aulock, W. H., Ed., *Handbook of Microwave Ferrite Materials*, Academic Press, New York, 1965.

[205] von Hippel, A. R., *Dielectrics and Waves*, Wiley, New York, 1954.

[206] Wait, J. R., *Electromagnetic Wave Theory*, Harper & Row, New York, 1985.

[207] Wait, J. R., *Electromagnetic Waves in Stratified Media*, Pergamon Press, Oxford, UK, 1970.

[208] Wait, J. R., *Introduction to Antennas and Propagation*, Peter Peregrinus, LTD, London, UK, 1986.

[209] Wang, C.-C., *Mathematical Principles of Mechanics and Electromagnetism; Part A: Analytical and Continuum Mechanics*, Plenum Press, New York, 1979.

[210] Webster, A. G., *Partial Differential Equations of Mathematical Physics*, Dover Publications, New York, 1966.

[211] Weigelhofer, W. S., and Hansen, S. O., *Faraday and Chiral Media Revisited–I: Fields and Sources*, IEEE Transactions on Antennas and Propagation, vol. 47, no. 5, May 1999.

[212] Weinberger, H., *A First Course in Partial Differential Equations with Complex Variables and Transform Methods*, Dover Publications, New York, 1965.

[213] Weyl, H., *Space–Time–Matter*, Dover Publications, New York, 1952.

[214] Whitaker, S., *Intoduction to Fluid Mechanics*, Prentice–Hall, Englewood Cliffs, NJ, 1968.

[215] Yaghjian, A. D., *Electric Dyadic Green's Functions in the Source Region*, Proceedings of the IEEE, vol. 68, pp. 248–263, 1980.

[216] Zierep, J., *Similarity Laws and Modeling*, Marcel Dekker, New York, 1971.

[217] Zwillinger, D., *Handbook of Differential Equations*, Academic Press, Boston, 1992.

Index